The Future of Amazonian Aquatic Biota

Samara Silva de Souza •
Susana Braz-Mota • Adalberto Luis Val
Editors

The Future of Amazonian Aquatic Biota

Springer

Editors
Samara Silva de Souza
Laboratory of Ecophysiology and Molecular Evolution
Brazilian National Institute for Research of the Amazon (INPA)
Manaus, Amazonas, Brazil

Susana Braz-Mota
Laboratory of Ecophysiology and Molecular Evolution
Brazilian National Institute for Research of the Amazon (INPA)
Manaus, Amazonas, Brazil

Adalberto Luis Val
Laboratory of Ecophysiology and Molecular Evolution
Brazilian National Institute for Research of the Amazon (INPA)
Manaus, Amazonas, Brazil

ISBN 978-3-031-66821-0 ISBN 978-3-031-66822-7 (eBook)
https://doi.org/10.1007/978-3-031-66822-7

This Springer imprint is published by the registered company Springer Nature Switzerland AG
The registered company address is: Gewerbestrasse 11, 6330 Cham, Switzerland

Acknowledgements

We would like to extend a special thanks to all the reviewers who contributed their valuable suggestions, which made each chapter even more special and unique.

List of Reviewers
Adalberto Luis Val
Ahana Maitra
Ana Paula Justino Faria
Andrea Viviana Waichman
Bianca Weiss Albuquerque
Carlos Henrique dos Santos
Célio Magalhães
Daiani Kochhann
Derek Felipe Campos
Emanuel Gloor
Ferdyanne Saboia Peixoto
Gabriel Costa Borba
Gelcirene de Albuquerque Costa
Giorgi Dal Pont
Grazyelle Sebrenski da Silva
Guacira de Figueiredo Eufrasio Pauly
Helen Agasild
Inácio Mateus Assane
Jaqueline Custodio da Costa
Jamerson Aguiar-Santos
James Albert
Jochen Schöngart
John Melack
Jonatas da Silva Castro
Larissa Corteletti da Costa
Maíra Rocha
Rafael Mendonça Duarte
Renan Diego Amanajás
Scott Smith
Samara Silva de Souza

Samuel Coelho de Faria
Susana Braz-Mota
Vanessa Kruth Verdade
Tyson James MacCormack
Waldir Heinrichs Caldas

Introduction

The Amazon biome contains a high diversity of organisms spread across its 7.76 million km^2, which accounts for over 10% of the planet's described biodiversity. It also exhibits a high rate of endemism (Zapata-Ríos et al., 2021; Val & Wood, 2022). This megadiversity is the result of a long history of tectonism and climate changes (Val et al., 2022), which have given rise to an unparalleled heterogeneity of landscapes and a complexity of habitats (Guayasamin et al., 2021). In addition, the Amazon is home to the world's largest river basin, which has unique characteristics that promote diversification: for example, periodic fluctuations in river levels lead to the availability of new and transitional habitats due to the flooding of forest areas; and the different types of water (white, black, and clear) have specific physicochemical characteristics and high interconnectivity (Sioli, 1984; Junk et al., 1989; Val & Almeida-Val, 1995).

Amazonian aquatic environments undergo natural daily and seasonal variations in physicochemical characteristics (Junk, 1984; Val & Almeida-Val, 1995). The organisms inhabiting these environments have adapted to cope with these natural events (Val & Almeida-Val, 1995). However, environmental changes caused by human activities threaten ecosystems and affect biodiversity at all levels of biological organization, causing drastic changes in climatic conditions on a shortened time scale (Val & Wood, 2022), leaving little time for organisms to adapt.

Despite the vast amount of information available today, knowledge about the biodiversity of the Amazon region remains underestimated due to the complexity of its ecosystems (Val, 2019). Understanding the Amazon is a significant challenge, not only because of the biome's vastness but also due to the difficulty of accessing its complex environments. Furthermore, human activity has been altering the biome's characteristics, leading to increased temperatures, changes in flood patterns, and altered levels of dissolved oxygen, which have compromised the aquatic biota (Val & Wood, 2022). In this sense, this book presents and discusses, based on currently available information, what the future scenario of biodiversity in the Amazon region will be in terms of natural and anthropogenic pressures. Divided into five parts, the book provides an unprecedented compilation of information on the future of different groups of organisms, including endogenous symbionts, insects, crustaceans, fish, plants, and amphibians.

The first part addresses the 'Impacts of anthropogenic pressures and climate change in the responses of Amazonian aquatic organisms'. The first chapter of this section deals with how climate change is altering the hydrological cycle in the Amazon, affecting floodplain lakes and the habitats of Amazonian fish, and brings the understanding that these impacts are crucial for the development of conservation strategies for aquatic biodiversity, as well as for the maintenance of fish stocks, which are important food resources for the people of the Amazon. After this general overview of the diversity and characteristics of Amazon aquatic environments, the responses of different organisms to climate change are addressed, starting with insect vectors (Chap. 2), focusing on biological aspects of the life cycle, molecular responses, and vector control mechanisms of the main mosquitoes of epidemiological importance in the region. Chapter 3 then presents relevant aspects of dissolved organic carbon (DOC), a major component of Amazonian blackwater rivers, and the effects of climate change on DOC quantity and quality. The impact of road construction on fish in forest streams is addressed in Chap. 4, where the authors examine the effects of climate change and local disturbances on stream fish communities in the Amazon. Chapter 5 looks at the response of Amazonian aquatic insects to environmental change, with approaches to the effects of different land-use practices and extreme events on this group of organisms. The changes that the Amazon's humid areas are undergoing are discussed in Chap. 6, highlighting the main species of aquatic herbaceous plants in Amazonian wetlands and their current challenges. Chapter 7 brings an interesting discussion on the unprecedented transformation of the Amazon basin due to anthropogenic disturbances and increasing hydroclimatic extremes; and finally, this part brilliantly ends with Chap. 8, which discusses the challenges facing aquatic food webs in the Amazon rainforest due to climate change and land-use change, emphasizing the interconnectedness of terrestrial and aquatic components and how perturbations such as habitat drying and changes in fish communities are affecting the ecological balance and productivity of the region, underscoring the need for conservation efforts.

In the scenario of increasing human influence on the Amazon, aquatic biota is not only affected by climate change (Jung et al., 2020; Campos et al., 2021; Val & Wood, 2022), but also by aquatic contamination (Rico et al., 2021, 2022; Val et al., 2022). Therefore, the second part of this book, titled 'Ecotoxicology: Fish and Amphibian Sensitivity', will address the responses of biota to contamination by different classes of pollutants. Chapter 9 will present information on how environmental pollution by different classes of pesticides and microplastics affects anuran amphibians, including the interactive effects of pollution and changes in temperature on this group of organisms. In Chap. 10, the responses and degree of sensitivity of Amazonian fish in relation to other tropical fish will be presented, with emphasis on contamination by pesticides and metals. Chapter 11 will present information on the effects of organophosphate pesticides on the cytogenomic responses and gene expression of Amazonian fish.

The aquatic biota of the Amazon will be significantly affected by the extreme climate events predicted by the Intergovernmental Panel on Climate Change (IPCC) for the next century (Campos et al., 2021; Val & Wood, 2022), with effects already observed even under current climate conditions (Ilha et al., 2018). Aquatic organisms, such as Amazonian fish, have adaptations to deal with changes in physical-chemical factors of water, particularly in dissolved oxygen concentrations (Val & Almeida-Val, 1995). Responses of aquatic biota to environmental variations are heterogeneous, presenting different degrees of sensitivity (Val & Wood, 2022). In the third part of this book, titled 'Phenotypic Plasticity and Adaptive Mechanisms: Change to Survive', we will briefly discuss the mechanisms that aquatic biota can adjust to survive the changes that have been occurring.

Chapter 12 will discuss the composition and function of the symbiotic microbiome of Amazonian fish and how it may be affected by anthropogenic pressures. Additionally, the role of endogenous symbionts will be addressed as a possible adaptive mechanism for fish to deal with future climate events. Chapter 13 discusses the applications of environmental nucleic acid techniques, eRNA and eDNA, in the Amazon region. These techniques are useful tools for increasing knowledge of the region's diversity. Understanding Amazon biodiversity is crucial for seeking mechanisms for sustainable use of its resources. Chapter 14 will review the effects of frequent hypoxic events on respiratory activity in Amazonian fish, particularly air-breathing fish. Therefore, this discussion will explore whether increased reliance on aerial respiration leads to a decrease in gill surface area and the resulting effects on acid–base balance, ionic homeostasis, nitrogen excretion, and predation risks.

Part IV of this book will present information on 'Physiological and Biochemical Response'. Chapter 15 will explore the biochemical, reproductive, morphological, and ecotoxicological responses of the shrimp *Macrobrachium amazonicum*, as well as the effects of climate change on ecological and economic aspects of the species. Chapter 16 presents a comparative approach to the physiological responses to temperature in animals from temperate and tropical zones. The chapter focuses on the activity of the electron transport system (ETS) to understand the responses of fish under different climatic contexts in a warming world. In Chap. 17, this review will cover the thermal limits and acclimatization capacity of Amazonian fish, as well as the application of this data in a prediction model that can provide information about the future of these organisms.

The final part of this book, titled 'Aquaculture as a Tool to Ensure Food Security in the Amazon', provides a brief overview of the risks that climate change poses to this sector, particularly in terms of parasitism and diseases. Chap. 18 will thus discuss the impact of climate change on aquaculture and provide an overview of strategies for the sustainable exploitation of pirarucu (*Arapaima gigas*) with a focus on bioeconomy, aligning society, environment, and economy. Chapter 19 will specifically address the effects of climate change on fish diseases. The discussion of the epidemiological triangle will consider the dynamics of the parasite–host–environment relationship, as well as potential mechanisms for controlling parasites using Amazonian plants.

Laboratory of Ecophysiology and Molecular Evolution, Brazilian National Institute for Research of the Amazon (INPA)
Manaus, Amazonas, Brazil

Samara Silva de Souza
Susana Braz-Mota
Adalberto Luis Val

References

Campos, D. F., Amanajás, R. D., Almeida-Val, V. M. F., & Val, A. L. (2021). Climate vulnerability of South American freshwater fish: Thermal tolerance and acclimation. *Journal of Experimental Zoology Part A, Ecological and Integrative Physiology, 335*, 723–734. https://doi.org/10.1002/jez.2452

Guayasamin, J. M., Ribas, C. C., Carnaval, A. C., et al. (2021). Evolution of Amazonian Biodiversity. In C. Nobre, A. Encalada, E. Anderson, et al. (Eds.), *Amazon Assessment Report 2021*. United Nations Sustainable Development Solutions Network.

Ilha P., Schiesari L, Yanagawa FI, Jankowski K, Navas C (2018) Deforestation and stream warming affect bodysize of Amazonian fishes. PLoS One 13, e0196560. https://doi.org/10.1371/journal.pone.0196560.

Jung, E. H., Brix, K. V., Richards, J. G., et al. (2020). Reduced hypoxia tolerance and survival at elevated temperatures may limit the ability of Amazonian fishes to survive in a warming world. *Science of the Total Environment, 748*, 141349. https://doi.org/10.1016/j.scitotenv.2020.141349

Junk WJ, Bayley PB, Sparks RE (1989) The flood-pulse concept in river-floodplain systems. In: Dodge D (ed.) Proceedings of the International Large River Symposium. Canadian Special Publications of Fisheries and Aquatic Science, p 110–127.

Rico, A., Oliveira, R., Nunes, G. S. S., Rizzi, C., et al. (2021). Pharmaceuticals and other urban contaminants threaten Amazonian freshwater ecosystems. *Environment International, 155*, 106702. https://doi.org/10.1016/j.envint.2021.106702

Rico, A., Oliveira, R., Nunes, G. S. S., Rizzi, C., et al. (2022). Ecological risk assessment of pesticides in urban streams of the Brazilian Amazon. *Chemosphere, 291*, 132821. https://doi.org/10.1016/j.chemosphere.2021.132821

Sioli H (1984) The Amazon and its main affluents: Hydrology, morphology of the rivers course, and river types. Junk Publishers, Netherlands. Val AL. (2019). Fishes of the Amazon: diversity and beyond. *Anais da Academia Brasileira de Ciências, 91*, e20190260. https://doi.org/10.1590/0001-3765201920190260

Val AL, Almeida-Val VMF (1995) Fishes of the Amazon and their environments. Springer Verlag, Heidelberg. Val AL, Duarte RM, Campos D, Almeida-Val VMF. (2022). Environmental stressors in Amazonian riverine systems. In N. A. Fangue, S. J. Cooke, A. P. Farrel, et al. (Eds.), *Conservation Physiology for the Anthropocene - Issues and Applications* (pp. 209–2017). Academic Press.

Val, A. L., & Wood, C. M. (2022). Global change and physiological challenges for fish of the Amazon today and in the near future. *The Journal of Experimental Biology, 225*(10), jeb216440. https://doi.org/10.1242/jeb.216440

Zapata-Ríos, G., Andreazzi, C. S., Carnaval, A. C., et al. (2021). Biological Diversity and Ecological Networks in the Amazon. In C. Nobre, A. Encalada, E. Anderson, et al. (Eds.), *Amazon Assessment Report 2021*. United Nations Sustainable Development Solutions Network.

Contents

Part I

Impacts of Anthropogenic Pressure and Climate Change in the Responses of Amazonian Aquatic Organisms

Anthropogenic Climate Change Is a Present and Future Stressor for Fish Communities in the Amazon Basin

Carlos E. C. Freitas, Lawrence E. Hurd, Bruna S. Andrade, Marcos A. Mereles, and Flávia K. Siqueira-Souza

Abstract

The main reported effect of global climate change on marine fish assemblages has been the movement of species to higher latitudes as ocean waters have warmed. In contrast, the Amazon basin is the largest freshwater system in the world and, here, climate change has created extremes in the annual hydrological cycle: floodplain lakes, which are crucial habitats for Amazonian fish, have now become more isolated from the river in periods of low water; and, in the periods of high water, lakes extend further into the adjacent forest. These events strongly affect the life cycle and reproduction of fish, especially those that require seasonal connectivity between lakes and river channels in order to migrate. Our earlier results indicated that responses are species-specific, with both winners and losers being based largely on trophic level and migratory behavior. This chapter is about predicting changes in fish assemblages based on characteristics of life history. Understanding these impacts caused by climate change is key to preserving aquatic biodiversity and conserving an important food resource

C. E. C. Freitas (✉) · F. K. Siqueira-Souza
Department of Fisheries Sciences, Federal University of Amazonas, Manaus, Brazil
e-mail: cefreitas@ufam.edu.br

L. E. Hurd
Department of Biology, Washington and Lee University, Lexington, VA, USA

B. S. Andrade
Graduate Program in Biology of Freshwater and Inland Fishing (PPGBADPI), Brazilian National Institute for Amazonian Research (INPA), Manaus, Brazil

M. A. Meireles
Graduate Program in Animal Science and Fisheries Resources, Federal University of Amazonas, Manaus, Brazil

S. S. de Souza et al. (eds.), *The Future of Amazonian Aquatic Biota*,
https://doi.org/10.1007/978-3-031-66822-7_1

for humans in the Amazon. We suggest the need for a more comprehensive, cooperative research program to investigate the abundance and distribution of fish species in floodplain lakes of different regions in the Amazon.

1 Preamble

The main issue regarding our general understanding of biodiversity, and its present endangerment due to anthropogenic disturbances, is our limited knowledge of adaptive tolerance that species have evolved to withstand changing environments. Changes in the global climate are essentially moving the goalposts that constitute the tolerance limits of aquatic species, including endangered ones, i.e., altering their environments beyond their evolved niches. We use niche in an ecological context: the conditions necessary to support the vital activities and fitness of a specific organism (Alley, 1982). In an operational sense, the niche consists of all biotic and abiotic environmental factors necessary for survival and continued existence in a habitat; therefore, we have to know what those tolerance limits are before we can understand how species will respond to alterations in temperature, water chemistry, and resource availability.

We know more about the niche requirements of temperate zone species than about tropical species, primarily because species diversity is so much higher near the equator than at latitudes above and below the tropics. This is known as the latitudinal gradient in species diversity, which is characteristic of most (but not all) animal taxa. There are lots of hypotheses about what causes this (e.g., see Table 15.4 in Lomolino et al., 2006), but it has become axiomatic these days that the niches of tropical species are, in general, more narrowly constructed than those of temperate zone species. This is one reason why tropical organisms are more easily susceptible to extirpation when their environment changes. Thus, there are far more tropical fish species than elsewhere, they have lower tolerance to environmental change, and we know less about them.

Knowing the precise limits of tolerance for physical niche requirements would require extensive testing in the laboratory, and for how many species do we have this information? Furthermore, beyond physical factors such as water chemistry, temperature, depth, and turbidity that often are measured in the field during sampling studies, what are the characteristics of life history such as feeding at different stages of the life cycle, migratory behavior, best food sources during reproductive season, or competitive and predatory interactions with other fish? We know a lot more about fish that we exploit for food than the tremendous variety that we do not normally eat, but how many fish do we know well enough to make predictions about their probability of surviving the environmental changes underway in places like the Amazon basin?

Nonetheless, we must try to envision, based on what we already know and can continue to discover, how the fish portion of the aquatic biota within the Amazon ecosystem will respond to climate change. All this, while also keeping in mind the effects of other anthropogenic stressors such as dam building, introduction of invasive species, and overharvesting of fish stocks.

2 Introduction

Fish ecologists working in the Amazon basin have a double challenge. Bearing in mind the lack of knowledge of species-specific answers of Amazonian fish species to climate change, they need to predict what will happen on the fish assemblage level. Also, they are faced with a spatially disproportionated knowledge of the fish diversity itself (Carvalho et al., 2023). As a consequence, our approach to predicting how anthropogenic climate change would affect Amazonian fish assemblages should be an exercise of layers overlap.

In this chapter, we try to summarize the environmental characteristics, discussing the main threats that can change it. Then we present our knowledge on fish diversity, with emphasis on functional diversity, which could be assumed as the component of biodiversity that represent the range of things that organisms do in communities or ecosystems (Petchey & Gaston, 2006), and have been recognized as a good tool to assess the impact of anthropogenic stressors on natural communities (Cantera et al., 2023).

3 Part I. The Environment: The Amazon Basin Now and the Predicted Changes

3.1 Main Basin Characteristics: Geology and Water Types

The Amazon basin is the largest hydrological basin on Earth, with an area of 5.83×10^6 km^2 (Hess et al., 2015). This huge ecosystem includes everything from a complex network of small streams to the biggest rivers on Earth (Melack & Coe, 2020). The dynamics of large Amazonian rivers are controlled by an annual mono-modal flood pulse, with an amplitude varying from 2 to 18 m, depending on their location in.the basin (Richey et al., 1986), and in the low water season large floodplains with numerous lakes form adjacent to the main channels of the rivers. The floodplains are covered by different vegetation types, such as floating herbaceous plants in the lakes, and forests that are flooded in the high water season (Melack & Hess, 2010), all of which provide spatially and temporally heterogeneous environments that help explain high fish species diversity in the basin (Boschman et al., 2023; Freitas et al., 2018).

According to the Flood Pulse Concept (FPC), the hydrologic cycle is the environmental driver that controls the flow of nutrients, primary and secondary production, presence and distribution of organisms, and life history traits (Junk et al., 1989). Largely due to the contribution of its aquatic environments, the ecosystem of the Amazon is important to the carbon cycle and climate on a global scale (Phillips et al., 2009). The flood pulse influences fish migration and spatial distribution (Freitas et al., 2010; Röpke et al., 2015) and also is responsible for the connectivity among lake habitats of the floodplains, including metapopulation dynamics within regional fish populations of this area (Hurd et al., 2016).

A gross classification of the rivers based on the color of their waters and the geology of the drainage area was initially proposed by Siol (1984) and has since been useful for ecological and conservation purposes.

(i) White-water rivers (e.g., Solimões and Madeira rivers) have their headwaters in the Andean and pre-Andean regions, carry large volumes of organic and inorganic materials, and have high dissolved and suspended loads, causing them to have a pH close to neutral. The floodplains adjacent to these rivers, known as *varzeas*, are the most productive areas in the basin (Junk et al., 2011; Fig. 1).
(ii) Blackwater rivers (e.g., Negro and Uatumã rivers) run through the ancient lands of the Guianas Shield. In areas with low sediment content and that are rich in humic acids, resulting in acidic water that is poor in nutrients (Junk et al., 2011; Siol, 1984; Fig. 1).

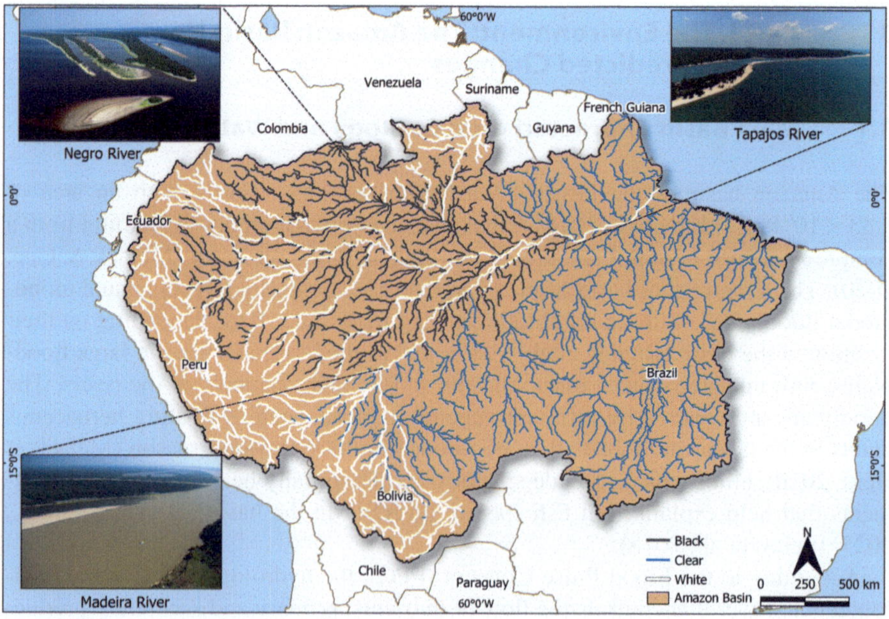

Fig. 1 Map of the Amazon basin showing the types of river water and their geographical distribution

(iii) Clearwater rivers (e.g., Tapajós and Xingu rivers) drain ancient and eroded areas, mainly in the Brazilian Shield. They have a low suspension load, high transparency, and a slightly acid to neutral pH (Siol, 1984; Fig. 1).

3.2 Anthropogenic Threats to the Amazon Basin

Human colonization of the Amazon basin was not historically uniform. Its eastern portion and the areas around the estuary of the Amazon river were firstly colonized relatively rapidly by the Portuguese (Castro & Campos, 2015). In contrast, the western portion of the Brazilian Amazon was colonized slowly by the Portuguese, and there are no cities with more than 100,000 inhabitants upstream beyond Manaus, the location of the confluence between the Negro and Solimões rivers. This process is reflected in the present land use and status of forest conservation in the Amazonian ecosystem, with more anthropic pressures acting from east to west.

In the last decades of the twentieth century, the Amazon Forest became the last wild frontier in South America due to the expansion of cattle ranching and soybean plantations from south to north (Prates & Bacha, 2011), which has resulted in large areas of deforestation, especially in the south of the Amazon basin. A large area known as the "Arc of Deforestation" comprises the main region under pressure of deforestation and environmental degradation (Pfaff et al., 2015; Silva et al., 2021). The effects of the reduction of forest cover on the aquatic ecosystem and its biota have been well documented (Arantes et al., 2017).

River impoundment is another important threat to the Amazon basin. The dams that have already been built in the basin have resulted in many negative impacts on the ecosystem and its biota (Castello et al., 2013; Santos et al., 2018). In the western pre-Andean and Andean areas, anthropogenic pressure on aquatic biota will be magnified by the building of several planned hydroelectric dams (Finer & Jenkins, 2012), with the potential to change the nutrient flow that is the key driver of high productivity in the floodplains of the central Amazon (Forsberg et al., 2017). If one considers the importance of the floodplains to the life cycle of many of the fish of the Amazon, especially the many migratory species, changes in this environment could be catastrophic for the maintenance of regional fish diversity. Unfortunately, important stakeholders such as fishers have not factored dam-building into the decision-making process about dam implementation in the Amazon basin (Doria et al., 2018).

Sousa et al. (2018, 2021) demonstrated that even the large number of small dams used for aquaculture could be disruptive to the environment, with negative effects on fish diversity. In addition, the synergic effect of thousands of small dams, built for either aquaculture and hydropower, negatively impacts streams that are essential to the nutrient flow and movements of fish in the Amazon basin (Freitas et al., 2022).

It is important to understand that the terrestrial and aquatic environments of the Amazon are completely integrated, and therefore the many described threats have impacts on the system as a whole (Castello et al., 2013). Moreover, these local and regional threats to fish diversity could interact synergistically with global climate changes, magnifying their consequences. For example, the reduction in precipitation predicted for the eastern portion of the Amazon basin (Melack & Coe, 2012) could intensify the disruptive effects of the Belo Monte dam on water availability immediately below the dam, where a reduction has already been provoked by the dam itself (Souza-Cruz-Buenaga et al., 2019).

3.3 Predicted Changes Due to Climate Change and Synergistic Effects

Many local and regional threats in the Amazon basin can be viewed as being part of the scenario of global climate change. Following the Technical Summary of the IPCC published in 2022, the most confidently predicted climate changes are a decrease in the number of cold days and nights and an increase in the number of warm days and nights, on a global scale. More relevant to our thesis, two effects have been noticeable with "very high confidence": (i) alterations in marine, terrestrial, and freshwater ecosystems of the world due to climate change and (ii) biological adaptations, including alterations in physiology, growth rates, abundance, geographic distribution, and seasonal timing, are often not fast enough to deal with the rate of climate change. The predicted alterations in the freshwater ecosystems of Central and South America are made with high or very high confidence, as are impacts on species distribution and fisheries.

Hydrological models of the Amazon basin predict an increase in the volume of river discharge for the large rivers that drain the Andean and pre-Andean areas, which could lead to higher inundation events on the floodplains in Peru and along the Solimões river (Fleischmann et al., 2022). An opposing pattern is predicted to occur in the middle and lower Amazon river and its tributaries (Langerwich et al., 2013; Sorribas et al., 2016). Langerwich et al. (2013) predicted that most changes will happen in the northwestern portion of the Amazon basin, mainly an increase in the flooded area, lengthening of inundation season, shifts in the month of the peak low, and high-water combined with alterations in the frequency and duration of extreme events. These predicted patterns have been corroborated by multi-decadal hydrological data that show an increase in minimum annual discharge associated with droughts in the southern and central-southern regions of the Amazon basin (Correa et al., 2017). These contrasting patterns in hydrological conditions along the axis of the Amazon basin could promote perceptible changes in the ecosystem and elicit distinct responses in fish, both at the level of individual species and community-level interactions among species and trophic levels.

Extreme events of drought and flood are the most apparent of these alterations in the climate-hydrological patterns of the Amazon basin. They have become more frequent in recent decades (Marengo et al., 2013) and their effects on fish assemblages are quite perceptible (Freitas et al., 2013; Röpke et al., 2017, 2022). In years of extreme droughts, the environmental conditions could be severely degraded by low levels of dissolved oxygen, or even complete hypoxia. As we will discuss ahead, extreme floods also can impact the physiological responses of aquatic biota.

4 Part II. The Fish

4.1 Taxonomic and Functional Fish Diversity at the Amazon Basin

The Amazon basin hosts the richest freshwater fish fauna on Earth, with more than 2700 species already described (Dagosta & Pinna, 2019), of which at least 1089 species are endemic (Reis et al., 2016). It is difficult to define which are the most diverse areas in this huge ecosystem since the number of described species seems to be dependent on the intensity of studies, and that varies across regions. However, all the studies on Amazonian fish assemblages show a common pattern, with the dominance of Characiformes, Siluriformes, and Cichliformes, in proportions that are dependent on the type of environment: river, lake, or stream (Lowe-McConnell, 1999).

The evolutionary processes driving fish speciation in the Amazon basin remain a subject for discussion, however, some prominent patterns have been proposed. Oberdorff et al. (2019) suggested that the center of Amazonian fish diversity is located westward with a negative gradient from upriver to downriver, which correlates with the known geological history of the Amazon basin. Siqueira-Souza et al. (2021) found striking similarities in pelagic fish diversity patterns between black- and whitewater river basins, such as (1) species richness was virtually equal (165 in whitewater and 168 in blackwater rivers); (2) both species richness and number of migratory species per lake increased toward the confluence of the rivers; and (3) the proportion of unique species was similar in whitewater lakes and blackwater lakes (41 and 43%, respectively), boosting total regional richness to 237 species. Nonetheless, species composition in whitewater lakes was more homogeneous, exhibiting lower beta diversity than in blackwater lakes.

Notwithstanding, patterns and drivers of Amazonian fish diversity are likely to be scale-dependent (Freitas et al., 2014). Some studies have tried to identify drivers of Amazonian fish diversity on different scales, such as the physical–chemical variables of the water measured in the sampling site (Morales & Deus, 2021), or landscape components (e.g., area of flooded forest or area of macrophyte meadows), which could be represented in buffers of different sizes around the sampling sites (Arantes et al., 2017; Freitas et al., 2018; Lobón-Cerviá et al., 2015). Moreover, it is possible that small differences in the environmental characteristics determine the fish species distribution in the same environment. Marshall et al. (2011) observed that small differences in depth and temperature determine the habitat dominance of *Paracheirodon axelroldi* and *P. simulans* in an interfluvial palm swamp of the Negro river basin.

4.2 Predicted Responses to Climate Change by Taxonomic and Functional Fish Groups

Impacts of climate change on fish distribution, with fishes moving toward higher latitudes when ocean waters are warmer (Heath et al., 2012), and mismatches between phytoplankton bloom and fish spawning (Asch et al., 2019), already have been observed in marine environments. The west-to-east flow of the Amazon river from the Andes highlands, almost parallel with the equator, makes it probable that the water temperature itself would be quite homogeneous over the mainstream of the Amazon river. However, the occurrence of extreme events, mainly droughts, could be catastrophic to the fish assemblages of the Amazonian floodplain lakes due to the combined effect of high temperature and low oxygen (or complete depletion of it) as the water evaporates and fish become crowded in shallow waters (Freitas et al., 2013; Fig. 2a, b).

The availability of many different habitats is one of the factors that is responsible for the high taxonomic diversity of fish in the Amazon basin (Albert et al., 2011). So

Fig. 2 Dead fish along the shore of the Manaquiri lake during the extreme drought of 2005, with a view of the dead fishes on the shore on the left (**a**), and a close-up showing the variety of fish species, on the right (**b**) (*Source*: Silva, C.A.)

far, few publications have addressed the impact of climate change on the fish assemblages of the Amazon. Most publications to date are reviews, perspective papers, or book chapters (Andrade et al., 2019; Val, 2011; Val & Almeida-Val, 2008; Zuanon, 2008). These papers have made predictions about the types of climate change impacts on fish assemblages, although they are primarily based on reported effects already observed in other parts of the planet.

Only a small proportion of these publications are data-based, with time intervals including exceptional years of droughts or floods, which allow us to evaluate the effects of extreme events associated with climate change. Freitas et al. (2013) analyzed a four-year database (from 2004 to 2007) of fish assemblages that were sampled quarterly in floodplain lakes of the Solimões river. This study showed that the extreme drought that occurred in 2005 provoked a marked reduction in beta diversity among lakes, but with a noticeable response related to functional traits, such as its trophic ecology. Two other studies demonstrated that the responses of fish to extreme droughts appear mainly to be specific to certain functional groups of fish. Correia et al. (2015) evaluated the intra- and inter-annual changes in the condition factor of three Curimatidae species in Amazonian floodplain lakes. These fish are detritivorous and could be classified as small seasonal in sensum of Winemiller and Rose (1992) by Röpke et al. (2017). They showed a similar pattern with better body condition during the period of high water and exhibited similar reductions in body condition during the extreme low water period of 2005. On the other hand, Tribuzy-Neto et al. (2018) employed a similar approach to evaluate the effect of the 2005 extreme flood on four species of Characidae. They observed distinct responses in body condition by omnivorous and carnivorous species. The latter were favored by the reduction in the aquatic environment, probably due to the improved ability to find prey under increasingly crowded conditions.

Röpke et al. (2017) analyzed a longer database (from October 1999 to October 2014) of fish assemblages that were sampled monthly in a floodplain lake at the confluence of the Negro and Solimões rivers, and developed another approach that analyzed the effects of extreme climatic events on the taxonomic and functional structure of the fish assemblages. They observed that the intense drought of 2005 acted as a trigger for changes in the relative abundance of fish species. In addition, they proposed that the anomalous hydrological conditions that happened in the following years were obstacles to the recovery of the assemblages.

Considering that the effect of climate change or other habitat alterations on fish assemblages may well be both species-specific and associated with functional traits of each species, we performed a search of the literature on Amazonian fish diversity in order to identify the proportion of species by three categories of functional groups: trophic guild, migratory behavior, and life history strategy following Winemiller and Rose (1992; see Box 1), who further classified the fishes as equilibrium, periodic, and opportunistic. The fish assemblages of these functional groups were discriminated in different types of water (white, black, and clear) as per Sioli (1984).

Box 1

Winemiller and Rose (1992) proposed a model encompassing three primary life history strategies, forming endpoints on a triangle, to describe general patterns observed in fishes. This classification has been useful for management and conservation purposes (Winemiller, 2005) and we will use them to identify the impacts of climate change:

a) Equilibrium: Moderate to long generation time, low reproductive effort, variable body size, low fecundity and high investment per offspring, and reproduction events evenly distributed throughout the year, such as *Cichla* spp., *Hoplias malabaricus*, *Osteoglossum bicirhosum*, *Arapaima gigas*, *Serrasalmus rhombeus*, and *Potamotrygon motoro*.
b) Seasonal or periodic: Long generation time, moderate reproductive effort, large body size, high fecundity, low investment per offspring, small oocytes, large fluctuations in local populations, and the predominance of adults in the dry season and juveniles in the rainy season, such as some small Curimatidae (*Curimata inornata*, *Potamorhina altamazonica*, *P. latior*, *Psectrogaster amazonica*, and *P. rutiloides*) and other very abundant fish in Amazonian rivers such as *Mesonauta festivus*, *Triportheus* spp., *Prochilodus nigricans*, *Rhytiodus microlepis*, and some Serrasalmids (*Colossoma macropomum*, *Mylossoma aureum*, and *Serrasalmus maculatus*).
c) Opportunistic: short generation time, high reproductive effort, repeated reproductive periods, small oocytes, small body size, low fecundity, and low investment per offspring such as *Anodus elongatus*, *Anchoviella jamesi*, *Hemigrammus levis*, *Moenkhausia lepidura*, *Nannostomus eques*, *Apistogramma agassizii*, *Semaprochilodus* spp., *Brycon amazonicus*, and *Acestrorhynchus microlepis*.

Omnivores comprise the largest trophic guild of Amazonian fishes, especially in whitewater systems, followed in abundance by carnivores, detritivores, and invertivores. Algivores and spongivores were found only in clearwater rivers, while mucophages, ovivores, and hematophages were representative only in whitewater rivers (Fig. 3).

By a small margin, most fish in the Amazon basin are in the equilibrium category of life histories, with about 400 classified species, and about 350 species having periodic strategies. Whitewater rivers have a greater number of species with opportunistic and equilibrium strategies, while clear and blackwater rivers have a greater number of species with periodic and equilibrium strategies (Fig. 4).

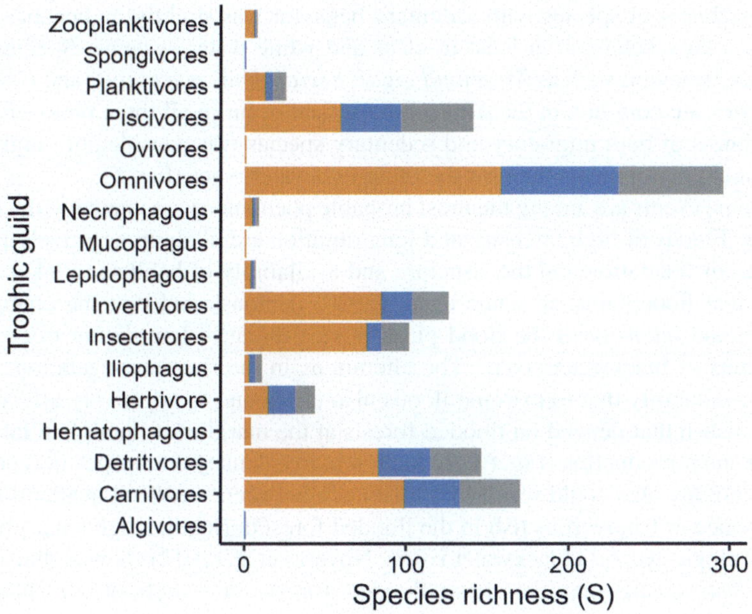

Fig. 3 Species richness (S) by trophic guilds sampled in different types of water in the Amazon basin. Orange = whitewater; blue = clearwater; gray = blackwater

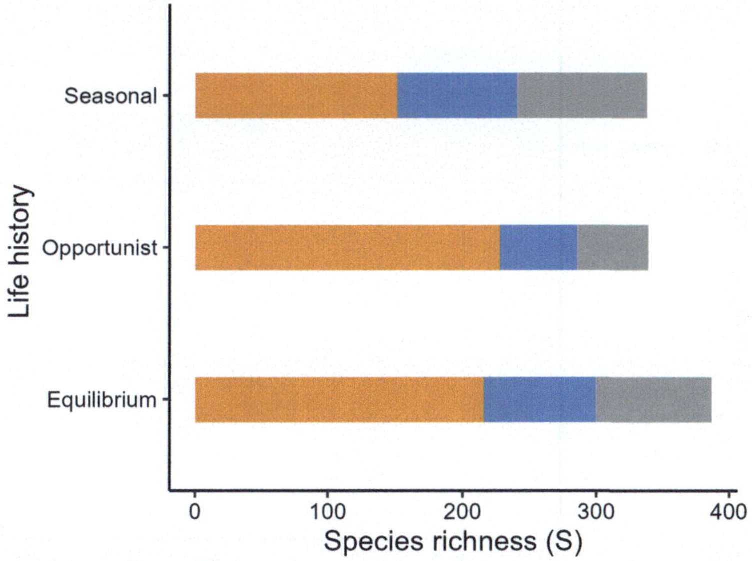

Fig. 4 Fish richness by life history pattern (sensu Winemiller & Rose, 1992), sampled in different types of water in the Amazon basin, where: orange = equilibrium, blue = opportunistic, and gray = periodic

The richness of species with sedentary behavior was slightly higher than those with migratory behavior, at least in clear and white water systems. Species with sedentary behavior, such as *Arapaima gigas*, *Serrasalmus maculatus,* and *Chalceus erythrurus,* are common in the Amazon basin and occur in all three types of water. The richness of both migratory and sedentary species was considerably higher for whitewater environments than for the other two water types (Fig. 5).

Extreme events are among the most probable phenomena associated with climate changes. Floods of high intensity and long duration are likely to negatively impact the riparian vegetation and the structure and availability of habitats in lakes of the Amazonian floodplains. Resende et al. (2020) demonstrated that the changes in duration and intensity of the flood pulse pose a threat to long-living trees in the floodplains of blackwater rivers. The alterations in the riparian vegetation and in habitat availability due to extreme floods also are bound to negatively affect some groups of fish that depend on flooded forests at the margins of floodplain lakes for feeding and reproduction (Fig. 6). Reduction in tree density or changes in reproductive phenology also could result in a mismatch between seed/fruit availability and the presence of frugivorous fish in the flooded forest during the highwater phase of the hydrologic cycle. In blackwater rivers, Noveras et al. (2012) showed that a number of fish species explore these flooded forests in search of allochthonous food items.

Extreme drought can exert catastrophic impacts on fish assemblages (Fig. 6). As happened in 2005 and 2019, thousands of fish can die due to total depletion of

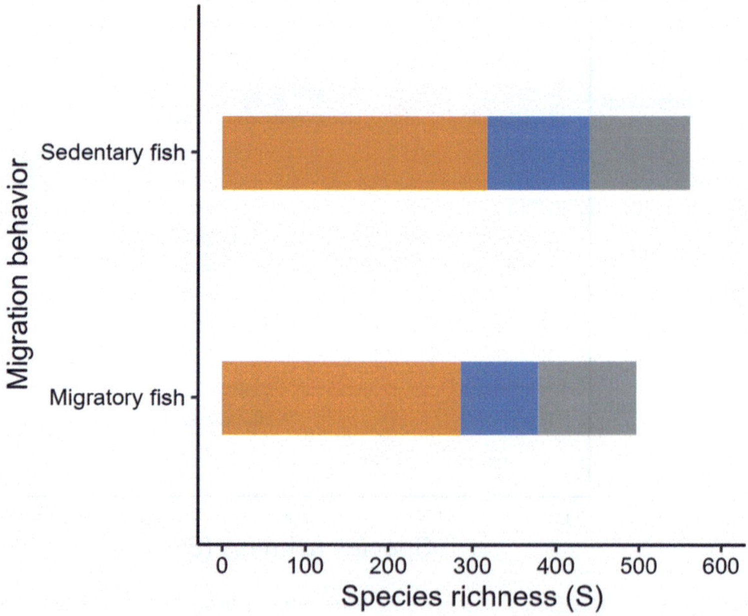

Fig. 5 Fish richness by migration behavior type, sampled in different types of water in the Amazon basin, where orange = whitewater, blue = clearwater, and gray = blackwater

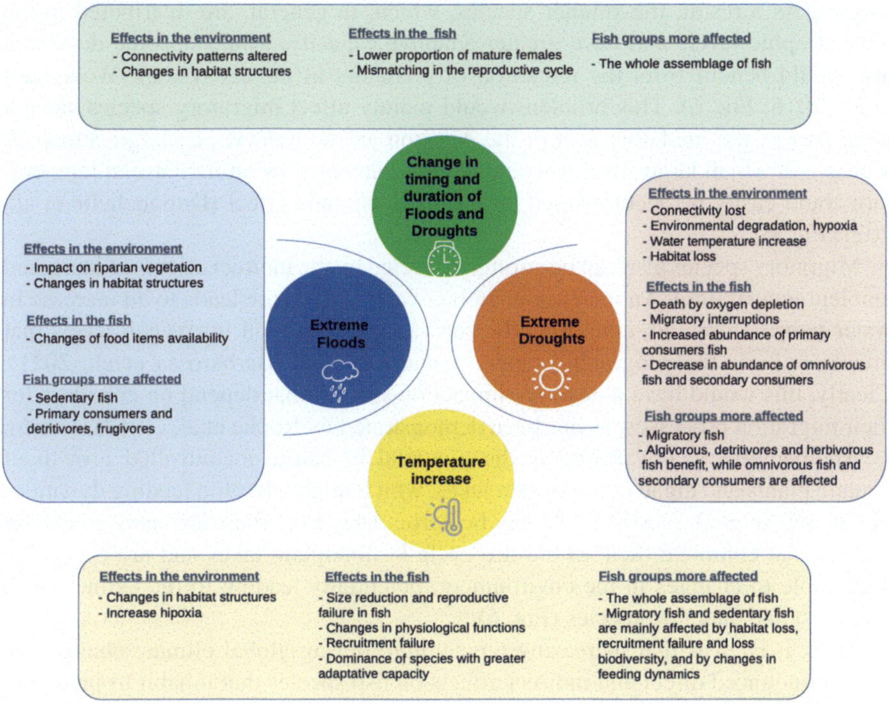

Fig. 6 Predicted effects on the environment and fishes related to the expected environmental changes

dissolved oxygen and/or an increased temperature as lakes become shallower and reduced in area. During the drought of 2019, the area of a huge floodplain lake system called as Lago do Rei, placed at the right margin of the Amazonas river, was substantially reduced and the banks of the small, shallow, residual aquatic habitats that remained became filled with dead fish (Pinheiro et al., 2022).

The timing and duration of the low-water season also can have a dramatic effect on some fish that have evolved reproductive strategies triggered by the timing of the hydrologic cycle (Duponchelle et al., 2021). For instance, Arantes et al. (2010) noted that species such as *Arapaima* spp., in which reproduction is biannual, build their nest in response to an optimal water depth (Fig. 6).

According to the Intergovernmental Panel on Climate Change (IPCC, 2007), the temperature increase in the Amazon region could be as much as 3 °C by the end of the present century. With this increase, it is expected that there will be negative consequences for the entire biota of the Amazon, both at a physiological and at ecological level since organisms are dependent on an optimal temperature range for their biological processes. Woodward et al. (2016) pointed out that at the physiological level, the increase in temperature could be a disadvantage for larger fish species due to their metabolic allometry. At the ecological level, this means that these species would be more susceptible to local extinction when compared to smaller species. As a result, the smaller species, which, in general, are distributed in the lower trophic levels and have greater adaptive capacity, would become dominant and would benefit from the reduction of predators in the ecosystem (Woodward et al., 2016; Fig. 6). This problem would mainly affect migratory species since a large part of the predatory fish in the Amazon are migratory, i.e., large catfish. A scenario in which large species would be disadvantaged by an increase in temperature could easily be accompanied by a trophic cascade effect (Duponchelle et al., 2021; Fig. 6).

Migratory species also can be strongly affected by the indirect effects of increased ambient temperature. An increase in ambient air temperature leads to an increase in water temperature and, consequently, can cause a reduction in oxygen levels that greatly reduces the geographic range of many species (Barbarossa et al., 2021). Clearly, this would have a negative impact on species that depend on corridors for their migration processes, as has been demonstrated by Röpke et al. (2017) (Fig. 6).

Another effect of warming is the potential to cause uncontrolled growth of aquatic plants and algae in floodplain lakes, which might alter the feeding dynamics of fish (Ficke et al., 2007). This may be particularly important for many sedentary species that complete their whole life cycle in floodplain lakes and are especially vulnerable to changes in the environment, potentially leading to the extinction of specialist and endemic species (Fig. 6).

Thus, it is clear that increasing temperature during global climate change can have pronounced direct and indirect effects on fish species that inhabit tropical lentic environments such as floodplain lakes, and that many of these effects are at present hard to predict with precision, but we already have some indications of what to expect. Extreme, catastrophic events, and their occurrence with increasing frequency can eliminate species in the long term, which can, in turn, reduce both alpha and beta diversities of fish assemblages (Freitas et al., 2013), and alter the physiology of aquatic organisms (Val & Wood, 2022; Fig. 6). Some studies have already shown some of the consequences of temperature increase, such as impacts on the growth of individuals (Amanjás & Val, 2022; Oliveira & Val, 2017), changes in physiological functions (Val & Wood, 2022), failures in the reproductive development of species (Castro et al., 2020), and biodiversity loss (Barbarossa et al., 2021; Fig. 6).

A recent study by Campos et al. (2023) has shown that the increase in temperature in the Amazon already has caused an increase in the uncertainty of natural mortality of species, owing to species-specific types of environment and life history strategies. Both freshwater and marine fish species exhibit similar patterns of increasing uncertainty of natural mortality with increasing temperature. The species that have an equilibrium strategy presented lower uncertainty in the estimates of natural mortality, while those that have a periodic strategy present mortality estimates with intermediate uncertainty values, and opportunistic species present the highest uncertainties in the estimated values of natural mortality (Campos et al., 2023).

4.3 Impacts on Amazonian Fisheries

Environmental changes that influence fish populations and their assemblages clearly will have a strong impact on important fisheries. The Amazon basin hosts four types of fisheries: small-scale commercial, subsistence, recreational, and ornamental (Freitas & Rivas, 2006). The first two are widespread in the basin and have great importance as the main source of animal protein and generation of income for riverine communities.

The relationship between landscape variables of the Amazon river and its adjacent floodplains, and the fishing yields of the small-scale fisheries has been demonstrated (Lopes et al., 2023). The negative impact of loss of forest cover on the floodplain and on fishery yields also has been established (Barros et al., 2020; Castello et al., 2018; Pereira et al., 2022), and alteration of the landscape due to climate change could exert interactive and disruptive effects due to local deforestation. The relationship between hydrological variability and Amazonian small-scale multispecies fisheries has also been demonstrated (Isaac et al., 2016) and was demonstrated for one unique species *Prochilodus nigricans* (Barros et al., 2021). As the hydrology of Amazonian rivers is closely related to the global climate phenomena of El Niño Southern Oscillation—ENSO and South Atlantic weather (Marengo & Espinoza, 2015)—it is highly likely that changes in hydrological patterns caused by these events will exert strong impacts on Amazonian fisheries.

Physiological responses of fishes to climate change would be reflected in population parameters which are essential to the assessment of fish stocks. Campos et al. (2023) demonstrated that uncertainties around estimates of natural mortality would increase with scenarios of temperature increase predicted by IPCC. Moreover, these authors demonstrated that freshwater fish would be even more sensitive than marine species to increasing temperature.

5 Part III. Final Considerations

Surely the most common phrase, with variation of wording, that appears at or near the end of a scientific report is "needs further research." So, it is here as well. There is much to be learned if we are to (1) understand and (2) preserve aquatic biodiversity in the Amazon river basin. In the face of the rapidly increasing impacts of global climate change, there is an unprecedented urgency to this endeavor.

A good subject upon which to focus that gives clues to both of these issues is research into what, if anything, regulates populations; and then by extension, what—again, if anything—regulates species diversity. This is an investigation into the evolutionary responses of native species to the environment in their natural habitat, and seeks to answer to the old and venerable question of why there are so many species in a given ecosystem (Hutchinson, 1959). This effort is already hampered by the interference of *Homo sapiens* with natural environmental variables under which species have evolved; however, it is not yet hopeless.

The consensus in ecology is that true regulation involves a relationship between population density (numbers of individuals/area) and population growth, such that an equilibrium in density is achieved by negative feedback of density on growth. Density-dependent regulation can be achieved a number of different ways, all of them involving biotic interactions, either among individuals within a species population (intraspecific), or among populations of different species (interspecific). In a classic book, Pielou (1974) divides these up into eight categories, which we here modify slightly to those that are apt to be most relevant to fish:

1. Intraspecific competition
2. Contagious disease
3. Interspecific competition
4. Predation
5. Parasitism

The first two of these belong under the heading, "self-regulation," wherein the negative effect of population density slows or halts growth by either (or both) reducing rates of reproductive output or increasing mortality. As a critical resource becomes scarcer, or rates of disease transmission are high, population growth slows and may reverse if overtakes specific limits. Conversely, as population density declines below the level of resource replacement, or to the point where the birth rate outpaces pathogenic mortality, positive growth can be restored. The ideal ratio between population density and growth in either case will be the density at which the birth rate is equal to the death rate, i.e., an equilibrium value known as carrying capacity. The difference between regulation by a resource and a disease is that we do not often expect the effects of a disease to last, while a resource is an ongoing requirement for a population.

The problem here is that we do not know the carrying capacity of most fish species, including those that are exploited for food by humans. We certainly can see when fish stocks become depleted by overfishing, but rarely do we really know what rate of fishing can be sustained so that reproduction does not fall below mortality. This is chiefly because environments can change unpredictably over time, so that, for instance, a critical food resource for a fish population can vary in abundance from year to year, causing carrying capacity to fluctuate.

Categories 3–5 are generally interspecific, and populations regulated by these interactions are by definition not self-regulated as described above, i.e., they are below the carrying capacity in the sense that, in the absence of their competitor, predator, or parasite, they could sustain a larger population density. The exception to this is in the category predation, which can include intraspecific predation, i.e., cannibalism. Cannibalism is common to many predators (Polis et al., 1989), and it would qualify as a self-regulating mechanism along with categories 1 and 2. Category 5 (parasitism) is the one of which we know least about, even though a wide variety of invertebrate parasites are found in Amazonian fish (Thatcher, 2006), and parasites have been shown to exert strong effects on the population dynamics of a wide variety of animals, including fish (Poulin, 2000).

The many anthropogenic stressors of Amazonian fish, including overfishing, dam-building for hydroelectric power and aquaculture, deforestation, droughts brought on by deforestation and climate warming, and invasive species, will adversely affect fish population dynamics and species diversity, often synergistically. Invasive species, for instance, can outcompete and/or or prey on native fish, and/or bring novel parasites and disease organisms with them (categories 2–5).

What all this boils down to is the need for a concerted effort by many people in many places to gather data that reveal reliable patterns in abundance and distribution. The Amazon basin is a complex meta-ecosystem with multiple regions of different water types, species assemblages, and combinations of stressors. The design of strategies to conserve aquatic biodiversity and sustainable fishery stocks will depend on our ability to recognize the differences and similarities among these regions. Much of the information gathered to date about fish diversity in floodplain lakes consists of data gathered from different groups of researchers, at different times, in one or few lakes, asking different questions of the data, and sometimes using different sampling methods. A more comprehensive, cooperative venture in which multiple teams of researchers from different institutions agree on how, when, where, and with what techniques fish assemblages are to be sampled might go a long way toward establishing patterns of species abundance and distribution that could be applied to questions of population regulation and community structure. Obviously, this will require sufficient funding as well as the will of governments to underwrite such an ambitious undertaking, but the time grows shorter the longer we delay.

References

Albert, J. S., Carvalho, T. P., Petry, P., Titular, M. A., Maxime, E. L., Espino, J., Corahua, I., Quispe, R., Rengifo, B., Ortega, H., & Reis, R. E. (2011). Aquatic biodiversity in the Amazon: Habitat specialization and geographic isolation promote species richness. *Animals, 1*, 205–241. https://doi.org/10.3390/ani1020205

Alley, T. R. (1982). Competition theory, evolution, and the concept of an ecological niche. *Acta Biotheoretica, 31*, 165–179. https://doi.org/10.1007/BF01857239

Amanjás, R. D., & Val, A. L. (2022). Thermal biology of tambaqui (*Colossoma macropomum*): General insights for aquaculture in a changing world. *Reviews in Aquaculture, 15*(2), 480–490. https://doi.org/10.1111/raq.12732

Andrade, B. S., Singh, C. L., Santos, J. Á., Gonçalves, V. V., Siqueira-Souza, F., & Freitas, C. E. (2019). Efeitos das mudanças climáticas sobre as comunidades de peixes na Bacia Amazônica. *Rev Ciênc Soc, 2*(4), 107–124.

Arantes, C. C., Castello, L., Stewart, D. J., Cetra, M., & Queiroz, H. L. (2010). Population-density, growth and reproduction of arapaima in na Amazonian river-floodplain. *Ecology of Freshwater Fish, 19*, 455–465. https://doi.org/10.1111/j.1600-0633.2010.00431.x

Arantes, C. C., Winemiller, K. O., Petrere, M., Castello, L., Hess, L. L., & Freitas, C. E. (2017). Relationships between forest cover and fish diversity in the Amazon River floodplain. *Journal of Applied Ecology, 55*, 386–395. https://doi.org/10.1111/1365-2664.12967

Asch, R. G., Stock, C. A., & Sarmiento, J. L. (2019). Climate change impacts on mismatches between phytoplankton blooms and fish spawning phenology. *Global Change Biology, 25*(8), 2544–2559. https://doi.org/10.1111/gcb.14650

Barbarossa, V., Bosmans, J., Wanders, N., King, H., Bierkens, M. F. P., Huijbregts, M. A. J., & Schipper, A. M. (2021). Threats of global warming to the world's freshwater fishes. *Nature Communications, 12*, 1701. https://doi.org/10.1038/s41467-021-21655-w

Barros, D. F., Petrere-Jr, M., Lecours, V., Betturi-Gomes, D., Castello, L., & Isaac, V. J. (2020). Effects of deforestation and other environmental variables on floodplain fish catch in the Amazon. *Fisheries Research, 230*, 105643. https://doi.org/10.1016/j.fishres.2020.105643

Barros, D., Petrere-Jr, M., Castello, L., Santos, P. B., Butturi-Gomes, D., & Isaac, V. J. (2021). Hydrologic variability effects on catches of Prochilodus nigricans in the lower Amazon. *Aquatic Sciences, 83*(22), 1–9. https://doi.org/10.1007/s00027-021-00782-y

Boschman, L. M., Carraro, L., Cassemiro, F. A. S., deVries, J., Altermatt, F., Hagen, O., Horn, C., & Pellisier, L. (2023). Freshwater fish diversity in the Western Amazon Basin shaped by Andean uplift since the late cretaceous. *Nature Ecology & Evolution*. https://doi.org/10.1038/s41559-023-02220-8

Campos, C. P., Bitar, S. D. B., & Freitas, C. (2023). Uncertainties regarding the natural mortality of fish can increase due global climate change. *PeerJ, 11*, e14989. https://doi.org/10.7717/peerj.14989

Cantera, I., Jésequel, C., Dejean, T., Murienne, J., Vigouroux, R., Valentini, A., & Brosse, S. (2023). Functional responses to deforestation in fish communities inhabiting neotropical streams and rivers. *Ecological Processes, 12*, 52. https://doi.org/10.1186/s13717-023-00463-8

Carvalho, et al. (2023). Pervasive gaps in Amazonian ecological research. *Current Biology*. https://doi.org/10.1016/j.cub.2023.06.077

Castello, L., McGrath, D. G., Hess, L. L., Coe, M. T., Lefebvre, P. A., Petry, P., Macedo, M. N., Renó, V. F., & Arantes, C. C. (2013). The vulnerability of Amazon freshwater ecosystems. *Conservation Letters, 6*(4), 217–229. https://doi.org/10.1111/conl.12008

Castello, L., Hess, L. L., Thapa, R., McGrath, D. G., Arantes, C. C., Renó, V., & Isaac, V. I. N. (2018). Fish yields vary with land cover on the Amazon river-floodplain. *Fish and Fisheries, 19*(3), 431–440.

Castro, E. R., & Campos, I. (2015). *Formação socioeconômica da Amazônia. Coleção Formação Regional da Amazônia vol 2* (p. 640). NAEA.

Castro, J. S., Braz-Mota, S., Campos, D. F., Souza, S. S., & Val, A. L. (2020). High temperature, pH, and hypoxia cause oxidative stress and impair the spermatic performance of the Amazon fish *Colossoma macropomum*. *Frontiers in Physiology, 11*, 772. https://doi.org/10.3389/fphys.2020.00772

Correa, S. W., Paiva, R. C. D., Espinoza, J. C., & Collishonn, W. (2017). Multi-decadal hydrological retrospective: Case study of Amazon floods and droughts. *Journal of Hydrology, 549*, 667–684. https://doi.org/10.1016/j.hydrol.2017.04.019

Correia, G. B., Siqueira-Souza, F. K., & Freitas, C. E. C. (2015). Intra- and inter-annual changes in the condition factors of three Curimatidae detritivores from the Amazonian floodplain lakes. *Biota Neotropica, 15*, 1–7. https://doi.org/10.1590/1676-0611-BN-2014-0001

Dagosta, F. C. P., & Pinna, M. (2019). The fishes of the Amazon: Distribution and biogeographical patterns, with a comprehensive list of species. *Bulletin of the American Museum of Natural History, 431*, 1–163.

Doria, C. R. C., Athayde, S., Marques, E. E., Lima, M. A. L., Dutka-Gianelli, J., Ruffino, M. L., Kaplan, D., Freitas, C. E. C., & Isaac, V. N. (2018). The invisibility of fisheries in the process of hydropower development across the Amazon. *Ambio, 47*, 453–465. https://doi.org/10.1007/s13280-017-0994-7

Duponchelle, F., Isaac, V. L., Doria, C. R. C., Van Damme, P. A., Herrera-R, G. A., Anderson, E. P., Cruz, R. E. A., Hauser, M., Hermann, T. W., Agudelo, E., Bonilla-Castillo, C., Barthem, R., Freitas, C. E. C., Garcia-Davila, C., Garcia-Vasquez, A., Renno, J. F., & Castello, L. (2021). Conservation of migratory fishes in the Amazon basin. *Aquatic Conservation Marine Freshwater Ecosystems, 31*(5), 1087–1105.

Ficke, A. D., Myrick, C. A., & Hansen, L. J. (2007). Potential impacts of global climate change on freshwater fisheries. *Reviews in Fish Biology and Fisheries, 17*, 581–613. https://doi.org/10.1007/s11160-007-9059-5

Finer, M., & Jenkins, C. N. (2012). Proliferation of hydroelectric dams in the Andean Amazon and implications for Andes-Amazon connectivity. *PLoS One, 7*(4), e35126. https://doi.org/10.1371/journal.pone.0035126

Fleischmann, A. S., Papa, F., Hamilton, S. K., Fassoni-Andrade, A., Wongchuig, S., Espinoza, J. C., Paiva, R. C. D., Melack, J. M., Fluet-Chouinard, E., Castello, L., Almeida, M., Bonnet, M. P., Alves, L. G., Moreira, D., Yamazaki, D., Revel, M., & Collischonn, W. (2022). Increased floodplain inundation in the Amazon since 1980. *Environmental Research Letters, 18*, 034024.

Forsberg, B. R., Melack, J. M., Dunne, T., Barthem, R. B., Goulding, M., et al. (2017). The potential impact of new Andean dams on Amazon fluvial ecosystems. *PloSONE, 12*(8), e0182254. https://doi.org/10.1371/journal.pone.0182254

Freitas, C. E. C., & Rivas, A. A. F. (2006). A pesca e os recursos pesqueiros na Amazônia Ocidental. *Ciênc Cult, 8*, 30–32.

Freitas, C. E. C., Siqueira-Souza, F. K., Guimarães, A. R., Santos, F. A., & Santos, I. L. A. (2010). Interconnectedness during high water maintains similarity in fish assemblages of Island floodplain lakes in the Amazonian basin. *Zoologia, 27*(6), 931–938. https://doi.org/10.1590/S1984-46702010000600014

Freitas, C. E. C., Siqueira-Souza, F. K., Humston, R., & Hurd, L. E. (2013). An initial assessment of drought sensitivity in Amazonia fish communities. *Hydrobiologia, 705*, 159–171.

Freitas, C. E. C., Siqueira-Souza, F. K., Florentino, A. C., & Hurd, L. E. (2014). The importance of spatial scale to analysis of fish diversity in Amazonian floodplain lakes and implications for conservation. *Ecology of Freshwater Fish, 23*, 470–477. https://doi.org/10.1111/eff.12099

Freitas, C. E. C., Laurenson, L., Yamamoto, K. C., Forsberg, B. R., Petrere, M. J., Arantes, C., & Siqueira-Souza, F. K. (2018). Fish species richness is associated with the availability of landscape components across seasons in the Amazonian floodplain. *PeerJ, 6*, e5080. https://doi.org/10.7717/peerj.5080

Freitas, C. E. C., Mereles, M. A., Pereira, D. V., Siqueira-Souza, F., Hurd, L., Kahn, J., Morais, G., & Sousa, R. G. C. (2022). Death by a thousand cuts: Small local dams can produce large regional impacts in the Brazilian legal Amazon. *Environmental Science & Policy, 136*, 447–452. https://doi.org/10.1016/j.envsci.2022.07.013

Heath, M. R., Neat, F. C., Pinnegar, J. K., Reid, D. G., Sims, D. W., & Wright, P. J. (2012). Review of climate change impacts on marine fish and shellfish around the UK and Ireland. *Aquatic Conservation Marine Freshwater Ecosystem, 22*(3), 337–367. https://doi.org/10.1002/aqc.2244

Hess, L. L., Melack, J. M., Affonso, A. G., Barbosa, C., Gastil-Buhl, M., & Novo, E. M. L. M. (2015). Wetlands of the lowland Amazon Basin: Extent, vegetative cover, and dual-season inundated area as mapped with JERS-1 synthetic aperture radar. *Wetlands, 35*(4), 745–756. https://doi.org/10.1007/s13157-015-0666-y

Hurd, L. E., Souza, R. G. S., Siqueira-Souza, F. K., Cooper, G. J., Kahn, J. R., & Freitas, C. E. C. (2016). Amazon floodplain fish communities: Habitat connectivity and conservation in a rapidly deteriorating environment. *Biological Conservation, 195*, 118–127. https://doi.org/10.1016/j.biocon.2016.01.005

Hutchinson, G. E. (1959). Homage to Santa Rosalia, or why are there so many kinds of animals? *The American Naturalist, 93*, 145–159.

IPCC. (2007). Climate change: Impacts, adaptation and vulnerability. In M. Parry, O. Canziani, J. Palutikof, P. V. D. Linden, & C. Hanson (Eds.), *Report of the intergovernmental panel on climate change* (p. 976). Cambridge University Press.

IPCC. (2022). Climate change 2022: Impacts, adaptation and vulnerability. In H. O. Pörtner, D. C. Roberts, M. Tignor, E. S. Poloczanska, K. Mintenbeck, A. Alegría, M. Craig, S. Langsdorf, S. Löschke, V. Möller, A. Okem, & A. Rama (Eds.), *Report of the intergovernmental panel on climate change*. Cambridge University Press. 3056 pp. https://doi.org/10.1017/9781009325844

Isaac, V. J., Castello, L., Santos, P. R. B., & Ruffino, M. (2016). Seasonal and interannual dynamics of river-floodplain multispecies fisheries in relation to flood pulses in the lower Amazon. *Fisheries Research, 183*, 352–359. https://doi.org/10.1016/j.fishres.2016.06.017

Junk, W. J., Bayley, P. B., & Sparks, R. E. (1989). The flood pulse concept in river–floodplain systems. In D. P. Dodge (Ed.), *Proceedings of the international large river symposium* (Vol. 106, pp. 110–127). Canadian Special Publications of Fish Aquatic Sci.

Junk, W. J., Piedade, M. T. F., Schöngart, J., Cohn-Haft, M., Adeney, J. M., & Wittmann, F. (2011). A classification of major naturally-occurring Amazonian lowland wetlands. *Wetlands, 31*, 623–640. https://doi.org/10.1007/s13157-011-0190-7

Langerwich, F., Rost, S., Gerten, D., Poulter, B., Ramming, A., & Cramer, W. (2013). Potential effects of climate change on inundation patterns in the Amazon Basin. *Hydrology and Earth System Sciences, 17*, 2247–2262. https://doi.org/10.5194/hess-17-2247-2013

Lobón-Cerviá, J., Hess, L. L., Melack, J. M., & Araújo-Lima, C. A. R. M. (2015). The importance of forest cover for fish richness and abundance on the Amazon floodplain. *Hydrobiologia, 750*(1), 245–255. https://doi.org/10.1007/s10750-014-2040-0

Lomolino, M. V., Riddle, B. R., & Brown, J. H. (2006). *Biogeography*. Sunderland.

Lopes, G. C. S., Matos, O. F., & Freitas, C. E. C. (2023). Spatial dynamics of Amazonian commercial fisheries: An analysis of landscape composition and fish landings. *Boletim do Instituto de Pesca, 48*, e664.

Lowe-McConnell, R. H. (1999). *Estudos ecológicos de comunidades de peixes tropicais*. São Paulo.

Marengo, J. A., & Espinoza, J. C. (2015). Extreme seasonal droughts and floods in Amazonian: Causes, trends and impacts. *International Journal of Climatology, 36*, 1033–1050.

Marengo, J. A., Borma, L. S., Rodriguez, D. A., Pinho, P., Soares, W. R., & Alves, L. M. (2013). Recent extremes of drought and flooding in Amazonia: Vulnerabilities and human adaptation. *American Journal of Climate Change, 2*, 87–96.

Marshall, B. G., Forsberg, B. R., Hess, L., & Freitas, C. E. C. (2011). Water temperature diferences in interfluvial palm swamp habitats of *Paracheirodon axelroldi* and *P. Simulans* (Osteichytyes: Characidae) in the middle Rio Negro. *Ichthyological Exploration of Freshwaters, 22*, 377–383.

Melack, J. M., & Coe, M. T. (2012). Climate change and the Floodplain Lakes of the Amazon Basin. In C. R. Golman, M. Kumagai, & R. D. Robarts (Eds.), *Climatic change and global warming of inland waters* (pp. 295–310). Wiley. https://doi.org/10.1002/9781118470596.ch17

Melack, J. M., & Coe, M. T. (2020). Amazon floodplain hydrology and implications for aquatic conservation. *Aquatic Conservation: Marine and Freshwater Ecosystems, 31*(5), 1029–1040. https://doi.org/10.1002/aqc.3558

Melack, J. M., & Hess, L. L. (2010). Remote sensing of the distribution and extent of wetlands in the Amazon basin. In W. J. Junk, M. T. F. Piedade, F. Wittmann, J. Schongart, & P. Parolin (Eds.), *Amazonian floodplain forests: Ecophysiology, biodiversity and sustainable management* (pp. 43–59). Springer.

Morales, B. F., & Deus, C. P. (2021). The role of fishery management and environmental variables on the fish fauna in floodplain lakes in the lower Purus River, Amazon Basin, Brazil. *Lakes & Reservoirs: Science, Policy and Management for Sustainable Use, 26*(4), e12385. https://doi.org/10.1111/lre.12385

Noveras, J., Yamamoto, K. C., & Freitas, C. E. (2012). Use of the flooded forest by fish assemblages in lakes of the National Park of Anavilhanas (Amazonas, Brazil). *Acta Amazonica, 42*(4), 561–566.

Oberdorff, T., Dias, M. S., Jézéquel, C., Albert, J. S., Arantes, C. C., et al. (2019). Unexpected fish diversity gradients in the Amazon Basin. *Science Advances, 5*(9), eaav8681. https://doi.org/10.1126/sciadv.aav8681

Oliveira, A. M., & Val, A. L. (2017). Effects of climate scenarios on the growth and physiologu of the Amazonian fish tambaqui (*Colossoma macropomum*)(Characiformes: Serrasalmidae). *Hydrobiologia, 789*(1), 167–178. https://doi.org/10.1007/s10750-016-2926-0

Pereira, D. V., Arantes, C. C., Sousa, K. N., & Freitas, C. E. C. (2022). Relationships between fishery catch rates and land cover along a longitudinal gradient in floodplains of the Amazon River. *Fisheries Research, 258*, 106521.

Petchey, I. L., & Gaston, K. J. (2006). Functional diversity: Back to basics and looking forward. *Ecology Letters, 9*, 741–758. https://doi.org/10.1111/j.1461-0248.2006.00924.x

Pfaff, A., Robalino, J., Sandoval, C., & Herrera, D. (2015). Protected are types, strategies and impacts in Brazil's Amazon: Public protected area strategies do not yield a consistent ranking of protected area types by impact. *Philosophical Transactions of the Royal Society B, 370*, 1681. https://doi.org/10.1098/rstb.2014.0273

Phillips, O. L., Aragão, L. E. O. C., Lewis, S. L., Fisher, J. B., Lloyd, J., et al. (2009). Drought sensitivity of the Amazon rainforest. *Science, 323*, 1344–1347. https://doi.org/10.1126/science.1164033

Pielou, E. C. (1974). *Population and community ecology: Principles and methods*. Gordon and Breach.

Pinheiro, J. A. C., Gonçalves, V. V. C., Pereira, H. S., Fraxe, T. J. P., Okac, J. M., Siqueira-Souza, F., & Freitas, C. E. C. (2022). Perception of Amazonian fishers regarding environmental changes as causes of drastic events of fish mortality. *Brazilian Journal of Biology, 82*, e263339. https://doi.org/10.1590/1519-6984.263339

Polis, G. A., Myers, C. A., & Holt, R. D. (1989). The ecology of intraguild predation. *Annual Review of Ecology and Systematics, 20*, 297–330.

Poulin, R. (2000). Variation in the intraspecific relationship between fish length and intensity of parasitic infection: Biological and statistical causes. *Journal of Fish Biology, 56*, 123–137.

Prates, R. C., & Bacha, C. J. C. (2011). Os processos de desenvolvimento e desmatamento da Amazônia. *Economies et Societes, 20*(3), 601–636. https://doi.org/10.1590/S0104-06182011000300006

Reis, R. E., Albert, J. S., Di Dario, F., Mincarone, M. M., Petry, P., & Rocha, L. A. (2016). Fish biodiversity and conservation in South America. *Journal of Fish Biology, 89*, 12–47. https://doi.org/10.1111/jfb.13016

Resende, A. F., Piedade, M. T. F., Feitosa, Y. O., Andrade, V. H. F., Trumbore, S. E., Durgante, F. M., et al. (2020). Flood-pulse disturbances as a threat for long-living Amazonian trees. *The New Phytologist, 227*, 1790–1803. https://doi.org/10.1111/nph.16665

Richey, J. E., Meade, R. H., Salati, E., Devol, A. H., Nordin, C. F., Jr., & Santos, U. (1986). Water discharge and suspended sediment concentrations in the Amazon River. *Water Resources Research, 22*, 756–764.

Röpke, C. P., Amadio, S. A., Winemiller, K. O., & Zuanon, J. (2015). Seasonal dynamics of the fish assemblage in a floodplain lake at the confluence of the negro and Amazon Rivers. *Journal of Fish Biology, 89*(1), 194–212. https://doi.org/10.1111/jfb.12791

Röpke, C. P., Amadio, S., Zuanon, J., Ferreira, E. J. G., Deus, C. P., Pires, T. H. S., & Winemiller, K. O. (2017). Simultaneous abrupt shifts in hydrology and fish assemblage structure in a floodplain lake in the Central Amazon. *Scientific Reports, 7*, 40170. https://doi.org/10.1038/srep40170

Röpke, C. P., Pires, T. H. S., Zuchi, N., Zuanon, J., & Amadio, S. (2022). Effects of climate-driven hydrological changes in the reproduction of Amazonian floodplain fishes. *Journal of Applied Ecology, 59*, 1134–1145. https://doi.org/10.1111/1365-2664.14126

Santos, R. E., Pinto-Coelho, R., Fonseca, R., Simões, N. R., & Zanchi, F. B. (2018). The decline of fisheries on the Madeira River, Brazil: The high cost of the hydropower dams in the Amazon Basin. *Fisheries Management and Ecology, 25*(5), 380–391. https://doi.org/10.1111/fme.12305

Silva, S. D. P., Santos, S. B., Pereira, P. C. G., Melo, M. R. S., & Eugenio, F. C. (2021). Landscape analysis in a municipality in the arc of deforestation of the Brazilian Amazon rainforest. *Ecological Engineering, 173*, 106417. https://doi.org/10.1016/j.ecoleng.2021.106417

Siol, H. (1984). The Amazon and its main affluents: Hydrography, morphology of the river courses, and river types. In H. Sioli (Ed.), *The Amazon, 56*, 127–165. Springer.

Siqueira-Souza, F. K., Hurd, L. E., Yamamoto, K. C., Soares, M. G. M., Cooper, G. J., Kahn, J. R., & Freitas, C. E. C. (2021). Patterns of pelagic fish diversity in floodplain lakes of whitewater and Blackwater drainage systems within the Central Amazon River basin of Brasil. *Frontiers in Ecology and Evolution, 9*, 602895. https://doi.org/10.3389/fevo.2021.602895

Sorribas, M. V., Paiva, R. C. D., Melack, J. M., Bravo, J. M., Jones, C., et al. (2016). Projections of climate change effects on discharge and inundation in the Amazon Basin. *Climatic Change, 136*, 555–570. https://doi.org/10.1007/s10584-016-1640-2

Sousa, R. G. C., Mereles, M. A., Siqueira-Souza, F. K., Hurd, L. E., & Freitas, C. E. C. (2018). Small dams for aquaculture negatively impact fish diversity in Amazonian streams. *Aquaculture Environment Interactions, 10*, 89–98.

Sousa, R. G. C., Freitas, H. C. P., Zacardi, D. M., & Faria-Junior, C. H. (2021). Effects of river dams on the fish guilds in the northwest region of the Brazilian Amazon. *Fisheries Research, 243*, 106091. https://doi.org/10.1016/j.fishres.2021.106091

Souza-Cruz-Buenaga, F. V. A., Espig, S. A., Castro, T. L. C., & Santos, M. A. (2019). Environmental impacts of a reduced flow stretch on hydropower plants. *Brazilian Journal of Biology, 79*(3), 470–487. https://doi.org/10.1590/1519-6984.183883

Thatcher, V. E. (2006). *Amazon fish parasites*. Pensoft Publishers.

Tribuzy-Neto, I. A., Conceição, K. G., Siqueira-Souza, F. K., Hurd, L. E., & Freitas, C. E. C. (2018). Condition factor variations over time and trophic position among four species of Characidae from Amazonian floodplain lakes: Effects of an anomalous drought. *Brazilian Journal of Biology, 78*, 337–344. https://doi.org/10.1590/1519-6984.166332

Val, A. L. (2011). As mudanças climáticas e os peixes de água doce. In *63ª Reunião anual da Sociedade Brasileira para o Progresso da Ciência (SBPC)*. Universidade Federal de Goiás.

Val, A. L., & Almeida-Val, V. M. (2008). Mudanças climáticas e biodiversidade na Amazônia. In *60ª Reunião anual da Sociedade Brasileira para o Progresso da Ciência (SBPC), Conferência biodiversidade na Amazônia x mudanças climáticas: causas e consequências*. Universidade de Campinas.

Val, A. L., & Wood, C. M. (2022). Global change and physiological challenges for fish of the Amazon today and in the near future. *The Journal of Experimental Biology, 225*(10), jeb216440. https://doi.org/10.1242/jeb.216440

Winemiller, K. O. (2005). Life history strategies, population regulation, and implications for fisheries management. *Canadian Journal of Fisheries and Aquatic Sciences, 62*(4), 872–885. https://doi.org/10.1139/F05-040

Winemiller, K. O., & Rose, K. A. (1992). Patterns of life-history diversification in north American fishes: Implications for population regulation. *Canadian Journal of Fisheries and Aquatic Sciences, 49*, 2196–2218.

Woodward, G., et al. (2016). The effects of climatic fluctuations and extreme events on running water ecosystems. *Philosophical Transactions of the Royal Society B, 371*(20150274), 1–15. https://doi.org/10.1098/rstb.2015.0274

Zuanon, J. (2008). Peixes, pesca e clima na Amazônia: um ensaio sobre os efeitos das mudanças climáticas globais sobre os recursos pesqueiros na região do Rio Negro, Amazonas, Brasil. In G. V. P. Crus & S. Andrade (Eds.), *Rio Negro, Manaus e as mudanças no clima*. Instituto Socioambiental.

Genetics, Toxicological Background, and Environmental Adaptation Processes of Disease: Vector Mosquitoes in the Amazon

Joaquim Ferreira do Nascimento Neto, Rosemary Aparecida Roque, Sabrina da Fonseca Meireles, Daniel Luís Viana Cruz, Junielson Soares da Silva, Neiva Silva de Vasconcelos, Jessica Cavalcante Barros, Bruna Felipe Olavo da Silva, Ricardo de Melo Katak, Adalberto Luis Val, Ana Cristina da Silva Pinto, Francisco Célio Maia Chaves, Wanderli Pedro Tadei, and Míriam Silva Rafael

Author "Wanderli Pedro Tadei" is deceased at the time of publication.

J. F. do Nascimento Neto · R. A. Roque · R. de Melo Katak · M. S. Rafael (✉)
Brazilian National Institute for Research of the Amazon (INPA), Coordination of Society, Environment, and Health (COSAS), Laboratory of Malaria and Dengue Vectors (LMD), Manaus, AM, Brazil

S. da Fonseca Meireles · D. L. V. Cruz · J. S. da Silva · N. S. de Vasconcelos · J. C. Barros · B. F. O. da Silva
Graduate Program in Genetics, Conservation and Evolutionary Biology – PPG-GCBEV, Brazilian National Institute for Research of the Amazon (INPA), Manaus, AM, Brazil

A. L. Val
Laboratory of Ecophysiology and Molecular Evolution, Brazilian National Institute for Research of the Amazon (INPA), Manaus, Amazonas, Brazil

A. C. da Silva Pinto
Federal University of Amazonas (UFAM), Manaus, AM, Brazil

F. C. M. Chaves
Brazilian Agricultural Research Corporation, Center for Agroforestry Research of the Western Amazon, Manaus, AM, Brazil

W. P. Tadei
Laboratory of Ecophysiology and Molecular Evolution (LEEM), Brazilian National Institute of Amazonian Research (INPA), Manaus, AM, Brazil

S. S. de Souza et al. (eds.), *The Future of Amazonian Aquatic Biota*,
https://doi.org/10.1007/978-3-031-66822-7_2

Abstract

The influence of climate change on Amazonian ecosystems has produced changes in the way of life of several mosquitoes of epidemiological importance. In this context, aspects of the biological and molecular cycle of mosquitoes in the Amazon were studied to develop new vector control strategies that do not cause damage to the associated fauna. In this study, the transcription of lipophorin, vitellogenin genes, and some biological parameters showed significant changes as the climate simulation environments became more severe in relation to the current environment. In addition, assays to determine the mortality, cytotoxicity, and genotoxicity of *Ae. aegypti* exposed to dillapiole derivatives were carried out as a new alternative to control of this mosquito resistant to certain types of insecticides in certain geographic areas. Mortality and cytotoxicity occurred in larvae exposed to the dillapiole piperidyl and dillapiole isopropyl ether substances and, furthermore, dillapiole ethyl ether caused a high frequency of damage to the DNA of larvae compared to the negative control. The screening of bacteria revealed three strains (End64LB, End82LB, and End84LB) that belong to the phyla Proteobacteria and Firmicutes and have larvicidal potential against *Ae. aegypti*, suggesting that dillapiole derivatives and these bacteria are promising for the biological control of *Ae. aegypti*.

1 Introduction

The Amazon is one of the most biodiverse areas on the planet and possesses many vertebrate animals such as the tambaqui (*Colossoma macropomum*) and plants such as the cupuaçu (*Theobroma grandiflorum*) (Lopes et al., 2018; Barbalho et al., 2022). The diversity of species also extends to large invertebrates such as the titan beetle (*Titanus giganteus*) and small ones such as mosquitoes, some of which are of great medical importance, such as *Anopheles darlingi*, the vector of malaria, *Aedes aegypti* and *Aedes albopictus*, primary and secondary vectors, respectively, of dengue, zika, chikungunya, and urban yellow fever viruses (Kraemer et al., 2015; Tadei et al., 2017; Leta et al., 2018; Dahmana et al., 2020). The population density of mosquitoes is intrinsically linked to the establishment of breeding sites that present areas of water in natural reservoirs, such as rivers and lakes, or artificial ones such as water tanks and trays for potted plants, which are used to carry out oviposition laying (Arcos et al., 2018; Flaibani et al., 2020). In this way, the rainfall regime in these places determines the establishment of breeding grounds for their offspring and, consequently, the increase in the population density of mosquitoes (Kubota et al., 2003; Lima-Camara, 2016).

Currently, sanitary surveillance agencies use the elimination or mitigation of breeding sites as the main strategy to control the population density of vectors, which reduces human contact with mosquitoes and, consequently, decreases the circulation of the etiological agents (Ernawati et al., 2021; Rahman et al., 2021). In addition to the control of breeding sites, chemical and biological insecticides are employed as a second form of controlling the immature and winged phases of the vectors (Silva et al., 2021; Raguvaran et al., 2022). The use of chemical insecticides, such as pyrethroids, organophosphates, and carbamates, causes several impacts on the environment and its associated fauna, such as mosquitoes some of which are of great medical importance (Malhotra et al., 2021; Raguvaran et al., 2022).

As a result of the intensification of anthropogenic actions after the industrial revolution, which includes the use of nonrenewable energy sources, burning of forests, and accelerated deforestation, the release of greenhouse gases into the atmosphere has intensified, contributing to the increase in global average temperature (Alves de Oliveira et al., 2021). The increase in greenhouse gases, including CO_2, contributes to the increase in the global average temperature, affecting the rainfall regime and consequently, the availability of natural and artificial breeding sites (Derkzen et al., 2017; Kriegler et al., 2017). Furthermore, the rainfall regime may intensify in some regions and the dry season may intensify in others (Rose et al., 2020; Wimalasiri-Yapa et al., 2021). Some ecosystems, such as those of the Amazon, are highly sensitive to climate change since they need a sensitive global climate balance for their maintenance. They may also be affected by the presence of the large urban centers (Cochrane & Barber, 2009; Menezes et al., 2018).

Due to the extensive territorial area of the Amazon, climate variations do not occur homogeneously, and therefore, affect the diverse biological community of mosquitoes in the Amazon in different ways (Tadei et al., 2017; Do Nascimento Neto et al., 2020). High temperatures significantly affect the development of culicids, making the biological cycle shorter, affecting their permanence and interaction with humans and, consequently, the spread of their arboviruses (Chaves et al., 2014; Ezeakacha & Yee, 2019). In this way, they influence the way vectors interact with humans and the number of disease cases with which they are associated (Majeed et al., 2014; Raji et al., 2019).

In Brazil, the prolonged use of synthetic chemical insecticides in vector control campaigns, especially organophosphates such as temephos, has proven resistance to chemical insecticides of these vectors (Chediak et al., 2016; Valle et al., 2019), in addition to causing damage to the environment (Hamid et al., 2018; Collins et al., 2019; World Health Organization—WHO, 2020).

The search for new alternatives for the control of mosquitoes of epidemiological importance, especially *Ae. aegypti*, has become urgent (Valle et al., 2019). In this sense, the genus *Piper* has stood out for the presence of active metabolites in their composition, that can be used to control mosquito vectors of diseases (Marques and Kaplan, 2014). *Piper marginatum* Jacq. (marigold pepper), *Piper arboretum* Aubl. (jaborandi-pepper), and *Piper aduncum* Vell. (spiked pepper) showed larvicidal positive activity against *Ae. aegypti* (Santana et al., 2015).

Bacteria produce active metabolites that are useful for the biological control of populations *Ae. aegypti*. It is important due to the resistance of the mosquitoes to the synthetic chemical insecticides temephos and deltamethrin Brazil (Bellinato et al., 2016; Valle et al., 2019).

There are substances of botanical origin, such as extracts, essential oils (EO), isolated molecules, and semisynthetic derivatives, that have been analyzed in regard to their effectiveness against mosquito vectors, principally in order to verify their ovicidal, larvicidal, and adulticidal effects (Rafael et al., 2008; Pinto et al., 2012; Domingos et al., 2014; Meireles et al., 2016; Silva et al., 2019; Santos et al., 2020; Silvério et al., 2020; Silva et al., 2021). Extracts of the leaves of *P. aduncum* Vell. showed toxic effects against the larvae of *Ae. aegypti* (Pinto et al., 2012). The essential oil of the species *Piper caninum*, *Piper longum*, *Piper montium*, and *Piper mutabile* showed excellent larvicidal activity against *Ae. aegypti* (Huong et al., 2019). According to França et al. (2020), *Piper capitarianum* can be used in the development of new insecticidal targets for the control of *Ae. aegypti* and *Ae. albopictus* in the Amazon. In this sense, Oliveira et al. (2022), after studying the essential oil of *Piper purusanum*, cited that it caused the mortality of larvae of the genera *Aedes* and *Anopheles*, and recommended this substance as an alternative for the natural control of these vectors.

The substance dillapiole, a major component (80–90%) of the essential oil of the plant *Piper aduncunm*, has in its composition methylenedioxyphenyl (Maia et al., 1998). Other molecules tested with the same structure have shown larvicidal activity against *Culex quinquefasciatus* (Desmarchelier & Fukuto, 1974), *Culex pipiens pallens* Coq. (Lee, 2000), and toxic effects in *Spodoptera frugiper* (Harmatha & Nawrot, 2002).

In addition to the evaluation of toxicity, genotoxicity tests (Rafael et al., 2008) and the comet assay (Viana Cruz et al., 2020, in press) in *Ae. aegypti* can reveal the genotoxic action of dillapiole substances and derivatives of this essential oil, respectively. The derivatives of dillapiole, dillapiole ethyl ether and dillapiole *n*-butyl ether in *Ae. aegypti* (Domingos et al., 2014) and *Ae. albopictus* (Meireles et al., 2016) and dillapiole methyl ether (Silva et al., 2019, 2021) and isodillapiole in *Ae. aegypti* (Santos et al., 2020) demonstrated a proven increase in larvicidal and genotoxic activity when compared to its precursor, dillapiole (Rafael et al., 2008).

Another potential alternative for combatting *Ae. aegypti* is its biological control by means of entomopathogenic microorganisms associated with aquatic plants. *Eichhornia crassipes* (Mart.) Solms-Laub. (Pontederiaceae) is an aquatic macrophyte that is native to South America (Gopal, 1987) and is usually found in stagnant or slow-moving waters (Edwards & Musil, 1975). The tissues of these plants are reservoirs of endophytic microorganisms (El-Deeb et al., 2012), such as prolific entomopathogenic filamentous bacteria and fungi of a variety of metabolites, which can be used in the control of economically important pests (Katak et al., 2021; Silva et al., 2022).

In the present study, the toxic effect of dillapiole-derived substances (dillapiole piperidyl and dillapiole isopropyl ether) on *Ae. aegypti* was evaluated for the first time by analyzing egg-laying rates and its larvicidal activity in this mosquito. In addition, the genotoxic effect of the substance dillapiole ethyl ether was studied, through DNA damage, in comparison to the larvicide temephos of the organophosphate class. Further, the endophytic bacteria from the roots of *Ei. crassipes* were isolated, screened, and identified to assess their larvicidal potential against *Ae. aegypti*. These approaches were aimed at finding new alternatives for effective control of *Ae. aegypti*.

In recent decades, the intensification of several factors such as anthropogenic actions, global climatic events, and increased greenhouse gases have influenced climate change, ecosystems, and the ecological dynamics of mosquitoes of epidemiological importance. In this sense, the present work seeks new perspectives for the of control of endemic vectors in the face of the impacts of the climate changes expected by 2100 and their implications for these mosquitoes.

2 Materials and Methods

The use of the animals in a vector colony was carried out in accordance with the standards of the National Council for Control of Animal Experimentation (CONCEA), and the study was previously approved by the Ethics Commission on the Use of Animals (CEUA) of the National Institute for Amazonian Research (INPA) and registered under No. 012/2021. Mosquito collections occurred under authorization for insect capture and were registered under No. 32,941 and 77,078, in the years 2012 and 2020, respectively, in the Biodiversity Information and Authorization System (SISBIO) of the Chico Mendes Institute for Biodiversity Conservation (ICMBio), Brazil. The studies involving blood from human beings were approved by the Ethics Committee in Research Involving Human Beings of the Federal University of Amazonas/UFAM—CAAE: 66467417.6.0000.5020. Access to genetic heritage was registered in the National System for the Management of Genetic Heritage and Associated Traditional Knowledge (SISGEN), under No. A949853.

2.1 Rearing Mosquitoes Under Climate Simulation Conditions

The Laboratory of Malaria and Dengue Mosquito Research (LMD) has performed studies involving the effect of climate change on malaria and dengue vectors since 2011. It has obtained significant results in relation to the biology and genetics of mosquito vectors in the Amazon as the climate faces changes that will be necessary to occur by the year 2100, according to the Intergovernmental Panel on Climate Change (IPCC) variation (Kriegler et al., 2017).

Investigations are carried out in four climatic chambers, called environmental rooms, in which the temperature, relative air humidity, and atmospheric CO_2 concentration in parts per million (ppm) are controlled by a computer. Environment 1 presents the temperature and CO_2 values taken as a reference from the values recorded in an INPA permanent forest reserve area in the city of Manaus, state of Amazonas, Brazil. The information regarding the climatic parameters reproduced within environment 1, referring to the control environment or current scenario, are used as a reference for increasing the simulated environments.

The remaining three climate environments present increments in temperature and CO_2 provided for each climate condition predicted by the IPCC until the year 2100, being: mild scenario (REMIND-MAGPIE SSP5-45) 2 °C and 200 ppm CO_2;

intermediate scenario (REMIND-MAGPIE SSP5-60) 3.5 °C and 300 ppm CO_2; and extreme scenario (REMIND-MAGPIE SSP5-Baseline) 5.0 °C and 700 ppm CO_2 (Kriegler et al., 2017, Do Nascimento Neto et al., 2020). The colonies of *Ae. aegypti* were reared under the climate simulation conditions (do Nascimento Neto et al., 2020) for more than 80 generations and the synergistic effects of increased temperature and CO_2 on parameters that could affect the epidemiology of the vector with the worsening of climatic events expected until the year 2100 are evaluated.

The creation of colonies in climate simulation rooms follows the creation protocol provided by do Nascimento Neto et al. (2020). In general, strips of filter paper containing eggs are placed to hatch in basins of dechlorinated well water, and the larvae are fed with cat and rodent food in a 1:1 ratio. The pupae are collected using a pipette and special cups for adult emergence in screened cages. Males and females are fed with sugar solution and females (*Ae. aegypti* and *An. darlingi*) receive additional blood nutrition in anesthetized hamsters *Mesocricetus auratus*. After the females have been fed blood, the eggs are laid on paper strips in laying cups filled with moistened, chlorine-free well water.

2.2 *An. darlingi* Survival Assessment

Collections of adults were carried out in peri-urban breeding sites in the city of Manaus, Amazonas. The adults were captured using a manual field vacuum cleaner, stored inside paraffin-coated cups, and were taken to the Malaria and Dengue Laboratory (LMD) at INPA. In the laboratory, the mosquitoes were identified by means of taxonomic keys, and the adult *An. darlingi* were separated and kept in the laboratory to obtain F1.

In the experimental infection laboratory at LMD, adult females of *An. darlingi* from three to five days old were fed with uninfected blood. The procedure was performed by means of artificial feeders consisting of a glass feeder with a closed tubular central cavity, the lower part being sealed by a membrane. The feeders are heated by a water circulation system (water bath) connected by hoses and kept at a constant temperature of 39 °C and then taken to the environmental rooms (Dinglasan et al., 2009; Bahia et al., 2013).

After artificial feeding inside the climatic simulation environments of the environmental rooms, the F_1 generation eggs were obtained by oviposition in 500 mL paraffin-coated cups maintained in the current, intermediate, and extreme environments of the environmental rooms. After oviposition, female *An. darlingi* fed uninfected human blood and females fed only 10% sugar solution were kept separately in screened cages for daily observation of survival time and mortality rate in the current, intermediate, and extreme environments. The total duration of the experiment was 31 days.

2.3 Analysis of Gene Expression

The *Ae. aegypti* mosquitoes from the pre-established breeding in the environmental rooms in the current scenario (46th generation), intermediate scenario (48th generation), and extreme scenario (65th generation) were used for gene expression analysis. Thirty males and thirty females were collected at random, in triplicate, from the current, intermediate, and extreme scenarios for total RNA extraction, quantification, and construction of cDNA libraries, following the protocol established by Do Nascimento Neto et al. (2020).

The qPCR assays were performed in triplicate and analyzed in Quant Studio 6 (Applied Biosystems), and the reactions were prepared using Power SYBR Green PCR Master Mix (Applied Biosystems) with ROX, following the manufacturer's recommendations. In the assays, primers were used for the detection of the protein-coding genes *Hsp26, Hsp83,* and *Hsc70* and normalizers ACTIN, α-tubulin, RPS17, RPL32, and eEF1α. The transcripts were analyzed using relative expression levels and the ΔΔCt method (Livak and Schmittgen, 2001) using multiple endogenous controls.

2.4 Capture and Breeding of *Aedes Aegypti*

Samples of eggs of *Ae. aegypti* were captured in households and their external areas in the following neighborhoods city of Manaus, capital of the state of Amazonas, Brazil: Center (03° 08′ 33,5″ S, 60° 01′ 13,5″ W) in March 2019; Coroado (03°09′34,17″ S, 59°98′30.08″ W) in January 2020; and at four points in the Praça 14 neighborhood: Avenida Japurá (3°07′17.1″ S 60°00′40.3″ W), Avenida Codajás (3°07′11.3″ S 60°00′40.5″ W), Avenida Nhamundá (3°07′12.6″ S 60°00′53.4″ W), and the Rua do Cruzeiro (3°07′09.9″ S 60°00′45.1″ W).

The *Ae. aegypti* eggs were taken to the Insectarium of the Laboratory of Cytogenetics, Genomics and Evolution of Mosquitoes (LCGEM) at the National Institute for Amazonian Research (INPA), Manaus, Amazonas. These were kept in a controlled environment (26 ± 2 °C; 70-80% humidity, photoperiod of 12 h light and 12 h dark), until the eggs hatched, and the adults emerged, whereby they were taxonomically identified (Forattini, 2002; Consoli & Oliveira, 1994).

Male and female of *Ae. aegypti* were transferred to cages for successive crossings (10 generations) to obtain individuals less affected by the possible effects of synthetic insecticide molecules. From the oviposition of these generations of *Ae. aegypti*, the greatest number of eggs (n=5,000) were used in ovicidal bioassays with dillapiole piperidyl, dillapiole isopropyl ether, and dillapiole ethyl ether substances.

2.5 Production of the Dillapiole Piperidyl, Dillapiole Isopropyl Ether, and Dillapiole Ethyl Ether Substances

Leaves, thin stems, and flowers of *Piper aduncum* were collected from plants grown at Empresa Brasileira de Pesquisa Agropecuária (Embrapa Amazônia Ocidental) at highway marker 23 of the state highway AM-010 (02° 24′ 52″ S, 54° 42′ 36″ W) in the municipality of Manaus, state of Amazonas, Brazil. The dry material (20 kg) from *P. aduncum* was dehydrated in an oven at 40 °C. In the extraction of essential oil (EO) from the plant, a total of 12.5 kg of dry material was used for the isolation, purification, and identification of the dillapiole molecule (Pinto et al., 2012), followed by the reaction and acquisition of dillapiole piperidyl, dillapiole isopropyl ether, and dillapiole ethyl ether.

The derivative dillapiole piperidyl was obtained from dillapiole via the Mannich reaction, as adapted from Kumar et al. The derivative dillapiole piperidyl was isolated using a silica gel 60 chromatographic column and was eluted with hexane:AcOEt (8:2, 7:3). The dillapiole isopropyl ether was obtained from dillapiole (52 mg) in THF (0.2 mL) and mercury acetate ($Hg(OAc)_2$) in isopropyl alcohol (0.8 mL) under stirring and in an N_2 atmosphere (0 °C), for 72 h, at room temperature. After this time, the organomercury intermediate was reduced via the addition of an aqueous $NaBH_4$/KOH solution. The solution was filtered and extracted with dichloromethane. The organic phase was washed with water, a saturated solution of NaCl, and dried with $MgSO_4$. After the removal of the solvent, an etheric derivative was obtained in crude form (Pinto, 2008). Dillapiole isopropyl ether was purified by flash column chromatography by diluting with Hex:AcOEt (95:5, 1:1) and MeOH to obtain a light yellow oily substance that was identified using 1H and 13C NMR. The substances ethyl dillapiole ether and isopropyl dillapiole were obtained by adding the isopropyl (-CH3-CH-CH3) and ethyl (-CH2-CH3) radicals after a chemical process of oxymercuration, according to Pinto et al. (2012).

2.6 *Aedes aegypti* Larvicidal Bioassays with Dillapiole Derivatives

Dillapiole piperidyl was diluted in concentrations of 12.5–200 μg/mL (v/v) for the toxicity bioassay. In the positive control (PC), temephos was used at a concentration of 0.012 μg/mL (v/v) and in the negative control (NC) distilled water and dimethyl sulfoxide (DMSO—$(CH_3)_2SO$—Sigma Aldrich) were used at a concentration of 0.5% (v/v). In the insectarium, the larvae were kept in enameled trays containing drinking water and were fed with TetraMin Tropical Flakes® fish feed until pupation. After reaching adulthood, males and females were transferred to cages for breeding. Males were fed with 10% (w/v) sugar solution. The females were fed with blood from 4-week-old hamsters (*Mesocricetus auratus*), twice a week, for maturation of the eggs, to obtain the F_1 generation, of which about 500 eggs were transferred to trays for hatching. Of the larvae that reached the third instar, 20 specimens were transferred to plastic cups (50 mL), containing either 10 mL of drinking water with dillapiole piperidyl diluted in DMSO or the positive and negative controls, in triplicate, which totaled 420 individuals (Consoli & Oliveira, 1994; World Health Organization –WHO, 2005). Larval mortality was measured 24 h after the start of the bioassay. Dead larvae and those that did not respond to mechanical stimuli were counted to determine the larvicidal effect of dillapiole piperidyl.

Dillapiole isopropyl ether was used in the bioassay for toxicity. Third instar larvae were added to containers with 10 mL of drinking water and dillapiole isopropyl ether (μg/mL) at concentrations of 12.5, 25, 50, 100, 200 μg/mL diluted in DMSO a (0.05%), with 5 replicates each, and the negative control (NC), for 24 h. Larval counts were taken at 4-h intervals to verify the mortality rate. The lethal concentrations (LC_{50} and LC_{90}) in 50% and 90% of the subjects were calculated using Probit analysis in the Polo-Plus 2.0 program.

The egg count for the oviposition rate of *Ae. aegypti* subjected to dillapiole isopropyl ether concentrations was carried out for four generations via the separation of 10 couples for each concentration and generation, one in each 500 mL plastic cup, which was covered with phyllo mesh, with females fed in the first blood meal. For 10 days, the couples were offered 10% sugar solution. Egg counting and analysis were performed with the naked eye and with the aid of a stereomicroscope (binocular magnifier, Carl Zeiss), respectively.

2.7 Comet Assay

In the comet assay, the substance DEE and the chemical insecticide temephos (TE) (Fersol 500 CE) were used on the third instar larvae of *Ae. aegypti*, in triplicate for each treatment, at concentrations of 40, 60, and 80 μg/mL of DEE and 0.002, 0.005, and 0.007 μg/mL, and the negative control (NC), which contained filtered water, for 4 h. The larvae (n = 10 – total of 210 larvae) were transferred to 100 mL plastic cups. The larvae of each group were transferred to cups containing 20 mL of Ringer's solution (Equiplex®).

2.8 Cytological Preparations of *Ae. aegypti* for the Comet Assay

The heads of the *Ae. aegypti* larvae were dissected and placed on a Kline plate containing 50 μL of Ringer's solution. The brain nodes were dissected and transferred to 2 mL microtubes containing 300 μL of Ringer's solution and then homogenized. The cell suspension (50 μL) was mixed with 120 μL of low melting point agarose (0.5%) and then spread on glass slides, coated with normal melting point agarose (1.5%), and kept at 4 °C for 5 min. Then, the slides were immersed in a lysis solution (2.5 M NaCl, 100 mM ethylenediaminetetraacetic acid (EDTA), 1% Triton X-100, and 10 mM Tris), pH 10, freshly prepared, for 60 minutes at 4 °C.

After lysis, the slides were transferred to a horizontal electrophoresis vessel containing 300 mM NaOH and 1 mM EDTA (pH > 13) for DNA denaturation, for 20 min. Electrophoresis was performed at 1 V/cm (25 V and 300 mA) for 15 minutes. The slides were subsequently immersed in neutralizing buffer (0.4 mM Tris-HCl, pH 7.5) for 10 min. The cytological preparations were stained with GelRed® (50 μL): 1μL in 10,000 μL of distilled water. After drying at room temperature, the slides were fixed in ethanol PA for 10 minutes and stored until further analysis.

2.9 Microphotographs

Immediately after the preparation, the slides were analyzed using the Cy^3 filter at a wavelength of 510–575 nm by means of a camera (AxiomCam MRc) coupled to an epifluorescence microscope (Axio Imager 2, Zeiss), with a 100× immersion objective and 1×, 1.25× and 1.6× optovar. A total of 100 comets were captured per slide and analyzed at 400x magnification, totaling 500 structures per treated group. The results were presented as the frequency of damage observed on the comet's tails, in accordance with Collins et al. (1995).

2.10 Collection of *Ei. crassipes*, Bacterial Processing and Isolation

Collections of *Ei. crassipes* (n = 3) were carried out in a flooded area of the Jirau Hydroelectric Plant, Nova Mutum Paraná, Porto Velho, RO, Brazil. The samples were packed in polyethylene plastic bags and sent to the Laboratory of Cytogenetics, Genomics and Evolution of Mosquitoes (LCGEM) at INPA, in Manaus, Amazonas state.

The samples were processed according to the protocol of El-Deeb et al. (2006) with modifications, and began with the superficial disinfection of plant tissues (roots and stems) with a neutral detergent and a sterile sponge under running water to remove the external microbiota. In a laminar flow chamber, only the roots were soaked in 70% alcohol for 1 min and 3% sodium hypochlorite solution for 1 min. Shortly after, the roots were washed again in 70% alcohol for 30 seconds and three times in sterile milli-Q water.

For the isolation of the endophytic bacteria, after asepsis, the roots were cut into 5 mm fragments with the aid of a scalpel blade and placed in Petri dishes containing Luria-Bertani (LB) agar culture medium and 20 mg/mL of fluconazole. The plates were incubated in a bacteriological oven at 30 °C for 24 h. The isolated bacterial colonies were purified in LB medium and cryopreserved in 30% glycerol at −80 °C. Subsequently, 15 colonies were reactivated for molecular identification and selective pathogenicity testing.

2.11 Molecular Identification of Bacterial Isolates

The genomic DNA of 15 endophytic bacteria was extracted according to the protocol of Govindarajan et al. (2007), and the 16S rDNA gene was amplified in the polymerase chain reaction (PCR) using the bacteria-specific primer 27 F (5′-GAGTTTGATCACTGGCTCAG-3′) and the universal primer 1492 R (5′-TACGGCTACCTTGTTACGACTT-3′) (Byers et al., 1998). In summary, using the GoTaq® DNA Polymerase reagent set (Promega, Madison, WI, USA), each 25 μL reaction consisted of 10× buffer B, 25 mM $MgCl_2$, 10 mM dNTPs, 10 mM of each primer, 10 ng of the extraction products, and 5 U/μL Taq DNA polymerase and ultrapure water q.s.p. Amplification was performed in a thermocycler (ProFlex-PCR, Applied Biosystems) with the initial denaturation programs at 95 °C for 2 min, followed by 30 cycles of 95 °C for 60 s, 59 °C for 40 s, 72 °C for 90 s, and final extension at 72 °C for 5 min. Shortly after, the amplicons were visualized on 0.8% agarose gel and purified with the Agencourt® Ampure® XP kit, following the manufacturer's recommendations. The sequences were determined in a genetic analyzer (ABI 3500, Applied Biosystems), edited in the SeqManPro program (DNASTAR Lasergene 12), according to Meireles et al. (2022), and then compared in nonredundant DNA databases (GenBank, DDBJ, EMBL) using BLASTN of the National Center for Biotechnology Information (NCBI) to identify the closest bacterial species (Altschul et al., 1990).

2.12 Selective Pathogenicity Test of Endophytic Bacteria

The selective pathogenicity test was performed following the parameters established by World Health Organization—WHO (2005), along with the protocols modified by Soares-Silva et al. (2017) and Katak et al. (2021), in two procedural steps:

In the first stage, a laboratory colony of *Ae. aegypti* was reactivated and 15 endophytic bacteria from the cryopreserved *Ei. crassipes* roots. The standard methodology for breeding and maintaining mosquito colonies in the LCGEM/COSAS/INPA Insectarium, described previously, was adopted to obtain third instar *Ae. aegypti* larvae. To reactivate the bacteria, 15 strains were seeded in Petri dishes with LB medium and incubated at 30 °C for 24 h. Then, each strain was pre-inoculated in a 50 mL Falcon tube containing 5 mL of LB medium, kept at 30 °C and 180 rpm for 24 h. Of these pre-inoculated strains, 20 μL were used to prepare the final bacterial culture in a 25 mL Falcon tube containing 15 mL of LB medium, which was kept at 30 °C and 180 rpm for 120 h.

In the second stage, the bioassays were performed in 100 mL plastic cups, in triplicate, for the 17 treatments. A total of 510 third instar *Ae. aegypti* larvae were used in the tests, with 30 larvae for each treatment. Immediately after separation of the larvae in cups containing 20 mL of distilled water, 1 mL of culture of the 15 bacterial strains (Table 1) and 1 mL of culture of the standard strain *Bacillus thuringiensis israelis*—BtiBR101NA in the positive control (PC) were added. The negative control (NC) contained only sterile distilled water. The number of dead larvae was checked at intervals of 24, 48, and 72 h.

2.13 Statistical Analysis

Gene Expression Normality and equality of variance tests were performed. The two-way ANOVA was used to analyze sex and climate scenarios for each transcript, additionally with a paired post-hoc multiple comparison test (Tukey) with a significance level of $p < 0.05$. The tests were expressed as means ± standard deviations. Statistical analyses were performed in SigmaStat v3.5 software, with graphs elaborated in SigmaPlot V. 11.0 software.

Table 1 Toxic profile of endophytic bacteria from *Eichhornia crassipes* against *Aedes aegypti* larvae at 24, 48, and 72 h

N°	Strains	24 h	48 h	72 h
C+	*BtiBR101*	*90%*	*100%*	*100%*
C−	*H_2O*	*0*	*0*	*0*
1	End20LB	0	0	0
2	End22LB	0	0	0
3	End35LB	0	0	0
4	End52LB	0	0	0
5	End53LB	0	0	0
6	*End64LB*	*24%*	*60%*	*70%*
7	End69LB	0	0	0
8	End75LB	0	0	0
9	End79LB	0	0	0
10	End81LB	0	0	0
11	*End82LB*	*17%*	*30%*	*60%*
12	*End84LB*	*30%*	*60%*	*70%*
13	End128LB	0	0	0
14	End130LB	0	0	0
15	End139LB	0	0	0

Toxicity Considering the egg count and mortality of larvae *Ae. aegypti* submitted to the dillapiole ethyl ether, dillapiole isopropyl ether, and dillapiole piperidyl substances, we performed the Probit analysis (Finney 1971) to obtain the values of LC_{50} and LC_{90}, with the respective confidence intervals (CI-95%).

Comet Assay The results of the comet assays were obtained via the mean ± SD for each group of animals and an analysis of variance (ANOVA) was applied to determine the level of variation between the groups, followed by a multiple comparison using the Tukey test. This analysis was performed in the software R, version 4.1.0, and $p < 0.05$ was considered significant in all cases.

3 Results and Discussion

3.1 Evaluation of Transcripts and Biology of Vector Mosquitoes Under Simulated Climatic Conditions

Mosquitoes are a group of invertebrates that have a notorious prominence regarding public health, either due to their high capacity for adaptation or due to their peculiarities regarding their biology. As in studies with other vectors of medical importance, investigations involving mosquitoes seek new efficient control and monitoring strategies via their biology and genetics, either under current or using future climatic conditions via studies in simulators.

In the studies of do Nascimento Neto et al. (2020) with male and female *Ae. aegypti*, the transcription of the heat shock protein genes *Hsp26*, *Hsp83*, and *Hsc70* under current, intermediate, and extreme climatic conditions was evaluated. The transcription of the genes for the females of the vector presented high levels in the intermediate scenario; while, in the extreme scenario, they had values that were similar to those of the control environment. In males, there was a reduction in the transcription of the *Hps26* gene in the intermediate scenario of up to three times in relation to the current scenario. However, for the *Hsp83* and *Hsc70* genes, the differences in the transcription of genes from the treatment scenarios in relation to the current one were not significant. The results of the transcription of the heat shock protein genes suggest the action of compensation mechanisms on the mechanisms of protein stability in the face of extreme climatic conditions. In studies carried out after shock exposure followed by acclimatization, elevations in gene transcription levels are observed; however, shock exposure is not stable when compared to long acclimatization. This can affect the stability of cellular protein systems, in addition to mechanisms associated with the replication of certain viruses such as dengue (Muturi et al., 2011; do Nascimento Neto et al., 2020). Under these same climatic conditions, the influence of two climate scenarios for the year 2100 on the diet with sugar and blood in *An. darlingi* was investigated. A total of 361 adult males and 190 adult females were kept together in cages, half of which were fed with a sugar solution and the other half with human blood. Females fed blood survived longer than individuals (both males and females) who were given the sugar solution. The biological cycle of the mosquitoes was shorter in the warmer scenarios (in particular, in the intermediate environment) when compared to the control. The higher temperatures and higher CO_2 concentrations in the most extreme conditions exerted a negative pressure on the *An. darlingi* colonies, which caused a decline in the survival of these mosquitoes, regardless of the type of feeding.

In this sense, the analyses performed by Azevedo et al. (2015) evaluated the influence of the IPCC climate scenarios on colonies of *Ae. aegypti* via climate simulations in environmental rooms (LEEM/INPA) for more than 25 generations. In the study, feeding, average weight, and reproductive success had significant increases in the intermediate and extreme scenarios of the IPCC, while oviposition, egg viability, and longevity of adults decreased proportionally. This demonstrates that mosquito populations in future climate scenarios can survive in extreme climatic environments and could continue to be vectors of important arboviruses. These results mainly highlight the presence of a faster life cycle, which contributes to a higher population density (Kraemer et al., 2015). Temperature also influences egg dormancy and competition of vector larvae in breeding sites, which can influence mosquito density and, consequently, the spread of arboviruses to humans (Ezeakacha & Yee, 2019; Garzón et al., 2021).

The impact on the survival of *Ae. aegypti* larvae and the risk of predation by larvae of *Toxorhynchites haemorrhoidalis* in climate simulation scenarios were also evaluated. The study demonstrated that predation risk and its interactions did not significantly affect the survival of *Ae. aegypti* and *To. haemorrhoidalis* larvae. The *Ae. aegypti* larvae showed high survival rates even in extreme scenarios. In relation to the survival rate, both increased proportionally; thus, the risk of predation was not influenced. This suggests that, although climate change affects the development of larvae, their accelerated development in extreme climatic environments may favor their survival when competing with larvae of other species for the environment.

Other studies conducted with *Ae. aegypti* and climate simulation (in press) evaluated the transcription of genes associated with lipid maintenance of the mosquito during oogenesis. Significant decreases in the transcription levels of some genes have been detected in extreme climatic conditions, which can compromise the ability of eggs to survive for long periods in extreme climatic environments. The presence of lipid material in the eggs and the good formation of vitellogenic constituents are fundamental for the long-term maintenance of *Ae. aegypti* embryos (Sun et al., 2000; Fruttero et al., 2011). As such, the decrease in transcription of the associated genes can affect the survival time of eggs in breeding sites, thus influencing the population density of the vector.

3.2 Toxicity and Genotoxicity of Dillapiole Derivatives in *Ae. Aegypti*

Larvae of *Ae. aegypti* exposed to the substance dillapiole piperidyl, for 24 h, presented mortality of 80% at the highest concentration (200 μg/mL) and 17% at the lowest concentration (12.5 μg/mL); while, in the positive and negative controls, mortality was 95% and 3%, respectively.

When testing dillapiole, the precursor molecule of dillapiole piperidyl, Rafael et al. (2008) observed high mortality of *Ae. aegypti* in concentrations of 200 and 400 μg/mL. Toxic effects of other dillapiole derivatives against *Ae. aegypti* were observed by Domingos et al. (2014), 24 h after exposure to dillapiole ethyl ether and dillapiole *n*-butyl ether, which caused 100% mortality of larvae at concentrations ranging from 12.5 to 80 μg/mL. Against *Aa. albopictus*, both substances showed similar larvicidal activity (Meireles et al., 2016). When testing dillapiole methyl ether (DME) against *Ae. aegypti*, Silva et al. (2019) observed 100% larval mortality at the concentration of 140 μg/mL after 24 h, and at the concentrations of 120 and 140 μg/mL after 48 h.

The literature reports that dillapiole and the phenylpropanoids safrole, eugenol, and cinnamic aldehydes have the methylenedioxyphenyl group in their composition, which have larvicidal activity against *Ae. aegypti* (Silva et al., 2008). Similarly, dillapiole derivatives, which have this group in their structure, have shown a toxic effect against this mosquito (Pinto et al., 2012). The structural characteristics of chemical molecules are important in determining their biological activity against insects (Pinto et al., 2012). Thus, it is necessary to explore new active compounds of phytochemical origin, with potential insecticidal effects.

Regarding the evaluation of the mortality rate of the *Ae. aegypti* larvae exposed to dillapiole isopropyl ether, mortality was 68% and 84% in the third instar larvae after exposure for 24 h at concentrations of 100 μg/mL and 120 μg/mL, respectively. There was no significant variation between concentrations of 100 and 120 μg/mL of dillapiole isopropyl ether and, when compared to the negative control (water and DMSO 0.05%), all the concentrations caused a significant increase in larval mortality. Using other semisynthetic of dillapiole, Meireles et al. (2016) reported 100% mortality of *Ae. albopictus* at 40 μg/mL using dillapiole *n*-butyl ether (DBE) and 80 μg/mL of dillapiole ethyl ether (DEE) after 24 h of exposure.

The LC_{50} of dillapiole isopropyl ether (DPE) in *Ae. aegypti* larvae was 88.01 μg/ml, while the LC_{90} was 196.84 μg/mL after 24 h of exposure. Domingos et al. (2014) demonstrated that the semisynthetic dillapiole *n*-butyl ether and dillapiole ethyl ether caused 100% mortality of *Ae. aegypti* larvae, after 24 h, with an LC_{50} of 18.6 ± 1.9 and LC_{90} of 27.1 ± 2.3 μg/mL for dillapiole *n*-butyl ether and an LC_{50} of 61.8 ± 1.9 and LC_{90} of 89.0 ± 1.7 μg/mL for dillapiole ethyl ether.

The oviposition rates of female *Ae. aegypti*, in generations F_1 to F_4, treated at a concentration of 100 μg/mL of dillapiole isopropyl ether were 39.6%, 26.6%, 25.8%, and 22%, respectively. However, at the concentration of 120 μg/mL, the oviposition rates of 28.8%; 18.8%, 14.8%, and 15.4% were obtained. On the other hand, in the negative control, they were 41%, 41.4%, 45.8%, and 32.6%. Thus, according to the data for the oviposition rate of the 100 and 120 μg/mL treatments, these were statistically lower ($p < 0.05$) when compared to the negative control. There was a gradual reduction in the oviposition rate that was inversely proportional to the increase in concentrations during generation formation (F_1 to F_4). Similarly, Silva et al. (2019) obtained comparable results when exposing *Ae. aegypti* to the semisynthetic dillapiole methyl ether at a concentration of 140 μg/mL for 24 h, with nonviability of 97% of the eggs.

The larvicide temephos, from the organophosphate class, is widely used in Brazil to combat *Ae. aegypti*; however, due to its constant use over about 30 years, this mosquito has acquired resistance to the insecticide temephos and deltamethrin (Valle et al., 2019). In the present study, temephos and dillapiole ethyl ether at concentrations of 0.005 and 0.007 μg/mL and 40 and 60 μg/mL, respectively, showed a statistically significant increase in the frequency of genomic lesions in *Ae. aegypti* (Fig. 1) via the comet assay analysis. Similar results were reported by Silva et al. (2021), who observed the increase of chromosomal alterations in neuroblasts of *Ae. aegypti* larvae exposed to temephos and dillapiole methyl ether at concentrations of 0.002, 0.005, and 0.007 μg/mL and 60, 80, and 100, respectively.

In the larvae of *Ae. aegypti* exposed for 4 h to dillapiole ethyl ether at the same concentrations used in the present study, Domingos et al. (2014) detected micronucleus in interphase, pro-metaphase, and metaphase of neuroblasts and oocytes. This same substance was tested in *Ae. albopictus*, with reports of chromosomal alterations (Meireles et al., 2016). These data corroborate the results of the present study regarding the increased frequency of DNA damage in *Ae. aegypti* neuroblasts tested with dillapiole ethyl ether.

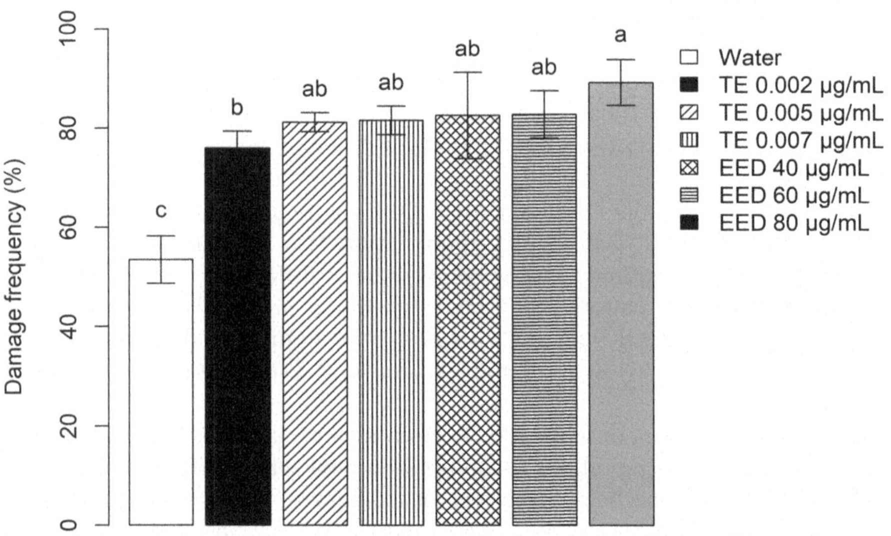

Fig. 1 Frequency of DNA damage to neuroblasts of *Aedes aegypti* larvae observed viathe comet assay, 4 h after the onset of exposure to dillapiole ethyl ether and temephos, at concentrations of 40, 60 and 80 μg/mL and 0.002, 0.005 and 0.007 μg/mL, respectively, and water (negative control). Statistical analysis was performed using ANOVA and Tukey's test, with a significance threshold of $p < 0.05$

3.3 Isolation and Identification of Entomopathogenic Bacteria

Regarding the search for entomopathogenic microorganisms of botanical origin, a total of 56 endophytic bacterial colonies were isolated from the roots of the aquatic plant *Ei. crassipes*. Of these, 15 strains were identified as belonging to the phyla Proteobacteria and Firmicutes, and three strains (End64LB, End82LB, and End84LB) showed larvicidal activity against third instar *Ae. aegypti* larvae (Table 1). Mortality caused by the End64LB strain was 24, 60, and 70%, at 24, 48, and 72 h, respectively. End82LB caused lethality of 17, 30, and 60% at 24, 48, and 72 h, respectively, while the End84LB strain caused lethality of 30, 60, and 70% at times of 24, 48, and 72 h, respectively. In the PC, the BtiBR101 strain caused mortality of 100% of the larvae, and in the NC, no mortality occurred in any of the intervals.

Endophytic bacteria have numerous functional characteristics that make them interesting targets for applications in biological pest control (Bensidhoum et al., 2016). Via the culture-dependent method used in this study, 56 endophytic bacterial colonies were isolated from the roots of *Ei. crassipes*. El-Deeb et al. (2012) isolated 95 bacterial strains from different parts of this plant also with this methodology. The selective pathogenicity test revealed the larvicidal potential of three bacterial strains against *Ae. aegypti* larvae, thus corroborating the results of Katak et al. (2021) who detected 19 strains of *Bacillus* sp. and two strains of *Brevibacillus halotolerans* from various Amazonian environments that showed larvicidal activity against *Ae.*

aegypti, and those of Dahmana et al. (2020) who demonstrated the potential of 14 entomopathogenic bacteria that were active against *Ae. Albopictus*.

In this work, molecular identification based on the 16S rRNA gene of 15 endophytic bacterial strains revealed that they belong to the phyla Proteobacteria and Firmicutes. The literature shows that most bacterial endophytes are grouped in the phyla Proteobacteria and Firmicutes, and to a lesser extent, Bacteroidetes and Actinobacteria (Liu et al., 2017; Ali et al., 2021). Of these, the genus *Bacillus* sp. (Firmicutes) has known metabolic diversity, such as insecticidal proteins (CRY, VIP, and SIP), hydrolytic enzymes (chitinases and beta-glucanases), and lipopeptides with mechanisms of pathogenicity against mosquitoes that are vectors of diseases (Valtierra-de-Luis et al., 2020).

4 Conclusions

Throughout the climate simulation analyses with *Ae. aegypti* and *An. darlingi*, impacts on key aspects of biology, such as food intake, lifespan, oviposition, and egg viability, were observed. The combined effects of temperature and CO_2, associated with biological data, point to a possible decrease in population density in warmer scenarios, despite the reproduction of the *Ae. aegypti* continuing to be efficient. However, climate change can negatively affect the population density of *An. darlingi*, which may be a factor that affects the vector capacity of the mosquito due to its lower ability to survive in the most severe scenarios. Nonetheless, more studies need to be conducted to better understand the long-term impact of climate change on the biology and genetics of *Ae. aegypti* and *An. darlingi*.

The tested substances dillapiole piperidyl and dillapiole isopropyl ether caused toxicity and mortality of *Ae. aegypti*. Dillapiole isopropyl ether caused a decrease in the oviposition rate in females, and dillapiole ethyl ether, even in low concentrations, caused DNA damage to larvae of this mosquito in a similar manner to the chemical larvicide temephos. In addition, the selective test of the three entomopathogenic strains (End64LB, End82LB, and End84LB) of *Ei. crassipes* against *Ae. aegypti* presented potential for the identification of promising active molecules for the biological control of this mosquito. These results suggest that dillapiole derivatives and the three new strains of entomopathogenic bacteria studied are promising for the control of *Ae. aegypti*.

Acknowledgments The authors would like to thank CNPq/FAPEAM/INCT—ADAPTAII/INPA, Process No. 465540/2014-7 for Adalberto Luis Val; Governo do Estado do Amazonas, Secretaria de Estado de Desenvolvimento Econômico, Ciência, Tecnologia e Inovação (SEDECTI), Fundação de Amparo à Pesquisa do Estado do Amazonas (FAPEAM)—Institutional program for the support of postgraduate studies Stricto Sensu (FAPEAM-POSGRAD 2020/2021) (T.O. 159/2020 and 186/2021 for Jacqueline da Silva Batista/PPG-GCBEv/INPA); the GCBEv post-graduate program at Instituto Nacional de Pesquisas da Amazônia (INPA), and the Laboratório Temático de Biologia Molecular (LTBM) at INPA, and Laboratório de Vetores da Malária e Dengue/INPA for logistical support and maintenance teams in the collection points of *Ae. aegypti* located in Manaus, capital of the state of Amazonas, Brazil.

Potential Competing Interest The authors declare that they have no conflicts of interest. This manuscript contains original data, which have not been published and are not being considered for publishing in any other journal.

Ethical Approval The use of animals in a vector colony occurred according to the National Council for the Control of Animal Experimentation (CONCEA) and the Ethics Commission for the Use of Animals (CEUA) of the National Institute for Amazonian Research (INPA), registration No. (012/2021). Mosquito collections were carried out under licenses No. 32,941 (2012) and 77,078 (2020), from the Biodiversity Information and Authorization System (SISBIO)/Chico Mendes Institute for Biodiversity Conservation (ICMBio), Brazil. Access to genetic heritage was registered in the National System for the Management of Genetic Heritage and Associated Traditional Knowledge (SISGEN), under No. A949853.

References

Ali, M., Ali, Q., Sohail, M. A., Ashraf, M. F., Saleem, M. H., Hussain, S., & Zhou, L. (2021). Diversity and taxonomic distribution of Endophytic bacterial Community in the Rice Plant and its Prospective. *International Journal of Molecular Sciences, 22*(18), 10165. https://doi.org/10.3390/ijms221810165

Altschul, S. F., Gish, W., Miller, W., Myers, E. W., & Lipman, D. J. (1990). Basic local alignment search tool. *Journal of Molecular Biology, 215*, 403–410. https://doi.org/10.1016/s0022-2836(05)80360-2

Alves de Oliveira, B. F., Bottino, M. J., Nobre, P., & Nobre, C. A. (2021). Deforestation and climate change are projected to increase heat stress risk in the Brazilian Amazon. *Communications Earth & Environment, 2*, 207. https://doi.org/10.1038/s43247-021-00275-8

Arcos, A. N., da Silva Ferreira, F. A., da Cunha, H. B., & Tadei, W. P. (2018). Characterization of artificial larval habitats of *Anopheles darlingi* (Diptera: Culicidae) in the Brazilian Central Amazon. *Revista Brasileira de Entomologia, 62*, 267–274. https://doi.org/10.1016/j.rbe.2018.07.006

Azevedo, J. B., Lima, J. B. P., & Tadei, W. P. (2015). *Análise do ciclo biológico do Aedes aegypti (Diptera: Culicidae) exposto a cenários de mudanças climáticas previstas pelo IPCC (Intergovernmental Panel on Climate Change)*. Instituto Nacional de Pesquisas da Amazônia. 53p.

Bahia, A. C., Oliveira, J. H. M., Kubota, M. S., Araújo, H. R. C., Lima, J. B. P., Ríos-Velásquez, C. M., et al. (2013). The role of reactive oxygen species in *Anopheles aquasalis* response to *plasmodium vivax* infection. *PLoS One, 8*, e57014.

Barbalho, G. N., Matos, B. N., da Silva Brito, G. F., da Cunha, M. T., Alencar-Silva, T., Sodré, F. F., et al. (2022). Skin regenerative potential of Cupuaçu seed extract (*Theobroma grandiflorum*), a native fruit from the Amazon: Development of a topical formulation based on chitosan-coated Nanocapsules. *Pharmaceutics, 14*, 207. https://doi.org/10.3390/pharmaceutics14010207

Bellinato, D. F., Viana-Medeiros, P. F., Araújo, S. C., Martins, A. J., Lima, J. B., & Valle, D. (2016). Resistance status to the insecticides temephos, deltamethrin, and diflubenzuron in Brazilian *Aedes aegypti* populations. *BioMed Research International, 2016*, 8603263. https://doi.org/10.1155/2016/8603263

Bensidhoum, L., Nabti, E., Tabli, N., Kupferschmied, P., Weiss, A., Rothballer, M., et al. (2016). Heavy metal tolerant *pseudomonas protegens* isolates from agricultural well water in northeastern Algeria with plant growth promoting, insecticidal and antifungal activities. *European Journal of Soil Biology, 75*, 38–46. https://doi.org/10.1016/j.ejsobi.2016.04.006

Byers, H. K., Stackebrandt, E., Hayward, C., & Blackall, L. L. (1998). Molecular investigation of a microbial mat associated with the great Artesian Basin. *FEMS Microbiology Ecology, 25*, 391–403. https://doi.org/10.1111/j.1574-6941.1998.tb00491.x

Chaves, L. F., Scott, T. W., Morrison, A. C., & Takada, T. (2014). Hot temperatures can force delayed mosquito outbreaks via sequential changes in *Aedes aegypti* demographic parameters in autocorrelated environments. *Acta Tropica, 129*, 15–24. https://doi.org/10.1016/j.actatropica.2013.02.025

Chediak, M., Pimenta, F. G., Jr., Coelho, G. E., et al. (2016). Spatial and temporal country-wide survey of temephos resistance in Brazilian populations of *Aedes aegypti*. *Memórias do Instituto Oswaldo Cruz, 111*, 311–321. https://doi.org/10.1590/0074-02760150409

Cochrane, M. A., & Barber, C. P. (2009). Climate change, human land use and future fires in the Amazon. *Global Change Biology, 15*, 601–612. https://doi.org/10.1111/j.1365-2486.2008.01786.x

Collins, A. R., Ai-Guo, M., & Duthie, S. J. (1995). The kinetics of repair of oxidative DNA damage (strand breaks and oxidised pyrimidines) in human cells. *Mutation Research, DNA Repair Reports, 336*(1), 69–77. https://doi.org/10.1016/0921-8777(94)00043-6

Collins, E., Vaselli, N. M., Sylla, M., et al. (2019). The relationship between insecticide resistance, mosquito age and malaria prevalence in *Anopheles gambiae* sl from Guinea. *Scientific Reports, 9*(1), 1–12. https://doi.org/10.1038/s41598-019-45261-5

Consoli, R. A. G. B., & Oliveira, R. L. (1994). *Principais Mosquitos de Importância Sanitária no Brasil* (1st ed., p. 228). Fiocruz.

Dahmana, H., Raoult, D., Fenollar, F., & Mediannikov, O. (2020). Insecticidal activity of bacteria from larvae breeding site with natural larvae mortality: Screening of separated supernatant and pellet fractions. *Pathogens, 9*, 486. https://doi.org/10.3390/pathogens9060486

Derkzen, M. L., van Teeffelen, A. J. A., & Verburg, P. H. (2017). Green infrastructure for urban climate adaptation: How do residents' views on climate impacts and green infrastructure shape adaptation preferences? *Landscape and Urban Planning, 157*, 106–130.

Desmarchelier, J. M., & Fukuto, T. R. (1974). Toxicological effects produced by some 1,3- benzodioxoles, catechols, and quinone in *Culex* mosquito larvae. *Journal of Economic Entomology, 67*, 153–158. https://doi.org/10.1093/jee/67.2.153

Dinglasan, R., Devenport, M., Florens, L., Johnson, J. R., McHugh, C., Donnelly-Doman, M., et al. (2009). The *Anopheles gambiae* adult midgut peritrophic matrix proteome. *Insect Biochemistry and Molecular Biology, 39*, 125–134.

do Nascimento Neto, J. F., da Mota, A. J., Roque, R. A., Heinrichs-Caldas, W., & Tadei, W. P. (2020). Analysis of the transcription of genes encoding heat shock proteins (hsp) in *Aedes aegypti* Linnaeus, 1762 (Diptera: Culicidae), maintained under climatic conditions provided by the IPCC (intergovernmental panel on climate change) for the year 2100. *Infection, Genetics and Evolution, 86*. https://doi.org/10.1016/j.meegid.2020.104626

Domingos, P. R. C., Pinto, A. C. S., Santos, J. M. M., & Rafael, M. S. (2014). Insecticidal and genotoxic potential of two semi-synthetic derivatives of dillapiole for the control of *Aedes (Stegomyia) aegypti* (Diptera: Culicidae). *Mutation Research, Genetic Toxicology and Environmental Mutagenesis, 772*, 42–54. https://doi.org/10.1016/j.mrgentox.2014.07.008

Edwards, D., & Musil, C. J. (1975). *Eichhornia crassipes* in South Africa—A general review. *Journal of the Limnological Society of Southern Africa, 1*, 23–27. https://doi.org/10.1080/03779688.1975.9632904

El-Deeb, B. A., El-Sharouny, H. M., & Fahmy, N. (2006). Plasmids incidence, antibiotic and heavy metal resistance patterns of endophytic bacteria isolated from aquatic plant, *Eichhornia crassipes*. *Journal of Botany, 33*, 151–171.

El-Deeb, B. A., Gherbawy, Y., & Hassan, S. (2012). Molecular characterization of Endophytic bacteria from metal Hyperaccumulator aquatic plant (*Eichhornia crassipes*) and its role in heavy metal removal. *Geomicrobiology Journal, 29*(10), 906–915. https://doi.org/10.1080/01490451.2011.635764-02760150409

Ernawati, K., Farras, R. M., Zakiyyah, A., et al. (2021). Community behavior in controlling *Aedes aegypti* Mosquito breeding places before and during the Covid-19 pandemic. *IOP Conference Series: Earth Environment Science, 940*. https://doi.org/10.1088/1755-1315/940/1/012081

Ezeakacha, N. F., & Yee, D. A. (2019). The role of temperature in affecting carry-over effects and larval competition in the globally invasive mosquito *Aedes albopictus*. *Parasites & Vectors, 12*, 123.

Flaibani, N., Pérez, A. A., Barbero, I. M., & Burroni, N. E. (2020). Different approaches to characterize artificial breeding sites of *Aedes aegypti* using generalized linear mixed models. *Infectious Diseases of Poverty, 9*, 107.

Forattini, O. P. (2002). *Culicidologia médica: identificação, biologia e epidemiologia* (Vol. 2). EDUSP.

França, L. P., Amaral, A. C. F., Ramos, A. S., et al. (2020). *Piper capitarianum* essential oil: A promising insecticidal agent for the management of *Aedes aegypti* and *Aedes albopictus*. *Environmental Science and Pollution Research International, 28*, 9760–9776. https://doi.org/10.1007/s11356-020-11148-6

Fruttero, L. L., Frede, S., Rubiolo, E. R., et al. (2011). The storage of nutritional resources during vitellogenesis of *Panstrongylus megistus* (Hemiptera: Reduviidae): The pathways of lipophorin in lipid delivery to developing oocytes. *Journal of Insect Physiology, 57*, 475–486. https://doi.org/10.1016/j.jinsphys.2011.01.009

Garzón, M., Maffey, J. L., Lizuain, A., Soto, D., Diaz, P. C., Leporace, M., Salomón, O. D., & Schweigmann, N. J. (2021). Temperature and photoperiod effects on dormancy status and life cycle parameters in *Aedes albopictus* and *Aedes aegypti* from subtropical Argentina. *Medical and Veterinary Entomology*. https://doi.org/10.1111/mve.12474

Gopal, B. (1987). *Water hyacinth. (aquatic. Plant studies 1)* (p. xii, 471). Elsevier. ISBN 0-444-42706-6.

Govindarajan, M., Kwon, S., & Weon, H. (2007). Isolation, molecular characterization and growth- promoting activities of endophytic sugarcane diazotroph *Klebsiella* sp. GR9. *World Microbialogy and Biotechnology, 23*, 997–1006. https://doi.org/10.1007/s11274-006-9326-y

Hamid, P. H., Ninditya, V. I., Prastowo, J., et al. (2018). Current status of *Aedes aegypti* insecticide resistance development from Banjarmasin, Kalimantan, Indonesia. *BioMed Research International*. https://doi.org/10.1155/2018/1735358

Harmatha, J., & Nawrot, J. (2002). Insect feeding deterrent activity of lignans and related phenylpropanoids with a methylenedioxyphenyl (piperonyl) structure moiety. *Entomologia Experimentalis et Applicata, 104*, 51–60. https://doi.org/10.1007/978-94-017-2776-1_6

Huong, L. T., Hung, N. H., Dai, D. N., et al. (2019). Chemical compositions and mosquito Larvicidal activities of essential oils from piper species growing wild in Central Vietnam. *Molecules, 24*, 3871. https://doi.org/10.3390/molecules24213871

Katak, R. M., Rocha, E. M., Oliveira, C., Muniz, V. A., Oliveira, M. R., et al. (2021). Larvicidal activities against *Aedes aegypti* of supernatant and pellet fractions from cultured *bacillus* spp. isolated from Amazonian microenvironments. *Tropical Medicine and Infectious Disease, 6*(2), 104. https://doi.org/10.3390/tropicalmed6020104

Kraemer, M. U., Sinka, M. E., Duda, K. A., et al. (2015). The global distribution of the arbovirus vectors *Aedes aegypti* and *Ae. Albopictus*. *eLife, 4*, e08347. https://doi.org/10.7554/eLife.08347

Kriegler, E., Bauer, N., Popp, A., Humpenöder, F., Leimbach, M., Strefler, J., et al. (2017). Fossil-fueled development (SSP5): An energy and resource intensive scenario for the 21st century. *Global Environmental Change, 42*, 297–315. https://doi.org/10.1016/j.gloenvcha.2016.05.015

Kubota, R. L., de Brito, M., & Voltolini, J. C. (2003). Método de varredura para exame de criadouros de vetores de dengue e febre amarela urbana. *Revista de Saúde Pública, 37*, 263–265. https://doi.org/10.1590/S0034-89102003000200017

Lee, S. E. (2000). Mosquito larvicidal activity of pipernonaline, a piperidine alkaloid derived from long pepper, *piper longum*. *Journal of the American Mosquito Control Association, 16*, 245–247.

Leta, S., Beyene, T. J., de Clercq, E. M., et al. (2018). Global risk mapping for major diseases transmitted by *Aedes aegypti* and *Aedes albopictus*. *International Journal of Infectious Diseases, 67*, 25–35. https://doi.org/10.1016/j.ijid.2017.11.026

Lima-Camara, T. N. (2016). Emerging arboviruses and public health challenges in Brazil. *Revista de Saúde Pública, 50*, 1–7.

Liu, H., Carvalhais, L. C., Crawford, M., et al. (2017). Inner plant values: Diversity, colonization and benefits from Endophytic bacteria. *Frontiers in Microbiology, 8*, 2552. https://doi.org/10.3389/fmicb.2017.02552

Lopes, I. G., Araújo-Dairiki, T. B., Kojima, J. T., Val, A. L., & Portella, M. C. (2018). Predicted 2100 climate scenarios affects growth and skeletal development of tambaqui (*Colossoma macropomum*) larvae. *Ecology and Evolution, 8*, 10039–10048. https://doi.org/10.1002/ece3.4429

Maia, J. G. S., Zohhbi, M. G. B., Andrade, E. H. A., et al. (1998). Constituents of the essential oil of *Piper aduncum* L. growing wild in the Amazon region. *Flavour and Fragrance Journal, 13*, 269–272. 10.1002/(SICI)1099-1026(1998070)13:4<269::AID-FFJ744>3.0.CO;2-A.

Majeed, S., Hill, S. R., & Ignell, R. (2014). Impact of elevated CO_2 background levels on the host-seeking behaviour of *Aedes aegypti*. *Journal of Experimental Biology, 217*, 598–604. https://doi.org/10.1242/jeb.092718

Malhotra, N. H. C. K., Chen, H. J., Lai, H., Uapipatanakul Roldan, B. M. J. M., Macabeo, A. P. G., Ger, T., & Hsiao, C. (2021). Physiological effects of Neonicotinoid insecticides on non-target aquatic animals—An updated review. *International Journal of Molecular Sciences, 22*(17), 9591. https://doi.org/10.3390/ijms22179591

Marques, A. C., & Kaplan, M. A. (2014). Active metabolites of the genus piper against *Aedes aegypti*: Natural alternative sources for dengue vector control. *Universitas Scientiarum, 20, 61.* https://doi.org/10.11144/javeriana.sc20-1.amgp

Meireles, S. F., Domingos, P. R. C., Pinto, A. C. S., & Rafael, M. S. (2016). Toxic effect and genotoxicity of the semisynthetic derivatives dillapiole ethyl ether and dillapiole n -butyl ether for control of *Aedes albopictus* (Diptera: Culicidae). *Mutation Research Genetic Toxicology Environment, 807*, 1–7. https://doi.org/10.1016/j.mrgentox.2016.07.003

Meireles, S. F., Santos, S. F., Rafael, M. S., Mota, A. J., & Nunes-Silva, C. G. (2022). Yeasts from the nests of two Amazonian stingless bees: Screening and PCR-RFLP molecular analysis. *Symbiosis, 87*, 153–163. https://doi.org/10.1007/s13199-022-00865-w

Menezes, J. A., Confalonieri, U., Madureira, A. P., de Duval, I. B., Santos, R. B., & Margonari, C. (2018). Mapping human vulnerability to climate change in the Brazilian Amazon: The construction of a municipal vulnerability index. *PLoS One, 13*, e0190808.

Muturi, E. J., Chang-Hyun, K., Alto, B. W., et al. (2011). Larval environmental stress alters *Aedes aegypti* competence for Sindbis virus. *Tropical Medicine & International Health, 16*(8), 955–964. https://doi.org/10.1111/j.1365-3156.2011.02796.x

Oliveira, A. C., Simões, R. C., Lima, C. A. P., et al. (2022). Essential oil of piper purusanum C.DC (Piperaceae) and its main sesquiterpenes: Biodefensives against malaria and dengue vectors, without lethal effect on non-target aquatic fauna. *Environmental Science and Pollution Research International, 29*, 47242–47253. https://doi.org/10.1007/s11356-022-19196-w

Pinto, A. C. S., Nogueira, K. L., Chaves, F. C. M., et al. (2012). Adulticidal activity of dillapiol and semi- synthetic derivatives of dillapiol against *Aedes aegypti* (L) (Culicidae). *Journal of Mosquito Research, 2*, 1–7. https://doi.org/10.5376/jmr.2012.01.0001

Rafael, M. S., Hereira-Rojas, W. J., Roper, J. J., et al. (2008). Potential control of *Aedes aegypti* (Diptera: Culicidae) with Piper aduncum L. (Piperaceae) extracts demonstrated by chromosomal biomarkers and toxic effects on interphase nuclei. *Genetics and Molecular Research, 7*, 772–781. https://doi.org/10.4238/vol7-3gmr481

Raguvaran, K., Kalpana, M., Manimegalai, T., et al. (2022). Larvicidal, antioxidant and biotoxicity assessment of (2-(((2-ethyl-2 methylhexyl)oxy)carbonyl))benzoic acid isolated from *Bacillus pumilus* against *Aedes aegypti*, *Anopheles stephensi* and *Culex quinquefasciatus*. *Archives of Microbiology, 204*(650). https://doi.org/10.1007/s00203-022-03264-3

Rahman, M. D. S., Faruk, M. D., Tanjila, O. S., Sabbir, N. M., Haider, N., & Chowdhury, S. (2021). Entomological survey for identification of *Aedes* larval breeding sites and their distribution in Chattogram, Bangladesh. *Journal of Basic and Applied Sciences, 10*, 32. https://doi.org/10.1186/s43088-021-00122-x

Raji, J. I., Melo, N., Castillo, J. S., GonzalezS, S. V., Stensmyr, M. C., et al. (2019). *Aedes aegypti* mosquitoes detect acidic volatiles found in human odor using the IR8a pathway. *Current Biology, 29*, 1253–1262.e7. https://doi.org/10.1016/j.cub.2019.02.045

Rose, N. H., Sylla, M., Badolo, A., Lutomiah, J., Ayala, D., Aribodor, O. B., et al. (2020). Climate and urbanization drive mosquito preference for humans. *Current Biology, 30*, 3570–3579.e6. https://doi.org/10.1016/j.cub.2020.06.092

Santana, H. T., Trindade, F. T. T., Stabeli, R. G., et al. (2015). Essential oils of leaves of piper species display larvicidal activity against the dengue vector, Aedes aegypti (Diptera: Culicidae). *Revista Brasileira de Plantas Medicinais, 17*, 105–111. https://doi.org/10.1590/1983-084X/13_052

Santos, L. H. F. D., Domingos, P. R. C., Meireles, S. D. F., et al. (2020). Genotoxic effects of semi-synthetic isodillapiole on oviposition in *Aedes aegypti* (Linnaeus, 1762) (Diptera: Culicidae). *Revista da Sociedade Brasileira de Medicina Tropical, 53*. https://doi.org/10.1590/0037-8682-0467-2020

Silva, H. H. G., Geris, R., Rodrigues Filho, E., et al. (2008). Diterpenoids from *Copaifera reticulata* ducke with larvicidal activity against *Aedes aegypti* (L.) (Diptera, Culicidae). *Revista do Instituto de Medicina Tropical de São Paulo, 50*, 25–28. https://doi.org/10.1590/s0036-46652008000100006

Silva, J. S., Pinto, A. C. S., Santos, L. H. F., et al. (2019). Efeito ovicida e larvicida do éter metil dilapiol (EMD) em *Aedes aegypti*, Manaus-AM. In Y. C. S. de Salgado (Ed.), *Patologias: Doenças Parasitárias* (1st ed., pp. 192–204). https://doi.org/10.22533/at.ed.97819180323

Silva, J. S., da Silva Pinto, A. C., Dos Santos, L. H. F., et al. (2021). Genotoxic and mutagenic effects of methyl ether dillapiole on the development of *Aedes aegypti* (Diptera: Culicidae). *Medical and Veterinary Entomology, 35*(4), 556–566. https://doi.org/10.1111/mve.12533

Silva, S. G. M., Melo, B. A., Santos, M. T., Rios, R. R. S., Santos, C. M. S., Júnior, K. A. L. R., Maranhão, F. C. A., Santos, T. M. C., & Fraga, A. B. (2022). Endophytic fungi: Benefits for plants and biotechnological potential. Res. *Social Development, 11*(4), e9211427008. https://doi.org/10.33448/rsd-v11i4.27008

Silvério, M. R. S., Espindola, L. S., Lopes, N. P., et al. (2020). Plant natural products for the control of *Aedes aegypti*: The Main vector of important Arboviruses. *Molecules, 25*, 3484. https://doi.org/10.3390/molecules25153484

Soares-da-Silva, J., Queirós, S. G., Aguiar, J. S., Viana, J. L., Neta, M. D. R. A. V., Silva, M. C., Pinheiro, V. C. S., Polanczyk, R. A., Carvalho-Zilse, G. A., & Tadei, W. P. (2017). Molecular characterization of the gene profile of *bacillus thuringiensis* Berliner isolated from Brazilian ecosystems and showing pathogenic activity against mosquito larvae of medical importance. *Acta Tropica, 176*, 197–205. https://doi.org/10.1016/j.actatropica.2017.08.006

Sun, J., Hiraoka, T., Dittmer, N. T., et al. (2000). Lipophorin as a yolk protein precursor in the mosquito, *Aedes aegypti*. *Insect Biochemistry and Molecular Biology, 30*, 1161–1171. https://doi.org/10.1016/S0965-1748(00)00093-X

Tadei, W. P., Rodrigues, I. B., Rafael, M. S., Sampaio, R. T. M., Mesquita, H. G., Pinheiro, V. C. S., et al. (2017). Adaptative processes, control measures, genetic background, and resilience of malaria vectors and environmental changes in the Amazon region. *Hydrobiologia, 789*, 179–196. https://doi.org/10.1007/s10750-016-2960-y

Valle, D., Bellinato, D. F., Viana-Medeiros, P. F., Lima, J. B. P., & Martins, J. A. J. (2019). Resistance to temephos and deltamethrin in *Aedes aegypti* from Brazil between 1985 and 2017. *Memórias do Instituto Oswaldo Cruz, 14*, e180544. https://doi.org/10.1590/0074-02760180544

Valtierra-de-Luis, D., Villanueva, M., Berry, C., & Caballero, P. (2020). Potential for *bacillus thuringiensis* and other bacterial toxins as biological control agents to combat dipteran pests of medical and agronomic importance. *Toxins (Basel), 12*(12), 773. https://doi.org/10.3390/toxins12120773

Viana Cruz, D. L., Sumita, T. C., Silva Leão Ferreira, M., et al. (2020). Histopathological, cytotoxicological, and genotoxic effects of the semi-synthetic compound dillapiole *n*-butyl ether in Balb/C mice. *Journal of Toxicology and Environmental Health, Part A: Current Issues, 83*(17–18), 604–615. https://doi.org/10.1080/15287394.2020.1804026

Wimalasiri-Yapa, B. M. C. R., Barrero, R. A., Stassen, L., Hafner, L. M., McGraw, E. A., Pyke, A. T., et al. (2021). Temperature modulates immune gene expression in mosquitoes during arbovirus infection: Temperature, mosquitoes and arboviruses. Open Biology 11.

World Health Organization - WHO. (2020). *Insecticide resistance*. World Health Organization. www.who.int/malaria/areas/vectors-control/insectide-resistance/en

World Health Organization -WHO. (2005). *Guidelines for laboratory and field testing of mosquito larvicides*. World Health Organization. https://apps.who.int/iris/handle/10665/69101

Dissolved Organic Carbon (DOC) of the Rio Negro, Amazonia: Importance, Processes and Potential Anthropogenic Impacts

Ora E. Johannsson and Aleicia Holland

Abstract

The Rio Negro watershed is unique among black water rivers, supporting extensive areas of white-sand ecosystems and peat deposits which influence the character of the dissolved organic carbon (DOC). The humic material in the DOC reaching the river is higher in aromaticity, mean molecular mass, and O/C ratio and lower in H/C ratio than those of other Amazonian black waters. The composition of the DOC changes with season having a higher percentage of aliphatic to aromatic structures during high water when much DOC enters the river from the land. These imply that the principle sources of DOC to the river change with the hydrological cycle.

DOC is the backbone of the Rio Negro aquatic ecosystem contributing to the acidic, stained nature of the system and supporting the microbial food-web which underlies the vast diversity and productivity of the fish community. Its secondary importance is the return of carbon to the atmosphere through photo-oxidation of DOC and microbial respiration which produce carbon dioxide. Thirdly, the Rio Negro is an important source of DOC to down river reaches where some DOC is sedimented out and the remainder is injected into the ocean.

Recognizing the importance of the hydrological cycle, temperature and climate variability on the functioning of this DOC-based ecosystem, we explore the

O. E. Johannsson (✉)
University of British Columbia, Department of Zoology, Vancouver, BC, Canada
e-mail: johannss@zoology.ubc.ca

A. Holland
La Trobe University, School of Life Science, Department of Ecology, Environment and Evolution, Murray Darling Freshwater Research Centre, Albury/Wodonga Campus, Wodonga, Vic, Australia

S. S. de Souza et al. (eds.), *The Future of Amazonian Aquatic Biota*,
https://doi.org/10.1007/978-3-031-66822-7_3

predictions for climate change in the region and the implications of deforestation, fire and dams on the production and character of DOC in the future.

1 Introduction

The Rio Negro encompasses a watershed of 716,604 km^2 (Trancoso et al., 2009), with its headwaters originating from the rainforests in Colombia and Venezuela. It snakes 2000 km east-southeast to the lowlands of the Amazon Basin near Manaus (Beltrão et al., 2019; Freitas et al., 2024; Schöngart et al., 2024). Its major tributary, the Rio Branco originates in the north in the Guiana Highlands. The Rio Negro is one of the largest and most important black-water rivers globally. The river is characterized by its soft waters (low buffering capacity), low pH and characteristic brown staining. The colour and pH of the water are due to the presence of dissolved organic carbon (DOC) high in humic acids which control the pH of the water due to the low buffering capacity of the system (Holland et al., 2012; Leenheer, 1980; Wood et al., 2003).

Dissolved organic carbon (DOC) is a product of the degradation of organic matter and is defined as the dissolved fraction of organic carbon which can be passed through a 0.45-um filter (Thurman, 1985). These molecules not only include carbon but also organic moieties of nitrogen, phosphorus and sulfur. Thus, DOC is also referred to as dissolved organic matter (DOM). The Rio Negro and the Amazon River process and export large quantities of DOC to the ocean, contributing greatly to the overall global carbon budget. The Amazon, itself, exports 11% of the global DOC that enters the ocean (Richey et al., 2022). The Rio Negro exports 5.2–6.7 Tg $C.y^{-1}$ as DOC; that is, 17–23% of the Amazon's total DOC annual export of 29.4 Tg $C.y^{-1}$ (Moreira-Turcq et al., 2003, Raymond & Spencer, 2015).

DOC is a major driver of both abiotic and biotic processes in freshwater ecosystems and is especially important within naturally acidic blackwaters such as the Rio Negro. In the river, DOC can follow several pathways. It can be sedimented out through adsorption to particles where it can enter the benthic microbial food chain or become buried. The water of the Rio Negro is very clear due to low levels of suspended particles (Leenheer, 1980), thus, sedimentation to the benthic regions would be less important than processes in the pelagic zone. The pelagic microbial food web consumes the labile components of the DOC, while photo-oxidation breaks bonds in the DOC molecules via use of light-energy, creating new and smaller molecules. Both photo-oxidation and microbial degradation of DOC break down the structure of DOC molecules making some components more labile (Kieber et al., 1989). Repeated or long-term exposure to light gradually releases additional labile compounds until only non-degradable DOC remains. Both microbial respiration and photo-oxidation release carbon dioxide (CO_2) back into the atmosphere—an important link in global carbon cycling (e.g. Johannsson et al., 2020). Due to the depth of the river, the CO_2 loss from microbial respiration greatly exceeds that lost through photo-oxidation (Amon & Benner, 1996).

The processing of DOC also influences the fluxes of nutrients and metals which can be constituents of the DOC molecules themselves, or complexed with the molecules (Bardy et al., 2011; Ertel et al., 1986; Küchler et al., 2000). These relationships between DOC and nutrients and metals influence the availability of these components to the biota which can affect their growth and health (Morris et al., 2021; Wood et al., 2011). Within the Rio Negro, DOC does more than feed biota and limit nutrients/metal availability: it also structures the aquatic habitat. Due to its dark colour, DOC absorbs light more readily than clear water, reducing the depth of light penetration and the spectrum of wavelengths present at depth. Light energy can be transformed to heat in molecules (kinetic energy): this heat warms the surface layer so that the greater the reduction in light penetration, the thinner and warmer the epilimnion (Read & Rose, 2013; Vähätalo & Vione, 2022). The lack of light penetration below the reduced photic zone leaves these waters cooler. In a blackwater lake within the Rio Negro basin, it was found that temperature differed between the surface waters and the bottom by up to 4.7 °C during flood periods and 4.5 °C during dry periods (Aprile & Darwich, 2009). Oxygen was also shown to differ between layers, with bottom layers being hypoxic (oxygen levels $\leq 2\%$ saturation) during both the wet and dry seasons, while surface concentrations ranged between 59 and 125% saturation (Aprile & Darwich, 2009). This restructuring of the light availability, the thermal profile and oxygen concentrations affects animal and plant distributions, their interactions and their productivities, especially in those species which rely on light as an energy resource (phytoplankton) or to find prey (e.g. zooplankton, fish).

Blackwaters, however, are at threat to changes in DOC concentration and composition due to predicted climate change, increased urbanization and deforestation and the possible development of dams and/or reservoirs along this river and its tributaries. This chapter will discuss the concentrations and characteristics of DOC in the Rio Negro and the possible threats these pressures may pose to carbon cycling within this unique, highly biodiverse, extreme waterway.

2 DOC in the Rio Negro

The Rio Negro contains high concentrations of DOC, generally ranging from 7 to 15 mg C L^{-1}, of terrestrially derived (allochthonous) material which is highly aromatic, of a high molecular weight and contains high amounts of humic substances enriched in carboxylic and phenolic groups (Ertel et al., 1986, Duarte et al., 2016, Holland et al., 2017, Johannsson et al., 2017, 2020). Values of DOC indices, which give an indication of aromaticity and molecular weight, such as specific absorbance at 254 nm and 340 nm ($SUVA_{254}$ and SAC_{340}) and spectral slopes, reported previously for the Rio Negro in the Anavilhanas Archipelago and further upstream are often higher than those recorded closer to Manaus (Johannsson et al., personal communication) and from other parts of the world (Duarte et al., 2016; Holland et al., 2017; Johannsson et al., 2020, 2017). For example, SAC_{340} and $SUVA_{254}$ values for the Rio Negro have been reported to range between 39–73 $cm^2.mg\ C^{-1}$ and

5.0 - 7.4 $cm^2.mg\ C^{-1}$ respectively. In contrast, samples collected from other parts of the world generally had SAC_{340} values which ranged between 2.3–50 $cm^2.mg\ C^{-1}$ and $SUVA_{254}$ values which ranged from 0.2–5.7 $cm^2.mg\ C^{-1}$ (Holland et al., 2017).

The structure of the humic portion of Rio Negro DOC was investigated by Ertel et al. (1986): DOC from the Rio Negro was unlike the DOC of other Amazonian rivers with (1) more extreme hydrogen and nitrogen deficits (low H/C [HA: H/C ratio was 0.79 compared with 0.83–1.10 for other rivers, FA: ratio was comparable with the other Amazonian rivers at 0.95], and low N/C ratios [HA: N/C ratio was 0.017 compared with 0.026–0.067 for the other rivers, FA: N/C ratio was 0.012 compared with 0.13–0.023 for the other rivers]), and (2) lower fulvic acid to humic acid ratios (and, therefore, lower O/C ratios)[FA/HA was 1.53, whereas, the ratios in all other rivers were > 2]. Degradation of lignin, a phenolic polymer unique to vascular plants and present in humic substances, can be followed through the ratio of the breakdown products vanillic acid/vanillin: the ratio increases with aerobic degradation (Ertel and Hedges, 1984). DOC degradation was greater in the Rio Negro than other black water Amazonia rivers, as evidenced by its lower total lignin concentration and higher vanillic acid/vanillin ratio, particularly of HA [2.0 as compared with < 1.2 for most other sites] (Ertel et al., 1986). Gonsior et al. (2016) and Li et al. (2019) built on this work, further linking these characteristics with the microbial oxidative processing of plant polyphenols. The source and character of the DOC in the Rio Negro are often attributed to the surrounding forested environment which undergoes annual wetting and drying, contributing large amounts of allochthonous carbon to the system (Leenheer, 1980; Thurman, 1985; Ertel et al., 1986; Bardy et al., 2011). Humic materials, such as humic acids, are often by-products of lignin and tannin degradation. Although Ertel et al. (1986) thought there was little if any further processing of DOC in the river, later studies showed the importance of photo-oxidation (Amaral et al., 2013, Johannsson et al., 2020) and particularly, microbial consumption (Amon & Benner, 1996) on in-river degradation.

3 The Land to River

Terrestrial processes regulate the input of DOM into the Rio Negro and thus drive not only the biogeochemical environment but also directly influence the biota through structuring the ecosystem (McClain & Elsenbeer, 2001). The white-sand ecosystems found throughout the Rio Negro watershed, as depressions in the upland regions surrounded by campina forests, along riparian stream edges, and across the broad lowland peneplain further upstream of the Anavilhanas Archipelago, are the source of the dark, humic DOC in the river (Bardy et al., 2011; Dubroeucq & Volkoff, 1998; Leenheer, 1980; Remington et al., 2007; Simon et al., 2021). White-sand ecosystems vary in their forest cover, the depth of their humic layer and their degree of wetness, ranging from systems which are dry to Giant Spodosols with Histosols (areas of peat formation) in their centres (Dubroeucq & Volkoff, 1998; Montes et al., 2011). Many of these white-sand ecosystems are contiguous and all are underlain by an unstratified layer of white sand (the albic layer) (Dubroeucq & Volkoff, 1998, Montes et al., 2011).

Direct litterfall contributes approximately 0.7 kg.m^{-2} year^{-1} to the organic matter pool in the Amazon Basin with this estimated to be similar across regions. Campina forest such as those in the headwaters of the Rio Negro are characterized by groundwater consisting of high concentrations of old refractory DOC, > 20 mg C L^{-1} which contribute to the high humified DOM in the Rio Negro throughout the basin (McClain & Elsenbeer, 2001; Wetlands also play a key role in the composition and concentration of DOC in the Rio Negro (McClain & Elsenbeer, 2001). Extensive peat deposits have also recently been mapped in the watershed of the Rio Negro above the confluence with the Rio Branco, within the Rio Branco basin and in the headwaters originating in Venezuela, and may be associated with wetlands (Gumbricht et al., 2017). The contribution of the various terrestrial sources of DOC to the Rio Negro DOC pool also varies depending on the wet or dry season and in relation to stormflow with increases in DOC in groundwater reported during the wet season (McClain et al., 1997).

Allochthonous DOC characteristics are determined by degradation processes on the land and its interactions with the soil before it enters the river. The degree of microbial processing of allochthonous DOC is governed by its environment: the seasonal and local hydrological and temperature regimes, oxygen availability, amount and type of litter, and soil properties, including permeability and clay content, and the microbial community present (Ågren et al., 2007, 2014; Bouillon et al., 2012; Simon et al., 2021). In white-sand ecosystems, the decay of litter in the humic layer produces humic substances (DOC) which leach into the albic layer and are transported laterally into the river (Leenheer, 1980). DOC composition changes little in the albic layer due to the relatively anaerobic conditions which inhibit oxidation, the lack of clay, which could selectively remove components of the DOC, to the lack of nutrients which may limit the activity of anaerobic bacteria and to the lack of light for photo-oxidation (Leenheer, 1980). Further, the lack or low concentrations of clay in the albic layer may allow more rapid movement of ground water toward the river. All these factors would tend to minimize further changes in the DOC. Indeed, the composition of DOC in the podzols is similar to that of the groundwater and local rivers (Bardy et al., 2011; Remington et al., 2007). The extent of DOC export from white-sand ecosystems is extensive. Simon et al. (2021) found that the molecular signature of DOC in the Rio Negro just upstream of Manaus and in the Anavilhanas Archipeligo (just downstream of the confluence with the Rio Branco), was more similar in composition to a dry, upland, white-sand ecosystem than to the DOC in the spodosols in the riparian valley suggesting white-sand ecosystems play a specific role in determining the characteristics of DOC in the Rio Negro (Gonsior et al., 2016; Simon et al., 2019). The character of Rio Negro DOC also resembles that of DOC from other temperate and tropical naturally acidic, humic waterways globally, such as, those draining peatlands (Holland et al., 2018; Thurman, 1985) and that of DOC from newly decomposed plant material (Seifert et al., 2016). Such DOM tends to be high in aromatic, high molecular weight humic acid material, H deficient and characterized by a higher O/C ratio due to the waterlogged conditions and slow decomposition rates of peatlands (Tfaily et al., 2013) or the short degradation time of new plant material. This suggests that much of the DOC reaching the river may come from more than one source, but that groundwater draining the humus-rich, peaty, hydromorphic spodozols along with the terrestrial input from the surrounding forested environment are important DOC sources for the Rio Negro.

DOC composition in the Rio Negro changes with the hydrological cycle. During high-water, aliphatic structures are more abundant than aromatic structures (86% to 14%), while the percentage of aliphatic structures declines and aromatic structures increases during the low-flow period (76% aliphatics, 23% aromatics) (Rodríguez-Zúñiga et al., 2008). Similarly, Holland et al. (2017) found that the $SUVA_{254}$ and $Ratio_{254/365}$, indices of aromaticity and molecular weight respectively, changed accordingly between the high water and low water seasons: as indicated by the lower SUVA values for the wet season (less aromaticity): 4.98 compared to 7.35, and $Ratio_{254/365}$ decreased from 3.63 to 3.43 between high-water and low-water periods (increased molecular weight). The δ C^{13} signature of the DOC, on the other hand, changes only slightly between seasons: −29.2 to −29.5%, indicating that the source material of the DOC is not changing (Albéric et al., 2017), suggesting that the microbial processing of some of the DOC entering the river is different between the two seasons. Higher aliphatic composition suggests more extensive microbial processing (Kurek et al., 2021). This could reflect high processing within the waterbody or of the DOC in the soils during the dry season and the release of these smaller molecules on flooding. However, to this time, no studies have been done on the DOC composition coming from the flooded forests (Simon et al., 2021).

The seasonal patterns shown in the Rio Negro DOC are not typical. In the Congo River, the second largest tropical river, DOC rich in aromatic compounds, enters the river throughout the year from the extensive, flooded forest in the Cuvette Central wetland (Lambert et al., 2016). As the river flows downstream, the character of the DOC becomes more aliphatic driven by photo-oxidation and bacterial consumption (Lambert et al., 2016). Other regions in the Congo basin possess different seasonal patterns in DOC sources. The DOC in the Epulu River, a pristine tropical rainforest river in the east of the Congo basin, has high aromatic characteristics during high water, but less aromatic characteristics during low water at the same sampling sites (Spencer et al., 2009). In the Oubangui River, Bouillon et al. (2014) needed three different DOC source pools to explain the seasonal patterns they observed in DOC and its $\delta 13$ C signatures. As Ågren et al. (2014) emphasized the character of DOC entering rivers and lakes is a product of geology, hydrography and biology.

4 DOC in the River

Once in the river, DOC structure and concentration are changed by competing forces of photo-oxidation, microbial degradation, complexation and sedimentation and a number of factors will influence these (Fig. 1). Factors which influence photo-oxidation are the concentration of DOC, water clarity, water depth, sufficient

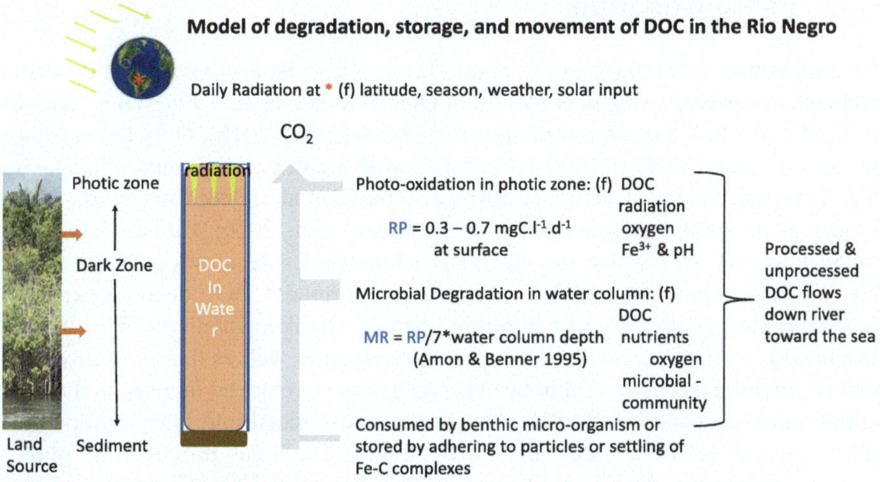

Fig. 1 Model of degradation, storage and movement of DOC in the Rio Negro. RP and MR are the rates of CO_2 production by photo-oxidation and microbial respiration, respectively. The estimates of RP were developed in this paper. The estimates of MR were taken from Amon and Benner (1996)

oxygen, weather together with latitude and time of the year as they determine the total radiation reaching the water's surface, and the concentration of iron (Fe^{+2}) which increases the rate of photo-oxidation (discussed in Johannsson et al., 2020, 2021). Factors which enhance the importance of microbial degradation are aeration (sufficient oxygen), higher temperatures, presence of nutrients in the DOC or water, breakdown of DOC by photo-oxidation or input of more labile DOC (priming effect) and time –time to interact with the volume of DOC. In deep, dark rivers, like the Rio Negro, the proportion of the water column (and DOC) available for photo-oxidation is small, and more DOC is converted to CO_2 by microbial degradation than by photo-oxidation (Amon & Benner, 1996). Sedimentation is important because it removes DOC from the pelagic environment either to the benthic environment or into long-term storage in the sediments. When blackwaters such as Rio Negro water mixes with water containing suspended particles such as kaolinite this can lead to the complexation of DOC onto the kaolinite causing flocculation and deposition into the sediments (McClain & Elsenbeer, 2001). Given the importance of photo-oxidation and microbial degradation in determining DOC concentrations and composition within river systems, these are discussed in more detail in the sections below.

5 Photo-oxidation

Photo-oxidation takes place in the photic layer, where both ultraviolet and visible irradiance contribute to the degradation of DOC (Granéli et al., 1998). DOC is composed of 100 s to 1000s of compounds (e.g. Riedel et al., 2016). Only the chromophoric (coloured) DOC (CDOC) is reactive with light, causing many changes in DOC composition and structure (Rodríguez-Zúñiga et al., 2008; Cory et al., 2009; Spencer et al., 2009; Remington et al., 2011; Riedel et al., 2016; Ward & Cory, 2016 amongst others). Principally, the mean size of molecules decreases as larger molecules are broken apart and smaller molecules are formed. CDOC decreases (photobleaching) along with a decline in lignin phenols. Breakdown/alteration of highly, unsaturated, aromatic compounds has been observed as well as that of highly oxygenated functional groups. Aliphatic side chains can be removed increasing the concentration of aromatic compounds. Oxygen is incorporated into many compounds during photo-oxidation, especially fulvic acids. The final products of photo-oxidation are carbon dioxide (CO_2) and carbon monoxide CO). The rate of photo-oxidation is measured by the production of CO_2 because it is easy to measure and it is a key component of the carbon cycle.

Photo-oxidation changes seasonally with light intensity and the character of the DOC. Light intensity is highest during the periods of rising (March) and falling (September) water because the Rio Negro is virtually on the equator (Fig. 2). Amaral et al. (2013) observed 'surface' rates of photo-oxidation of 0.47 mg $C.l^{-1}.d^{-1}$ (2008, DOC = 8.8 mg $C.l^{-1}$) and 2.40 mg $C.l^{-1}.d^{-1}$ (2009, DOC = 9.9 mg $C.l^{-1}$) during falling water. 'Surface' refers to samples held just below the water's surface. Their observed rates of photo-oxidation over the 6 h during the middle of the day were converted to daily estimates by assuming 75% of total daily photo-oxidation occurred during those 6 h (Granéli et al., 1998). Johannsson et al. (2020) reported similar 'surface' rates of 0.41 mg $C.l^{-1}$ to 0.68 mg $C.l^{-1}$ for September 2018 (N = 5, DOC = 8.7 mg $C.l^{-1}$). Variation in daily irradiance accounted for the variation in photo-oxidation rates (Johannsson et al., 2020, Fig. 9). During low water (December–January), two estimates of photo-oxidation exist: 0.31 mg $C.l^{-1}.d^{-1}$ (Amaral et al., 2013) and 0.93 mg $C.l^{-1}.d^{-1}$ (Granéli et al., 1998 as calculated by Johannsson et al., 2020). Per discussion by Johannsson et al. (2020), the December surface measures in Johannsson et al. (2017) are considered overestimates and not included here. The estimate by Granéli et al. (1998) is not directly comparable because the experimental tubes were held half-in and half-out of the water, exposing the DOC to greater irradiance, and thus produce higher rates of photo-oxidation. Amaral et al. (2013) observed high water (March–April) rates of photo-oxidation of 0.67 mg $C.l^{-1}.d^{-1}$, similar to that of Amon and Benner (1996) of 0.65 mg $C.l^{-1}.d^{-1}$ (as calculated in Johannsson et al., 2020), however, lower than in the falling water period (2.40 mg $C.l^{-1}.d^{-1}$). Amon and Benner's rate is not directly comparable because they did not filter the water (all other studies filtered the water through a 0.45 μm or 0.22 μm filters) which may have over-estimated DOC photo-oxidation. At the same time, their container was larger (more depth) and made of pyrex which did not transmit UVB and UVA radiation as well as quartz. These factors would underestimate photo-oxidation. Overall, most 'surface' rates of photo-oxidation for the Rio Negro fall between 0.3 and 0.7 mg $C.l^{-1}.d^{-1}$.

Data in Spencer et al. (2009) allowed calculation of a rate of photo-oxidation of DOC in the surface waters of the Congo River: total DOC was approximately 11.2 mg $C.l^{-1}$ (their Fig. 2) and the daily rate of decay of DOC in their long-term irradiation experiment was given as 0.0474. On the first day of the experiment, the most comparable to all the above data, photo-oxidation was 11.2 * 0.0474 or 0.53 mgC. $l^{-1}.d^{-1}$. This value is well within the range observed in the Rio Negro. Although photo-oxidation of another large tropical river, the Mekong, was measured, rates of photo-oxidation could not be calculated due to the experimental design (Riedel et al., 2016). Daily rates in the Amazon and Congo during periods of maximum solar radiation are within the range of rates observed at latitudes of ˜50 in Canada and higher than maximum summer rates in high latitude lakes in Sweden and Finland (Johannsson et al., 2020).

Photo-oxidation is not limited to the surface. Rates of photo-oxidation decline exponentially with depth following the extinction of light through the water column. Johannsson et al. (2020) measured areal, photo-oxidation directly: it was in the range of 28.8 mg $C.m^{-2}.d^{-2}$ and 39.3 mg $C.m^{-2}.d^{-2}$ for September 2018. They noted that the nearshore water in the experimental pool was originally quite dark suggesting that these were underestimates of offshore rates of photo-oxidation and that the second estimate of areal photo-oxidation would be closer to offshore rates. Remington et al. (2011) calculated the annual average rate of areal photo-oxidation as 24 mg $C.m^{-2}.d^{-2}$ (200 nm cm^{-2} d^{-1}) (as calculated by Johannsson et al., 2020). They considered this estimate conservative. Amaral et al. (2013) estimated areal

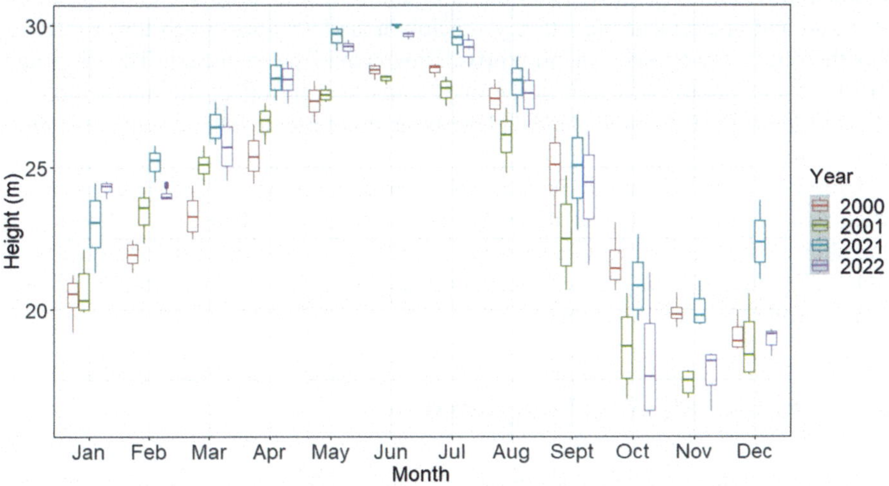

Fig. 2 Height in metres of the Rio Negro at the Porto de Manaus during the last 2 years (2021–2022) and 20 years ago (years 2000–2001) presented as a box and whiskers plot. Daily height (m) was used and obtained from https://www.portodemanaus.com.br/?pagina=nivel-do-rio-negro-hoje

rates of photo-oxidation across the seasonal cycle in 2008 and 2009. Their rates of areal photo-oxidation ranged from 69.6 mg $C.m^{-2}.d^{-2}$ (fall of 2008) to 334.8 mg $C.m^{-2}.d^{-2}$ (5.8 to 27.9 $mmol.m^{-2}.d^{-1}$) (fall of 2009). They did not measure photo-oxidation at depth but calculated it based on satellite irradiance estimates, literature estimates of the relative contribution of UVB, UVA and UVIS to surface photo-oxidation and calculations of attenuation of each set of wavelengths with depth. As the 'surface' photo-oxidative rates of Johannsson et al. (2020) and Amaral et al. (2013) (fall 2008) are similar, the differences in areal estimates (39.9 mg $C.m^{-2}.d^{-2}$ vs 69.6 mg $C.m^{-2}.d^{-2}$) are related primarily to the deep colour of the pool water (Johannsson) and differences in the methodology of measurement of photo-oxidation at depth. In summary, the rate of 'surface' photo-oxidation of the river has been shown to vary seasonally and in response to levels of irradiance over a range of 0.3 to 0.7 mg $C.l^{-1}.d^{-1}$. Areal production estimates should be revisited. Amaral et al. (2013) data indicate that areal photo-oxidation can vary by fivefold.

The Rio Negro basin also includes Ria Lakes which are formed in submerged valleys at the mouths of tributaries (World Landforms https://worldlandforms.com/landforms/ria/) and experience annual flooding. Photo-oxidation rates in Lake Tupé followed the pattern in the main river, but tended to be lower, except in high water, when they were the same. 'Surface' photo-oxidation ranged from 0.2 $mgC.l^{-1}.d^{-1}$ to 1.7 $mgC.l^{-1}.d^{-1}$, and areal photo-oxidation from 56.4 to 250.8 $mgC.l^{-1}.d^{-1}$ (Amaral et al., 2013). Examination of the Rio Negro on Google Earth found a small number of ria lakes along the Rio Negro. Their impacts would be local. During the low water period when the river does not influence these lakes, bacterial turnover times are longer and phytoplankton layers are observed in the upper water column (Rai & Hill, 1980). At this time of year, DOC in these lakes would be a combination of retained allochthonous DOC from the land and autochthonous DOC from phytoplankton and bacterial degradation. Autochthonous DOC has much lower CDOC, and therefore, lower rates of photo-oxidation would be expected. The high and equal rates of photo-oxidation in the high-water season reflect the major input of allochthonous DOC from the land and washing out of previously present autochthonous DOC.

Climate change, changes in land use (forests/logging, grasslands, cropland, urban) and dams are likely to affect the photo-oxidation of DOC in the Rio Negro and alter these rates. Many of these changes will affect the quantity and quality of DOC. The impact of these potential changes are discussed in more detail in the section on Anthropogenic Stressors.

6 Role of Microbial Degradation

The bioavailability of DOC depends on the type of DOC present which is dictated by its origin, previous degree of microbial degradation and photo-oxidation. Previously it was thought that smaller DOC molecules, such as proteins, were more labile than the more refractory, humic material. Recently, however, it has been suggested that allochthonous, humic DOC may be more bioavailable to some microbes, especially bacteria, than lower-molecular weight DOC of microbial origin, with some taxa able to break down humic DOC and actually preferring terrestrially derived, humic DOC over more labile autochthonous, microbially derived, protein-like DOC (Roiha et al., 2016). The degradation and consumption of humic DOM by bacteria leads to a change in DOC composition. Bacteria, fungi and mixotrophic algae release more labile, autochthonous DOC which can then be taken up by other members of the microbial community (Holland et al., 2020). Thus, the microbial loop is intricately linked to the type and quantity of the DOC, and DOC linked to the microbial loop. The degradation of humic DOC, like that found in the Rio Negro, is often associated with Proteobacteria and Actinobacteria which are found in freshwaters globally (Jezberová et al., 2010, Ghai et al., 2011). Proteobacteria and Actinobacteria respond quickly to an enhanced availability of DOC (Burkert et al., 2003), and some species contain biochemical pathways involved in the degradation of humic compounds (Tikhonov et al., 2010; Hoetzinger et al., 2017; Kamimura et al., 2017). A high presence of Proteobacteria and Actinobacteria is present in the Rio Negro and other blackwaters globally (Sylvain et al., 2023). Recent work by Sylvain et al. (2023) has shown that the bacteria within these phyla in the water of the Rio Negro are strongly linked with humic DOC and were correlated with the upregulated genes involved in degradation of terrestrially derived lignin compounds into humic acids and smaller organic molecules, thus playing an important role in the degradation of plant-derived DOC and converting it into the Rio Negro DOC pool. The degradation of humic DOC might also involve members of other kingdoms present within the microbial community. Fungi also contain the ability to degrade terrestrial, humic carbon. Increased concentrations of allocthonous terrestrial DOC have also been associated with higher abundance of some chlorophytes and diatoms (Arvola & Tulcnen, 1998).

Naturally, acidic waters such as the Rio Negro are dominated by bacteria that tend to outnumber algae by several folds unlike other waters where algae dominate (Holland et al., 2012). Mixotrophic flagellates and heterotrophs have also been reported to increase in blackwater as they feed off DOC (Holland et al., 2012). The decrease autotrophic production is a result of the low light availability, nutrient binding of the DOC, herbicidal effect of DOC and competition with bacteria and mixotrophs. Changes to climate, land use, river regulation and urbanization may alter DOC quantity and quality entering the Rio Negro, switching the DOC from humic-dominated to more labile proteinaceous material, affecting microbial communities and processing, overall nutrient concentrations, light availability and other key abiotic processes within the river and thus the overall function of the Rio Negro.

7 Anthropogenic Stressors

The effects of Anthropogenic Stressors on DOC within the Rio Negro can be viewed through the lens of their impacts on the forcing variables which influence the production and degradation of DOC: (1) climate change through increases in temperature and atmospheric CO_2, and effects on precipitation and basin hydrography; (2)

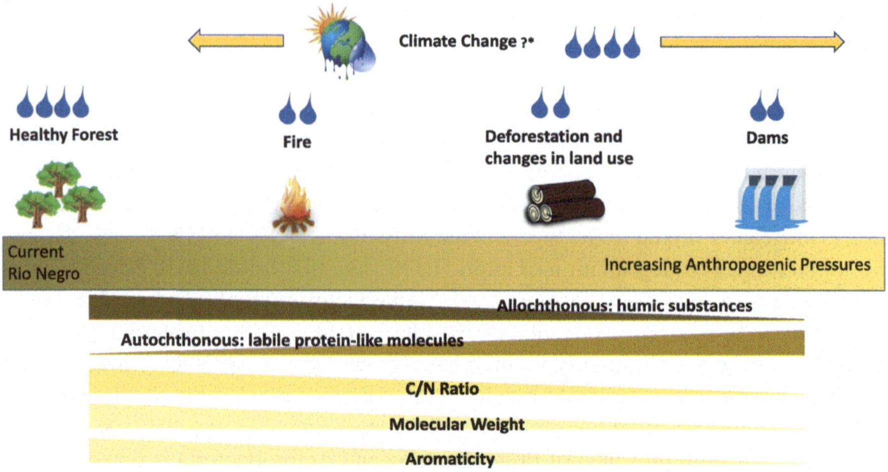

Fig. 3 Model of the impact of Anthropogenic Stressors on the tropical rainforest environment of the Rio Negro, and on the quality and concentrations of DOC then feed back to the river. Fire, precipitation, the question mark beside climate change precipitation indicates that the models are not clear on whether the rainfall in this region of the Amazon is going to decline. The * indicates that whether it declines or not, the pattern and amount of rainfall will become more variable across the years as is already seen

deforestation through changes in land use from forest to grasslands, to cropland to cities which effect water movement, susceptibility to fires, carbon incorporation, litter concentration and type, and soil DOC and nutrient retention; and (3) dams through creating lake-like systems with high temperatures/low oxygen environments and changing flood patterns in the surrounding forest (Fig. 3). Each of these stressors with the addition of fire, will be discussed in more detail in the following sections.

8 Climate Change and Increasing Temperature and CO_2

Much remains to be understood concerning the impacts of increasing CO_2 and temperature on moist tropical forests and nutrient cycling in those ecosystems (Cusack et al., 2016). These forests have been successful carbon sinks, at least until 2010, as assessed by a network of experimental plots (Phillips et al., 2017). Yet higher temperatures reduce net primary productivity while increased CO_2 promotes growth, suggesting that decomposition of these trees & litter will produce compounds with higher C/nutrient ratios (Cusack et al., 2016) which should affect DOC composition similarly. Cusack et al. (2016) laid out many research areas with respect to climate change and forest, soil and nutrient dynamics, in their comprehensive review of the topic. Only when these are better understood, will their impacts on DOC composition and concentration be understood.

9 Climate Change and Hydrology of the Rio Negro

Climate change will affect the Amazonian rainforest through changes in temperature, carbon dioxide (CO_2) concentrations, and levels of precipitation. While climate change will affect all parts of the watershed, it will not do so equally.

Climate change, assisted by anthropogenic behaviour, has the potential to alter the hydrological cycles in the Amazon. These cycles drive the production and degradation of litter and movement of DOC (Cusack et al., 2016). The hydrological cycle in the Rio Negro watershed is characterized by an annual cycle of flood and drought (Fig. 2). The period of intense rains from January through April cause the river water level to rise until the waters overflow their banks and flood the surrounding forests. The rains subside in May and June reaching particularly low levels from July to September. During this time, water flows from the land back into the river, and subsequently, the water level in the river falls, reaching minimum levels in November/December The austral rains gradually return through October to December building towards the return of the intense rains and increasing water levels. Disruption of this hydrological cycle could alter the production and movement of DOC to the river as DOC flows to the river both by base ground flow and overland transport McClain et al., 1997, Vidon et al., 2008). Both climate change and land use changes are predicted to alter this hydrological cycle. Changes are already evident, as outlined in Marengo et al. (2011) and Espinoza et al. (2022, Fig. 2), with the height of the Rio Negro increasing since 2001 with 2021 recording the largest historical flood in this system. Waters were also 1–4 m higher during 2022 compared to 2000–2001; however, this year also reported the lowest levels in October and November (Fig. 2). While revising this paper, the worst drought in Amazonian history has occurred with a lack of autumnal (August to November) rains causing water levels at Manaus to drop to 12.7 m by October 27 (NOAA Climate.gov, November 23, 2023). These changes are in line with climate change predictions of more variability in precipitation (floods and droughts) between years, and a sift to longer periods of consecutive dry days and shorter periods of consecutive wet days in Brazil (Avila-Diaz et al., 2020).

The base, hydrological cycle depends on the annual pattern of rainfall which is influenced by the El Nino Southern Oscillation (ENSO), the tropical Atlantic Dipole Mode (temperatures in the northern versus southern tropical Atlantic Ocean), the South American Monsoon System and forest evapotranspiration (ET) recycling of water across the jungle (Espinoza et al., 2022; Gloor et al., 2013; Lean & Warrilow, 1989; Makarieva & Gorshkov, 2007; Marques et al., 1977; Reyer et al., 2017; Shukla et al., 1990; Spracklen et al., 2012; Wang et al., 2018). Climate change models are continuously under further development and are not conclusive in their predictions of precipitation changes in the Amazon Basin. Duffy et al. (2015) led an inter-model comparison of 35 CMIP5 climate models (Coupled Model Intercomparison Project). Some model runs showed increased rainfall for the Rio Negro while others predicted increases in the head waters and decreases further downstream. On average, these models predicted that precipitation would increase in the north-west near the Andes and gradually become drier towards the lower reaches of the river by 2100. Sorribas et al. (2016) examined five of what they considered the most realistic of these models for South America and found that the precipitation projections were similar to those of Duffy et al. (2015) for the Amazon. They also saw increases in the length of the dry season as climate change progressed. Reyer et al. (2017) assessed changes in precipitation at a 2 °C and 4 °C rise in average world temperature. With a 2 °C rise in temperature, they observed a small change in precipitation (±10%), but the direction of change was unstable. More distinct changes were observed with a 4 °C rise in temperature, although regions of uncertainly still existed throughout much of the Amazon Basin during the austral summer (November, December, and January), and near the periphery of the Basin on an annual basis. In their analyses, the Rio Negro watershed clearly increased in precipitation during the austral summer. This was determined by overlaying the course of the Rio Negro on the panels of Fig. 2. In the annual models, the uncertainty was pushed to the periphery of the Amazon Basin (Reyer et al., 2017, Fig. 2). Changes in the Rio Negro and its tributaries were still uncertain.

Wang et al. (2018) have documented increased precipitation during the wet season in the Amazon, including the watershed of the Rio Negro, from 1976 to 2015, 'during the satellite era': 1975 is considered the tipping year when climate changes became noticeable. More than half of the increased precipitation can be attributed to the warming of the tropical Atlantic Ocean and a lesser portion to the colder temperatures in the eastern Pacific Ocean. The Rio Negro water-level data from the monitoring station at Manaus reflect some of this change and indicate that the Rio Negro is presently experiencing more periods of extreme flood and drought than in the past (Espinoza et al., 2022, Figs. 1b, 3). The high-water levels have been noticeably increasing: the lower high-water levels since the mid-1990s and the high high-waters since 2009–2011. The frequency and duration of these more extreme high-water level periods (major flood events with water levels > 29 m) have increased (Espinoza et al., 2022). Between 1903 and 1975, major flood events occurred once every 20 years: the time between flood events has gradually declined to every 4 years by 2014 (Barichivich et al., 2018). This reduced cycling time appears to persist up to 2021 (Espinoza et al., 2022, Fig. 1b). Six of these flood events occurred between 2009 and 2021, and lasted from 40 to 90 days (Espinoza et al., 2022, Fig. 1c). In addition, high high-water levels in 2009, 2013, 2015, 2019, 2021 and 2022 are higher than previous extreme peaks in 1953 and 1975.

The low-water trend declined through the early to mid-1990s (Espinoza et al., 2022, Fig. 1b). More extreme low-water conditions were observed in a cluster: 1995, 1997, 1998, 2005 and 2010. The difference between the high and low water levels (the amplitude) rebounded from a low in the 1970s, and then increased again in the mid-1990s. Eight years between 1995 and 2021 had amplitudes greater than observed prior to this date: 1996, 1998, 1999, 2006, 2011, 2012, 2013 and 2021, indicating that periods of more intense seasonal 'drought' are increasingly affecting the Rio Negro ecosystem. The drought in 2023 has been the most severe to date due to the high temperatures associate with climate change and the lack of August–October rains associated with a strong El Niño (Fernside & Araújo Silva, 2023).

Differences in the amount of carbon exported during the wet and dry seasons and between wet and dry years and extended periods have also been reported. Carbon exported in the wet season has been reported to account for approximately 70% of the total carbon within areas of the Rio Negro Basin (Waterloo et al., 2006), with higher in-river DOC concentrations reported during the wet season. DOC in runoff during low flow and high flow events also differ with daily averages of 8 mg C L^{-1} reported in low flow conditions and 27 mg C L^{-1} under high (Waterloo et al., 2006). Previously published organic carbon export rates range from 12.6 g C m^{-2} y^{-1} for the period 1982–1984 (Richey et al., 1990) 18.1 g C m^{-2} y^{-1} over period 1994–2000 (Moreira-Turcq et al., 2003) and 17.8 g C m^{-2} y^{-1} for the period 2002–2003 (Waterloo et al., 2006).

Gatti et al. (2014) compared the carbon balance above intact forest in the Amazon during the 2010 drought and 2011 extreme flood conditions. During the wet period, the forest took up carbon, while during the dry period, the forest normally reverted to a neutral state where community respiration equalled carbon uptake. However, during prolonged or severe drought, evapotranspiration, gross primary production and ecosystem respiration declined (Corlett, 2016; Granier et al., 2007). These changes lead to reduced rainfall further into the forest as well as reduced tree growth and litter production. DOC production would decrease under these conditions commensurate with the decrease in production of new plant material and litter. The decrease would be felt over a number of years as the lost production would not all have decayed at the same time. In severe drought, the trees weaken, leading to tree mortality (Berenguera et al., 2021; Corlett, 2016; Cusack et al., 2016). Extensive tree mortality and opening of the forest leads to higher local temperatures which increase rates of degradation (Cusack et al., 2016) and the production of DOC, exacerbates drought conditions which would reduce the amount of DOC production but gives more access to oxygen, affecting the composition of the DOC molecules, and leaves the region vulnerable to forest fires (Cochrane, 2011; Corlett, 2016). The effect of forest fires on DOC dynamics is considered in the section on Anthropogenic Stressors.

Soil water levels influence the effect of drought conditions (Granier et al., 2007). Where the water table is low, the surface of the soil can dry and crack, allowing oxygen to reach deeper carbon deposits. This is particularly true in peat soils which are common in the Rio Negro basin (Gumbricht et al., 2017). The oxygen allows microbial degradation of these previously buried reserves, releasing old DOC into the environment when the rains return (Ekblad & Bastviken, 2019; Gandois et al., 2013). This DOC is more biolabile and has a higher microbial signature than the normal DOC produced during the wet season (Drake et al., 2019; Ekblad & Bastviken, 2019; Gandois et al., 2013; Wilson & Xenopoulos, 2013).

More frequent and extensive periods of flooding may lead to (1) changes in forest species composition in newly flooded areas, (2) changes in amount of leaf litter, (3) longer periods of time for litter decay and for leaching of DOC into surface water, (4) more microbial-like DOC given the longer processing times (5) shorter dry periods before the next rains and (6) changes in overall amounts of carbon exported to the basin. The larger flooded area may mean that the pulse of DOC returning to the main river is larger. The longer periods of falling water observed at Manaus suggest that September rains are delayed and that the forest soils may have more time to dry out. Where the water table is deep, the forest likely experiences drought-like conditions, as discussed above. There is less water movement and less DOC would make it into the rivers at this time. However, when the rains do come, the rainfalls recently have generally been more intense likely to lead to higher DOC export.

Since 1995, droughts and floods have become more variable, more frequent and more intense (Espinoza et al., 2022), meaning that the forest has had to cope with a wider range of environmental conditions, and that DOC loads to the river and the character of that DOC have likely become more variable. The character of the DOC may show more seasonal variation between microbially processed and weakly processed composition and greater peaks and troughs of DOC concentrations within the rivers of the Rio Negro basin.

10 Land Use Change

In the shorter term, local environmental changes have more rapid and localized impacts on the production and movement of DOC than climate change (Roland et al., 2012). Changes in land use, including alterations in water flow (damming, diversions), alter landscape characteristics and deforestation affect the fate of carbon in the ecosystem (e.g. Wilson & Xenopoulos, 2013). Deforestation is the first step. To date, the Rio Negro has been spared the severe deforestation which has occurred along the southern boundary of the Amazon rainforest (Trancoso et al., 2009; Global Forest Watch, accessed March 17, 2023, https://www.globalforestwatch.org › country › BRA). In contrast, the Rio Negro watershed had lost 1.6% (11,294 km^2) of its forests by 2008, while at the same time, 50.6% of the watershed (362,766 km^2) had come under a protective status (Trancoso et al., 2009). Land use changes have continued to occur and the future is never certain.

Deforestation and subsequent land uses are major disturbances to a forested ecosystem, creating ecologically unstable conditions. The system takes time to move to a new equilibrium. For instance, in a meta-analysis, Don et al. (2011) found that 3–10 years were required to reach a new equilibrium in soil organic carbon in newly created crop lands. Other reports suggest that carbon stores and fluxes after deforestation need approximately 25 years to recover 70% of its original biomass and another 50 years to accumulate the other 30% after selective logging or agricultural use has stopped (Cochrane, 2011; Poorter et al., 2016) This suggests that DOC production would take a similar time to come into equilibrium. The largest impacts of deforestation on future DOC loads are the decline in litter with the loss of the forest (Könönen et al., 2018), disturbance to soil carbon stocks and increases in velocity and amount of runoff (Langerwisch et al., 2016). In line with this, Drake et al. (2019) found lower concentrations of DOC in the Congo River in regions which had greater deforestation. Langerwisch et al. (2016) predicted that by the end of the current century, riverine DOC concentrations will decrease by 90% in the Amazon region under high deforestation rates. Increased discharge on top of this will likely keep concentrations stable to 2050 due to higher amounts of DOC in runoff if a sufficient amount of forest remain; however, DOC concentrations are then predicted to drastically decrease (Langerwisch et al., 2016). The quality of the DOC will be determined partly by the new vegetation and secondly by the conditions of degradation. Decomposition of the litter left behind by logging can give a pulse of nutrients to the soil (and to the water draining from the soil) (Kreutzweiser et al., 2008). Immediately after deforestation, rates of DOC degradation can initially be 'high' as factors favouring microbial degradation are increased; i.e., higher temperatures due to increased irradiance, more available oxygen (especially if the land is drained), and increased nutrients. This likely leads to increased autochthonous production and this has been previously reported from numerous other catchments affected by changes in land use (Lambert et al., 2015, 2017). Forest soils will also influence the pattern of DOC production after deforestation. In organically rich soils, including peat, exposure of the soil surface can open the soil, letting oxygen enter which promotes degradation of old, newly exposed, stored carbon. The 'old' DOC is more biodegradable with lower molecular weight, less aromaticity, and higher protein (N) content (Drake et al., 2019; Ekblad & Bastviken, 2019; Gandois et al., 2013; Wilson & Xenopoulos, 2013). The watersheds of the Rio Negro and its tributary, the Rio Branco, contain extensive peat beds (Gumbricht et al., 2017) and white sands with large humic deposits, with large carbon stores which if disturbed could export similarly degraded DOC.

Several other effects of deforestation also affect DOC production and movement. Deforestation increases erosion, releasing DOC with the soil. Where soils have a higher adsorptive capacity, they will trap DOC and slow its movement towards the rivers, perhaps increasing the time it undergoes degradation. Fire is frequently used to clear the land and is a source of black carbon which can become part of the DOC pool (Drake et al., 2020). For instance, black dissolved carbon forms 7.5% of the DOC pool in the Congo (Drake et al., 2020). Extensive deforestation can also disrupt the forest water pump which moves precipitation across the forest via evapotranspiration, resulting in less precipitation further into the forest (Marquess et al., 1977; Lean & Warrilow, 1989; Makarieva & Gorshkov, 2007). Less precipitation can lead to drought conditions as discussed above. Extensive deforestation can be very difficult to reverse given the change in precipitation patterns (Shukla et al., 1990). Reduced precipitation will affect DOC discharge to the rivers.

The most important distinction among land use changes, is whether the use destroys or repeatedly breaks the surface of the soil or whether the soil surface is rapidly mended after deforestation and not further damaged. If the integrity of the soil is maintained (pasture, orchards, afforested regions), the decrease in litter results in less production of DOC, and if the litter is woody in nature, the DOC maintains the higher molecular weight and aromaticity of forest DOC. When the soil surface is continually broken as in crop lands, the DOC released comes from degradation of both new litter and old carbon stored in the soil. The ratio of the two will determine the character of the DOC. As the soil organic carbon is the most depleted by this use (Don et al., 2011), the DOC returned to the river should also be the most decreased. Therefore, any changes in land use in the Rio Negro basin are likely to have great affects on the DOC within this system.

11 Fire

Intact, tropical rainforests are very resistant to fire due to the high moisture levels and cooler conditions beneath their unbroken canopies: only extreme drought which initiates forest mortality, can open the forest to fire under natural situations (Cochrane, 2011; Nepstd et al., 2007). However, deforestation through clearcutting and selective logging, associated building of roads and slash and burn agriculture, can open the forest canopy, drying the forest and leaving it more susceptible to drought, mortality and forest fires (Cochrane, 2011). The Amazon has experienced a number of years of drought and extensive fires over the past two decades (Aragaõ et al., 2007; Berenguera et al., 2021). The probability of fire also increases with climate change driven increases in temperature: a 1% increase in temperature is projected to increase lightning strikes by 10% (Price, 2013). Fire has become a major threat to the Amazonian rainforest and its ecosystem.

These rainforests may have little resilience to survive fire or replace their species composition due to their lack of evolutionary experience with fire (Ciemer et al., 2019; Cochrane, 2011; Kukla et al., 2021). Regions which have experience with fire normally have higher resilience. Fire in the Rio Negro basin is particularly destructive because it causes the loss of critical nutrients which are not readily replaced due to a lack of underlying bedrock (Jordan & Herrera, 1981). The soils of the Rio Negro watershed do not hold nutrients well—it is the forest complex which has developed the capacity to hold nutrients within the ecosystem (Jordan & Herrera, 1981). When nutrients are lost to the fluvial system or to the atmosphere they are not readily replaced which will affect rates of replacement of these forest ecosystems and the development and flow of DOC.

The impact of forest fire on DOC depends foremost on the extent and severity of the fire (Rhoades et al., 2016). Recovery of the ecosystem depends on a wide variety of variables related to landscape, climate, soils, new nutrient supplies and sources of new seeds, microorganism and soil biota. An example of the effects of moderate and severe fire damage on DOC dynamics is the Hayman wildfire in Colorado in 2002. That fire burnt across numerous watersheds to various extents allowing comparison of the results of different levels of fire damage. The effects on carbon and nutrient cycling were extensive where over 60% of a watershed was burnt down to the soil with the loss of the O-layer at the top of the soil (Rhoades et al., 2016). With intense fire, much of the carbon on the land and in surface soils goes up in smoke removing vast quantities of 'future' litter which could have been decomposed in future years producing and contributing carbon in all forms, including DOC, to the ecosystem. The loss of DOC will extend for decades. Similarly, nutrients are lost to the system. Erosion further reduces soil capacity to heal the system. The lack of vegetation to capture nutrients that enter through atmospheric deposition prolongs the period of recovery. Such severe fires also destroy the seed bank, ensuring that the forest composition will change in the future which may or may not affect future DOC. DOC concentrations in streams running through these catchments were very low.

DOC in stream waters were highest in catchments with a moderate burn extent (30% to < 60%)—higher even than in unburnt areas (Rhoades et al., 2016), reflecting DOC losses from these catchments due to degradation of fire-damaged vegetation (Rhoades et al., 2016). This pattern was still evident 13 years after the fire when they were sampled at the end of the study. Burnt vegetation is higher in aromatics shifting the composition of the DOC to include black carbon (Santos et al., 2019; Wang et al., 2015). Streams in burnt catchments were hotter, suggesting that any DOC present would suffer an increased rate of microbial degradation. Thus, fire can alter DOC dynamics for many decades. Similar studies are needed in the Rio Negro basin.

12 Dams

Dams regulate the river flow through impoundment affecting flow regimes, links with the terrestrial environment and transport of organic matter and sediment downstream, trapping terrestrial carbon for extended periods of time (Ward & Stanford, 1995a, b; Winton et al., 2019). DOC concentrations within regulated systems are also generally lower due to increased processing and sedimentation (Acharya et al., 2023). DOC processing via microbial and/or photodegradation/photo-oxidation mechanisms is also increased and stimulate the production of microbial DOC (Larson et al., 2007). River regulation also lowers the magnitude of the peak flow by limiting the amount of rain-associated flows and organic matter pulses to downstream environments (Acharya et al., 2023; Goodman et al., 2011; Rohlfs et al., 2016). Therefore, all above processes are likely to modify DOC (in terms of concentration and composition) in the Rio Negro. The Rio Negro is likely to be more greatly affected due to its unique DOM composition (high amounts of aromatic, high molecular weight humic material) compared to other watersheds in the Amazon Basin. If the Rio Negro or Rio Branco were dammed, it would lead to increases in the amount of microbially produced and photo-oxidized DOM in this system, as well as decreasing inputs of terrestrially derived humic material from the surrounding forests and soils. This would have cascading affects not only on the DOM pool but also in regards to overall water quality within this system with acidity likely to decrease (and pH increase). This puts biota within this system at significant risk given their adaptation to acidic conditions.

The effects of regulation on downstream regulated rivers can be decreased via the input of unregulated tributaries (Acharya et al., 2023). In contrast to regulated rivers, unregulated tributaries are mainly driven by catchment properties and generally dominated by terrestrial organic matter (Acharya et al., 2023). Unregulated tributaries can recover certain water quality gradients (e.g. organic materials, sediments) of regulated rivers and may ameliorate the impact of upstream impoundments on a main stem (Acharya et al., 2023; Kiffney et al., 2006). However, the magnitude to which tributaries inflow restore main-stem water quality is dependent on tributaries size and their distance from the dam (Benda et al., 2004; Kiffney et al., 2006). Rohlfs et al. (2016) and Acharya et al. (2023) observed distinct DOC pulses below confluences with tributaries in regulated rivers. The Rio Negro has numerous tributaries which join its main stem with the main tributary being the Rio Branco, a whitewater river. White water rivers contain lower concentrations of DOC and a different type of DOC compared to blackwater rivers (Holland et al., 2017; Sylvain et al., 2023), as such, in flows from whitewater or clearwater tributaries into the blackwater Rio Negro may not restore the adverse effects of upstream regulation and may even increase the impact of dams on DOC concentrations and characteristics within this system.

13 Other Anthropogenic Impacts

Other anthropogenic activities (urbanization, mines and feedlots) will also impact the concentration and nature of DOC, and therefore, the microbial and broader food-web. However, they are beyond the scope of the present review.

14 Conclusion

The Rio Negro contains a unique DOC pool dominated by humic material of high aromaticity and molecular weight. The DOC within this system not only forms the basis of the food web by acting as an essential energy resource for microbes, it also determines the water quality and chemistry of the system. Climate change and anthropogenic pressures are significant threats to DOM quantity and quality in the Rio Negro. Deforestation, in particular, whether selective logging or clear cutting, is the most serious threat to the wellbeing and persistence of the forest and its ability to produce its unique DOC. Any changes will likely have cascading effects across the ecosystem and cause catastrophic changes to the biodiversity and functioning of these highly biodiverse, unique systems.

Acknowledgements The authors thank Samara Silva de Souza, Susana Braz-Mota and Adalberto Luis Val for organizing this volume on the Rio Negro, and two thoughtful reviewers who improved the manuscript. OJ also thanks Chris Wood and AL Val for encouraging her to study the properties and function of DOC in the Rio Negro system.

OJ is retired without direct scientific support. AH is a senior Lecturer at La Trobe University.

References

Acharya, S., Holland, A., Rees, G., Brooks, A., Coleman, D., Hepplewhite, C., et al. (2023). Relevance of tributary inflows for driving molecular composition of dissolved organic matter (DOM) in a regulated river system. *Water Research, 237*, 119975. https://doi.org/10.1016/j.watres.2023.119975

Ågren, A., Buffam, L., Jansson, M., & Laudon, H. (2007). Importance of seasonality and small streams for the landscape regulation of dissolved organic carbon export. *Journal of Geophysical Research, 112*, G03003.

Ågren, A. M., Buffam, I., Cooper, D. M., Tiwari, T., Evans, C. D., & Laudon, H. (2014). Can the heterogeneity in stream dissolved organic carbon be explained by contributing landscape elements? *Biogeosciences, 11*, 1199–1213.

Albéric, P., Pérez, M. A. P., Moreira-Turcq, P., Benedetti, M. F., Boullion, S., & Abril, G. (2017). Variation of the isotopic composition of dissolved organic carbon during the runoff cycle in the Amazon River and the floodplains. *Comptes Rendus Geoscience, 50*, 65–75.

Amaral, J. H. F., Suhett, A. L., Melo, S., & Farjalla, V. F. (2013). Seasonal variation and interaction of photodegradation and microbial metabolism of DOC in black water Amazonian ecosystems. *Aquatic Microbial Ecology, 70*, 157–168.

Amon, R. M. W., & Benner, R. (1996). Photochemical and microbial consumption of dissolved organic carbon and dissolved oxygen in the Amazon River system. *Geochimica et Cosmochimica Acta, 60*, 1783–1792.

Aprile, F., & Darwich, A. (2009). Regime térmico e a dinâmica do oxigênio em um lago meromítico de águas pretas da região amazônica. *Brazilian Journal of Aquatic Science and Technology., 13*, 37–43.

Aragaõ, L. E. O. C., Malhi, Y., Roman-Cuesta, R. M., Saatchi, S., Anderson, L. O., & Shimabukuro, Y. E. (2007). Spatial patterns and fire response of recent Amazonian droughts. *Geophysical Research Letters, 34*, L07701.

Arvola, L., & Tulcnen, T. (1998). Effects of allochthonous dissolved organic matter and inorganic nutrients on the growth of bacteria and algae from a highly humic lake. *Environment International, 24*, 509–520.

Avila-Diaz, A., Benezoli, V., Justino, F., Torres, R., & Wilson, A. (2020). Assessing current and future trends in climate extremes across Brazil based on reanalyses of earth system model projections. *Climate Dynamics, 55*, 1403–1426.

Bardy, M., Derenne, S., Allard, T., Benedetti, M. F., & Fritsch, E. (2011). Podzolisation and exportation of organic matter in black waters of the Rio Negro (upper Amazon basin, Brazil). *Biogeochemistry, 106*, 71–88.

Barichivich, J., Gloor, E., Peylin, P., Brienen, R. J. W., Schöngart, J., Espinoza, J. C., & Pattnayak, K. C. (2018). Recent intensification of Amazon flooding extremes driven by strengthened Walker circulation. *Science Advances, 4*, eaat8785.

Beltrão, H., Zuanon, J., & Ferreira, E. (2019). Checklist of the ichthyofauna of the Rio Negro basin in the Brazilian Amazon. *ZooKeys, 881*, 53–89.

Benda, L., Poff, N. L., Miller, D., Dunne, T., Reeves, G., Pess, G., et al. (2004). The network dynamics hypothesis: How channel networks structure riverine habitats. *Bioscience, 54*(5), 413–427, 415.

Berenguera, E., Lennox, G. D., Ferreira, J., Malhia, Y., Aragãoe, L. E. O. C., Barretog, J. R., et al. (2021). Tracking the impacts of El Niño drought and fire in human-modified Amazonian forests. *PNAS, 118*(30), e2019377118. https://www.pnas.org/doi/full/10.1073/pnas.2019377118

Bouillon, S., Yambélé, A., Spencer, R. G. M., Gillikin, D. P., Hernes, P. J., Six, J., et al. (2012). Organic matter sources, fluxes and greenhouse gas exchange in the Oubangui River (Congo River basin). *Biogeosciences, 9*, 2045–2062.

Bouillon, S., Yambélé, A., Gillikin, D. P., Teodoru, C., Darchambeau, F., Lambert, T., et al. (2014). Contrasting biogeochemical characteristics of the Oubangui River and tributaries (Congo River basin). *Scientific Reports, 4*, 5402–5411.

Burkert, U., Warnecke, F., Babenzien, D., Zwirnmann, E., & Pernthaler, J. (2003). Members of a readily enriched β-Proteobacterial clade are common in surface waters of a humic lake. *Applied Environmental Microbiology, 69*. https://doi.org/10.1128/AEM.69.11.6550-6559.2003

Ciemer, C., Boers, N., Hirota, M., Kurths, J., Müller-Hansen, F., Oliveira, R. S., et al. (2019). Higher resilience to climatic disturbances in tropical vegetation exposed to more variable rainfall. *Nature Geoscience, 121*, 174–179.

Cochrane, M. A. (2011). The past, present and future importance of fire in tropical rainforests. In M. B. Bush, J. R. Flenley, & W. D. Gosling (Eds.), *Topical rainforest responses to climate change* (2nd ed., pp. 213–240). Springer-Praxis Books in Environmental Sciences.

Corlett, R. T. (2016). Impacts of drought in tropical forests. *Trends in Plant Science, 21*, 584–593.

Cory, R., Cotner, J., & McNeill, K. (2009). Quantifying interactions between singlet oxygen and aquatic fulvic acids. *Environmental Science and Technology, 43*, 718–723.

Cusack, D. F., Karpman, J., Ashdown, D., Cao, Q., Ciochina, M., Halterman, S., et al. (2016). Global change effects on humid tropical forests: Evidence for biogeochemical and biodiversity shifts at an ecosystem scale. *Reviews of Geophysics, 54*, 523–610. https://doi.org/10.1002/2015RG000510

Don, A., Schumacher, J., & Freibauer, A. (2011). Impact of tropical land-use change on soil organic carbon stocks—A meta-analysis. *Global Change Biology, 17*, 1658–1670.

Drake, T. W., Van Oost, K., Barthel, M., Bauters, M., Hoyt, A. M., Podgorski, D. C., et al. (2019). Mobilization of aged and biolabile soil carbon by tropical deforestation. *Nature Geoscience, 12*, 541–546.

Drake, T. W., Wagner, S., Stubbins, A., Wabakanghanzi, J. N., Six, J., & Spencer, R. G. M. (2020). Du feu à l'eau: Source and flux of dissolved black carbon from The Congo River. *Global Biogeochemical Cycles*. https://doi.org/10.1029/2020GB006560

Duarte, R. M., Smith, D. S., Val, A. L., & Wood, C. M. (2016). Dissolved organic carbon from the upper Rio Negro protects zebrafish (*Danio rerio*) against ionoregulatory disturbances caused by low pH exposure. *Scientific Reports, 6*, 20377. https://doi.org/10.1038/srep20377

Dubroeucq, D., & Volkoff, B. (1998). From Oxisols to Spodosols and Histosols: Evolution of the soil mantles in the Rio Negro basin Amazonia. *Catena, 32*, 245–280.

Duffy, P. B., Brandoa, P., Asnerb, G. P., & Field, C. B. (2015). Projections of future meteorological drought and wet periods in the Amazon. *The Proceedings of the National Academy of Sciences, 112*, 13172–13177.

Ekblad, A., & Bastviken, D. (2019). Deforestation releases old carbon. *Nature Geoscience, 12*, 499–500.

Ertel, J. R., Hedges, J. I., Devol, A. H., Richey, J. E., & de Nazaré Góes Ribero, M. (1986). Dissolved humic substances of the Amazon River system. *Limnology and Oceanography, 31*, 739–754.

Espinoza, J. C., Marengo, J. A., Schongart, J., & Jimenez, J. C. (2022). The new historical flood of 2021 in the Amazon River compared to major floods of the 21st century: Atmospheric features in the context of the intensification of floods. *Weather and Climate Extremes, 35*, 100406.

Fernside, P., & Araújo Silva, R. (2023). Amazon hit by a trio of droughts in a snapshot of the century to come. The Conversation, November 23 2023.

Freitas, C., Hurd, L., Bruna Andrade, B., Marcos Mereles, M., & Flávia Siqueira-Souza, F. (2024). Anthropogenic climate change is a present and future stressor for fish communities in the Amazon Basin. In S. S. de Souza, S. Braz-Mota, & A. L. Val (Eds.), *The future of Amazonian aquatic biota*. Springer.

Gandois, L., Cobb, A. R., Hei, C., Lim, L. B. L., Abu Salim, K., & Harvey, C. F. (2013). Impact of deforestation on solid and dissolved organic matter characteristics of tropical peat forests: Implications for carbon release. *Biogeochemistry, 114*, 183–199.

Gatti, L. V., Gloor, M., Miller, J. B., Doughty, C. E., Malhi, Y., Domingues, L. G., et al. (2014). Drought sensitivity of Amazonian carbon balance revealed by atmospheric measurements. *Nature, 506*, 76–80.

Ghai, R., Rodríguez-Valera, F., McMahon, K. D., Toyama, D., Rinke, R., Cristina Souza de Oliveira, T., et al. (2011). Metagenomics of the water column in the pristine upper course of the Amazon River. *PLoS One, 6*, e23785.

Gloor, M., Brienen, R. J. W., Galbraith, D., Feldpausch, T. R., Schöngart, J., Guyot, J.-L., et al. (2013). Intensification of the Amazon hydrological cycle over the last two decades. *Geophysical Research Letters, 40*, 1729–1733.

Gonsior, M., Valle, J., Schmitt-Kopplin, P., Hertkorn, N., Bastviken, D., Luek, J., et al. (2016). Chemodiversity of dissolved organic matter in the Amazon Basin. *Biogeosciences, 13*, 4279–4290.

Goodman, K., Baker, M. A., & Wurtsbaugh, W. A. (2011). Lakes as buffers of stream dissolved organic matter (DOM) variability: Temporal patterns of DOM characteristics in mountain stream-lake systems. *Journal of Geophysical Research-Biogeosciences, 116*. https://doi.org/10.1029/2011jg001709

Granéli, W., Lindell, M., de Faria, B. M., & Esteves, F. (1998). Photo-production of dissolved inorganic carbon in temperate and tropical lakes—Dependence on wavelength band and dissolved organic carbon concentration. *Biogeochemistry, 43*, 175–195.

Granier, A., Reichstein, M., Bréda, N., Janssens, I. A., Falge, E., Ciaisf, P., et al. (2007). Evidence for soil water control on carbon and water dynamics in European forests during the extremely dry year: 2003. *Agricultural and Forest Meteorology, 143*, 123–145.

Gumbricht, T., Roman-Cuesta, R. M., Verchot, L., Herold, M., Wittmann, F., Householder, E., et al. (2017). Expert system model for mapping tropical wetlands and peatlands reveals South America as the largest contributor. *Global Change Biology, 23*, 3581–3599.

Hoetzinger, M., Schmidt, J., Jezberová, J., Koll, U., & Hahn, M. W. (2017). Microdiversification of a pelagic polynucleobacter species is mainly driven by acquisition of genomic islands from a partially interspecific gene pool. *Applied and Environmental Microbiology, 83*, e02266–e02216.

Holland, A., Duivenvoorden, L. J., & Kinnear, S. H. W. (2012). Naturally acidic waterways: Conceptual food webs for better management and understanding of ecological functioning. *Aquatic Conservation: Marine and Freshwater Ecosysterms, 22*(2012), 836–847.

Holland, A., Wood, C. M., Smith, D. S., Correia, T. G., & Val, A. L. (2017). Nickel toxicity to cardinal tetra (*Paracheirodon axelrodi*) differs seasonally and among the black, white and clear river waters of the Amazon basin. *Water Research, 123*, 21–29.

Holland, A., Stauber, J., Wood, C. M., Trenfield, M., & Jolley, D. F. (2018). Dissolved organic matter signatures vary between naturally acidic, circumneutral and groundwater-fed freshwaters in Australia. *Water Research, 137*, 184–192.

Holland, A., McInerney, P. J., Shackleton, M. E., Rees, G. N., Bond, N. R., & Silvester, E. (2020). Dissolved organic matter597and metabolic dynamicsin dryland lowland rivers. *Spectrochimica Acta, Part A: Molecular and Biomolecular Spectroscopy, 229*, 117871.

Jezberová, J., Jezbera, J., Brandt, U., Lindström, E. S., & Langenheder, S. W. (2010). Ubiquity of Polynucleobacter necessarius ssp. asymbioticus in lentic freshwater habitats of a heterogenous 2000 km area. *Environmental Microbiology, 12*, 658–669. https://doi.org/10.1111/j.1462-2920.2009.02106.x

Johannsson, O. E., Smith, D. S., Sadauskas-Henrique, H., Cimprich, G., Wood, C. M., & Val, A. L. (2017). Photo-oxidation processes, properties of DOC, reactive oxygen species (ROS), and their potential impacts on native biota and carbon cycling in the Rio Negro (Amazonia, Brazil). *Hydrobiologia, 789*, 7–29.

Johannsson, O. E., Ferreira, M. S., Smith, D. S., Crémazy, A., Wood, C. M., & Val, A. L. (2020). Effects of natural light and depth on rates of photo-oxidation of dissolved organic carbon in a major black-water river, the Rio Negro, Brazil. *Science of the Total Environment, 733*, 139193. https://doi.org/10.1016/j.scitotenv.2020.139193

Johannsson, O. E., Ferreira, M. S., Smith, D. S., Wood, C. M., & Val, A. L. (2021). Interplay of oxygen and light in the photooxidation of dissolved organic carbon. *Water Research, 201*, 117332.

Jordan, C. F., & Herrera, R. (1981). Tropical rain forests: Are nutrients really critical? *The American Naturalist, 117*, 167–180.

Kamimura, N., Takahashi, K., Mori, K., Araki, T., Fujita, M., Higuchi, Y., & Masai, E. (2017). Bacterial catabolism of ligninderived aromatics: New findings in a recent decade. *Environmental Microbiology Reports, 9*, 679–705.

Kieber, D. J., McDaniel, J., & Mopper, K. (1989). Photochemical source of biological substrates in sea water: Implications for carbon cycling. *Nature, 341*, 637–639.

Kiffney, P. M., Greene, C. M., Hall, J. E., & Davies, J. R. (2006). Tributary streams create spatial discontinuities in habitat, biological productivity, and diversity in mainstem rivers. *Canadian Journal of Fisheries and Aquatic Sciences, 63*, 2518–2530.

Könönen, M., Jauhiainen, J., Straková, P., Heinonsalo, J., Laiho, R., Kusi, K., et al. (2018). Deforested and drained tropical peatland sites show poorer peat substrate quality and lower microbial biomass and activity than unmanaged swamp forest. *Soil Biology and Biochemistry, 123*, 229–241.

Kreutzweiser, D., Hazlet, P. W., & Gunn, J. M. (2008). Logging impacts on the biogeochemistry of boreal forest soils and nutrient export to aquatic systems: A review. *Environmental Reviews, 16*, 157–179.

Küchler, I. L., Miekeley, N., & Forsberg, B. R. (2000). A contribution to the chemical characterization of rivers in the Rio Negro basin, Brazil. *Journal of the Brazilian Chemical Society, 11*, 286–292.

Kukla, T., Ahlström, A., Maezumi, S. Y., Chevalier, M., Lu, Z., Winnick, M. J., et al. (2021). The resilience of Amazon tree cover to past and present drying. *Global and Planetary Change, 202*, 103520.

Kurek, M. R., Stubbins, A., Drake, T. W., Moura, J. M. S., Holmes, R. M., Osterholz, H., et al. (2021). Drivers of organic molecular signatures in the Amazon River. *Global Biogeochemical Cycles, 35*, e2021GB006938. https://doi.org/10.1029/2021GB006938

Lambert, T., Darchambeau, F., Bouillon, S., Alhou, B., Mbega, J. D., Teodoru, C., et al. (2015). Landscape control on the spatial and temporal variability of Chromophoric dissolved organic matter and dissolved organic carbon in large African Rivers. *Ecosystems, 18*, 1224–1239. https://doi.org/10.1007/s10021-015-9894-5

Lambert, T., Bouillon, S., Darchambeau, F., Massicotte, P., & Borges, A. V. (2016). Shift in the chemical composition of dissolved organic matter in The Congo River network. *Biogeosciences, 13*, 5405–5420.

Lambert, T., Bouillon, S., Darchambeau, F., Morana, C., Roland, F., Descy, J. P., et al. (2017). Effects of human land use on the terrestrial and aquatic sources of fluvial organic matter in a temperate river basin (the Meuse River, Belgium). *Biogeochemistry, 136*, 191–211. https://doi.org/10.1007/s10533-017-0387-9

Langerwisch, F., Walz, A., Rammig, A., Tietjen, B., Thonickel, K., & Cramer, W. (2016). Deforestation in Amazonia impacts riverine carbon dynamics. *Earth System Dynamics, 7*, 953–968.

Larson, J. H., Frost, P. C., Zheng, Z., Johnston, C. A., Bridgham, S. D., Lodge, D. M., et al. (2007). Effects of upstream lakes on dissolved organic matter in streams. *Limnology and Oceanography, 52*, 60–69. https://doi.org/10.4319/lo.2007.52.1.0060

Lean, J., & Warrilow, D. A. (1989). Simulation of the regional climatic impact of Amazon deforestation. *Nature, 342*, 411–413.

Leenheer, J. A. (1980). Origin and nature of humic substances of the Amazon River basin. *Acta Amazonica, 10*, 513–526.

Li, Y., Nie, C., Liu, Y., Du, W., & He, P. (2019). Soil microbial community composition closely associates with specific enzyme activities and soil carbon chemistry in a long-term nitrogen fertilized grassland. *Science of the Total Environment, 654*, 264–274.

Ling, L., Fu, Y., Jeewani, P. H., Tang, C., Pan, S., Reid, B. D., et al. (2021). Organic matter chemistry and bacterial community structure regulate decomposition processes in post-fire forest soils. *Soil Biology and Biochemistry, 160*, 108311.

Makarieva, A. M., & Gorshkov, V. G. (2007). Biotic pump of atmospheric moisture as driver of the hydrological cycle on land. *Hydrology and Earth System Sciences, 11*, 1013–1033.

Marengo, J. A., Tomasella, J., Alves, L. M., Soares, W. R., & Rodriguez, D. A. (2011). The drought of 2010 in the context of historical droughts in the Amazon region. *Geophysical Research Letters, 38*, L12703. https://doi.org/10.1029/2011GL047436

Marques, J., Marden dos Santos, J., Villa Nova, N. A., & Salati, E. (1977). Precipitable water and water vapor flux between Belém and Manaus. *Acta Amazonica, 7*, 355–362.

McClain, M. E., & Elsenbeer, H. (2001). Terrestrial inputs to Amazon streams and internal biogeochemical processing. In M. E. McClain, R. L. Victoria, & J. E. Richey (Eds.), *The biogeochemistry of the Amazon Basin* (pp. 185–208).

McClain, M. E., Richey, J. E., Brandes, J. A., & Pimentel, T. P. (1997). Dissolved organic matter and terrestrial-lotic linkages in the Central Amazon Basin of Brazil. *Global Biogeochemical Cycles, 11*, 295–311.

Montes, C. R., Lucas, Y., Pereira, O. J. R., Achard, R., Grimaldi, M., & Melfi, A. J. (2011). Deep plant-derived carbon storage in Amazonian podzols. *Biogeosciences, 8*, 113–120.

Moreira-Turcq, P., Seyler, P., Guyot, J. L., & Etcheber, H. (2003). Exportation of organic carbon from the Amazon river and its main tributaries. *Hydrological Processes, 17*, 1329–1344.

Morris, C., Val, A. L., Brauner, C. J., & Wood, C. M. (2021). The physiology of fish in acidic waters rich in dissolved organic carbon, with specific reference to the Amazon basin: Ionoregulation, acid–base regulation, ammonia excretion, and metal toxicity. *Journal of Experimental Zoology A, 335*, 843–863.

Nepstd, D. C., Tohver, I. M., Ray, D., Moutinho, P., & Cardinot, G. (2007). Mortality of large trees and lianas following experimental drought in an Amazon forest. *Ecology, 88*, 2259–2269.

NOAA. (2023, Dec). https://www.climate.gov/news-features/blogs/enso/december-2023-el-nino-update-adventure#:~:text=El%20Ni%C3%B1o%20is%20zipping%20along,the%20top%20 5%20on%20record. Accessed 16.01.24.

Phillips, O. L., Brienen, R. J. W., & RAINFOR collaboration. (2017). Carbon uptake by mature Amazon forests has mitigated Amazon nations' carbon emissions. *Carbon Balance and Management, 12*, 1. https://doi.org/10.1186/s13021-016-0069-2

Poorter, L., Bongers, F., Aide, T., et al. (2016). Biomass resilience of Neotropical secondary forests. *Nature, 530*, 211–214. https://doi.org/10.1038/nature16512

Price, C. G. (2013). Lightning applications in weather and climate research. *Survey of the Geophysical Society, 34*, 755–767.

Rai, H., & Hill, G. (1980). Classification of central amazon lakes on the basis of their microbiological and physico-chemical characteristics. *Hydrobiologia, 72*, 85–99.

Raymond, P. A., & Spencer, R. G. M. (2015). Riverine DOM. In D. A. Hansell & C. A. Carlson (Eds.), *Biogeochemistry of marine dissolved organic matter* (2nd ed., pp. 509–535). Academic Press.

Read, J. S., & Rose, K. C. (2013). Physical responses of small temperate lakes to variation in dissolved organic carbon concentrations. *Limnology and Oceanography, 58*, 921–931.

Remington, S. M., Strahm, B. D., Neu, V., Richey, J. E., & DaCunha, H. B. (2007). The role of sorption in control of riverine dissolved organic carbon concentrations by riparian zone soils in the Amazon basin. *Soil Science, 172*, 279–291.

Remington, S., Krusche, A., & Richey, J. (2011). Effects of DOM photochemistry on bacterial metabolism and CO_2 evasion during falling water in a humic and a whitewater river in the Brazilian Amazon. *Biogeochemistry, 105*, 185–200.

Reyer, C. P. O., Adams, S., Albrecht, T., Baarsch, F., Boit, A., Trujillo, N. C., et al. (2017). Climate change impacts in Latin America and the Caribbean and their implications for development. *Regional Environmental Change, C-F17*, 1601–1621.

Rhoades, C. C., Chow, A. T., Covino, T. P., Fegel, T. S., Pierson, D. N., & Rhea, A. E. (2016). The legacy of a severe wildfire on stream nitrogen and carbon in headwater catchments. *Ecosystems, 22*, 643–657. https://doi.org/10.1007/s10021-018-0293-6

Richey, J. E., Hedges, J. I., Devol, A. H., Quay, P. D., Victoria, R. L., Martinelli, L. A., et al. (1990). Biogeochemistry of carbon in the Amazon River. *Limnology and Oceanography, 35*, 352–371.

Richey, J. E., Spencer, R. G. M., Drake, T. W., & Ward, N. D. (2022). Chapter 20: Fluvial carbon dynamics across the land to ocean continuum of great tropical rivers: The Amazon and Congo. In R. M. Tshimanga, G. D. M. N'kaya, & D. Alsdorf (Eds.), *Congo Basin hydrology, climate, and biogeochemistry: A foundation for the future, geophysical monograph* (Vol. 269, First ed., English Version). https://doi.org/10.1002/9781119657002.ch20

Riedel, T., Zark, M., Vahatalo, A. V., Niggeman, J., Spencer, R. G. M., Hernes, P. J., et al. (2016). Molecular signatures in biogeochemical transformation of dissolved organic matter from ten world rivers. *Frontiers in Earth Science, 4*. https://doi.org/10.3389/feart.2016.00085

Rodríguez-Zúñiga, U. F., Pereira Milori, D. M. B., Lopez da Silva, W. T., Martin-Neto, L., Camargo Oliveira, L., & Rocha, J. C. (2008). Changes in optical properties caused by UV-irradiation of aquatic humic substances from the Amazon River basin: Seasonal variability evaluation. *Environmental Science and Technology, 42*, 1948–1953.

Rohlfs, A. M., Mitrovic, S. M., Williams, S., Hitchcock, J. N., & Rees, G. N. (2016). Dissolved organic carbon delivery from managed flow releases in a montane snowmelt river. *Aquatic Sciences, 78*, 793–807.

Roiha, T., Peura, S., Cusson, M., & Rautio, M. (2016). Allochthonous carbon is a major regulator to bacterial growth and community composition in subarctic freshwaters. *Scientific Reports, 6*, 34456.

Roland, F., Huszar, V. L. M., Farjalla, V. F., Enrich-Prast, A., Amado, A. M., & Ometto, J. P. H. B. (2012). Climate change in Brazil: Perspective on the biogeochemistry of inland waters. *Brazilian Journal of Biology, 72*, 709–722.

Santos, F., Wymore, A. S., Jackson, B. K., Mažeika, S., Sullivan, P., McDowell, W. H., et al. (2019). Fire severity, time since fire, and site-level characteristics influence stream water chemistry at baseflow conditions in catchments of the Sierra Nevada, California, USA. *Fire Ecology, 15*, 3. https://doi.org/10.1186/s42408-018-0022-8

Schöngart, J., Wittmann, F., Junk, W. J., Lopes, A., Demarchi, L. O., Biem Mori, G., et al. (2024). The fate of floodplain forests in a changing Amazon. In S. S. de Souza, S. Braz-Mota, & A. L. Val (Eds.), *The future of Amazonian aquatic biota*.

Seifert, A.-G., Roth, V.-N., Dittmar, T., Gleixner, G., Breuer, L., Houska, T., et al. (2016). Comparing molecular composition of dissolved organic matter in soil and stream water: Influence of land use and chemical characteristics. *Science of the Total Environment, 571*, 142–152.

Shukla, J., Nobre, C., & Sellers, P. (1990). Amazon deforestation and climate change. *Science, 247*, 1322–1325.

Simon, C., Osterholz, H., Koschinsky, A., & Dittmar, T. (2019). Riverine mixing at the molecular scale—An ultrahigh-resolution mass spectrometry study on dissolved organic matter and selected metals in the Amazon confluence zone (Manaus, Brazil). *Organic Geochemistry, 129*, 45–62.

Simon, C., Pimentel, T. P., Monteiro, M. Y. F., Candido, L. A., Gastmans, D., Geilmann, H., et al. (2021). Molecular links between whitesand ecosystems and Blackwater formation in the Rio Negro watershed. *Geochimica et Cosmochimica Acta, 311*, 274–291.

Sorribas, M. V., Paiva, R. C. D., Melack, J. M., Bravo, J. M., Jones, C., Carvalho, C., Beighley, E., Forsberg, B., & Costa, M. H. (2016). Projections of climate change effects on discharge and inundation in the Amazon basin. *Climatic Change, 136*, 555–570. https://doi.org/10.1007/s10584-016-1640-2

Spencer, R. G. M., Stubbins, A., Hernes, P. J., Baker, A., Mopper, K., Afdenkampe, A. K., et al. (2009). Photochemical degradation of dissolved organic matter and dissolved lignin phenols from The Congo. *Journal of Geophysical Research, 114*, G03010. https://doi.org/10.1029/2009JG000968

Spracklen, D. V., Arnold, S. R., & Taylor, C. M. (2012). Observations of increased tropical rainfall preceded by air passage over forests. *Nature, 489*, 282–285.

Sylvain, F. É., Bouslama, S., Holland, A., Leroux, N., Mercier, P.-L., Val, A. L., et al. (2023). Bacterioplankton communities in dissolved organic carbon-rich Amazonian black water. *Microbiology Spectrum*. https://doi.org/10.1128/spectrum.04793-22

Tfaily, M. M., Hamdan, R., Corbett, J. E., Chanton, J. P., Glaser, P. H., & Cooper, W. T. (2013). Investigating dissolved organic matter decomposition in northern peatlands using complimentary analytical techniques. *Geochimica et Cosmochimica Acta, 112*, 116–129.

Thurman, E. M. (1985). *Organic geochemistry of natural waters*. Junk Publishers, Kluwer Academic Publishers Group, Dordrecht/Boston/Lancaster. 499 pp.

Tikhonov, V. V., Yakushev, A. V., Zavgorodnyaya, Y. A., Byzov, B. R., & Demin, V. V. (2010). Effects of humic acids on the growth of bacteria. *Eurasian Soil Science, 43*, 305–313.

Trancoso, R., Carneiro Filho, A., Tomasella, J., Schietti, J., Forsberg, B. R., & Miller, R. P. (2009). Deforestation and conservation in major watersheds of the Brazilian Amazon. *Environmental Conservation, 36*, 277–288.

Vähätalo, A. V., & Vione, C. L. (2022). *Photochemical Reactions in Inland Waters. Encyclopedia of Inland Waters* (2nd ed.). https://doi.org/10.1016/B978-0-12-819166-8.00047-5

Vidon, P., Wagner, L. E., & Soyeux, E. (2008). Changes in the character of DOC in streams during storms in two Midwestern watersheds with contrasting land uses. *Biogeochemistry, 88*, 257–270.

Wang, J. J., Dahlgren, R. A., Ersan, M. S., Karanfil, T., & Chow, A. T. (2015). Wildfire altering terrestrial precursors of disinfection byproducts in forest detritus. *Environmental Science & Technology, 49*, 5921–5929.

Wang, X. Y. L., Zhu, J., & Tanajura, C. A. S. (2018). The strengthening of Amazonian precipitation during the wet season driven by tropical sea surface temperature forcing. *Environmental Research Letters, 13*, 094015.

Ward, J. V., & Stanford, J. A. (1995a). Ecological connectivity in alluvial river ecosystems and its disruption by flow regulation. *Regulated Rivers: Research & Management, 11*, 105–119.

Ward, J. V., & Stanford, J. A. (1995b). The serial discontinuity concept: Extending the model to floodplain rivers. *Regulated Rivers: Research & Management, 10*, 159–168.

Ward, C. P., & Cory, R. M. (2016). Complete and partial photo-oxidation of dissolved organic matter draining permafrost soils. *Environmental Science Technology, 50*, 3545–3553. https://doi.org/10.1021/acs.est.5b05354

Waterloo, M. J., de Oliveira, M., Drucker, S., Nobre, D., Cuartas, A., Hodnett, L., et al. (2006). Export of organic carbon in runoff from an Amazonian Blackwater catchment. *Hydrological Processes, 20*, 2581–2597. https://doi.org/10.1002/hyp.6217

Wilson, H. F., & Xenopoulos, M. A. (2013). Diel changes of dissolved organic matter in streams of varying watershed land use. *River Research and Applications, 29*, 1330–1339.

Winton, R. S., Calamita, E., & Wehrli, B. (2019). Reviews and syntheses: Dams, water quality and tropical reservoir stratification. *Biogeosciences, 16*, 1657–1167.

Wood, C. M., Matsuo, A. Y. O., Wilson, R. W., Gonzalez, R. J., Patrick, M. L., Playle, R. C., et al. (2003). Protection by natural Blackwater against disturbances in ion fluxes caused by low pH exposure in freshwater stingrays endemic to the Rio Negro. *Physiological Biochemical Zoology, 76*, 12–27.

Wood, C. M., Al-Reasi, H. A., & Smith, D. S. (2011). The two faces of DOC. *Aquatic Toxicolocy, 105S*, 3–8.

World Land Forms. https://worldlandforms.com/landforms/ria/. Accessed 28.01.23.

Impacts of Climate Change and Local Disturbance on Stream Fish Assemblages in the Amazon

Gabriel Gazzana Barros, Barbara Mascarenhas, Jansen Zuanon, and Cláudia Pereira de Deus

Abstract

Freshwater environments are among the most threatened by anthropogenic impacts in the world. Among these environments, streams are especially vulnerable to structural changes, water pollution, alterations in food chains, and loss of biodiversity, resulting from deforestation, agricultural land use, as well as small dams and road construction. Changes in rainfall patterns and increased temperature resulting from climate change may boost the negative effects of local habitat disturbances and disrupt physiological, biological and behavioral responses of the aquatic biota, including the instability of seasonal predictability. Here, we evaluate the current knowledge regarding the effects of local anthropogenic disturbances and those from ongoing planetary climate changes on the aquatic environment and fish fauna of small streams worldwide, with a special focus on Amazonian forest streams. To do so, we conducted a (non-exhaustive) literature review searching for information regarding direct and indirect impacts of climate change on the structure and hydrological dynamics of headwater streams around the world. Then, we present information on the effects of such environmental changes in species distributions, on the taxonomic and functional structure and composition of stream fish assemblages in the Amazon and in other biomes. Despite the scarcity of published references dealing with combined/synergic

G. G. Barros (✉) · B. Mascarenhas · J. Zuanon
Graduate Program in Biology of Freshwater and Inland Fishing (PPGBADPI), Brazilian National Institute for Amazonian Research (INPA), Manaus, Amazonas State, Brazil
e-mail: gabriel.barros.bolsista@icmbio.gov.br

C. P. de Deus
Graduate Program in Biology of Freshwater and Inland Fishing (PPGBADPI), Brazilian National Institute for Amazonian Research (INPA), Manaus, Amazonas State, Brazil

Biodiversity Coordination, Brazilian National Institute for Amazonian Research (INPA), Manaus, Amazonas State, Brazil

S. S. de Souza et al. (eds.), *The Future of Amazonian Aquatic Biota*,
https://doi.org/10.1007/978-3-031-66822-7_4

effects of climate change and local anthropogenic impacts on the biology and ecology of stream fishes worldwide, we bring some insights about these emerging environmental problems affecting Amazon stream fishes.

1 Introduction

Thousands of animal and plant species have been directly exploited by humans for millennia, generating varying degrees of negative impacts on the populations of these organisms. Recently, Darimont et al. (2023) estimated the extent of the impacts resulting from the use of the majority of the world's vertebrate fauna (approximately 47,000 species assessed). This study revealed humans as the planet's top predator, exploiting 300 times more species than comparable non-human predators, with records of the use of approximately 15,000 species of vertebrates; about 42% of bony fish species (Actinopterygii) and 35% of freshwater fish species are exploited by humans as food and/or pets.

The detrimental effects of unsustainable human use of natural resources and of habitat loss due to anthropogenic environmental degradation are evident, widely distributed across the planet's biomes, and present among different biological groups. However, these direct negative impacts have been exacerbated by the effects of ongoing climate change, which can directly affect species (by surpassing certain physiological limits, for example) and/or indirectly (by altering the environmental conditions necessary for species to successfully complete their life cycles). In fact, signs of global climate change have been observed for decades, especially after the identification of the role played by the greenhouse effect, resulting from the significant increase in CO_2 and CH_4 emissions since the Industrial Revolution in the second half of the nineteenth century (IPCC, 2023). Although well-supported in the scientific literature, the direct and indirect effects of these phenomena on species can be challenging to ascertain based on empirical ecological data, mainly due to the scarcity of long-term, controlled ecological observations. Additionally, the effects of climate change on global temperature rise are subtle and incremental, making it difficult to perceive their multiple and complex effects on ecosystem functioning, species biology, and biodiversity richness in real-time, especially when considering the prominence of other threats.

Despite the eloquence of these facts, other factors have been identified as major drivers of biodiversity loss on the planet. For example, Jaureguiberry et al. (2022) showed that changes in land use have been the most significant direct drivers of recent freshwater biodiversity loss. The second most important factor identified by these authors was pollution, followed by direct use of natural resources, and in fourth place, the detrimental effects of invasion by non-native species; the negative effects of climate change only ranked fifth, the last among the drivers analyzed in that study. Additionally, the study by Jaureguiberry et al. (2022) suggests that ongoing climate change has had a stronger impact on the composition of biological communities than on the population structure of the species that make up those communities. This apparent lower relative importance of climate change in biodiversity loss probably represents only a historical artifact, as the identification of the climate change process and its anthropogenic origin are relatively recent compared to the accumulated records of other anthropogenic drivers' effects.

Despite our recent awareness regarding the ongoing climate change, global greenhouse gas emissions have been increasing uninterrupted, resulting from the unwise use of energy, increasing deforestation and land use changes, unsustainable lifestyles and, consumption and production patterns, factors that vary among regions and countries (IPCC, 2023). In recent decades (mainly from 1980 onwards), an increase in precipitation has been recorded in most of the Amazon Basin (with the exception of the southern and southwestern portions), with a significant increase during the rainy season and a slight reduction in historical precipitation averages during the dry season (Gloor et al., 2015). Extreme floods and severe droughts have also been recorded during this period (Gloor et al., 2015; Marengo et al., 2008; Marengo & Espinoza, 2016), causing disruptions in the annual flood pulse of large river-floodplain systems and changes in the hydrological dynamics of small upland forest streams, which immediately respond to the effects of local rainfall (e.g., Trancoso, 2006). Furthermore, climate models predict significant changes in rainfall patterns throughout the Amazon Basin, with an increase in the western portion and a reduction in the eastern portion (Sorribas et al., 2016), which may exacerbate the effects of local disturbances in aquatic systems.

Several authors have documented the high vulnerability of freshwater ecosystems to anthropogenic impacts worldwide (e.g., Tickner et al., 2020; van Rees et al., 2021), both due to direct alterations to water courses like dams, road construction, pollution, excessive water withdrawal, and introduction of non-native species, as well as indirect ones resulting from deforestation and land use changes (e.g., Trancoso et al., 2010). Various characteristics of freshwater systems make them particularly vulnerable to the effects of climate change, such as the limited ability of many species to disperse between fragmented environments in response to environmental changes; the strong influence of climate on water availability and temperature; and the simultaneous exposure to various other types of anthropogenic impacts at different levels of organization, from individuals and species to trophic chains (Perkins et al., 2010) and ecosystems (Woodward et al., 2010). In tropical regions, Taniwaki et al. (2017) conducted a meta-analysis of studies on the effects of climate change (rainfall increase or decrease, and temperature increase) and environmental stressors related to increase in land use intensity in small streams, where they summarized the knowledge about these anthropogenic impacts on aquatic organisms (algae, macroinvertebrates, and fish) and ecological processes (stream metabolism and decomposition rates). Biological responses detected by Taniwaki et al. (2017) include reduction in species richness and abundance/density of aquatic macroinvertebrates, reduction in algal and epilithic biomass, increase in dominance of certain species, shifts in fish assemblage structure, and increase in the respiration rates of stream whole metabolism. Despite the relevance of the study to the understanding of the effects of climate change in tropical streams, the majority of the articles analyzed by Taniwaki et al. (2017) do not address the effects of climate change on stream fish fauna in Brazil. Regarding temperature, only two of the articles evaluated by these authors refer to Brazilian streams (Pinto et al., 2009; Rolla et al., 2009), both conducted in the Southeast region of the country (in the Atlantic Forest biome), with no published reference on this type of impact in Amazonian streams.

Despite being the biome that harbors the greatest diversity of freshwater fish species on the planet, there are few studies presenting empirical data on the effects of climate change on Amazonian fish assemblages. These studies focus on the composition and trophic structure of fish assemblages of large rivers and lake systems after an extreme drought event and have recorded changes in species distributions and reduction in reproductive activity (e.g., Freitas et al., 2013; Herrera-R et al., 2020; Röpke et al., 2017, Röpke 2022), with fewer studies in fish assemblages of small headwater streams (Borba et al., 2020; Costa et al., 2020; Frederico et al., 2021; Martins et al., 2022). This scarcity of information is especially concerning for Amazonian fish species already threatened with extinction due to other types of anthropogenic impacts, which may have their risk increased by additive or synergistic effects generated by climate change (e.g., Frederico et al., 2016). In this sense, there is a need to evaluate the current state of knowledge on the topic and advance in the analysis of synergistic (actual or potential) effects of ongoing climate change on the planet simultaneously with the disturbances caused by multiple forms of anthropogenic impacts on the fish fauna of small Amazonian streams.

In the present study, we evaluate the current knowledge on the effects of climate change on fish in upland forest streams in the Amazon and the possible synergistic negative effects of these changes with other ongoing anthropogenic impacts in the biome. To do so, we initially discuss the effects of climate change (represented mainly by temperature and rainfall changes) on the structure and hydrological dynamics of headwater streams around the world. Then, we present information on the effects of climate change on fish species and assemblages worldwide, and follow with a discussion about the effects of these changes specifically for Amazonian stream fishes in pristine environments. Considering that climate changes are expected to affect pristine and disturbed environments differently, we then present a compilation of results from studies on the effects of the most frequent anthropogenic impacts in Amazonian streams. Finally, we discuss the combined and synergistic effects of environmental alterations caused by climate change together with the impacts caused by habitat loss resulting from deforestation, infrastructure construction (including roads), water pollution, invasion by non-native species, and modifications in the parasite-host relationships of Amazonian stream fish.

2 Methods

This study was conducted based on the compilation of bibliographic references obtained from the internet (Google Scholar) and personal files of the authors, using different combinations of keywords selected according to the themes addressed in the article: "climate change", "temperature change", "rainfall change", "forest streams", "Amazonian streams", "headwater streams", "stream fishes", "stream fish fauna", "stream ichthyofauna", combined with "anthropogenic impacts", "anthropogenic changes", "land use changes", "deforestation", "damming" and "road construction". The time range employed in the literature search was from 1990 to 2023. References explicitly mentioning the occurrence of climate change effects on the structure and hydrological dynamics of small headwater streams worldwide were sought as a way to identify widespread effects on these systems through studies developed with historical data series and ecological modeling. In addition, references on the effects of climate change on stream fishes worldwide were sought in order to obtain a general perspective on the types of impacts specifically affecting this group of organisms and type of aquatic environment. Then, in order to gather specific information about the Amazonian ichthyofauna, references on the effects of climate change on stream fishes in the biome were searched. Finally, publications on the effects of other anthropogenic impacts on Amazonian stream fishes were researched, including those resulting from direct disturbances, such as deforestation, dams, pollution, etc., as well as indirect and diffuse impacts resulting from land use changes in the drainage basin.

3 Effects of Climate Change on the Structure and Hydrological Dynamics of Headwater Streams Across the World

A fundamental step in understanding the effects of climate change on the ichthyofauna is to evaluate or predict the main impacts of these environmental transformations on fish habitats. For this, historical data series are needed on variations in air temperature (and water temperature, when available), rainfall (which directly or indirectly affects water flow), and hydrodynamics of these systems (in the form of flow data and information on water flow permanence). The analysis of these factors, individually or combined, allows us to estimate how aquatic environments have been (or will be) affected by climate change, and how these transformations may affect various aspects of ecological system functioning, including changes in species distribution, taxonomic and functional structure of assemblages, availability of natural resources for human populations, and various types of environmental services.

A fundamental obstacle to the development of predictions about the biological and ecological responses of the ichthyofauna to climate change lies precisely in the lack of empirical information about species and biological communities. Historical series on biological data are rare worldwide, which makes predicting the impacts of global temperature increase and changes in rainfall an extremely complex task, particularly in hyperdiverse ecosystems like the Amazon. Thus, the choice to model climatic and hydrological variables lies in the existence of many long-term time series of environmental data, obtained for various historical reasons and made widely available mainly through the advances in informatics (i.e., high data storage and analysis capacity) and the internet (which connected institutions, databases, and researchers).

The main methodology used to evaluate the impacts of climate change on the structural, limnological, and hydrological dynamics of stream systems consists of statistical modeling of databases, which can involve historical series of a river system, sets of watersheds, or even entire countries' watersheds. This happens either due to countries having their own specific methods of measurement and monitoring of parameters or due to the absence of databases that allow for the direct analysis of the effects of climate change over time. Most of the available articles analyzed by us have mainly focused on assessing the effects of climate change on water quality, flow alteration, and sediment transport in rivers and streams (Chang et al., 2001; Gain et al., 2011; Gao et al., 2013; Fazel et al., 2017; Phan et al., 2011; Xu et al., 2013; Al-Mukhtar et al., 2014; Woznicki et al., 2016; Wang et al., 2018; Givati et al., 2019; Lyu et al., 2019; Oliveira et al., 2019; Rogers et al., 2020), change in precipitation patterns (Maghsood et al., 2019; Sun et al., 2013), and water temperature increases (Adem et al., 2016; Du et al., 2019; Rogers et al., 2020; Stefan & Sinokrot, 1993), without inferences about the impacts of these environmental alterations on biological communities. Among the analytical methods employed in these studies, the most used tool has been the SWAT—Soil and Water Assessment Tool (Arnold et al., 1998), which allows for projecting the impacts of climate change and its effects on water flow and sediment transport.

Among the above-mentioned studies, Sun et al. (2013) and Xu et al. (2013) employed environmental modeling to evaluate the possible effects of different scenarios of greenhouse gas emissions in varying time windows (periods of 30–90 years in the future) in China, and have pointed to a significant reduction in rainfall and streamflow, with serious negative consequences for the functioning of these systems and for the use of water resources. Another study conducted in China by Wang et al. (2018) analyzed temperature, rainfall, and effluent pollution data in the drainage basin that supplies Beijing's main drinking water reservoir. The models showed a significant increase in rainfall and consequently in streamflow (about 15% in the period 2021–2050, compared to the period 1961–1990), with negative effects on water quality worsened by the predicted increase in total nitrogen and total phosphorus concentration. In Vietnam, in a similar approach, Phan et al. (2011) found that higher greenhouse gas emissions could change precipitation patterns and cause an increase in streamflow by 11.4% and sedimentation rates by 15.3% in the rainy season, and by 8.8% and 12.6% in the dry season, respectively. These results clearly point to significant environmental alterations of rivers and streams, with probable negative impacts on the biota. On the other hand, a recently conducted study in Japan by Suzuki et al. (2022) showed that streams heavily dependent on groundwater are likely to have smaller annual temperature fluctuations and lower summer temperatures in the future, compared to streams with less groundwater influence. Thus, the negative effects on the environmental characteristics of streams and on the biota could be markedly different in type and intensity in a given region, which makes predictions about the ecological consequences of climate change in lotic systems even more complex and context-dependent.

On a much larger spatial scale, Punzet et al. (2012) addressed the effects of climate change on stream water temperature worldwide (despite the low sample coverage in some regions and continents). According to the study's projections, there would be a widespread increase of 1–4 °C in stream water temperature by the year 2070. This global modeling also predicted a 25% temperature increase in December, January, and February in the equatorial zone, and in June, July, and August in the northern and southern zones of the planet. This global approach demonstrated that the effects of climate change on stream temperature are not uniform, combining regional and seasonal differences in the rate of water warming.

4 Effects of Climate Change on Taxonomic and Functional Composition of Stream Fish Assemblages Worldwide

The effects of climate change on water temperature, rainfall regime, streamflow, and sediment load in streams can directly and indirectly affect the distribution and phenology of stream fish species, leading to changes in the taxonomic and functional structure of fish assemblages. From a global perspective, Xenopoulos et al. (2005) combined two scenarios from the Intergovernmental Panel on Climate Change (IPCC) with a hydrological model to estimate the combined effects of reduced river flow and increased freshwater withdrawal for human use on fish species richness: A2 emissions scenario assumes economic and population trends consistent with a regionalized and economically oriented world, where the fertility patterns across regions converge very slowly, resulting in continuously increasing global population. The future levels of global greenhouse gas emissions are among the highest of the IPCC scenarios; B2 emissions scenario also assumes economic and population trends consistent with a regionalized world but with a stronger environmental focus. It is a scenario with a continuously increasing global population at a rate lower than A2. Based on a relationship between flow and local fish species richness in over 200 rivers across the world, the authors demonstrated that in rivers with reduced flow, up to 75% of local fish biodiversity could be at risk of extinction by 2070 due to the combined effects of climate change and increased freshwater consumption. This projection was shown to be even more severe in less developed countries, which concentrate high rates of species richness and endemism.

Temperature increase resulting from climate change is especially worrisome in freshwater systems, since one of the variables that most limits fish distribution is water temperature (Kirk & Rahel, 2022). Fish species inhabiting streams in colder regions or higher elevation streams in a given watershed are more likely to lose their suitable habitat characteristics (e.g., water temperature ideal for reproduction) due to increased water temperature. On the other hand, species occurring in lower elevation streams tend to gain more suitable habitats due to the warming of water from higher regions and the consequent increase in the probability of suitable habitats for adult individuals and for fry survival (Rogers et al., 2020).

Impacts of climate change on local fish assemblages at a regional scale have been also studied in France, where initial studies did not indicate a clear pattern of change in the structure and composition of fish assemblages in response to climate change. Buisson et al. (2008) estimated the intensity of turnover rates in the composition of fish assemblage, where headwater stream species experienced a greater detrimental effect (attributed to their lower tolerance to higher temperatures), while downstream species expanded their distribution to intermediate or upstream locations in the basin. Subsequently, Buisson and Grenouillet (2009) modeled the distribution of 35 common fish species in French streams based on predicted temperature and precipitation changes, focusing on both the taxonomic and functional structure of assemblages. The authors predicted an increase in functional and taxonomic diversity along the upstream-downstream gradient of streams in response to climate change (temperature and precipitation regime), but with different effects in its upper, middle, and lower reaches. Assemblages occupying the upper reaches (headwaters) are expected to become more differentiated from each other, mainly due to local extinctions, while assemblages in the lower reaches are likely to become more similar to each other due to increased colonization rates by allochthonous species.

Buisson et al. (2013) also examined the effects of climate change on the distribution areas of stream fish species in France based on the uniqueness and functional singularity of species. Contrary to the authors' initial hypothesis, the most unique and singular species (with lower functional redundancy in assemblages) did not have the highest potential reductions in distribution range. Despite this unexpected result, Buisson et al. (2013) state that the predicted changes in species distribution ranges could cause a drastic reduction in the functional diversity of assemblages, together with its homogenization. Indeed, an analysis of occurrence data of 32 common fish species in France in two periods (1980–1992 and 2003–2009) conducted by Comte and Grenouillet (2013) demonstrated changes in species distribution ranges due to temperature increase. Some species dispersed to upstream and higher altitude sections, but also showed a tendency toward reduced distribution ranges near the lower limits of occurrence areas. However, the observed changes in fish species distribution range do not seem to be fast enough to compensate for the temperature increase resulting from climate change. Furthermore, it is important to carefully assess the results of studies on species' response speed in changing environments, since analyses of species richness and population data may be biased by the occurrence of temporal autocorrelations, especially regarding population trends of long-living species (Kuczynski et al., 2023).

5 Impacts of Climate Change on Amazonian Stream Fish Assemblages

The literature on the impacts of climate change on fish in the Amazon Basin is scarce and predominantly focuses on the effects of extreme events (abnormally strong droughts) on the structure of floodplain lake fish assemblages (e.g., Freitas et al., 2013; Röpke et al., 2017), changes in the flood pulse and the effects on the reproduction of commercial fish species (Röpke et al., 2019), and disturbances in the reproductive migrations of large catfish (Feng et al., 2020). One of the earliest studies to model the probability of Amazonian fish extinction due to climate change was conducted by Oberdorff et al. (2015), which used the pirarucu *Arapaima gigas* as an example and concluded that direct anthropogenic threats resulting from the expansion of human infrastructure and economic activities have a much greater potential impact on fish communities than ongoing climate changes. The predominant focus of those studies on large Amazonian rivers and lakes are likely due to higher interest in studying commercially valuable species, as well as the greater availability of occurrence and abundance data resulting from historical monitoring of fish landings. As a result, small headwater streams and small-sized fish fauna (with low commercial interest) have historically been neglected in ecological monitoring studies.

Only recently have empirical data been used to analyze the effect of changes in rainfall patterns on stream habitat structure and the taxonomic and functional composition of fish assemblages. As far as we could verify, the work of Borba et al. (2020) was the first to assess the effects of climate change on Amazonian stream assemblages based on empirical data collected over a period of about 20 years of periodic monitoring. These authors demonstrated that the increase in the frequency of torrential rains (more than 20 mm of rain/day) observed over the last decades in the Central Brazilian Amazon (Fig. 1) is associated with changes in the predominant substrate type in upland streams, transitioning from litter banks to sand. The taxonomic and functional composition of fish assemblages also underwent changes related to substrate modification, with a shift from a predominance of open-water species (laterally compressed body with less-developed fins and terminal mouth) to species adapted to lentic environments in lateral pools and shallow marginal areas (fusiform body, more developed fins, and superior mouth). Over the whole sampling years, *Pyrrhulina* aff. *brevis*, *Microcharacidium eleotrioides*, *Laimosemion kirovskyi*, *Bryconops giacopinii*, *Iguanodectes geisleri* and *Hemigrammus* cf. *pretoensis* increased in relative abundance, while *Hyphessobrycon* aff. *melazonatus*, *Gymnotus pedanopterus*, and *Hypopygus lepturus* decreased. Therefore, depending on the magnitude, direction, frequency, and spatial extent of the impact that climate changes cause on the habitat, certain species that better adapt to the new predominant environmental conditions in the area (or are able to seek refuge from direct impacts of habitat modification) are favored. In fact, Costa et al. (2020) demonstrated simultaneous changes in the taxonomic and functional composition of fish, ants, birds, trees, palms, and lianas assemblages in a forest reserve in Central Amazon, involving reductions in species abundance and richness, as well as changes in functional composition. Despite some differences in the direction of observed changes, the results indicate that forests and micro basins have become more dynamic in response to changes in the intensity of rainy and dry seasons, and that the new predominant environmental conditions have acted as environmental filters for certain groups of aquatic and terrestrial species (Costa et al., 2020).

The lack of observational studies on the effects of climate change on stream fish assemblages is accompanied by a scarcity of research exploring the physiological mechanisms that underlie the observed responses in the environment. One of the rare studies on the physiological responses of Amazonian stream fish to the stressful effects of climate change was conducted by Campos et al. (2019). They compared

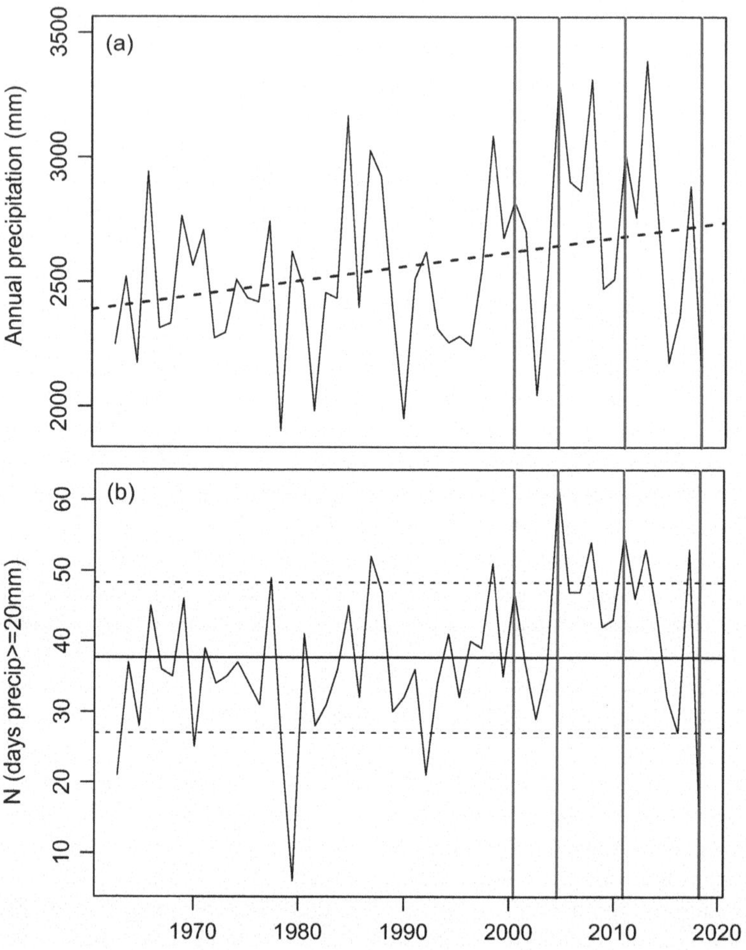

Fig. 1 Example of ongoing climate change in the Amazon, illustrated by temporal rainfall trends at Ducke Forest Reserve, Brazilian Central Amazon, from 1966 to 2018, with red vertical lines indicating surveyed years (2001, 2005, 2011, and 2018) in the study of Borba et al. (2020). Variation in annual precipitation over time (**a**); the historical mean annual precipitation was 2563 mm ± 351 mm, and there is a significant linear regression slope represented by the full line (increase of 6.08 ± 2.98 mm/year along the 53-year period). Variation in the frequency of stormy days (i.e., precipitation higher than 20 mm) over time (**b**); the historical mean was 38 ± 10 storm days, and there is a non-significant regression slope with time represented by the full line. (*Source*: Borba et al., 2020, Freshwater Biology 66:753–764. DOI: 10.1111/fwb.13675)

three fish species in a laboratory experiment within a microcosm environment simulating climate change effects (4.5 °C and 900 ppm CO_2 above current levels).

In general, the fish showed an increase in metabolic rate and oxidative stress, but there were variations in the responses among the species analyzed. The small-sized tetra *Hyphessobrycon aff. Melazonatus* (a rheophilic species occupying the stream channels) showed lipid peroxidation, lower thermal tolerance (36 °C), and low survival rate. On the other hand, the dwarf cichlid *Apistogramma agassizii* and the pencil fish *Pyrrhulina brevis*, which inhabit marginal environments and have more sedentary habits, were more tolerant to climate change; they demonstrated higher survival rates, greater thermal tolerance (40 °C and 39 °C, respectively), and lower values of physiological stress response biomarkers. These experimental results indicate that Amazonian stream fish exhibit distinct responses to increased temperature and other environmental disturbances associated with climate change, and some of them may adapt (at least temporarily) to the new environmental conditions.

6 Direct Anthropogenic Disturbances in Amazonian Streams and Their Impacts on the Fish Fauna

Some effects of global climate change, such as the net temperature increase, are subtle and really challenging to be directly perceived in real-time. However, local anthropogenic disturbances are far more conspicuous and their consequences for stream fish have been frequently recorded in the Amazon, not only in the Brazilian region but also in neighboring countries (Table 1). A pioneering study on the effects of deforestation on stream fish assemblages in the Amazon was conducted by Bojsen and Barriga (2002) in Ecuador. In that study, the authors found that species richness was not correlated with deforestation, but Beta diversity was higher among streams with preserved forest cover. On the other hand, fish density was higher in deforested streams, but there was a shift in the trophic composition of the assemblages, transitioning from dominated by omnivores and insectivores (mainly Characiformes) in preserved environments to being dominated by periphytivores (Loricariidae catfish) in deforested streams. Changes in stream substrate cover, particularly the presence of litter banks, were the environmental variable most closely related to the local fish assemblage structure. The importance of litter banks as an explanatory variable for fish assemblage composition was also observed by Borba et al. (2020) in the Central Brazilian Amazon, in streams affected by an increase in the precipitation and in the frequency of stormy days.

In addition to local factors that can generate changes in the structure of fish assemblages, as observed by the authors cited above, several other stressors, such as changes in land use, pollution, increased temperature, alterations in substrate composition, siltation, and reduced water flow, can occur in association and generate combined or synergistic effects on stream fishes. In the Amazon, extensive cattle ranching and large-scale agriculture are some of the main drivers of deforestation in current times (Macedo et al., 2018). These stressors contribute to a high level of environmental degradation, modifying the physical quality of the habitat, altering light and nutrient input (and the energy flow of the aquatic system), reducing connectivity between water bodies (and fragmenting fish populations), and thus altering the biological communities of streams. Even pollution by microplastics has already been detected in Amazonian stream fishes. Ribeiro-Brasil et al. (2020) analyzed the presence of nano, micro, and mesoplastics in the digestive tract and gills of stream fish sampled in the eastern Amazon, State of Pará, where cattle ranching,

Table 1 Environmental and biological/ecological responses of Amazonian stream fish assemblages to the effects of direct anthropogenic disturbances (deforestation, pollution, damming, road construction) and of climate change (mostly resulting from temperature and rainfall pattern changes)

Main driver of change	Effect/Variable	Environmental responses	Biological/Ecological responses	Site/River basin/Region	Reference
Direct anthropogenic impacts	Contamination by plastic waste	N.A.	Presence of nano, micro and mesoplastics in 98% of analyzed individuals	Guamá, Acará and Capim river basins, Eastern Brazilian Amazon	Ribeiro-Brasil et al. (2020)
Direct anthropogenic impacts	Small dams construction	N.A.	Decrease of species richness, diversity and evenness of assemblages; increase of taxonomical dominance upstream of dams, especially for equilibrium and periodic species	Western Brazilian Amazon (Rondônia state)	Sousa et al. (2018)
Direct anthropogenic impacts	Impacts of road crossings	Increase of stream depth; reduction of current velocity in stretches dammed by road crossing	Change of taxonomical and functional composition of assemblages between upstream and downstream of dams; change in Beta diversity patterns (nestedness and turnover)	Marapanim and Maracanã river basins, Northeastern Pará state, Eastern Brazilian Amazon	Brejão et al. (2020)
Direct anthropogenic impacts	Deforestation	Reduction of canopy cover; increase of macroinvertebrate density and periphyton biomass	Change of relative abundance of species; changes in taxonomical and trophic composition of assemblages	Ecuadorian Amazon	Bojsen and Barriga (2002)
Direct anthropogenic impacts	Deforestation (age and time since impact)	Simplification and homogenization of substrate (mostly sand) and marginal vegetation (predominance of grasses)	Species' threshold of negative responses at low deforestation levels (<20%) and soon after impact (<10 years); positive responses less common and with higher deforestation levels (>70%) and later (>10 years)	Machado River basin, Rondônia state, Western Brazilian Amazon	Brejão et al. (2017)
Direct anthropogenic impacts	Deforestation of riparian zone	Reduction of physical habitat integrity (Habitat Integrity Index)	Decrease of functional diversity; changes in Functional Trophic Groups composition	Amazon forest-savannah transition, Central-Western Brazil	Freitas et al. (2021)

Direct anthropogenic impacts	Deforestation; conversion of forest to pasture	Change of substrate structure and composition; reduction of canopy cover	Overall increase of abundance; a decrease in functional divergence	Eastern Brazilian Amazon	Cantanhêde et al. (2021)
Direct anthropogenic impacts	Impacts of Reduced Impact Logging	Canopy cover loss; reductions of mean thalweg depth; increase of water temperature and conductivity and dissolved oxygen	Without significant changes in taxonomical structure and composition of assemblages compared to non-logged streams	Capim River basin, Eastern Brazilian Amazon	Prudente et al. (2017)
Direct anthropogenic impacts	Deforestation and land use changes	N.A.	Common and rare species contribute similarly to the functional richness of assemblages, but respond differently to landscape changes	Capim and Acará river basins, Eastern Brazilian Amazon	Benone et al. (2020)
Direct anthropogenic impacts	Deforestation and land use changes	Change in substrate structure and composition; reduction of canopy cover; increase in water temperature and electrical conductivity	Overall increase in fish abundance; changes in assemblage species composition	Xingu River headwaters, Southeastern Brazilian Amazon	Ilha et al. (2019)
Direct anthropogenic impacts	Deforestation and land use changes	Small direct influence of the environmental gradient	Dispersal limitation and environmental filters affect the co-occurrence patterns of fish species, without direct influence of environmental or spatial gradients	Acará River basin, Eastern Brazilian Amazon	Lobato et al. (2022)
Direct anthropogenic impacts	Deforestation and land use changes	Reduction of canopy cover; decrease of dissolved oxygen; changes in stream channel structure	Change of Functional Trophic Groups of assemblages	Capim River basin, Eastern Brazilian Amazon	Maia et al. (2022)
Direct anthropogenic impacts	Deforestation and land use changes	Reduction of canopy cover and large woody debris; increase of fine sediments	Change in Functional Beta Diversity; change in the pattern of functional turnover of assemblages	Capim River basin, Eastern Brazilian Amazon	Seabra et al. (2022)

(continued)

Table 1 (continued)

Main driver of change	Effect/Variable	Environmental responses	Biological/Ecological responses	Site/River basin/Region	Reference
Direct anthropogenic impacts	Deforestation and land use changes	Loss of catchment and local riparian forest cover; increase of fine sediments, water temperature and abundance of aquatic macrophytes	Changes in macroinvertebrates and fish incidence and abundance in disturbed streams, but with different tolerances to the impacts; congruency between the two biological groups, but with slightly different responses to anthropogenic stressors	Amazon, Tapajós, Curuá-Una, Capim and Gurupi River basins, in Paragominas and Santarém–Belterra municipalities, Pará state, Eastern Brazilian Amazon	Martins et al. (2022)
Direct anthropogenic impacts	Deforestation, land use changes and small dams construction	Reduction of riparian cover; increase of water temperature	N.A.	Xingu River headwaters, Southeastern Brazilian Amazon	Macedo et al. (2018)
Direct anthropogenic impacts	Deforestation, land use changes, and road crossings	Changes in stream channel structure; changes in substrate structure and composition	Decrease in species richness, evenness, and functional divergence of assemblages	Santarém and Paragominas municipalities, Pará state, Eastern Brazilian Amazon	Leitão et al. (2018)
Direct anthropogenic impacts	Different logging regimes	Reduction of canopy cover; increase of fine sediments	Change of ecomorphological structure of assemblages compared to non-logged streams; different effects related to logging regimes	Capim River basin, Eastern Brazilian Amazon	Jacob et al. (2021)
Direct anthropogenic impacts	Different logging regimes and small-scale gold mining	Increase of coarse substrate in mined (current or historical) streams	Small-scale gold mining strongly changes the taxonomical richness and functional trophic structure of assemblages; Reduced Impact Logging shows smaller negative effects and restricted to a single functional group	French Guiana	Allard et al. (2016)
Direct anthropogenic impacts	Substitution of forest cover for oil palm plantation	Increase of fine sediments and ORP	Changes in assemblage composition	Moju and Acará river basins, Eastern Brazilian Amazon	Ferreira et al. (2018)

Direct anthropogenic impacts	Substitution of forest cover for oil palm plantation	Reduction of coarse litter; increase of large instream cover features; higher substrate embeddedness at stream channel and margins	Changes in substrate composition, in the amount of instream cover, and amount of large woody debris condition species richness of fish and insects (Heteroptera and Odonata)	Moju and Acará river basins, Tailândia municipality, Northeastern Pará state, Eastern Brazilian Amazon	Juen et al. (2016)
Climate change	Changes in rainfall; temperature increase	Increase of water temperature; reduction of dissolved oxygen; change in stream discharge	Increased extinction risk of species already threatened by other environmental disturbances (e.g. dams, deforestation, pollution)	Brazilian Amazon basin	Frederico et al. (2016)
Climate change	Increase in rainfall	Temporary flow interruption	Persistent (several years) changes in the composition of assemblages	Xingu River basin, Eastern Brazilian Amazon	Martins et al. (2022)
Climate change	Increase in rainfall (frequency of stormy days with >20 mm rain)	Change of substrate composition (from litter-dominated to free sand)	Changes in the functional composition of assemblages	Brazilian Central Amazon	Borba et al. (2020)
Climate change	Increase in water temperature	N.A.	Increased metabolic rate and oxidative stress	Laboratorial experiment	Campos et al. (2019)
Combined effects of climate change and direct anthropogenic impacts	Increase in water temperature; small dams construction	River fragmentation and loss of hydrological connectivity	Distribution range contraction; decrease in the long-run probability of species persistence in the habitat	Andean rivers of Peru, Bolivia, Colombia and Ecuador	Herrera-R et al. (2020)

N.A. = Not Applicable; *ORP* = Oxidation Reduction Potential

agriculture, and deforestation are intense. Only one of the 68 specimens evaluated (a small sand-dwelling catfish, *Mastiglanis* cf. *asopos*) contained no plastic particles, and no difference was found in the contamination of the gills and digestive tract, which means that plastics are being effectively ingested. Other species, such as *Hemigrammus unilineatus* presented less contamination of both the gills and the digestive tract than the other species, while *Polycentrus schomburgkii* had less plastic in the gastrointestinal tract, and *Crenicichla regani* and *Pimelodella gerii* both had a larger quantity of plastic adhered to their gills in comparison with the other species. Nanoplastics and microplastics adhered most to the gills, while plastic fibers were the most common type of material overall. This was the first study to analyze plastic contamination in fish from Amazonian streams and revealed high levels of contamination, and some species were shown to possibly be more susceptible than others. The combination of different sources of environmental disturbance (land use change, deforestation, pollution) reinforces the potential synergistic effects of several stressors impacting the biological community.

There is a reasonable number of empirical studies on the impacts of anthropogenic disturbances on stream fish assemblages in the Brazilian Amazon. These studies have been mainly conducted in the Eastern portion of the basin, in the state of Pará, in a region known as the "Arc of deforestation," where there has been a rapid process of occupation and expansion of the agricultural frontier in recent decades. There is significant concern and interest in understanding, monitoring, and mitigating the effects of deforestation and changes in land use on the stream ecosystems in that region of the Brazilian Amazon. Many of these studies are aimed at analyzing the loss of taxonomic and functional diversity of fish assemblages, as well as modifications in the community assembling processes and patterns of species associations in the streams (Benone et al., 2020; Cantanhêde et al., 2021; Freitas et al., 2021; Lobato et al., 2022; Maia et al., 2022). According to these studies, the level of alteration in fish assemblages depends on the intensity of the disturbance, the time of occurrence of the alterations, the local environmental characteristics (e.g., canopy cover, diversity of substrates water temperature), and the original fish fauna.

Fish assemblages are formed by sets of species with varying abundances and ecological characteristics, shaped by biotic interactions and environmental filters (e.g., Mouillot et al., 2013). In the Neotropics, the taxonomic composition of fish assemblages is characterized by a higher species richness of the orders Characiformes, Siluriformes, and Cichliformes (Dagosta & De Pinna, 2019; Lowe-McConnell, 1987). However, the abundance of individuals in the local assemblages varies greatly among species, where few are abundant and most have intermediate to low abundances (Magurran & MacGill, 2011). Both common and rare species contribute to the overall functional richness in an assemblage, but may respond differently to environmental alterations imposed by deforestation or other anthropogenic disturbances. Rare species often show higher habitat specialization and thus become more vulnerable to landscape changes than common species (e.g., Leitão et al., 2016). These observations were particularly striking in studies conducted in streams of the Eastern Brazilian Amazon. The conversion of forest into pasture alters the functional structure of fish assemblages, making them less diverse due to the reduction in niche differentiation among species (Cantanhêde et al., 2021; Maia et al., 2022). In some cases, the alterations observed in the composition of functional trophic groups were not necessarily accompanied by changes in taxonomic richness or total fish biomass in the assemblages. This seems to occur when environmental conditions minimally suitable for the species are maintained in the streams, despite the impacts resulting from changes in land use occurring in the drainage basin (Cantanhêde et al., 2021; Ilha et al., 2019; Maia et al., 2022). On the other hand, the apparent lack of changes in these parameters may represent only a temporal delay in the responses of assemblages to environmental disturbances, as new colonization events are usually recorded earlier than local extinctions (Kuczynski et al., 2023).

In the formation of stream fish assemblages, the most abundant species are generally few, but they are important because they can ensure greater functional redundancy of the assemblages. For example, Castanho et al. (2019) analyzed the functional structure of fish assemblages in tributary streams of the Rio Negro in the Brazilian Amazon, where there is selective fishing of small-sized fish species for the ornamental fish trade. The authors simulated scenarios of selective removal of ornamental species (overfishing) and demonstrated that the exclusion of these species from the assemblages would result in a 24.6% reduction in functional diversity and up to 12% reduction in functional redundancy, which could compromise the ecological resilience of the system. On the other hand, rare species are also ecologically important as they contribute significantly to the functional richness, specialization, and uniqueness of the assemblages (Leitão et al., 2018). These elements (functional redundancy and uniqueness) add functional complexity to the assemblages and thus increase resilience to environmental disturbances.

Among the various drivers of environmental changes in streams (Fig. 2), the construction of roads and highways is considered one of the most widespread stressors. Roads are essential for social integration and have high economic importance, facilitating the transportation of goods between regions. Regarding environmental issues, in some cases, roads can have a few positive effects when they indirectly act as habitat boundaries for some animal or plant species, or even as limiting zones for the geographic distribution of terrestrial animals (Daigle, 2010). However, the negative effects are much more frequent and strongly impact aquatic systems (especially headwater streams) and their biota. Most of the impacts of road construction on streams fish result from the impoundment of large sections (and environmental disturbances caused by the drowning of riparian vegetation, limnological and hydrodynamic alterations upstream the road) and the obstacles or blockage of fish movement along streams (due to inadequate water passage structures under the roads).

According to Trombulak and Fressel (2000), roads cause seven main types of environmental impacts to the fauna in general: mortality resulting from road construction, mortality due to collisions with vehicles, changes in animal behavior, changes in the physical environment, alterations in the chemical environment, spread of non-native species, and increased human use of the area (including the creation of new vectors of deforestation and environmental disturbances). More specifically, road construction generates disturbances in the structure and dynamics

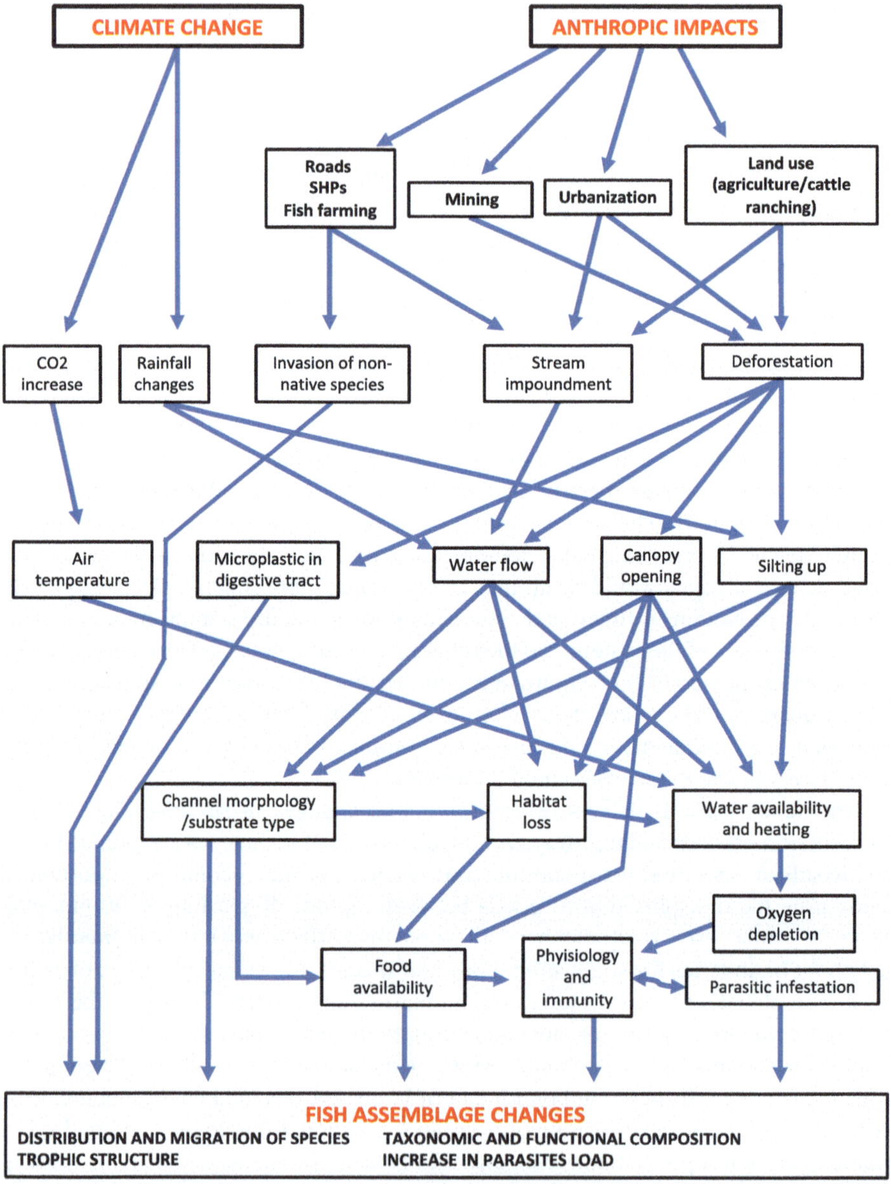

Fig. 2 Flowchart of expected pathways for the effects of climate change and other direct anthropogenic disturbances (deforestation, disturbances on riparian vegetation, land use change in the watershed, pollution, impoundment, road crossings) on the channel structure, limnological characteristics, hydrodynamics, and substrate composition of Amazon streams. Arrows indicate the effects (actual and/or predicted) of each disturbance driver on fish assemblages, including changes in assemblage structure and composition, fish reproduction disruption, food availability, and parasite load. Taxonomic structure comprises species richness, diversity, abundance, and evenness. Functional structure comprises functional richness, evenness, divergence, originality, and identity. *SHPs* = Small Hydroelectric Plants

of the landscape, such as the suppression of riparian vegetation, increased sediment load in rivers and streams, alterations in the physical and chemical conditions of the water, increased water temperature, changes in flow and channel morphology, bed clogging due to erosion, substrate homogenization (which consequently reduces microhabitat diversity), and restriction of organism movement (e.g., Daigle, 2010; Forman & Alexander, 1998). Additionally, road construction modifies the composition of invertebrate species (aquatic insects and zooplankton; Khan & Colbo, 2008) and the composition and functional structure of fish assemblages (Brejão et al., 2020; Smith et al., 2018; Yuhara, 2012).

The effects of road construction and operation on the environment and animal species have been extensively studied in various regions of the world, especially in North America, Europe, and Australia, where comprehensive reviews on the subject are available (e.g., Daigle, 2010; Forman & Alexander, 1998; Trombulak & Fressel, 2000). However, these reviews provide general analyses of road impacts at different spatial scales without specifying the deleterious effects on specific groups of organisms. In Brazil, for example, the review of Smith et al. (2018) presents a synthesis of the impacts of roads and highways in different river basins and biomes, discussing the disturbances resulting from deforestation, impoundment, grounding of springs and streams, and channel erosion and siltation on aquatic organisms, mainly on fish. According to these authors, the construction of roads causes the fragmentation of streams and rivers through landfills or diversions from natural courses. This leads to changes in the morphology of rivers (width, depth, flow), making access to spawning sites difficult, or even reducing the diversity of microhabitats to be explored by the fish. The Amazon region, which hosts an enormous and complex hydrographic system, has received some attention regarding the deleterious effects of road construction, but the focus has mainly been on the role of these infrastructure projects as vectors of deforestation and other large-scale environmental disturbances (e.g., Fearnside & Graça, 2006; Graça et al., 2014).

Regarding the specific impacts of roads on fish assemblages in Amazonian streams, few studies have been conducted, such as those by Mascarenhas (2019), Brejão et al. (2020) and Barros (2022). Brejão et al. (2020) demonstrated the effects of stream impoundment caused by road construction as a limiting factor for fish dispersal, and the changes in the functional and taxonomic structure of fish assemblages resulting from the impact of transforming lotic stretches into lentic habitats. Mascarenhas (2019), in a more specific approach, assessed the effects of stream impoundment caused by the construction of the BR-174 highway in the Brazilian Central Amazon on the populations of the nektonic fish *Bryconops giacopinii* (Iguanodectidae), a locally abundant species with a wide distribution in the region. Individuals from impounded streams exhibited poorer body condition (Fig. 3), as measured by the Condition Factor (cf. Froese, 2006), and a higher incidence of parasites when compared to individuals sampled in pristine streams (Fig. 4) likely compromising the fitness of the individuals (Barros, 2022). Stream impoundment can increase water temperature and create favorable conditions for the occurrence of opportunistic parasites, such as the observed high prevalence of the black-spot disease (subcutaneous infestations by digenean trematode larvae) in individuals of

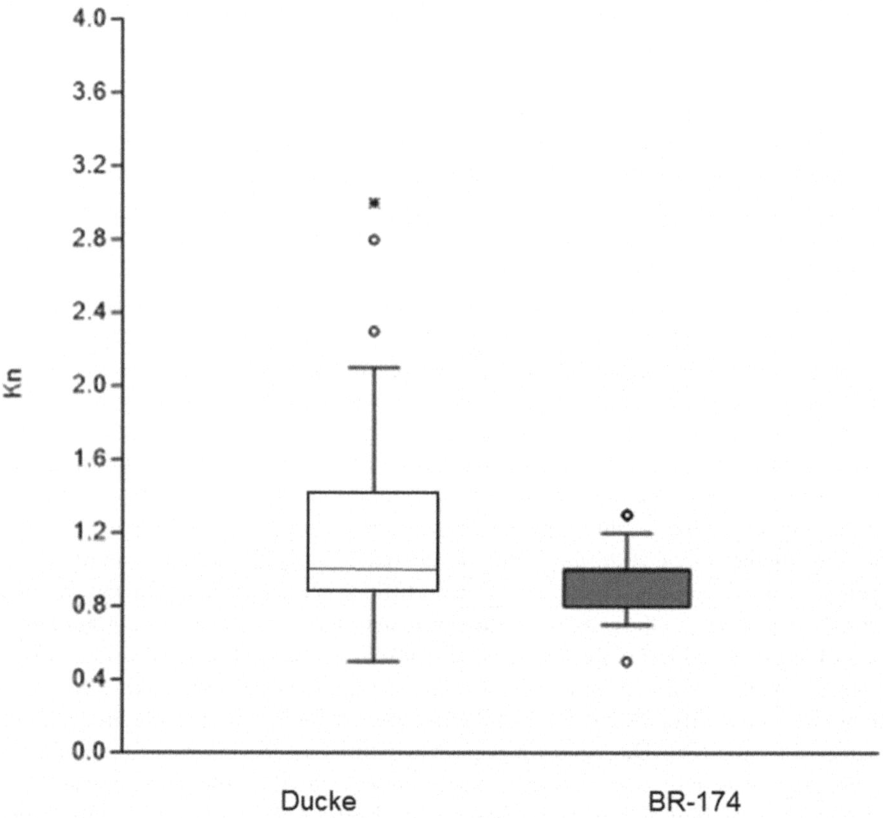

Fig. 3 Condition Factor (K) of *Bryconops giacopinii* specimens sampled in pristine (Ducke, empty box) and dammed (BR-174, gray-filled box) streams. Ducke = Adolpho Ducke Forest Reserve; BR-174 = Manaus-Boa Vista highway). Upper and lower limits of the boxes = interquartiles distance; horizontal lines outside boxes = highest and lowest values excluding outliers; circles = outliers

Bryconops giacopinii (Iguanodectidae) in streams disturbed by the construction of a highway in the Brazilian Central Amazon (Barros, 2022; Mascarenhas, 2019).

The elevation of the water temperature due to canopy opening facilitates the proliferation of these parasites and increases infestation rates in fish, especially through the creation of favorable environmental conditions for the proliferation of mollusks, which act as the first intermediate hosts of these trematodes and sustain the complex life cycle of these organisms (Flores-Lopes, 2014). The transformation of lotic environments in lentic impoundments probably increases the encounter rate of cercariae and host fishes (Marcogliese, 2001). In fact, Cairns et al. (2005), Flores-Lopes and Thomaz (2011), and Flores-Lopes (2014) observed that the highest intensity of infestation occurs at higher temperatures and in lentic environments. High infection by the black spot parasites leads to deleterious effects for the individual, affecting energy reserves and consequently weakening the fish's immune

Fig. 4 Individuals of *Bryconops giacopinii* (Characiformes: Iguanodectidae) collected in a pristine stream and in an impounded stream by the BR-174 highway crossing. All streams belong to the Tarumã-Açu River basin, itself a tributary of the lower Negro River in Brazilian Central Amazon. The black dots on the body surface of the lower fish are metacercariae of a trematode worm provisionally identified as *Neascus* sp. (Diplostomatidae: Crassiphialine). Individuals collected in pristine streams were almost completely free from such parasites

system (Bush et al., 2001), as well as impairing the individual's swimming ability (Barros, 2022). Additionally, individuals affected by black spot disease become more susceptible to predation, as the metacercariae encystment results in the formation of black marks that make the individual more visible to piscivorous birds (the final hosts of the trematodes) (Flores-Lopes, 2014).

An unexpected impact of stream impoundment by roads on Amazonian fish was observed by Barros (2022). Working in Central Amazonian streams, the author observed a higher degree of trophic specialization (expressed both in diet composition and Carbon isotopic signatures of muscle, liver, and fin tissues) of *B. giacopinii* individuals affected by stream impoundment. Individuals sampled in the impacted streams consumed a higher proportion of items of vegetable origin (algae, fruits, and seeds), whereas individuals from pristine streams strongly relied on allochthonous invertebrates as the main food items (Fig. 5). Furthermore, changes in the body morphology and lower swimming performance of *B. giacopinii* individuals sampled in streams impacted by the road indicate the possible occurrence of a process of selection of morphotypes adapted to the semi-lentic environments, with loss of deleterious consequences for the ecological performance of the individuals (Fig. 6).

Various engineering alternatives and remediation or mitigation solutions for the environmental problems generated by the disturbances resulting from roads construction can be implemented, such as the restoration of riparian vegetation and establishment of a buffer zone adjacent to the highway (Casatti et al., 2006; Daigle, 2010; Melo et al., 2003), avoidance and control of waste discharge into rivers/creeks (Daigle, 2010; Fonseca et al., 2014), installation of more suitable devices (such as culverts and gabions) for allowing fish passage and improving access to upstream

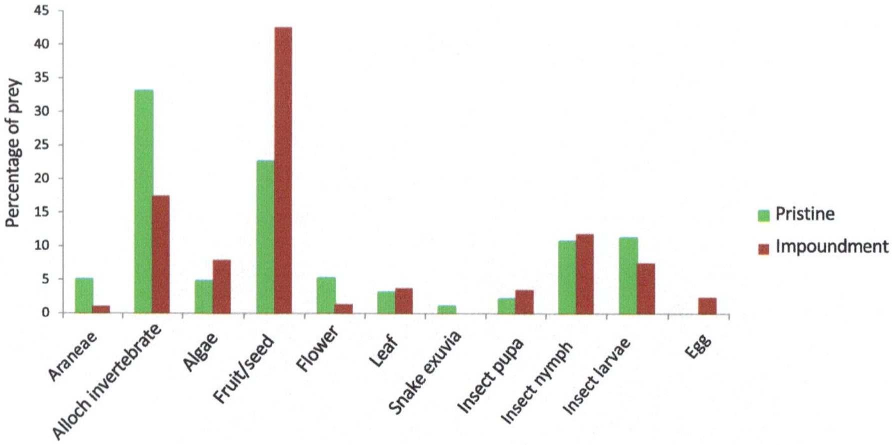

Fig. 5 Proportion of prey by food item category found in 172 stomachs of *Bryconops giacopinii* from pristine ($N = 91$) and impounded streams ($N = 81$). Alloch invert = allocthonous invertebrates

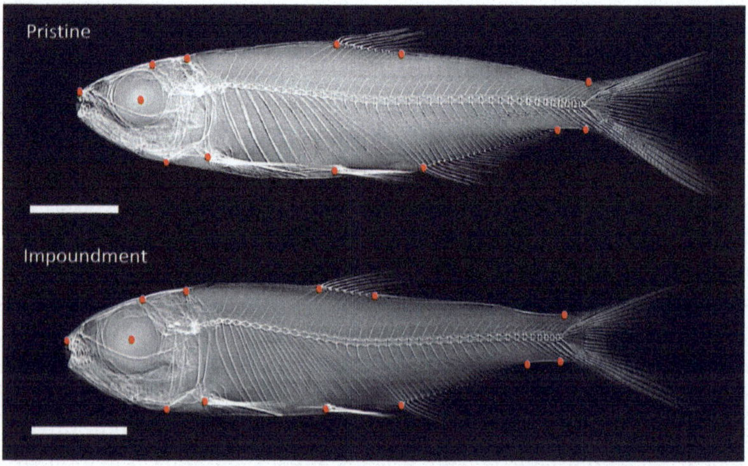

Fig. 6 X-ray images of specimens of *Bryconops giacopinii* representing the extremes of morphological variation between pristine (lotic) and impacted (impounded) streams. Red dots indicate the landmarks used to capture shape changes. White bars represent the one-centimeter scale

habitats (Baker & Votapka, 1990; Daigle, 2010; Oliveira & Bennemann, 2005), banks stabilization with selected plants and restoration of riparian vegetation (Daigle, 2010; Melo et al., 2003). Such mitigation measures can help buffering the impacts of the roads and may help recovering the impounded stream stretch, and possibly increasing the resilience of the system as a whole. This, in turn, tends to reduce the likelihood of population extirpation and supports the maintenance of species within the environment (Daigle, 2010; Smith et al., 2018).

Another dramatic anthropogenic impact on small streams is the widespread environmental degradation resulting from unplanned urbanization processes. This problem, common worldwide but especially concerning in developing countries, involves a set of simultaneous impacts arising from deforestation, degradation of riparian zones, physical alteration of streambeds, dumping of waste, and pollution from domestic and industrial effluents (Brown et al., 2005). In addition to the well-known loss of biodiversity resulting from these impacts, the process of urbanization is often accompanied by the introduction of non-native species that have been increasingly frequent in the Amazon (Doria et al., 2021) which significantly exacerbates the impacts on stream ecosystems. The combined effects of environmental degradation and biological invasions also manifest in different ecological relationships within local biological communities (e.g., Francis et al., 2019), including parasite-host relationships (Lacerda et al., 2018; Moreira et al., 2015). In fact, the health of an ecosystem can be assessed through the interpretation of parasite-host relationships (e.g., Hudson et al., 2006), which can constitute reliable indicators and early sentinels of ecological disturbances.

In urbanized streams, pollution typically occurs simultaneously with the removal of native riparian vegetation and an increase in water temperature caused by higher solar radiation on the water surface. In the Brazilian Central Amazon, Gomes (2020) assessed the effects of environmental degradation resulting from the urbanization process on the diversity, richness, and composition of endoparasite species in fish from upland stream ecosystems in the municipality of Manaus. The author analyzed the intensity of endoparasite infestations in native and non-native fish species along a gradient of environmental degradation, and recorded the presence of 21 morphospecies of parasites in 21 fish species. Parasite diversity and richness were affected by the interaction of environmental degradation with the diversity and richness of host fish, as well as a replacement of endoparasite species between pristine and degraded streams. Additionally, the authors did not find a preference of parasites for host species along the gradient, nor between native and non-native species. These results illustrate the combined effects of various stressors on stream ecosystems, including local extirpations of species and loss of biodiversity, biotic homogenization, and modification of complex parasite-host ecological relationships.

7 Combined and Synergistic Impacts of Climate Change Plus Direct Anthropogenic Disturbances on Amazonian Stream Fishes

Various types of anthropogenic impacts (river impoundment, pollution, urbanization, agriculture, land use etc.) can act additively or synergistically with the effects of climate change, leading to significant alterations in fish communities, such as decreased species richness, changes in trophic groups, shifts in species distribution and abundance, reduced catches, and alterations in functional diversity of assemblages, among others (Fazel et al., 2017; Herrera-R et al., 2020; Murdoch et al., 2020; Nelson et al., 2009). Some studies investigating the combined effects of climate change and other direct anthropogenic disturbances in lotic systems, particularly in the Neartic and Neotropical regions, have used ecological models based on long-term observation databases, which enable the identification of environmental stressors. Nelson et al. (2009) studied the possible synergistic effects of urbanization and climate change on North American headwater stream fishes. The authors used models to predict changes in growth and reproduction of headwater stream fishes in the Chesapeake Bay, based on ten scenarios that combined characteristics related to urbanization and future climate changes. The effects of urbanization alone would reduce the growth and reproduction of 8 out of 39 fish species present in the streams, while the effect of climate change alone would affect 22 to 29 of the 39 species; however, the combined effects of these two factors would be more severe for a larger number of species, altering composition and leading to significant diversity loss. Nelson et al. (2009) also found that among the 10 most common fish species, six would experience an impact of 40 to 90% on their growth and reproduction when affected by the combined effects of urbanization and climate change, resulting in an overall ecological reconfiguration of streams and loss of ecosystem services.

Murdoch et al. (2020) also support the occurrence of synergistic effects of climate stressors and those resulting from land use in boreal stream fishes, based on predictive studies of 300 watersheds in Canada. Species richness and biological productivity were not negatively related to changes in land use, but a decline in the number of sensitive species (species less resistant and more vulnerable to changes in environmental conditions, including salmonids) was observed, which varied depending on the interactions between local climate, land use, and stream type. The number of sensitive species decreased in regions with greater land use intensity and higher warming, while in colder regions, the number of sensitive species increased.

For the Neotropical region, despite being considered one of the richest in biodiversity, empirical studies on the synergistic effects of climate change and other anthropogenic impacts on stream fish are rare. Herrera-R et al. (2020) evaluated the potential impacts of climate change on fish species ranges in rivers and streams of the Andean Amazon when combined with the effects of river fragmentation by dams. The authors demonstrated an overall tendency of range contraction for most of the fish species, particularly those inhabiting highlands. It was also shown that some of these dams could prevent upstream migration for many species, which can reduce fish diversity in some basins. The synergistic effects of climate change and dam construction were found to promote a decrease in the probability of species persisting in the long term due to the isolation of sub-populations in smaller habitat fragments. Moreover, Herrera-R et al. (2020) demonstrated that the negative impacts of the synergistic effects of climate change and habitat fragmentation would not be similar among species, with strongly rheophilic and specialized species being the most vulnerable to severe range contractions and an increased risk of local extirpation.

The extreme scarcity of empirical data on the combined effects of multiple stressors on stream habitats and aquatic organisms, including the transformations resulting from ongoing climate change on the planet, highlights the need for the maintenance of long-term data series of environmental/ecological data, so allowing more in-depth analyses and the increasing our ability to predict long-term alterations in stream ecosystems.

8 Amazonian Stream Fishes Under Climate Change: Concluding Remarks and Future Perspectives for Research and Conservation

The present review provides a general perspective of the effects of climate change on the ichthyofauna of Amazonian streams and its interactions with other direct anthropogenic impacts. Many studies have been conducted on the impacts of global climate change on aquatic systems, seeking to predict the effects of increased water temperature and changes in rainfall patterns on the physical habitat structure and hydrological dynamics of these systems (Xenopoulos et al., 2005; Punzet et al., 2012; Sanchez et al., 2019; Borba et al., 2020; Kirk & Rahel, 2022). Similarly, various studies on the effects of climate change on stream fishes have used Species Distribution Models (SDM models) to predict changes in species distribution ranges, as well as in the taxonomic and functional structure of fish assemblages (Comte & Grenouillet, 2013; Buisson et al., 2013; Kwon et al., 2015; Wang et al., 2018). However, published references focusing on direct and indirect effects of climate change on fish species and assemblages through historical series of empirical data are scarce, both in the Amazon and globally.

Studies on various types of direct anthropogenic impacts on stream environments and their ichthyofauna in the Brazilian Amazon are somewhat more numerous but are geographically concentrated in the Central and Eastern portions of the basin, close to the two largest urban centers (Manaus and Belém). Consequently, significant knowledge gaps remain, and not all adverse effects of stressors acting on the biome are well understood or even recognized (see Carvalho et al., 2023 for a recent analysis of research gaps in the Amazon). Additionally, the lack of information on fish's natural history and the scarcity of well-resolved phylogenies hinders the establishment of more appropriate and effective conservation strategies for Amazonian stream ichthyofauna (e.g., García-Girón et al., 2023). For example, various studies in temperate regions have shown that animal and plant species can respond to global warming and habitat changes by shifting their distribution ranges toward higher elevation (or higher latitude) areas (Comte & Grenouillet, 2013; Woodward et al., 2010). This suggests that spatial planning of protected areas should be revisited, as species may shift, contract, or expand their ranges inside or outside conservation units, as noted by Struebig et al. (2015) for Bornean mammals. In this sense, formally protected area systems worldwide should be revised to accommodate these foreseen distributional changes, which may require expanding existing protected areas or creating new ones. However, considering that most of the Amazon Basin is situated in lowlands, the potential role of high-altitude areas as refuges for fish fauna is probably low.

An alternative approach to seeking microclimatic refuges for Amazonian stream fishes in a scenario of global warming would involve occupying more suitable habitats in the surrounding landscape. For example, in areas with naturally lower canopy cover and higher stream water temperature, such as Amazonian savannas, the effects of a warmer climate could drive range shifts of more sensitive species toward the milder conditions found in adjacent forested areas, where streams are protected by a denser canopy cover (see Mendes Jr, 2013). However, such species range shifts would likely result in perturbations to the forested stream fish assemblages due to the limits to the carrying capacity of strongly oligotrophic Amazon headwater streams (Fittkau et al., 1975; Junk, 1983; Schmidt, 1972; Sioli, 1975). Given the immense drainage network of small headwater streams in the Amazon, the rich biodiversity present in these environments, and their vulnerability to the deleterious effects of direct and indirect anthropogenic impacts, it is crucial that more studies be conducted to enable its effective conservation.

Freshwater systems support a disproportionately higher diversity of fishes than the area occupied by this type of environment on the planet, yet they have been neglected in conservation strategies employed worldwide, which heavily rely on terrestrial flora and fauna to create protected areas (e.g., van Rees et al., 2021, Frederico et al., 2021). Several factors contribute to this situation, from the physiography of lotic systems (i.e., linear and dendritic systems that are often used as boundaries for terrestrial protected areas, and not as focal points for conservation strategies targeting aquatic environments and their biota), to the relative invisibility of freshwater fishes compared to marine and more charismatic terrestrial species. This is particularly concerning because freshwater systems are vital for the ecological functioning of terrestrial and marine ecosystems and for human existence (Tickner et al., 2020; van Rees et al., 2021). Moreover, it has been shown that Neotropical freshwater fishes provide highly valuable ecosystem services, including provisioning (fisheries resources, species for aquaculture, medicinal products), regulating (seed dispersal in river basins, decomposition, and top-down control of biological communities), supporting (nutrient cycling, ecosystem engineering), and cultural services (recreational fishing and tourism, education, generation of scientific knowledge about ecological and evolutionary processes, as well as values related to cosmologies of traditional communities) (Pelicice et al., 2022). Therefore, the conservation of freshwater systems in the Amazon should be urgently prioritized, both due to the intrinsic benefits of the conservation of aquatic biodiversity and the quality of life of human populations in the region (e.g., van Rees et al., 2021), and because of the predicted high benefits and low costs of this change in the conservation strategies employed so far (see Leal et al., 2020).

The greatest limiting factor for advancing studies on the effects of climate change on Amazonian stream fishes keeps being in the lack of a historical series of empirical data collected through standardized sampling methods. The scarcity of such databases makes it challenging to disentangle the effects of ecological modifications caused by climate change from those generated by other types of simultaneous anthropogenic impacts in the watershed and/or landscape. Other vertebrate groups better known from a taxonomic and ecological perspective, such as birds, have long data series that allow subtle temporal changes in species composition of assemblages to be detected, even in pristine areas historically employed as reference conditions (e.g., Stouffer et al., 2021). As far as we could verify, only the results of Borba et al. (2020) were generated with empirical long-term data series from stream fishes, and using a standardized sampling protocol. Fortunately, the periodic standardized samplings in that area (Adolpho Ducke Forest Reserve, in Manaus, Amazonas state, Brazil) keep running as part of a Long Term Ecological Research (LTER) site, which should allow for new analyses of the process of taxonomic and functional changes in local fish assemblages. In this regard, long-term ecological studies should be encouraged and expanded through specific calls and adequate funding, which would allow verifying the validity of the findings of this study for other areas of the Amazon.

Acknowledgments We thank the ADAPTA project for providing support to field and laboratory activities related to our research with stream fishes, and especially during the development of the Doctoral Thesis (GGB) and the Master's Dissertation (MBCB). This study was financed in part by the Coordenação de Aperfeiçoamento de Pessoal de Nível Superior— Brasil (CAPES)—Finance Code 001, and Fundação de Amparo à Pesquisa do Estado do Amazonas (FAPEAM) through the Post-Graduate Program Biologia de Água Doce e Pesca Interior at INPA. We are also grateful to CNPq, FAPEAM and CAPES for long-term financial support to the Igarapés project; INPA provided logistic support to field and laboratory activities, as well as office space and other facilities, for which we sincerely thank. Camila S. dos Anjos helped with fish parasites identification, Eurizângela Dary with analysis of fish stomach contents, and José da Silva Lopes ("Seu Zé") provided invaluable help during the field trips. We also thank Dr. Gilvan Yogui from the Laboratory of Organic Compounds in Coastal and Marine Ecosystems, Oceanography Department of Federal University of Pernambuco in Recife, Brazil, for the support with stable isotope analyses, and Dr. Márcio Silva Araújo from Universidade Estadual Paulista (UNESP-Rio Claro) for the support with stable isotope, geometric morphometric and individual specialization analyses. Fish specimen collections were authorized by SISBIO (permit # 10199-5).

Competing Interests The authors have no conflicts of interest to declare that are relevant to the content of this chapter.

Ethics Approval Research activities with fishes at INPA were authorized by the institutional Ethics Committee on the Use of Animals in Research (permit # 052/2012 granted to Projeto Igarapés/JZ).

References

Adem, A. A., Tilahun, S. A., Ayana, E. K., Worqlul, A. W., Assefa, T. T., Dessu, S. B., & Melesse, A. M. (2016). Climate change impact on stream flow in the upper Gilgel Abay catchment, Blue Nile Basin, Ethiopia. In *Landsc Dynam Soils and Hydrol Proc in Varied Climates* (pp. 645–673). Springer.

Allard, L., Popée, M., Vigouroux, R., & Brosse, S. (2016). Effect of reduced impact logging and small-scale mining disturbances on Neotropical stream fish assemblages. *Aquatic Sciences, 78*, 315–325.

Al-Mukhtar, M., Dunger, V., & Merkel, B. (2014). Assessing the impacts of climate change on hydrology of the upper reach of the spree river: Germany. *Water Resources Management, 28*, 2731–2749.

Arnold, J. G., Srinivasan, R., Muttiah, R. S., & Williams, J. R. (1998). Large area hydrologic modeling and assessment part I: Model development. *JAWRA, 84*(1), 73–89.

Baker, C. O., & Votapka, F. E. (1990). *Fish passage through culverts* (Vol. 990). US Department of Transportation, Federal Highway Administration.

Barros, G. G. (2022). *Respostas de um peixe de riacho à construção de uma rodovia na Amazônia: variação fenotípica inter e intrapopulacional e especialização individual.* Thesis, National Institute of Amazonian Research.

Benone, N. L., Soares, B. E., Lobato, C. M. C., Seabra, L. B., Bauman, D., & Montag, L. F. S. (2020). How modified landscapes filter rare species and modulate the regional pool of ecological traits. *Hydrobiology, 849*, 4499–4514.

Bojsen, B. H., & Barriga, R. (2002). Effects of deforestation on fish community structure in Ecuadorian Amazon streams. *Freshwater Biology, 47*, 2246–2260.

Borba, G. C., Costa, F. R. C., Espírito-Santo, H. M. V., Leitão, R. P., Dias, M. S., & Zuanon, J. (2020). Temporal changes in rainfall affect taxonomic and functional composition of stream fish assemblages in Central Amazonia. *Freshwater Biology, 66*(4), 753–764. https://doi.org/10.1111/fwb.13675

Brejão, G. L., Hoeinghaus, D. J., Perez-Mayorga, M. A., Ferraz, S. F. B., & Casatti, L. (2017). Threshold responses of Amazonian stream fishes to timing and extent of deforestation. *Conservation Biology, 32*(4), 860–871.

Brejão, G. L., Teresa, F. B., & Gerhard, P. (2020). When roads cross streams: Fish assemblage responses to fluvial fragmentation in lowland Amazonian streams. *Neotropical Ichthyology, 18*(3), e200031.

Brown, L. R., Gray, R. H., Hughes, R. M., & Meador, M. R. (2005). Introduction to effects of urbanization on stream ecosystems. *American Fisheries Society Symposium, 47*, 1–8.

Buisson, L., & Grenouillet, G. (2009). Contrasted impacts of climate change on stream fish assemblages along an environmental gradient. *Diversity and Distributions, 15*(4), 613–626.

Buisson, L., Thuiller, W., Lek, S., Lim, P. U. Y., & Grenouillet, G. (2008). Climate change hastens the turnover of stream fish assemblages. *Global Change Biology, 14*(10), 2232–2248.

Buisson, L., Grenouillet, G., Villéger, S., Canal, J., & Laffaille, P. (2013). Toward a loss of functional diversity in stream fish assemblages under climate change. *Global Change Biology, 19*(2), 387–400.

Bush, A. O., Fernandez, J. C., Esch, G. W., & Seed, J. R. (2001). *Parasitism: The diversity and ecology of animal parasites*. Cambridge University Press. 312p.

Cairns, M. A., Ebersole, J. L., Baker, J. P., Wigington, P. J., Jr., Lavigne, H. R., & Davis, S. M. (2005). Influence of summer stream temperatures on black spot infestation of juvenile Coho Salmon in the Oregon coast range. *Transactions of the American Fisheries Society, 134*, 1471–1479. https://doi.org/10.1577/T04-151.1

Campos, D. F., Braz-Mota, S., Val, A. L., & Almeida-Val, V. M. F. (2019). Predicting thermal sensitivity of three Amazon fishes exposed to climate change scenarios. *Ecological Indicators, 101*, 533–540. https://doi.org/10.1016/j.ecolind.2019.01.051

Cantanhêde, L. G., Andrade, A. L., Leão, H., & Montag, L. F. A. (2021). How does conversion from forest to pasture affect the taxonomic and functional structure of the fish assemblages in Amazonian streams? *Ecology of Freshwater Fish, 30*, 334–346.

Carvalho, R. L., Resende, A. F., Barlow, J., França, F. M., Moura, M. R., Maciel, R., et al. (2023). Pervasive gaps in Amazonian ecological research. *Current Biology, 33*, 1–9. https://doi.org/10.1016/j.cub.2023.06.077

Casatti, L., Langeani, F., & Ferreira, C. P. (2006). Effects of physical habitat degradation on the stream fish assemblage structure in a pasture region. *Environmental Management, 38*, 974–982.

Castanho, D. G. R., Deus, C. P., Zuanon, J., Santorelli, S., Leitão, R. P., & Teresa, F. B. S. (2019). Simulation of over-exploitation of ornamental fish and its consequences for the functional structure of assemblages of Amazonian streams. *Ecology of Freshwater Fish*. https://doi.org/10.1111/eff.12524

Chang, H., Evans, B. M., & Easterling, D. R. (2001). The effects of climate change on stream flow and nutrient loading 1. *JAWRA, 37*(4), 973–985.

Comte, L., & Grenouillet, G. (2013). Do stream fish track climate change? Assessing distribution shifts in recent decades. *Ecography, 36*(11), 1236–1246.
Costa, F., Zuanon, J., Baccaro, F., Schietti, J., Menger, J., Souza, J. L. P., et al. (2020). Effects of climate change on central Amazonian forests: A two decades synthesis of monitoring tropical biodiversity. *Oecol Australis, 24*, 317–335.
Dagosta, F. C. P., & De Pinna, M. (2019). The fishes of the Amazon: Distribution and biogeographical patterns, with a comprehensive list of species. *Bulletin of the American Museum of Natural History, 431*. 163 p.
Daigle, P. A. (2010). Summary of the environmental impacts of roads management responses and research gaps: A literature review. *BC Journal of Ecosystems and Management, 10*(3), 65–89.
Darimont, C. T., Cooke, R., Bourbonnais, M. L., Bryan, H. M., Carlson, S. M., Estes, J. A., et al. (2023). Humanity's diverse predatory niche and its ecological consequences. *Communications Biology, 6*, 609. https://doi.org/10.1038/s42003-023-04940-w
Doria, C. R. C., Agudelo, E., Akama, A., Barros, B., Bonfim, M., Carneiro, L., et al. (2021). The silent threat of non-native fish in the Amazon: ANNF database and review. *Frontiers in Ecology and Evolution, 9*, 646–702. https://doi.org/10.3389/fevo.2021.646702
Du, X., Shrestha, N. K., & Wang, J. (2019). Assessing climate change impacts on stream temperature in the Athabasca River basin using SWAT equilibrium temperature model and its potential impacts on stream ecosystem. *Science of the Total Environment, 650*, 1872–1881.
Fazel, N., Haghighi, A. T., & Klove, B. (2017). Analysis of land use and climate change impacts by comparing river flow records for headwaters and lowland reaches. *Global and Planetary Change*. https://doi.org/10.1016/j.gloplacha.2017.09.014
Fearnside, P., & Graça, P. M. L. A. (2006). BR-319: Brazil's Manaus-Porto Velho highway and the potential impact of linking the arc of deforestation to Central Amazonia. *Environmental Management, 38*(5), 705–716.
Feng, D., Raoufi, R., Beighley, E., Melack, J. M., Goulding, M., Barthem, R. B., et al. (2020). Future climate impacts on the hydrology of headwater streams in the Amazon River basin: Implications for migratory goliath catfishes. *Hydrological Processes, 34*(26), 5402–5416. https://doi.org/10.1002/hyp.13952
Ferreira, M. C., Begot, T. O., da Silveira, P. B., Juen, L., & Montag, L. F. A. (2018). Effects of oil palm plantations on habitat structure and fish assemblages in Amazon streams. *Environmental Biology of Fishes, 101*, 547–562.
Fittkau, E. J., Irmler, U., Junk, W. J., Reiss, F., & Schmidt, G. W. (1975). Productivity, biomass, and population dynamics in Amazonian water bodies. In *Tropical ecological systems: Trends in terrestrial and aquatic research* (pp. 289–311). Springer.
Flores-Lopes, F. (2014). The occurrence of black spot disease in *Astyanax* aff. *Fasciatus* (Characiformes: Characidae) in the Guaíba Lake basin, RS, Brazil. *Brazilian Journal of Biology, 74*(3), 127–134.
Flores-Lopes, F., & Thomaz, A. T. (2011). Assessment of environmental quality through analysis of frequency of the black spot disease in an assemblage of fish, Guaíba Lake, RS, Brazil. *Brazilian Journal of Biology, 71*, 915–923.
Fonseca, A., Botelho, C., Boaventura, R. A., & Vilar, V. J. (2014). Integrated hydrological and water quality model for river management: A case study on Lena River. *Science of the Total Environment, 485*, 474–489.
Forman, R. T. T., & Alexander, L. E. (1998). Roads and their major ecological effects. *Annual Review of Ecology, Evolution, and Systematics, 29*, 207–231.
Francis, R. A., Chadwick, M. A., & Turbelin, A. J. (2019). An overview of non-native species invasions in urban river corridors. *River Research and Applications, 35*(8), 1269–1278. https://doi.org/10.1002/rra.3513
Frederico, R. G., Olden, J. D., & Zuanon, J. (2016). Climate change sensitivity of threatened, and largely unprotected, Amazonian fishes. *Aquatic Conservation: Marine and Freshwater Ecosystems, 26*, 91–102.
Frederico, R. G., Dias, M. S., Jezequel, C., Tedesco, P. A., Hugueny, B., Zuanon, J., et al. (2021). The representativeness of protected areas for Amazonian fish diversity under climate change.

Aquatic Conservation: Marine and Freshwater Ecosystems, 31(5), 1158–1166. https://doi.org/10.1002/aqc.3528

Freitas, C. E. C., Siqueira-Souza, F. K., Humston, R., & Hurd, L. E. (2013). An initial assessment of drought sensitivity in Amazonian fish communities. *Hydrobiology, 705*(536), 159–171. https://doi.org/10.1007/s10750-012-1394-4

Freitas, P. V., Montag, L. F. A., Ilha, P., Torres, N. R., Maia, C., Deegan, L., et al. (2021). Local effects of deforestation on stream fish assemblages in the Amazon-Savannah transitional area. *Ichthyos, 19*(3), e210098.

Froese, R. (2006). Cube law, condition factor and weight–length relationships: History, meta-analysis and recommendations. *Journal of Applied Ichthyology, 22*(4), 241–253.

Gain, A. K., Immerzeel, W. W., Sperna Weiland, F. C., & Bierkens, M. F. P. (2011). Impact of climate change on the stream flow of the lower Brahmaputra: Trends in high and low flows based on discharge-weighted ensemble modeling. *Hydrology and Earth System Sciences, 15*(5), 1537–1545.

Gao, P., Geissen, V., Ritsema, C. J., Mu, M. X., & Wang, F. (2013). Impact of climate change and anthropogenic activities on stream flow and sediment discharge in the Wei River basin, China. *Hydrology and Earth System Sciences, 17*, 961–972.

García-Girón, J., Bini, L. M., & Heino, J. (2023). Shortfalls in our understanding of the causes and consequences of functional and phylogenetic variation of freshwater communities across continents. *Biological Conservation, 282*, 110082. https://doi.org/10.1016/j.biocon.2023.110082

Givati, A., Thirel, G., Rosenfeld, D., & Paz, D. (2019). Climate change impacts on streamflow at the upper Jordan River based on an ensemble of regional climate models. *Journal of Hydrology: Regional Studies, 21*, 92–109.

Gloor, M., Barichivich, J., Ziv, G., Brienen, R., Schöngart, J., Peylin, P., et al. (2015). Recent Amazon climate as background for possible ongoing and future changes of Amazon humid forests. *Global Biogeochemical Cycles, 29*(9), 1384–1399. https://doi.org/10.1002/2014GB005080

Gomes, E. S. (2020). *Efeitos da degradação ambiental sobre relações parasito-hospedeiro em riachos urbanos na Amazônia*. Unpublished Masters Dissertation, INPA, Manaus, 43p.

Graça, P. M. L. A., Dos Santos Jr, M. A., Rocha, V. M., Fearnside, P. M., Emilio, T., Menger, J. S., et al. (2014). Cenários de desmatamento para região de influência da rodovia BR-319: perda potencial de habitats, status de proteção e ameaça para a biodiversidade. In T. Emilio & F. Luizão (Eds.), *Cenários para a Amazônia: Clima, Biodiversidade e Uso da Terra* (pp. 91–101). Editora INPA.

Herrera-R, G. A., Oberdorff, T., Anderson, E. P., Brosse, S., Carvajal-Vallejos, F. M., Frederico, R., et al. (2020). The combined effects of climate change and river fragmentation on the distribution of Andean Amazon fishes. *Global Change Biology, 26*, 5509–5523.

Hudson, P. J., Dobson, A. P., & Lafferty, K. D. (2006). Is a healthy ecosystem one that is rich in parasites? *TREE, 21*(7), 381–385. https://doi.org/10.1016/j.tree.2006.04.007

Ilha, P., Rosso, S., & Schiesari, L. (2019). Effects of deforestation on headwater stream fish assemblages in the upper Xingu River basin, southeastern Amazonia. *Neotropical Ichthyology, 17*, e180099.

IPCC. (2023). In Core Writing Team, H. Lee, & J. Romero (Eds.), *Climate change 2023: Synthesis report, summary for policymakers. Contribution of working groups I, II and III to the sixth assessment report of the intergovernmental panel on climate change*. IPCC.

Jacob, L. L., Prudente, B. S., Montag, A. L. F., & Silva, R. R. (2021). The effect of different logging regimes on the ecomorphological structure of stream fish assemblages in the Brazilian Amazon. *Hydrobiology, 848*, 1027–1039.

Jaureguiberry, P., Titeux, N., Wiemers, M., Bowler, D. E., Coscieme, L., Golden, A. S., et al. (2022). The direct drivers of recent global anthropogenic biodiversity loss. *Science Advances, 8*(45), eabm9982. https://doi.org/10.1126/sciadv.abm9982

Juen, L., Cunha, E. J., Carvalho, F. G., Ferreira, M. C., Begot, T. O., Andrade, A. L., et al. (2016). Effects of oil palm plantations on the habitat structure and biota of streams in eastern Amazon. *River Research and Applications, 32*(10), 2081–2094.

Junk, W. J. (1983). As águas da Bacia Amazônica. In E. Salati, W. J. Junk, S. HOR, & A. E. Oliveira (Eds.), *Amazônia: desenvolvimento, integração e ecologia* (pp. 45–100). São Paulo.

Khan, B., & Colbo, M. H. (2008). The impact of physical disturbance on stream communities: Lessons from road culverts. *Hydrobiology, 600*, 229–235.

Kirk, M. A., & Rahel, F. J. (2022). Air temperatures over-predict changes to stream fish assemblages with climate warming compared with water temperatures. *Ecological Applications, 32*(1), e02465.

Kuczynski, L., Ontiveros, V. J., & Hillebrand, H. (2023). Biodiversity time series are biased towards increasing species richness in changing environments. *Nature Ecology & Evolution, 7*, 994–1001. https://doi.org/10.1038/s41559-023-02078-w

Kwon, Y. S., Bae, M. J., Hwang, S. J., Kim, S. H., & Park, Y. S. (2015). Predicting potential impacts of climate change on freshwater fish in Korea. *Ecological Informatics, 29*, 156–165.

Lacerda, A. C. F., Roumbedakis, K., Junior, J. B., Nuñer, A. P. O., Petrucio, M. M., & Martins, M. L. (2018). Fish parasites as indicators of organic pollution in southern Brazil. *Journal of Helminthology, 92*(3), 322–331.

Leal, C. G., Lennox, G. D., Ferraz, S. F. B., Ferreira, J., Gardner, T. A., Thomson, J. R., et al. (2020). Integrated terrestrial-freshwater planning doubles conservation of tropical aquatic species. *Science, 370*, 117–121.

Leitão, R. P., Zuanon, J., Villéger, S., Williams, S. E., Baraloto, C., Fortunel, C., et al. (2016). Rare species contribute disproportionately to the functional structure of species assemblages. *Proceedings of the Royal Society B, 20160084*. https://doi.org/10.1098/rspb.2016.0084

Leitão, R. P., Zuanon, J., Mouillot, D., Leal, C. G., Hughes, R. M., Kaufmann, P. R., et al. (2018). Disentangling the pathways of land use impacts on the functional structure of fish assemblages in Amazon streams. *Ecography, 4*, 219–232.

Lobato, C. M. C., Benone, N. L., Brasil, L. S., & Montag, L. F. A. (2022). Land use effects on the co-occurrence patterns of streams ichthyofauna in the eastern Amazon. *Ecological Indicators, 145*, 109659.

Lowe-McConnell, R. H. (1987). *Ecological studies in tropical fish communities*. Cambridge University Press.

Lyu, J., Mo, S., Luo, P., Zhou, M., Shen, B., & Nover, D. (2019). A quantitative assessment of hydrological responses to climate change and human activities at spatiotemporal within a typical catchment on the loess plateau, China. *Quaternary International, 527*, 1–11.

Macedo, N. M. N., Coe, M. T., DeFries, R., Uriarte, M., Brando, P. M., Neill, C., et al. (2018). Land-use-driven stream warming in southeastern Amazonia. *Philosophical Transactions of the Royal Society B, 368*, 20120153.

Maghsood, F. F., Moradi, H., MassahBavani, A. R., Panahi, M., Berndtsson, R., & Hashemi, H. (2019). Climate change impact on flood frequency and source area in northern Iran under CMIP5 scenarios. *Water, 11*(2), 1–21.

Magurran, A., & MacGill, B. J. (2011). *Biological diversity: Frontiers in measurement and assessment*. Oxford University Press.

Maia, C., Salvador, G. N., Begot, T. O., Freitas, P. V., Nonato, F. A. S., Torres, N. R., et al. (2022). Fish functional responses to local habitat variation in streams within multiple land uses areas in the Amazon. *Neotropical Ichthyology, 20*(4), e220091.

Marcogliese, D. J. (2001). Implications of climate change for parasitism of animals in the aquatic environment. *Canadian Journal of Zoology, 79*(8), 1331–1352.

Marengo, J., & Espinoza, J. C. (2016). Extreme seasonal droughts and floods in Amazonia: Causes, trends and impacts. *International Journal of Climatology, 36*(3), 1033–1050. https://doi.org/10.1002/joc.4420

Marengo, J., Nobre, C. A., Tomasella, J., Oyama, M. D., de Oliveira, G. S., de Oliveira, R., et al. (2008). The drought of Amazonia in 2005. *Journal of Climate, 21*(3), 495–516. https://doi.org/10.1002/joc.4420

Martins, A. R., Bastos, D. A., Sousa, L. M., Giarizzo, T., Vieira, T. B., & Zuanon, J. (2022). Intense droughts affect temporal stability of Amazonian stream fish assemblages. *Freshwater Biology, 67*(9), 1656–1667. https://doi.org/10.1111/fwb.13970

Mascarenhas, M. B. C. (2019). *Efeitos da alteração ambiental sobre Bryconops giacopinii (Characidae) em igarapés de terra firme: dieta, fator de condição, parasitismo e reprodução.* Unpublished Masters Dissertation, National Institute for Amazonian Research, Manaus, Brazil.

Melo, C. E., Machado, F. A., & Pinto-Silva, V. (2003). Diversidade de peixes em um córrego de cerrado no Brasil Central. *Brazilian Journal of Ecology, 1*(2), 17–23.

Mendes Jr, R. N. G. (2013). *Composição e estrutura das assembleias de peixes de pequenos riachos afluentes do Lago Ajuruxi, Mazagão-AP.* Unpublished Masters Dissertation, National Institute for Amazonian Research., Manaus, Brazil.

Moreira, L. H. A., Takemoto, R. M., & Pavanelli, G. C. (2015). Urbanization effects on the host/parasite relationship in fishes from tributary streams of Pirapó River, Paraná state: Assessment of potential environmental bioindicators. *Acta Scientiarum Biological Sciences, 37*(3), 319–326.

Mouillot, D., Graham, N. A. J., Villéger, S., Mason, N. W. H., & Bellwood, D. R. (2013). A functional approach reveals community responses to disturbances. *Trends in Ecology & Evolution, 28*(3), 167–177. https://doi.org/10.1016/j.tree.2012.10.004

Murdoch, A., Mantyka-Pringle, C., & Sharma, S. (2020). The interactive effects of climate change and land use on boreal stream fish communities. *Science of the Total Environment.* https://doi.org/10.1016/j.scitotenv.2019.134518

Nelson, K. C., Palmer, M. A., Pizzuto, J. E., Moglen, G. E., Angermeier, P. L., Hilderbrand, R. H., et al. (2009). Forecasting the combined effects of urbanization and climate change on stream ecosystems: From impacts to management options. *Journal of Applied Ecology, 46*, 154–163.

Oberdorff, T., Jézéquel, C., Campero, M., Carvajal-Vallejos, F., Cornu, J. F., Dias, M. S., et al. (2015). Opinion paper: How vulnerable are Amazonian freshwater fishes to ongoing climate change? *Journal of Applied Ichthyology, 31*, 4–9.

Oliveira, D. C. D., & Bennemann, S. T. (2005). Ictiofauna, recursos alimentares e relações com as interferências antrópicas em um riacho urbano no sul do Brasil. *Biota Neotropica, 5*, 95–107.

Oliveira, V. A., de Mello, C. R., Beskow, S., Viola, M. R., & Srinivasan, R. (2019). Modeling the effects of climate change on hydrology and sediment load in a headwater basin in the Brazilian Cerrado biome. *Ecological Engineering, 133*, 20–31.

Pelicice, F. M., Agostinho, A. A., Azevedo-Santos, V. M., Bessa, E., Casatti, L., Garrone-Neto, D., et al. (2022). *Ecosystem services generated by Neotropical freshwater fishes* (pp. 1–24). Hydrobiology.

Perkins, D. M., Reiss, J., Yvon-Durocher, G., & Woodward, G. (2010). Global change and food webs in running waters. *Hydrobiologia, 657*, 181–198. https://doi.org/10.1007/s10750-009-0080-7

Phan, D. B., Wu, C. C., & Hsieh, S. C. (2011). Impact of climate change on stream discharge and sediment yield in North Viet Nam. *Water Resources, 38*, 827–836.

Pinto, B. C. T., Araújo, F. G., Rodrigues, V. D., & Hughes, R. M. (2009). Local and ecoregion effects on fish assemblage structure in tributaries of the Rio Paraiba do Sul, Brazil. *Freshwater Biology, 54*, 2600–2615.

Prudente, B. S., Pompeu, P. S., Juen, L., & Montag, L. F. A. (2017). Effects of reduced-impact logging on physical habitat and fish assemblages in streams of eastern Amazonia. *Freshwater Biology, 62*(2), 303–316.

Punzet, M., Voß, F., Voß, A., Kynast, E., & Bärlund, I. (2012). A global approach to assess the potential impact of climate change on stream water temperatures and related in-stream first-order decay rates. *Journal of Hydrometeorology, 13*(3), 1052–1065.

Ribeiro-Brasil, D. R. G., Torres, N. R., Picanço, A. B., Sousa, D. S., Ribeiro, V. S., Brasil, L. S., & Montag, L. F. A. (2020). Contamination of stream fish by plastic waste in the Brazilian Amazon. *Environmental Pollution, 266*, 115241.

Rogers, J. B., Stein, E. D., Beck, M. W., & Ambrose, R. F. (2020). The impact of climate change induced alterations of streamflow and stream temperature on the distribution of riparian species. *PLoS One, 15*(11), e0242682.

Rolla, A., Esteves, K. E., & Avila-da-Silva, A. O. (2009). Feeding ecology of a stream fish assemblage in an Atlantic Forest remnant (Serra do Japi, SP, Brazil). *Neotropical Ichthyology, 7*, 65–76.

Röpke, C. P., Amadio, S., Zuanon, J., Ferreira, F., Deus, C. P., Pires, T. H. S., & Winemiller, K. O. (2017). Simultaneous abrupt shifts in hydrology and fish assemblage structure in a floodplain lake in the Central Amazon. *Scientific Reports, 7*, 40170. https://doi.org/10.1038/srep40170

Röpke, C. P., Pires, T. H., Winemiller, K. O., de Fex, W. D., Deus, C. P., & Amadio, S. (2019). Reproductive allocation by Amazon fishes in relation to feeding strategy and hydrology. *Hydrobiology, 826*, 291–305.

Röpke, C. P., Pires, T. H. S., Zuchi, N., Zuanon, J., & Amadio, S. (2022). Effects of climate-driven hydrological changes in the reproduction of Amazonian floodplain fishes. *Journal of Applied Ecology, 59*(4), 1134–1145. https://doi.org/10.1111/1365-2664.14126

Sanchez, J. D. J. S., de Jesús M. J., Carranza, M., Hernández, R., & del Rocío, M. (2019). *Distribución potencial de peces dulceacuícolas tropicales bajo escenarios de cambio climático*. Tesis Maestro. El Colegio de la Frontera Sur. EcoSur.

Schmidt, G. W. (1972). Chemical properties of some waters in the tropical rain - forest region of Central - Amazonia along the new road Manaus - Caracarai. *Amazoniana, 3*(2), 199–207.

Seabra, L. B., Benone, N. L., & Montag, L. F. A. (2022). Assessing the effects of multiple land uses on the functional beta diversity of stream fishes in the Amazon region. *Hydrobiology, 849*(20), 4515–4527.

Sioli, H. (1975). Tropical Rivers: The Amazon. In B. A. Whitton (Ed.), *River ecology* (pp. 461–488). University of California Press.

Smith, W. S., Lima, C. R., Silva, L. C. M., Corrêa, C. S., Teodoro, C. C., Vaz, A. A., et al. (2018). A duplicação de rodovias no Brasil sob o olhar da Ictiofauna. *Boletim Sociedade Brasileria de Ictiologia, 125*, 16–23.

Sorribas, M. V., Paiva, R. C., Melack, J. M., Bravo, J. M., Jones, C., Carvalho, L., et al. (2016). Projections of climate change effects on discharge and inundation in the Amazon basin. *Climatic Change, 136*(3), 555–570.

Sousa, R. G. C., Almeida Mereles, M., Siqueira-Souza, F. K., Hurd, L. E., & Carvalho Freitas, C. E. (2018). Small dams for aquaculture negatively impact fish diversity in Amazonian streams. *Aquaculture Environment Interactions, 10*, 89–98.

Stefan, H. G., & Sinokrot, B. A. (1993). Projected global climate change impact on water temperatures in five north central US streams. *Climatic Change, 24*(4), 353–381.

Stouffer, P. C., Jirinec, V., Rutt, C. L., Bierregaard, R. O., Jr., Hernández-Palma, A., Johnson, E. I., et al. (2021). Long-term change in the avifauna of undisturbed Amazonian rainforest: Ground-foraging birds disappear and the baseline shifts. *Ecology Letters, 24*(2), 186–195. https://doi.org/10.1111/ele.13628

Struebig, M. J., Wilting, A., Gaveau, D. L. A., Meijaard, E., Smith, R. J., The Borneo Mammal Distribution Consortium, et al. (2015). Targeted conservation to safeguard a biodiversity hotspot from climate and land-cover change. *Current Biology, 25*, 372–378. https://doi.org/10.1016/j.cub.2014.11.067

Sun, J., Lei, X., Tian, Y., Liao, W., & Wang, Y. (2013). Hydrological impacts of climate change in the upper reaches of the Yangtze River Basin. *Quaternary International, 304*, 62–74.

Suzuki, H., Nakatsugawa, M., & Ishiyama, N. (2022). Climate change impacts on stream water temperatures in a snowy cold region according to geological conditions. *Water, 14*(14), 2166.

Taniwaki, R. H., Cassiano, C. C., Filoso, S., de Barros Ferraz, S. F., de Camargo, P. B., & Martinelli, L. A. (2017). Impacts of converting low-intensity pastureland to high-intensity bioenergy cropland on the water quality of tropical streams in Brazil. *Science of the Total Environment, 584*, 339–347.

Tickner, D., Opperman, J. J., Abell, R., Acreman, M., Arthington, A. H., Bunn, S. E., et al. (2020). Bending the curve of global freshwater biodiversity loss: An emergency recovery plan. *Bioscience, 70*(4), 330–342. https://doi.org/10.1093/biosci/biaa002

Trancoso, R. (2006). *Mudanças na cobertura da terra e alterações na resposta hidrológica de bacias hidrográficas na Amazônia*. Dissertation, National Institute of Amazonian Research.

Trancoso, R., Carneiro Filho, A., Tomasella, J., Schietti, J., Forsberg, B. R., & Miller, R. P. (2010). Deforestation and conservation in major watersheds of the Brazilian Amazon. *Environmental Conservation, 36*(4), 277–288. https://doi.org/10.1017/S0376892909990373

Trombulak, S. C., & Fressel, C. A. (2000). Review of ecological effects of roads on terrestrial and aquatic communities. *Conservation Biology, 14*(1), 18–30.

van Rees, C. B., Waylen, K. A., Schmidt-Kloiber, A., Thackeray, S. J., Kalinkat, G., Martens, K., et al. (2021). Safeguarding freshwater life beyond 2020: Recommendations for the new global biodiversity framework from the European experience. *Conservation Letters, 14*(1), e12771.

Wang, X., Li, Z., & Li, M. (2018). Impacts of climate change on stream flow and water quality in a drinking water source area, northern China. *Environment and Earth Science, 77*, 1–14.

Woodward, G., Perkins, D. M., & Brown, L. E. (2010). Climate change and freshwater ecosystems: Impacts across multiple levels of organization. *Philosophical Transactions. Royal Society of London, 365*, 2093–2106. https://doi.org/10.1098/rstb.2010.0055

Woznicki, S. A., Nejadhashemi, A. P., Tang, Y., & Wang, L. (2016). Large-scale climate change vulnerability assessment of stream health. *Ecological Indicators, 69*, 578–594.

Xenopoulos, M. A., Lodge, D. M., Alcamo, J., Märker, M., Schulze, K., & Van Vuuren, D. P. (2005). Scenarios of freshwater fish extinctions from climate change and water withdrawal. *Global Change Biology, 11*(10), 1557–1564.

Xu, Y. P., Zhang, X., Ran, Q., & Tian, Y. (2013). Impact of climate change on hydrology of upper reaches of Qiantang River Basin, East China. *Journal of Hydrology, 483*, 51–60.

Yuhara, T. Y. (2012). *Avaliação do impacto da utilização de manilhas na construção de estradas sobre ambientes aquáticos e na ictiofauna em duas rodovias na região sul de Minas Gerais*. Dissertation, Universidade Federal de Lavras.

Amazonian Aquatic Insects in a Changing Landscape

Renato Tavares Martins, Viviane Caetano Firmino, and Neusa Hamada

Abstract

Multiple local anthropogenic pressures and climate changes threaten the biodiversity of aquatic insects in the Amazon. In this chapter, we review the impacts of various land use practices (e.g., urbanization, pasture, agriculture, mining, and dams) and climate change on aquatic insect assemblages in the Amazon region. We explore how these factors might influence the composition, richness, abundance, diversity, and ecosystem processes (e.g., leaf decomposition) of these insect communities. Finally, we outline future perspectives for studies of insects and aquatic environments in the Amazon.

1 Temporal Trends in Studies on Anthropogenic Activity and the Diversity of Aquatic Insects

The Amazon harbors not only the largest tropical forest but also contains approximately 10–15% of known global biodiversity, including endemic species and endangered fauna and flora (Marengo et al., 2018; Nobre et al., 2016). It comprises 6.7 million km of extension in eight countries of South America (Brazil, Bolivia, Peru, Ecuador, Colombia, Venezuela, Guyana, and Suriname) and the territory of French Guiana. Amazon forests regulate temperature and humidity and are intrinsically associated with the water cycle and regional and even global climate patterns, given

R. T. Martins (✉) · N. Hamada
Biodiversity Coordination, Brazilian National Institute for Amazonian Research (INPA), Manaus, AM, Brazil
e-mail: renato.martins@inpa.gov.br

V. C. Firmino
Graduate Program in Zoology, Institute of Biological Sciences, Federal University of Pará, Belém, Pará, Brazil

S. S. de Souza et al. (eds.), *The Future of Amazonian Aquatic Biota*,
https://doi.org/10.1007/978-3-031-66822-7_5

the large amount of carbon stored in the Amazon, about 150 to 200 billion tons (Marengo et al., 2018; Nepstad et al., 2008; Nobre et al., 2016). The discharge of the Amazon River ranges from 210,000 to 220,000 m^3s^{-1} which represents approximately 15% of the freshwater input into the oceans (Callede et al., 2012; Marengo et al., 2018; Nobre et al., 2016).

Aquatic insects are important elements in the ecological dynamics of aquatic ecosystems, developing key functions in the cycling of organic matter, nutrients, and energy flow in the food web of these ecosystems (Bispo et al., 2006; Cummins et al., 1989; Hynes, 1970; Vannote et al., 1980). In addition, these organisms are important indicators of environmental quality and are often used in the biomonitoring of aquatic environments, as they respond quickly to changes in local environmental conditions and exhibit different levels of tolerance (Martins et al., 2014; Rosenberg & Resh, 1993). However, Amazonian aquatic ecosystems suffer from multiple human pressures such as global warming, deforestation, habitat fragmentation, land use change, increasing urbanization, and the construction of dams and reservoirs, which occur around or directly in these environments, altering the physical and chemical structure of these aquatic ecosystems and, consequently, aquatic communities (Couceiro et al., 2007; Juen et al., 2016; Martins et al., 2017a), endangering global climate and hydrologic security, reducing biodiversity, and impairing ecosystem functions (Duffy et al., 2015; Lovejoy & Nobre, 2018; Marengo et al., 2018; Nobre et al., 2016).

We conducted a literature review to understand the impacts of various factors on aquatic insects in the Brazilian Amazon. We searched for published articles in the Scopus, Web of Science, and Google Scholar databases. We employed the following keywords: Amazon*, aquatic insect*, agriculture, climatic change*, dam*, lotic*, mining, pasture, pollution, river*, stream*, urbanization, and, and all Orders of aquatic insects (e.g., Diptera, Ephemeroptera). In this search, we used keywords in English, Spanish and Portuguese. After excluding papers due to their lack of relevance to our question, 74 studies remained. Most of the studies were carried out by research groups from the National Institute for Amazon Research (*Laboratório de Citotaxonomia e Insetos Aquáticos* – LACIA) and the Federal University of Pará (*Laboratório de Ecologia e Conservação* – LABECO) in the states of Amazonas and Pará. The majority of studies (49%) focused on assessing the effects of pasture and agriculture on aquatic insects. The other impacts evaluated on Amazonian aquatic insects were non-agricultural deforestation (14%), urbanization (12%), mining (12%), climatic change (9%), and dams (3%).

In our search, the first study evaluating the effects of human impacts on aquatic insects in Amazonian streams was recorded in 1998 (Fig. 1). The study focuses on the impact of various factors on aquatic insects and ecosystem processes. In 1998, there was a significant increase in publications discussing the effects of mining on aquatic insects. However, along the time, the interest in this topic has a decline. Since 2014, there has been a continuous increase in studies evaluating the impacts of pasture and agriculture on aquatic insects. From 2010 to 2021, non-agricultural deforestation effects on aquatic insects appeared to maintain a relatively constant number of publications. Similarly, during the period 2006 to 2022, urbanization

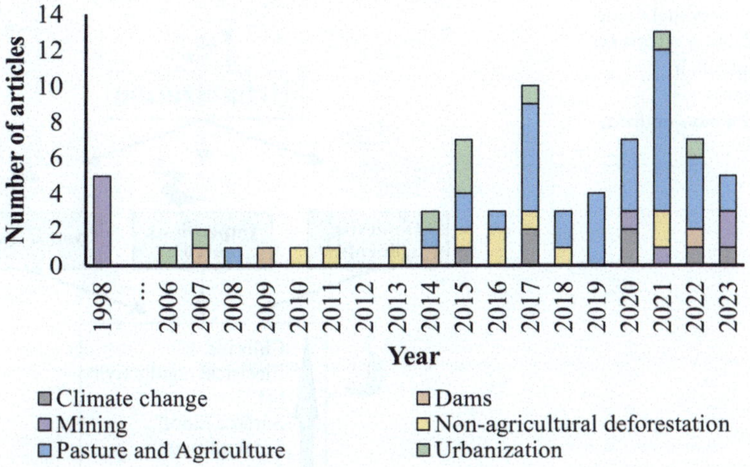

Fig. 1 Trend in publications on the effects of human impacts on aquatic insects in Amazonian streams over time

effects on aquatic insects showed a consistent level of interest among researchers, as reflected in the relatively steady number of publications. The effects of climate change on aquatic insects and ecosystem processes gained greater interest in 2015, keeping publications on this theme constant until 2023. On the other hand, the impact of dams on aquatic insects seemed to generate sporadic interest, resulting in temporally spaced publications rather than a consistent trend (Fig. 1).

2 What Are the Effects of Urbanization on Aquatic Insects?

Although the Amazon is renowned for its vast forests, it is important to highlight the presence of sizable urban centers within this region. Cities like Manaus (~ 2.2 million inhabitants) and Belém (~ 1.5 million inhabitants) stand out among the 11 most populous cities in Brazil (IBGE, 2022). The rapid urbanization of these two cities can be directly attributed to rubber exploitation in the 1970s and industrialization in the 1990s, leading to significant impacts on aquatic environments (Barham & Coomes, 1994; Bentes, 2005). For instance, between 1970 and 2010, the populations of Manaus and Belém experienced a staggering growth of 600% and 219% respectively, showcasing the magnitude of their demographic expansion.

In general, the urban areas of the Amazon have witnessed the removal of riparian vegetation and the discharge of domestic effluents, resulting in significant alterations to aquatic environments (Couceiro et al., 2007; Martins et al., 2017a; Fig. 2). Urban streams across the Amazon have shown a notable increase in water temperature, chloride concentration, pH, electrical conductivity, total nitrogen, total phosphorus, and impervious soil area. On the other hand, the influence of these human

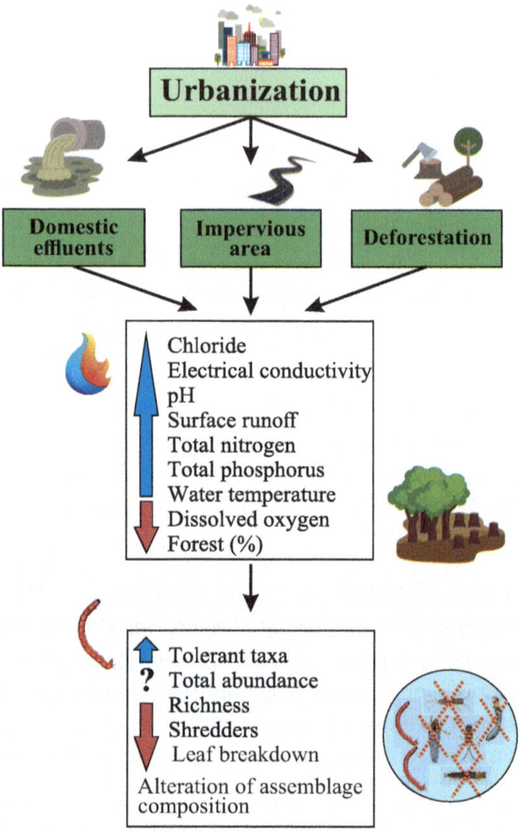

Fig. 2 Conceptual model of urbanization effects on aquatic insects in Amazonian streams.? = non-uniform response

impacts has led to a decrease in dissolved oxygen concentration (Couceiro et al., 2007; Ferreira et al., 2021; Martins et al., 2017a). Additionally, during the rainy season, the increased input of domestic sewage, fertilizers, and dissolved organic matter through surface runoff can result in the elevated concentration of various contaminants in urban streams of the Amazon (Nascimento et al., 2022).

The environmental changes in urban streams of the Amazon have led to alterations in the composition of aquatic insect assemblages, including a reduction in species richness due to the exclusion of sensitive taxa such as Ephemeroptera, Plecoptera, and Trichoptera—EPT (e.g., Couceiro et al., 2012; Martins et al., 2015; Martins et al., 2017a). Additionally, there has been an increase in the abundance of tolerant organisms such as *Chironomus* (Chironomidae), Ephydridae, and Psychodidae (Cleto-Filho & Walker, 2001; Couceiro et al., 2007), which have adaptations such as pigments and respiratory siphons that allow them to survive in low-oxygen environments. Overall, no significant relationship has been observed between insect abundance (entire assemblage) and the degree of urbanization in Amazonian streams (Couceiro et al., 2007; Martins et al., 2017a). In addition to spatial variation in aquatic insects, temporal variation has also been examined, with no significant differences found in terms of abundance, species richness, and assemblage composition between collection periods (Couceiro et al., 2021).

However, urbanization has consistently had a negative impact on species richness and composition across all collection periods (Couceiro et al., 2021). A study investigating interannual variation (2004 and 2010) of invertebrates in streams near Manaus revealed an increased negative effect of urbanization on insect species richness, along with an increase in the dominance of tolerant taxa over time (Martins et al., 2017a). Unfortunately, due to Manaus population growth (2003: ~1.5 million; 2010: ~1.8 million; 2023: ~2.2 million; IBGE) and low sewage treatment rates (22%; Trata Brasil, 2023), combined with the absence of effective public policies, this situation is not expected to improve in the short term.

Studies have also been conducted on the effects of urbanization on specific groups of insects. For instance, in relation to adult Odonata, a positive correlation was observed between the richness of the Zygoptera suborder and environmental integrity, while Anisoptera showed a negative correlation (Monteiro-Júnior et al., 2014, 2015). In urban and industrialized areas affected by kaolin and bauxite mining in Barcarena, Pará, a decrease in richness and alterations in the composition of Gerromorpha (Heteroptera) fauna have been recorded, highlighting the importance to preserve riparian forests even in urban and industrialized areas to maintain habitats, food sources, and breeding sites for these organisms (Cunha et al., 2022). In addition to changes in insect assemblages, the increase in deforestation and organic pollution in urban streams can have a negative impact on ecosystem processes, such as leaf decomposition, resulting from a reduction in biomass or even the exclusion of shredder insects (Martins et al., 2015).

In addition to organic pollution and deforestation, urban streams can be affected by other negative impacts. For example, in 1999, in the urban area of Manaus, a submerged oil pipeline ruptured in the Cururu stream, leading to the release of petroleum-derived oil. Despite containment measures being implemented, a significant amount of oil remained in the stream bed, resulting in a decrease in the richness and abundance of aquatic insects compared to a preserved area. This decline can be attributed to reduced dissolved oxygen levels and increased concentrations of total phosphorus and total nitrogen (Couceiro et al., 2006). Furthermore, many of the collected individuals had their mouthparts covered in oil, which impeded their feeding abilities (Couceiro, personal observation).

3 What Are the Effects of Pasture and Agriculture on Aquatic Insects?

Throughout history, the Amazon rainforest has witnessed the expansion of various agricultural practices, including rubber, livestock, soybeans, and more recently, oil palm (Fearnside, 2006; Homma, 2001). The expansion of agriculture and cattle grazing in the Pan-Amazonia has resulted in the loss of ~14 of the native forest by 2018 (Berenguer et al., 2021). Large-scale agricultural activities bring about significant changes to the landscape, characterized by the conversion, loss, and fragmentation of native vegetation (Laurance et al., 2014; Fig. 3). The conversion of forests to agriculture often leads to altered water flow, increased sedimentation in aquatic ecosystems (Nakamura & Yamada, 2005), intensified sunlight penetration, and higher concentrations of nutrients and agricultural pesticides (Macdonald et al., 2003). Furthermore, it can cause the homogenization of substrates due to a decrease in streambed debris and organic matter (Price et al., 2003).

In Amazon, it is estimated that 30 million hectares are suitable for monocultures of *Elaeis guineensis* Jacq. (oil palm; Ramalho-Filho et al., 2010). In the state of Pará alone, there are over 256,000 hectares of oil palm plantations. These areas can exhibit reduced structural complexity and canopy cover, resulting in increased light penetration compared to forested areas (Carvalho et al., 2018; Cunha et al., 2015). In these areas, there is also an increase in roads, which can lead to greater sediment input into streams (Paiva et al., 2017). Additionally, the use of fertilizers and pesticides may occur in oil palm plantation areas, which can be carried into aquatic environments (Ferreira et al., 2018). As a result, oil palm monocultures can cause

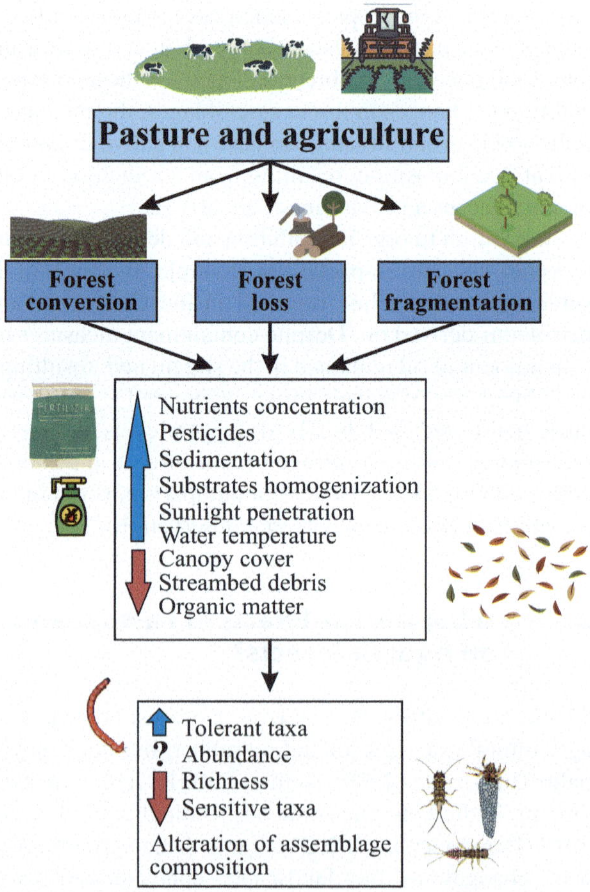

Fig. 3 Conceptual model of agriculture and pasture effects on aquatic insects in Amazonian streams. Note:? = non-uniform response

physical alterations (e.g., woody debris decrease) and abiotic changes (e.g., higher pH and temperature) in streams (Carvalho et al., 2018; Cunha et al., 2020; Juen et al., 2016).

Studies have been conducted with different insect groups in oil palm plantations, and they have found a decrease in taxonomic and/or functional richness in Heteroptera (Cunha et al., 2015; Cunha & Juen, 2017), Ephemeroptera, Plecoptera, and Trichoptera (EPT; Shimano & Juen, 2016; Luiza-Andrade et al., 2017; Paiva et al., 2017) compared to forested areas. Additionally, forested areas have shown a higher abundance of winged Heteroptera individuals (Cunha et al., 2020) and greater morphological diversity of Odonata larvae (Mendes et al., 2020). On the other hand, average Odonata richness was higher in oil palm areas (Mendes et al., 2019), and these areas also exhibited higher β-diversity of aquatic Heteroptera (Cunha & Juen, 2017). Overall, the abundance of aquatic insects was similar in forested and oil palm areas (Cunha et al., 2015; Luiza-Andrade et al., 2017; Shimano & Juen, 2016).

Regarding the taxonomic composition of the assemblage, the elimination of specialist taxa in forested areas was observed for Odonata (Carvalho et al., 2018; Mendes et al., 2019), EPT (Luiza-Andrade et al., 2017), and Heteroptera (Cunha & Juen, 2017), resulting in differences compared to oil palm areas. On the other hand, studies evaluating the taxonomic composition of Ephemeroptera (Shimano & Juen, 2016) and Plecoptera and Trichoptera (Paiva et al., 2017), as well as the functional composition of EPT (Luiza-Andrade et al., 2017), did not find differences between the two areas. In addition to differences in the aquatic insect assemblage, oil palm plantations can result in alterations in ecosystem processes, such as organic matter processing, due to their slower leaf decomposition rate (Firmino et al., 2021).

Nessimian et al. (2008) conducted a study in the Biological Dynamics of Forest Fragments Project (BDFFP) and recorded a reduction of up to 34% in species richness and a shift in the composition of aquatic insects in streams when comparing pasture areas to forested areas (both primary and secondary). The study also recorded the exclusion of sensitive insect taxa in pasture areas, leading to a decrease in the richness of EPT genera. Furthermore, the intensity of the impact on aquatic insects was found to be influenced by the heterogeneity of the area, distance from forest fragments, presence of secondary forest areas, and the dispersal capacity of organisms (Nessimian et al., 2008). Lima et al. (2022) observed changes in the functional composition of EPT insects in streams within pasture areas, including a decrease in the proportion of shredders and collector-gatherers, and an increase in the dominance of collector-filtering insects. In the eastern Amazon, alterations in both, taxonomic and functional composition of EPT assemblage were observed between streams in pasture and forest areas. However, no significant differences were found in terms of abundance, taxonomic richness, and functional richness (Luiza-Andrade et al., 2023). Regarding semiaquatic bugs (Gerromorpha), the richness and abundance in streams within pasture areas in the eastern Amazon were similar to forested streams. However, the composition of the semiaquatic bug assemblage in pasture streams was mainly composed of tolerant species, resulting in compositional differences compared to forested streams (Guterres et al., 2021).

Streams in agricultural areas are primarily influenced by deforestation and the introduction of fertilizers and agricultural pesticides (Mateo-Sagasta et al., 2017). Changes in the Zygoptera assemblage were observed in streams located within a soybean plantation matrix in the "arc of deforestation" region of the Amazon, and these were linked to a decline in environmental integrity (Brasil et al., 2014a). Similarly, in this region, alterations in the assemblage of aquatic and semiaquatic Heteroptera were observed between streams in agricultural and forested areas, and these differences were attributed to variations in riparian vegetation (Silva-Costa et al., 2023).

In the northeastern region of Pará State, similar richness of Chironomidae (Diptera) genera was observed in streams located in agricultural areas and forested areas, but richness was higher in these two areas compared to Capoeira (secondary forest patches resulting from slash-and-burn agriculture; Sonoda et al., 2018). In Sonoda's study, the abundance in agricultural streams was 2.2 to 13.5 times higher than in forested and Capoeira streams. The taxonomic composition in forested and agricultural streams differed from Capoeira streams, while the functional composition differed only between forested and agricultural streams. Forested streams predominantly featured collector-gatherers, Capoeira streams had collector-filterers as the dominant group, and agricultural streams exhibited a combination of predators, collector-gatherers, and collector-filterers.

In the southeastern agricultural frontier of the Amazon in Mato Grosso, it was observed that the removal of riparian vegetation resulted in a change in the food source for EPTs. Forested watersheds exhibited a predominance of shredders, while grazers were more abundant in cropland watersheds with riparian forests (CropRip), and collector-filterers dominated in cropland watersheds without riparian forest buffers (CropNoRip; Marques et al., 2021). However, the functional composition was similar across all streams. Abundance and taxonomic richness did not differ among the sampled streams. The average EPT functional diversity was higher in forested and CropRip streams compared to CropNoRip streams. The taxonomic composition of the EPT assemblage differed among the stream categories.

In the Eastern Amazon, a reduction in the number of insect families and EPT was observed in streams located in Capoeira and agricultural areas compared to forested streams (Monteles et al., 2021). Simpson's diversity (both taxonomic and functional), the percentage of Chironomidae, and the number of functional feeding groups did not show significant differences among the stream categories. The functional and taxonomic composition of the aquatic insect assemblage was influenced by the overall land use and land cover in the catchment area, the riparian corridor, and the local reach, resulting in variations among the three stream categories. Luiza-Andrade et al. (2023), who conducted research in streams within forested, conventionally logged, and pasture areas, found similar taxonomic richness, taxonomic abundance, and functional richness of EPT among the stream categories. However, changes in the taxonomic and functional composition of the insect assemblage were observed, particularly in streams located in pasture areas, which differed from the other two categories.

Also in the Eastern Amazon, a series of studies were carried out in the municipalities of Santarém, Belterra, and Paragominas to evaluate the effects of anthropogenic impacts on aquatic insects. The regions presented a land-use gradient, encompassing primary and secondary forests, pastures for livestock, mechanized annual crops such as rice and soybeans (*Oryza sativa* L.; *Glycine max* L.), as well as forested areas with planted species such as teak (*Tectona grandis* L.f.), *Eucalyptus* spp. or fern tree (*Schizolobium parahyba* var. *amazonicum* (Huber ex Ducke) Barneby) (Gardner et al., 2013). The most frequently studied order in these investigations was Odonata, and the response of the assemblage varied depending on the chosen taxonomic metric and the stream categorization. Oliveira Junior et al. (2015) documented higher taxonomic richness of Odonata in altered streams (mean deforestation = 48%) and degraded streams (mean deforestation = 56%) compared to preserved streams (mean deforestation = 2%). They also observed differences in the assemblage composition among the stream categories. Odonata also exhibited variations in taxonomic diversity and distinctness, with higher values in altered streams and lower values in degraded streams (Miguel et al., 2017).

Additionally, the gradient of anthropogenic impact affected the phylogenetic and morphological metrics of Odonata species in the Eastern Amazon (Bastos et al., 2021; Carvalho et al., 2021). On the other hand, some studies did not find differences in taxonomic richness as the impact increased (Oliveira-Junior et al., 2017; Oliveira-Junior & Juen, 2019). The observed change in taxonomic composition may be attributed to the increased occurrence of species with a wider thorax and endophytic oviposition, resulting in assemblage homogeneity, loss of biodiversity, and functional decline in impacted streams (Pereira et al., 2019). When considering the suborders separately, the taxonomic richness of Anisoptera was higher in degraded streams, while Zygoptera had a higher richness in altered streams (Oliveira Junior et al., 2015). Other studies reported greater richness and relative abundance of Anisoptera in preserved streams, whereas Zygoptera showed higher richness and relative abundance in preserved streams (Oliveira-Junior et al., 2017, 2021; Oliveira-Junior & Juen, 2019). An index for stream classification based on relative abundance was proposed by Oliveira-Junior and Juen (2019), indicating that streams with an Odonata assemblage composed of over 54% Zygoptera are associated with preserved areas, while those with over 59% Anisoptera are associated with impacted areas. Mendoza-Penagos et al. (2021) observed a negative relationship between the degree of impact and the Zygoptera assemblage, except for Libellulidae.

The impact of deforestation for agriculture and/or livestock on other groups of aquatic insects was also documented for Santarém, Belterra, and Paragominas. Chen et al. (2017) utilized various metrics for insect fauna and observed that the taxonomic richness of Anisoptera and the functional structure of EPT increased with the intensity of the impact gradient. Conversely, the percentage of sensitive Zygoptera, the taxonomic richness of EPT and Heteroptera, EPT functional metrics (diversity and richness) exhibited higher values in less impacted areas. It was noted that streams with reduced vegetation cover limit the occurrence of EPT sensitive genera, while in impacted streams, environmental changes contribute to the persistence of locally tolerant EPT genera (Luiza-Andrade et al., 2022). For Heteroptera, a decline in taxonomic richness and alteration of the assemblage due to the impact were recorded (Moy et al., 2022). Martins et al. (2022) observed that disturbances in the catchment and riparian areas led to increased dissimilarity in macroinvertebrate assemblages based on abundance and incidence data. Studies evaluating the nonlinear response of aquatic invertebrate assemblages to the impact gradient revealed that even small rates of local riparian forest deforestation (1–11%) or forest loss in the catchment (1–9%) result in significant alterations in the assemblage, leading to the exclusion of several sensitive taxa (Brito et al., 2020; DalaCorte et al., 2020; Martins et al., 2021).

4 What Is the Effect of Non-agricultural Deforestation on the Aquatic Insects?

In the Amazon, the construction of roads can have significant impacts on aquatic environments (Fig. 4). For instance, the clearance of forest strips for road and well construction related to oil and gas exploration in the Urucu Petroleum Province (Coari-Amazonas) has resulted in an increased influx of suspended inorganic sediments and a decrease in sediment color index. These changes have led to a reduction in density and richness (both taxonomic and functional) and a modification of the assemblage composition in streams impacted by these activities compared to reference streams (Couceiro et al., 2010). The density of shredder organisms has declined by 40% in impacted streams due to the burial of leaf litter, which serves as food and shelter for these organisms (Couceiro et al., 2011). In Manaus, the removal of vegetation near streams for road construction has caused a substitution of specialist species with generalists, resulting in alterations in the composition of adult Odonata assemblages, while the overall richness of this insect order remains unaffected (Monteiro-Júnior et al., 2013).

In addition to conventional logging, which has a significant environmental impact, studies in the Amazon region have examined the effects of reduced-impact logging (RIL) in forest areas. RIL aims to preserve the habitat and maintain the biodiversity present in the area (Almeida et al., 2012). Conventional logging (timber extraction) involves the complete removal of the forest, while reduced-impact logging focuses on selectively harvesting a limited number of target species, leaving most of the forest intact (Calvão et al., 2016). In general, due to the absence or minimal variation in environmental variables, habitat availability, and food resources, the richness, abundance, and composition of different insect groups are similar between streams in RIL areas and those without vegetation removal (Calvão et al., 2016; Cardoso et al., 2018; Guterres et al., 2021; Nogueira et al., 2016; Roque et al., 2015). However, Mendes et al. (2021) observed greater environmental differences between sampling areas, resulting in intermediate richness values for the Anisoptera and Zygoptera suborders (Odonata) compared to forested areas and areas subjected to conventional logging.

In the eastern Amazon, researchers evaluated the impact of traditional populations on EPT (Faria et al., 2017). The study area includes the cultivation of açaí

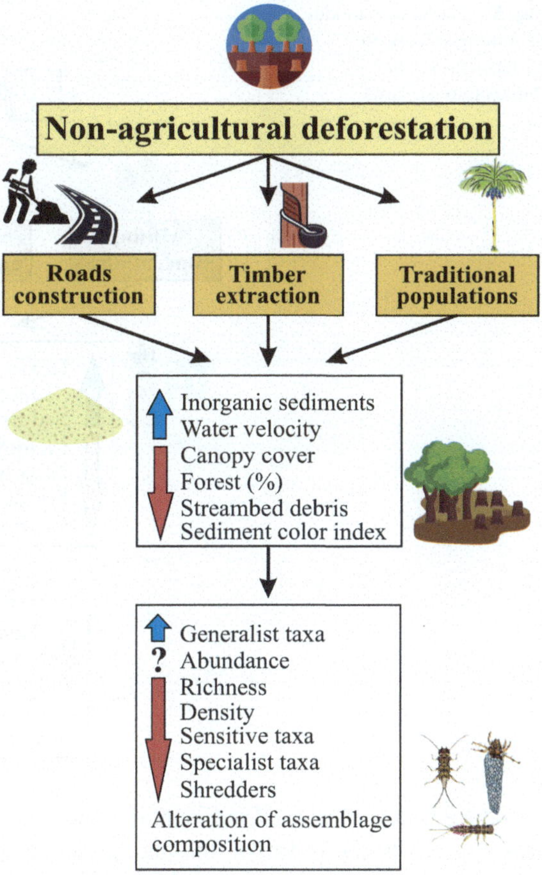

Fig. 4 Conceptual model of non-agricultural deforestation effects on aquatic insects in Amazonian streams. Note:? = non-uniform response

berry (*Euterpe oleracea* Mart) and cassava (*Manihot esculenta* Crantz), as well as the harvesting of natural resources. The riverside communities also engage in debris removal from the water channels, which can alter the morphology of the streams. It was observed that the streams affected by traditional populations exhibited increased water velocity and reduced amounts of woody debris, leaf litter, and riparian vegetation cover compared to forested streams. These changes in habitat characteristics resulted in alterations to the EPT assemblage, as more sensitive taxa were excluded (Faria et al., 2017). However, no significant differences were found in the abundance and richness of EPT between the sampled areas.

5 What Are the Effects of Mining on Aquatic Insects?

Mining activities can cause alterations in the abiotic factors of streams, including a decrease in the availability of organic matter and dissolved oxygen, as well as changes in sediment grain size (Callisto et al., 1998a; Fig. 5). Additionally, they can

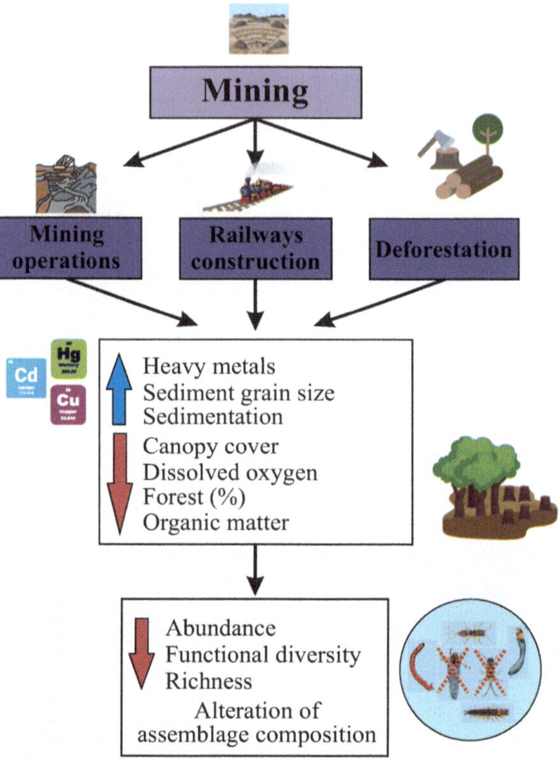

Fig. 5 Conceptual model of mining effects on aquatic insects in Amazonian streams

introduce heavy metals and sediment into the water (Enríquez-Espinosa et al., 2020; Kusin et al., 2017). In Brazil, mining operations are governed by distinct environmental legislation (Federal Law 12.561/2012), which mandates the safeguarding of riparian vegetation in areas impacted by such activities (Soares-filho et al., 2014). However, extensive deforestation is often associated with mining operations (Sonter et al., 2017). Furthermore, in certain regions of the Amazon, such as the Porto Trombetas region in Pará state, the construction of railways for ore transportation has led to the damming of multiple streams, resulting in changes in hydrodynamics and abiotic variables in these aquatic environments (Callisto et al., 1998a). These changes include increased sedimentation, such as silt and clay resulting from erosion, as well as the accumulation of organic matter and humic substances from decomposing flooded forests (Callisto et al., 1998a, b).

In a series of studies conducted on the effects of bauxite mining on the assemblage of aquatic invertebrates, three different categories were sampled (reference, dammed for railway construction, and impacted by bauxite tailings) in the Porto Trombetas region. The areas impacted by bauxite tailings showed changes in the composition of functional groups and a decrease in species richness. During the rainy periods of 2004 and 2005, no invertebrates were found, probably due to low organic matter concentration and alterations in sediment grain size caused by the deposition of large bauxite particles on the substrate (Callisto et al., 1998a, b, c; Callisto & Esteves, 1998). Moreover, the effects of bauxite deposition can be long-lasting. For instance, even 6 years after the cessation of bauxite deposition in a section of the Caranã stream, no benthic invertebrates had recolonized the area (Callisto et al., 1998a).

In the Serra dos Carajás region, ~14% of the area is dedicated to the extraction of iron, gold, and copper, with mining operations that have been ongoing for over 50 years (Ibama, 2003). Generally, the streams' integrity tends to decline as the proximity to mining operations increases (Enríquez-Espinosa et al., 2020), and a reduction in abundance and changes in the composition of the Odonata assemblage have been observed in relation to the proximity of the mines (Brito et al., 2021). There is also a decrease in the abundance and richness of EPT in streams impacted by mining activities (Rivera-Pérez et al., 2023). The functional diversity of Ephemeroptera is higher among impacted streams, indicating that the assemblage structure in this insect order, in mining areas, has low similarity (Enríquez-Espinosa et al., 2023). However, according to Enríquez-Espinosa et al. (2020), the taxonomic composition of the Ephemeroptera genera assemblage is similar between reference streams and those impacted by mining, perhaps if the identification had been carried out at specific levels differences could be detected since species in the same genus can present different levels of sensitivity to an impact (Buss & Salles, 2007; Heino, 2014).

6 What Is the Effect of Dam on the Aquatic Insects?

Dams are one of the oldest man-made infrastructures, having been an integral part of economic development in the history of humanity. These infrastructures can be used to control floods, irrigate crops, provide water, generate electricity, and facilitate navigation (Zhang & Gu, 2023). The implications of building dams and reservoirs transcend the socioeconomic and geopolitical spheres, also having broad environmental consequences, such as changes in aquatic and terrestrial habitats, loss of biodiversity, and impairment of the functioning of aquatic environments (Anderson et al., 2018; Forsberg et al., 2017; Zhang & Gu, 2023). Among the environmental changes, the fragmentation of rivers and loss of connectivity, flooding of habitats, interruption of sediment and nutrient transport, alteration of the flow regime and blockage of migration routes, and dispersion of numerous species stand out (Edmonds et al., 2020; Thieme et al., 2021; Fig. 6).

In the Amazon, the impacts of dams are often associated with the generation of electricity, water supply for agricultural activities, and the construction of roads (Nimer, 1977; Junk & Mello, 1990; Teyssier, 2002; Brasil et al., 2014b). Although hydroelectric dams are seen as a source of renewable energy and a promising way to generate low-carbon electricity (Thieme et al., 2021; Zhang & Gu, 2023), recent studies have found results that contradict the clean electricity proposal from dams and reservoirs in tropical forests (Faria et al., 2015; Fearnside, 2015a, b; Thieme et al., 2021). Especially in the Amazon basin, dams and/or reservoirs of

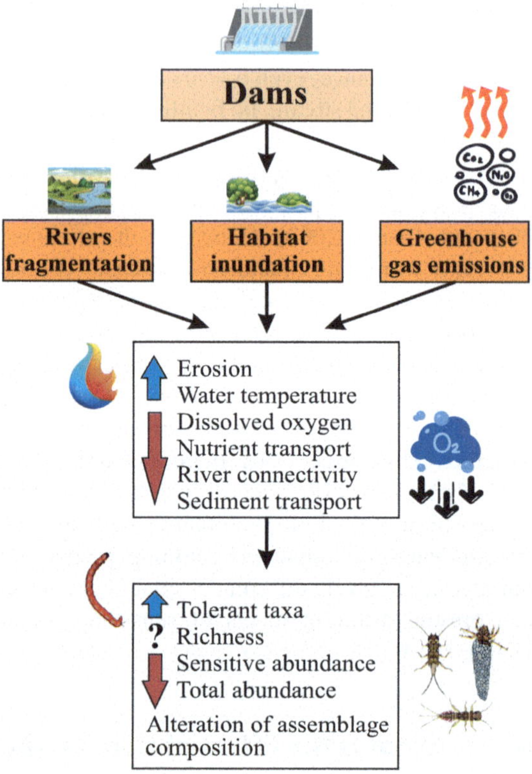

Fig. 6 Conceptual model of dam effects on aquatic insects in Amazonian streams. Note:? = non-uniform response

hydroelectric plants produced greenhouse gases, mainly carbon dioxide and methane, in the first 10 years of operation. In addition, specific dams and/or reservoirs in the Amazon basin (e.g., Balbina, Tucuruí, and Samuel dams; Santos et al., 2006; Kemenes et al., 2011; Fearnside & Pueyo, 2012) have emitted greenhouse gases in the same order of magnitude as thermal sources of electrical energy (Faria et al., 2015; Fearnside, 2015a). Another aggravating factor is that a survey carried out in the Andean Amazon identified 142 hydroelectric project dams in operation and/or construction (Anderson et al., 2018). In that same study, over 160 proposals for building dams in this region at different planning stages were counted (Anderson et al., 2018).

The effects of small hydroelectric dams on aquatic insects were evaluated in the Juruena River before and after dams within the Cerrado – Amazon transition (Arcifa et al., 2023). The abundance, richness, and diversity of insect groups were reduced in the post-dam phase with the predominance of Diptera, especially Chironomidae. While in the pre-dam phase, the assemblage was more diverse, containing especially, Trichoptera, Ephemeroptera, Coleoptera, and Diptera (Arcifa et al., 2023). In another study performed in the Cerrado–Amazon transition, changes in Odonata assemblages were recorded between the upstream and downstream sections of dams in the Xingu River basin (Brasil et al., 2014b). In this study, it was observed that six of the 18 species of Odonata collected occurred only in the upstream section of the dam (Brasil et al., 2014b).

7 What Is the Effect of Climate Change on Aquatic Insects?

Land use, deforestation, and the burning of fossil fuels have been key drivers of climate change (Canadell et al., 2007; Silvério et al., 2015). In the Amazon, it is projected that by 2100, there could be a temperature increase of up to 8 °C compared to the average from 1995 to 2014, accompanied by CO_2 levels approaching 1000 parts per million by volume (ppmv; IPCC, 2021; Fig. 7). The frequency of extreme droughts and floods is also expected to rise (Fearnside, 1995; Nobre et al., 2016; Sorribas et al., 2016). These climate changes can lead to modifications in the composition and biodiversity of riparian vegetation (Piedade et al., 2013) and may even contribute to the process of Amazon rainforest savannization (Gomes et al., 2019). Furthermore, physiological changes in vegetation can result in leaf litter with higher concentrations of structural and secondary compounds (Coviella & Trumble, 1999). The elevated temperature and CO_2 levels can directly impact the metabolism and survival of aquatic insects by increasing their respiratory rates (Rumbos et al., 2010; Villanueva et al., 2011), potentially leading to local extinctions (Wiens, 2016).

In order to evaluate the impact of projected climate change on various aquatic insects in the Amazon region, laboratory studies were conducted using the microcosm system of the National Institute of Science and Technology (INCT) Adapta-II,

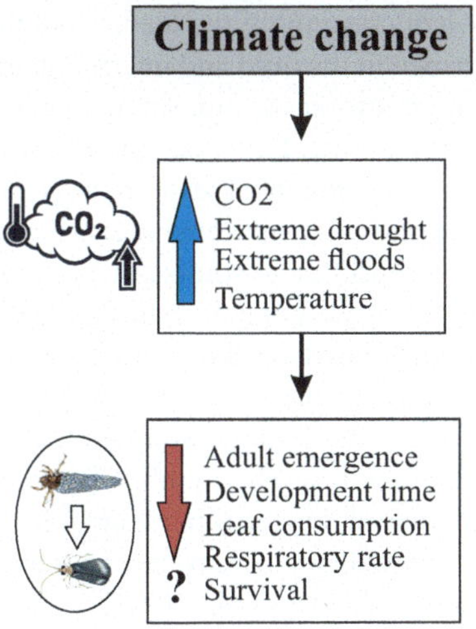

Fig. 7 Conceptual model of climate change effects on aquatic insects in Amazonian streams. Note:? = non-uniform response

located within the Instituto Nacional de Pesquisas da Amazônia (INPA). This microcosm consists of four chambers that replicate real-time scenarios proposed by the IPCC (2007, 2021), incorporating variations in temperature and CO_2 levels, resulting in temperature increases of up to 4.5 °C and CO_2 levels of approximately 900 ppmv compared to the control condition. Assessing the effects of climate change on the interactions among larvae of Diptera species *Aedes aegypti* (L.) and *Culex quinquefasciatus* (Say), similar survival was recorded between the two species in all different climatic conditions. However, the developmental time of the larvae decreased with temperature and CO_2 increase. Additionally, climate change has the potential to influence the pattern of interaction between these two species (Leonel et al., 2015). In another study involving Diptera, Piovezan-Borges et al. (2020) examined the effects of predation risk and climate change on *A. aegypti*. The survival of the larvae was found to be unaffected by predation risk or the increase in temperature and CO_2 levels. However, adult emergence was higher in the intermediate chamber (28 °C; 885 ppmv) compared to the other chambers of the microcosm, indicating that biological control of *A. aegypti* may not be a viable strategy.

Studies were also conducted in the Adapta microcosm to understand the effects of increased temperature and CO_2 on shredder insects (Trichoptera: *Phylloicus elektoros* Prather, 2003) and the leaf decomposition process, which is essential for the availability of fine particulate organic matter in preserved streams. In a 21-day study, a decrease in survival and leaf consumption by the shredders was observed with increasing temperature and CO_2, but the growth rate was similar across the climatic conditions (Martins et al., 2017b). However, in a shorter study duration (7 days), survival and leaf consumption by the shredders did not differ among the climatic conditions (Martins et al., 2017c). Martins et al. (2020) evaluated the effect of increased temperature and CO_2 on the chemical composition of *Montrichardia arborescens* (L.) Schott leaves and the consumption and survival of shredders. No difference in larval survival was recorded among the different climatic conditions when consuming the leaves, but the leaf consumption rate was higher in the debris grown in the mild and intermediate chambers. Recently, the effect of climate change and microplastics on shredder insects was assessed, and both stressors reduced the survival rate. Leaf consumption was affected by the interaction between microplastics and climate change, being lower in conditions with higher levels of microplastics and under extreme scenarios (Firmino et al., 2023).

In addition to studies in the Adapta microcosm, a 10-day study was carried out in BOD incubators to examine the impact of temperature elevation (23.2 °C and 27.5 °C) on two plant species. In treatments conducted at higher temperature conditions, especially with lower quality leaves, larval survival was found to be lower. The effect of temperature on leaf consumption was dependent on plant species. For plants with higher-quality leaves, leaf consumption decreased as the temperature increased. On the other hand, plants with lower-quality leaves exhibited an opposite trend, with an increase in consumption rate in response to higher temperature conditions (Martins et al., 2022).

8 Future Perspectives for Insects and Aquatic Environments in the Amazon

The Amazon biome is of global importance, in addition to harboring a great diversity of aquatic and terrestrial organisms, it plays a crucial role in global carbon sequestration, water cycling, and climate regulation (Nobre et al., 2016). Despite its ecological importance, its position in preserving biodiversity and mitigating climate change remains uncertain and largely neglected in regional development plans. These plans have led to increased population mobility, disorderly urbanization, and environmental degradation, which have significant impacts on the delicate Amazonian ecosystem (Fearnside et al., 2021). Anthropogenic impacts may even result in the extinction of species not discovered by science, as certain areas of the Amazon remain unexplored due to accessibility challenges and the limited presence of taxonomists and ecologists specialized in the study of different groups of tropical organisms.

Despite the progress made in understanding human impacts on Amazonian aquatic insects, several crucial questions remain largely unknown and require further investigation. For example, little is known about the effects of emerging pollutants and contaminants on aquatic insects in Amazonian streams. Specifically, the interactions between polystyrene microbeads and climate change, and their potential to impact the consumption and survival of Amazonian shredder insects, are still not fully understood (Firmino et al., 2023). A study that evaluated the ecological risks of pesticides in the Amazon region showed that urban streams with higher concentrations of malathion and chlorpyrifos could lead to the decline of aquatic insects of the order Ephemeroptera (Rico et al., 2022). However, the impacts of other micro(nano)materials (e.g., organic and inorganic polymers and heavy metals), fungicides, herbicides, insecticides, antibiotics, bactericides, plant extracts, and their derivatives remain largely unknown. Furthermore, the effects of drugs (e.g., synthetic hormones) and other endocrine disruptors on aquatic insect are not known and require thorough investigation as these drugs can reach rivers and streams. Another important point to be considered is the interaction of multiple stressors and how these interactions affect biodiversity and ecosystem functions in Amazonian aquatic systems. Changes in climate and land use can interact in complex ways, influencing aquatic ecosystems, their biodiversity, and ecosystem functions. These interactions can result in synergistic, additive, or antagonistic effects (Birk et al., 2020; Capon et al., 2021; Dudgeon, 2012).

Moreover, nutrient enrichment, hydrological changes, habitat loss, and pollution often have synergistic effects, leading to increased biodiversity extinction rates in freshwater ecosystems (Vescovi et al., 2009). Consequently, for a more realistic understanding of the Amazonia stream dynamics, future studies must consider the interactions of multiple stressors, as well as their potential impacts on biodiversity and ecosystem functions. A comprehensive approach that considers the cumulative effects of these stressors is essential in developing effective conservation strategies and management plans for these valuable and vulnerable environments.

References

Almeida, L. S. D., Gama, J. R. V., Oliveira, F. D. A., Carvalho, J. O. P. D., Gonçalves, D. C. M., & Araújo, G. C. (2012). Fitossociologia e uso múltiplo de espécies arbóreas em floresta manejada, Comunidade Santo Antônio, município de Santarém, Estado do Pará. *Acta Amazonica, 42*, 185–194. https://doi.org/10.1590/S0044-59672012000200002

Anderson, E. P., Jenkins, C. N., Heilpern, S., et al. (2018). Fragmentation of Andes-to-Amazon connectivity by hydropower dams. *Science Advances, 4*, eaao1642. https://doi.org/10.1126/sciadv.aao16

Arcifa, M. S., Coelho, P. N., Silva, L. H. S., & Righi-Cavallaro, K. O. (2023). Impacts of small dams on the phytoplankton and invertebrates in an Amazonian river. *River Research and Applications, 39*, 661–674. https://doi.org/10.1002/rra.4103

Barham, B. L., & Coomes, O. T. (1994). Reinterpreting the Amazon rubber boom: Investment, the state, and Dutch disease. *Latin American Research Review, 29*, 73–109.

Bastos, R. C., Brasil, L. S., Oliveira-Junior, J. M. B., Carvalho, F. G., Lennox, G. D., Barlow, J., et al. (2021). Morphological and phylogenetic factors structure the distribution of damselfly and dragonfly species (Odonata) along an environmental gradient in Amazonian streams. *Ecological Indicators, 122*, 107257. https://doi.org/10.1016/j.ecolind.2020.107257

Bentes, N. (2005). *Manaus Realidade E Contrastes Sociais* (1st ed., p. 182). Editora Valer.

Berenguer, E., Armenteras, D., Lees, A. C., Fearnside, P. M., Smith, C. C., Alencar, A., et al. (2021). Drivers and ecological impacts of deforestation and forest degradation. In C. Nobre, A. Encalada, et al. (Eds.), *Amazon assessment report 2021. Science panel for the Amazon (SPA)*. United Nations Sustainable Development Solutions Network. 10.55161/AIZJ1133.

Birk, S., Chapman, D., Carvalho, L., et al. (2020). Impacts of multiple stressors on freshwater biota across spatial scales and ecosystems. *Nature Ecology & Evolution, 4*, 1060–1068. https://doi.org/10.1038/s41559-020-1216-4

Bispo, P. D. C., Oliveira, L. G., Bini, L. M., & Sousa, K. G. D. (2006). Ephemeroptera, Plecoptera and Trichoptera assemblages from riffles in mountain streams of Central Brazil: Environmental factors influencing the distribution and abundance of immatures. *Brazilian Journal of Biology, 66*, 611–622. https://doi.org/10.1590/S1519-69842006000400005

Brasil, L. S., Batista, J. D., Giehl, N. F. D. S., Valadão, M. B. X., Santos, J. O. D., & Dias-Silva, K. (2014a). Environmental integrity and damselfly species composition in Amazonian streams at the" arc of deforestation" region, Mato Grosso, Brazil. *Acta Limnologica Brasiliensia, 26*, 278–287. https://doi.org/10.1590/S2179-975X2014000300007

Brasil, L. S., Giehl, N. F. D. S., Almeida, S. M., Valadão, M. B. X., Santos, J. O., Pinto, N. S., et al. (2014b). Does the damming of streams in the southern Amazon basin affect dragonfly and damselfly assemblages (Odonata: Insecta)? A preliminary study. *International Journal of Odonatology, 17*, 187–197. https://doi.org/10.1080/13887890.2014.963712

Brito, J. G., Roque, F. O., Martins, R. T., Nessimian, J. L., Oliveira, V. C., Hughes, R. M., et al. (2020). Small forest losses degrade stream macroinvertebrate assemblages in the eastern Brazilian Amazon. *Biological Conservation, 241*, 108263. https://doi.org/10.1016/j.biocon.2019.108263

Brito, J., Calvão, L., Cunha, E., Maioli, L., Barbirato, M., Rolim, S., et al. (2021). Environmental variables affect the diversity of adult damselflies (Odonata: Zygoptera) in western Amazonia. *International Journal of Odonatology, 24*, 108–121. https://doi.org/10.23797/2159-6719_24_8

Buss, D. F., & Salles, F. F. (2007). Using Baetidae species as biological indicators of environmental degradation in a Brazilian river basin. *Environmental Monitoring and Assessment, 130*, 365–372.

Callede, J., Guyot, J. L., Ronchail, J., Molinier, M., & Oliveira, E. (2012). L'Amazone à Óbidos (Brésil): étude statistique des débits et bilan hydrologique. *Hydrological Sciences Journal, 47*, 321–333. https://doi.org/10.1080/02626660209492933

Callisto, M., & Esteves, F. A. (1998). Caracterização funcional dos macroinvertebrados bentônicos em quatro ecossistemas lóticos sob influência das atividades de uma mineração de bauxita na Amazônia central (Brasil). *Oecologia Brasiliensis, 5*, 23–234.

Callisto, M., Esteves, F. A., Gonçalves, J. F., & Leal, J. J. F. (1998c). Impact of bauxite tailings on the distribution of benthic macrofauna in a Small River ('Igarapé') in Central Amazonia. *Brazilian Journal of Kansas Entomological Society, 71*, 447–455. http://www.jstor.org/stable/25085861

Callisto, M., Esteves, F. A., Gonçalves, J. F., & Fonseca, J. J. L. (1998a). Benthic macroinvertebrates as indicators of ecological fragility of small rivers ('igarapés') in a bauxite mining region of Brazilian Amazonia. *Amazoniana, 15*, 1–9. http://labs.icb.ufmg.br/benthos/index_arquivos/pdfs_pagina/Callisto.etal-1998d.pdf

Callisto, M., Gonçalves, J. F., & Fonseca, J. J. L. (1998b). Benthic macroinvertebrates of four Amazonian streams influenced by bauxite mining (Brazil). *Verhandlungen der Internationalischen Vereinigung für Theoretische und Angewandte Limnologie, 26*, 983–985. https://doi.org/10.1080/03680770.1995.11900865

Calvão, L. B., Nogueira, D. S., Montag, L. F. A., Lopes, M. A., & Juen, L. (2016). Are Odonata communities impacted by conventional or reduced impact logging? *Forest Ecology and Management, 382*, 143–150. https://doi.org/10.1016/j.foreco.2016.10.013

Canadell, J. G., Quéré, C., Raupach, M. R., Field, C. B., Buitenhuis, E. T., Ciais, P., et al. (2007). Contributions to accelerating atmospheric CO2 growth from economic activity, carbon intensity, and efficiency of natural sinks. *Proceedings of the National Academy of Sciences, 104*, 18866–18870.

Capon, S. J., Stewart-Koster, B., & Bunn, S. E. (2021). Future of freshwater ecosystems in a 1.5 C warmer world. Front. *Environmental Sciences, 9*, 596. https://doi.org/10.3389/fenvs.2021.784642

Cardoso, M. N., Calvão, L. B., Montag, L. F. A., Godoy, B. S., & Juen, L. (2018). Reducing the deleterious effects of logging on Ephemeroptera communities through reduced impact management. *Hydrobiologia, 823*, 191–203. https://doi.org/10.1007/s10750-018-3705-x

Carvalho, F. G., Roque, F., Barbosa, L., Montag, L. F., & Juen, L. (2018). Oil palm plantation is not a suitable environment for most forest specialist species of Odonata in Amazonia. *Animal Conservation, 21*, 526–533. https://doi.org/10.1111/acv.12427

Carvalho, F. G., Duarte, L., Nakamura, G., Seger, G. D. S., & Juen, L. (2021). Changes of phylogenetic and taxonomic diversity of Odonata (Insecta) in response to land use in Amazônia. *Forests, 12*, 1061. https://doi.org/10.3390/f12081061

Chen, K., Hughes, R. M., Brito, J. G., Leal, C. G., Leitão, R. P., Oliveira-Júnior, J. M., et al. (2017). A multi-assemblage, multi-metric biological condition index for eastern Amazonia streams. *Ecological Indicators, 78*, 48–61.

Cleto-Filho, S. E. N., & Walker, I. (2001). Efeitos da ocupação urbana sobre a macrofauna de invertebrados aquáticos de um igarapé da cidade de Manaus/AM-Amazônia Central. *Acta Amazonica, 31*, 69–89. https://doi.org/10.1590/1809-43922001311089

Couceiro, S. R., Forsberg, B. R., Hamada, N., & Ferreira, R. L. M. (2006). Effects of an oil spill and discharge of domestic sewage on the insect fauna of Cururu stream, Manaus, AM, Brazil. *Brazilian Journal of Biology, 66*, 35–44.

Couceiro, S. R. M., Hamada, N., Luz, S. L. B., Forsberg, B. R., & Pimentel, T. P. (2007). Deforestation and sewage effects on aquatic macroinvertebrates in urban streams in Manaus, Amazonas, Brazil. *Hydrobiologia, 575*, 271–284. https://doi.org/10.1007/s10750-006-0373-z

Couceiro, S. R. M., Hamada, N., Forsberg, B. R., & Padovesi-Fonseca, C. (2010). Effects of anthropogenic silt on aquatic macroinvertebrates and abiotic variables in streams in the Brazilian Amazon. *Journal of Soils and Sediments, 10*, 89–103. https://doi.org/10.1007/s11368-009-0148-z

Couceiro, S. R., Hamada, N., Forsberg, B. R., & Padovesi-Fonseca, C. (2011). Trophic structure of macroinvertebrates in Amazonian streams impacted by anthropogenic siltation. *Austral Ecology, 36*, 628–637.

Couceiro, S. R., Hamada, N., Forsberg, B. R., Pimentel, T. P., & Luz, S. L. B. (2012). A macroinvertebrate multimetric index to evaluate the biological condition of streams in the Central Amazon region of Brazil. *Ecological Indicators, 18*, 118–125. https://doi.org/10.1016/j.ecolind.2011.11.001

Couceiro, S. R. M., Dias-Silva, K., & Hamada, N. (2021). Influence of climate seasonality on the effectiveness of the use of aquatic macroinvertebrates in urban impact evaluation in Central Amazonia. *Limnology, 22*, 237–244. https://doi.org/10.1007/s10201-020-00648-6

Coviella, C. E., & Trumble, J. T. (1999). Effects of elevated atmospheric carbon dioxide on insect-plant interactions. *Conservation Biology, 13*, 700–712.

Cummins, K. W., Wilzbach, M. A., Gates, D. M., Perry, J. B., & Taliaferro, W. B. (1989). Shedders and riparian vegetation: Leaf litter that falls into streams influences communities of stream invertebrates. *Bioscience, 39*, 24–30. https://doi.org/10.2307/1310804

Cunha, E. J., & Juen, L. (2017). Impacts of oil palm plantations on changes in environmental heterogeneity and Heteroptera (Gerromorpha and Nepomorpha) diversity. *Journal of Insect Conservation, 21*, 111–119. https://doi.org/10.1007/s10841-017-9959-1

Cunha, E. J., Montag, L. F. A., & Juen, L. (2015). Oil palm crops effects on environmental integrity of Amazonian streams and Heteropteran (Hemiptera) species diversity. *Ecological Indicators, 52*, 422–429. https://doi.org/10.1016/j.ecolind.2014.12.024

Cunha, E. J., Guterres, A. P. M., Godoy, B. S., & Juen, L. (2020). Wing dimorphism in semiaquatic bugs (Hemiptera, Heteroptera, Gerromorpha) as a tool for monitoring streams altered by oil palm plantation in the Amazon. *Ecological Indicators, 117*, 106707. https://doi.org/10.1016/j.ecolind.2020.106707

Cunha, E. J., Cruz, G. M., Faria, A. P. J., Oliveira, J. N., & Juen, L. (2022). Urban development and industrialization impacts on semiaquatic bugs diversity: A case study in eastern Amazonian streams. *Water Biology and Security, 1*, 100061. https://doi.org/10.1016/j.watbs.2022.100061

DalaCorte, R. B., Melo, A. S., Siqueira, T., et al. (2020). Thresholds of freshwater biodiversity in response to riparian vegetation loss in the Neotropical region. *Journal of Applied Ecology, 57*, 1391–1402. https://doi.org/10.1111/1365-2664.13657

Dudgeon, D. (2012). Threats to freshwater biodiversity globally and in the indo-Burma biodiversity hotspot. In D. J. Allen, K. G. Smith, & W. R. T. Darwall (Eds.), *The status and distribution of freshwater biodiversity in indo-Burma* (pp. 1–28). IUCN (International Union for the Conservation of Nature).

Duffy, P. B., Brando, P. M., Asner, G. P., & Field, C. B. (2015). Projections of future meteorological drought and wet periods in the Amazon. *Proceedings of the National Academy of Sciences, USA112*, 13172–13177. https://doi.org/10.1073/pnas.1421010112

Edmonds, D. A., Caldwell, R. L., Brondizio, E. S., Brondizio, E. S., & Siani, S. M. O. (2020). Coastal flooding will disproportionately impact people on river deltas. *Nature Communications, 11*, 4741. https://doi.org/10.1038/s41467-020-18531-4

Enríquez-Espinosa, A. C., Shimano, Y., Rolim, S., Maioli, L., Juen, L., & Dunck, B. (2020). Effects of mining and reduced turnover of Ephemeroptera (Insecta) in streams of the eastern Brazilian Amazon. *Journal of Insect Conservation, 24*, 1061–1072. https://doi.org/10.1007/s10841-020-00275-7

Enríquez-Espinosa, A. C. E., Cunha, E. J., Shimano, Y., Rolim, S., Mioli, L., Juen, L., et al. (2023). Functional diversity of mayflies (Ephemeroptera, Insecta) in streams in mining areas located in the eastern Amazon. *Hydrobiologia, 850*, 929–945. https://doi.org/10.1007/s10750-022-05134-x

Faria, F. A., Jaramillo, P., Sawakuchi, H. O., Richey, J. E., & Barros, N. (2015). Estimating greenhouse gas emissions from future Amazonian hydroelectric reservoirs. *Environmental Research Letters, 10*, 124019. https://doi.org/10.1088/1748-9326/10/12/124019

Faria, A. P. J., Ligeiro, R., Callisto, M., & Juen, L. (2017). Response of aquatic insect assemblages to the activities of traditional populations in eastern Amazonia. *Hydrobiologia, 802*, 39–51. https://doi.org/10.1007/s10750-017-3238-8

Fearnside, P. M. (1995). Queimadas e desmatamentos na Amazônia. pp. 2127. In: M.F.S. Faria (ed.) Meio Ambiente e Sociedade (Série Estudos Contemporâneos 1). Serviço Nacional de Aprendizagem Comercial (SENAC), Rio de Janeiro, Brasil. 68 pp.
Fearnside, P. M. (2006). Desmatamento na Amazônia: dinâmica, impactos e controle. *Acta Amazonica, 36*, 395–400.
Fearnside, P. M. (2015a). Emissions from tropical hydropower and the IPCC. Environ Sci. *Policy, 50*, 225–239. https://doi.org/10.1016/j.envsci.2015.03.002
Fearnside, P. M. (2015b). Tropical hydropower in the clean development mechanism: Brazil's Santo Antônio dam as an example of the need for change. *Climatic Change, 131*, 575–589. https://doi.org/10.1007/s10584-015-1393-3
Fearnside, P. M., & Pueyo, S. (2012). Greenhouse-gas emissions from tropical dams. *Nature Climate Change, 2*, 382–384. https://doi.org/10.1038/nclimate1540
Fearnside, P. M., Berenguer, E., Armenteras, D., Duponchelle, F., Guerra, F. M., Jenkins, C. N., et al. (2021). Drivers and impacts of changes in aquatic ecosystems. Chapter 20. In C. Nobre, A. Encalada, et al. (Eds.), *Amazon assessment report 2021. Science panel for the Amazon (SPA)* (pp. 305–343). United Nations Sustainable Development Solutions Network. Part II. https://doi.org/10.55161/IDMB5770
Ferreira, M. C., Begot, T. O., Prudente, B. S., Juen, L., & Montag, L. F. A. (2018). Effects of oil palm plantations on habitat structure and fish assemblages in Amazon streams. *Environmental Biology of Fishes, 101*, 547–562.
Ferreira, S. J. F., Pinel, S., Ríos-Villamizar, E. A., et al. (2021). Impact of rapid urbanization on stream water quality in the Brazilian Amazon. *Environment and Earth Science, 80*, 316. https://doi.org/10.1007/s12665-021-09621-7
Firmino, V. C., Brasil, L. S., Martins, R. T., Ligeiro, R., Tonin, A., Gonçalves, J. F., et al. (2021). Litter decomposition of exotic and native plant species of agricultural importance in Amazonian streams. *Limnology, 22*, 289–297. https://doi.org/10.1007/s10201-021-00655-1
Firmino, V. C., Martins, R. T., Brasil, L. S., Cunha, E. J., Pinedo, R. B., Hamada, N., et al. (2023). Do microplastics and climate change negatively affect shredder invertebrates from an amazon stream? An ecosystem functioning perspective. *Environmental Pollution, 321*, 121184. https://doi.org/10.1016/j.envpol.2023.121184
Forsberg, B. R., Melack, J. M., Dunne, T., Barthem, R. B., Goulding, M., Paiva, R. C., et al. (2017). The potential impact of new Andean dams on Amazon fluvial ecosystems. *PLoS One, 12*, e0182254. https://doi.org/10.1371/journal.pone.0182254
Gardner, T. A., Ferreira, J., Barlow, J., Lees, A. C., Parry, L., Vieira, I. C. G., et al. (2013). A social and ecological assessment of tropical land uses at multiple scales: The sustainable Amazon network. *Philosophical Transactions of the Royal Society B: Biological Sciences, 368*, 20120166.
Gomes, V. H., Vieira, I. C., Salomão, R. P., & ter Steege, H. (2019). Amazonian tree species threatened by deforestation and climate change. *Nature Climate Change, 9*, 547–553.
Guterres, A. P. M., Cunha, E. J., & Juen, L. (2021). Tolerant semiaquatic bugs species (Heteroptera: Gerromorpha) are associated to pasture and conventional logging in the eastern Amazon. *Journal of Insect Conservation, 25*, 555–567. https://doi.org/10.1007/s10841-021-00316-9
Heino, J. (2014). Taxonomic surrogacy, numerical resolution and responses of stream macroinvertebrate communities to ecological gradients: Are the inferences transferable among regions? *Ecological Indicators, 36*, 186–194.
Homma, A. K. O. (2001). O desenvolvimento da agroindústria no estado do Pará. Embrapa Amazônia Oriental-Artigo em periódico indexado (ALICE). https://www.alice.cnptia.embrapa.br/bitstream/doc/403795/1/4725.pdf. Accessed 29.05.20.
Hynes, H. B. N. (1970). *The ecology of running waters*. Liverpool University Press.
IBAMA. (2003). Instituto Brasileiro do Meio Ambiente e dos Recursos Naturais. PLANO DE MANEJO PARA USO MÚLTIPLO DA FLORESTA NACIONAL DE CARAJÁS. Brasília. https://www.gov.br/icmbio/pt-br/assuntos/biodiversidade/unidade-de-conservacao/unidades-de-biomas/amazonia/lista-de-ucs/flona-de-carajas/arquivos/dcom_icmbio_plano_de_manejo_flona_carajas_volume_ii.pdf. Acessed 12.06.23.

IBGE. (2022). INSTITUTO BRASILEIRO DE GEOGRAFIA E ESTATÍSTICA. Censo Brasileiro. https://www.ibge.gov.br/pt/inicio.html. Acessed 12.06.23.

IPCC. (2021). In V. Masson Delmotte, P. Zhai, A. Pirani, S. L. Connors, C. Péan, S. Berger, N. Caud, Y. Chen, L. Goldfarb, M. I. Gomis, M. Huang, K. Leitzell, E. Lonnoy, M. JBR, T. K. Maycock, T. Waterfield, O. Yelekçi, R. Yu, & B. Zhou (Eds.), *Summary for policymakers in: Climate change 2021: The physical science basis. Contribution of working group I to the sixth assessment report of the intergovernmental panel on climate change*. Cambridge University Press. https://www.ipcc.ch/report/ar6/wg1/downloads/report/IPCC_AR6_WGI_SPM_final.pdf. Accessed 23.04.22.

IPCC – Intergovernmental Panel on Climate Change. (2007). *Climate change 2007: The physical science basis. Contribution of the working group I to the fourth assessment report of the intergovernmental panel on climate change*. Cambridge University Press.

Juen, L., Cunha, E. J., Carvalho, F. G., Ferreira, M. C., Begot, T. O., Andrade, A. L., et al. (2016). Effects of oil palm plantations on the habitat structure and biota of streams in eastern Amazon. *River Research and Applications, 32*, 2081–2094. https://doi.org/10.1002/rra.3050

Junk, W. J., & Mello, J. A. S. (1990). Impactos ecológicos das represas hidrelétricas na bacia amazônica brasileira. *Estudos avançados, 4*, 126–143.

Kemenes, A., Forsberg, B. R., & Melack, J. M. (2011). CO2 emissions from a tropical hydroelectric reservoir (Balbina, Brazil). *Journal of Geophysical Research—Biogeosciences, 116*. https://doi.org/10.1029/2010JG001465

Kusin, F. M., Rahman, M. S. A., Madzin, Z., Jusop, S., Mohamat-Yusuff, F., Ariffin, M., et al. (2017). The occurrence and potential ecological risk assessment of bauxite mine-impacted water and sediments in Kuantan, Pahang, Malaysia. *Environmental Science and Pollution Research, 24*, 1306–1321. https://doi.org/10.1007/s11356-016-7814-7

Laurance, W. F., Sayer, J., & Cassman, K. G. (2014). Agricultural expansion and its impacts on tropical nature. *Trends in Ecology & Evolution, 29*, 107–116. https://doi.org/10.1016/j.tree.2013.12.001

Leonel, B. F., Koroiva, R., Hamada, N., Ferreira-Keppler, R. L., & Roque, F. O. (2015). Potential effects of climate change on ecological interaction outcomes between two disease-vector mosquitoes: A mesocosm experimental study. *Journal of Medical Entomology, 52*, 866–872.

Lima, M., Firmino, V. C., Paiva, C. K. S., Juen, L., & Brasil, L. S. (2022). Land use changes disrupt streams and affect the functional feeding groups of aquatic insects in the Amazon. *Journal of Insect Conservation, 26*, 137–148. https://doi.org/10.1007/s10841-022-00375-6

Lovejoy, T. E., & Nobre, C. A. (2018). Amazon tipping point. *Science Advances, 4*, 2340. https://doi.org/10.1126/sciadv.aat2340

Luiza-Andrade, A., Brasil, L. S., Benone, N. L., Shimano, Y., Farias, A. P. J., Montag, L. F. A., Dolédec, S., & Juen, L. (2017). Influence of oil palm monoculture on the taxonomic and functional composition of aquatic insect communities in eastern Brazilian Amazonia. *Ecological Indicators, 82*, 478–483. https://doi.org/10.1016/j.ecolind.2017.07.006

Luiza-Andrade, A., Silva, R. R., Shimano, Y., Faria, A. P. J., Cardoso, M. N., Brasil, L. S., et al. (2022). Niche breadth and habitat preference of Ephemeroptera, Plecoptera, and Trichoptera (Insecta) in streams in the Brazilian Amazon. *Hydrobiologia, 849*, 4287–4306. https://doi.org/10.1007/s10750-022-04987-6

Luiza-Andrade, A., Silva, R. R., & Juen, L. (2023). Contribution of rare genera of aquatic insects to functional diversity in streams with multiple land use in the Amazon. *Hydrobiologia, 850*, 21–38. https://doi.org/10.1007/s10750-022-05035-z

Macdonald, J. S., Beaudry, P. G., Macisaac, E. A., & Herunter, H. E. (2003). The effects of forest harvesting and best management practices on stream flow and suspended sediment concentrations during snowmelt in headwater streams in sub-boreal forests of British Columbia, Canada. *Canadian Journal of Forest Research, 1407*, 1397–1407. https://doi.org/10.1139/X03-110

Marengo, J. A., Souza, C. M., Jr., Thonicke, K., Burton, C., Halladay, K., Betts, R. A., et al. (2018). Changes in climate and land use over the Amazon region: Current and future variability and trends. *Frontiers in Earth Science, 228*. https://doi.org/10.3389/feart.2018.00228

Marques, N. C. S., Jankowski, K. J., Macedo, M. N., Juen, L., Luiza-Andrade, A., & Deegan, L. A. (2021). Riparian forests buffer the negative effects of cropland on macroinvertebrate diversity in lowland Amazonian streams. *Hydrobiologia, 848*, 3503–3520. https://doi.org/10.1007/s10750-021-04604-y

Martins, R. T., Oliveira, V. C., & Salcedo, A. K. M. (2014). Uso de insetos aquáticos na avaliação de impactos antrópicos em ecossistemas aquáticos. In N. Hamada, J. L. Nessimian, & R. B. Querino (Eds.), *Insetos aquáticos na Amazônia brasileira: taxonomia, biologia e ecologia* (pp. 117–128). Editora do Inpa.

Martins, R. T., Melo, A. S., Gonçalves, J. F., & Hamada, N. (2015). Leaf-litter breakdown in urban streams of Central Amazonia: Direct and indirect effects of physical, chemical, and biological factors. *Freshwater Science, 34*, 716–726. https://doi.org/10.1086/681086

Martins, R. T., Couceiro, S. R., Melo, A. S., Moreira, M. P., & Hamada, N. (2017a). Effects of urbanization on stream benthic invertebrate communities in Central Amazon. *Ecological Indicators, 73*, 480–491. https://doi.org/10.1016/j.ecolind.2016.10.013

Martins, R. T., Melo, A. S., Gonçalves, J. F., Campos, C. M., & Hamada, N. (2017b). Effects of climate change on leaf breakdown by microorganisms and the shredder *Phylloicus elektoros* (Trichoptera: Calamoceratidae). *Hydrobiologia, 789*, 31–44. https://doi.org/10.1007/s10750-016-2689-7

Martins, R. T., Rezende, R. S., Gonçalves Júnior, J. F., Lopes, A., Piedade, M. T. F., Cavalcante, H. L., et al. (2017c). Effects of increasing temperature and, CO2 on quality of litter, shredders, and microorganisms in Amazonian aquatic systems. *PLoS One, 12*, e0188791. https://doi.org/10.1371/journal.pone.0188791

Martins, R. T., Gonçalves, J. F., Campos, C. M., Lopes, A., Piedade, M. T. F., & Hamada, N. (2020). Leaf consumption by invertebrate aquatic shredders in the Amazon: effects of climate change and microbial conditioning. *Limnology, 21*, 257–266.

Martins, R. T., Brito, J., Dias-Silva, K., et al. (2021). Low forest-loss thresholds threaten Amazonian fish and macroinvertebrate assemblage integrity. *Ecological Indicators, 127*, 107773. https://doi.org/10.1016/j.ecolind.2021.107773

Martins, R. T., Brito, J., Dias-Silva, K., et al. (2022). Congruence and responsiveness in the taxonomic compositions of Amazonian aquatic macroinvertebrate and fish assemblages. *Hydrobiologia, 849*, 2281–2298. https://doi.org/10.1007/s10750-022-04867-z

Mateo-Sagasta, J., Zadeh, S. M., Turral, H., & Burke, J. (2017). Water pollution from agriculture: A global review. *Executive Summary*. https://www.fao.org/3/i7754e/i7754e.pdf

Mendes, T. P., Benone, N. L., & Juen, L. (2019). To what extent can oil palm plantations in the Amazon support assemblages of Odonata larvae? *Insect Conservation and Diversity, 12*, 448–458. https://doi.org/10.1111/icad.12357

Mendes, T. P., Amado, L. L., Ribeiro, R. A. B., & Juen, L. (2020). Morphological diversity of Odonata larvae (Insecta) and abiotic variables in oil palm plantation areas in the eastern Amazon. *Hydrobiologia, 847*, 161–175. https://doi.org/10.1007/s10750-019-04079-y

Mendes, T. P., Montag, L. F. A., Alvarado, S. T., & Juen, L. (2021). Assessing habitat quality on alpha and beta diversity of Odonata larvae (insect) in logging areas in Amazon forest. *Hydrobiologia, 848*, 1147–1161. https://doi.org/10.1007/s10750-021-04524-x

Mendoza-Penagos, C. C., Calvao, L. B., & Juen, L. (2021). A new biomonitoring method using taxonomic families as substitutes for the suborders of the Odonata (Insecta) in Amazonian streams. *Ecological Indicators, 124*, 107388. https://doi.org/10.1016/j.ecolind.2021.107388

Miguel, T. B., Oliveira-Junior, J. M. B., Ligeiro, R., & Juen, L. (2017). Odonata (Insecta) as a tool for the biomonitoring of environmental quality. *Ecological Indicators, 81*, 555–566.

Monteiro-Júnior, C. S., Couceiro, S. R. M., Hamada, N., & Juen, L. (2013). Effect of vegetation removal for road building on richness and composition of Odonata communities in Amazonia, Brazil. *International Journal of Odonatology, 16*, 135–144. https://doi.org/10.1080/13887890.2013.764798

Monteiro-Júnior, C. S., Juen, L., & Hamada, N. (2014). Effects of urbanization on stream habitats and associated adult dragonfly and damselfly communities in central Brazilian Amazonia. *Landscape and Urban Planning, 127*, 28–40. https://doi.org/10.1016/j.landurbplan.2014.03.006

Monteiro-Júnior, C. S., Juen, L., & Hamada, N. (2015). Analysis of urban impacts on aquatic habitats in the Central Amazon basin: Adult odonates as bioindicators of environmental quality. *Ecological Indicators, 48*, 303–311. https://doi.org/10.1016/j.ecolind.2014.08.021

Monteles, J. S., Gerhard, P., Ferreira, A., & Sonoda, K. C. (2021). Agriculture impacts benthic insects on multiple scales in the eastern Amazon. *Biological Conservation, 255*, 108998. https://doi.org/10.1016/j.biocon.2021.108998

Moy, K. M., Brasil, L. S., Oliveira-Junior, J. M. B., Juen, L., Vieira, T. B., & Dias-Silva, K. (2022). Effects of environmental changes on Gerromorpha (Heteroptera: Hemiptera) communities from Amazonian streams. *Hydrobiology, 1*, 111–121. https://doi.org/10.3390/hydrobiology1010008

Nakamura, F., & Yamada, H. (2005). Effects of pasture development on the ecological functions of riparian forests in Hokkaido in northern Japan. *Ecological Engineering, 24*, 539–550. https://doi.org/10.1016/j.ecoleng.2005.01.010

Nascimento, T. S. R., Nascimento Monte, C., Corrêa, E. S., et al. (2022). The seasonality of contaminants in an urbanized microbasin in the Brazilian Amazon. *Water, Air, and Soil Pollution, 233*, 412. https://doi.org/10.1007/s11270-022-05879-0

Nepstad, D. C., Stickler, C. M., Filho, B. S., & Merry, F. (2008). Interactions among Amazon land use, forests and climate: Prospects for a near-term forest tipping point. *Philosophical Transactions of the Royal Society B: Biological Sciences, 363*, 1737–1746.

Nessimian, J. L., Venticinque, E. M., Zuanon, J., De Marco, P., Gordo, M., Fidelis, L., et al. (2008). Land use, habitat integrity, and aquatic insect assemblages in central Amazonian streams. *Hydrobiologia, 614*, 117–131. https://doi.org/10.1007/s10750-008-9441-x

Nimer, E. (1977). Clima. In IBGE (Ed.), *Geografia do Brasil* (Vol. 3, pp. 51–89). IBGE.

Nobre, C. A., Sampaio, G., Borma, L. S., Castilla-Rubio, J. C., Silva, J. S., & Cardoso, M. (2016). Land-use and climate change risks in the Amazon and the need of a novel sustainable development paradigm. *PNAS, 113*, 10759–10768. https://doi.org/10.1073/pnas.1605516113

Nogueira, D. S., Calvão, L. B., Montag, L. F. A., Juen, L., & De Marco, P. (2016). Little effects of reduced-impact logging on insect communities in eastern Amazonia. *Environmental Monitoring and Assessment, 188*, 441. https://doi.org/10.1007/s10661-016-5431-z

Oliveira Junior, J. M. B., Shimano, Y., Gardner, T. A., Hughes, R. M., De Marco, P., & Juen, L. (2015). Neotropical dragonflies (Insecta: Odonata) as indicators of ecological condition of small streams in the eastern Amazon. *Austral Ecology, 40*, 733–744. https://doi.org/10.1111/aec.12242

Oliveira Junior, J. M. B., Teodosio, M. A., & Juen, L. (2021). Patterns of co-occurrence and body size in dragonflies and damselflies (Insecta: Odonata) in preserved and altered Amazonian streams. *Austral Entomology, 60*, 436–450. https://doi.org/10.1111/aen.12535

Oliveira-Junior, J. M. B., & Juen, L. (2019). The Zygoptera/Anisoptera ratio (Insecta: Odonata): A new tool for habitat alterations assessment in Amazonian streams. *Neotropical Entomology, 48*, 552–560. https://doi.org/10.1007/s13744-019-00672-x

Oliveira-Junior, J. M. B., De Marco, P., Dias-Silva, K., Leitão, R. P., Leal, C. G., Pompeu, P. S., et al. (2017). Effects of human disturbance and riparian conditions on Odonata (Insecta) assemblages in eastern Amazon basin streams. *Limnologica, 66*, 31–39. https://doi.org/10.1016/j.limno.2017.04.007

Paiva, C. K. S., Faria, A. P. J., Calvão, L. B., & Juen, L. (2017). Effect of oil palm on the Plecoptera and Trichoptera (Insecta) assemblages in streams of eastern Amazon. *Environmental Monitoring and Assessment, 189*, 393. https://doi.org/10.1007/s10661-017-6116-y

Pereira, D. F. G., Oliveira-Junior, J. M. B., & Juen, L. (2019). Environmental changes promote larger species of Odonata (Insecta) in Amazonian streams. *Ecological Indicators, 98*, 179–192. https://doi.org/10.1016/j.ecolind.2018.09.020

Piedade, M. T. F., Schöngart, J., Wittmann, F., Parolin, P., & Junk, W. (2013). Impactos da inundação e seca na vegetação de áreas alagáveis amazônicas. In L. S. Borma & C. Nobre (Eds.), *Secas na Amazônia: causas e consequências* (pp. 268–305). Oficina de Textos.

Piovezan-Borges, A. C., Valente-Neto, F., Tadei, W. P., Hamada, N., & Roque, F. O. (2020). Simulated climate change, but not predation risk, accelerates *Aedes aegypti* emergence in a microcosm experiment in western Amazonia. *PLoS One, 15*, e0241070.

Prather, A. L. (2003). *Revision of the neotropical caddisfly genus Phylloicus (Trichoptera: Calamoceratidae), 275*, 1–214. Magnolia Press.

Price, K., Suski, A., Mcgarvie, J., Beasley, B., & Richardson, J. S. (2003). Communities of aquatic insects of old-growth and clearcut coastal headwater streams of varying flow persistence. *Canadian Journal of Forest Research, 1432*, 1416–1432. https://doi.org/10.1139/X03-089

Ramalho-Filho, A., Motta, P. E. F., Freitas, P. L., & Teixeira, W. G. (2010). *Zoneamento agro-ecológico, produção e manejo da cultura de palma de óleo na Amazônia*. Embrapa Solos.

Rico, A., Oliveira, R., Souza-Nunes, G. S., Rizzi, C., Villa, S., Vizioli, B. D. C., et al. (2022). Ecological risk assessment of pesticides in urban streams of the Brazilian Amazon. *Chemosphere, 291*, 132821. https://doi.org/10.1016/j.chemosphere.2021.132821

Rivera-Pérez, J. M., Shimano, Y., Luiza-Andrade, A., Pinto, N. S., Dias, L. G., Ferreira, K. S., et al. (2023). Effect of mining on the EPT (Ephemeroptera, Plecoptera and Trichoptera) assemblage of Amazonian streams based on their environmental specificity. *Hydrobiologia, 850*, 645–664. https://doi.org/10.1007/s10750-022-05111-4

Roque, F. O., Escarpinati, S. C., Valente-Neto, F., & Hamada, N. (2015). Responses of aquatic Saproxylic macroinvertebrates to reduced-impact logging in Central Amazonia. *Neotropical Entomology, 44*, 345–350. https://doi.org/10.1007/s13744-015-0295-4

Rosenberg, D. M., & Resh, V. H. (1993). Introduction to freshwater biomonitoring and benthic macroinvertebrates. In D. M. Rosenberg & V. H. Resh (Eds.), *Freshwater biomonitoring and benthic macroinvertebrates* (pp. 1–9). Chapman/Hall.

Rumbos, C. I., Stamopoulos, D., Georgoulas, G., & Nikolopoulou, E. (2010). Factors affecting leaf litter decomposition by *Micropterna sequax* (Trichoptera: Limnephilidae). *International Review of Hydrobiology, 95*, 383–394.

Santos, M. A., Rosa, L. P., Sikar, B., Sikar, E., & Santos, E. O. (2006). Gross greenhouse gas fluxes from hydro-power reservoir compared to thermo-power plants. *Energy Policy, 34*, 481–488. https://doi.org/10.1016/j.enpol.2004.06.015

Shimano, Y., & Juen, L. (2016). How oil palm cultivation is affecting mayfly assemblages in Amazon streams. *Annales de Limnologie, 52*, 35–45. https://doi.org/10.1051/limn/2016004

Silva-Costa, I., Giehl, N. F. S., Pozzobom, U. M., Carvalho-Soares, A. A., Oliveira-Junior, J. M. B., Cabette, H. S. R., et al. (2023). Aquatic and semiaquatic Heteroptera (Hemiptera: Insecta) distribution in streams on the Cerrado–Amazon Ecotone in headwaters of Xingu River. *Art, 1*, 13–24. https://doi.org/10.3390/arthropoda1010004

Silvério, D. V., Brando, P. M., Macedo, M. N., Beck, P. S., Bustamante, M., & Coe, M. T. (2015). Agricultural expansion dominates climate changes in southeastern Amazonia: The overlooked non-GHG forcing. *Environmental Research Letters, 10*, 104015.

Soares-filho, B., Rajão, R., Macedo, M., Carneiro, A., Costa, W., Coe, M., et al. (2014). Cracking Brazil's Forest code. *Science, 344*, 363–364. https://doi.org/10.1126/science.1246663

Sonoda, K. C., Monteles, J. S., Ferreira, A., & Gerhard, P. (2018). Chironomidae from eastern Amazon: Understanding the differences of land-use on functional feeding groups. *Journal of Limnology, 77*. https://doi.org/10.4081/jlimnol.2018.1799

Sonter, L. J., Herrera, D., Barrett, D. J., Galford, G. L., Moran, C. J., & Soares-Filho, B. S. (2017). Mining drives extensive deforestation in the Brazilian Amazon. *Nature Communications, 8*, 1013. https://doi.org/10.1038/s41467-017-00557-w

Sorribas, M. V., Paiva, R. C., Melack, J. M., Bravo, J. M., Jones, C., Carvalho, L., et al. (2016). Projections of climate change effects on discharge and inundation in the Amazon basin. *Climatic Change, 136*, 555–570.

Teyssier, A. (2002). *Criação e manejo de pontos de água para o gado da aldeia: um guia designado aos extensionistas rurais na zona sudano-saheliana*. Agromisa.

Thieme, M. L., Tickner, D., Grill, G., et al. (2021). Navigating trade-offs between dams and river conservation. *Global Sustain, 4*, e17. https://doi.org/10.1017/sus.2021.15

Trata Brasil. (2023). Ranking do Saneamento 2023 - Trata Brasil. https://tratabrasil.org.br/ranking-do-saneamento-2023/. Accessed 20.06.23.

Vannote, R. L., Minshall, G. W., Cummins, K. W. L., Sedell, J. R., & Cushing, C. E. (1980). The river continuum concept. *Canadian Journal of Fisheries and Aquatic Sciences, 37*, 130–137. https://doi.org/10.1139/f80-017

Vescovi, L., Berteaux, D. B., & Blois, S. D. (2009). Freshwater Biodiversity versus Anthropogenic Climate Change: The United Nations World Water Development Report–N° 3–2009. https://unesdoc.unesco.org/ark:/48223/pf0000185075

Villanueva, V. D., Albariño, R., & Canhoto, C. (2011). Detritivores feeding on poor quality food are more sensitive to increased temperatures. *Hydrobiologia, 678*, 155–165.

Wiens, J. J. (2016). Climate-related local extinctions are already widespread among plant and animal species. *PLoS Biology, 14*, e2001104.

Zhang, A. T., & Gu, V. X. (2023). Global dam tracker: A database of more than 35,000 dams with location, catchment, and attribute information. *Scientific Data, 10*, 111. https://doi.org/10.1038/s41597-023-02008-2

Aquatic Herbaceous Plants of the Amazonian Wetlands: Key Species Under Pressure

Aline Lopes, Jochen Schöngart, Gisele Biem Mori, Adriano Quaresma, Layon Oreste Demarchi, Sthefanie Gomes Paes, Giuliette Barbosa Mano, Michelle Gil Guterres-Pazin, Viviane Pagnussat Klein, and Maria Teresa Fernandez Piedade

Abstract

Aquatic herbaceous plants play a crucial role in Amazonian wetland ecosystems which are characterized by their unique hydrological cycles and diverse plant life. However, the fragile balance of these ecosystems is facing unprecedented challenges due to anthropogenic activities and the impacts of climate change. In this chapter, we explore the profound changes that Amazonian wetlands are

A. Lopes (✉)
Graduate Program in Clean Technologies at Cesumar University – Unicesumar. Productivity Fellow at the Cesumar Institute of Science, Technology and Innovation – ICETI, Parana, Brazil

Ecology, Monitoring and Sustainable use of Wetlands - MAUA, Brazilian National Institute of Amazonian Research, Manaus, Brazil
e-mail: aline.llopes@unicesumar.edu.br

J. Schöngart · A. Quaresma · L. O. Demarchi · S. G. Paes · G. B. Mano · V. P. Klein · M. T. F. Piedade
Ecology, Monitoring and Sustainable use of Wetlands - MAUA, Brazilian National Institute of Amazonian Research, Manaus, Brazil

G. B. Mori
Ecology, Monitoring and Sustainable use of Wetlands - MAUA, Brazilian National Institute of Amazonian Research, Manaus, Brazil

State University of Mato Grosso – UNEMAT, Nova Xavantina, Brazil

M. G. Guterres-Pazin
Ecology, Monitoring and Sustainable use of Wetlands - MAUA, Brazilian National Institute of Amazonian Research, Manaus, Brazil

Amazonian Aquatic Mammals Research Group (GPMAA), Mamirauá Institute for Sustainable Development (IDSM), Tefé, Brazil

S. S. de Souza et al. (eds.), *The Future of Amazonian Aquatic Biota*,
https://doi.org/10.1007/978-3-031-66822-7_6

undergoing and the direct effects of these changes on aquatic herbaceous vegetation. Specifically, we focused on this group of plants that inhabit three types of floodplains: the fertile white-water (várzea) and the oligotrophic black and clear-water (igapó). These distinct ecosystems have unique characteristics and constraints resulting from their natural peculiarities in synergy with anthropogenic disturbances. Combining scientific studies with field observations, we offer a comprehensive view of the characteristics and challenges faced by aquatic herbaceous plants in the Amazonian floodplains. In addition, we propose the potential impacts that projected climate change to the year 2100 will have on key aquatic macrophyte species in these environments. By highlighting the ecological importance and the vulnerabilities and threats faced by these species, we offer solutions for developing effective management plans focused on the conservation of the Amazon wetlands and their unique biodiversity.

1 Amazonian Wetlands Characteristics and Main Typologies

The Amazon wetlands are extensive low-lying areas that suffer periodic flooding due to the region's distinct hydrological cycles (Junk et al., 1989) resulting in the seasonal flooding of vast areas. Wetlands correspond to about 30% of the Amazon region, and among them, the floodplains along the large rivers are the most expressive, covering an area of about 750,000 km^2 (Wittmann & Junk, 2016).

The floodplains of the Amazon play a crucial role in maintaining the overall health and function of the rainforest ecosystem. They act as vital connectors between terrestrial and aquatic habitats (Junk et al., 2011). The main driving force in floodplains is the flood pulse which leads to one predictable aquatic and one terrestrial phases per year (Junk et al., 1989). Such dynamic is of major importance to the biota and also influences abiotic cycles in the associated plains and river systems, creating unique and diverse ecosystems along the river margins that support a wide array of plant and animal life (Junk, 2013). As a result, the floodplains serve as important feeding and nesting grounds for numerous animals, among them fish species, providing a source of sustenance for both aquatic and terrestrial predators. They also offer shelter and foraging opportunities for migratory birds and other wildlife, making them crucial stopover points during their long-distance journeys (Junk et al., 2011). Additionally, floodplains are colonized historically and currently by indigenous and traditional populations that develop low-impact activities in a multiple-use system vital for their lives and cultures (Harris, 2005; Junk et al., 2000).

According to their physical and chemical properties, the Amazonian floodplains can be divided into different typologies (Prance, 1979; Junk et al., 2011; Fig. 1). The white-water fertile floodplains (várzeas) drain geologically more recent areas from the Andean and pre-Andean region; the associated rivers have a muddy color because they carry a high load of nutrient-rich sediments. Their pH is close to neutrality, and they present a very pronounced sedimentation and erosion processes associated with the dynamic of the river waters (Junk et al., 2011). Among others,

Fig. 1 Amazon wetlands: (**a**) clear-water igapó, Tapajos River with *Luziola spruceana* Benth. ex Döll dominance; (**b**) black-water igapó, Aracá River with emergent *Oryza latifolia* Desv. dominance; (**c**) white-water várzea, Amazonas River with rooted with floating-leaves *Victoria amazonica* (Poepp.) J.C. Sowerby; (**d**) flooded white sand Campinarana, near Uatuma River with *Sobralia granitica* G.A. Romero & Carnevali and *Actinostachys pennula* (Sw.) Hook. Photos: Aline Lopes (**a**, **c**), Layon Demarchi (**b**), and Viviane Klein (**d**)

the main white-water rivers are the Solimões/Amazonas, Madeira, Purus, Japurá, Juruá, Javari, and Içá. The igapó ecosystems, on the other hand, drain very old and eroded areas, are poor in sediments and minerals, and originate mainly from the Guiana and Central Brazil shields (Junk et al., 2015); the water might be black (black-water igapós) or clear (clear-water igapós). Black-water igapós are poorer and more acidic (pH ~4–6) than the clear-water igapós, which may present highly varying pH values and intermediate mineral levels between the várzea and black-water igapós. The main black-water rivers that drain these areas are the Negro, Jaú, Vaupés, Içana, and Uatumã, among others, while the main clear-water rivers are the Tapajós, Xingú, Trombetas, and Branco. Despite their contrasting properties, várzeas and igapós share a comparable annual, predictable, long-lasting, and high-amplitude flood pulse (Junk et al., 1989). On the other hand, water transparency is limited in the várzea (<0.5 m), increases in the black-water igapós (1.3–2.9 m), and may reach higher values (1.1–4.3 m) in the clear-water igapós (Sioli, 1984).

In the várzea floodplains, during the aquatic phase, the waters carry a substantial amount of sediments and nutrients. Toward the dry season, as the waters recede, the enriched fertile soils are exposed providing a fertile substrate for plant growth. Consequently, the várzea floodplains harbor high levels of primary productivity, contributing to the overall carbon sequestration capacity of the Amazon rainforest

(Piedade et al., 1991; Ribeiro et al., 2021; Schöngart & Wittmann, 2011). Due to their pronounced sediment dynamics, the várzeas are composed of a mosaic of habitats, including forests, lakes, dikes, and other aquatic or semi-aquatic environments, each exhibiting specific ecological features (Junk et al., 2012). The duration and depth of flooding vary across the floodplain, leading to the formation of different vegetation zones (Wittmann et al., 2022). Due to the temporal fluctuations in the water level of rivers, four distinct periods can be delineated during the seasonal cycle, the receding and rising water periods, and the two extreme phases, the dry season (low water) and the flood season (high water) (Bittencourt & Amadio, 2007). During the dry season, lakes typically have a lower volume of water and may lose the connection to the associated river. Conversely, during the flood season, characteristics usually opposite to those of the dry season are observed, with a significant influx of water from rivers into lacustrine environments. This can lead to more homogeneous limnological characteristics (Thomaz et al., 2007).

In association with the degree of connectivity to the adjacent rivers, significant changes in the ecological processes of floodplain lakes can be observed throughout the seasonal cycle (Aprile & Darwich, 2023). Among these changes, particular attention can be given to those related to resource availability and the water mass mixing pattern, which is influenced, among other factors, by the thermal structure of the water column. In these environments, in addition to possible diurnal variations, which in tropical lakes may be comparable to seasonal variations in temperate lakes (Aprile & Darwich, 2023), seasonal variations become important due to the influence of the flood pulse (Junk & Piedade, 1993a). These zones are essential for understanding the distribution and dynamics of herbaceous plants in the várzea floodplains (Piedade et al., 2018).

Black and particularly clear-water igapós are still relatively poorly studied compared to the várzeas ecosystems. However, the knowledge gathered so far indicates that because they have modest sediment loads and lower minerals, aquatic herbaceous plants are reduced in both ecosystems. Contrasting between them is the presence of more subaquatic species in the clear-water where light penetration is higher (Crema et al., 2019). Altogether, it is evident that the dynamics and ecological characteristics of the three floodplains are different and understanding them is crucial to comprehend the complex interactions, adaptations, and responses of herbaceous plants to the unique conditions of these habitats.

Besides the floodplains, some wetlands harbor a few aquatic herbaceous species (Lopes et al., 2021). This is the case of the campinaranas or white-sand ecosystems (Fig. 1), with predominantly sandy, acidic, and heavily leached soils of very low fertility, and, in some areas, with toxic levels of aluminum (Adeney et al., 2016; Anderson, 1981; Mendonça et al., 2015). During the dry season, the campinarana plant community may experience physiological drought owing to the low water retention capacity of the sandy soils; during the rainy season, the root system may experience waterlogging due to high water levels, blockage of water drainage by a cemented soil layer (hardpan), or impermeable rock and clay layers (Franco & Dezzeo, 1994). Campinaranas that experience flooding due to the overflow of the water table are considered wetlands (Demarchi et al., 2022; Junk et al., 2011) and are partially addressed in this chapter. Although being annual, the flood pulse of campinaranas is of low to medium predictability, medium to long duration, and low amplitude (Junk et al., 2011). Due to their different characteristics, the richness of aquatic herbaceous species differs between the four cited ecosystems, although some species may be shared between some of them (Lopes et al., 2021).

2 Adaptations, Composition, and the Role of Common Aquatic Herbaceous Plants in Amazonian Wetlands

Aquatic herbaceous plants vary widely in their physiological and morphological characteristics. The diversity of leaves and roots with specialized structures that allow their floating or submersion in the water are particularly distinctive for their success within different aquatic habitats. For instance, floating species have special structures, such as stems that are hollow or filled with spongy tissues (aerenchyma), giving plants buoyancy capacity and allowing their aerial parts to stay above the water surface (Schulthorpe, 1967).

Amazonian aquatic herbaceous plants are specialists in the aquatic and terrestrial phases of these environments and can occupy specific ecological niches thanks to their unique physiological and morphological functions (Junk & Piedade, 1997). The species developed a series of adaptations to deal with variable water conditions, for instance, specialized aerial tissues, leaves with thick cuticles, adapted roots, floating structures, and water storage mechanisms (Coan et al., 2002; Piedade et al., 2018; Pierini & Thomaz, 2004; Rocha & Martins, 2011). Among these adaptive structures, the aerial, adventitious, and tuberous roots are distinguishing, because they act on osmotic and oxygen control during periods of drought and flood (Lopes et al., 2016a; Piedade et al., 2005). In general, to avoid transpiration, aquatic herbaceous species use different physiological mechanisms, such as the opening and closing of stomata; development of morphological traits, such as large leaves and/or leaves with protective waxy layers; and changes in the energy regulatory metabolism along the year (Schulthorpe, 1967).

Considering the prevalence of flooding in this habitat, adaptations of species to flooding are more conspicuous than those for fire (Gagnon et al., 2021; Rocha et al., 2015). Unlike adjacent upland areas, where fire plays a fundamental ecological role (Gagnon et al., 2021), it becomes a significant factor during droughts in floodplain forests (Mitsch et al., 2010). Research in the Pantanal has demonstrated that fire induces changes in floodplain floristic composition due to annual species variation. However, some species exhibit the capacity to colonize gaps and regenerate from underground organs and seeds following a fire event (Rocha et al., 2015).

Common species in the Amazonian várzea, such as the free-floating *Pistia stratiotes* L. (Araceae) and *Ludwigia helminthorrhiza* (Mart.) H. Hara (Onagraceae), have pendulous roots, hydrophobic, and aerenchymatous leaves with a large amount of trichomes, which act in osmotic control and protection against leaf deterioration (Cook et al., 1974; Costa et al., 2016; Junk & Piedade, 1997). The species *L. helminthorrhiza* is also characterized by the presence of aerenchyma in the roots (Junk,

1986; Pott & Pott, 2000). Another species with hydrophobic leaves that act by repelling water is *Salvinia auriculata* Aubl. (Salviniaceae), a floating macrophyte that grows fast and forms a dense cover on the surface of the water (Piedade et al., 2010).

Eichhornia crassipes (Mart.) and *Eichhornia azurea* (Sw.) Kunth (Pontederiaceae) have large, rounded leaves that form a dense covering on the surface of the water that efficiently captures sunlight, even in environments with high competition for light. While *E. crassipes* is more commonly found in the white-water rivers of Central Amazon, *E. azurea* occurs in higher density in clear-water rivers, such as the Tapajós. The root system of *E. crassipes* is dense and can make up to 40–50% of the total plant's biomass (Piedade et al., 2010), acting as structural support and allowing the plant to obtain nutrients directly from the water column and both the leaf and the petiole have aerenchyma, responsible for the plant's floatation (Barrett & Seaman, 1980; Cook et al., 1974; Gopal & Sharma, 1981; Junk & Piedade, 1993b; Kissmann, 1991; Ritter et al., 2017; Santos & Thomaz, 2007). The species *Echinochloa polystachya* (Kth) Hitchc (Poaceae) is an aquatic grass with high phenotypic plasticity that has its morphology modified according to the environment (Fig. 2). It has characteristics such as leaves and spikelets with trichomes and a blade filled with aerenchymatous tissue that facilitates floating. *E. polystachya* has the incredible characteristic of following the annual rise of water with its growth, where plants over 15 m long have been recorded (Piedade et al., 1991).

Differently, submerged species have leaves adapted to carry out photosynthesis even when completely immersed in water. In this case, the leaves tend to be thinner and more flexible, allowing them to move with the currents and adapt to changes in the depth of the water body. Specific structural characteristics can be found in the carnivorous submerged herbaceous *Utricularia foliosa* L (Lentibulariaceae), which has gelatinous-looking leaves when young and robust aerchymatous stolons (Fig. 2). The species uses utricles to feed on zooplankton and phytoplankton (Albuquerque, 1981; Junk & Howard-Williams, 1984; Junk & Piedade, 1993b; Oliveira Mota & Zappi, 2018; Santos & Thomaz, 2007; Walker, 2004).

The famous Amazonian giant water-lily *Victoria amazonica* (Poepp.) J.C. Sowerby (Nymphaeaceae) has adapted leaves, roots, and limbs (Fig. 2). Their leaves have a layer of wax that helps to reduce water loss by transpiration; the leaf blade has drainage pores capable of acting in the drainage of water when in excess, and the petiole has thorns and air channels that facilitate floating. The species has a rhizome and long roots attached to the substrate from which nutrients are obtained (Cook et al., 1974; Junk, 1986; Junk & Piedade, 1997; Pott & Pott, 2000; Prance, 1974; Ritter et al., 2017; Rosa-Osman et al., 2011; Schulthorpe, 1967).

The number of aquatic herbaceous plants recorded so far in the Amazonian várzea floodplains is comparatively high, reaching approximately 400 species (Junk & Piedade, 1993a, 1997). Numbers in black and clear-water igapós and campinaranas are much more modest (33; 49; 11, respectively; Table 1; Fig. 3) and derive from fewer inventories. By comparing data from the black-water igapós and campinaranas of the Jaú National Park (JNP), the Uatumã Sustainable Development Reserve (USDR), and herbaceous wetland plants inventories of clear-water igapó (Crema et al., 2019) and várzea (Junk & Piedade, 1993a), Lopes et al. (2021) found that the

Fig. 2 Common aquatic macrophytes species at flooded amazon wetlands: (**a**) *Salvinia auriculata*, (**b**) *Montrichardia arborescens* (aerenchyma and hypertrophic stem lenticels), (**c**) *Cabomba aquatica*, (**d**) *Eichhornia azurea*, (**e**) *Alternanthera aquatica*, (**f**) *Victoria amazonica (fecunde flower)*, (**g**) *Utricularia foliosa*, (**h**) *Echihoclhoa polystachya*. Photos: Aline Lopes (**a**, **b**, **c**, **d**, **e**, **g** and **h**); Jochen Schöngart (**f**)

flora composition of the JNP igapó is closely related with the Central Amazonian várzea with eight species in common. However, the campinarana of the JNP only shared three species with the same ecosystem at the Viruá National Park (VNP) (Costa et al., 2016). The black-water igapó at USDR had 19 species found in the várzea and eight in the campinarana of the VNP, while the campinarana of the USDR shared only two out of ten species with the same ecosystem at the VNP. In

Table 1 Number of species and water characteristics of four different Amazon wetlands

Environment	Number of species	pH	Conductivity (μS/cm)	Turbidity (NTU)	Authors
Igapó (clear water), Tapajós River	49	6.56	14.33	8.47	Crema (2017)
Igapó (black water), Jaú River	33	4.42	14.50	3.64	Lopes et al. (2021)
Várzea, Solimões River	388	6.8	65.97	65.78	Junk and Piedade (1993a, 1997)
Campinarana, Jaú National Park	11				Lopes et al. (2021)

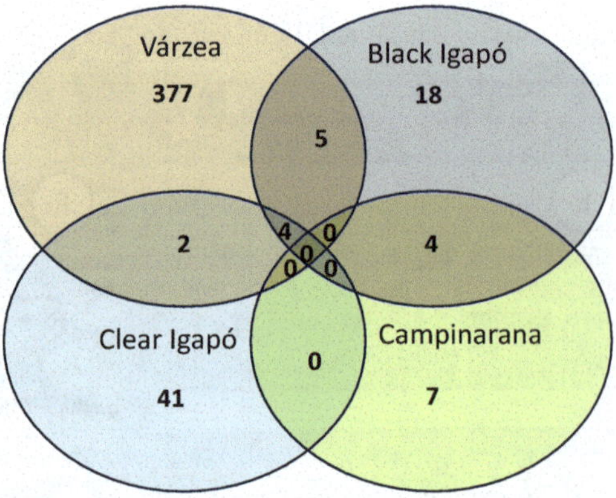

Fig. 3 Venn diagram of aquatic herbaceous composition from Solimões/Amazonas white-water várzea, black-water igapó and campinarana from Jaú National Park (Lopes et al., 2021), and clear-water igapó from Tapajós (Crema, 2017)

the clear-water igapó of the Tapajós River (Crema, 2017), they found 49 species, of which only eight species were found in the black-water counterpart igapós analyzed (Fig. 3; Table 1). Most black-water igapó species had ample distribution across the Neotropics, while most campinaranas species were restricted to the Amazon Basin (Lopes et al., 2021).

2.1 Food Chains and the Aquatic Herbaceous Plants

In the fertile várzea floodplains, during periods of growth, herbaceous aquatic species, mainly grasses, retain nutrients in biomass, releasing them during decomposition; this turnover affects the electrolytic balance and increases the productivity in Central Amazonia lakes (Furch et al., 1983). The large extension occupied and their massive production of biomass result in the key position of the aquatic herbaceous plants as habitats for aquatic and terrestrial fauna. Therefore, these communities are considered the most densely populated habitats of the várzea water bodies (Junk, 1973); when oxygen is not limiting and the amounts of inorganic particles in suspension is low, an impressive number of invertebrates (780,000 ind. m^{-2}) are found associated to stands of aquatic plants in the várzea (Junk & Robertson, 1997). A large number of terrestrial invertebrates also colonize herbaceous plants associated with the terrestrial phase of the hydrological cycle, as resumed in Junk and Piedade (1997) and Piedade et al. (2010). To the black and clear-water igapós, and also to the campinaranas, there are no comparable studies analyzing the interactions with aquatic herbaceous plants and invertebrates.

Among vertebrates, fish, especially the young stages, constitute the most important group associated with aquatic herbaceous plants. They use them as habitat, as shelter to avoid predation, and to gather food in roots, leaves, and stems. Some fish species exhibit feeding flexibility in response to spatial and temporal patterns in food source availability (Mortillaro et al., 2015). While floodplains support a high biomass of C_4 grasses, relatively little carbon from this source seems to be assimilated by fishes (Jepsen & Winemiller, 2007). In the Amazon River, C_3 plants serve as the primary carbon source for 34 fish species, accounting for an average of 82.4% to 97.5% of the carbon in all species (Forsberg et al., 1993). However, in the Pantanal, C_4 aquatic macrophytes are the primary food source for fish, such as *Prochilodus lineatus* (Prochilodontidae, Characiformes), particularly during the flood season (Lopes et al., 2007). For birds, the environmental heterogeneity created by different water bodies is an important determinant of waterfowl community composition, where many species feed and breed in aquatic herbaceous vegetation (Cintra et al., 2007).

The most important mammals consuming large amounts of aquatic herbaceous plants are the Manatee (*Trichechus inunguis* Natterer, 1883) and capybara (*Hydrochoerus hydrochaeris* L., 1766), although the present populations of these animals have declined as a result of intensive hunting in the Amazon Basin. Also, the large river turtles (*Podocnemis* spp.) are subject to intensive hunting, and the eggs are predated by animals and humans (Junk & da Silva, 1997). The manatee deserves a deeper analysis because this aquatic mammal which lives its entire life in the water colonizes floodplains with distinct physical and chemical properties: black and clear-water (igapós) and white-water (várzeas). In these three ecosystems, this herbivore consumes macrophytes, roots, and vegetation from floodplains (Crema, 2017).

3 Vulnerability of Wetlands to Natural and Anthropogenic Disturbances

Wetlands are susceptible to natural disturbances, including extreme hydroclimatic events such as droughts and floods (Barichivich et al., 2018; Piedade et al., 2013), which disrupt its ecological balance with consequently cascading effects on the herbaceous plant communities. Extreme meteorological events can cause physical damage to the forests, leading to tree falls, increasing mortality rates, and susceptibility to fire (Carvalho et al., 2021). Floods, especially those of higher magnitude or altered timing, can submerge the vegetation for extended periods, affecting plant growth and survival (Piedade et al., 2013; Resende et al., 2020). Moreover, natural disturbances can impact the reproductive cycles and breeding patterns of aquatic plant species that rely on the dynamics of the várzea and igapó ecosystem, potentially affecting their populations and overall biodiversity. Although less pronounced in amplitude and duration, flooding in campinaranas is an important factor, influencing the richness, structure, and composition of its plant species (Damasco et al., 2013; Pivello, 2011; Targhetta et al., 2015; Veldman & Putz, 2011). The intensity and duration of flooding can be affected by the climate regime, altitude, soil depth, and proximity to a watercourse.

Deforestation in the Amazon region has reached alarming rates in recent decades due to many human activities such as agriculture, mining, construction of hydroelectric plants, fire, logging, and urban expansion. The clearing of vast areas of forest has profound consequences for the floodplains and their herbaceous plant communities. When forests are removed, the natural vegetation that helps regulate the hydrological cycles of the region is lost, resulting in altered patterns of rainfall and flooding, disrupting the intricate balance between the floodplains and the surrounding forests, and affecting the availability of water, the nutrient cycling, and increased sediment deposition (see Schöngart et al., 2025, Chapter 7). The risk of erosion increases, leading to augmented sedimentation in rivers and floodplain areas, leading to habitat loss and fragmentation in the wetland areas.

Mining activities, particularly illegal gold mining, pose a severe threat to the igapó ecosystems (Goulding et al., 1996), for instance, soil erosion, sedimentation in water bodies, and pollution by mercury and other toxic chemicals. These activities not only directly impact herbaceous plant communities but also contaminate water sources, affecting other aquatic species and the overall ecological integrity of the ecosystem (Castello & Macedo, 2016).

Anthropogenic disturbances also contribute to the introduction and spread of invasive species that in the várzea floodplains may lead to outcompete native vegetation and disrupt ecological interactions, potentially interfering in the successional processes of the wood vegetation (Mano et al., 2023). Invasive species can alter nutrient cycling, reduce biodiversity, and impact the habitat suitability for herbaceous plants adapted to the natural conditions of the wetlands (Lopes et al., 2023).

3.1 Changes in Wetlands Affecting Aquatic Herbaceous Plants

Natural and anthropogenic disturbances in wetlands lead to habitat degradation and loss, directly affecting the distribution and abundance of aquatic herbaceous plants adapted to flooding conditions which play key functions within these ecosystems (Esteves & Camargo, 1986; Junk & Piedade, 1993a). The decline of key species can have cascading effects on the entire ecosystem. Many herbaceous plants, particularly in the várzea floodplains, provide important ecological services, such as nutrient cycling, soil stabilization, and habitat provisioning for other organisms (Piedade et al., 2000). The loss of these species can disrupt ecosystem services and have implications for the survival of associated wildlife including, among others, birds, amphibians, and fish. Furthermore, changes in species composition and ecosystem structure can alter ecological interactions, such as pollination and seed dispersal, potentially leading to further declines in plant diversity and hindering the natural regeneration processes within the várzea, igapó, and campinarana ecosystems.

Habitat alteration, particularly through deforestation and land use changes, is a primary driver of the loss of species diversity in the Amazonian floodplains (Latrubesse et al., 2021; Fig. 4). The conversion of natural habitats to agricultural fields, urban areas, and infrastructure disrupts the intricate ecological balance of the floodplain ecosystems, leading to the decline and local extinction of many plant species. Habitat alteration fragments and isolates populations, reducing gene flow and increasing the risk of inbreeding depression. Fragmented habitats also limit the available area for species to establish and expand their populations, further exacerbating the loss of genetic diversity (Martins et al., 2016).

The deposition of sediments can alter the composition of the floodplain soil and impact the growth and survival of herbaceous plants adapted to specific nutrient conditions (Bleich et al., 2014). Deforestation reduces the connectivity between the floodplains and the upland areas, disrupting ecological corridors for plant and animal dispersal. This fragmentation can hinder the natural regeneration of floodplain vegetation and limit the ability of species to adapt to changing environmental conditions.

Fig. 4 (**a**) Burnt area in igapós of Jaú National Park, (**b**) deforestation in the igapós of Tapajós River, (**c**) cattle in the várzea of the Amazon River, (**d**) Cururu River, affluent of Negro River, contaminated with oil. Photos: Jochen Schöngart (**a**, **b**), Layon Demarchi (**c**), and Aline Lopes (**d**)

More specifically for campinarana environments, the illegal extraction of sand for civil construction has caused many impacts in several areas, mainly close to large urban areas and large rivers (Ferreira et al., 2013). After removing the first horizons of soil, only white sand remains with almost no organic matter, completely changing the composition and dynamics of these ecosystems, making recovering very difficult.

Amazonia is experiencing an increase in the frequency of extreme droughts and wildfires (Fearnside et al., 2021). Fires are another significant consequence of human activities in the Amazon, particularly during the dry season when vegetation is more susceptible to ignition. These fires, often set intentionally for land clearance or agricultural practices (Barni et al., 2021), can spread into the floodplain areas and have detrimental effects on herbaceous plant communities (Silveira et al., 2022). Fires in the floodplains can cause immediate loss of vegetation, including both aquatic and terrestrial species. In the Pantanal variation in flood, duration and the occurrence of fires interact synergistically to decrease stem numbers during regeneration and alter species composition and distribution (Oliveira et al., 2014). In the Amazon, more than 90% of the area burned in igapós of the JNP was associated with extreme hydroclimatic conditions during El Niño years (Carvalho et al., 2021). Around 79% of the 254 fires detected in this igapó area occurred close (<10 km away) to human settlements, suggesting that human activities are the main source of ignition (Carvalho et al., 2021).

The intense heat can destroy seeds, seed banks, and underground rhizomes, severely impacting the ability of herbaceous plants to regenerate. Fire together with floods impacts seed bank richness, abundance, and diversity, influencing seasonal regeneration and plant community structure in the Pantanal (Kohagura et al., 2020). Recent burns reduce seed abundance due to plant mortality, although they can break seed dormancy, particularly in legumes. Flooding and fires lower seed bank abundance and species richness, but their interaction does not affect these factors (Kohagura et al., 2020). Additionally, fires release large amounts of carbon dioxide into the atmosphere, contributing to greenhouse gas emissions and exacerbating climate change. The ecological effects of fires extend beyond the immediate destruction of vegetation. The altered fire regimes can lead to shifts in species composition, favoring fire-adapted plants and potentially reducing the diversity and abundance of herbaceous species that are not adapted to fire (Feng et al., 2021; van Marle et al., 2017). The loss of vegetation also affects the habitat suitability for wildlife, including aquatic organisms that depend on the floodplains for breeding and foraging (see Chapter 7, Schöngart et al., 2025).

Urbanization in the Amazon, while concentrated in specific regions, has already begun to manifest in observable changes in aquatic ecosystems, notably an increase in aquatic plant proliferation due to eutrophication. Eutrophication is a prevalent consequence of the urbanization process, particularly evident in stagnant water environments like reservoirs designated for public water supply. In these aquatic settings, algae naturally extend their reach, driven by the presence of nitrogen, phosphorus, carbon dioxide, and various organic decomposition byproducts. However, this natural phenomenon can rapidly transform into an uncontrolled process when excessive nutrient discharges into the water lead to the overgrowth of aquatic vegetation, both within the water column and on the water surface. This is the case of the Amazonian species *E. crassipes*, considered the most invasive aquatic plant in the world (Gopal & Sharma, 1981). This transformation is commonly termed anthropic (or cultural) eutrophication, representing a disruption of the natural environmental characteristics, which in turn affects multiple water uses and, consequently, the ecosystem services they provide. These services encompass vital functions such as public water supply and the preservation of aquatic life. Human activities hold the potential to significantly influence the global distribution of macrophyte species and subsequently impact freshwater habitats worldwide. Therefore, a comprehensive evaluation of these influences should consider their distinctive attributes, geographical locations, management techniques, and global significance (Pompêo, 2008).

The construction of hydroelectric dams in the Amazon region has had significant impacts on the floodplains and the herbaceous plant communities that depend on the natural hydrological cycles. Dams alter the flow patterns of rivers, interrupting the natural flooding and receding cycles that are essential for the maintenance of the integral functioning of floodplain ecosystems (Schöngart et al., 2021). The regulation of river flow by dams can lead to reduced or prolonged periods of inundation in the floodplains, disrupting the ecological processes and habitat availability for aquatic herbaceous plants. Changes in water levels can affect seed germination, plant growth, and nutrient availability, leading to shifts in species composition and reduced biodiversity (Lobo et al., 2019; Schöngart et al., 2021). Furthermore, the construction of dams often involves the flooding of large areas of land, resulting in the direct loss of floodplain vegetation and habitats. The inundation of previously terrestrial areas can also lead to the decomposition of organic matter, releasing greenhouse gases into the atmosphere and contributing to climate change (Fearnside, 1995).

Extraction and transportation of petroleum in the Amazon region is another important problem that may affect directly and indirectly the floodplain ecosystem and its herbaceous plant communities (Lopes et al., 2009; Lopes & Piedade, 2014). Petroleum is transported from the Urucu River province in central Amazonia to the refinery in Manaus by ship (about 650 km fluvial distance). Oil spills from extraction and transportation activities can contaminate water bodies, soil, and vegetation in the floodplains, leading to long-term ecological impacts. Because the richness of aquatic herbaceous plants recorded in the Amazonian várzea floodplains is high, reaching approximately 400 species (Junk & Piedade, 1993a; Junk & Robertson, 1997), in the case of accidents involving oil spills, several of these species could suffer significant negative impacts. Therefore, it is evident that studies evaluating the effects of Urucu oil on other unstudied species of aquatic herbaceous plants are essential. However, considering the wide distribution, biomass production, and ecological role of the species already assessed, particularly *Echinochloa polystachya* (H.B.K) Hitchcock and *Eichhornia crassipes*, some scenarios regarding potential massive contamination of these plants can be discussed.

A potential oil spill in the Amazonian floodplains, particularly in the Solimões and Urucu River várzeas, would have significant impacts due to the importance of these environments for nutrient cycling and economic activities such as fishing, agriculture, and tourism (Lopes et al., 2020; Fig. 5). Studies carried out on aquatic plants *Echinochloa polystachya*, *Eichhornia crassipes,* and *Pistia stratiotes* have shown the high toxicity of Urucu oil, affecting their reproduction, growth, and survival (Lopes et al., 2009; Lopes & Piedade, 2014; Silva & Camargo, 2007). The sensitivity of these plants varied depending on their habits and characteristics, such as the presence of calcium oxalate crystals in *P. stratiotes*, which may influence the absorption of chemical elements. Compared to other types of oil, Urucu oil has proven to be highly toxic to the aquatic plants in the region (Lopes et al., 2020). Also, the period of contamination may play an important role in the toxic effects of Urucu's crude oil. For instance, Lopes and Piedade (2010) evaluated in a greenhouse the growth of *Echinochloa polystachya* planted in 2 L pots of floodplain soil contaminated with increasing dosages of Urucu oil. Two experiments were carried out in a period of the year with lower temperatures (July–September) and higher ones (September–November). Planting time influenced plant response, with more negative effects at higher temperatures, indicating that a spill in the period of higher temperatures may be more detrimental to the species and the ecosystem. Studies on the effects of oil spills in black and clear-water ecosystems are not available.

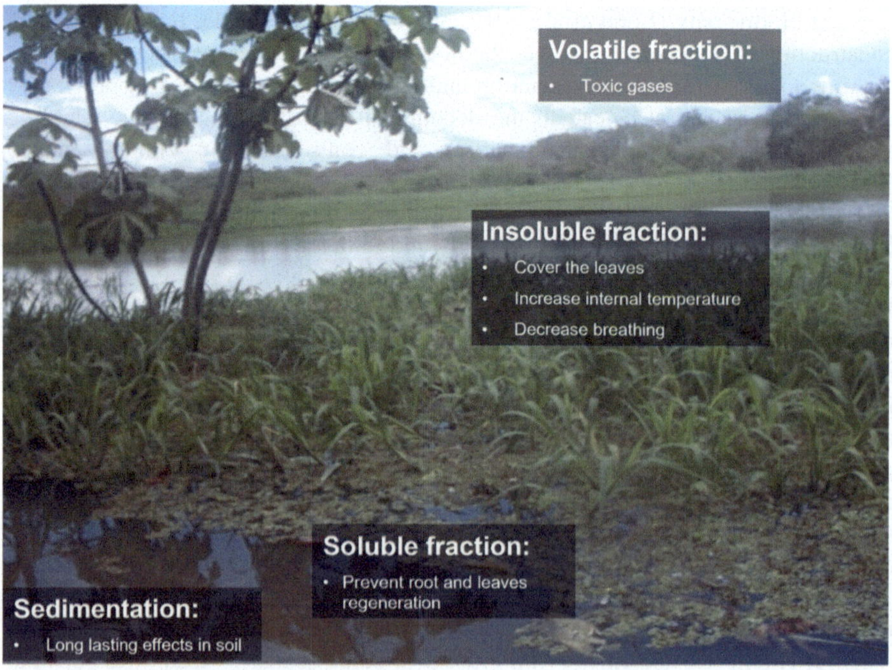

Fig. 5 Simplified scheme of oil contamination in different components of a várzea floodplain in Central Amazonia. Photo: Aline Lopes

3.2 Genetic Diversity and Adaptability of Aquatic Herbaceous Species

Additionally, to morphological diversity, aquatic herbs have great genetic diversity that increases their capacity to adapt to a wide range of different environmental conditions and extreme events, such as droughts and floods (Junk & Piedade, 1997; Piedade et al., 2013; Vieira et al., 2007). Genetic variations can lead to the formation of subspecies or ecotypes, which have species specific to certain habitats within the Amazon region (Piedade et al., 2013, 2020). High genetic variability within populations of Amazonian aquatic plants was observed for the species *Eichhornia crassipes*, *Victoria amazonica*, *Pistia stratiotes,* and *Cabomba* spp. (Cardoso et al., 2002, 2005; Martins et al., 2003).

Although Amazonian aquatic herbaceous plants developed a diversity of adaptive mechanisms to occupy different habitats, climate change may negatively affect their growth and development by influencing their photosynthetic metabolism, leaf length, and biomass production. In controlled experiments with elevated CO_2 and temperature, the species *Montrichardia arborescens* (L.) Schott showed reduced photosynthesis rates and inhibited biomass production (Lopes et al., 2016b). Similar results were also observed for the species *P. stratiotes* and *E. crassipes* (Lopes et al., 2018; Souza et al., 2021).

4 Climate Change Impacts on Aquatic Herbaceous Species in Wetlands

4.1 Climate Change Projections for the Amazon Region

The Amazon region is expected to experience significant climate changes in the coming decades. Climate models predict a warming trend in the Amazon with rising average temperatures and more frequent heatwaves (Castellanos et al., 2022). Marengo and Souza Jr (2018) point out an elevation in temperature of 0.6 to 0.7 °C for the region between the years 1949 and 2017. Through complex climate models presented by the Intergovernmental Panel on Climate Change (IPCC), an increase of more than 4 °C in the average air temperature in the Amazon is expected by the end of the twenty-first century, but rainfall is expected to vary geographically, with an increase in precipitation for the western Amazon region and a decrease in the eastern portion (Alves & Marengo, 2010; IPCC, 2022; Marengo & Souza Jr, 2018). Rising atmospheric CO_2 levels due to global warming can influence plant physiology and ecosystem dynamics (Castello & Macedo, 2016; Ribeiro et al., 2021).

Several studies have shown an increase in the frequency of extreme events along the Amazon Basin, such as increased or reduced rainfall in recent decades (Nobre et al., 2007, 2016; Espinoza et al., 2009, 2014; Marengo et al., 2012, 2013; Aleixo et al., 2019; Chevuturi et al., 2023). The impacts resulting from the growing frequency of these events can range from increased risk of fires, extreme temperature increases, as well as droughts and floods, causing major impacts both for human communities and for local flora and fauna (Marengo & Espinoza, 2016; Nobre et al., 2007; Piedade et al., 2013). In fact, the droughts in the Amazon in 1997/1998, 2005, 2010, and 2015/2016 were so intense that, due to water stress, large trees died, and there was also a rise in the number of outbreaks of forest fires, in addition to the release of large amounts of carbon into the atmosphere (Barichivich et al., 2018; Marengo & Espinoza, 2016; Nobre et al., 2016). In 2012, 2014, 2015, and 2021, the intense increase in rainfall caused large floods in several states, such as Acre and Rondônia, flooding cities, roads, and farms, resulting in several damages to homes and to the livelihood of the population (Barichivich et al., 2018; Espinoza et al., 2022; Marengo et al., 2013). Hydrological data show a trend toward an increase in extreme events in the Amazon region; a fact consistent with models developed by the IPCC, which demonstrate the same trend for the future (Nobre et al., 2007; Marengo & Espinoza, 2016; Marengo & Souza Jr, 2018, see more details in Chapter 7, Schöngart et al., 2025).

These climate changes are likely to have profound impacts specifically in floodplains, because they affect the regular pattern of seasonal flooding and water dynamics for their ecological functioning (Castello & Macedo, 2016). Alterations in temperature and precipitation patterns can disrupt the natural flood pulse, affecting the timing, duration, and intensity of floods. Such changes can have cascading effects on the vegetation, including aquatic macrophyte species, and the overall structure and functioning of the wetland's ecosystem (Dhir, 2015; Hossain et al., 2017), including their complex and vital food webs. Aquatic ecosystems are among the most vulnerable to climate change, and an increase in the temperature of rivers is expected as a result of the increase in air temperature and the balance between these two environment variables (Hossain et al., 2017).

4.2 Key-Species of Herbaceous Aquatic Species and Their Vulnerability

Herbaceous aquatic species play a fundamental role in structuring, functioning, and offering ecosystem services for the aquatic environment (O'Hare et al., 2018). The Amazonian wetlands are home to a diverse array of aquatic macrophyte species, each with unique ecological traits and adaptations (Junk & Piedade, 1997). Aquatic macrophytes can influence both the hydrology and the geomorphology of aquatic ecosystems, in addition to influencing physical processes such as the transport of solutes, sediment deposition, and penetration of light into the water, among others (Junk & Piedade, 1997; Piedade et al., 2000; Worbes et al., 1992), often considered engineering plants. Certain aquatic macrophyte species may be particularly susceptible to climate change due to their precise habitat requirements, life cycle characteristics, and physiological limitations (Nilsson & Berggren, 2000). Species that depend on specific flood regimes, temperature ranges, or nutrient availability may face challenges as these conditions shift under climate change (Salimi et al., 2021).

Lopes et al. (2021) studying aquatic herbs from igapó and campinarana showed that the potential distribution range of most campinaranas species was restricted to the Amazon Basin compared to the igapó that presented a wider potential distribution range in the Neotropics. These results are congruent with studies that described similar distribution patterns for tree species, which emphasizes a high degree of endemism in Amazonian campinarana (Adeney et al., 2016; Demarchi et al., 2022). In their study, Souza et al. (2021) emphasized the adverse effects of climate change on two abundant macrophyte species, *Pistia stratiotes* and *Eicchornia crassipes*. These species exhibited a reduced growth and survival rate when subjected to high concentrations of CO_2 and elevated temperatures; a pattern that has been documented for other macrophyte species as well along the Amazonian wetlands (Lopes et al., 2017). Furthermore, climate change has the potential to induce shifts in the geographical range and alterations in the habitats of macrophyte species in the floodplains (Dang et al., 2021). Alterations in temperature and precipitation patterns can cause suitable habitats to relocate, thereby forcing certain species to migrate to new areas or experience range contractions. This phenomenon has been observed for two native grasses within the Amazon basin, *Echinochloa polystachya* and *Paspalum fasciculatum* Willd. ex. Fluegge, which have exhibited a decline in their suitability area when facing climate change for the short (2040) and long (2100) term under different scenarios of greenhouse gas emission (Mano et al., 2023). Such shifts can have significant implications for the overall structure and functioning of floodplain ecosystems, affecting nutrient cycling, sediment and succession dynamics, and the availability of habitats for other organisms (Mano et al., 2023; Piedade et al., 1991).

The expansion of invasive species also poses a significant threat to the Amazon region. Alien species, known for their competitive abilities, have the potential to eliminate native species, and it has already been documented that they can benefit from global warming (Damasceno & Fidelis, 2020). In a recent study with ten invasive grass species, Lopes et al. (2023) found that, although their responses to climate change may vary, all species are capable of effectively establishing themselves in various Neotropical regions, overlapping their occurrence with those of native species, thus representing a threat to the natural environment. Moreover, their remarkable adaptability, rapid growth, and high reproductive rates increase their chances of thriving amidst changing climatic conditions (Xavier et al., 2017).

5 Conservation and Management Strategies

Considering the potential impacts of climate change on aquatic macrophyte species in the Amazonian wetlands, effective conservation and management strategies are essential. Protecting and restoring habitats for aquatic herbaceous species is crucial for their long-term survival (Piedade et al., 2010). Conservation efforts should focus on maintaining the connectivity and integrity of wetland habitats considering the ecosystem characteristics, as well as identifying and protecting key areas that are likely to remain suitable under future climate scenarios (Nielsen et al., 2020). Restoration initiatives can also play a vital role in recovering degraded habitats and providing additional opportunities for herbaceous colonization (Piedade et al., 2010).

Monitoring and early warning systems are essential for detecting changes in macrophyte species and their habitats (Fares et al., 2021). Regular monitoring of species distributions, population sizes, functional diversity, and ecological responses can provide valuable data for understanding the impacts of climate change and informing management responsible for allowing science-based decisions (Bellard et al., 2012). Early warning systems can help identify critical thresholds or indicators of change, allowing for timely intervention and adaptation measures (Silva et al., 2010).

In summary, addressing the challenges posed by climate change on macrophyte species in the Amazonian floodplains requires a combination of proactive conservation efforts, adaptive management strategies, and ongoing monitoring. By implementing these approaches, we can enhance the resilience of macrophyte species and the overall ecological integrity of the floodplain ecosystem in the face of a climate change.

6 Conclusions and Final Remarks

In this chapter, we have delved into the intricate world of the Amazonian wetlands and explored the various aspects that shape the dynamics and vulnerability of the ecosystem. Through a comprehensive examination of aquatic herbaceous plants, anthropogenic actions, hydrological changes, loss of diversity, climate change impacts, and conservation strategies, we have gained valuable insights into the challenges and opportunities for safeguarding the future of the wetlands. One of the key findings of this chapter is the profound influence of human activities on the wetlands. Deforestation, fires, mining, hydropower dams, and other anthropogenic actions as well as synergies and feedback loops with climate change have significantly altered the hydrological cycles, vegetation patterns, and overall ecological functioning of the wetlands with a high spatiotemporal variation across the Amazon Basin. These changes have led to the loss of biodiversity, disruption of ecological services, and increased vulnerability of key species.

Understanding and addressing these challenges requires interdisciplinary research and interinstitutional collaboration. The complexity of the wetlands' ecosystem requires the integration of ecological, hydrological, climatological, and socioeconomic perspectives. By bringing together scientists, policymakers, and local communities, holistic approaches that balance human needs with the preservation of the wetlands' ecological integrity can be developed. Policymakers have a crucial role to play in enacting measures that promote sustainable land use practices, protect critical habitats, and mitigate the impacts of climate change. This requires informed decision-making based on scientific evidence and a long-term vision for the preservation of the wetlands' unique biodiversity and ecosystem services.

Research in wetlands must be encouraged as a priority to improve the knowledge of the diversity of these ecosystems and their response to the increasing environmental changes. By conducting innovative studies, monitoring the health of key species, and assessing the effectiveness of conservation strategies, we can provide the necessary information for evidence-based decision-making and adaptive management.

Local communities living in and around the wetlands are vital partners in conservation efforts. Their traditional knowledge, practices, and close relationship with the environment can contribute significantly to the sustainable management of the wetlands. Engaging and empowering local communities in conservation initiatives not only ensures their well-being but also enhances the resilience of the ecosystem.

The Amazonian wetlands are a unique and valuable ecosystem that faces numerous challenges related to human activities and climate change. However, by understanding the intricacies of the system, promoting interdisciplinary collaboration, and taking collective action, we can strive toward a sustainable future for the wetlands. The preservation of this remarkable ecosystem is not only essential for the plants, animals, and local communities that depend on it but also for the global ecological balance. Let us join hands in protecting and conserving the Amazonian wetlands for generations to come.

Acknowledgements This study has been conducted in the framework of the Long-Term Ecological Research Program (PELD) of the Brazilian National Council for Scientific and Technological Development (CNPq), performed since 2012 by the research group MAUA (Ecology, Monitoring and Sustainable Use of Wetlands) of the Brazilian National Institute for Amazon Research (INPA). We acknowledge funding support from the Long-term Ecological Research Program (PELD) by the Brazilian National Council for Scientific and Technological Development (CNPq; grant numbers 403792/2012-6, 441590/2016-0, 441811/2020-5) and the Amazonas Research Foundation (FAPEAM; grant numbers 062.00466/2014; 062.01357/2017). We further acknowledge support by the National Institute of Science and Technologies in Adaptations of the Amazon Aquatic Biology (INCT-ADAPTA) during the first (CNPq grant number 465540/2014-7; FAPEAM grant number 062.1187/2017) and second (CNPq grant number 465540/2014-7) phases. This study was financed in part by the Coordination for the Improvement of Higher Education Personnel—Brazil (CAPES; Finance Code 001). We further received funding and support from the Technical/Scientific Cooperation between INPA and the Max-Planck Society with a focus on floodplain forests.

References

Adeney, J. M., Christensen, N. L., Vicentini, A., & Cohn-Haft, M. (2016). White-sand ecosystems in Amazonia. *Biotropica, 48*, 7–23. https://doi.org/10.1111/btp.12293

Albuquerque, B. W. P. (1981). Plantas forrageiras da Amazônia. *Acta Amaz, 11*, 457–471.

Aleixo, I., Norris, D., Hemerik, L., et al. (2019). Amazonian rainforest tree mortality driven by climate and functional traits. *Nature Climate Change, 9*, 384–388. https://doi.org/10.1038/s41558-019-0458-0

Alves, L. M., & Marengo, J. (2010). Assessment of regional seasonal predictability using the PRECIS regional climate modeling system over South America. *Theoretical and Applied Climatology, 100*, 337–350. https://doi.org/10.1007/s00704-009-0165-2

Anderson, A. B. (1981). White-sand vegetation of Brazilian Amazonia. *Biotropica, 13*, 199–210. http://www.jstor.org/stable/2388125

Aprile, F., & Darwich, A. J. (2023). Lake stratification in the Brazilian Amazon: Types and the first description of permanent meromixia. *Journal of South American Earth Sciences, 124*, 104224. https://doi.org/10.1016/j.jsames.2023.104224

Barichivich, J., Gloor, E., Peylin, P., et al. (2018). Recent intensification of Amazon flooding extremes driven by strengthened Walker circulation. *Science Advances, 4*, eaat8785. https://doi.org/10.1126/sciadv.aat8785

Barni, P. E., Rego, A. C. M., Silva, F. C. F., et al. (2021). Logging Amazon forest increased the severity and spread of fires during the 2015–2016 El Niño. *Forest Ecology and Management, 500*, 119652. https://doi.org/10.1016/j.foreco.2021.119652

Barrett, S. C. H., & Seaman, D. E. (1980). The weed flora of Californian rice fields. *Aquatic Botany, 9*, 351–376. https://doi.org/10.1016/0304-3770(80)90036-4

Bellard, C., Bertelsmeier, C., Leadley, P., et al. (2012). Impacts of climate change on the future of biodiversity. *Ecology Letters, 15*, 365–377. https://doi.org/10.1111/j.1461-0248.2011.01736.x

Bittencourt, M. M., & Amadio, S. A. (2007). Proposta para a identificação rápida dos períodos hidrológicos em áreas de várzea do rio Solimões—Amazonas nas proximidades de Manaus. *Acta Amaz, 37*, 303–308. https://doi.org/10.1590/S0044-59672007000200019

Bleich, M. E., Mortati, A. F., André, T., & Piedade, M. T. F. (2014). Riparian deforestation affects the structural dynamics of headwater streams in southern Brazilian Amazonia. *Tropical Conservation Science, 7*, 657–676.

Cardoso, L. R., Martins, D., Kuramae, E. E., et al. (2002). Variabilidade genética de acessos de aguapé coletados no Estado de São Paulo. *Planta Daninha, 20*, 1–5.

Cardoso, L. R., Martins, D., Mori, E. S., & Terra, M. A. (2005). Variabilidade genética entre populações de *Pistia stratiotes*. *Planta Daninha, 23*, 181–185.

Carvalho, T. C., Wittmann, F., Piedade, M. T. F., et al. (2021). Fires in Amazonian Blackwater floodplain forests: Causes, human dimension, and implications for conservation. *Frontiers in Forests and Global Change, 4*. https://doi.org/10.3389/ffgc.2021.755441

Castellanos, E., Lemos, M. F., Astigarrada, L., et al. (2022). Central and South America. In H.-O. Portner, D. C. Roberts, M. Tignor, et al. (Eds.), *Climate change 2022: Impacts, adaptation and vulnerability. Contribution of working group II to the sixth assessment report of the intergovernmental panel on climate change* (pp. 1689–1816). Cambridge University Press.

Castello, L., & Macedo, M. N. (2016). Large-scale degradation of Amazonian freshwater ecosystems. *Global Change Biology, 22*, 990–1007. https://doi.org/10.1111/gcb.13173

Chevuturi, A., Klingaman, N. P., Woolnough, S. J., et al. (2023). Forecasting annual maximum water level for the Negro River at Manaus using dynamical seasonal predictions. *Climate Services, 1*, e18. https://doi.org/10.1016/j.cliser.2023.100342

Cintra, R., Santos, P. M. R. S., & Leite, C. B. (2007). Composition and structure of the lacustrine bird communities of seasonally flooded wetlands of western Brazilian Amazonia at high water. *Waterbirds, 30*, 521–540. https://doi.org/10.1675/1524-4695(2007)030[0521:CASOTL]2.0.CO;2

Coan, A. I., Scatena, V. L., & Giulietti, A. M. (2002). Anatomy of some aquatic species of the Brazilian Eriocaulaceae. *Acta Botânica Brasílica, 16*, 371–384. https://doi.org/10.1590/S0102-33062002000400001

Cook, C. D. K., Gut, B. J., Rix, E. M., et al. (1974). *Water plants of the world*. Dr. W. Junk.

Costa, S. M., Barbosa, T. D. M., Bittrich, V., & Amaral, M. C. E. (2016). Floristic survey of herbaceous and subshrubby aquatic and palustrine angiosperms of Viruá National Park, Roraima, Brazil. *PhytoKeys, 58*, 21–48. https://doi.org/10.3897/phytokeys.58.5178

Crema, L. C. (2017). *Caracterização de igapós de águas claras e pretas e suas disponibilidades alimentares para o peixe-boi-da-Amazônia (Trichechus inunguis)*. Thesis, Instituto Nacional de Pesquisas da Amazônia.

Crema, L. C., da Silva, V. M. F., Botta, S., et al. (2019). Does water type influence diet composition in Amazonian manatee (*Trichechus inunguis*)? A case study comparing black and Clearwater rivers. *Hydrobiologia, 835*, 1–19. https://doi.org/10.1007/s10750-019-3900-4

Damasceno, G., & Fidelis, A. (2020). Abundance of invasive grasses is dependent on fire regime and climatic conditions in tropical savannas. *Journal of Environmental Management, 271*, 111016. https://doi.org/10.1016/j.jenvman.2020.111016

Damasco, G., Vicentini, A., Castilho, C. V., et al. (2013). Disentangling the role of edaphic variability, flooding regime and topography of Amazonian white-sand vegetation. *Journal of Vegetation Science, 24*, 384–394. https://doi.org/10.1111/j.1654-1103.2012.01464.x

Dang, A. T. N., Kumar, L., Reid, M., & Anh, L. N. T. (2021). Modelling the susceptibility of wetland plant species under climate change in the Mekong Delta, Vietnam. *Ecological Informatics, 64*, 101358. https://doi.org/10.1016/j.ecoinf.2021.101358

Demarchi, L. O., Klein, V. P., Aguiar, D. P. P., et al. (2022). The specialized white-sand flora of the Uatumã sustainable development reserve, Central Amazon, Brazil. *Check List, 18*, 187–217. https://doi.org/10.15560/18.1.187

Dhir, B. (2015). Status of aquatic macrophytes in changing climate: A perspective. *Journal of Environmental Science and Technology, 8*, 139–148. https://doi.org/10.3923/jest.2015.139.148

Espinoza, J. C., Ronchail, J., Guyot, J. L., et al. (2009). Spatio-temporal rainfall variability in the Amazon basin countries (Brazil, Peru, Bolivia, Colombia, and Ecuador). *International Journal of Climatology, 29*, 1574–1594. https://doi.org/10.1002/joc.1791

Espinoza, J. C., Marengo, J. A., Ronchail, J., et al. (2014). The extreme 2014 flood in South-Western Amazon basin: The role of tropical-subtropical South Atlantic SST gradient. *Environmental Research Letters, 9*, 124007. https://doi.org/10.1088/1748-9326/9/12/124007

Espinoza, J. C., Marengo, J. A., Schöngart, J., & Jimenez, J. C. (2022). The new historical flood of 2021 in the Amazon River compared to major floods of the 21st century: Atmospheric features in the context of the intensification of floods. *Weather Climate Extremes, 35*, 100406. https://doi.org/10.1016/j.wace.2021.100406

Esteves, F. A., & Camargo, A. F. M. (1986). Sobre o papel das macrófitas aquáticas na estocagem e ciclagem de nutrientes. *Acta Limnologica Brasiliensia, 1*, 273–298.

Fares, A. L. B., Sousa, R. L., Gurgel, E. C., et al. (2021). Diversity of macrophytes in the Amazon deforestation arc: Information on their distribution, life-forms and habits. *Rodriguésia, 72*, e00312020. https://doi.org/10.1590/2175-7860202172117

Fearnside, P. M. (1995). Potential impacts of climatic change on natural forests and forestry in Brazilian Amazonia. *Forest Ecology and Management, 78*, 51–70. https://doi.org/10.1016/0378-1127(95)03597-3

Fearnside, P., Berenguer, E., Armenteras, D., et al. (2021). Drivers and impacts of changes in aquatic ecosystems. Amazon Assessment Report 2021.

Feng, X., Merow, C., Liu, Z., et al. (2021). How deregulation, drought and increasing fire impact Amazonian biodiversity. *Nature, 597*, 516–521. https://doi.org/10.1038/s41586-021-03876-7

Ferreira, L. V., Chaves, P. P., Cunha, D. A., et al. (2013). A extração ilegal de areia como causa do desaparecimento de campinas e campinaranas no estado do Pará, Brasil. *Pesquisas. Botânica, 64*, 157–173.

Forsberg, B. R., Araújo-Lima, C. A. R. M., Martinelli, L. A., et al. (1993). Autotrophic carbon sources for fish of the Central Amazon. *Ecology, 74*, 643–652. https://www.jstor.org/stable/1940793

Franco, W., & Dezzeo, N. (1994). Soil and water regime in the terra firme-caatinga forest complex near San Carlos de Rio Negro, Amazonas, Venezuela. *Interciencia, 19*, 305–316.

Furch, K., Junk, W. J., Dieterich, J., & Kochert, N. (1983). Seasonal variation in the major cation (Na, K, mg, and Ca) content of the water of Lago Camaleão, an Amazonian floodplain-lake near Manaus, Brazil. *Amazoniana*, 75–89.

Gagnon, P. R., Battaglia, L. L., Hanberry, B. B., Conner, W. H., & King, S. L. (2021). Fire in floodplain forests of the southeastern USA. Fire Ecology and Management: Past, Present, and Future of US Forested Ecosystems, 201–242.

Gopal, B., Sharma, K. P. (1981). Water-hyacinth (*Eichornia crassipes*) the most troublesome weed of the world. Hindasia.

Goulding, M., Smith, N. J. H., & Maghar, D. J. (1996). *Floods of fortune. Ecology and economy along the Amazon*. Columbia University Press.

Harris, M. (2005). Nature makes them lazy: Contested perceptions of the place and knowledge in the lower Amazon floodplain of Brazil. *Conservation and Society, 3*, 461–478. https://www.jstor.org/stable/26396587

Hossain, K., Yadav, S., Quaik, S., et al. (2017). Vulnerabilities of macrophytes distribution due to climate change. *Theoretical and Applied Climatology, 129*, 1123–1132. https://doi.org/10.1007/s00704-016-1837-3

IPCC. (2022). The Intergovernmental Panel on Climate Change.

Jepsen, D. B., & Winemiller, K. O. (2007). Basin geochemistry and isotopic ratios of fishes and basal production sources in four neotropical rivers. *Ecology of Freshwater Fish, 16*(3), 267–281.

Junk, W. J. (1973). Investigations on the ecology and production-biology of the floating meadows on the middle Amazon. *Amazoniana, IV*, 9–102.

Junk, W. J. (1986). Aquatic plants of the Amazon system. In *The ecology of river systems* (pp. 318–328). Springer Science and Business Media.

Junk, W. J. (2013). *The Central Amazon floodplain: Ecology of a pulsing system*. Springer Science and Business Media.

Junk, W. J., & da Silva, V. M. (1997). Mammals, reptiles and amphibians. In *The Central Amazon floodplain: Ecology of a pulsing system* (pp. 409–417). Springer.

Junk, W. J., & Howard-Williams, C. (1984). Ecology of aquatic macrophytes in Amazonia. In *The Amazon: Limnology and landscape ecology of mighty tropical river and its basin. Dordrecht* (pp. 269–293).

Junk, W. J., & Piedade, M. T. F. (1993a). Herbaceous plants of the Amazon floodplain near Manaus: Species diversity and adaptations to the flood pulse. *Amazoniana, 12*, 467–484.

Junk, W. J., & Piedade, M. T. F. (1993b). Biomass and primary-production of herbaceous plant communities in the Amazon floodplain. *Hydrobiologia, 263*, 155–162.

Junk, W. J., & Piedade, M. T. F. (1997). Plant life in the floodplain with special reference to herbaceous plants. In W. J. Junk (Ed.), *The Central Amazon floodplain: Ecological studies*. Elsevier.

Junk, W. J., & Robertson, B. A. (1997). Aquatic invertebrates. In *The Central Amazon floodplain* (pp. 279–296). Springer-Verlag.

Junk, W. J., Bayley, P. B., & Sparks, R. E. (1989). The flood pulse concept in river-floodplain systems. In *Spec Publ Can J Fish Aquat Sci* (Vol. 106, pp. 110–127).

Junk, W. J., Ohly, J. J., Piedade, M. T. F., & Soares, M. G. M. (2000). *Central Amazon floodplain: Actual use and options for a sustainable management*. Backhuys Publishers.

Junk, W. J., Piedade, M. T. F., Schöngart, J., et al. (2011). A classification of major naturally-occurring Amazonian lowland wetlands. *Wetlands, 31*, 623–640. https://doi.org/10.1007/s13157-011-0190-7

Junk, W. J., Piedade, M. T. F., Schöngart, J., & Wittmann, F. (2012). A classification of major natural habitats of Amazonian white-water river floodplains (várzeas). *Wetlands Ecology and Management, 20*, 461–475. https://doi.org/10.1007/s11273-012-9268-0

Junk, W. J., Wittmann, F., Schöngart, J., & Piedade, M. T. F. (2015). A classification of the major habitats of Amazonian black-water river floodplains and a comparison with their white-water counterparts. *Wetlands Ecology and Management, 23*, 677–693. https://doi.org/10.1007/s11273-015-9412-8

Kissmann, K. G. (1991). *Plantas infestantes e nocivas*. Basf.

Kohagura, T. D. C., Souza, E. B. D., Bao, F., Ferreira, F. A., & Pott, A. (2020). Flood and fire affect the soil seed bank of riparian forest in the Pantanal wetland. *Rodriguésia, 71*, e00052018. https://doi.org/10.1590/2175-7860202071013

Latrubesse, E. M., D'Horta, F. M., Ribas, C. C., et al. (2021). Vulnerability of the biota in riverine and seasonally flooded habitats to damming of Amazonian rivers. *Aquatic Conservation: Marine and Freshwater Ecosystems, 31*, 1136–1149. https://doi.org/10.1002/aqc.3424

Lobo, G. S., Wittmann, F., & Piedade, M. T. F. (2019). Response of black-water floodplain (igapó) forests to flood pulse regulation in a dammed Amazonian river. *Forest Ecology and Management, 434*, 110–118. https://doi.org/10.1016/j.foreco.2018.12.001

Lopes, A., & Piedade, M. T. F. (2010). O período da contaminação com petróleo influencia a rebrota de *Echinochloa polystachya* (H.B.K.) Hitchcock em solo de várzea da Amazônia Central? *Biota Neotropica, 10*, 143–148. https://doi.org/10.1590/s1676-06032010000400019

Lopes, A., & Piedade, M. T. F. (2014). Experimental study on the survival of the water hyacinth (*Eichhornia crassipes* (Mart.) Solms—Pontederiaceae) under different oil doses and times of exposure. *Environmental Science and Pollution Research, 21*, 13503–13511. https://doi.org/10.1007/s11356-014-3307-8

Lopes, C. A., Benedito-Cecilio, E., & Martinelli, L. A. (2007). Variability in the carbon isotope signature of *Prochilodus lineatus* (Prochilodontidae, Characiformes) a bottom-feeding fish of the Neotropical region. *Journal of Fish Biology, 70*(6), 1649–1659. https://doi.org/10.1111/j.1095-8649.2007.01388.x

Lopes, A., Rosa-Osman, S. M., & Piedade, M. T. F. (2009). Effects of crude oil on survival, morphology, and anatomy of two aquatic macrophytes from the Amazon floodplains. *Hydrobiology, 636*, 295–305. https://doi.org/10.1007/s10750-009-9959-6

Lopes, A., Parolin, P., & Piedade, M. T. F. (2016a). Morphological and physiological traits of aquatic macrophytes respond to water chemistry in the Amazon Basin: An example of the genus *Montrichardia* Crueg (Araceae). *Hydrobiology, 766*, 1–15. https://doi.org/10.1007/s10750-015-2431-x

Lopes, A., Wittmann, F., Schöngart, J., et al. (2016b). Modeling of regional- and local-scale distribution of the genus Montrichardia Crueg. (Araceae). *Hydrobiology, 789*, 45–57. https://doi.org/10.1007/s10750-016-2721-y

Lopes, A., Wittmann, F., Schöngart, J., et al. (2017). Modeling of regional- and local-scale distribution of the genus *Montrichardia* Crueg. (Araceae). *Hydrobiology, 789*, 45–57. https://doi.org/10.1007/s10750-016-2721-y

Lopes, A., Ferreira, A. B., Pantoja, P. O., et al. (2018). Combined effect of elevated CO_2 level and temperature on germination and initial growth of *Montrichardia arborescens* (L.) Schott (Araceae): A microcosm experiment. *Hydrobiology, 814*, 19–30. https://doi.org/10.1007/s10750-015-2598-1

Lopes, A., Uchoa, K., Demarchi, L. O., et al. (2020). Impactos antrópicos nas várzeas amazônicas e seus efeitos sobre as macrófitas aquáticas. *Várzeas Amazônicas desafios para um manejo sustentável*, 106–121.

Lopes, A., Demarchi, L. O., Franco, A. C., et al. (2021). Predicting the potentioal distribution of aquatic herbaceous plants in oligotrophic central Amazonian wetland ecosystems. *Acta Botânica Brasílica, 35*, 22–36. https://doi.org/10.1590/0102-33062020abb0188

Lopes, A., Demarchi, L. O., Piedade, M. T. F., et al. (2023). Predicting the range expansion of invasive alien grasses under climate change in the Neotropics. *Perspectives in Ecology and Conservation, 21*, 128–135. https://doi.org/10.1016/j.pecon.2023.02.005

Mano, G., Lopes, A., & Piedade, M. T. F. P. (2023). Will climate change favor exotic grasses over native ecosystem engineer species in the Amazon basin. *Ecological Informatics, 75*, 102102. https://doi.org/10.1016/j.ecoinf.2023.102102

Marengo, J. A., & Espinoza, J. C. (2016). Extreme seasonal droughts and floods in Amazonia: Causes, trends and impacts. *International Journal of Climatology, 36*, 1033–1050. https://doi.org/10.1002/joc.4420

Marengo, J. A., & Souza, Jr C. (2018). *Mudanças Climáticas: impactos e cenários para a Amazônia*. São Paulo.

Marengo, J. A., Tomasella, J., Soares, W. R., et al. (2012). Extreme climatic events in the Amazon basin. *Theoretical and Applied Climatology, 107*, 73–85. https://doi.org/10.1007/s00704-011-0465-1

Marengo, J. A., Borma, L. S., Rodriguez, D. A., et al. (2013). Recent extremes of drought and flooding in Amazonia: Vulnerabilities and human adaptation. *American Journal of Climate Change, 2*, 87–96. https://doi.org/10.4236/ajcc.2013.22009

Martins, D., Cardoso, L. R., Mori, E. S., & Tanaka, R. H. (2003). Caracterização genética de acesso de egéria (*Egeria* spp.) coletados no Estado de São Paulo RAPD1. *Planta Daninha, 21*, 1–6.

Martins, K., Kimura, R. K., Francisconi, A. F., et al. (2016). The role of very small fragments in conserving genetic diversity of a common tree in a hyper fragmented Brazilian Atlantic Forest landscape. *Conservation Genetics, 17*, 509–520. https://doi.org/10.1007/s10592-015-0800-7

Mendonça, B. A. F., Filho, E. I. F., Schaefer, C. E. G. R., et al. (2015). Os solos das campinaranas na Amazônia Brasileira: Ecossistemas arenícolas oligotróficos. *Cienc Florest, 25*, 827–839. https://doi.org/10.5902/1980509820581

Mitsch, W. J., Nahlik, A., Wolski, P., Bernal, B., Zhang, L., & Ramberg, L. (2010). Tropical wetlands: Seasonal hydrologic pulsing, carbon sequestration, and methane emissions. *Wetlands Ecology and Management, 18*, 573–586. https://doi.org/10.1007/s11273-009-9164-4

Mortillaro, J. M., Pouilly, M., Wach, M., et al. (2015). Trophic opportunism of Central Amazon floodplain fish. *Freshwater Biology, 60*, 1659–1670. https://doi.org/10.1111/fwb.12598

Nielsen, D. L., Merrin, L. E., Pollino, C. A., et al. (2020). Climate change and dam development: Effects on wetland connectivity and ecological habitat in tropical wetlands. *Ecohydrology, 13*, e2228. https://doi.org/10.1002/eco.2228

Nilsson, C., & Berggren, K. (2000). Alterations of riparian ecosystems caused by river regulation. *Bioscience, 50*, 783–792. https://doi.org/10.1641/0006-3568(2000)050[0783:AORECB]2.0.CO;2

Nobre, C., Sampaio, G., & Salazar, L. (2007). Mudanças climáticas e Amazônia. *Ciência e Cultura, 53*, 6.

Nobre, C. A., Sampaio, G., Borma, L. S., et al. (2016). Land-use and climate change risks in the Amazon and the need of a novel sustainable development paradigm. *Proceedings of the National Academy of Sciences of the United States of America, 113*, 10759–10768. https://doi.org/10.1073/pnas.1605516113

O'Hare, M. T., Aguiar, F. C., Asaeda, T., et al. (2018). Plants in aquatic ecosystems: Current trends and future directions. *Hydrobiologia, 812*, 1–11. https://doi.org/10.1007/s10750-017-3190-7

Oliveira Mota, N. F., & Zappi, D. C. (2018). Flora das cangas da Serra dos Carajás, Pará, Brasil: Lentibulariaceae. *Rodriguésia, 69*, 119–132. https://doi.org/10.1590/2175-7860201869110

Oliveira, M. T., Damasceno-Junior, G. A., Pott, A., Paranhos Filho, A. C., Suarez, Y. R., & Parolin, P. (2014). Regeneration of riparian forests of the Brazilian Pantanal under flood and fire influence. *Forest Ecology and Management, 331*, 256–263. https://doi.org/10.1016/j.foreco.2014.08.011

Piedade, M. T. F., Junk, W. J., & Long, S. P. (1991). The productivity of the C4 grass Echinochloa polystachya on the Amazon floodplain. *Ecology, 72*, 1456–1463. https://doi.org/10.2307/1941118

Piedade, M. T. F., Junk, W. J., & Parolin, P. (2000). The flood pulse and photosynthetic response of trees in a white water floodplain (várzea) of the Central Amazon, Brazil. *Verhandlungen des Internationalen Verein Limnologie, 27*, 1–7.

Piedade, M. T. F., Schöngart, J., & Junk, W. J. (2005). O manejo sustentável das áreas alagáveis da Amazônia Central e as comunidades de herbáceas aquáticas. *Uakari, 1*, 29–38.

Piedade, M. T. F., Junk, W., DÂngelo, S. A., et al. (2010). Aquatic herbaceous plants of the Amazon floodplains: State of the art and research needed. *Acta Limnologica Brasiliensia, 22*, 165–178. https://doi.org/10.4322/actalb.02202006

Piedade, M. T. F., Schöngart, J., Wittmann, F., et al. (2013). Impactos ecológicos da inundação e seca na vegetação das áreas alagáveis amazônicas. In L. S. Borma & C. Nobre (Eds.), *Secas na Amazônia: causas e conseqüências* (pp. 268–305). Oficina de Textos.

Piedade, M. T. F., Lopes, A., Demarchi, L. O., et al. (2018). *Guia de campo de herbáceas aquáticas: várzea Amazônica*. Editora INPA.

Piedade, M. T. F., Lopes, A., Demarchi, L. O., et al. (2020). Os campos naturais e as plantas herbáceas na planície de inundação amazônica e sua utilização. In *Várzeas Amazônicas: Desafios para um Manejo Sustentável* (pp. 86–105).

Pierini, S. A., & Thomaz, M. (2004). Adaptações de plantas submersas à absorção do carbono inorgânico. *Acta Botânica Brasílica, 18*, 629–641. https://doi.org/10.1590/S0102-33062004000300020

Pivello, V. R. (2011). The use of fire in the Cerrado and Amazonian rainforests of Brazil: Past and present. *Fire Ecology, 7*, 24–39. https://doi.org/10.4996/fireecology.0701024

Pompêo, M. (2008). Monitoramento e manejo de macrófitas aquáticas. *Oecologia Brasiliensis, 12*, 5.

Pott, V. J., & Pott, A. (2000). *Plantas aquáticas do Pantanal*. Embrapa.

Prance, G. T. (1974). *Victoria amazonica* ou *Victoria regia*? *Acta Amaz, 4*, 5–8. https://doi.org/10.1590/1809-43921974043005

Prance, G. T. (1979). Notes on the vegetation of Amazonia. 3. The terminology of Amazonian forest types subject to inundation. *Brittonia, 31*, 26–38. https://doi.org/10.2307/2806669

Resende, A. F., Piedade, M. T. F., Feitosa, Y. O., et al. (2020). Flood-pulse disturbances as a threat for long-living Amazonian trees. *The New Phytologist, 227*, 1790–1803. https://doi.org/10.1111/nph.16665

Ribeiro, G. G., Anderson, L. O., Barretos, N. J. C., et al. (2021). Attributing the 2015/2016 Amazon basin drought to anthropogenic influence. *Climate, Resilience & Sustainability, 1*, 1–10. https://doi.org/10.1002/cli2.25

Ritter, C. D., McCrate, G., Nilsson, R. H., et al. (2017). Environmental impact assessment in Brazilian Amazonia: Challenges and prospects to assess biodiversity. *Biological Conservation, 206*, 161–168. https://doi.org/10.1016/j.biocon.2016.12.031

Rocha, D., & Martins, D. (2011). Levantamento de plantas daninhas aquáticas no reservatório de Alagados, Ponta Grossa-PR. *Planta Daninha, 29*, 237–246. https://doi.org/10.1590/s0100-83582011000200001

Rocha, M., Santos Júnior, C. C., Damasceno-Junior, G. A., Pott, V. J., & Pott, A. (2015). Effect of fire on a monodominant floating mat of Cyperus giganteus Vahl in a neotropical wetland. *Brazilian Journal of Biology, 75*, 114–124. https://doi.org/10.1590/1519-6984.08613

Rosa-Osman, S. M., Rodrigues, R., Mendonça, M. S., et al. (2011). Morfologia da flor, fruto e plântula de *Victoria amazonica* (Poepp.) J.C. Sowerby (Nymphaeaceae). *Acta Amaz, 41*, 21–28. https://doi.org/10.1590/S0044-59672011000100003

Salimi, S., Almuktar, S. A. A. A. N., & Scholz, M. (2021). Impact of climate change on wetland ecosystems: A critical review of experimental wetlands. *Journal of Environmental Management, 286*, 112160. https://doi.org/10.1016/j.jenvman.2021.112160

Santos, A. M., & Thomaz, S. M. (2007). Aquatic macrophytes diversity in lagoons of a tropical floodplain: The role of connectivity and water level. *Austral Ecology, 32*, 177–190. https://doi.org/10.1111/j.1442-9993.2007.01665.x

Schöngart, J., & Wittmann, F. K. (2011). Biomass and net primary production of central Amazonian floodplain forests. In *Ecophysiology, biodiversity and sustainable management* (pp. 347–388).

Schöngart, J., Wittmann, F., Resende, A. F., et al. (2021). The shadow of the Balbina dam: A synthesis of over 35 years of downstream impacts on floodplain forests in Central Amazonia. *Aquatic Conservation: Marine and Freshwater Ecosystems, 31*, 1117–1135. https://doi.org/10.1002/aqc.3526

Schöngart, J., Wittmann, F., Junk, W. J., Lopes, A., Demarchi, L. O., Biem Mori, G., et al. (2025). The fate of floodplain forests in a changing Amazon. In S. S. de Souza, S. Braz-Mota, & A. L. Val (Eds.), The future of Amazonian aquatic biota.

Schulthorpe, C. D. (1967). *The biology of aquatic plants*. Edward Arnold.

Silva, R. S., & Camargo, A. F. M. (2007). Effect of Urucu crude oil on the aquatic macrophyte *Pistia stratiotes*. *Acta Limnologica Brasiliensia, 19*, 155–161.

Silva, T. S. F., Costa, M. P. F., & Melack, J. M. (2010). Spatial and temporal variability of macrophyte cover and productivity in the eastern Amazon floodplain: A remote sensing approach. *Remote Sensing of Environment, 114*, 1998–2010. https://doi.org/10.1016/j.rse.2010.04.007

Silveira, M. V. F., Silva-Junior, C. H. L., Anderson, L. O., & Aragão, L. E. O. C. (2022). Amazon fires in the 21st century: The year of 2020 in evidence. *Global Ecology and Biogeography, 31*, 2026–2040. https://doi.org/10.1111/geb.13577

Sioli, H. (1984). The Amazon and its main affluents: Hydrography, morphology of the river courses and river types. In H. Sioli (Ed.), *The Amazon: Limnolgy and landscape ecology of mighty tropical river and its basin, first* (pp. 127–165). DW Junk.

Souza, S. D. N. G., Piedade, M. T. F., Demarchi, L. O., & Lopes, A. (2021). Implications of global climate change for the development and ecological interactions between two key Amazonian aquatic macrophytes. *Acta Botânica Brasílica, 35*, 111–121. https://doi.org/10.1590/0102-33062020abb0138

Targhetta, N., Kesselmeier, J., & Wittmann, F. (2015). Effects of the hydroedaphic gradient on tree species composition and aboveground wood biomass of oligotrophic forest ecosystems in the Central Amazon basin. *Folia Geobotanica, 50*, 185–205. https://doi.org/10.1007/s12224-015-9225-9

Thomaz, S. M., Bini, L. M., & Bozelli, R. L. (2007). Floods increase similarity among aquatic habitats in river-floodplain systems. *Hydrobiology, 579*, 1–13. https://doi.org/10.1007/s10750-006-0285-y

van Marle, M. J., Field, R. D., & van der Werf, G. (2017). Fire and deforestation dynamics in Amazonia (1973–2014). *Global Biogeochemical Cycles, 31*, 24–38. https://doi.org/10.1002/2016GB005445

Veldman, J. W., & Putz, F. E. (2011). Grass-dominated vegetation, not species-diverse natural savanna, replaces degraded tropical forests on the southern edge of the Amazon Basin. *Biological Conservation, 144*, 1419–1429. https://doi.org/10.1016/j.biocon.2011.01.011

Vieira, V. C., Alves, P. L. C. A., Lemos, M. V. F., et al. (2007). Variabilidade genética em acessos de trapoeraba (*Commelina benghalensis* L.). *Arquivos do Instituto Biológico, 74*, 315–320. https://doi.org/10.1590/1808-1657v74p3152007

Walker, I. (2004). Trophic interactions within the *Utricularia* habitat in the reservoir of the Balbina hydroelectric powerplant (Amazonas, Brazil). *Acta Limnologica Brasiliensia, 16*, 183–191.

Wittmann, F., & Junk, W. J. (2016). Amazon River basin. In *The wetland book*. Springer Science.

Wittmann, F., Householder, J. E., Piedade, M. T. F., et al. (2022). A review of the ecological and biogeographic differences of Amazonian floodplain forests. *Water, 14*, 1–28. https://doi.org/10.3390/w14213360

Worbes, M., Klinge, H., Revilla, J. D., & Martius, C. (1992). On the dynamics, floristic subdivision and geographical distribution of várzea forests in Central Amazonia. *Journal of Vegetation Science, 3*, 553–564. https://doi.org/10.2307/3235812

Xavier, R. O., Leite, M. B., & Silva-Matos, D. M. (2017). Stress responses of native and exotic grasses in a Neotropical savanna predict impacts of global change on invasion spread. *Austral Ecology, 42*, 562–576. https://doi.org/10.1111/aec.12475

Fate of Floodplain Forests in a Changing Amazon

Jochen Schöngart, Florian Wittmann, Wolfgang Johannes Junk, Aline Lopes, Layon Oreste Demarchi, Gisele Biem Mori, Angelica Faria de Resende, Flávia Machado Durgante, Adriano da Silva Costa Quaresma, John Ethan Householder, Gilvan Costa, Bianca Weiss, Maíra Rocha, Yanka Laryssa Almeida Alves, Elizabeth Rebouças, Augusto Cesar Franco, Cristiane Silva Ferreira, Pia Parolin, and Maria Teresa Fernandez Piedade

Abstract

The Amazon basin is undergoing an unprecedented transformation during recent decades driven by anthropogenic-dominated disturbance regimes and increasing hydroclimatic extremes. This affects the large Amazonian river-floodplains influenced by annual, regular and predictable flood pulses with high amplitudes, the *várzeas* and *igapós*, which cover an area of more than 750,000 km^2. These ecosystems are densely forested with varying environmental conditions, distinct

J. Schöngart (✉) · L. O. Demarchi · B. Weiss · M. Rocha · Y. L. A. Alves · E. Rebouças M. T. F. Piedade
Working Group Ecology, Monitoring and Sustainable Use of Wetlands (MAUA), Brazilian National Institute for Amazon Research (INPA), Manaus, Brazil
e-mail: jochen.schongart@inpa.gov.br

F. Wittmann · J. E. Householder
Department of Wetland Ecology, Institute for Geography and Geoecology, Karlsruhe Institute for Technology (KIT), Rastatt, Germany
e-mail: florian.wittmann@kit.edu; john.householder@kit.edu

W. J. Junk
National Institute for Science and Technology in Wetlands (INCT-INAU), Federal University of Mato Grosso (UFMT), Cuiabá, Brazil

S. S. de Souza et al. (eds.), *The Future of Amazonian Aquatic Biota*,
https://doi.org/10.1007/978-3-031-66822-7_7

species composition, and different levels of resilience and vulnerability which we summarize and discuss in this chapter. We provide an overview of the influence of large-scale coupled ocean-atmosphere phenomena from the Pacific and Atlantic Oceans at high (interannual) and low (multidecadal) frequencies on the river-floodplains. The impact of hydroclimatic variability and extreme events on the large river-floodplains and their human populations are discussed. Disturbances caused by land-use changes on large river-floodplains are shown in case studies highlighting hydropower dams and fires. We finalize with an overall synthesis about the fate of floodplains in the changing Amazon basin indicating recommendations for conservation.

A. Lopes
Working Group Ecology, Monitoring and Sustainable Use of Wetlands (MAUA), National Institute for Amazon Research (INPA), Manaus, Brazil

Cesumar Institute of Science, Technology, and Innovation, Graduate Program in Clean Technologies, Cesumar University (Unicesumar), Maringá, Brazil

G. B. Mori
Working Group Ecology, Monitoring and Sustainable Use of Wetlands (MAUA), National Institute for Amazon Research (INPA), Manaus, Brazil

Graduate Program of Ecology and Conservation, Mato Grosso State University (UNEMAT), Nova Xavantina, Brazil

A. F. de Resende
Biological and Environmental Sciences, University of Stirling, Stirling, UK

Department of Forest Sciences, Luiz de Queiroz College of Agriculture, University of São Paulo, São Paulo, Brazil

F. M. Durgante · A. C. Quaresma
Working Group Ecology, Monitoring and Sustainable Use of Wetlands (MAUA), National Institute for Amazon Research (INPA), Manaus, Brazil

Department of Wetland Ecology, Institute for Geography and Geoecology, Karlsruhe Institute for Technology (KIT), Rastatt, Germany

G. da Silva Costa Quaresma
Working Group Ecology, Monitoring and Sustainable Use of Wetlands (MAUA), National Institute for Amazon Research (INPA), Manaus, Brazil

Biology of Continental Aquatic Environments, Federal University of Rio Grande (FURG), Rio Grande, Brazil

A. C. Franco · C. S. Ferreira
Department of Botany, University of Brasilia (UnB), Brasilia, Brazil
e-mail: acfranco@unb.br

P. Parolin
Max Planck Institute of Animal Behavior, Radolfzell, Germany
e-mail: piaparolin@gmx.de

1 Introduction

The Amazon basin is the largest watershed (~6.87 million km^2) on the planet providing about 16–22% of the global freshwater discharge to the oceans (Costa et al., 2021). It hosts approximately 40% of the remaining tropical rainforest on Earth, harbours about 10% of the known global terrestrial biodiversity, contributes about 16% of all terrestrial photosynthetic productivity and stores 150–200 billion tons of carbon in the biomass and soils (Brienen et al., 2015; Guayasamin et al., 2021; Malhi et al., 2021). The diverse rainforest is a major centre of deep atmospheric convection in the global climate system, coupled with large-scale atmospheric circulations and promoted by complex biosphere–atmosphere interactions, cooling the atmosphere near the surface via evapotranspiration, generating humidity and maintaining biodiversity and biogeochemical cycles (Costa et al., 2021). About 14% of the global rainfall on land is concentrated in the Amazon basin, representing less than 5% of the Earth's continental area. These numbers emphasize the global relevance of this biome for the Earth system and humanity. During the last decades, however, the Amazon has undergone an unprecedented transition mainly driven by anthropogenic impacts resulting from land-use (deforestation, forest fragmentation, logging, fires, hydropower dams, mining, road networks and urban footprints) and climate changes (increasing temperature and extreme hydroclimatic events) as well as feedback loops and synergies of these disturbance regimes (Albert et al., 2023; Nobre et al., 2016) with the risk to transition the biome towards its tipping point (Flores et al., 2024; Lovejoy & Nobre, 2018). These scenarios are discussed for the non-flooded *terra firme* forests (Lapola et al., 2023).

However, the Amazon basin has a diverse composition of ecosystems that developed through evolutionary time spanning millions of years (Hoorn et al., 2010). More than 30% of the Amazon corresponds to different typologies of wetlands (> 2.3 million km^2) (Fleischmann et al., 2022; Junk et al., 2011; Melack & Hess, 2010; Wittmann & Junk, 2016), which have been classified by a hierarchical system considering climate, hydrology, edaphic (water and soil) and plant communities as environmental indicators (Junk et al., 2011). Wetlands occur in the aquatic–terrestrial transition zone (ATTZ) (Junk et al., 1989), exposed to permanent or seasonal flooding or water logging in coastal and continental regions (natural and anthropogenic environments) by waters of different physiochemical characteristics with plant and animal species adapted to the prevailing environmental dynamics (Junk et al., 2014). The dominant wetland types in the Amazon basin are those with seasonal inundations shaped by flood pulses of varying frequency, predictability and amplitude (Junk et al., 2011). Their flood dynamics and seasonal water storage depend mainly on the catchment runoff, the amount of exchanged water with the river, precipitation, evaporation and groundwater seepage which influences hydrology, biogeochemical processes and sediment transport (Lesack & Melack, 1995, Paiva et al., 2013, Rudorf et al., 2014). In this chapter, we focus on the seasonally inundated wetlands along the large Amazonian rivers, generally denominated as floodplains.

2 Amazonian River-Floodplains

About 750,000 km^2 of the Amazon basin is covered by Amazonian river-floodplains (Wittmann & Junk, 2016). These ecosystems are influenced by monomodal flood pulses with high amplitudes and one regular and predictable high-water (aquatic phase) and one low-water period (terrestrial phase) over the year (Bonnet et al., 2008; Junk et al., 1989; Rudorff et al., 2014). The flood pulse is the main driver of geomorphological processes, biogeochemical and life cycles as well as growth rhythms, reproduction periods and interactions of its adapted biota (Junk, 1997; McClain et al., 2001; Melack & Coe, 2021). Due to the regular seasonal shift between terrestrial and aquatic phases, floodplains are key elements in the vast Amazonian landscapes. These ecosystems offer seasonal habitats for numerous migrating species of aquatic and terrestrial fauna (Wittmann & Junk, 2016). They further harbour a huge diversity of partially endemic flora and fauna which are important elements of diversification and speciation processes (Junk & Piedade, 1993; Thom et al., 2020; Wittmann et al., 2010, 2013). In addition, they provide fundamental ecosystem services, such as water storage and purification, which is essential to buffer river discharge, mitigate extreme hydroclimatic events and recharge groundwater (Paiva et al., 2013). They contribute to sediment retention and microclimate regulation and trigger biogeochemical and nutrient cycles distinct from the non-flooded environments (Junk et al., 2018; Melack et al., 2009). Amazonian floodplains have been occupied and used by indigenous populations for thousands of years and by traditional riverine populations for hundreds of years, providing shelter, water and multiple natural resources for subsistence and trade (Junk et al., 2000, 2020), contributing to their cultural safeguarding (Junk et al., 2014).

The Amazonian large river-floodplains present varying edaphic conditions and three major systems have been classified based on hydrochemical characteristics, that in turn depend mainly on the geology of the catchment area and physiochemical features of the draining rivers (Sioli, 1954), as well as on botanical criteria (Kubitzki, 1989; Prance, 1979): the *várzea* (white-water floodplains) and the black-water and clear-water *igapós* (Junk et al., 2011). However, some rivers can have intermediate physiochemical conditions in their waters, and in some rivers, the physiochemical conditions of the water change along their course caused by environmental or anthropogenic alterations (Marinho et al., 2020; Rios-Villamizar et al., 2020; Santos et al., 1985). A review by Wittmann et al. (2022a) provides an actual overview of the main general characteristics of these ecosystems, which are summarized in Table 1.

Table 1 Main characteristics of large Amazonian river-floodplains

Parameter	White-water	Black-water	Clear-water	References
Major rivers	Marañon-Solimões-Amazon mainstem, Madeira, Caquetá-Japurá, Purus, Juruá, Javari, Putumayo-Içá and Ucayali	Negro and the majority of its affluents, Uatumã, Urubu and Nhamundá	Branco, Tapajós, Xingú, Tocantins-Araguaia, Trombetas, Jarí, Paru and Araguari	Junk et al. (2011) and Wittmann and Junk (2016)
Watersheds	Andes and Andean foothills (NW to SW-Amazon)	Guiana Shield and podzolic formations (NW-Amazon)	Savanna complexes of Guiana and Central Brazilian Shields (NE and SE-Amazon)	Wittmann and Junk (2016)
Water colour	Beige (like milk coffee)	Brown-blackish (like black coffee)	Clear-greenish	Wittmann et al. (2022a)
Water transparency (Secchi disk)	<0.5 m	1.3–2.9 m	1.1–4.3 m	Sioli (1984) and Rios-Villamizar et al. (2020)
Water pH	6.2–7.6	3.8–4.9	4.5–7.8	Sioli (1984), Silva et al. (2019), Küchler et al. (2000) and Rios-Villamizar et al. (2020)
Suspended sediments	100–400 mg/l	3–27 mg/l	12–45 mg/l	Santos et al. (1985) and Richey et al. (1986)
Electric conductivity	40–140 μS/cm	<20 μS/cm	5–40 μS/cm	Santos et al. (1985), Junk et al. (2011), and Rios-Villamizar et al. (2020)
Organic substances	Intermediate	High	Low	Junk et al. (2011)
Floodplain extension	>450,000 km^2	~140,000 km^2	~160,000 km^2	Wittmann and Junk (2016)

(continued)

Table 1 (continued)

Parameter	White-water	Black-water	Clear-water	References
Geomorphologic stability	Low	High	Varying	Wittmann et al. (2022a)
Dominating soils	Fluvisols (Entisols), Gleysols (Inceptsols, Entisols and Ultisols)	Arenosols, Podzols and Gleysols	Fluvisols, Gleysols	Quesada et al. (2011) and Wittmann et al. (2022a)
Soil fertility	High	Low	Intermediate	Furch (1997)
Annual precipitation (mm)	1200- > 4000	1800–3600	1200–2400	Sombroek (2001)
Number of consecutive months <100 mm precipitation	0–7	0–3	3–8	Sombroek (2001)
Amplitude flood-pulse (m) (high river order)	5–15	4–10	3–7	Fassoni-Andrade et al. (2021)
Vegetation	*Várzea*	*Igapó*	*Igapó*	Prance (1979)
Forest cover (%)	61–93	85–93	78–96	Melack and Hess (2010)
Tree species	>1000	>600	>150	Wittmann et al. (2006a, 2010, 2022a), Ferreira et al. (2013), Scudeller and Vegas-Vilarrúbia (2018), Householder et al. (2021) and MAUA database (*unpubl.*)
Herbaceous species	>388	>178	>49	Junk and Piedade (1993), Crema (2017) and Lopes et al. (2019)
Density of human populations	High	Low	Intermediate	Junk et al. (2020) and Piedade et al. (2024)

(continued)

Table 1 (continued)

Parameter	White-water	Black-water	Clear-water	References
Land-use changes	Hydropower dams (catchments), local deforestation, fires and mining (small-scale)	Local hydropower dams, small-scale deforestation, fires and mining (small-scale)	Large-scale deforestation (headwaters), hydropower dams, mining (small-scale and concessions) and fires	Castello et al. (2013), Latrubesse et al. (2017) and Fearnside (2019)
Overall threats	High (Madeira), intermediate (Marañon and Ucayali) and low (Japurá, Purus, Juruá and Solimões-Amazon)	Low (Negro and Urubu) and intermediate (Uatumã)	High (Tapajós), intermediate (Branco and Xingú) and low (Trombetas, Jari, Paru and Araguarí)	Castello et al. (2013) and Latrubesse et al. (2017)
Overall conservation status	High (Caquetá-Japurá), intermediate (Purus, Juruá and Madeira) and low (Solimões-Amazon mainstem)	High (Negro and Nhamundá), intermediate (Uatumã) and low (Urubu)	High (Trombetas, Jari, Paru and Araguari), intermediate (Branco, Tapajós and Xingú) and low (Tocantins-Araguaia)	RAISG (2019)
Overall scientific knowledge	Good	Intermediate	Poor	Wittmann and Junk (2016) and Junk et al. (2020)

2.1 Várzeas

Floodplains seasonally inundated by large white-water rivers with headwaters in the Andean Cordillera and pre-Andean slopes are called *várzea* covering more than 450,000 km^2 (Junk et al., 2012; Wittmann & Junk, 2016). They occur along the mainstream of the Amazon River and other major rivers (Ucayali, Napo, Putumayo-Içá, Juruá, Purus and Javari), passing through vast landscapes of river terrace systems formed by Andean sediments in western Amazonia limited by the Caquetá-Japurá (northwest) and Madeira (southwest) river basins (Fig. 1). These rivers transport large amounts of suspended sediments (100–400 mg/l) of multilayered clay minerals, such as smectite, illite and montmorillonite (Irion et al., 2010;

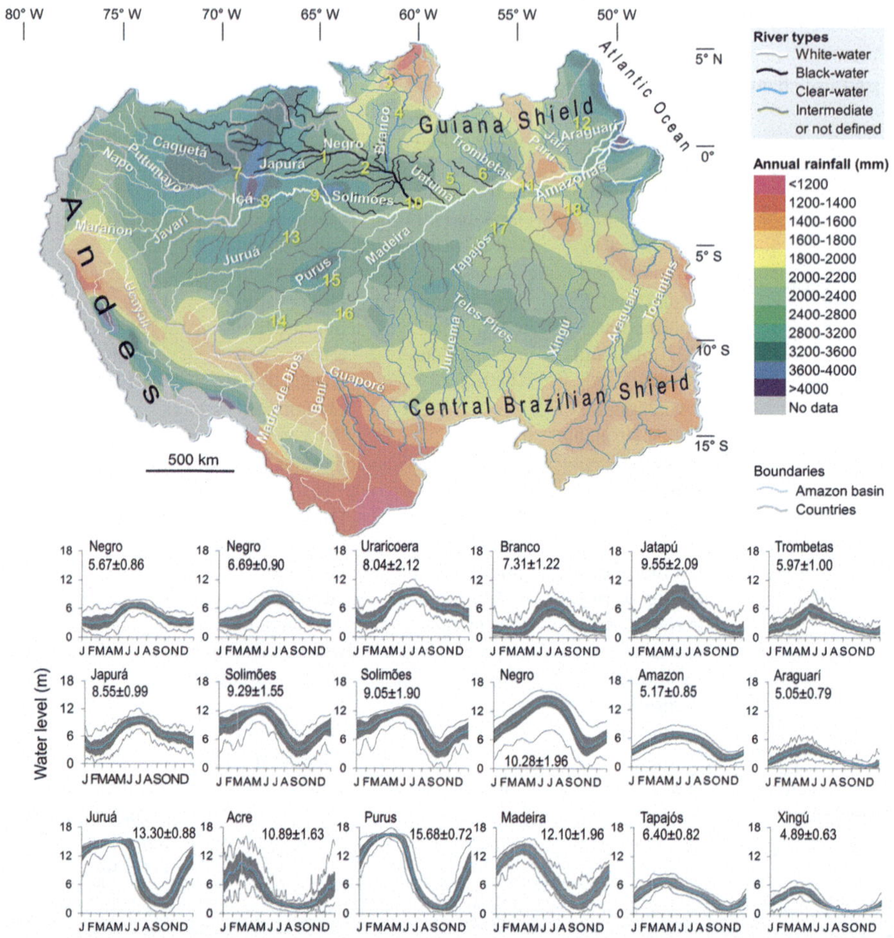

Fig. 1 Main rivers draining the Amazon basin (including the Tocantins-Araguaia basin), classified into white-water, black-water and clear-water rivers and those with intermediate or unknown hydrochemical and morphological characteristics (Junk et al., 2011, Rios-Villamizar et al., 2020). Annual mean precipitation (30-year average) has been obtained from Sombroek (2001). Flood pulses (locations of stations indicated in the map by numbers: (1) Sta. Isabel do Rio Negro; (2) Barcelos; (3) Maracá; (4) Caracaraí; (5) Base Siderama-Jusante; (6) Caramucho; (7) Vila Bittencourt; (8) São Paulo de Olivença; (9) Fonte Boa; (10) Manaus; (11) Santarem; (12) Capivara; (13) Gavião; (14) Rio Branco; (15) Lábrea; (16) Altamira) are shown by hydrographs indicating the mean water level (blue), standard deviation (grey area), as well as maximum and minimum data during the year [data obtained from National Water Resources Information–SNIRH operated by the Brazilian National Water and Sanitation Agency–ANA for the period 1970–2010 (station numbers 8, 11, 13–18), 1978–2019 (1), 1967–2019 (2, 4), 1981–2022 (3, 7), 1971–2019 (5), 1986–2019 (6), 1983–2005 (9), 1903–2022 (10), 1982–2019 (12)]. Decimal numbers in hydrographs indicate the average (±standard deviation) of the annual amplitude of the flood pulse for the given period in metres

Richey et al., 1986). Their waters have low transparency (<0.5 m based on Secchi depth), an almost neutral pH (6–7) and high electric conductivity (40–140 μS/cm), which tends to decrease along the Amazon mainstem from the west towards the east (Wittmann et al., 2022a). The deposits of the sediments during flooding and the weathering of the clay minerals lead to alluvial substrates with elevated cation exchange capacity and soil fertility (Furch, 1997). The intense hydrogeomorphological dynamics of erosion and sedimentation processes of the river-floodplain system are reflected by distinct successional stages of highly productive herbaceous plant communities (Engle et al., 2008; Piedade et al., 1991; Silva et al., 2009; 2013) and diverse forests (Wittmann et al., 2004; Worbes et al., 1992), leading to the formation of a heterogeneous alluvial landscape (Junk et al., 2012). Since the human occupation of the Amazon basin about 13,000 years ago, *várzeas* and adjacent areas have been populated for thousands of years by indigenous people and for the last 300 years by traditional communities (Denevan, 1996). The livelihood of the traditional populations is intrinsically related to the flood pulse through economic activities such as fishing, hunting, agriculture, small-scale livestock production and extraction of timber and non-timber forest products according to the water level conditions (Junk et al., 2023). In many regions, *várzea* forests upon the higher and less flooded elevations are transformed into lands for agriculture, extensive livestock production, and settlements and have been degraded by the unsustainable use of natural resources, such as timber and non-timber forest products, during the last decades and centuries (colonization period) (Renó & Novo, 2019). Only in some regions, *várzeas* are integrated with indigenous lands and conservation units, mainly along the tributaries of the Solimões River. Overall, the scientific knowledge about this ecosystem is high compared to the *igapós* (Gash et al., 2009; Junk, 1997; Junk et al., 2000, 2010, 2020; McClain et al., 2001; Nagy et al., 2016; Piedade et al., 2019; Sioli, 1984; Wittmann et al., 2010).

2.2 Black-Water *Igapós*

The oligotrophic *igapó* covers about 119,000 km^2 in the Negro River basin (i.e., Negro, Jáu, Unini, Içana and Uapés rivers) (Melack & Hess, 2010) and also occurs along smaller black-water rivers (Urubu, Uatumã, Abacaxí and Nhamundá rivers) outside this basin (total estimate is about 140,000 km^2) (Fig. 1). These rivers drain deeply weathered tertiary sediments of Pre-Cambrian origin and large oligotrophic wetland complexes (*campinaranas*) on mainly Podzols in the northwestern portion of the Amazon basin (Quesada et al., 2011). The brown-blackish colour of its water results from large amounts of organic substances leading to a low pH (4–5). The water transparency of these rivers is higher (1–3 m based on Secchi depth) compared to the white-water counterparts as a result of the low amount of suspended sediments (< 30 mg/l) and, therefore, the electric conductivity is low (< 20 μS/cm) (Junk et al., 2015). The geomorphology of the black-water *igapós* is relatively stable (Wittmann et al., 2022a), and soils are nutrient-poor, often developing on sandy bedrocks (Latrubesse & Franzinelli, 2005). The vegetation is dominated by forests (> 85% coverage) (Melack & Hess, 2010) with slow dynamics, low regeneration capacity and high vulnerability to anthropogenic disturbances such as logging, hydropower dams and fires (Carvalho et al., 2021; Flores et al., 2014; Resende et al., 2019; Scabin et al., 2012; Schöngart et al., 2021). The density of human populations is comparatively low but was regionally much higher in the past during the pre-Columbian periods and European colonization than today (Leonardi, 1999; Sampaio, 2003). Nowadays, although black-water *igapós* are inserted into huge mosaics of protected areas (indigenous lands and different categories of conservation units), they are locally impacted by deforestation, fires, mining activities (*garimpo*) and hydropower generation (Piedade et al., 2024).

2.3 Clear-Water *Igapós*

Clear-water *igapós* occur along rivers that drain the Precambrian Shields in the north (Guiana Shield) and south (Central Brazilian Shield) (Fig. 1). Although Prance (1979) classified its vegetation as *igapó*, there is growing evidence that this ecosystem also hosts many typical *várzea* species (Wittmann & Junk, 2016). The catchments of these rivers integrate large savanna complexes in the north (Branco, Jatapú, Trombetas, Jarí, Paru and Araguarí rivers) and south (Guaporé, Tapajós, Xingú and Araguaia-Tocantins rivers). Their water has a greenish colour with high transparency of up to 4.5 m (Secchi depth) and highly varying sediment loads (12–45 mg/l), pHs (4.5–8) and electric conductivities (5–40 μS/cm) (Junk et al., 2011). Clear-water *igapós* cover about 160,000 km^2 and have soils of varying and general intermediate fertility compared to the oligotrophic *igapó* and eutrophic *várzeas* (Wittmann et al., 2022a). Overall, this ecosystem is less studied compared to its black-water and white-water counterparts, but regionally highly threatened by many hydropower dams, road networks, deforestation, fires and mining activities on small (*garimpo*) and large (concessions) scales (Conde et al., 2024; Wittmann & Junk, 2016). Most of the northern affluents have their headwater regions inserted into large complexes of protected areas, while those in the south have only some protected river sections integrating floodplains.

3 Adaptations of Tree Species to the Flood Pulse

The flood pulse is the main environmental filter for the tree species (Junk, 1989). All tree species can survive the extreme hydric conditions due to their different adaptations (for herbaceous species, see Lopes et al., Chapter 6). Seed dispersal of many tree species in the floodplains is adapted to the predictable seasonal flooding. Most species fruit during the aquatic phase (Kubitzki & Ziburski, 1994; Parolin et al., 2010; Schöngart et al., 2002) and seeds developed adaptations (i.e., air-filled tissues, hairs that retain air) to float on the water for several weeks (Waldhoff et al., 1996) or even to germinate underwater (De Melo et al., 2015). Generally, water dispersal (hydrochory) is unilateral (downstream); however, in floodplain lakes, water movements can occur in different directions due to changes in river inundation and local runoff, altering flow directions (Lesack & Melack, 1995). Seeds are also distributed by the fauna, where fishes (ictyochory) play an important role as they disperse diaspores upstream and between floodplain systems (Correa et al., 2015; Gottsberger, 1978; Weiss et al., 2016). These specific dispersal mechanisms contrast with those from the non-flooded upland (*terra firme*) forests (Weiss et al., 2023).

Seasonal floods induce anoxic conditions as the remaining oxygen is consumed within days by microorganisms and fine roots, reducing the soil's redox potential and leading to the formation of phytotoxic compounds such as Mn^{2+}, Fe^{2+} and H_2S (Gambrel & Patrick, 1978, Crawford, 1992). To survive these harsh seasonal environmental conditions, tree species have developed and combined different adaptive mechanisms (Fig. 2) at morpho-anatomical (i.e., hypertrophic lenticels, adventitious roots, aerenchyma formation and suberization of roots) and physiological

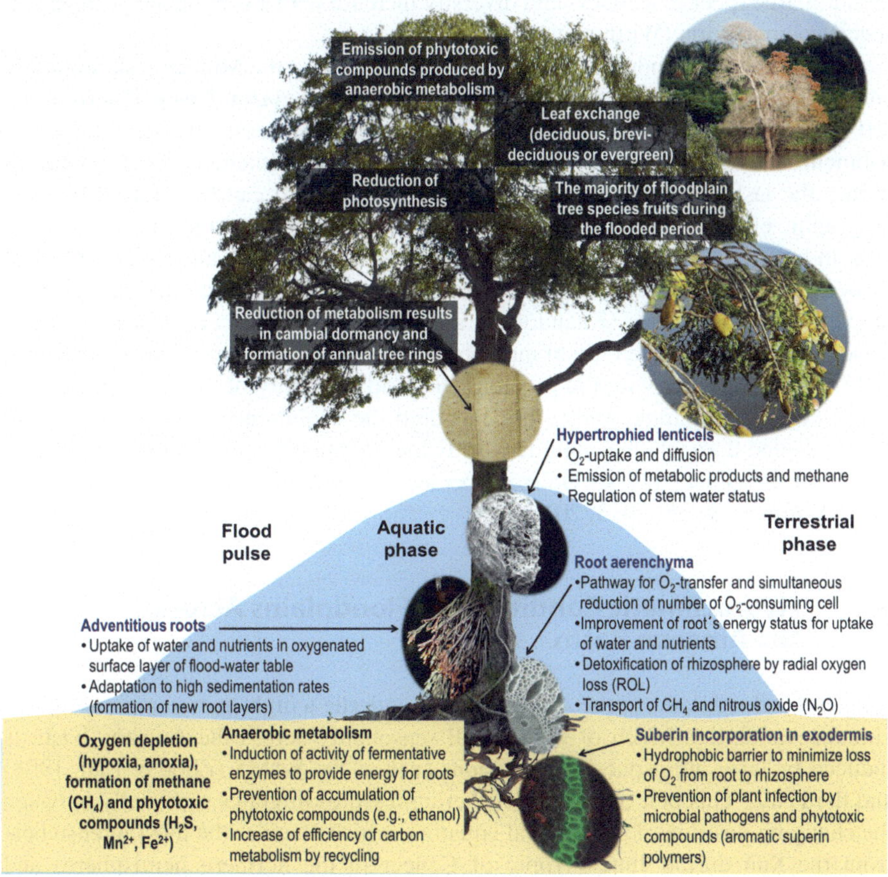

Fig. 2 Adaptive mechanisms of adaptations to tolerate floodings of Amazonian floodplain tree species and their functions

levels (i.e., anaerobic metabolism, elimination of phytotoxic compounds and phytohormones) (De Simone et al., 2002; Ferreira et al., 2009; Haase & Rätsch, 2010; Junk, 1989; Parolin et al., 2004; Piedade et al., 2010; Schlüter & Furch, 1992). This allowed them to colonize the floodplains over evolutionary periods (Wittmann et al., 2010) up to the maximum flood-induced tree line, which is lower in the central Amazonian *igapó* (up to 10 months per year on average corresponding to maximum flood heights of up to 9 m) compared to the *várzea* (up to 8 months per year corresponding to maximum flood heights of 7 m on average) (Wittmann et al., 2010; Wittmann et al., 2022a). Possible explanations for this are higher light availability and oxygen levels of the bottom water in the *igapó* compared to the *várzea* due to the higher water transparency and slower decomposition rate of organic material (Wittmann et al., 2022a) and/or more sophisticated, so-far unknown, adaptations of the *igapó* tree species to long-term floodings. The level of specific flood adaptation is reflected by a high turnover of tree species along topographical elevation in the floodplains, as many of them have narrow ranges of flood tolerance (Householder et al., 2021; Wittmann et al., 2010; Worbes, 1997). With decreasing flood height and duration, tree species' richness and diversity increase with a simultaneous change in species composition (Wittmann et al., 2010).

The uplift of the Andean mountains, primarily during the Miocene, significantly altered the quantity and seasonality of rainfall in the Amazon basin (Hoorn et al., 2010). The formation of the modern transcontinental Amazon River and the establishment of riverine connections among the principal sedimentary basins occurred during the last ten million years (Hoorn et al., 2017). It can be inferred that the orographic effect of the Andean uplift, combined with the established monsoon climate induced by the South American Monsoon System (SAMS) and the Intertropical Convergence Zone (ITCZ), resulted in a distinctive seasonal precipitation regime in the vast river catchments (Kaandorp et al., 2005), generating a regular and annual flood pulse. The high diversity of macrohabitats in the floodplains (Junk et al., 2012, 2015, 2018), together with the predictability and regularity of the flood pulse over long time periods (Junk, 1989), has favoured the evolution of a species-rich and flood-adapted flora which is unique worldwide (Wittmann et al., 2006a) with a high degree of endemism and specialized species (Householder et al., 2024; Kubitzki, 1989; Wittmann et al., 2013, 2017).

4 Hydroclimatic Conditions in Floodplains Across the Amazon Basin

Large river-floodplains in the Amazon basin underlie a high spatiotemporal hydroclimatic variability. Much of the rainfall seasonality across the Amazon basin is induced by large atmospheric circulations such as the SAMS (Zhou & Lau, 1998) and the ITCZ. This belt of low atmospheric pressures migrates seasonally between the changing position of the thermal equators, which receives the most intense heat from the Sun during June (Tropic of Cancer in the northern hemisphere) and December (Tropic of Capricorn in the southern hemisphere) solstices. The seasonal, regular and predictable oscillation of the ITCZ induces a distinct rainfall seasonality across the Amazon basin, which presents a shift of more or less 6 months between both hemispheres. Coupled with the ITCZ and the Hadley Cells, southeast and northeast trade winds resulting from the Coriolis force and acting on the near-surface branch of the Hadley Cells import large amounts of water vapour generated by the warm surface waters of the tropical Atlantic into the Amazon basin. The westwards shifting air masses across the Amazonian rainforests concentrate increasing water vapour by biosphere–atmosphere interactions providing an effective water recycling by evapotranspiration processes (Salati & Vose, 1984) and by biogenic emissions (pollen, spores, volatile organic compounds, biogenic salts) from the forests which pass through complex processes in the atmospheric boundary layer forming organic aerosols functioning as cloud condensation nuclei (Pöhlker et al., 2012; Pöschl et al., 2010). Due to the Andean Cordillera in the west (orographic effect), annual rainfall increases from about 1200–1600 mm in the southeast towards 3600–4000 mm in the northwest (Fig. 1), accompanied by a declining dry season length (monthly rainfall <100 mm) from 5 to 7 months (southeastern section) towards zero (northwestern section) (Sombroek, 2001). Reaching the Andean Cordillera, the concentrated humid air masses drift southward during austral summer (wet season) along this physical barrier forming so-called aerial rivers providing humidity for other regions of Brazil and South America (Arraut et al., 2012).

The spatiotemporal variations of rainfall in the watersheds shape the flood pulses of the large Amazonian rivers, which present different periods of annual water level extremes (flood and drought) and amplitudes (difference between minimum and maximum water levels) (Fassoni-Andrade et al., 2021; Fig. 1). Rivers with catchment situated in the northern hemisphere generally have the minimum water level in January–March period, while those located in the southern hemisphere have the minimum stages around August–November. Along the Solimões-Amazon mainstem in the central portion of the basin, the annual minimum water level occurs in the September–November period with a shift of about 2 months from the west (September) towards the east (November) (Fassoni-Andrade et al., 2021). This time lag is explained by the slow progression of the generated flood wave along the transcontinental mainstem and the water storage in the connected floodplains (Brêda et al., 2023; Paiva et al., 2013). The maximum water levels present a shift of about 3–4 months from the southern (February–April), across the central (May–July) and towards the northern (June–August) section of the Amazon basin (Fig. 1).

The annual amplitude of the flood pulse depends mainly on the amount and seasonality of rainfall, catchment size and geomorphological characteristics of the watershed. However, alterations can occur due to land-use changes, such as large-scale deforestation of the headwater regions and hydropower dams (e.g. Costa et al., 2003; Timpe & Kaplan, 2017). With the onset of the wet season, the increasing frequency and magnitude of rainfall events in the headwater regions result in unpredictable, irregular and frequent (polymodal) flood pulses with low amplitude in the lower-order rivers (creeks) (Strahler, 1957). As the downslope moving water masses integrate the discharge from headwater areas, the flood pulse gradually transforms along the drainage system to a predictable, regular and monomodal shape, generally increasing the annual amplitude, characteristic of the high-order rivers. Nevertheless, also so-called backwater effects, which are a result of the hydraulic barrier created by a river with higher order upon a river of lower order, influence the hydrological regime in the upstream direction of the low-order river (Meade et al., 1991; Paiva et al., 2013). Especially, in the flat central Amazon basin, where several large rivers (Juruá, Japurá-Caquetá, Purus, Negro and Madeira rivers) have their confluence with the Solimões-Amazon mainstem, backwater effects can reach large extensions (Junk et al., 2015; Meade et al., 1991). This central part of the Amazon basin has a flood-pulse regime with an annual amplitude of 9–11 m, with maximum water levels in May–July and minimum during September–November (Fassoni-Andrade et al., 2021). From the central region towards the east and west, the amplitude of the Solimões-Amazon mainstem decreases. White-water rivers draining the southwestern part of the Amazon region (e.g. Juruá, Purus and Madeira) have in their middle stretch annual amplitudes of 12–15 m on average as their catchments integrate large regions with a distinct rainfall seasonality and along the extensive river courses annual precipitation increases (Fig. 1). The other rivers generally present lower flood amplitudes which tend to increase from the headwater regions towards the middle stretches and then either increase (tributaries from the northwestern and northern Amazon basin) or decrease (tributaries from the southern and eastern Amazon basin) towards the river mouth (Fassoni-Andrade et al., 2021) because of the backwater effects induced by the Solimões-Amazon mainstem with the highest river order (Meade et al., 1991).

The high spatiotemporal variability of flood-pulse patterns and rainfall regimes (Fig. 1), together with varying soil textures, plant communities and distinct microclimatic features, results in complex hydroclimatic conditions which are distinct from the non-flooded *terra firme* forests (Wittmann et al., 2022a). Rainfall regimes and flood pulses have a synchronized pattern in the headwater regions. The low amount of rainfall, the distinct seasonality and extensive dry season (monthly rainfall <100 mm for 5–7 months) in the huge savanna complexes of the north (*Lavrado* and the central region of the Guiana Shield), south (*Cerrado*) and southwest (*Chaco*) of the Amazon basin generate a severe cumulative hydroclimatic water deficit in the exposed river-floodplains of predominantly white-water (southwestern Amazon basin) and clear-water (northern and southern Amazon basin) rivers (Fleischmann et al., 2023a). Towards the equatorial region, the climatological water deficit in these river-floodplains tends to decline as rainfall increases with a simultaneous decrease in the dry-season length. In the case of extensive large rivers with small altitudinal gradients, the slow movement of the flood wave from the headwater region towards its mouths induces a delay of the flood pulse for several weeks in comparison to the rainfall regime, as, for instance, along the Solimões-Amazon mainstem in the central and eastern Amazon regions (Fassoni-Andrade et al., 2021; Schöngart et al., 2002).

Consequently, the climatological precipitation water deficit during the first part of the dry season gets buffered as the large areas of the central Amazonian floodplains are still flooded, and the minimum water level coincides with the transition from the dry to the wet season. In river-floodplains in the eastern Amazon, along the mainstem (tidal *várzea*) and lower sections of the northeastern and southeastern affluents (mainly clear-water *igapós*), the climate-induced water deficit is further buffered by tidal inundations originating from the Atlantic Ocean, as measurable up to the region of the city of Santarém, located approximately 600 km upstream the Amazon mouth to the Atlantic (Meade et al., 1991). Black-water *igapós* occur mainly in the northwestern Amazon basin with generally high annual precipitation and a short or even absent dry season (Sombroek, 2001). The *várzea* vegetation occurs along a gradient from perhumid conditions in the northwest (i.e. Caquetá-Japurá River basin) to annual precipitation regimes within the savanna biome with a dry season length of up to 7 months (i.e. Beni River basin) in the southwest of the Amazon basin. Regarding the hydroclimatic conditions, *várzeas* as such occupy a much larger hydroclimatic gradient than the black-water *igapós*.

5 Microclimatic Conditions of Large River-Floodplains

Large Amazonian river-floodplains comprise a mosaic of open water (lakes, rivers and channels), bare soils (e.g. point bars) and habitats covered by various plant communities (aquatic, semi-terrestrial and terrestrial herbaceous species, shrub and tree species) (Melack & Hess, 2010; Wittmann et al., 2022a). The flood-pulsing rivers lead to changing extension of the inundated area over the years (Fleischmann et al., 2022; Hess et al., 2015). In the case of the sediment-loaded white-water rivers, the fluvial dynamics results further in new arrangements of the diverse habitats as a consequence of the intense sedimentation and erosion processes (Kalliola et al., 1991; Peixoto et al., 2009; Salo et al., 1986). This results in a high spatiotemporal variability of microclimatic conditions across the alluvial landscape as albedos (fraction of solar radiation that is reflected) vary between the water surface, bare soils and habitats covered by different plant communities (Ponce et al., 1997). A forest canopy generally has a low albedo leading to a high absorption of solar radiation. This provides energy for trees for photosynthesis and transpiration leading to a high latent heat loss cooling the surface. Bare soils, especially sandy soils, have a higher albedo, absorbing less energy at the surface. As the excess energy cannot be removed by plant transpiration, the latent heat loss is reduced, and the increased sensible heat loss warms the surface (Foley et al., 2003). While exposed substrates without vegetation present an extreme thermal oscillation during the day, aquatic habitats have attenuated diurnal temperature variations as water acts as a buffer because of its high specific heat capacity (Tonolla et al., 2010). Minimum temperatures at night, for instance, can be up to 4 °C higher during the aquatic phase compared to the terrestrial phase (Irmler, 1986) and adjacent *terra firme* forests. These factors together produce a complex and spatiotemporally varying microclimate across the large river-floodplain landscape which is subjected to higher ranges of temperature and less rainfall compared to *terra-firme* forests (Fleischmann et al., 2023a; Ribeiro & Adis, 1984; Wittmann et al., 2022a).

Várzea floodplains generally have a lower forest coverage than the *igapós* (Melack & Hess, 2010) because of the lower flood-induced tree line combined with constantly deposited, unvegetated substrates (Wittmann et al., 2022a). Due to the constant hydrogeomorphological disturbances, the *várzea* has heterogeneous soil textures with highly varying water retention capacities associated with topographical elevations and flooding conditions. This is reflected by different successional stages of herbaceous plant communities and forests with varying structures, net primary productivities and dynamics (Piedade et al., 1991; Wittmann et al., 2004). While the main growing season of the floodplain trees is the terrestrial phase (Schöngart et al., 2002; Worbes, 1997), the highly productive semi-aquatic and aquatic herbaceous vegetation grows mainly during the flooded period (Piedade et al., 2001). Compared to the *várzea*, *igapós* are floodplains with a high cover of century-old forest communities with an overall lower net primary production (NPP) due to the oligotrophic conditions leading to the absence of high-productive herbaceous plant communities (Junk et al., 2015). *Igapó* forests have a reduced water retention capacity, occasionally resulting in a strong water deficit and turning this ecosystem vulnerable to fire, especially during El Niño events (Flores & Holmgren, 2021, Carvalho et al., 2021; see also Sect. 9.1). The silty and clayish soils of *várzea* forests can store water for much longer periods, buffering the effect of the climatological water deficit and turning this ecosystem more resilient against fire (Wittmann et al., 2022a). Overall, this suggests distinct microclimates within and between the main floodplain ecosystems.

6 Resilience and Vulnerability of the Amazonian Floodplain Vegetation Against Disturbances

Flood tolerance, natural disturbance regimes and the resulting patterns of ecological succession shape tree species composition, diversity, forest structure and dynamics in floodplains which vary considerably between the *várzea* and the *igapó*. In the *várzea*, the high fluvial dynamics lead to continuous hydrogeomorphological disturbances driven by sedimentation and erosion processes shaping the alluvial landscape (Salo et al., 1986). This leads to the formation of diverse habitats such as point bars, islands, swales, ridges, lakes and secondary river channels (Wittmann et al., 2022a). In general, the fluvial dynamics are more intense in the western Amazon basin compared to the central and eastern parts (Kalliola et al., 1991, Peixoto et al., 2009). Over areas created by sedimentation processes, primary succession initiates and passes through different stages along the development (Wittmann et al., 2004; Worbes et al., 1992; Fig. 3). The successional dynamics in the *várzea* result from changes in coupled environment-vegetation processes. Plants of a respective successional stage in the floodplain alter the environmental conditions that allow the subsequent plant community to establish, as described by the concept of facilitation, tolerance and inhibition (Connell & Slatyer, 1977). New areas formed by sedimentation deposits in the *várzea* are first colonized by herbaceous plants (mainly C4 grasses such as *Paspulum* and *Echinochloa*), characterized by a high tolerance of solar radiation (up to 100%) and NPP of 30–100 Mg/ha/year dry biomass with a high turnover rate due to their annual life cycle (Engle et al., 2008; Piedade et al.,

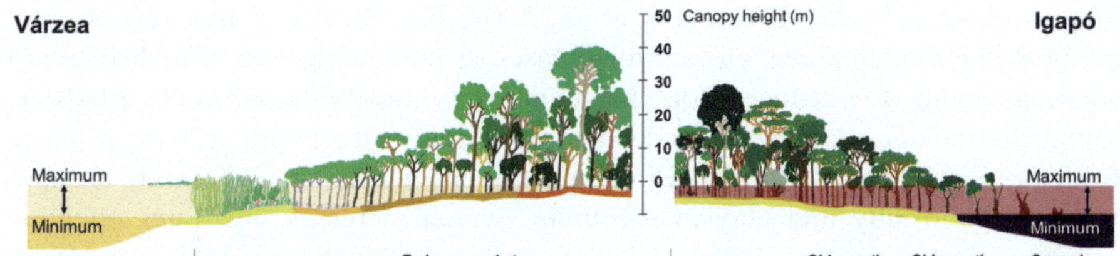

Environmental factors	Herbaceous communities	Pioneer stages	Early secondary stages	Late secondary stages	Late stages	Old-growth forests	Old-growth forests	Old-growth forests	Opened monodominant forests
Habitat stability[1,2]	Unstable through sedimentation and erosion processes and channel migration					Stable during centuries to millennia			
Habitat diversity[3]	High through small-scale hydro-geomorphic dynamics, heterogenous microclimate and different flood levels					Low, mostly defined by flood-levels			
Average inundation height[4,5]	8-10 m	5-7 m	4-6 m	3-5 m	<3 m	<3 m	3-5 m	5-7 m	8-10 m
Average inundation length[4,5] (days)	260-310	180-240	150-200	100-180	<100	<100	100-180	180-240	260-310
Sedimentation rate[6]	dm-m	cm-dm	mm-cm	mm	mm	<mm			
Soil texture[4,6,7]	sand	fine sand	silt		clay	silt-clay upon sandy soil bed			
Soil fertilty[8]	high					low			
rPAR(%)[9,10]	70-100	30-70	20-30	5-10	<5	<5	5-10	10-25	20-70
Soil water retention capacity[11]		low	very low	intermediate	high	low	low	very low	low
Drought vulnerability[11]		high	intermediate	low	low	intermediate	high	very high	very high
Vulnerability to fire[9,11]		low	low	low	low	intermediate	high	very high	intermediate
Life history traits									
Leaves[12,13]	Larger leaf area, higher specific leaf area, leaf nitrogen, phosphorus and potassium concentrations, and larger proportion of deciduous tree species					Sclerophyllic, higher leaf dry matter content, leaf chlorophyll content, lenticel density, and a larger proportion of evergreen tree species			
Fine roots[14]	Roots with depth >40 cm					Root mats on top of soil surface and concentrated up to 30 cm depth			
Mean wood density[15,16]		0.3-0.4	0.4-0.5	0.5-0.6	0.5-0.6	0.6-0.7	0.6-0.7	0.6-0.7	0.7-0.8
Mean diameter growth rates (mm/year)[17,18]		10-20	5-10	2-5	2-5	2.5-3.5	2-3	2-2.5	<2
Stand age (max. tree age) (years)[18-20]		<30	50-150	>200	>200	>400	>400	>500	>800
Overall strategy[12]	Resource acquisition					Resource conservation and tissue conservation			
Vegetation structure									
Tree density/ha (>10 cm DBH)[5,6]		100-200	600-1000	400-600	400-600	400-700	600-1000	600-1000	100-200
Tree species richness[5,6]		1-3	30-50	60-90	>100	50-70	40-50	20-40	1-3
Upper canopy height (m)[18,20]		<20	20-30	25-35	30-40	20-30	15-25	15-20	10-15
Biomass (Mg/ha)[16,20]	60-80	<100	100-200	200-250	200-300	150-300	150-300	150-300	50-150
NPP (Mg/ha/year)[16,20]	30-100	20-30		15-20		<15			?

Fig. 3 Main environmental characteristics and biotic features of central Amazonian floodplain forests (nutrient-rich *várzea* and nutrient-poor *igapó*) along the flood gradient. References: [1]Junk et al. (2015), [2]Kalliola et al. (1991), [3]Junk et al., 2018, [4]Wittmann et al. (2004), [5]Aguiar (2015), [6]Wittmann et al. (2010), [7]Latrubesse and Franzinelli (2005), [8]Furch (1997), [9]Wittmann and Junk (2003), [10]Carvalho (2019), [11]Wittmann et al. (2022a), [12]Mori et al. (2021), [13]Parolin (2001), [14]Meyer et al. (2010), [15]Wittmann et al. (2006b), [16]Schöngart et al. (2010), [17]Worbes et al. (1992), [18]Schöngart (2003), [19]Resende et al. (2020) and [20]Batista (2015)

1991, 2001). This favours sedimentation allowing the colonization of fast-growing shrub and tree pioneer genera (e.g. *Salix*, *Alchornea* and *Tessaria*) with short longevity (<10–15 years). Forming opened canopies with low heights, they start to shade out herbaceous plants and create favourable conditions for moderately shade-tolerant genera forming early (e.g. *Cecropia*, *Nectandra* and *Ficus*; up to 40 years old) and late (e.g. *Pseudobombax* and *Luehea*; up to 100 years old) secondary stages (Schöngart, 2003; Worbes et al., 1992). Over time (decades to centuries), these finally develop to species-rich late successional stages (>200 years old) at lower (e.g. *Piranhea*, *Oxandra* and *Handroanthus*) and higher elevations (e.g. *Ceiba*,

Maquira and *Sterculia*) (Wittmann et al., 2004; Fig. 3). Along this successional gradient, the topographical elevation increases several metres over time because of siltation, leading to a reduced flood height and duration (Wittmann et al., 2010). As sedimentation rates decline (from decimetres to millimetres) with growing distance from the main channel and increasing topographical elevation, soil texture changes from sandy to silty and clayish substrates, which increases the water retention capacity (Wittmann et al., 2022a). This favours the formation of higher and more stratified canopies reducing solar radiation on the forest floor (from 70–100% to < 5% relative photosynthetically active radiation—rPAR) changing the microclimate (Wittmann et al., 2022a).

The oligotrophic black-water *igapó* has, in general, a more stable geomorphology due to the neglectable sediment load (Richey et al., 1986), especially in the river-floodplains. Mainly, islands in anabranching and braided (Mariuá and Anavilhanas archipelagos of the Negro River) river channels present visible small-scale successional patterns as a consequence of slowly moving channel bedloads of sandy substrates (Wittmann et al., 2022a). The zonation of forest vegetation in this ecosystem is primarily the result of the adaptive capacity to tolerate flooding (Householder et al., 2021) as the variation of soil texture is relatively low (Wittmann et al., 2022a). This results in more homogenous landscapes dominated by century-old forests down to the treeline defined by the maximum flood tolerance (Junk et al., 2015; Resende et al., 2020; Fig. 3). For the clear-water *igapós*, not much information is available so far about vegetation structure, species composition and dynamics associated with natural disturbance regimes indicating the urgent need for further studies. Therefore, the following sections will focus mainly on the *várzea* and the black-water *igapós*, which have contrasting environmental features.

Although the central Amazonian *várzea* and *igapó* are influenced by similar flood pulse regimes which require complex mechanisms of adaptations against inundation (anoxic conditions), the tree species similarity between the two floodplain systems is low with about 10% (Wittmann et al., 2022a), and overall, *igapós* have a lower tree species richness (Montero et al., 2014; Wittmann, 2012). Therefore, it can be suggested that the contrasting nutrient levels and microclimatic conditions are additionally important and effective environmental filters leading to the selection of a distinct tree species flora along evolutionary time scales in both floodplain ecosystems. Tree species in *igapós* adapted to the low nutrient status which selects for traits favouring resource conservations and high investment in persistent tissues (Mori et al., 2021; Fig. 3). This is reflected by generally higher leaf dry matter contents (especially carbon), higher seed masses and wood densities, lower diameter growth rates and biomass production as well as higher longevities of trees (Mori et al., 2019; Parolin, 2000; Parolin et al., 1998; Resende et al., 2020; Schöngart et al., 2005, 2010) compared to the eutrophic *várzea*. The high nutrient status in the *várzea*, maintained by the regular deposits of sediments by the flood pulse, allows tree species to invest in functional traits for resource acquisition. Tree species in this environment tend to have leaves with higher nutrient contents (especially N and P), a higher photosynthetic capacity with shorter lifespans (a large proportion of the species are deciduous), higher growth rates in diameter and biomass and shorter longevities (Furch, 1997; Mori et al., 2019; Parolin, 2009; Schöngart, 2010; Schöngart et al., 2005; Wittmann et al., 2022a; Worbes, 1997; Fig. 3). With different setups of adaptations to the changing environmental conditions and filters along the successional gradient, the plant communities of the *várzea* have effective mechanisms for resource allocation to increase biomass stocks up to 250 Mg/ha (dry weight) within 50 years (Schöngart et al., 2010). Fine roots of trees in the eutrophic *várzea* have a depth of several decimetres, and some species (i.e. *Salix martiana*) even form new horizontal layers of adventitious roots due to high sedimentation rates (Wittmann & Parolin, 2005; Worbes, 1997). The oligotrophic *igapó* stocks on predominantly sandy soil beds and trees form a dense system of fine roots with intensive branching near or even on top of the soil surface indicating an efficient nutrient uptake that derives primarily from the litter (Furch, 1997; Meyer et al., 2010).

In summary, it can be concluded that the primary environmental filter in the *igapó* is the adaptive capacity to tolerate flooding (Householder et al., 2021), as the geomorphology is more stable, and soil texture and microclimatic conditions seem to be more homogeneous compared to the *várzea* (Wittmann et al., 2022a). A second important filter is the low nutrient status requiring different setups of functional characteristics to adapt to the harsh edaphic conditions in the *igapó* (Mori et al., 2021). The eutrophic and dynamic *várzea* has a high spatiotemporal variability of hydroclimatic and microclimatic conditions on small and large scales which generate a set of simultaneously acting environmental filters. We can assume that these environmental filters probably resulted in selecting a large and distinct pool of tree species with specific adaptations to its complex and high environmental dynamics. Therefore, we hypothesize that the *várzea* has a higher adaptive capacity to cope with changing conditions and, consequently, resilience when compared to the oligotrophic *igapó*, which is vulnerable to disturbance regimes as shown in the subsequent sections. For the clear-water *igapós*, almost no data and information is available to discuss the capacity of resilience, adaptations and vulnerability. Future studies are necessary to understand the resilience of the floodplain vegetation from different systems and regions with contrasting environmental conditions across the Amazon basin, integrating analyses of phylogenetics and functional characteristics. Understanding the evolutionary history of the tree flora in the floodplains will provide insights into their vulnerability and resilience against the intensifying hydrological regime, increasing temperatures and alterations of land-uses in the Amazon basin, which are discussed in the following sections.

7 Role of Coupled Ocean-Atmosphere Variability on Hydroclimatic Conditions in the Amazonian Floodplains

Although the regular periodicity of the flood pulse and precipitation regimes across the Amazon basin are mainly defined by the ITCZ, large-scale climate phenomena resulting from coupled atmosphere-ocean variability at interannual (high frequency) and decadal (low frequency) time scales influence the hydrological regimes. This can generate extreme floods and meteorological and hydrological droughts in the Amazonian floodplains. Many studies have shown that the seasonal and interannual variability of precipitation regimes and hydrological cycles of large rivers in the Amazon region can be partly attributed to variations in sea surface temperature (SST) anomalies across the equatorial Pacific (El Niño-Southern Oscillation—ENSO) and the tropical Atlantic (Atlantic Meridional Mode—AMM) (e.g. Marengo & Espinoza, 2016; Yoon & Zeng, 2010).

During a positive AMM phase with an anomalously warm tropical North Atlantic (TNA), the ITCZ is displaced northwards (Chiang & Vimont, 2004), and air humidity generated by the warmed water surface does not reach large parts of the Amazon basin, generating severe meteorological drought conditions, mainly in the southern and southwestern regions of the basin during the dry season (Lewis et al., 2011; Marengo et al., 2008; Marengo et al., 2021). The accumulating climatic water deficit during the dry season of the southern Amazon basin induced by the AMM gets integrated by the drainage system at large scales. This generates hydrological droughts in the downstream areas that affect the large river-floodplains in regions that can even be out of the spatial range of the meteorological droughts with a time lag of up to several weeks or even months. For instance, the generated meteorological droughts during the dry season (July–September) in southwest Amazonia in 2005 resulted in extreme hydrological drought conditions in the floodplains along the central Amazon mainstem towards the end of the dry season (October–December) (Marengo et al., 2008).

Severe hydroclimatic events influencing the floodplains in the northern, central and eastern regions of the Amazon basin are generally influenced by the ENSO, characterized by warm (El Niño) and cold phases (La Niña). El Niño events decrease rainfall and increase air temperature in those regions, while La Niña events have the opposite pattern (Cai et al., 2020; Foley et al., 2002; Marengo & Espinoza, 2016). The cumulative negative rainfall anomalies in the affected Amazonian river catchments during El Niño events result in a decrease of the streamflow regime (hydrological drought), extending the terrestrial phase in the large river-floodplain and leading to a reduced maximum water level and shortened aquatic phase (Schöngart et al., 2004; Schöngart & Junk, 2007). Especially, El Niños with the magnitude of increased SSTs in the eastern equatorial Pacific (EP) causes a larger and longer persisting footprint of reduced precipitation and streamflow compared to those located in the central equatorial Pacific (CP) (Cai et al., 2020). Strong EP El Niños on record are the episodes 1982/1983, 1991/1992, 1997/1998, 2023/2024 and partially 2015/2016 (mix of CP and EP), which had severe impacts in their footprints within the Amazon basin (Marengo & Espinoza, 2016; Marengo et al., 2024), generating huge cumulative water deficits turning forests vulnerable to fires (Aragão et al., 2018; Nepstad et al., 1999). For the La Niña episodes, CP events have stronger and longer impacts than EP events, increasing rainfall in the Amazon basin with a similar footprint as the EP El Niños (Cai et al., 2020).

In addition to the interannual variability of coupled ocean-atmosphere interactions, low-frequency oscillations strongly influence the precipitation and streamflow regime in the Amazon basin. The Atlantic Multidecadal Oscillation (AMO) is the dominating mode influencing the rainfall and streamflow regimes, mainly in the south and southwest of the Amazon basin (Aragão et al., 2018). The AMO is characterized by cold (negative) and warm (positive) phases with cycles of approximately 60–80 years duration (Enfield et al., 2001; Kerr, 2000; Trenberth & Shea, 2006) that result from an ocean-atmosphere interaction related to low-frequency fluctuations in the thermohaline circulation (Knight et al., 2005). Cold phases of the AMO occurred in the periods 1902–1925 and 1965–1994. The periods from 1930 to 1964 and since the mid-1990s are characterized by warm phases (Parker et al., 2007; Sutton & Dong, 2012; Trenberth & Shea, 2006). The Interdecadal Pacific Oscillation (IPO) has a tripolar pattern across the Pacific basin (equatorial region, northern and southern hemispheres). The IPO has SST anomaly patterns in the tropical Pacific similar to ENSO but in a decadal mode of changing warm and cold phases (Salinger et al., 2001) with cycles (warm+cold phase) of 40–60 years. Warm phases persisted in 1924–1944 and from 1977 to the end of the 1990s, while cold phases occurred in 1945–1976 and from the 2000s onwards (Henley et al., 2015). The influence of warm ENSO phases (El Niño) on hydroclimatic regimes of Amazonia is strengthened when they occur in-phase with the decadal oscillations (warm IPO phase) (Wang et al., 2014). The same holds for La Niña events which have stronger impacts on rainfall regimesduring cold phases of the IPO (Campozano et al., 2020; Nguyen et al., 2021).

Low-frequency oscillations of the Atlantic (AMO) and Pacific (IPO) have only recently been described as large-scale phenomena in the global climate system (Mantua & Hare, 2002, Kerr, 2000, Salinger et al., 2001), but their influence on hydroclimatic regimes in the Amazon basin is difficult to explore, as most instrumental time series hardly have more than 50 years on record (Marengo et al., 2021). However, exactly dated climate-sensitive tree-ring records and derived stable oxygen time series from the Amazon basin spanning over 100 years highlight the low-frequency mode (Brienen et al., 2012; Granato-Souza et al., 2020; Schöngart et al., 2004) suggesting that multidecadal oscillations are important drivers of hydroclimatic regimes and consequently of tree growth and forest dynamics in the Amazonian floodplains.

In the following, we analyze the over 120-year-long time series of daily water level measurements at the Port of Manaus (available since September 1902), which is the longest continuous instrumental record in the Brazilian Amazon (Fig. 4). This record is a robust hydroclimatic indicator, as rainfall anomalies and associated large-scale ocean-atmospheric factors responsible for much of its variability are integrated across the catchments of the Solimões and Negro rivers with an area of approximately three million square kilometres (Schöngart & Junk, 2007). The annual maximum levels (floods) occur between the second half of May and the first half of August with the highest frequency during June (75%). The temporal variability of the minimum water level is higher and about 60% appears between the second half of October and the first half of November (Schöngart & Junk, 2020). The average (±standard deviation) of the maximum water level is 27.87 ± 1.15 m and attains in 69% of the years (1903–2015) at stages between 27 and 29 m with reference to the Port of Manaus. The average minimum water level is 17.66 ± 1.84 m

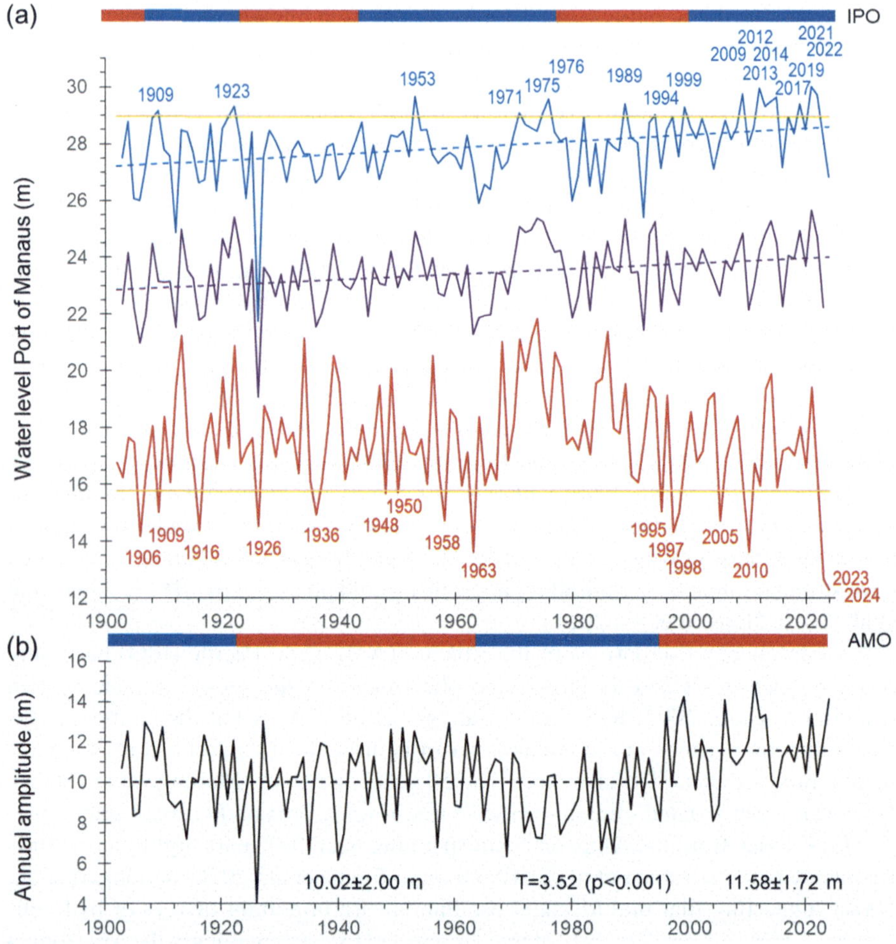

Fig. 4 Intensification of the hydrological cycle in the central Amazon basin. (**a**) Over 120-year long record (Set. 1902—Oct. 2024) of annual maximum (blue; calendar years represent severe flood events corresponding to emergency threshold ≥ 29.0 m indicated in orange), mean (violet) and minimum (red; calendar years represent extreme hydrological drought events corresponding to emergency threshold < 15.8) water levels. Dashed lines indicate significant trends at a 99.9% confidence level evidenced by the Mann-Kendall trend test. (**b**) Annual flood amplitude (difference between the annual maximum and minimum water level of the previous year). A T-test indicates the significant difference in the flood amplitude between this and the previous century (Water level data: National Information System on Water Resources—SNIRH of the Brazilian National Water and Sanitation Agency—ANA). Warm (red) and cold (blue) phases of the Interdecadal Pacific Oscillation (IPO; Henley et al., 2015) and the Atlantic Multidecadal Oscillation (AMO; Sutton & Dong, 2012) are indicated

and occurs in 46% of the cases at stages between 16 and 18 m. The annual mean amplitude (difference between the minimum and subsequent maximum water level) is about 10.2 m (Fig. 4). The absolute maximum value is 30.02 m (16 June 2021), and the lowest water level on record is 12.13 m (01 November 2024; data: Hidroweb/ Brazilian National Water and Sanitation Agency—ANA; fluviometric station code 14990000). Both flood and drought regimes, have a huge inter-annual variation, and extreme values of both regimes can vary by more than 8 m. The highest minimum water level attained 21.84 m (1974) and was even higher than the lowest maximum level on record at 21.76 m (1926). So far, 16 severe droughts and 18 extreme flood events have been recorded (1902–2024). Extreme events are those attaining the threshold of 29.0 m (floods) and 15.8 m (droughts), leading to the declaration of an emergency (Maciel et al., 2020).

Almost 89% of the 18 severe flood events on record are associated with the cold IPO-phase (Fig. 4). Many of the severe flood events (1971, 1975, 1976, 1989, 1999 and 2021) are caused by CP La Niña conditions (Cai et al., 2020, Espinoza et al., 2022), while others are associated with warm SST anomalies of the tropical South Atlantic (2009) or both oceanographic regions (2012) (Marengo et al., 2021). However, the severe flooding in 2014 in the Madeira River basin has been associated with warm SST anomalies in the subtropical South Atlantic Ocean and western Pacific-Indian Ocean (Espinoza et al., 2014). A remarkable pluriannual event is the climate shift in the 1970s (Marengo et al., 2021), characterized by consecutive CP La Niña conditions (Cai et al., 2020) at the end of an intense cold IPO phase (Henley et al., 2015). This provoked extremely high minimum, mean and maximum water levels in the central Amazonian floodplains over four consecutive years (1971–1974) (Fig. 4). The water level remained 1767 days above the 20-m stage with reference to the Port of Manaus which is representative for the lower Negro River due to backwater effects (Carvalho et al., 2021; Meade et al., 1991). The generated permanent flooding conditions at the low topographical elevations of the central Amazonian *igapó* and *várzea* resulted in massive large-scale mortality of shrubs and trees (Piedade et al., 2013), exceeding the adaptive capacity to cope with long-term flooding of these species (Assahira et al., 2017; Resende et al., 2020). Especially, the slow dynamics of the oligotrophic *igapó* seem to respond to the multidecadal oscillations, particularly to the IPO, which has a strong influence on the hydroclimatic conditions in *igapó* regions (Aragão et al., 2018). Growth and mortality patterns of *Eschweilera tenuifolia* (Lecythidaceae), forming monodominant stands at the flood-induced tree line of the central Amazonian *igapó* (Fig. 3), are exposed to an average aquatic phase of more than 10 months per year (Resende et al., 2020). Analysis of exactly dated basal area increments by dendrochronological techniques indicates distinct clusters corresponding with the phase changes of the IPO around 1945, 1975 and at the end of the 1990s (Henley et al., 2015). Exploring the year of death by radio-carbon dating, Resende et al. (2020) indicate that mortality of *E. tenuifolia* occurs solely within the cold IPO phases close to La Niña events which are favoured by the in-phase constellation (Cai et al., 2020).

About 87% (13 from 16 events on record) of extreme hydrological droughts occurred during warm AMO phases (Fig. 4). Some of these severe drought events (2010, 1963 and 2005) have been mainly caused by warming of the TNA (positive AMM phase) (Marengo et al., 2021), while others are mainly favoured by EP El Niños during warm IPO phases, especially the episodes 1982/1983, 1991/1992 and 1997/1998 (Cai et al., 2020). Remarkable are droughts that occurred in periods of synchronized warm phases of the IPO and AMO, which corresponds to 19% of the 120-year-old record, such as in 1926, 1936, 1995, 1997 and 1998. In the high-frequency mode, extreme drought events are generated by synergisms of warm AMM and ENSO (El Niño) phases (Marengo et al., 2021). This is the case of the historical droughts of 2023/2024 which caused a huge climatological water deficit almost across the entire Amazon basin resulted from a simultaneously occurring EP El Niño and an anomalously warmed TNA during a warm AMO phase. River levels of the 2023/2024 period reached record lows during two consecutive years (Fig. 4) combined with heat waves and anomalous warm water in the shallow floodplain lakes causing mass mortality of river dolphins (*Inia geoffrensis* and *Sotalia fluviatilis*), fish and other aquatic organisms (Marengo et al., 2024; Marmontel et al., 2024). Another outstanding drought event is the episode 1925/1926 with the lowest flood amplitude and maximum water level (1926) on the instrumental record (Fig. 4). This event was provoked by the strong El Niño in 1925/1926 (Williams et al., 2005) during an intense warm IPO phase (Henley et al., 2015). The extreme meteorological drought conditions (1925/1926) are reflected by a flood amplitude of only 4.1 m resulting in the low maximum water level in 1926, followed by an extreme hydrological drought in 1926 caused by a warmed TNA and a warm AMO phase. While this severe drought episode favoured tree growth in the central Amazonian *várzea* (Schöngart et al., 2004), it caused large-scale fires in *igapós* of the Negro River basin (Williams et al., 2005). This episode is also an outstanding value of a 100-year-long stable oxygen isotope chronology (Brienen et al., 2012) in Bolivia, receiving the wind trajectories across the Amazon basin which integrate the meteorological drought conditions by intense biosphere–atmosphere interactions (Baker et al., 2016).

8 Intensification of the Hydrological Cycle and Its Impact on Amazonian Floodplains

The hydrological cycle of the largest hydrological basin on Earth is intensifying (Gloor et al., 2013). Over the last decades, extreme floods and droughts increased in frequency and magnitude (Fig. 4). The four highest maximum (2021, 2012, 2009 and 2022) and three lowest minimum (2024, 2023 and 2010) annual water levels occurred during the last 15 years, condensed on the last 13% of the long-term record. The annual flood amplitude increased by about 1.6 m in this century compared to the previous one (Fig. 4), and over the entire 120-year period, the mean water level significantly increased by about 1 m (Barichivich et al., 2018). The intensification of the hydrological cycle is mainly a result of a significant increase in the maximum water levels (Fig. 4). In contrast, no significant trends and changes in the return period of severe hydrological droughts can be observed along the 120-year period (Barichivich et al., 2018). This raises the question of whether the intensification of the hydrological cycle is already a result of anthropogenic climate change or if it still can be explained by its natural climate variability and associated large-scale climate anomalies responsible for much of this variation as shown in Sect. 7.

8.1 Mechanisms Beyond the Intensifying Hydrological Cycle

During the last decades, a persistent formation of a climate dipole can be observed along a north-south gradient of the Amazon basin. This dipole is characterized by increasing wet-season rainfall and maximum streamflow regimes in the northern Amazon (Barichivich et al., 2018; Gloor et al., 2013; Wang et al., 2018), contrasting increasing dry-season length and dry-day frequency (< 1 mm/day), with a simultaneous decrease in dry-season precipitation and minimum river discharge in the southern and southwestern Amazon region (Espinoza Villar et al., 2009, Espinoza et al., 2019, Marengo et al., 2018, Haghtalab et al., 2020, Heerspink et al., 2020). The formation of the dipole probably started with the climate shift in 1976/1977 after consecutive years of severe flood conditions in the northern and central Amazon basin and severe pluriannual droughts in the southern Amazon basin associated with a phase change in the IPO (Marengo et al., 2021). Reduced precipitation, together with an increased frequency of dry days during the dry season and delayed onset of the wet season, directly impacts the floodplains, especially in the headwater regions of white-water rivers in the southwestern Amazon basin and the clear-water rivers in the southern Amazon region (Fig. 1) already experiencing long dry seasons (Espinoza Villar et al., 2009, Heerspink et al., 2020, Marengo et al., 2021). In addition to rainfall trends observed by satellites and in-situ monitoring, time series of stable oxygen isotopes obtained from dated tree rings of the floodplain species *Macrolobium acaciifolium* from the Marañon River (Peru), reflect the trend of increasing drought conditions since the 1970s (Cintra et al., 2022), which is consistent with the observed trends in rainfall and discharge (Lavado Casimiro et al., 2013).

The underlying mechanisms of the increasing drought conditions during the dry season are complex. One of the attributed mechanisms is related to trends of large-scale climate anomalies over the last 40–50 years, mainly the observed increase of SST in the TNA (Gloor et al., 2013, 2015), favoured by the warm AMO phase since the 2000s among other atmospheric changes (Marengo et al., 2021). Other mechanisms are related to land-use changes, such as large-scale deforestation and associated fires to expand areas for industry, agriculture and livestock production in the southern Amazon basin. Wright et al. (2017) show, based on isotopic analyses, that the dry-season length also depends on the evapotranspiration of the rainforests in the southern Amazon basin, which might initiate atmospheric changes favouring the onset of the wet season before the arrival of the ITCZ. However, increasing large-scale deforestation and forest degradation in the southern Amazon basin increases temperatures by the enhanced sensible heat loss and decreases evapotranspiration, leading to less air humidity and water recycling in the atmosphere (Covey et al., 2021; Foley et al., 2003) and delaying the onset of the rainy season (Marengo et al., 2018). Frequent biomass burning associated with deforestation in this region (Brando et al., 2020) causes further large emissions of aerosols polluting the atmosphere, leading to an upward shift of convection and increasing lifetime of clouds, which probably contribute to the delay of the onset of the rainy season (Wright et al., 2017). These coupled feedback loops of increasing drought conditions induced by climate and land-use changes are considered the main mechanism responsible for much of the observed increasing drought conditions during the dry season in the southern Amazon basin, which likely affects tree growth and mortality in floodplains (Cintra et al., 2019; Resende et al., 2020). However, no information is available on the responses of different tree species of clear-water *igapó* and *várzea* floodplains to the increased drought conditions of these regions, requiring more studies.

Recent studies have brought new insights into the complex mechanism beyond the intensification of severe floods in the Amazon basin. The warming tropical Atlantic (Gloor et al., 2013) and the cooling central-eastern region of the equatorial Pacific over the last decades intensified the zonal Walker Circulation, leading to increased cloud convection and precipitation over the northwestern, northern and central Amazon basin (Barichivich et al., 2018; Espinoza et al., 2022). The warming of the tropical Atlantic is further enhanced by the drift of warm surface waters from the Indian Ocean into the Atlantic (Agulhas Leakage) and the northward transport by the Atlantic Meridional Overturning Circulation (AMOC) (Beal et al., 2011). This mechanism is sustained by the poleward shift of the Southern Westerlies over several decades caused by the increasing ozone hole over the Antarctic and atmospheric greenhouse gas concentrations and is a mechanism linked to anthropogenic climate change (Barichivich et al., 2018; Biastoch et al., 2009). The increasing freshwater discharge into the tropical Atlantic decreased the salinity of the Amazon River plume (ARp) by about 3.5% per year in the 2002–2016 period (Gouveia et al., 2019). This favoured the formation of a barrier layer reducing the mixing of the warmer, lighter surface water with the cooler, dense bottom water resulting in an eastward extension of the ARp by several hundred kilometres. These complex mechanisms together have resulted in a warming of the tropical Atlantic's surface waters over the last decades (Gloor et al., 2013), which is the main source of the water vapour imported into the Amazon basin during the wet season by trade winds (Gloor et al., 2015; Salati & Vose, 1984).

The formation of the climate dipole resulting in the intensification of the hydrological cycle in the Amazon basin is a consequence of mechanisms related to natural climate variability, anthropogenic climate change and land-use changes with complex interactions, synergies and feedback loops. This raises concerns about the impacts on biodiversity, ecosystem services, food webs and the livelihood and food security of hundreds of thousands of people depending on the floodplains and the society in general.

8.2 Impacts of the Intensifying Hydrological Cycle in Amazonian Floodplains Including Human Dimensions

Severe hydrological drought conditions lead to the isolation of communities of traditional and indigenous populations and municipalities, which historically have established along rivers and floodplain lakes and channels. Besides the risk for fluvial navigation due to the appearance of point bars and rock beds, the isolation causes massive problems in supplying the population with drinking water, food, medicine and essential public services (e.g. health and education). Riverine populations in isolated regions must face long walkways under extremely hot conditions to access essential goods and services. The limited fluvial navigation causes further huge economic losses for commerce and many industrial sectors, as goods and materials are not timely delivered. The hot climate and extreme meteorological drought conditions result in losses of agricultural crops and livestock production. The extremely warm water in the floodplains causes massive mortality of fishes, aquatic mammals (manatees and river dolphins) and other aquatic organisms, as observed during the historical drought in 2023 (Marmontel et al., 2023). Health conditions of the affected populations are impacted by air pollution in urban and rural regions caused by large-scale fires, extremely hot weather conditions and, especially in the floodplains, illness caused by the consumption of contaminated water and food. Floodings over several weeks cause damages and losses to private and public properties and goods, restrict essential public services (e.g. health and education), impact socioeconomic sectors (e.g. commerce and trade) and increase the risk of diseases as vulnerable populations are exposed to polluted wastewater due to missing basic sanitation and the dumping of several categories of industrial and domestic garbage (Espinoza et al., 2022).

Based on the long-term water level record in Manaus, the emergency situation caused by severe hydrological droughts (< 15.8 m), extreme floods (≥ 29.0 m) and both extreme events have been calculated as cumulative values for the period 1903–2023 (Fig. 5a). Until the 1970s, the emergency situation in Manaus, representative for the hydrological regime in Central Amazonia, has been dominated by severe hydrological droughts. This scenario changed with the occurrence of nine severe flood events within a 13-year period (2009–2022). The duration of the emergency period provoked by extreme flood events in Manaus is in this century already 29% higher compared to the entire past century (Fig. 5a). Overall, the emergency situation increased potentially during the last three decades caused by the increasing frequency and magnitude of severe hydroclimatic events. About 7.7% (every 13th day) of this century is characterized by an emergency situation caused by severe hydroclimatic events.

Along the Solimões-Amazon mainstem (from Iquitos in Peru to Gurupá in the eastern Brazilian Amazon), the annual maximum inundation extent of flooded areas increased by 26% in the period 1980–2014, evidenced by hydrological-hydrodynamic models and remote sensing data (Fleischmann et al., 2023b). In rural regions, especially in the *várzea*, the increasing flood conditions (Fig. 5b) cause losses of annual crops (mainly varieties of *Manihot esculenta*—manioc, which is

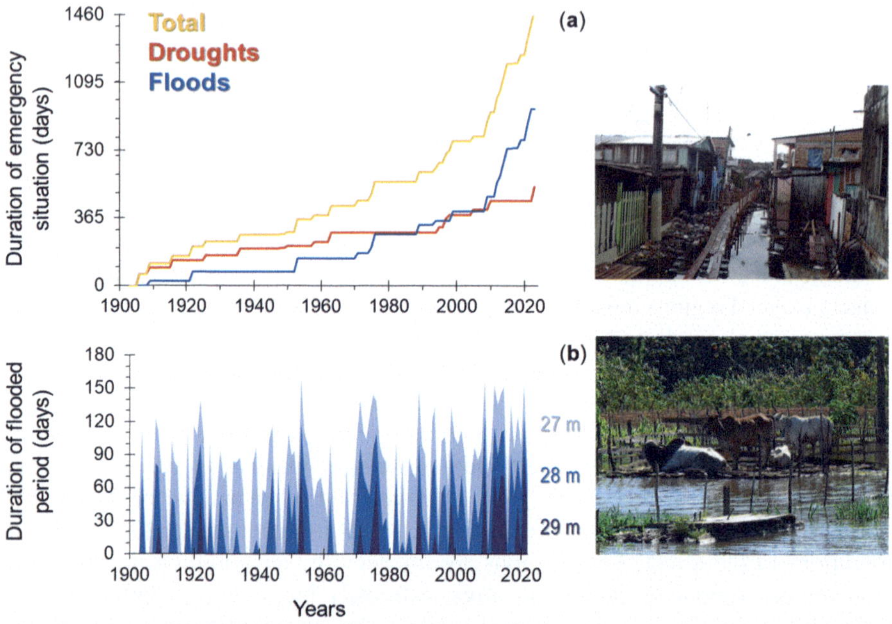

Fig. 5 (**a**) Cumulative period of the emergency situation caused by severe hydrological droughts (< 15.8 m), extreme flood events (≥ 29.0 m) and the total emergency period in the city of Manaus (period 1903–2023). (**b**) Duration of flooded period of topographies (27, 28 and 29 m) relevant for agriculture and livestock production in central Amazonian floodplains (Data: National Information System on Water Resources—SNIRH of the Brazilian National Water and Sanitation Agency—ANA). Photos: flooded suburb in Manaus during the historical flood in 2021 (left); cattle on a floating meadow (*maromba*) to pass the severe flood conditions of the year 2009 in the central Amazon region (right) (Credits: Jochen Schöngart)

the main provider of carbohydrates for the riverine populations) which do not attain maturity due to the shortened terrestrial phase or mortality of perennial crops because of the prolonged anoxic conditions during flooding (Junk et al., 2023). Cattle must be removed from the floodplains and transferred to non-flooded areas or floating meadows (*marombas*), involving high costs for the riverine people (Schöngart & Junk, 2007). The observed increase in the flood amplitude during the last three decades might accelerate hydrogeomorphological disturbance regimes and turn-over rates of erosion and sedimentation processes in the *várzea* floodplains along the main river channels. Many traditional populations, villages and even municipalities get isolated during severe hydrological droughts in areas undergoing increased sedimentation rates driven by the frequent extreme flood events (Fleischmann et al., 2023b). The extremely low water levels destabilize the alluvial soils and riverbanks, leading to landslides (regionally called *terra caída*) resulting in death and loss of houses, communities and large infrastructures (i.e., ports).

The increase in the maximum extent of flooded areas is likely to induce increased tree mortality on sites not exposed to longer floodings before (Herraiz et al., 2017).

The increase in the duration, extent and connectivity of floodplains likely enhances methane emissions under the extended period of anaerobic soil conditions (Fleischmann et al., 2023b; Pangala et al., 2017). The increase in the annual mean water level in the central Amazonian floodplains possibly leads to changes in plant species composition and vertical shifts in the species zonation over time, propagating to food webs and ecosystem functioning. Increasing flood events induce higher rates of sediment deposits in the *várzea* (Aalto et al., 2003), which buffers at least along the main channel of the white-water rivers the prolonged flooding conditions. The highly dynamic vegetation of engineering herbaceous plants and forest succession has sophisticated mechanisms to adapt to rapidly changing environmental conditions. In *várzea*, floodplains distant from the main river channels, sedimentation rates, however, are lower (Wittmann et al., 2010), and it can be expected that the increase in flood duration over the last decades causes shifts in the species composition and zonation. As discussed earlier, it can be assumed that the *várzea* tree species have an overall higher adaptive capacity to environmental disturbance regimes than the oligotrophic *igapós*. As this ecosystem has a relatively stable geomorphology, increasing flood levels over time provoke tree mortality (Assahira et al., 2017; Resende et al., 2020), and a shift of the tree species zonation towards upper elevations can be expected. The question is to what extent the oligotrophic *igapó* is able to respond and adapt to the rapidly intensifying hydrological cycle, as this ecosystem has a slow dynamic, high vulnerability and low adaptive capacity.

We assume that the impacts of severe droughts may be more pronounced in the *igapó* than in the *várzea*. Trees from the *igapó* have larger vessels and higher xylem hydraulic conductivity and are presumably more vulnerable to hydraulic failure caused by water deficit when compared to congeneric tree species from other central Amazonian ecosystems (Fontes et al., 2020). Fine roots of trees growing in the *igapó* are concentrated near or on top of the soil surface forming dense root mats (Santos & Nelson, 2013), while in the *várzea*, fine roots are observed also at soil depths of several decimetres (Meyer et al., 2010). The clay fraction of soil substrates in the *várzea* increases at higher topographical elevations (Wittmann et al., 2010), characterized by a high and dense canopy which reduces solar radiation (rPAR) at the soil (Fig. 3) and probably avoids desiccation even under severe and long-lasting dry periods. In the *igapó*, the dominating soil fraction is fine sand (> 80%) with a low water retention capacity (Wittmann et al., 2022a). The forest canopy in the *igapó* is generally less stratified and low, resulting in lower relative air humidity on the forest floor (Almeida et al., 2016; Resende et al., 2014).

Not much data is available to evaluate the impact of severe droughts on the floodplain vegetation. Some insights into the vulnerability of *igapó* trees against drought can be provided by historical extreme events. Due to the damming of the Uatumã River (central Amazonia) to fill the reservoir of the Balbina dam (1987–1989), severe artificial drought conditions were induced in the downstream *igapó* floodplain (Schöngart et al., 2021; see also Sect. 9.2). Resende et al. (2020) showed, based on radiocarbon-dating, increased mortality of *E. tenuifolia* trees during this period of severe and prolonged drought conditions. Secondary forests occurring along the Uatumã River downstream of the Balbin dam for several dozen kilometres, probably developed after the artificially caused prolonged drought conditions leading to mass mortality of the former *igapó* forests (Neves et al., 2019; Salerno et al., 2022).

Granato-Souza et al. (2019, 2020) used tree-ring chronologies of *Cedrela odorata* from the eastern Amazonian *terra firme* (Paru River basin) to reconstruct wet season precipitation totals from 1759 to 2016 within the ENSO-IPO footprint. The climate reconstruction revealed pluriannual drought conditions during the period 1864–1881, which also included the severe El Niño event from 1878/1879, which had severe socioenvironmental impacts around the tropics (Aceituno et al., 2009). This period of pluriannual drought conditions coincides with the occurrence of a warm IPO phase, as highlighted by a 700-year-long IPO-reconstruction exploring large climate-sensitive tree-ring networks from Thailand and North America (Buckley et al., 2019). The year with the highest reconstructed IPO-index on the record is 1865 (Buckley et al., 2019), which is the outstanding negative value of the entire period of reconstructed Amazonian rainfall (Granato-Souza et al., 2020). Many historical documents point out, that the 'Forgotten Drought' of 1865 (Granato Souza & Stahle, 2023) had dimensions similar to the recently observed extreme events. The prolonged drought conditions in the past likely caused increased mortality in the oligotrophic floodplains within the ENSO-IPO footprint (Cai et al., 2020). Analyses of the age structures of *igapó* forests in the affected regions (Neves et al., 2019; Reis, 2021) suggest significant age cohorts of trees with mean ages of 120–130 years which probably established after the prolonged drought period. *Igapó* forests are also vulnerable to disturbances caused by land-use changes, which will be demonstrated in the following section.

9 Impacts of Land-Use Changes on Central Amazonian Floodplains

9.1 Vulnerability of Floodplains to Fire

In general, the humid climate in the Amazon biome does not favour the propagation of wildfires caused by lightning, and the forest vegetation is in general not adapted to fire regimes as, for instance, the savanna vegetation (Pivello, 2011). Most fire ignition sources in Amazonia are anthropogenic, applied to clear lands after large-scale deforestation (industrial agriculture, livestock production and large infrastructure projects), to maintain pastures or to perform slash-and-burn activities at small scales for subsistence. In severe drought years, fires can escape, accidentally or intentionally, affecting intact *terra-firme* forests, both outside and within protected areas (Nepstad et al., 2006). However, the impacts caused by fires vary considerably, depending on the fire dynamics, environmental conditions (climate, hydrology and soils) and vegetation (resilience and recovery) (Barlow et al., 2020). The increasing population in the Amazon basin, the growing demand for natural resources from this biome by national and global markets, the weak presence of governance in this continental-size region and the growing feedback loops and synergisms between land-use and climate change turned fires into unprecedented drivers of disturbances in the Amazon region during the last decades (Aragão et al., 2018; Barlow et al., 2020; Davidson et al., 2012).

Fires are also reported for Amazonian *igapós* and *várzeas* (Flores et al., 2017). However, the causes and impacts of fires in the large river-floodplains vary considerably, as well as their vulnerability and resilience to fires. In the fertile *várzea*, fires are mainly intended to clear areas for agriculture and pasture after deforestation at small scales which are commonly practised land-uses by the riverine population (Junk et al., 2023; Reis et al., 2021). The oligotrophic *igapó* is unattractive for these land-uses (Junk et al., 2015) and has a relatively low human population density (Piedade et al., 2024; Wittmann et al., 2022b). Nevertheless, sources of ignition in *igapó* floodplains are mainly anthropogenic, such as accidental fire escapes originating from slash-and-burn activities in adjacent *terra firme* forests near human settlements or not extinguished campfires on the river margin, also in remote areas (Carvalho et al., 2021; Ritter et al., 2012). Although the *igapó* in the Negro River basin is integrated into large frameworks of protected areas and indigenous lands and there are no major deforestations in this ecosystem, fires can attain huge dimensions in these regions killing thousands of hectares of floodplain forests (Carvalho et al., 2021; Flores & Holmgren, 2021).

Most of the oligotrophic *igapós* occur in regions with, on average, high amounts of annual rainfall with short or even no distinct dry seasons. However, severe hydroclimatic droughts occasionally occur in these regions generated by El Niño events (Aragão et al., 2018; Carvalho et al., 2021; Flores et al., 2014; Labat et al., 2012). Especially, EP El Niños during warm IPO phases can provoke severe hydroclimatic drought conditions in the *igapós*, such as the 1925/1926, 1982/1983, 1997/1998 and 2015/2016 events. For these severe El Niño episodes, large-scale fires in the *igapó* are reported or observed (Carvalho et al., 2021; Flores et al., 2014; Williams et al., 2005). The occurrence of El Niños coincides with the non-flooded period of the *igapó*, generating a huge climatological water deficit as rainfall reduces and temperature increases which enhances potential evapotranspiration (Carvalho et al., 2021). These conditions can extend into the period of the rainy season. Due to the low precipitation in the headwater region, the flood pulse is weaker than usual, leading to an extension of the terrestrial phase (Schöngart & Junk, 2007). Especially, the *igapós* of the middle and upper Negro River basin are highly vulnerable, as the minimum water level in these regions occurs in the period from January to March (Fassoni-Andrade et al., 2021; Junk et al., 2015), which coincides with the generated maximum cumulative water deficit. The strong El Niño episodes in 1982/1983, 1991/1992, 1997/1998 and 2015/2016 induced high cumulative water deficits in the *igapós* and large-scale fires which comprise 78% of the total burnt floodplain area of the Jaú National Park monitored during a 35-year period (1982/1983–2016/2017) (Carvalho et al., 2021). Although the source of fire ignition is mainly caused by humans (Ritter et al., 2012), natural fire ignition due to high lightning frequency, which occasionally might attain large canopy trees with trapped methane in trunk hollows, could be a mechanism of natural ignition during severe droughts (Carvalho et al., 2021), an issue which deserves more studies.

The severe hydroclimatic drought conditions induced by El Niño are coupled with specific characteristics of the *igapó* turning this ecosystem highly vulnerable to fires (Fig. 6). As the soil bedding of *igapó* floodplain forests is predominantly sandy (Latrubesse & Franzinelli, 2005), the soil water retention capacity is low. Because of the oligotrophic conditions, trees form dense mats of fine roots on top or near the soil surface (Meyer et al., 2010; Santos & Nelson, 2013). The nutrient-poor conditions of the *igapó* are reflected by sclerophyllous leaves formed by many tree species with high levels of structural carbohydrates, low nutrient contents and a high leaf dry matter content (Furch & Junk, 1997; Irmler & Furch, 1980; Mori et al., 2019; Waldhoff & Furch, 2002). Due to a combination of these structural features and the anoxic conditions during the long inundation periods, which limits the activities of soil invertebrates (Adis et al., 2010; Irmler & Furch, 1980; Wantzen et al., 2008), the decomposition of organic material is slow, accumulating fine litter

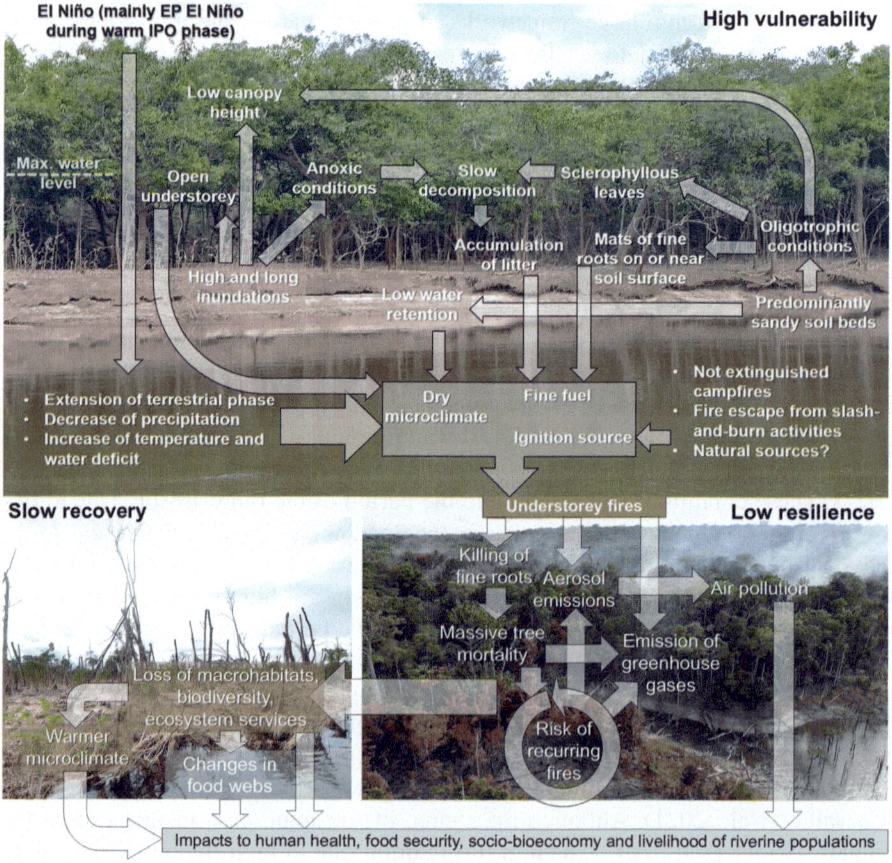

Fig. 6 Main underlying mechanisms leading to the high vulnerability, low resilience and slow recovery of oligotrophic *igapós* in Amazonia regarding fire regimes. Photos: Jochen Schöngart and Nikolai da Silva Espinoza (aerial photo of burning *igapó*)

on the forest floor (Kauffman et al., 1988). Together with the root mats, this forms a considerable layer of fine fuel, much higher than the non-flooded *terra firme* forests (Santos & Nelson, 2013). Particularly, forests growing on low topographic elevations have low canopies and open understories, generating a dry microclimate with relative air humidity below 65%, which favours the ignition of fine fuel (Almeida et al., 2016; Carvalho, 2019; Resende et al., 2014; Uhl et al., 1988). The dry hydrometeorological conditions generated by El Niño, together with the structural features of the *igapó* forests, favours the spreading of understorey fires along the soil surface, leading to massive tree mortality as fine roots on top of the soil or near the surface are killed by the high combustion heat (Resende et al., 2014; Ritter et al., 2012).

For the oligotrophic *igapó* with its high vulnerability, low resilience and slow recovery, fire is a severe disturbance vector. *Igapós* present long recovery periods that keep burnt areas open for several years, further exposing them to recurring fires (Flores et al., 2014; Flores & Holmgren, 2021) and generating a hot microclimate. Especially, large-scale fires in the *igapó* cause losses of ecosystem services and biodiversity and alter food webs, as recently reported for fishes (Carvajal-Lugo et al., 2023), directly affecting traditional and indigenous populations. In general, fires further increase atmospheric emissions of greenhouse gases, pollutants and aerosols which affect human health, change atmospheric chemistry and physics and alter local and regional climate conditions and hydrological cycles (Fig. 6; Urrutia-Pereira et al., 2021; Wright et al., 2017).

Although the population density nowadays is low in the *igapó* floodplains, Carvalho et al. (2021) provided a historical panorama of the last 400 years, considering political, sociocultural and economic aspects based on literature surveys (Leonardi, 1999). Cross-dating the historical panorama with reconstructions of El Niño events (Cook et al., 2018) and warm IPO phases (Buckley et al., 2019; Porter et al., 2021) for this period brought new insights about fire disturbances in the *igapós*, which probably were much frequent in the past with higher magnitudes than nowadays. This holds especially for those periods and regions with a high human occupation, coinciding with warm IPO phases and severe El Niño conditions. This constellation is evident for the period of the rubber boom (around 1880–1920), when thousands of immigrants, mainly from the Brazilian Northeast, settled in *igapós* of the lower and middle Negro River and lower Branco River to explore rubber (*Hevea* spp.) and other forest products (e.g. *Couma* spp. and *Chrysophyllum* spp.) (Leonardi, 1999). This human occupation brought a deep sociocultural transformation to the Amazonian floodplains in general, as the immigrants from the Brazilian Northeast had no intrinsic relationship with the different Amazonian forest ecosystems as the former indigenous and descendant traditional populations which were decimated or even extinct by slavery, epidemics and conflicts along the period of colonization by the Portuguese Crown and the Brazilian Empire (Carvalho et al., 2021; Leonardi, 1999). Large amounts of firewood for the vulcanization of rubber and the increasing steamboat navigation along the Negro River were extracted from the *igapó* during the rubber boom (Leonardi, 1999). The predominate warm IPO phases during this period and frequent El Niño conditions turned the *igapó* highly vulnerable to fires (Carvalho et al., 2021). At least in those regions which were economically strategic during the rubber boom (Airão, Moura, Carvoeiro, Barcelos, Santa Isabel do Rio Negro, among others) (Leonardi, 1999; Sampaio, 2003), we can assume that many of the *igapó* forests nowadays are a mosaic of secondary forests originating from the intense human occupation during a period of frequent and intense hydroclimatic drought conditions.

These findings have substantial implications for the conservation of this ecosystem. Nowadays, most black-water *igapós* are integrated into a mosaic of indigenous lands and conservation units of different categories (Piedade et al., 2024). However, these huge efforts are insufficient, unless coupled with efficient public policies to prevent and mitigate fires in this vulnerable ecosystem. Such measures comprise socioenvironmental education campaigns not only for traditional populations, which rely on fire for their subsistence, but also for other stakeholders, such as tourist agencies and municipalities, among others (Carvalho et al., 2021). Management plans of protected areas designed for sustainable resource management in this region need a revision, prohibiting the use of fire in the *igapó* in general. As the evolution of El Niño years generally starts during the second half of the year when the *igapó* floodplains are still flooded, this time interval allows the development and implementation of early warning systems with efficient mechanisms of widespread dissemination of alerts. Still, large areas of oligotrophic *igapós* are not integrated into the Brazilian National System of Conservation Units (SNUC). This holds, in particular, for the Mariuá Archipelago along the middle stretch of the Negro River (since 2018 integrated into the Regional Ramsar Site Rio Negro with more than 12 million hectares).

9.2 Impacts of Hydropower Dams on Floodplain Forests

While many countries in Europe, North America and Australia have removed several of their hydropower dams due to growing socioenvironmental awareness in the last decades, other countries, especially in tropical regions, have undergone an unprecedented boom of hydropower dam installations in the last decades (Finlayson et al., 2018; Piedade et al., 2023). This also holds for the Amazon basin, which still is one of the largest remaining networks of free-flowing rivers on Earth (Grill et al., 2019). More than 400 hydropower dams (> 1 Megawatt—MW) are currently in the phase of operation, construction and planning. More than half of the hydropower dams will be located in the Andean headwater regions at higher elevations dominated by white-water rivers (N = 224), others in rivers draining the cratonic formations of the Guiana and Central Brazilian Shields dominated by clear-water rivers with attenuated reliefs and waterfalls in the headwater regions (N = 200) or in the Amazonian lowlands (black-water and white-water rivers) (N = 4) (Latrubesse et al., 2017). Many studies in the Amazon basin report the various and severe socioenvironmental problems caused by hydropower dams such as the removal of traditional and indigenous populations, emissions of greenhouse gases (mainly methane), the obstruction of migration routes of many aquatic organisms, among others

(Fearnside, 2019). However, the main focuses of these studies are the areas directly affected by the creation of the dams and reservoirs. In contrast, only a few studies consider the downstream impacts on floodplains, which have been extensively studied over the last ten years in the case of the Uatumã River (see the map of Fig. 1) in the central Amazon region (Schöngart et al., 2021). In the following, we will synthesize the main findings of downstream impacts in the *igapós* and discuss how these can be extrapolated to other regions in the Amazon basin and the tropics.

The Uatumã River is a black-water river that had, in periods before the construction of the Balbina hydropower plant, a monomodal flood pulse, similar to the Jatapú River, its major clear-water tributary on the left margin (Fig. 1). The Balbina dam was constructed in the period 1983–1987 in the middle stretch of the Uatumã River and is one of the most emblematic hydropower plants due to its disastrous socioenvironmental problems (Fearnside, 1989). The river was dammed from October 1987 to February 1989 to fill the reservoir with almost 3000 km^2 surface. The damming of the river induced an artificial drought in *igapós* downstream of the dam as river discharge was reduced to about 5–20 m^3/s (mean discharge of more than 650 m^3/s under normal conditions). Simultaneously, El Niño episodes from the second half of 1986 until the beginning of 1988 during a warm IPO phase generated meteorological droughts in this region. Schöngart et al. (2021) originally hypothesized that large-scale fires probably affected the *igapós* during severe hydroclimatic drought conditions favouring the establishment of pioneer tree species and the formation of the secondary forests after fire disturbance. These homogeneous secondary forests have an almost continuous occurrence, dominating the intermediate topographical elevations of the *igapó* along the main river course extending over 120 km downstream of the dam. However, although burning occurred occasionally and locally during dam operation, it cannot explain the large occurrence of these homogeneous secondary forests. The synergized drought condition likely caused a massive tree mortality in the *igapós* during river damming (Resende et al., 2020), which affected the downstream floodplains along the main river course. We can assume that the artificially induced hydrological drought in the *igapó* downstream of the hydropower dam was mitigated towards the headwaters of smaller affluents as they received water from the drainage system of the adjacent *terra-firme* forests in this period. With increasing fluvial distance from the hydropower dam, the impact of the artificial hydrological drought in the *igapós* was probably buffered, receiving more and more water from the increasing drainage area of upland forests. Along the mainstem of the Uatumã River, however, most trees likely had no access to soil water under the severe artificial hydrological drought conditions and the hot and dry microclimate with high atmospheric vapour-pressure deficit. As fine roots of tree species in the *igapó* are concentrated at the soil surface (Meyer et al., 2010) and functional traits indicate a high vulnerability to hydraulic failures (Fontes et al., 2020; Resende et al., 2020; Wittmann et al., 2022a), the mass mortality was facilitated (Fig. 7). This large-scale disturbance was followed by the establishment of homogenous and species-poor secondary forests, mainly dominated by the pioneer *Nectandra amazonum* (Lauraceae) and the palm species *Astrocaryum jauari* (Arecaceae) (Rocha

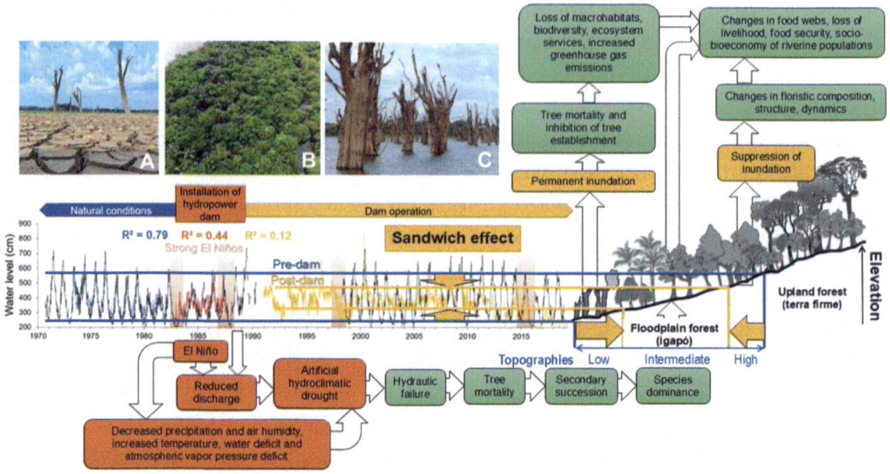

Fig. 7 Spatiotemporal disturbances in the *igapó* floodplains downstream of the Balbina dam over a 35-year period. Shown are three periods comparing the flood-pulse regime of the Uatumã River downstream of the Balbina dam (coloured) with that of the Jatapú River (black), the largest undisturbed affluent in the river basin. Square coherencies (R^2) between both hydrological regimes are indicated for the period before dam construction (blue), dam implementation (red) and dam operation (yellow). During the installation of the Balbina dam and the reservoir filled by river damming, in combination with El Niño conditions (in 1982/1983 and 1986–1988), severe dry conditions were generated in the downstream floodplains (red boxes). This led to hydraulic failure and massive tree mortality (Photo A), resulting in forest succession and the establishment of dominating pioneer tree species forming continuous, species-poor and homogeneous secondary forests along the main river course (Photo B). Hydrological alterations during the period of dam operation generated the typical 'sandwich effect' (yellow arrows), characterized by the suppression of the terrestrial phase at lower elevations and suppression of the aquatic phase at higher elevations of the *igapó* as a consequence of higher minimum and lower maximum water levels, respectively (yellow boxes). This caused mass mortality of tree species mainly in connected floodplain lakes of the Uatumã River (low topographical elevations) (Photo C), while the upper topographies are getting more and more colonized by secondary tree species originating from the adjacent *terra firme*. (Adapted from Schöngart et al., 2021; photos: Jochen Schöngart and Nikolai da Silva Espinoza)

et al., 2020; Schöngart et al., 2021; Targhetta et al., 2015). Dendroecological studies in these secondary forests demonstrate tree cohorts of *N. amazonum* whose mean age, with a narrow range of a few years, matches the dam implementation period (Neves et al., 2019). Applying remote-sensing techniques, Salerno et al. (2022) indicate that about 77 km^2 of the downstream *igapós* have undergone a vegetation gain during the post-dam period, which are mainly secondary forests along the main river channel. The hot microclimate in the opened areas after large-scale tree mortality likely favoured the germination of *A. jauari* (Ferreira, 2022) and mainly pioneer tree species adapted to these harsh environmental conditions, such as *N. amazonum*, a common pioneer tree species of the *várzea* floodplains (Gonçalves et al., 2021). Seeds of *N. amazonum* probably originated from the *várzeas* of the Amazon River and were dispersed upstream by migrating fish species, mainly *Brycon amazonicus* (matrinxã) (Weiss et al., 2023).

The period of dam operation is characterized by the sandwich effect (Schöngart et al., 2021; Wittmann et al., 2019; Fig. 7). As the operation of the power plant aims to generate uniform hydroelectric energy during the year, the cumulating upstream water excess during the rainy season is stored in the reservoir leading to a lower maximum water level in the downstream area due to the controlled release of water from the reservoir to generate power. The stored water volume in the reservoir is released during the dry season increasing the downstream minimum annual water level. This caused, throughout dam operation, the suppression of the terrestrial phase at the low topographic elevations. In contrast, the aquatic phase at the upper elevations of the *igapó* was more or less extinct (Assahira et al., 2017; Lobo et al., 2019). This large-scale disturbance induced a slow process of mass mortality of *igapó* tree species as the permanent inundation over consecutive years exceeded their adaptative capacity to tolerate seasonal floodings (Assahira et al., 2017; Resende et al., 2020). Compared to the overall and sudden tree mortality of mainly upland forests (*terra firme*) induced by the reservoir fill in the upstream region, the downstream mortality in the *igapó* is a long process extending over many years and even decades. So far, 12% of the *igapó* downstream of the Balbina dam died at the low topographies (mainly monodominant formations of *E. tenuifolia*) along a downstream fluvial distance of about 125 km and is expected to extend in time and space with ongoing dam operation (Resende et al., 2019; Salerno et al., 2022). However, this area of mapped dead *igapó* forests by remote-sensing techniques does not account for those forests killed during the period of river damming, which are represented by secondary forests mainly on intermediate topographic elevations of the *igapó* (Neves et al., 2019; Targhetta et al., 2015), suggesting a much larger forest degradation than previously estimated (Salerno et al., 2022). On the high elevations of the *igapó*, the suppression of the flooded period led to an encroachment of mainly secondary species from the adjacent *terra firme*, changing the species composition and structure of the *igapó* forests (Lobo et al., 2019; Rocha et al., 2019). In summary, the entire *igapó* ecosystem along a fluvial distance of more than 125 km downstream of the hydropower dam was affected by a complex spatiotemporal and persistent disturbance regime over decades (Schöngart et al., 2021; Fig. 7). This caused a loss of macrohabitats, mass mortality of tree species with cascading impacts on food webs affecting the entire ecosystem functioning and provision of ecosystem services and livelihood of the resident traditional population. Furthermore, greenhouse gas (mainly methane and carbon dioxide) emissions are likely to increase due to the decomposition of large amounts of dead organic material and the enhanced period of flooding and anoxic conditions (Resende et al., 2019; Schöngart et al., 2021).

The findings on the downstream impacts on floodplains caused by a hydropower dam, exemplifying the Balbina dam, are unique for the tropical regions. However, their extrapolation to other regions and ecosystems which are targets of hydropower dams in the Amazon and other tropical regions is limited, as no other studies are available for comparisons (Schöngart et al., 2021). Downstream impacts caused by hydropower dams vary depending on many factors such as the size and elevation of the reservoir, time since dam construction, technical aspects of the impoundment and synergies with other land-use forms and intensities (Altahyde et al., 2019; Latrubesse et al., 2017; Rufin et al., 2019; Timpe & Kaplan, 2017). Furthermore, downstream impacts differ depending on the geomorphology, hydrochemical conditions, sediment load, biogeography, biodiversity, dynamics, vulnerability and resilience of the floodplain vegetation (Schöngart et al., 2021). The large amount of hydropower dams in operation, under construction or planning of the cratonic clear-water rivers draining the Guiana and Central Brazilian shields, are likely to cause severe impacts. This holds especially for multiple dams which are designed for the Tocantins-Araguaia, Xingú, Tapajós and tributaries, Araguari, Parú, Jarí and Trombetas rivers (Fearnside, 2019), as they induce cascading and cumulative effects in the downstream area (Timpe & Kaplan, 2017). However, no information on the vulnerability and resilience of clear-water *igapós* and *várzeas* is available, emphasizing the urgent need for more studies with a transdisciplinary approach considering the multiple factors, including social and economic dimensions. The most critical reservoirs are those planned in the Negro River basin operating with large reservoirs at low altitudes (Fearnside, 2019; Timpe & Kaplan, 2017) as the oligotrophic *igapó* has a high vulnerability and low resilience as shown in Fig. 8. But also,

Fig. 8 Disturbances of the oligotrophic *igapó* in the Uatumã Sustainable Development Reserve (central Amazonia) about 90 km downstream of the Balbina dam (fluvial distance) showing massive tree mortality caused by hydrological alterations as a consequence of hydropower generation (flooded forests on low topographical elevations) and anthropogenic fire (forests on intermediate topographical elevations) which occurred during the severe El Niño 2015/2016 (Credit: Nikolai da Silva Espinoza)

the headwater regions of the white-water rivers in the Andes are the focus of the hydropower expansion in the Amazon basin (Latrubesse et al., 2017). Large dams in this region likely cause reservoir siltation, greenhouse gas emissions and mercury contamination, changes of downstream hydrological regimes and reduction of downstream sediment and consequently nutrient transports (e.g. phosphorous) (Forsberg et al., 2017). Minimizing impacts from future hydropower expansion in the Amazon basin requires, therefore, an integrated assessment of multiple environmental, social and economic criteria and a better coordination among the Amazonian nations (Flecker et al., 2022; Latrubesse et al., 2017). Furthermore, environmental impact studies and reports should integrate the downstream areas of floodplains in their assessments, at least until the next confluence with an undisturbed river of the same order (Schöngart et al., 2021).

Tropical floodplains along large flood-pulsing rivers such as the Tapajós, Orinoco, Magdalena, Congo, Mekong and many others (Wittmann et al., 2022b), which undergo an unprecedented boom of hydropower dam installations are likely to be impacted by the sandwich effect in the downstream areas (Latrubesse et al., 2019, 2021; Winemiller et al., 2016), as it has been already evidenced for the Tocantins, Paraná and São Francisco rivers in Brazil (Agostinho et al., 2004; Bustamante et al., 2019; Junk et al., 2021). This threatens the integrity of the last remaining large networks of free-flowing rivers and their connected large floodplain systems over thousands of square kilometres in the Amazon and other tropical regions (Grill et al., 2019). The potential loss of the integral functioning of one of the most valuable ecosystems (Davidson et al., 2019) must be considered in the cost-benefit analyses of hydropower dam projects involving a consortium that should integrate governments, scientists, NGOs, stakeholders from indigenous and traditional populations, industry and financial agencies. This is vital for the conservation of the unique biodiversity of floodplains, the maintenance of their ecosystem functioning, the food security of indigenous and traditional populations and their following generations and the sustainable development of a socio-bioeconomy. Overall, the findings discourage the installation of further hydropower dams in the Amazon basin.

10 Synthesis and Conclusions Concerning Conservation of Floodplains in the Changing Amazon

The Amazon basin has undergone an unprecedented transition over the last 50 years, characterized by growing anthropogenic impacts due to land-use and climate change and resulting synergies and feedback loops among these disturbance regimes which approximate the biome to its critical tipping points (Lovejoy & Nobre, 2018). A sophisticated modelling approach by Flores et al. (2024) projected that by 2050, 10–47% of Amazonian forests might approach critical transition trajectories. However, these are still uncertain regarding the large-river floodplains, as these ecosystems respond to different environmental drivers compared to the *terra-firme* forests. Simultaneously, the persistent formation of the climate dipole between the northern and southern regions of the Amazon basin since the climate shift in the 1970s (Marengo et al., 2021) has intensified the hydrological cycle. Since then, wet-season rainfall and maximum discharge have increased in the northern Amazon basin, while the southern part is facing an increasing dry-season length and dry-day frequency with a simultaneous decrease in precipitation and minimum river discharge. These trends are reflected by the increasing annual amplitude in the central Amazon basin over the last decades. Compared to the northern part, where the trend is mainly driven by climate variability and change, the intensifying drought conditions in the south are mainly a result of synergies of climate variability, climate change and massive land-use changes, as the environmental degradation turns the remaining ecosystems more vulnerable to extreme droughts which are mainly driven by the tropical Atlantic Ocean. Large impacts of these persistent changes can be expected for the large Amazonian river-floodplains which, however, have different species assemblies adapted to distinct environmental conditions and natural disturbance regimes with varying levels of resilience and vulnerability.

We suggest that the *várzea* vegetation along the nutrient-rich white-water floodplains has a higher resilience against disturbances as the *igapó* owing to its higher adaptive capacity responding to intense sedimentation and erosion processes which create a heterogeneous landscape with a high spatiotemporal variability of hydroclimatic and microclimatic conditions. Overall, functional characteristics (roots, leaves and wood) of plant species in the *várzea* have arrangements for resource acquisition resulting in fast growth and turnover rates. Additionally, *várzeas* occur in Amazonian regions with a high variability of different hydroclimatic configurations from almost perhumid conditions in the northwest to severe hydroclimatic drought conditions in the southwest of the Amazon basin. The black-water *igapó*, on the other hand, occurs mainly in the northwest and central Amazon region with a short or indistinct dry season. Due to the oligotrophic conditions, the vegetation of this ecosystem has slow dynamics as most tree species have functional traits (roots, leaves and wood) with arrangements for resource conservation resulting in slow growth and turnover rates. This turns the oligotrophic *igapó* highly vulnerable to disturbance and abrupt changes in environmental conditions exceeding the resilience of this ecosystem. We highlighted this by the examples of abrupt alterations of the hydrological cycles caused by a hydropower dam and anthropogenic fires favoured by severe hydroclimatic drought conditions generated by El Niño events (Fig. 8). In the case of the clear-water *igapós*, not sufficient information is available regarding data on tree species composition, functional traits, growth and turnover rates and analyses of their resilience against disturbance regimes. The generation of this knowledge is of high priority as these regions are undergoing an unprecedented transition driven by anthropogenic pressure. We can assume that the highly varying hydroclimatic and soil conditions between different regions of clear-water *igapós* also result in different environmental filters to which tree species adapted over evolutionary time scales. This is reflected by the tree species assemblies which have floristic components of both, the eutrophic *várzea* and the oligotrophic *igapó* (Conde et al., 2024; Wittmann & Junk, 2016).

Large river-floodplains have a key role in the conservation of the Amazon basin as they occur at the land-water boundary, harbour specific flora and fauna and provide essential ecosystem services (Wittmann & Junk, 2016). In general, the design of protected areas in the large Amazonian river-floodplains should be based on ecological-economic zoning that considers the natural capacity of the ecosystem for a sustainable use of its natural resources, aspects of the biodiversity, environmental services and socioeconomic factors such as population demographics, traditional uses and management of natural resources, the history of human occupation, among other aspects. Ideally, such an approach should be delineated on the level of macrohabitats (functional units in the landscape) which allows the development of specific sustainable natural resource managements, conservation strategies and environmental legislation (Junk et al., 2018).

Várzea floodplains allow, due to the high fertility of their alluvial soils, a sustainable use of their natural resources. Considering the high population density in these rural regions, scientists have proposed several strategies to improve the management of the floodplain's natural resources, such as the multiple-use concept (Junk et al., 2000, 2020, 2023), the concept of participatory management of fishery resources (McGrath et al., 2020) and the concept of growth-oriented logging (GOL) for timber resources (Schöngart, 2008, 2020). Protected areas in the *várzea* should, therefore, favour categories that allow the sustainable use of natural resources, based on existing scientific knowledge and participatory management that integrates traditional knowledge of riverine populations, ensuring their food security, promoting a sustainable socio-bioeconomy and safeguarding their livelihoods.

In the case of oligotrophic black-water *igapós*, the situation is different. The population density in this ecosystem is low, and its vulnerability limits the sustainable use of natural resources due to the low regenerative capacity and slow dynamics (Carvalho et al., 2021; Piedade et al., 2024; Schöngart et al., 2021). Conservation units should favour the integral protection of biodiversity; however, considering regions with historical human occupation and traditional extractive uses of non-timber forest products and offering economic alternatives such as ecotourism, sport and ornamental fisheries through participatory management integrating resident traditional populations under an efficient governance regime. For clear-water *igapós*, science lacks robust information to contribute to the conservation of these ecosystems which currently suffer the greatest threats from land-use and climate changes and the synergies and feedback loops that result from these disturbance regimes. It is of outmost importance to increase research efforts in clear-water *igapós* by interdisciplinary approaches, considering these actual scenarios.

However, the establishment of protected areas and the development of conservation strategies in Amazonian floodplains must be sustained by efficient public policies based on scientific knowledge, integrating different governmental levels, non-governmental organizations, traditional and indigenous populations and other relevant stakeholder groups mainly from private sectors. This is of vital importance to prevent or at least mitigate the complex impacts caused by increasing disturbance regimes on floodplains and the human populations which depend on the functional integrity of these ecosystems to maintain their unique biodiversity, vital ecosystem services and food webs. The increasing inundation length and impacts caused by severe droughts threaten the traditional management systems and consequently the socio-bioeconomy, food and water security and livelihood of traditional and indigenous populations. Large investments are, therefore, necessary to develop specific innovative technologies to adapt traditional resource management and livelihoods, especially in the *várzea* based on interdisciplinary research to guarantee water and food security for traditional populations in the floodplains. In the background of this major threat, it is absurd to suggest the expansion of farming in the *várzea* to prevent the disturbance of upland forests (Coomes et al., 2022).

The intensification of the flood-pulse regime in the central Amazon region also threatens human populations in vulnerable locations of urban zones and affects diverse socioeconomic sectors and public services. Seasonal forecast models of the maximum water level of the Negro River (with reference to the Port of Manaus) with a lead time of more than 3 months have been developed (Chevuturi et al., 2022, 2023; Maciel et al., 2020; Schöngart & Junk, 2007, 2020). Such modelling approaches should be extended in time and space to provide reliable forecasts of minimum, maximum and monthly water levels at strategic locations. Seasonal forecasts of the hydrological and climatic conditions together with adaptations of the traditional management systems and livelihoods of the vulnerable populations are efficient measures to mitigate socioeconomic impacts in urban zones and rural regions if they are sustained by efficient public policies and the society. The intensification of the hydrological regime and the resulting complex environmental, social and economic impacts are some of the biggest challenges faced by, nowadays, public policies at different governmental levels and society requiring urgent actions to guarantee the water and food security, socio-bioeconomy and livelihood of vulnerable human populations in the Amazon basin.

While it is relatively clear that large-scale hydrological change is underway in the Amazon, relevant biological data for assessing the consequences on flora and fauna are far too scarce, especially at basin-level spatial scales that are increasingly needed to keep pace with and mitigate the growing threats. Key challenges include not only an improved understanding of floodplain occupancy patterns of many thousands of species but also how this links to the wide variation in hydrological patterns across Amazonia. Householder et al. (2024) provide an interesting example of what this might look like on regional scales for floodplain tree species. Collating nearly 2000 forest inventory plots across Amazonia, Householder et al. (2024) used a spatially explicit approach to look at differences in tree species composition (or species turnover) between floodplain and *terra-firme* forests to convincingly demonstrate the crucial role that flooding patterns have on sustaining the unique tree species compositions of these ecosystems, and what this means for safeguarding landscape biodiversity and anticipating floodplain vulnerability in the face of future threats arising from climate and land-use changes.

The results indicate that species turnover of floodplains with surrounding *terra-firme* is exceptionally high in the central Amazonian region, precisely where floods reach their highest amplitude and longest duration (Householder et al., 2024). In other words, tree species distributions are strongly constrained to either floodplain or *terra-firme*, but not both. Other regions with high species turnover are *várzeas* in the western Amazon basin, black-water *igapós* of the Negro River basin and clear-water *igapós* in the Tapajós River basin (Fig. 9). Further analyses indicated that *várzeas* and *igapós* with a high tree species turnover are mainly composed of specialist species with narrow preferences for floodplain habitats. This group of 301 identified tree species (154 spp. in *várzea*, 115 spp. in *igapó* and 32 spp. occurring in both floodplain types) from a total of 1450 taxonomically identified tree species occurring in Amazonian floodplains comprise in regions with high species turnover more than 70% of all individuals (Householder et al., 2024). These specialists

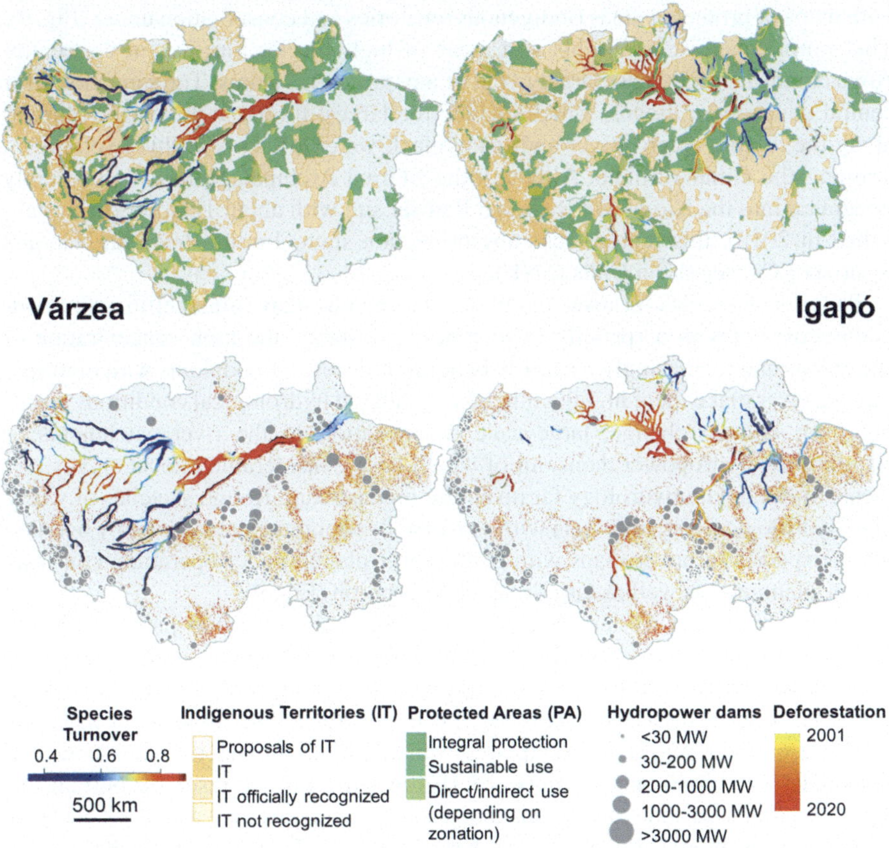

Fig. 9 Tree-species turnover in *várzea* (left) and *igapó* (right) floodplains as a measure of the level of floristic dissimilarity with adjacent upland (*terra firme*) forests (Householder et al., 2024). Areas in red indicate a high species turnover caused by the dominance of specialist tree species. The modelled species turnover is overlapped with protected areas and indigenous territories (upper panels) indicating regions that lack protection status (Data: Rede Amazônica de Informações Socioambiental—RAISG with reference to 2022). Threats for Amazonian floodplains arise from hundreds of hydropower plants (>1 MW) under operation, construction and planning and large-scale deforestation and forest degradation in the catchment areas (lower panels) (Data: RAISG, Fearnside, 2019, Fearnside et al., 2021)

represent a unique genetic patrimonial value due to their sophisticated adaptations to tolerate flooding and are key elements in the floodplain ecosystem to maintain high levels of biodiversity through interactions with other plant and animal species. Furthermore, they provide critical ecosystem services maintaining the forest structure and sustaining food webs. Floodplain regions with a high species turnover are thus indicators for conservation priorities, but also sentinels of hydrological changes.

Because the degree of species turnover reflects the level of overlap between the floodplain and *terra-firme* species compositions, prioritizing floodplain regions with high species turnover can potentially maximize benefits for biodiversity conservation in these ecosystems. We might more precisely target these floodplains by overlaying maps of *várzea* and *igapó* species turnover (Householder et al., 2024) with maps of protected areas (indigenous territories and conservation units) (Fig. 9). This simple exercise spotlights vast areas of high species turnover in the central Amazonian *várzea* that are largely void of any protective status. The same holds for central Amazonian *igapós,* which include the largest fluvial archipelago (Mariuá) along the middle Negro River and many black-water affluents in this river stretch draining the Guiana Shield. Although the Mariuá Archipelago has been recently integrated into the Rio Negro Regional Ramsar site with about 120,000 km^2 (established in 2018), this region lacks any protection status by the Brazilian National System of Conservation Units (SNUC).

Patterns of species turnover might also be used to map vulnerabilities to floodplain ecosystems in a spatially explicit way. Arguably, the high concentration of specialists and increasingly narrow habitat preferences in floodplains with high species turnover make them highly vulnerable to altered hydrological conditions resulting from climate change, large-scale deforestation of the river catchments or proliferating hydropower dams, all of which have considerable effects on key elements of floodplain hydrology far up and/or downstream. Rough sketches of where high hydrological risk intersects with sensitive floodplains can be drawn by overlaying the species turnover maps with hydropower plants (under operation, construction or planning) as well as large-scale deforestation (Fig. 9).

The *várzea* in the western Amazon basin, for example, seems particularly vulnerable, where the hydrological regimes of extensive floodplains with high species turnover will be affected by dozens of upstream hydropower plants. The large number of dams might also lead to considerable downstream reductions in the sediment load and nutrient supply to floodplains, which can also impact species compositions. Dissolved inorganic mercury (MeHg) produced in the reservoir bottom further can affect downstream biodiversity, species interactions and food webs in floodplains by bioaccumulation (Forsberg et al., 2017). In another example, the *várzea* along the lower Madeira River with a high species turnover is likewise under greater risk due to large hydropower plants in the upstream regions as well as surrounding deforestation (Latrubesse et al., 2017). The same holds for the clear-water *igapós* along the Branco and Tapajós rivers, which are threatened by cumulative hydrological alterations that might be expected from large-scale deforestation in the catchment areas, persistent climate dipole formation and the multiplying effects of a chain of successive hydropower plants (Flores et al., 2024; Latrubesse et al., 2017;

Timpe & Kaplan, 2017). The drivers of hydrological alteration to floodplains vary regionally and the synergisms of multiple disturbance vectors impacting these ecosystems remain unexplored. These arising threats underpin the need for a watershed-based conservation strategy to mitigate the impacts of hydrological alterations on floodplains caused by land-use and climate change and their complex synergetic effects to prevent massive losses of a unique floodplain biodiversity (Householder et al., 2024).

The transboundary character of the Amazon basin implies a basin-wide management and sustainable development to mitigate impacts on floodplains caused by hydrological alterations. For instance, alterations of the hydrological regime, sediment load and contaminants caused by large-scale deforestation, hydropower dams, mining, oil extraction and other land-use changes in catchments of the western Amazon basin (e.g. Bolivia, Peru and Colombia) likely affect downstream floodplains located in neighbouring countries, such as Brazil (Forsberg et al., 2017; Latrubesse et al., 2017). This holds especially for the *várzea* floodplains along white-water rivers, but also for the black-water *igapós* of the Negro River. On the other hand, Brazil comprises more than 60% of the Amazon forest biome which generates water vapour transported westwards by the trade winds sustaining rainfall regimes of other Amazonian countries (Flores et al., 2024). Land-use changes altering the regional hydrological cycle can thus act in both transboundary directions, upstream and downstream. This connectivity implies efficient international policies and increased efforts among the Amazonian countries to develop watershed-based conservation strategies which is a major challenge for the Amazon Cooperation Treaty Organization (ACTO).

The conservation status and the impacts of accumulated anthropogenic disturbances along land-use history vary considerably among the large Amazonian river basins (Latrubesse et al., 2017, Soares-Filho et al., 2023). The northern portion comprises several oligotrophic and vulnerable wetland ecosystems and landscapes (e.g. white-sand ecosystems, black-water and clear-water floodplains and hydromorphic savannas) with a high partially endemic biodiversity and sociocultural diversity (Science Panel for the Amazon, 2021). The majority of watersheds of the river basins in the north are partially or totally integrated into a large framework of protected areas, indigenous and *quilombola* (descendants of Afro-Brazilian slaves) lands with an overall high conservation status (Soares-Filho et al., 2023; Piedade et al., 2024; Fig. 9). Efforts to increase effective conservation measures and to avoid large-scale infrastructural projects must be a priority for the maintenance of its unique biodiversity, ecosystem services, sociocultural diversity and integrity of the traditional and indigenous populations. In contrast, the southern regions undergo increasing drought conditions and an unprecedented transition mainly driven by unsustainable logging, large-scale deforestation and associated fires as well as forest fragmentation to expand agriculture and livestock production on mainly undesignated public forests (Brando et al., 2020; Covey et al., 2021). This turns the region into a net carbon source (Gatti et al., 2021; Kruid et al., 2021) associated with losses of biodiversity and ecosystem services which have negative feedback on climate change approximating these areas to the predicted tipping points (Flores et al.,

2024; Lovejoy & Nobre, 2018). Large-scale deforestation, mining activities and dozens of operating, under construction or planned hydropower dams, mainly in watersheds of the Tocantins-Araguaia, Xingú, Tapajós and Madeira river basins are expected to cause severe environmental impacts on the river connectivity (Latrubesse et al., 2017; Timpe & Kaplan, 2017). These unprecedented land-use changes and simultaneous increasing temperature and drought conditions are likely to affect the floodplains of these rivers altering streamflow and physiochemical characteristics, increasing the emission of greenhouse gases and causing losses of biodiversity (Costa et al., 2003; Latrubesse et al., 2017, 2021; Timpe & Kaplan, 2017). Finally, this will affect the integral functioning of floodplains, as well as the water and food security of indigenous and traditional human populations depending on these ecosystems. It is a paradox to expand hydropower generation into a region with a persistent trend of increasing length and intensity of the dry season during the last decades (Marengo et al., 2021), which is expected to continue as a result of the synergies of anthropogenic land-use and climate change approaching, especially, this region to a critical transition (Flores et al., 2024).

In the southern portion, immediate actions must include adequate ecological restoration of the basins with native species considering the varying vectors of environmental degradation and future climate scenarios, integrating the existing protected areas and sociocultural diversity into a watershed-based conservation strategy that is missing in this region. The adoption of new and innovative land-use forms, such as agroforestry, agrosilvopastoral and low-carbon agricultural systems sustained by socioenvironmental education, are effective measures to transform undesignated public areas of this region from a net carbon source into a long-term net carbon sink (Torres et al., 2017). This holds also for the floodplains and other wetlands in these regions which need specific measures of protection, sustainable resource management and conservation (Junk et al., 2014). Overall, restoring degraded areas, developing a sustainable socio-bioeconomy and innovative land-use forms and avoiding new deforestation and fires could restore essential ecosystem services at regional scales to avoid the approaching tipping points (Flores et al., 2024, Lovejoy & Nobre, 2018) and future large-scale migration of resident populations as a consequence of the increasing drought conditions and heatwaves. Simultaneously, global efforts must decrease greenhouse gas emissions to avoid temperature increasing beyond the internationally established thresholds by the UN Climate Change Conferences (COP).

Much of this recent development in the Amazon basin is driven and intensified by national and global economic interests. However, this also implies more support from the international community to achieve the Sustainable Development Goals formulated by the United Nations for the Amazon region, considering the United Nations Framework Convention on Climate Change (UNFCC; Conferences-of-the-Parties—COP; REDD and REDD+ conventions), the Ramsar Convention and other international treaties. This is a major challenge for the Amazonian countries, particularly Brazil, the ACTO and the international community. The Amazonian countries have a unique chance to promote a differentiated development for the largest hydrobasin on Earth hosting a unique network of free-flowing rivers and associated floodplains (Grill et al., 2019; Householder et al., 2024). This can be achieved by avoiding well-known errors of historical development in many countries and investing in specific and adapted technologies and innovations based on transdisciplinary sciences. Until today, more than 50% of the Amazon basin is under protection by indigenous lands and different categories of conservation units with varying governmental levels. This is an unprecedented effort made by the Amazonian countries with international support (e.g. SNUC, Amazon Protected Areas Programme—ARPA, Ramsar Convention, World Heritage Convention, Biosphere Reserves, Amazon Fund, among others) (Science Panel for the Amazon, 2021), which deserve more international recognition. Mankind has the power to counteract climate change by increasing measures of mitigation, adaptation and resilience. This can only be promoted by education to increase awareness and change the acting of individuals, communities and societies at national and international levels. It is our responsibility to guarantee the livelihood of vulnerable human populations, but also regarding our following generations. The fate of the large and unique Amazonian floodplains will depend on this.

Acknowledgements This study has been conducted in the framework of the Long-Term Ecological Research Programme (PELD) of the Brazilian National Council for Scientific and Technological Development (CNPq), performed since 2012 by the research group MAUA (Ecology, Monitoring and Sustainable Use of Wetlands) of the Brazilian National Institute for Amazon Research (INPA). We acknowledge funding support from the Long-term Ecological Research Programme (PELD) by the Brazilian National Council for Scientific and Technological Development (CNPq; grant numbers 403792/2012-6, 441590/2016-0, 441811/2020-5) and the Amazonas Research Foundation (FAPEAM; grant numbers 062.00466/2014; 062.01357/2017; 01.02.016301.02630/2022-76). We further acknowledge support by the National Institute of Science and Technologies in Adaptations of the Amazon Aquatic Biology (INCT-ADAPTA) during the first (CNPq grant number 465540/2014-7; FAPEAM grant number 062.1187/2017) and second (CNPq grant number 465540/2014-7) phase. This study was financed in part by the Coordination for the Improvement of Higher Education Personnel—Brazil (CAPES; Finance Code 001). We acknowledge the support of the BiodivERsA 2019–2020 Joint COFUND Call on 'Biodiversity and Climate Change' (German Federal Ministry of Education and Research—BMBF: 16LC2025A). We further received funding and support from the Technical-Scientific Cooperation between INPA and the Max-Planck Society with a focus on floodplain forests.

References

Aalto, R., Maurice-Bourgoin, L., Dunne, T., et al. (2003). Episodic sediment accumulation on Amazonian floodplains influenced by El Niño/southern oscillation. *Nature, 425*, 493–497.

Aceituno, P., Prieto, M. R., Solari, M. E., et al. (2009). The 1877–1878 El Niño episode: Associated impacts in South America. *Climatic Change, 92*, 389–416.

Adis, J. A., Erwin, T. L., Battirola, L. D., et al. (2010). The importance of floodplain forests for animal biodiversity: Beetles in canopies of floodplain and upland forests. In W. J. Junk, M. T. F. Piedade, F. Wittmann, et al. (Eds.), *Amazonian floodplain forests. Ecophysiology, biodiversity and sustainable management, ecological studies* (Vol. 210, pp. 313–328). Springer.

Agostinho, A. A., Tomaz, S. M., & Gomes, L. C. (2004). Threats for biodiversity in the floodplain of the upper Paraná River: Effects of hydrological regulation by dams. *Ecohydrology and Hydrobiology, 4*, 255–268.

Aguiar, D. P. P. (2015). *Influência dos fatores hidro-edáficos na diversidade, composição florística e estrutura da comunidade arbórea de igapó no Parque Nacional do Jaú, Amazônia Central*. Dissertation, Instituto Nacional de Pesquisas da Amazônia.

Albert, J. S., Carnaval, A. C., Flantua, S. G. A., et al. (2023). Human impacts outpace natural processes in the Amazon. *Science, 379*, 348.

Almeida, D. R. A., Nelson, B. W., Schietti, J., et al. (2016). Contrasting fire damage and fire susceptibility between seasonally flooded forest and upland forest in the Central Amazon using portable profiling Lidar. *Remote Sensing of Environment, 184*, 153–160.

Altahyde, S., Mathews, M., Bohlman, S., et al. (2019). Mapping research on hydropower and sustainability in the Brazilian Amazon: Advances, gaps in knowledge and future directions. *Current Opinion in Environment Sustainability, 37*, 50–69.

Aragão, L. E. O. C., Anderson, L. O., Fonseca, M. G., et al. (2018). 21st century drought-related fires counteract the decline of Amazon deforestation carbon emissions. *Nature Communications, 9*, 536.

Arraut, J. M., Nobre, C., Barbosa, H. M. J., et al. (2012). Aerial rivers and lakes: Looking at large-scale moisture transport and its relation to Amazonia and to subtropical rainfall in South America. *Journal of Climate, 25*, 543–556.

Assahira, C., Piedade, M. T. F., Trumbore, S. E., et al. (2017). Tree mortality of a flood-adapted species in response of hydrographic changes caused by an Amazonian river dam. *Forest Ecology and Management, 396*, 113–123.

Baker, J. C. A., Gloor, M., Spracklen, D. V., et al. (2016). What drives interannual variation in tree ring oxygen isotopes in the Amazon? *Geophysical Research Letters, 43*, 11831–11840.

Barichivich, J., Gloor, E., Peylin, P., et al. (2018). Recent intensification of Amazon flooding extremes driven by strengthened Walker circulation. *Science Advances, 4*, eaat8785.

Barlow, J., Berenguer, E., Carmenta, R., et al. (2020). Clarifying Amazonia's burning crisis. *Global Change Biology, 26*, 319–321.

Batista, E. S. (2015). *Estimativas de produtividade de biomassa lenhosa ao longo de gradientes ambientais em florestas alagáveis na Amazônia Central*. Dissertation, Instituto Nacional de Pesquisas da Amazônia.

Beal, L. M., De Ruijter, W. P. M., Biastoch, A., et al. (2011). On the role of the Agulhas system in ocean circulation and climate. *Nature, 472*, 429–436.

Biastoch, A., Böning, C. W., & Lutjeharms, J. R. E. (2009). Increase in Agulhas leakage due to poleward shift of southern hemisphere westerlies. *Nature, 462*, 495–498.

Bonnet, M. P., Barroux, G., Martinez, J. M., et al. (2008). Floodplain hydrology in an Amazon floodplain lake (Lago Grande de Curuai). *Journal of Hydrology, 349*, 18–30.

Brando, P., Macedo, M., Silvério, D., et al. (2020). Amazon wildfires: Scenes from a foreseeable disaster. *Flora, 268*, 151609.

Brêda, J. P. L. F., Paiva, R. C. D., Siqueira, V. A., et al. (2023). Assessing climate change impact on flood discharge in South America and the influence of its main drivers. *Journal of Hydrology, 619*, 129284.

Brienen, R. J. W., Helle, G., Pons, T. L., et al. (2012). Oxygen isotopes in tree rings are a good proxy for Amazon precipitation and El Niño-southern oscillation variability. *Proceedings of the National Academy of Sciences, 109*, 16957–16962.

Brienen, R. J. W., Phillips, O. L., Feldpausch, T. R., et al. (2015). Long-term decline of the Amazon carbon sink. *Nature, 519*, 344–348.

Buckley, B. M., Ummenhofer, C. C., D'Arrigo, R. D., et al. (2019). Interdecadal Pacific oscillation reconstructed from trans-Pacific tree rings: 1350–2004 CE. *Climate Dynamics, 53*, 3181–3196.

Bustamante, M. M. C., Metzger, J. P., Scariot, A., et al. (2019). Tendências e impactos dos vetores de degradação e restauração da biodiversidade e dos serviços ecossistêmicos. In C. A. Joly, F. R. Scarano, C. S. Seixas, et al. (Eds.), *1° Diagnóstico Brasileiro de Biodiversidade e Serviços Ecossistêmicos, capítulo 3*. Editora Cubo.

Cai, W., McPhaden, M. J., Grimm, A. M., et al. (2020). Climate impacts of the El Niño–southern oscillation on South America. *Nature Reviews Earth and Environment, 1*, 215–231.

Campozano, L., Robaina, L., & Samaniego, E. (2020). The Pacific decadal oscillation modulates the relation of ENSO with the rainfall variability in coast of Ecuador. *International Journal of Climatology, 40*, 5801–5812.

Carvajal-Lugo, A., Holmgren, M., Zuanon, J., et al. (2023). Fish on fire: Shifts in Amazonian fish communities after floodplain forest fires. *Journal of Applied Ecology*. https://doi.org/10.1111/1365-2664.14434

Carvalho, T. C. (2019). *Susceptibilidade ao fogo de florestas de igapó de águas pretas no Parque Nacional do Jaú, Amazônia Central*. Dissertation, Instituto Nacional de Pesquisas da Amazônia, Manaus, Brazil.

Carvalho, T. C., Wittmann, F., Piedade, M. T. F., et al. (2021). Fires in Amazonian Blackwater floodplain forests: Causes, human dimension, and implications for conservation. *Frontiers in Forests and Global Change, 4*, 75544.

Castello, L., McGrath, D. G., Hess, L. L., et al. (2013). The vulnerability of Amazon freshwater ecosystems. *Conservation Letters, 6*, 217–229.

Chevuturi, A., Klingaman, N. P., Rudorff, C. M., et al. (2022). Forecasting annual maximum water level for the Negro River at Manaus. *Climate Resilience and Sustainability, 1*, e18.

Chevuturi, A., Klingaman, N. P., Woolnough, S. J., et al. (2023). Forecasting annual maximum water level for the Negro River at Manaus using dynamical seasonal predictions. *Climate Services, 30*, 100342.

Chiang, J. C. H., & Vimont, D. J. (2004). Analogous Pacific and Atlantic meridional modes of tropical atmosphere–ocean variability. *Journal of Climate, 17*, 4143–4158.

Cintra, B. B. L., Gloor, M., Boom, J., et al. (2019). Contrasting controls on tree ring isotope variation for Amazon floodplain and terra firme trees. *Tree Physiology, 39*(5), 845–860.

Cintra, B. B. L., Gloor, M., Boom, A., et al. (2022). Tree-ring oxygen isotopes record a decrease in Amazon dry season rainfall over the past 40 years. *Climate Dynamics, 59*, 1401–1414.

Conde, M. L. G., Piedade, M. T. F., Wittmann, F., et al. (2024). Evaluation of the management potential of timber resources in Clearwater floodplain forests in the Amazon using growth models. *Journal of Environmental Management, 729*, 119781.

Connell, J. H., & Slatyer, R. O. (1977). Mechanisms of succession in natural communities and their role in community stability and organization. *The American Naturalist, 111*, 1119–1144.

Cook, B. I., Williams, A. P., Smerdon, J. E., et al. (2018). Cold tropical Pacific Sea surface temperatures during the late sixteenth-century north American megadrought. *Journal of Geophysical Research—Atmospheres, 123*, 11307–11320.

Coomes, O. T., Takasaki, Y., & Abizaid, C. (2022). Sparing of Amazonian old-growth forests with floodplain access. *Nature Sustainability, 5*, 965–972.

Correa, S. B., Costa-Pereira, R., Fleming, T., et al. (2015). Neotropical fish–fruit interactions: Eco-evolutionary dynamics and conservation. *Biological Reviews, 90*, 1263–1278.

Costa, M. H., Botta, A., & Cardille, J. A. (2003). Effects of large-scale changes in land cover on the discharge of the Tocantins River, southeastern Amazonia. *Journal of Hydrology, 283*, 206–217.

Costa, M. H., Borma, L. S., Espinoza, J. C., et al. (2021). Chapter 5: The physical hydroclimate system of the Amazon. In C. Nobre, A. Encalada, E. Anderson, et al. (Eds.), *Amazon assessment report 2021*. United Nations Sustainable Development Solutions Network.

Covey, K., Soper, F., Pangala, S., et al. (2021). Carbon and beyond: The biogeochemistry of climate in a rapidly changing Amazon. *Frontiers in Forests and Global Change, 4*, 618401.

Crawford, R. M. M. (1992). Oxygen availability as an ecological limit to plant distribution. *Advances in Ecological Research, 23*, 93–185.

Crema, L. C. (2017). *Caracterização de igapós de águas claras e pretas e suas disponibilidades alimentares para o peixe-boi-da- Amazônia (Trichechus inunguis)*. Dissertation, Instituto Nacional de Pesquisas da Amazônia, Manaus, Brazil.

Davidson, E. A., Araújo, A. C., Artaxo, P., et al. (2012). The Amazon basin in transition. *Nature, 481*, 321–328.

Davidson, N. C., van Dam, A. A., Finlayson, C. M. et al. (2019). Worth of wetlands: Revised global monetary values of coastal and inland wetland ecosystem services. *Marine and Freshwater Research, 70*, 1189–1194.

De Melo, R. B., Franco, A. C., Silva, C. O., et al. (2015). Seed germination and seedling development in response to submergence in tree species of the central Amazonian floodplains. *AoB Plants, 7*, plv041.

De Simone, O., Haase, K., Müller, E., et al. (2002). Impact of root morphology on metabolism and oxygen distribution in roots and rhizosphere from two Central Amazon floodplain tree species. *Functional Plant Biology, 29*, 1025–1035.

Denevan, W. M. (1996). A bluff model of riverine settlement in prehistoric Amazonia. *Annals of the Association of American Geographers, 86*, 654–681.

Enfield, D. B., Mestas-Nuñez, A. M., & Trimble, P. J. (2001). The Atlantic multidecadal oscillation and its relation to rainfall and river flows in the continental U.S. *Geophysical Research Letters, 28*, 2077–2080.

Engle, D. L., Melack, J. M., Doyle, R. D., et al. (2008). High rates of net primary productivity and turnover for floating grasses on the Amazon floodplain: Implications for aquatic respiration and regional CO_2 flux. *Global Change Biology, 14*, 369–381.

Espinoza Villar, J. C., Guyot, J. L., Ronchail, J., et al. (2009). Contrasting regional discharge evolutions in the Amazon basin (1974-2004). *Journal of Hydrology, 375*, 297–311.

Espinoza, J. C., Marengo, J. A., Ronchail, J., et al. (2014). The extreme 2014 flood in South-Western Amazon basin: The role of tropical-subtropical South Atlantic SST gradient. *Environmental Research Letters, 9*, 124007.

Espinoza, J. C., Ronchail, J., Marengo, J. A., et al. (2019). Contrasting north–south changes in Amazon wet-day and dry-day frequency and related atmospheric features (1981–2017). *Climate Dynamics, 52*, 5413–5430.

Espinoza, J. C., Marengo, J. A., Schöngart, J., et al. (2022). The new historical flood of 2021 in the Amazon River compared to major floods of the 21st century: Atmospheric features in the context of the intensification of floods. *Weather and Climate Extremes, 35*, 100406.

Fassoni-Andrade, A. C., Fleischmann, A. S., Papa, F., et al. (2021). Amazon hydrology from space: Scientific advances and future challenges. *Reviews of Geophysics, 59*, e2020RG000728.

Fearnside, P. M. (1989). Brazil's Balbina dam: Environment versus the legacy of the pharaohs in Amazonia. *Environmental Management, 13*, 401–423.

Fearnside, P. M. (Ed.). (2019). *Hidrelétricas na Amazônia: Impactos Ambientais e Sociais na Tomada de Decisões sobre Grandes Obras* (Vol. 3). Editora INPA.

Fearnside, P. M., Berenguer, E., Armenteras, D., et al. (2021). Chapter 20: Drivers and impacts of changes in aquatic ecosystems. In C. Nobre, A. Encalada, E. Anderson, et al. (Eds.), *Amazon assessment report 2021*. United Nations Sustainable Development Solutions Network.

Ferreira, A. B. (2022). *Efeito do fogo induzido na germinação de diásporos de Astrocaryum jauari (Mart.) em cenários de mudanças climáticas*. Dissertation, Instituto Nacional de Pesquisas da Amazônia.

Ferreira, C. S., Piedade, M. T. F., Tiné, M. A. S., et al. (2009). The role of carbohydrates in seed germination and seedling establishment of *Himatanthus sucuuba*, an Amazonian tree with populations adapted to flooded and non-flooded conditions. *Annals of Botany, 104*, 1111–1119.

Ferreira, L. V., Cunha, D. A., & Chaves, P. P. (2013). Impacts of hydroelectric dams on alluvial riparian plant communities in eastern Brazilian Amazonia. *Anais da Academia Brasileira de Ciências, 85*, 1013–1023.

Finlayson, C. M., Milton, R., Prentice, C., et al. (Eds.). (2018). *The wetland book. Vol 2: Distribution, description, and conservation*. Springer.

Flecker, A. S., Shi, Q., Almeida, R. M., et al. (2022). Reducing adverse impacts of Amazon hydropower expansion. *Science, 375*, 753–760.

Fleischmann, A. S., Papa, F., Fassoni-Andrade, A., et al. (2022). How much inundation occurs in the Amazon River basin? *Remote Sensing of Environment, 278*, 113099.

Fleischmann, A. S., Laipelt, L., Papa, F., et al. (2023a). Patterns and drivers of evapotranspiration in south American wetlands. *Nature Communications, 14*, 6656.

Fleischmann, A. S., Papa, F., Hamilton, S. K., et al. (2023b). Increased floodplain inundation in the Amazon since 1980. *Environmental Research Letters, 18*, 034024.

Flores, B. M., & Holmgren, M. (2021). White-sand savannas expand at the core of the Amazon after forest wildfires. *Ecosystems, 24*, 1624–1637.

Flores, B. M., Piedade, M. T. F., & Nelson, B. W. (2014). Fire disturbance in Amazonian Blackwater floodplain forests. *Plant Ecology & Diversity, 7*, 319–327.

Flores, B. M., Holmgren, M., Xu, C., et al. (2017). Floodplains as an Achilles' heel of Amazonian forest resilience. *Proceedings of the National Academy of Sciences, 114*, 4442–4446.

Flores, B. M., Montoya, E., Sakschewski, B., et al. (2024). Critical transitions in the Amazon forest system. *Nature, 626*, 555–564.

Foley, J. A., Botta, A., Coe, M. T., et al. (2002). El Niño-southern oscillation and the climate, ecosystems and rivers of Amazonia. *Global Biogeochem Cycles, 16*(4), 1132.

Foley, J. A., Costa, M. H., Delire, C., et al. (2003). Green surprise? How terrestrial ecosystems could affect Earth's climate. *Frontiers in Ecology and the Environment, 1*, 38–44.

Fontes, C. G., Fine, P. V. A., Wittmann, F., et al. (2020). Convergent evolution of tree hydraulic traits in Amazonian habitats: Implications for community assemblage and vulnerability to drought. *The New Phytologist, 228*, 106–120.

Forsberg, B. R., Dunne, T., Melack, J. M., et al. (2017). Potential impact of new Andean dams on the Amazon fluvial ecosystems. *PLoS One, 12*(8), e0182254.

Furch, K. (1997). Chemistry of várzea and igapó soils and nutrient inventory of their flood-plain forests. In W. J. Junk (Ed.), *The Central Amazon floodplain: Ecology of a pulsing system, ecological studies* (Vol. 126, pp. 47–67). Springer.

Furch, K., & Junk, W. J. (1997). Physiochemical conditions in floodplains. In W. J. Junk (Ed.), *The Central Amazon floodplain: Ecology of a pulsing system, ecological studies* (Vol. 126, pp. 69–108). Springer.

Gambrel, R. P., & Patrick, W. H. (1978). Chemical and microbiological properties of anaerobic soils and sediments. In D. D. Hook & R. M. M. Crawford (Eds.), *Plant life in anaerobic environments* (pp. 233–247). Ann Arbor Science Publication.

Gash, J., Keller, M., & Silva-Dias, P. (Eds.). (2009). *Amazonia and global change. Geophysical monograph series 186*. American Geophysical Union.

Gatti, L. V., Basso, L. S., Miller, J. B., et al. (2021). Amazonia as a carbon source linked to deforestation and climate change. *Nature, 595*, 388–393.

Gloor, M., Brienen, R. J. W., Galbraith, D., et al. (2013). Intensification of the Amazon hydrological cycle over the last two decades. *Geophysical Research Letters, 40*, 1729–1733.

Gloor, M., Barichivich, J., Ziv, G., et al. (2015). Recent Amazon climate as background for possible ongoing and future changes of Amazon humid forests. *Global Biogeochem Cycles, 29*, 1384–1399.

Gonçalves, J. Q., Durgante, F. M., Wittmann, F., et al. (2021). Minimum temperature and evapotranspiration in central Amazonian floodplains limit tree growth of *Nectandra amazonum* (Lauraceae). *Trees, 35*, 1367–1384.

Gottsberger, G. (1978). Seed dispersal by fish in the inundated regions of Humaitá, Amazonia. *Biotropica, 10*, 170–183.

Gouveia, N. A., Gherardi, D. F. M., & Aragão, L. E. O. C. (2019). The role of the Amazon River plume on the intensification of the hydrological cycle. *Geophysical Research Letters, 46*, 12221–12229.

Granato Souza, D., & Stahle, D. W. (2023). Drought and flood extremes on the Amazon River and in Northeast Brazil, 1790-1900. *Journal of Climate, 36*, 7213–7229.

Granato-Souza, D., Stahle, D. W., Barbosa, A. C. M. C., et al. (2019). Tree rings and rainfall in the equatorial Amazon. *Climate Dynamics, 52*, 1857–1869.

Granato-Souza, D., Stahle, D. W., Torbenson, M. C. A., et al. (2020). Multidecadal changes in wet season precipitation totals over the eastern Amazon. *Geophysical Research Letters, 47*, e2020GL087478.

Grill, G., Lehner, B., Thieme, M., et al. (2019). Mapping the world's free-flowing rivers. *Nature, 569*, 215–221.

Guayasamin, J. M., Ribas, C. C., Carnaval, A. C., et al. (2021). Chapter 2: Evolution of Amazonian biodiversity. In C. Nobre, A. Encalada, E. Anderson, et al. (Eds.), *Amazon assessment report 2021*. United Nations Sustainable Development Solutions Network.

Haase, K., & Rätsch, G. (2010). The morphology and anatomy of tree roots and their aeration strategies. In W. J. Junk, M. T. F. Piedade, F. Wittmann, et al. (Eds.), *Amazonian floodplain forests. Ecophysiology, biodiversity and sustainable management, ecological studies* (Vol. 210, pp. 141–161). Springer.

Haghtalab, N., Moore, N., Heerspink, B. P., et al. (2020). Evaluating spatial patterns in precipitation trends across the Amazon basin driven by land cover and global scale forcings. *Theoretical and Applied Climatology, 140*, 411–427.

Heerspink, B. P., Kendall, A. D., Coe, M. T., et al. (2020). Trends in streamflow, evapotranspiration, and groundwater storage across the Amazon Basin linked to changing precipitation and land cover. *Journal of Hydrology: Regional Studies, 32*, 100755.

Henley, B. J., Gergis, J., Karoly, D. J., et al. (2015). A Tripole index for the Interdecadal Pacific oscillation. *Climate Dynamics, 45*, 3077–3090.

Herraiz, A. D., Graca, P. M. L. D., & Fearnside, P. M. (2017). Amazonian flood impacts on managed Brazilnut stands along Brazil's Madeira River: A sustainable forest management system threatened by climate change. *Forest Ecology and Management, 406*, 46–52.

Hess, L. L., Melack, J. M., Affonso, A. G., et al. (2015). Amazonian wetlands: Extent, vegetative cover, and dual season inundation area. *Wetlands, 35*, 745–756.

Hoorn, C., Wesselingh, F. P., ter Steege, H., et al. (2010). Amazonia through time: Andean uplift, climate change, landscape evolution, and biodiversity. *Science, 330*, 927–931.

Hoorn, C. M., Bogotá-A, G. R., Romero-Baez, M., et al. (2017). The Amazon at sea: Onset and stages of the Amazon River from a marine record, with special reference to Neogene plant turnover in the drainage basin. *Global and Planetary Change, 153*, 51–65.

Householder, J. E., Schöngart, J., Piedade, M. T. F., et al. (2021). Modeling the ecological responses of tree species to the flood pulse of the Amazon Negro River floodplains. *Frontiers in Ecology and Evolution, 9*, 628606.

Householder, J. E., Wittmann, F., Schöngart, J., et al. (2024). One sixth of Amazonian tree diversity is dependent on river floodplains. *Nature Ecology & Evolution*. https://doi.org/10.1038/s41559-024-02364-1

Irion, G., De Mello, J. A. S. N., Morais, J., et al. (2010). Development of the Amazon valley during the middle to late quaternary: Sedimentological and climatological observations. In W. J. Junk, M. T. F. Piedade, F. Wittmann, et al. (Eds.), *Amazonian floodplain forests. Ecophysiology, biodiversity and sustainable management, ecological studies* (Vol. 210, pp. 313–328). Springer.

Irmler, U. (1986). Temperature dependent generation cycle for the cicindelid beetle *Pentacomia egregia* Chaud. (Coleoptera, Carabidae, Cicindelidae) of the Amazon valley. *Amazoniana, 9*, 431–439.

Irmler, U., & Furch, K. (1980). Weight, energy and nutrient changes during decomposition of leaves in the emersion phase of Central-Amazonia inundation forests. *Pedobiologia, 20*, 118–130.

Junk, W. J. (1989). Flood tolerance and tree distribution in central Amazonian floodplains. In L. B. Holm-Nielsen, I. C. Nielsen, & H. Balslev (Eds.), *Tropical forests: Botanical dynamics, speciation and diversity* (pp. 47–64). Academic Press.

Junk, W. J. (Ed.). (1997). *The Central Amazon floodplain: Ecology of a pulsing system* (Ecological Studies, vol) (Vol. 126). Springer Verlag.

Junk, W. J., & Piedade, M. T. F. (1993). Herbaceous plants of the Amazon floodplains near Manaus: Species diversity and adaptations to the flood pulse. *Amazoniana, 12*, 467–484.

Junk, W. J., Bayley, P. B., & Sparks, R. E. (1989). The flood pulse concept in river-floodplain systems. Proceedings of the international large river symposium, Ottawa. *Canadian Journal of Fisheries and Aquatic Sciences, 106*, 110–127.

Junk, W. J., Ohly, J. J., Piedade, M. T. F., et al. (Eds.). (2000). *The Central Amazon floodplain: Actual use and options for a sustainable management*. Backhuys Publishers b.V.

Junk, W. J., Piedade, M. T. F., Wittmann, F., et al. (Eds.). (2010). *Central Amazonian floodplain forests: Ecophysiology, biodiversity and sustainable management* (Ecological Studies) (Vol. 210). Springer.

Junk, W. J., Piedade, M. T. F., Schöngart, J., et al. (2011). A classification of major Amazonian wetlands. *Wetlands, 31*, 623–640.

Junk, W. J., Piedade, M. T. F., Schöngart, J., et al. (2012). A classification of major natural habitats of Amazonian white water river floodplains (várzeas). *Wetlands Ecology and Management, 20*, 461–475.

Junk, W. J., Piedade, M. T. F., Lourival, R., et al. (2014). Brazilian wetlands: Definition, delineation, and classification for research, sustainable management, and protection. *Aquatic Conservation: Marine and Freshwater Ecosystems, 24*, 5–22.

Junk, W. J., Wittmann, F., Schöngart, J., et al. (2015). A classification of the major habitats of Amazonian black-water river floodplains and a comparison with their white-water counterparts. *Wetlands Ecology and Management, 23*, 677–693.

Junk, W. J., Piedade, M. T. F., Nunes da Cunha, C., et al. (2018). Macrohabitat studies in large Brazilian floodplains to support sustainable development in the face of climate change. *Ecohydrology and Hydrobiology, 18*, 334–344.

Junk, W. J., Piedade, M. T. F., Wittmann, F., et al. (Eds.). (2020). *Várzeas Amazônicas: Desafios para um Manejo Sustentável*. Editora INPA.

Junk, W. J., Nunes da Cunha, C., Thomaz, S. M., et al. (2021). Macrohabitat classification of wetlands as a powerful tool for management and protection: The example of the Paraná River floodplain, Brazil. *Ecohydrology and Hydrobiology, 21*, 411–424.

Junk, W. J., Schöngart, J., Piedade, M. T. F., et al. (2023). The Várzeas of the Brazilian Amazon River: Living with the flood pulse. In K. M. Wantzen (Ed.), *River culture: Life as a dance to the rhythm of the waters* (pp. 469–495). UNESCO Publishing.

Kaandorp, R. J. G., Vonhof, H. B., Wesselingh, F. W., et al. (2005). Seasonal Amazonian rainfall variation in the Miocene climate optimum. *Palaeo3, 221*, 1–6.

Kalliola, R., Puhakka, M., Salo, J., et al. (1991). The dynamics, distribution and classification of swamp vegetation in Peruvian Amazonia. *Annales Botanici Fennici, 28*, 225–239.

Kauffman, J. B., Uhl, C., & Cummings, D. L. (1988). Fire in the Venezuelan Amazon. 1. Fuel biomass and fire chemistry in the evergreen rainforest of Venezuela. *Oikos, 53*, 167–175.

Kerr, R. A. (2000). A North Atlantic climate pacemaker for the centuries. *Science, 288*, 1984–1986.

Knight, J. R., Allan, R. J., Folland, C. K., et al. (2005). A signature of persistent natural thermohaline circulation cycles in observed climate. *Geophysical Research Letters, 32*, L20708.

Kruid, S., Macedo, M. N., Gorelik, S. R., et al. (2021). Beyond deforestation: Carbon emissions from land grabbing and forest degradation in the Brazilian Amazon. *Frontiers in Forests and Global Change, 4*, 645282.

Kubitzki, K. (1989). The ecogeographical differentiation of Amazonian inundation forests. *Plant Systematics and Evolution, 162*, 285–304.

Kubitzki, K., & Ziburski, A. (1994). Seed dispersal in flood-plain forests of Amazonia. *Biotropica, 26*, 30–43.

Küchler, I. L., Miekely, N., & Forsberg, B. R. (2000). A contribution to the chemical characterization of rivers in the Rio Negro basin. *Journal of the Brazilian Chemical Society, 11*, 286–292.

Labat, D., Espinoza, J. C., Ronchail, J., et al. (2012). Fluctuations in the monthly discharge of Guyana shield rivers, related to Pacific and Atlantic climate variability. *Hydrological Sciences Journal, 57*, 1081–1091.

Lapola, D. M., Pinho, P., Barlow, J., et al. (2023). The drivers and impacts of Amazon forest degradation. *Science, 379*, eabp8622.

Latrubesse, E. M., & Franzinelli, E. (2005). The late quaternary evolution of the Negro River, Amazon, Brazil: Implications for Island and floodplain formation in large anabranching tropical systems. *Geomorphology, 70*, 372–397.

Latrubesse, E. M., Arima, E. Y., Dunne, T., et al. (2017). Damming the rivers of the Amazon basin. *Nature, 546*, 363–369.

Latrubesse, E. M., Arima, E., Ferreira, M. E., et al. (2019). Fostering water resource governance and conservation in the Brazilian Cerrado biome. *Conservation Science and Practice, 1*, e77.

Latrubesse, E. M., DHorta, F. M., Ribas, C. C., et al. (2021). Vulnerability of the biota in riverine and seasonally flooded habitats to damming of Amazonian rivers. *Aquatic Conservation: Marine and Freshwater Ecosystems, 31*, 1136–1149.

Lavado Casimiro, W. S., Labat, D., Ronchail, J., et al. (2013). Trends in rainfall and temperature in the Peruvian Amazon-Andes basin over the last 40 years (1965-2007). *Hydrological Processes, 27*, 2944–2957.

Leonardi, V. (1999). *Os historiadores e os rios: Natureza e ruína na Amazônia brasileira. Paralelo15*. Editora da Universidade de Brasília.

Lesack, L. F. W., & Melack, J. M. (1995). Flooding hydrology and mixture dynamics of lake water derived from multiple sources in an Amazon floodplain lake. *Water Resources Research, 31*, 329–345.

Lewis, S. L., Brando, P. M., Phillips, O. L., et al. (2011). The 2010 Amazon drought. *Science, 331*, 554.

Lobo, G. S., Wittmann, F., & Piedade, M. T. F. (2019). Response of black-water floodplain (igapó) forests to flood pulse regulation in a dammed Amazonian river. *Forest Ecology and Management, 434*, 110–118.

Lopes, A., Crema, L. C., Demarchi, L. O., et al. (2019). Conservação de herbáceas aquáticas em igapós de água preta dentro e fora de Unidades de Conservação no Estado do Amazonas. *Biodiversidade Brasileira, 9*, 45–62.

Lovejoy, T. E., & Nobre, C. A. (2018). Amazon tipping point. *Science Advances, 4*, eaat2340.

Maciel, J. S. C., Alves, L. G. S., Corrêa, B. G. S., et al. (2020). Flood forecast in Manaus, Amazonas, Brazil. In S. Mambretti (Ed.), *Urban water systems & floods III, WIT transactions on the built environment, vol 194. Polytechnic of Milan, D. Proverbs* (pp. 63–72). Birmingham City University.

Malhi, Y., Melack, J., Gatti, L. V., et al. (2021). Chapter 6: Biogeochemical cycles of the Amazon. In C. Nobre, A. Encalada, E. Anderson, et al. (Eds.), *Amazon Assessment Report 2021*. United Nations Sustainable Development Solutions Network.

Mantua, N. J., & Hare, S. R. (2002). The Pacific decadal oscillation. *Journal of Oceanography, 58*, 35–44.

Marengo, J. A., & Espinoza, J. C. (2016). Extreme seasonal droughts and floods in Amazonia: Causes, trends and impacts. *International Journal of Climatology, 36*, 1033–1050.

Marengo, J. A., Espinoza, J. C., Fu, R., et al. (2021). Chapter 22: Long-term variability, extremes and changes in temperature and hydro meteorology in the Amazon region. In C. Nobre, A. Encalada, E. Anderson, et al. (Eds.), *Amazon Assessment Report 2021*. United Nations Sustainable Development Solutions Network.

Marengo, J. A., Nobre, C. A., Tomasella, J., et al. (2008). Hydro-climatic and ecological behaviour of the drought of Amazonia in 2005. *Philosophical Transactions of the Royal Society B, 363*, 1773–1778.

Marengo, J. A., Souza, C. M., Jr., Thonicke, K., et al. (2018). Changes in climate and land use over the Amazon region: Current and future variability and trends. *Frontiers in Earth Science, 6*, 228.

Marengo, J. A., Cunha, A. P., Espinoza, J.-C. et al. (2024). The drought of Amazonia in 2023-2024. *American Journal of Climate Change, 13*, 567–597.

Marinho, R. R., Filizola Junior, N. P., & Cremon, E. H. (2020). Analysis of suspended sediment in the Anavilhanas archipelago, Rio Negro, Amazon Basin. *Water, 12*, 1073.

Marmontel, M., Fleischmann, A., Val, A., et al. (2024). Safeguard Amazon's aquatic fauna against climate change. Nature, 625, 450.

McClain, M. E., Victoria, R. L., & Richey, J. E. (Eds.). (2001). *The biogeochemistry of the Amazon basin*. Oxford University Press.

McGrath, D. G., Arantes, C., Gama, A. S. P., et al. (2020). Manejo comunitário da pesca no Baix Amazonas, Estado do Pará. In W. J. Junk, M. T. F. Piedade, F. Wittmann, et al. (Eds.), *Várzeas amazônicas: desafios para um manejo sustentável* (pp. 226–241). Editora INPA.

Meade, R. H., Rayol, J. M., Da Conceicão, S. C., et al. (1991). Backwater effects in the Amazon River basin of Brazil. *Environmental Geology and Water Sciences, 18*, 105–114.

Melack, J. M., & Coe, M. T. (2021). Amazon floodplain hydrology and implications for aquatic conservation. *Aquatic Conservation: Marine and Freshwater Ecosystems, 31*, 1029–1040.

Melack, J. M., & Hess, L. L. (2010). Remote sensing of the distribution and extent of wetlands in the Amazon basin. In W. J. Junk, M. T. F. Piedade, F. Wittmann, et al. (Eds.), *Amazonian floodplain forests. Ecophysiology, biodiversity and sustainable management, ecological studies* (Vol. 210, pp. 163–178). Springer.

Melack, J. M., Novo, E. M. L. M., Forsberg, B. R., et al. (2009). Floodplain ecosystem processes. In J. Gash, M. Keller, & P. Silva-Dias (Eds.), *Amazonia and global change. Geophysical monograph series 186* (pp. 525–541). American Geophysical Union.

Meyer, U., Junk, W. J., & Linck, C. (2010). Fine root systems and mycorrhizal associations in two central Amazonian inundation forests: igapó and várzea. In W. J. Junk, M. T. F. Piedade, F. Wittmann, et al. (Eds.), *Amazonian floodplain forests. Ecophysiology, biodiversity and sustainable management, ecological studies* (Vol. 210, pp. 163–178). Springer.

Montero, J. C., Piedade, M. T. F., & Wittmann, F. (2014). Floristic variation across 600 km of inundation forests (Igapó) along the Negro River, Central Amazonia. *Hydrobiologia, 729*, 229–246.

Mori, G. B., Schietti, J., Poorter, L., et al. (2019). Trait divergence and habitat specialization in tropical floodplain forests trees. *PLoS One, 14*, e0212232.

Mori, G. B., Poorter, L., Schietti, J., et al. (2021). Edaphic characteristics drive functional traits distribution in Amazonian floodplain forests. *Plant Ecology, 222*, 349–360.

Nagy, L., Forsberg, B. R., & Artaxo, P. (Eds.). (2016). *Interactions between biosphere, atmosphere and human land use in the Amazon Basin, Ecological Studies* (Vol. 227). Springer.

Nepstad, D., Verssimo, A., Alencar, A., et al. (1999). Large-scale impoverishment of Amazonian forests by logging and fire. *Nature, 398*, 505–508.

Nepstad, D., Schwartzman, S., Bamberger, B., et al. (2006). Inhibition of Amazon deforestation and fire by parks and indigenous lands. *Conservation Biology, 20*, 65–73.

Neves, J. N. R., Piedade, M. T. F., Resende, A. F., et al. (2019). Impact of climatic and hydrological disturbances on Blackwater floodplain forests in Central Amazonia. *Biotropica, 51*, 484–489.

Nguyen, P. L., Min, S. K., & Kim, Y. H. (2021). Combined impacts of the El Niño-southern oscillation and Pacific decadal oscillation on global droughts assessed using the standardized precipitation evapotranspiration index. *International Journal of Climatology, 41*, E1645–E1662.

Nobre, C. A., Sampaio, G., Borma, L. S., et al. (2016). Land-use and climate change risks in the Amazon and the need of a novel sustainable development paradigm. *Proceedings of the National Academy of Sciences, 113*, 10759–10768.

Paiva, R. C. D., Buarque, D. C., Collischonn, W., et al. (2013). Large-scale hydrologic and hydrodynamic modeling of the Amazon River basin. *Water Resource Research, 49*, 1226–1243.

Pangala, S., Enrich-Prast, A., Basso, L., et al. (2017). Large emissions from floodplain trees close the Amazon methane budget. *Nature, 552*, 230–234.

Parker, D., Folland, C., Scaife, A., et al. (2007). Decadal to multidecadal variability and the climate change background. *Journal of Geophysical Research, 112*, D18115.

Parolin, P. (2000). Seed mass in Amazonian floodplains forest with contrasting nutrients supplies. *Journal of Tropical Ecology, 16*, 417–428.

Parolin, P. (2001). Morphological and physiological adjustments to waterlogging and drought in seedlings of Amazonian floodplain trees. *Oecologia, 128*, 326–335.

Parolin, P. (2009). Submerged in darkness: Adaptations to prolonged submergence by woody species of the Amazonian floodplains. *Annals of Botany, 103*, 359–376.

Parolin, P., Ferreira, L. V., & Junk, W. J. (1998). Central Amazonian floodplains: Effect of two water types on the wood density of trees. *Verhandlungen des Internationalen Verein Limnologie, 26*, 1106–1112.

Parolin, P., De Simone, O., Haase, K., et al. (2004). Central Amazon floodplain forests: Tree survival in a pulsing system. *The Botanical Review, 70*, 357–380.

Parolin, P., Wittmann, F., & Schöngart, J. (2010). Tree phenology in Amazonian floodplain forests. In W. J. Junk, M. T. F. Piedade, F. Wittmann, et al. (Eds.), *Amazonian floodplain for-*

ests. Ecophysiology, biodiversity and sustainable management, ecological studies (Vol. 210, pp. 105–126). Springer.

Peixoto, J. M. A., Nelson, B. W., & Wittmann, F. (2009). Spatial and temporal dynamics of river channel migration and vegetation in central Amazonian white-water floodplains by remote-sensing techniques. *Remote Sensing of Environment, 113*, 2258–2268.

Piedade, M. T. F., Junk, W. J., & Long, S. P. (1991). The productivity of the C_4 grass *Echinochloa polystachya* on the Amazon floodplain. *Ecology, 72*, 1456–1463.

Piedade, M. T. F., Worbes, M., & Junk, W. J. (2001). Geo-ecological controls on elemental fluxes in communities of higher plants in Amazonian floodplains. In M. E. McClain, R. L. Victoria, & J. E. Richey (Eds.), *The biogeochemistry of the Amazon basin and its role in a changing world* (pp. 209–234). Oxford University Press.

Piedade, M. T. F., Ferreira, C. S., Oliveira Wittmann, A., et al. (2010). Biochemistry of Amazonian floodplain trees. In W. J. Junk, M. T. F. Piedade, F. Wittmann, et al. (Eds.), *Amazonian floodplain forests. Ecophysiology, biodiversity and sustainable management, ecological studies* (Vol. 210, pp. 127–140). Springer.

Piedade, M. T. F., Schöngart, J., Wittmann, F., et al. (2013). Impactos ecológicos da inundação e seca a vegetação das áreas alagáveis amazônicas. In C. A. Nobre & L. S. Borma (Eds.), *Eventos Climáticos Extremos na Amazônia: Causas e Conseqüências* (pp. 268–305). Oficina de Textos.

Piedade, M. T. F., Lopes, A., Demarchi, L. O., et al. (Eds.). (2019). *Guia de campo de herbáceas aquáticas: várzea Amazônica*. Editora INPA.

Piedade, M. T. F., Wittmann, F., Schöngart, J., et al. (2023). Wetlands ecosystems. In S. M. Scheiner (Ed.), *Encyclopaedia of biodiversity* (3rd ed.). Academic Press.

Piedade, M. T. F., Parolin, P., Lopes, A., et al. (2024). Chapter 7: Amazonian Rivers from the Guiana Shield. In: Graca, M. A. S., Callisto, M., Mello, F. T., et al. (Eds.), *Rivers of South America* (1st ed.). Elsevier.

Pivello, V. R. (2011). The use of fire in the Cerrado and Amazonian rainforests of Brazil: Past and present. *Fire Ecology, 7*, 24–39.

Pöhlker, C., Wiedemann, K. T., Sinha, B., et al. (2012). Biogenic potassium salt particles as seeds for secondary organic aerosol in the Amazon. *Science, 337*, 1075–1078.

Ponce, V. M., Lohani, A. K., & Huston, P. T. (1997). Surface albedo and water resources: Hydroclimatological impact of human activities. *Journal of Hydrologic Engineering, 2*, 197–203.

Porter, S. E., Mosley-Thompson, E., Thompson, L. G., et al. (2021). Reconstructing an Interdecadal Pacific oscillation index from a Pacific basin–wide collection of ice core records. *Journal of Climate, 34*, 3839–3852.

Pöschl, U., Martin, S. T., Sinha, B., et al. (2010). Rainforest aerosols as biogenic nuclei of clouds and precipitation in the Amazon. *Science, 329*, 1513–1516.

Prance, G. T. (1979). Notes on the vegetation of Amazonia III. The terminology of Amazonian forest types subject to inundation. *Brittonia, 3*, 26–38.

Quesada, C. A., Lloyd, J., Anderson, L. O., et al. (2011). Soils of Amazonia with particular reference to the RAINFOR sites. *Biogeosciences, 8*, 1415–1440.

RAISG. (2019). *Amazônia 2019: Áreas protegidas e Territórios indígenas. Rede Amazônica de Informação Socioambiental Georreferenciada (RAISG).* https://acervo.socioambiental.org/acervo/mapas-e-cartas-topograficas/amazonia/amazonia-2019-areas-protegidas-e-territorios-indigenas. Accessed 29.05.02.03.

Reis, A. A. (2021). *Relação do incremento em diâmetro e biomassa de florestas alagáveis com fatores hidro- edáficos na bacia Amazônica*. Dissertation, Instituto Nacional de Pesquisas da Amazônia.

Reis, M., Graça, P. M. L. A., Yanai, A. M., et al. (2021). Forest fires and deforestation in the Central Amazon: Effects of landscape and climate on spatial and temporal dynamics. *Journal of Environmental Management, 88*, 112310.

Renó, V., & Novo, E. (2019). Forest depletion gradient along the Amazon floodplain. *Ecological Indicators, 98*, 409–419.

Resende, A. F., Nelson, B. W., Flores, B. M., et al. (2014). Fire damage in seasonally flooded and upland forests of the Central Amazon. *Biotropica, 46*, 643–646.

Resende, A. F., Schöngart, J., Streher, A. S., et al. (2019). Massive tree mortality from flood pulse disturbances in Amazonian floodplain forests: The collateral effects of hydropower production. *Science of the Total Environment, 659*, 587–598.

Resende, A. F., Piedade, M. T. F., Feitosa, Y. O., et al. (2020). Flood-pulse disturbances as a threat for long-living Amazonian trees. *The New Phytologist, 227*, 1790–1803.

Ribeiro, M. N., & Adis, J. (1984). Local rainfall variability - a potential bias for bioecological studies in the Central Amazon. *Acta Amaz, 14*, 159–174.

Richey, J. E., Meade, R. H., Salati, E., et al. (1986). Water discharge and suspended sediment concentrations in the Amazon river: 1982-1984. *Water Resources Research, 22*, 756–764.

Rios-Villamizar, E. A., Adeney, J. M., Piedade, M. T. F., et al. (2020). Hydrochemical classification of Amazonian rivers: A systematic review and meta-analysis. *Caminhos de Geografia, 21*, 211–226.

Ritter, C. D., Andretti, C. B., & Nelson, B. W. (2012). Impact of past forest fires on bird populations in flooded forests of the Cuini River in the lowland Amazon. *Biotropica, 44*, 449–453.

Rocha, M., Assis, R. L., Piedade, M. T. F., et al. (2019). Thirty years after Balbina dam: Diversity and floristic composition of the downstream floodplain forest, Central Amazon, Brazil. *Ecohydrology, 12*, e2144.

Rocha, M., Feitosa, Y. O., Wittmann, F., et al. (2020). River damming affects the seedling communities of a floodplain forest in the Central Amazon. *Acta Botânica Brasílica, 34*, 192–203.

Rudorff, C. M., Melack, J. M., & Bates, P. D. (2014). Flooding dynamics on the lower Amazon floodplain: 2. Seasonal and interannual hydrological variability. *Water Resources Research, 50*, 635–649.

Rufin, P., Gollnow, F., Müller, D., et al. (2019). Synthesizing dam-induced land system change. *Ambio, 48*, 1183–1194.

Salati, E., & Vose, P. B. (1984). Amazon basin: A system in equilibrium. *Science, 225*, 129–138.

Salerno, L., Moreno-Martínez, Á., Izquierdo-Verdiguier, E., et al. (2022). Satellite analyses unravel the multi-decadal impact of dam management on tropical floodplain vegetation. *Frontiers in Environmental Science, 10*, 871530.

Salinger, M. J., Renwick, J. A., & Mullan, A. B. (2001). Interdecadal Pacific oscillation and South Pacific climate. *International Journal of Climatology, 21*, 1705–1721.

Salo, J., Kalliola, R., Häkkinen, I., et al. (1986). River dynamics and the diversity of Amazon lowland forest. *Nature, 322*, 254–258.

Sampaio, P. M. (2003). Cidades desaparecidas na Amazônia Portuguesa. Poiares, séculos XVIII e XIX. *Historia Social, 10*, 73–100.

Santos, A. R., & Nelson, B. W. (2013). Leaf decomposition and fine fuels in floodplain forests of the Rio Negro in the Brazilian Amazon. *Journal of Tropical Ecology, 29*, 455–458.

Santos, U. D. M., Bringel, S. R. B., Ribeiro, M. D. N. G., et al. (1985). Rios da bacia amazônica II. Os afluentes do rio Branco. *Acta Amazonica, 15*, 147–156.

Scabin, A., Costa, F., & Schöngart, J. (2012). The spatial distribution of illegal logging in the Anavilhanas archipelago (Central Amazonia) and logging impacts on the primary timber species. *Environmental Conservation, 39*, 111–121.

Schlüter, U. B., & Furch, B. (1992). Morphologische, anatomische und physiologische Untersuchungen zur Überflutungstoleranz des Baumes *Macrolobium acaciaefolium*, charakteristisch für die Weiß- und Schwarzwasser-Überschwemmungswälder bei Manaus, Amazonas. *Amazoniana, 12*, 51–69.

Schöngart, J. (2003). Dendrochronologische Untersuchungen in Überschwemmungswäldern der Várzea Zentralamazoniens. In *Göttinger Beiträge zur Land- und Forstwirtschaft in den Tropen und Subtropen* (Vol. 149). Erich Goltze Verlag.

Schöngart. (2008). *G*rowth-*O*riented *L*ogging (GOL): A new concept towards sustainable forest management in central Amazonian várzea floodplains. *Forest Ecology and Management, 256*, 46–58.

Schöngart, J. (2010). Growth-oriented logging (GOL): The use of species-specific growth information for forest management in in central Amazonian floodplain forests. In W. J. Junk, M. T. F. Piedade, F. Wittmann, et al. (Eds.), *Amazonian floodplain forests. Ecophysiology, biodiversity and sustainable management, ecological studies* (Vol. 210, pp. 437–462). Springer.

Schöngart, J., & Junk, W. J. (2007). Forecasting the flood-pulse in Central Amazonia by ENSO-indices. *Journal of Hydrology, 335*, 124–132.

Schöngart, J. (2020). O manejo florestal na várzea da Amazônia Central: História, desenvolvimento e futuros desafios. In W. J. Junk, M. T. F. Piedade, F. Wittmann, et al. (Eds.), *Várzeas amazônicas: desafios para um manejo sustentável* (pp. 178–206). Editora INPA.

Schöngart, J., & Junk, W. J. (2020). Clima e hidrologia nas várzeas da Amazônia Central. In W. J. Junk, M. T. F. Piedade, F. Wittmann, et al. (Eds.), *Várzeas amazônicas: desafios para um manejo sustentável* (pp. 44–65). Editora INPA.

Schöngart, J., Piedade, M. T. F., Ludwigshausen, S., et al. (2002). Phenology and stem-growth periodicity of tree species in Amazonian floodplain forests. *Journal of Tropical Ecology, 18*, 581–597.

Schöngart, J., Junk, W. J., Piedade, M. T. F., et al. (2004). Teleconnection between tree growth in the Amazonian floodplains and the El Niño-southern oscillation effect. *Global Change Biology, 10*, 683–692.

Schöngart, J., Piedade, M. T. F., Wittmann, F., et al. (2005). Wood growth patterns of *Macrolobium acaciifolium* (Benth.) Benth. (Fabaceae) in Amazonian black-water and white-water floodplain forests. *Oecologia, 145*, 454–461.

Schöngart, J., Wittmann, F., & Worbes, M. (2010). Biomass and NPP of central Amazonian floodplain forests. In W. J. Junk, M. T. F. Piedade, F. Wittmann, et al. (Eds.), *Amazonian floodplain forests. Ecophysiology, biodiversity and sustainable management, ecological studies* (Vol. 210, pp. 347–388). Springer.

Schöngart, J., Wittmann, F., Resende, A. F., et al. (2021). The shadow of the Balbina dam—A synthesis of over 35 years of downstream impacts on floodplain forests in Central Amazonia. *Aquatic Conservation: Marine and Freshwater Ecosystems, 31*, 1117–1135.

Science Panel for the Amazon. (2021). *Executive summary of the Amazon assessment report*. United Nations Sustainable Development Solutions Network.

Scudeller, V. V., & Vegas-Vilarrúbia, T. (2018). Distribution and β-diversity of tree species in igapó forests (Negro River basin, Brazilian Amazon). *Journal of Vegetation Science, 29*, 1052–1064.

Silva, T. S. F., Cost, M. P. F., & Melack, J. M. (2009). Annual net primary production of macrophytes in the eastern Amazon floodplain. *Wetlands, 29*, 747–758.

Silva, T. S. F., Melack, J. M., & Novo, E. M. L. (2013). Responses of aquatic macrophyte cover and productivity to flooding variability on the Amazon floodplain. *Global Change Biology, 19*, 3379–3389.

Silva, M. D. S. R., Ríos-Villamizar, E. A., da Cunha, H. B., et al. (2019). A contribution to the hydrochemistry and water typology of the Amazon River and its tributaries. *Caminhos de Geografia, 20*, 360–374.

Sioli, H. (1954). Beiträge zur regionalen Limnologie des Amazonasgebietes. *Archiv für Hydrobiologie, 45*, 267–283.

Sioli, H. (Ed.). (1984). *The Amazon. Limnology and landscape ecology of a mighty tropical river and its basin*. Dr W Junk Publisher.

Soares-Filho, B. S., Oliveira, U., Ferreira, M. N. et al. (2023). Contribution of the Amazon protected areas program to forest conservation. *Biological Conservation, 279*, 109928.

Sombroek, W. (2001). Spatial and temporal patterns of Amazon rainfall. *Ambio, 30*, 388–396.

Strahler, A. N. (1957). Quantitative analysis of watershed geomorphology. *Transactions of the American Geophysical Union, 38*, 913–920.

Sutton, R. T., & Dong, B. (2012). Atlantic Ocean influence on a shift in European climate in the 1990s. *Nature Geoscience, 5*, 788–792.

Targhetta, N., Kesselmeier, J., & Wittmann, F. (2015). Effects of the hydroedaphic gradient on tree species composition and aboveground wood biomass of oligotrophic forest ecosystems in the Central Amazon basin. *Folia Geobotanica, 50*, 185–205.

Thom, G., Xue, A. T., Sawakuchi, A. O., et al. (2020). Quaternary climate changes as speciation drivers in the Amazon floodplains. *Science Advances, 6*, 4718.

Timpe, K., & Kaplan, D. (2017). The changing hydrology of a dammed Amazon. *Science Advances, 3*, e1700611.

Tonolla, D., Acuña, V., Uehlinger, U., et al. (2010). Thermal heterogeneity in river floodplains. *Ecosystems, 13*, 727–740.

Torres, C. M. M. E., Jacovine, L. A. G., de Olivera, N., Neto, S., et al. (2017). Greenhouse gas emissions and carbon sequestration by agroforestry systems in southeastern Brazil. *Scientific Reports, 7*, 16738.

Trenberth, K. E., & Shea, D. J. (2006). Atlantic hurricanes and natural variability in 2005. *Geophysical Research Letters, 33*, L12704.

Uhl, C., Kauffman, J. B., & Cummings, D. L. (1988). Fire in the Venezuelan Amazon 2: Environmental conditions necessary for forest fires in the evergreen rainforest of Venezuela. *Oikos, 53*, 176–184.

Urrutia-Pereira, M., Rizzo, L. V., Chong-Neto, H. J., et al. (2021). Impact of exposure to smoke from biomass burning in the Amazon rain forest on human health. *Jornal Brasileiro de Pneumologia, 47*, e20210219.

Waldhoff, D., & Furch, B. (2002). Leaf morphology and anatomy in eleven tree species from central Amazonian floodplains (Brazil). *Amazoniana, 17*, 79–94.

Waldhoff, D., Saint-Paul, U., & Furch, B. (1996). Value of fruits and seeds from the floodplain forests of Central Amazonia as food resource for fish. *Ecotropica, 2*, 143–156.

Wang, S., Huang, J., He, Y., et al. (2014). Combined effects of the Pacific decadal oscillation and El Niño-southern oscillation on global land dry–wet changes. *Scientific Reports, 4*, 6651.

Wang, X. Y., Li, X., Zhu, J., et al. (2018). The strengthening of Amazonian precipitation during the wet season driven by tropical sea surface temperature forcing. *Environmental Research Letters, 13*, 94015.

Wantzen, K. M., Yule, C. M., Mathooko, J. M., et al. (2008). Organic matter processing in tropical streams. In D. Dudgeon (Ed.), *Aquatic ecology, tropical stream ecology* (pp. 43–64). Academic Press.

Weiss, B. W., Zuanon, J. A. S., & Piedade, M. T. F. (2016). Viability of seeds consumed by fishes in a lowland Forest in the Brazilian Central Amazon. *Tropical Conservation Science, 9*, 1–10.

Weiss, B. A., Santana, F. D., Calvi, P., et al. (2023). Effectiveness of fish assemblage as seed dispersers in Amazon oligotrophic flooded forests. *Austral Ecology*. https://doi.org/10.1111/aec.13330

Williams, E., Antonia, A. D., Antonia, V. D., et al. (2005). The drought of the century in the Amazon basin: An analysis of the regional variation of rainfall in South America in 1926. *Acta Amazonica, 35*, 231–238.

Winemiller, K. O., McIntyre, P. B., Castello, L., et al. (2016). Balancing hydropower and biodiversity in the Amazon, Congo, and Mekong. *Science, 351*, 128–129.

Wittmann, F. (2012). Tree species composition and diversity in Brazilian freshwater floodplains. In M. Pagano (Ed.), *Mycorrhiza: Occurrence and role in aquatic and riparian environments* (pp. 223–263). Nova Science.

Wittmann, F., & Junk, W. J. (2003). Sapling communities in Amazonian white-water forests. *Journal of Biogeography, 30*, 1533–1544.

Wittmann, F., & Junk, W. J. (2016). The Amazon River basin. In C. M. Finlayson, G. R. Milton, R. C. Prentice, et al. (Eds.), *The wetland book II: Distribution, description and conservation* (pp. 1–20). Springer.

Wittmann, F., & Parolin, P. (2005). Aboveground roots in Amazonian floodplain trees. *Biotropica, 37*, 609–619.

Wittmann, F., Junk, W. J., & Piedade, M. T. F. (2004). The várzea forests in Amazonia: Flooding and the highly dynamic geomorphology interact with natural forest succession. *Forest Ecology and Management, 196*, 199–212.

Wittmann, F., Schöngart, J., Montero, J. C., et al. (2006a). Tree species composition and diversity gradients in white-water forests across the Amazon Basin. *Journal of Biogeography, 33*, 1334–1347.

Wittmann, F., Schöngart, J., Parolin, P., et al. (2006b). Wood specific gravity of trees in Amazonian white-water forests in relation to flooding. *IAWA Journal, 27*, 255–266.

Wittmann, F., Schöngart, J., & Junk, W. J. (2010). Phytogeography, species diversity, community structure and dynamics of central Amazonian floodplain forests. In W. J. Junk, M. T. F. Piedade, F. Wittmann, et al. (Eds.), *Amazonian floodplain forests. Ecophysiology, biodiversity and sustainable management, ecological studies* (Vol. 210, pp. 61–102). Springer.

Wittmann, F., Householder, E., Piedade, M. T. F., et al. (2013). Habitat specifity, endemism and the neotropical distribution of Amazonian white-water floodplain trees. *Ecography, 36*, 690–707.

Wittmann, F., Marques, M. C. M., Damasceno Júnior, G., et al. (2017). The Brazilian freshwater wetscape: Changes in tree community diversity and composition on climatic and geographic gradients. *PLoS One, 12*(4), e0175003.

Wittmann, F., Damm, C., & Schöngart, J. (2019). Der Sandwich-Effekt: Einengung von Habitaten durch Staudämme gefährdet die grösstenund artenreichsten Flussauen der Erde. *Auenmagazin, 15*, 49–53.

Wittmann, F., Householder, J. E., Piedade, M. T. F., et al. (2022a). A review of the ecological and biogeographic differences of Amazonian floodplain forests. *Water, 14*, 3360.

Wittmann, F., Schöngart, J., Piedade, M. T. F., et al. (2022b). Tropical large river wetlands. In T. Mehner & K. Tockner (Eds.), *Encyclopedia of inland waters* (Vol. 3, 2nd ed., pp. 90–104). Elsevier.

Worbes, M. (1997). The forest ecosystem of the floodplains. In W. J. Junk (Ed.), *The Central Amazon floodplain: Ecology of a pulsing system, ecological studies* (Vol. 126, pp. 223–265). Springer.

Worbes, M., Klinge, H., Revilla, J. D., et al. (1992). On the dynamics, floristic subdivision and geographical distribution of várzea forests in Central Amazonia. *Journal of Vegetation Science, 3*, 553–564.

Wright, J. S., Fu, R., Worden, J. R., et al. (2017). Rainforest-initiated wet season onset over the southern Amazon. *Proceedings of the National Academy of Sciences, 114*, 8481–8486.

Yoon, J. H., & Zeng, N. (2010). An Atlantic influence on Amazon rainfall. *Climate Dynamics, 34*, 249–264.

Zhou, J., & Lau, K. M. (1998). Does a monsoon system exists over South America? *Journal of Climate, 11*, 1020–1040.

Amazonian Aquatic Food Webs, Their Variation, and Vulnerability in Regard to Climate and Land-Use Changes

Priit Zingel, Helen Agasild, and Arvo Tuvikene

Abstract

The Amazon rainforest's aquatic food webs are facing unprecedented challenges due to climate and land-use change. A diverse array of organisms is supported by these intricate and interconnected ecosystems, but the increasing threat of droughts and deforestation is disrupting the delicate balance of life. The Amazon food webs derive their energy from two primary sources: an autochthonous source, dependent on the productivity of the aquatic ecosystem (such as phytoplankton and aquatic macrophytes), and an allochthonous source, mainly from seasonally flooded forests (comprising fallen fruits and seeds that enter the water). The Amazonian food webs are currently influenced by various factors, including climate change and flooding patterns. These factors impact the structure, function, and energy sources of the food webs, affecting the abundance and distribution of organisms such as phytoplankton, bacteria, macrophytes, and invertebrates. The flood pulses have a pivotal role in shaping these food webs, affecting nutrient availability and impacting the abundance and biomass of fish. The interactions between the terrestrial and aquatic components are vital for the operation of Amazonian food webs, where floodplains serve significant functions as both nursery and feeding areas for numerous fish species. The disruption of food chains in the Amazon as a result of climate and land-use change is exemplified by the drying of aquatic habitats and changes in the fish community of floodplain lakes. This event generates a new scenario that has significant consequences for the ecological balance and productivity of the region, highlighting the urgent need for conservation and protection efforts.

P. Zingel (✉) · H. Agasild · A. Tuvikene
Chair of Hydrobiology and Fisheries, Institute of Agricultural and Environmental Science, Estonian University of Life Sciences, Tartu, Estonia
e-mail: priit.zingel@emu.ee

S. S. de Souza et al. (eds.), *The Future of Amazonian Aquatic Biota*,
https://doi.org/10.1007/978-3-031-66822-7_8

1 Variety of Amazonian Aquatic Environments

The Amazon rainforest is known for its incredible biodiversity, housing a vast array of plant and animal species. The aquatic food webs within this ecosystem are essential for maintaining the delicate balance of life. However, these food webs are now facing unprecedented challenges due to the increasing threat of climate change, land-use changes, and synergetic effects between both disturbance regimes (Zapata-Ríos et al., 2021). The Amazon basin contains a variety of rivers, lakes, and wetlands, each with its own unique set of species and ecological dynamics; however, the connection between the different habitats is extremely important for the maintenance of biodiversity. These diverse aquatic habitats support an extensive variety of organisms, from tiny plankton to large predators such as caimans and river dolphins. The food webs in these ecosystems are complex and interconnected, with each species having a specific role in the transfer of energy and nutrients (Moraes et al., 2021).

The Amazon's aquatic systems are known for their environmental diversity, presenting numerous distinctive obstacles for aquatic organisms. One of the most significant challenges is the varied habitats, which stem from the different types of rivers found in the region, including white-, black-, and clear-water (Sioli, 1954; Sioli, 1984; Zapata-Ríos et al., 2021). White-water rivers, such as the Amazon main stem, Caquetá-Japurá, Purus, Juruá, Ucayali, and Madeira, originate in the Andes (Moraes et al., 2021). These rivers are influenced by the Andean mountains, which supply most of the terrestrial sediments and mineral nutrients that affect floodplain ecology and productivity (Encalada et al., 2019; Filizola & Guyot, 2009; McClain & Naiman, 2008). The large sediment loads derived from the Andes control downstream channel erosion and width, impacting the connectivity between river channels and floodplains, as well as spatial patterns of inundation and floodplain productivity (Constantine et al., 2014; Forsberg et al., 2017). White-water rivers are turbid with water transparency ranging between 20 and 60 cm, and they have near-neutral pH (6.5–7.5) and relatively high electric conductivity (40–300 μS/cm) (Bogotá-Gregory et al., 2020; McClain & Naiman, 2008; Moraes et al., 2021). Diverse floodplain ecosystems surrounding white-water rivers are known as várzea (Wittmann et al., 2011) (Fig. 1). Black-water rivers, such as Negro in Brazil, originate in lowlands and drain the deeply weathered tertiary sediments of the pre-Cambrian origin and large oligotrophic wetland complexes (campinaranas) on mainly podzolic formations in the north-western portion of the Amazon basin. These rivers are characterized by high dissolved organic carbon concentrations but are low in nutrients (Moraes et al., 2021). They have low quantities of suspended matter but high amounts of humic acids, giving the water a brownish-reddish colour. Black-water rivers are also characterized by transparent water (transparency ranging between 60 cm and 4 m). The pH values of black-water rivers range from 3.5 to 6, and their electrical conductivity is rather low (5–20 μS/cm; Bogotá-Gregory et al., 2020). Ecosystems surrounding black-water rivers are known as igapó (Junk et al., 2015; Moraes et al., 2021) (Fig. 1). Clear-water rivers have their upper catchments in the cerrado region of the ancient Brazilian shield (such as the Tapajós and

Fig. 1 Diverse white-water floodplains known as várzea (**a**; Lake Janauaca) and flooded forests surrounding black-water rivers known as igapó (**b**; upper Tumbira River basin). Photos by P. Zingel

Xingu Rivers), or in the in large savanna complexes of the Guiana Shield (Branco, Jatapú, Trombetas, Jarí, Paru and Araguarí Rivers) (Sioli 1984; Moraes et al., 2021).

These rivers have high water transparency (1–3 m), low levels of suspended and dissolved solids, and varying pH levels (5.5–8.0) (Moraes et al., 2021). Their electrical conductivity ranges between 5 and 40 μS/cm (Bogotá-Gregory et al., 2020).

2 Amazonian Biodiversity

The Amazon rainforest stands as the most biologically diverse region on our planet. Covering an expansive area of approximately 5,800,000 square kilometres, the biodiversity within the Amazon's biogeographic province is immeasurable. Remarkably, over one-tenth of the world's species (marine life forms included) find their home in this extraordinary ecosystem (Bass et al., 2010; Mittermeier et al., 2002; Zapata-Ríos et al., 2021). The highly diverse ecosystems of the Amazonia form the central core of the Neotropical realm (Guayasamin et al., 2021; Jenkins et al., 2013; Raven et al., 2020; Reis et al., 2016; Stork, 2018). Endemism thrives in the Amazon basin, where a substantial number of mammals (40.5%), reptiles (70.1%), and amphibians (85.7%) exist exclusively within this region, without counterparts elsewhere. Quantifying the exact number of species inhabiting the Amazon remains an elusive challenge for scientists. Ongoing research continually unveils novel species, even within well-studied taxonomic groups such as mammals and birds (Milá et al., 2012; Patterson, 2001; Ribas & Aleixo, 2019). Remarkably, for numerous invertebrate taxa, the extent of undescribed biodiversity is so substantial that only a fraction of the species occurring in this region has been formally documented (Zapata-Ríos et al., 2021).

While there is no definitive list of vascular plants in the Amazon basin, estimates for seed plants occurring below 1000 metres vary significantly. These estimates range from 14,000 to 50,000 species (Cardoso et al., 2017; Gentry, 1997; Lewinsohn & Prado, 2005) (see also Chap. 7, Schöngart et al., 2024 and Chap. 6, Lopes et al., 2024). Concerning the diversity of lowland trees, the estimates span between 6000 and 16,000 species (Cardoso et al., 2017; Hubbell et al., 2008; Junk & Piedade, 1993; ter Steege et al., 2020). These figures underscore our incomplete understanding of the true richness of Amazonian plant life. Insufficiently studied species face the risk of extinction, even without being documented (Humphreys et al., 2019). Despite the extensive historical use of numerous local plants, the potential advantages offered by the majority of these species remain unexplored (Alcantara-Rodriguez et al., 2019; Antonelli et al., 2019; Zapata-Ríos et al., 2021).

While insects are dominant in terrestrial ecosystems in terms of species number or total biomass, the true richness of insects in the Amazonia remains largely unknown (Adis, 2007; Hanson, 2016; Zapata-Ríos et al., 2021). Understanding the distribution patterns of insects in the various aquatic ecosystems of the Amazon is crucial, as many small and medium fish species rely on aquatic insect larvae as their primary food source (see also Chap. 5, Martins et al., 2024). Among aquatic insects, the order Diptera, particularly the Chironomidae family, stands out, comprising half of the known aquatic insects (Trivinho-Strixino, 2019). It is noted that preserving riparian forests is essential for preventing the loss of aquatic insect species (Dala-Corte et al., 2020; Dias-Silva et al., 2020), but there is still much to be learned. Closing this knowledge gap will require the training of taxonomists, strengthening of collections, and an ongoing increase in the rate of new species descriptions (Rafael et al., 2009; Zapata-Ríos et al., 2021).

The Amazon basin is home to the world's most diverse freshwater fish fauna (see also Chap. 1, Freitas et al., 2024 and Chap. 4, Barros et al., 2024), with 2406 valid species across 56 families, 514 genera, and 18 orders described so far (Jézéquel et al., 2020; Tedesco et al., 2017; Zapata-Ríos et al., 2021). This exceptional diversity, representing about 15% of the world's freshwater fishes, includes 58% of species that are found exclusively in the Amazon basin (1402 endemics; Jézéquel et al., 2020). Unlike many other river basins where species richness increases downstream along fluvial gradients (Muneepeerakul et al., 2008), species diversity and endemism in the Amazon basin show decreasing west–east gradients, suggesting that the Amazonian fish fauna originated contemporary from the western portion of the basin (Oberdorff et al., 2019; Zapata-Ríos et al., 2021). This pattern of fish diversity also suggests that the colonization of the eastern portion of the basin is still ongoing and is consistent with the establishment of the modern Amazon River (Oberdorff et al., 2019; Zapata-Ríos et al., 2021).

The Amazon basin exhibits the highest density of amphibian species in the world—it is home to more than 600 amphibian species (Godinho & Da Silva, 2018; Mayer et al., 2019; Vacher et al., 2020). The region also hosts a significant number of endangered species (AmphibiaWeb, 2023; Bass et al., 2010; Scheele et al., 2019). Concerning reptilians, there are 371 species registered, inhabiting a wide range of terrestrial and aquatic environments (Avila-Pires & Ramalho, 2019; Mittermeier et al., 2003). However, the diversity of both Amazonian amphibians and reptiles is far from being fully documented (Zapata-Ríos et al., 2021).

The Amazon rainforest boasts the greatest diversity of bird species globally. It is home to over 1300 bird species. Among these avian inhabitants, around 265 species are endemic to the Amazon basin (Jetz et al., 2012; Mittermeier et al., 2003; Nores, 2000).

The Amazonian region is also home to a rich mammalian fauna, with around 140 genera and 425 species, constituting approximately one-third of all South American mammalian diversity (Mittermeier et al., 2003; Zapata-Ríos et al., 2021). Additionally, there are seven species of Amazonian aquatic mammals (*Inia geoffrensis*, *I. boliviensis*, *I. araguaiaensis*, *Sotalia fluviatilis*, *Trichechus inunguis*, *Pteronura brasiliensis*, and *Lontra longicaudis*), each playing crucial roles in their ecosystems and facing endangerment due to various human activities such as natural ecosystem modifications, energy production, mining, and climate change (Brum et al., 2021).

3 Food Webs in the Amazon Aquatic Ecosystems

A food web comprises species or groups and their direct interactions, such as consumer–prey relationships, which regulate the flow of nutrients (or energy) from decomposing material, recycling nutrients for primary (plant and algal growth), and secondary (animal growth) production (Angelini, 2019; Bartley et al., 2019). Various factors, including climate change and land-use changes, are concurrently impacting aquatic food webs and their components, leading to numerous synergistic interactions. Consequently, it is challenging to isolate the influence of each individual driver. Despite this complexity, scientists have dedicated significant effort to characterizing and evaluating the intricate and distinctive Amazonian river-floodplain food webs, including their structure, function, energy sources, pathways, and predicting future changes (e.g. Andrade et al., 2023; Capitani et al., 2021; Junk et al., 2010; Tregidgo et al., 2023; Zapata-Ríos et al., 2021). Our understanding of the interactions' complexity and their impact on ecosystem processes in the Amazonian region is only just beginning to unfold. However, the present environmental challenges posed by climate change and human-induced stressors are reshaping this knowledge and driving further research towards finding solutions for conservation and protection (Moraes et al., 2021).

Amazon food webs are fuelled by two main sources: an autochthonous source, which relies on the productivity of the aquatic ecosystem (such as phytoplankton and aquatic macrophytes), and an allochthonous source, primarily from seasonally flooded forests (including fallen fruits and seeds that enter the water) (González-Bergonzoni et al., 2019). White-water ecosystems are considered to be much more autochthonous than their black-water counterparts and also usually show higher productivity due to the higher nutrient concentration (Moraes et al., 2021). This production is channelled through zooplankton and other aquatic invertebrates to the fish (Fig. 2). Higher primary production is also reflected in respective higher fish productivity and biomasses (Bogotá-Gregory et al., 2020) (Fig. 3).

The yearly flood pulse plays a crucial role in shaping the food webs of Amazonian waters, as it supplies essential nutrients for all organisms within the food chains. The flood-pulse concept was first formulated by Junk et al. (1989), and it states that the pulsing of river discharge is the major driving force responsible for the existence, productivity, and interactions of biota in river-floodplain systems. For instance, lower water levels often lead to a high abundance of phytoplankton and bacteria (Casatti et al., 2003; Castillo, 2000; Chaparro et al., 2014; Izaguirre et al., 2010). However, during these periods, the prevalence of large cyanobacteria (Bouvy et al., 2003; O'Farrell et al., 2011; Paerl & Huisman, 2009; Unrein et al., 2010) may result in a decrease in the food resource quality (Deng et al., 2010). Changes in flooding patterns can also impact macrophyte communities (Boschilia et al., 2008; Gomes et al., 2020), altering further environmental conditions and affecting invertebrate populations (Burks et al., 2006; Chaparro et al., 2014; Scheffer et al., 2003). Flood pulses regulate zooplankton abundance through dilution, washing-out, or concentration processes (Baranyi et al., 2002; Chaparro et al., 2011; Chaparro et al., 2014; Jose, 1993). Different zooplankton groups respond differently to flushing

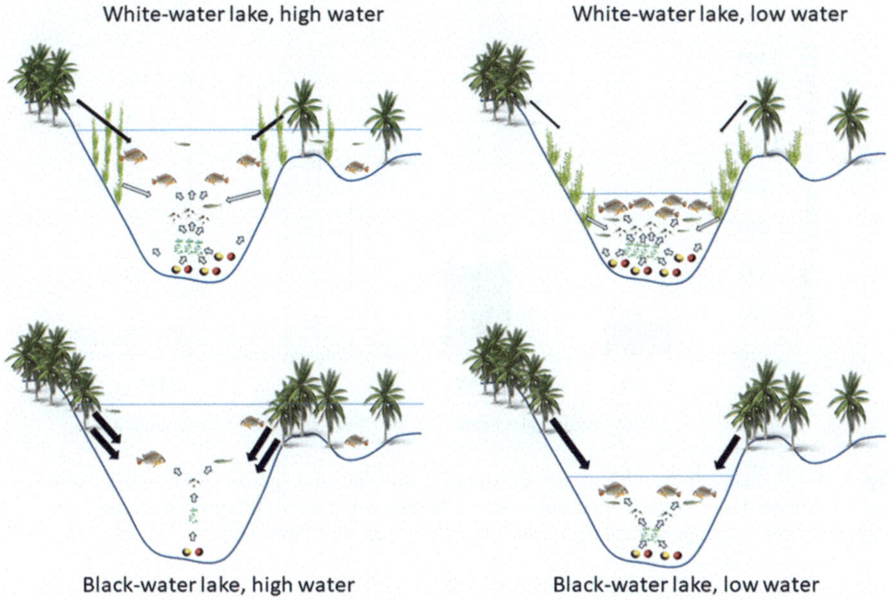

Fig. 2 Schematic representation of the Amazonian food webs in the white- and black-water floodplain lakes, both in high- and low-water periods. Black arrows represent the flow of nutrients and energy from terrestrial sources. Grey arrows represent the usage of nutrients for primary production (algae and aquatic macrophytes) and channelling this production through zooplankton and other aquatic invertebrates to the fish. Higher primary production is also reflected in respective higher secondary production (aquatic invertebrates and fish). The number of organisms shown in the figures corresponds conditionally to the number of organisms in the corresponding water volume and at the corresponding time. The most autochthonous are white-water ecosystems during low-water phases, whereas the most allochthonous are black-waters during flooding. Extreme droughts will lead to decreasing water levels and further disruption of hydrological connectivity

effects based on their growth rates. Rotifers, with short development times, recover quickly from flushing effects, while microcrustaceans, with longer growth rates, are more negatively impacted by floods (Costa Bonecker et al., 2005; Paidere, 2009). As floods favour more the small-sized opportunistic zooplankton species (rotifers and protozoa), the higher zooplankton abundances are often found during decreased water levels (Simões et al., 2012). As water levels rise, the physical, chemical, and biological conditions in aquatic environments tend to become more uniform (Thomaz et al., 2007). Specifically, for fish, the increase in water level expands the available area for dispersal, resulting in a relative decline in their biomass per area (Arrington et al., 2005; Fernandes et al., 2009; Chaparro et al., 2014) (Fig. 2). Typically, the most autochthonous are white-water ecosystems during low-water phases, whereas the most allochthonous are black-waters during flooding. This difference is also reflected in the total fish biomass, with white-waters at low-water levels exhibiting the highest values and black-waters at high-water levels showing the lowest (Fig. 3).

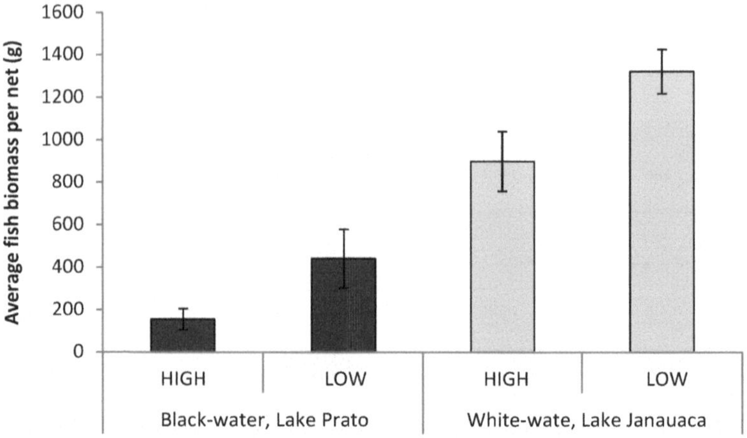

Fig. 3 Fish catches from white-water floodplain Lake Janauacá (connected to River Solimões) and blackwater Lake Prato (connected to River Negro) using NORDIC gillnets during low- and high-water season (unpublished data from Zingel, Agasild, and Tuvikene)

4 Impacts of Land-Use and Climate Change on Amazonian Aquatic Ecosystems

The interactions between terrestrial and aquatic components play a crucial role in Amazonian ecosystems. When floodwaters invade the floodplains, they carry terrestrial primary production, organic matter, and nutrients, which become the foundation of the aquatic food chain (Junk, 1984; Melack & Forsberg, 2001). Changes in energy and nutrient flows between terrestrial and aquatic habitats have significant implications for food web structures, biological cosecmmunities, and productivity in Amazonian floodplains. Understanding how human impacts will affect the functioning of floodplains and river ecosystems is essential for preserving fish biomass in the Amazonian region.

Multiple food web assessments have indicated that aquatic food webs, particularly those in tropical regions, have been significantly impacted over the past three decades (Angelini, 2019). Deforestation greatly affects the physical and chemical characteristics of water bodies. When agricultural practices take the place of forests, it leads to the introduction of fertilizers, herbicides, and other pollutants into the water, as well as sediments from soil erosion (see also Chap. 4, Barros et al., 2024 and Chap. 7, Schöngart et al., 2024). Approximately 12% of the rainforest has already been lost due to logging and agriculture (Stropp et al., 2020). The relationship between vegetation cover and moisture recycling is closely linked to the water, energy, and carbon cycles in the region (Marengo et al., 2018).

The impact of deforestation and the construction of hydropower dams has already disrupted the natural flow regime of certain Amazonian rivers (Almeida et al., 2020; Castello et al., 2013; Schöngart et al., 2021; Stropp et al., 2020; Timpe & Kaplan, 2017), potentially leading to disturbances in flood-pulse patterns across various river basins. This disruption of hydrological connectivity affects species migration, alters floodplain areas crucial for fish nutrition and spawning, and changes biogeochemical cycling in river floodplains. Changes in hydrology, together with land-use alterations and dam constructions, also disturb sediment transport and habitat heterogeneity, posing a significant threat to fish diversity in the Amazonian region (Flecker et al., 2022; Latrubesse et al., 2017; Santos et al., 2020; Sousa et al., 2021).

Deforestation of floodplains impacts aquatic primary production by diminishing C_3- and C_4-plant communities, which serve as the foundation for food webs and are important sources of food for numerous biological communities, especially in white-water rivers. For example, native C_4-grasses such as *Echinochloa polystachya* and *Paspalum fasciculatum*, which are well-adapted, are facing increasing threats due to human-induced disturbances. These disturbances include deforestation and land degradation, often for the purpose of expanding agricultural activities and cattle ranching. These activities frequently lead to the introduction of foreign grass species like *Urochloa* spp. (Mano et al., 2023). Grasses from this genus have demonstrated a significant invasive potential and are currently a cause for concern in various environments. These include the Pantanal (Bao et al., 2015), the Cerrado (Guimarães Silva et al., 2020; Pivello, 2011), and ecosystems in the Amazon (Fares et al., 2020; Ferreira et al., 2016). Furthermore, the construction of reservoirs often leads to the flooding of large forest areas, resulting in the decay of trees and the production of substantial amounts of methane (CH_4). This can lead to the development of significant anoxic zones in tropical reservoirs, reducing suitable habitats for aquatic life and causing high concentrations of CH_4 and CO_2 to be emitted into the atmosphere (Kemenes et al., 2007, 2011).

Climate change and land-use alterations often work together. Regional deforestation seems to exacerbate the intensity of droughts, making them drier and more severe compared to areas with full forest cover (Bagley et al., 2014). There is substantial evidence indicating that the Earth's climate is changing at a faster pace than in the past (IPCC, 2022; Wuebbles, 2018). Despite uncertainties at the local level, it is highly probable that extreme climatic events will occur more frequently (Franzke et al., 2023; IPCC, 2022). The majority of global climate models suggest significant drying in the Amazon region throughout the twenty-first century (Joetzjer et al., 2013; Sampaio et al., 2019). Based on current carbon emissions models, global temperatures are expected to increase by 6 °C by 2100 (IPCC, 2022). Ongoing climate change has already caused a rise in the annual mean temperature in the Amazon rainforest (Marengo et al., 2021), and projections indicate a potential further increase in the range of 2–8 °C in certain Amazon areas by the end of the century (IPCC, 2014, 2022).

Projections for the period 2041–2070 suggest that the northeastern region of the state of Amazonas may experience the most significant warming, with a potential increase of up to 5 °C. In contrast, the central region of the state of Amazonas and the upper Rio Negro are expected to undergo smaller thermal changes, with an increase of up to 3.5 °C (Menezes et al., 2018). Over the next century, water temperatures in Amazonia, currently ranging from 28 to 32 °C, are anticipated to rise by 2.2–7.0 °C (IPCC, 2014; Val & Wood, 2022). Without rapid evolutionary adaptation, the outlook for fish species appears grim—it is projected that even a slight long-term increase in temperature above current river temperatures may lead to the loss of a significant number of Amazonian fish species (Val & Wood, 2022). Temperature tolerance ranges vary by species, from those supporting only a narrow thermal range (stenothermal species) to those able to live in a wide thermal range (eurythermal species) (de Carvalho Freitas et al., 2012). Fish populations subject to changing thermal conditions may see fluctuations in abundance, experience range expansions or contractions, or face the threat of extinction (Ficke et al., 2007).

Due to the geographic and geological diversity of Amazonia, climate change will have varying impacts across its different regions, resulting in a complex mosaic of effects (Val & Wood, 2022). The Amazon rainforest itself contains various different ecosystems and macro habitats, each imposing unique selective pressures on resident organisms through their specific temperature and precipitation patterns (Adeney et al., 2016; Hess et al., 2015; Val & Wood, 2022). As for changes in the hydrological regime, recent years have witnessed regional trends in the Amazon, which, however, vary across the Amazon basin.

In the northern part of the Amazon basin, there has been a rise in rainfall during the wet season, leading to more frequent and severe flooding events (Barichivich et al., 2018; Espinoza et al., 2019; Gloor et al., 2013; Heerspink et al., 2020; Schöngart & Junk, 2020). Extensive research has been conducted to understand the factors behind these increasing floods, which are attributed to natural variations of large-scale climate anomalies at low frequencies, climate change, and various feedback loops (Barichivich et al., 2018; Fleischmann et al., 2023; Gouveia et al., 2019). Conversely, the southern Amazon basin has seen an increase in the frequency of dry days (less than 1 mm daily precipitation) and the length of the dry season (Marengo et al., 2018), likely due to a combination of factors such as land-use changes, including deforestation and fires, as well as rising sea surface temperatures in the tropical northern Atlantic (Marengo et al., 2021). This simultaneous increase in wet conditions in the northern part and drought conditions in the southern region is also reflected in rising trends of maximum and minimum discharge in their respective catchment areas, ultimately leading to increasing hydrological amplitude in the Central Amazon basin (Chaudhari et al., 2019; Espinoza Villar et al., 2009; Marengo et al., 2021).

The rapid intensification of the north-south sea surface temperature gradient in the tropical Atlantic (Hibbert et al., 2023) is expected to exacerbate the duration and severity of the dry season in much of southern and eastern Amazonia (de Carvalho Freitas et al., 2012). Research suggests that the most severe droughts in the Amazon are linked to strong El Niño-Southern Oscillation events (Berenguer et al., 2021; Jiménez-Muñoz et al., 2016; Nobre et al., 2009), significant warming of the sea surface in the Tropical North Atlantic, or a combination of these factors (Marengo et al., 2008; Marengo & Espinoza, 2016; Resende et al., 2020). The flood regime is the primary determining factor in Neotropical rivers (McConnell & Lowe-McConnell, 1987), and changes in precipitation during the dry season are likely the most critical determinant of the Amazon's climatic fate (IPCC, 2022). These extreme droughts can have devastating effects on aquatic organisms due to significant reductions in aquatic environments, particularly impacting floodplain lakes, which can shrink in size by several orders of magnitude (de Carvalho Freitas et al., 2012). For example, the extreme drought of 2023 led to unprecedented shrinkage of many floodplain lakes (Figs. 4 and 5). While some fish species can migrate to the main river channel during the dry season, resident species may face the risk of perishing. Some studies have shown that fish assemblages can recover rapidly from

Fig. 4 Black-water Lake Prato (Anavilhanas archipelago) during the normal low-water period (**a**; November 2019) and the exceptionally low-water period (**b**; November 2023). The photos are taken from exactly the same place. Photos by P. Zingel

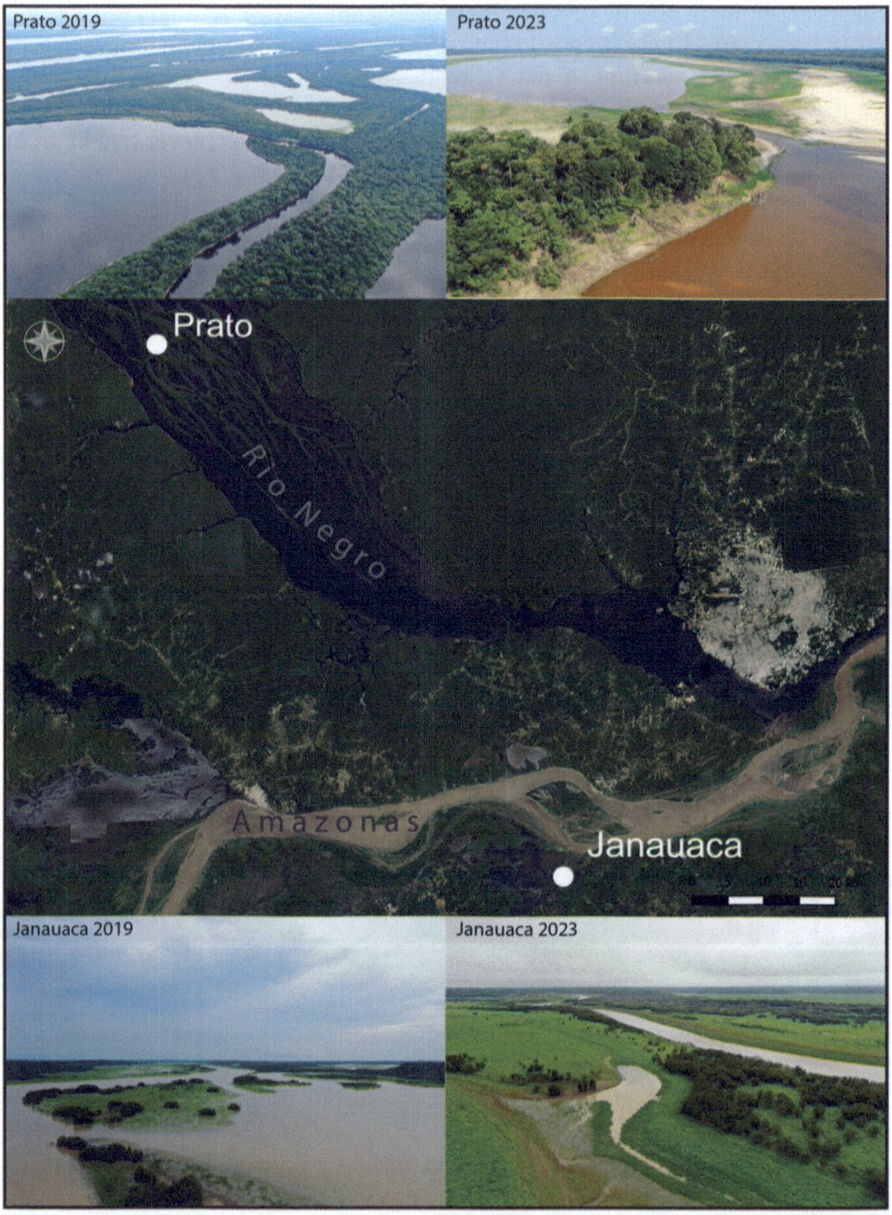

Fig. 5 Black-water Lake Prato (Anavilhanas archipelago) and white-water Lake Janauacá during the normal low-water period (November 2019) and the exceptionally low-water period (November 2023). Photos by A. Tuvikene and Google Maps (2024)

normal drought seasons, but there are indications that extreme droughts can alter fish populations and lead to the disappearance of certain species at a local level (de Carvalho Freitas et al., 2012; Humphries & Baldwin, 2003) (see also Chap. 1, Freitas et al., 2024).

Researchers have documented three significant changes in fish responses to climate warming: (1) shifts in the geographical distribution of species towards the poles (Comte et al., 2013; Perry et al., 2005); (2) changes in the timing of important life events such as migrations and spawning (Myers et al., 2017; Peer & Miller, 2014); and (3) a decrease in average body size, resulting in fewer large adults and a higher proportion of smaller, younger individuals in populations (Gardner et al., 2011; Solokas et al., 2023). Rising temperatures due to climate change can impact fish populations by reducing their performance when temperatures exceed their optimal range, leading to greater energy use and the need for additional energy supply. Many fish species rely on other fish or invertebrates as their primary food source, and changes in temperature can lead to mismatches in the timing of predator and prey activities, influencing trophic interactions (Domenici et al., 2019). These changes can have negative effects on predator–prey relationships in addition to direct temperature effects (Monaco et al., 2016).

Changes in temperature can have complex interactions with various physiological processes and responses. For instance, the chemical energy utilized by organisms for cellular maintenance may increase at a different rate as temperature rises (McKenzie, 2011). Inefficiencies also arise in the energy transfer from one trophic level to another. The cost of growth, defined as the energy expended in metabolism to produce a specific amount of biomass, can escalate with increasing temperature (Barneche et al., 2019; Barneche & Allen, 2018). Consequently, for a given primary production, the biomass at higher trophic levels fluctuates with temperature, impacting fish stocks significantly (Barneche & Allen, 2018).

The plasticity of physiological and molecular mechanisms across different time scales plays a crucial role in moderating the impact of temperature on individual fish. This plasticity varies among species and populations within species. Tropical fish are perceived to be more susceptible to warming due to some species living at temperatures near their upper thermal limits (Illing et al., 2020) and displaying limited capacity for acclimation of critical thermal maxima (Vinagre et al., 2016).

One of the most striking examples of food web disruption due to the impact of climate change is found in the central Amazon floodplains. In 2005 and 2010, the region experienced severe hydrological draughts, leading to the drying out of approximately 70% of aquatic habitats. This caused abrupt changes in the structure of fish communities in several floodplain lakes, with the new structure persisting over the following decade, indicating a regional shift in fish ecology (Freitas et al., 2013; Röpke et al., 2017). After the extreme hydrological drought in 2005 (Marengo et al., 2008), there were significant declines in abundance for species of all sizes with equilibrium life history strategies (characterized by moderate to long generation times, low fecundity, and high investment per offspring), as well as for large species with periodic life history strategies (characterized by long generation times, high fecundity, and low investment per offspring (Winemiller & Rose, 1992), and for all trophic levels except primary consumers. These findings suggest that the extreme drought triggered changes in the fish community, and subsequent abnormal hydrological conditions hindered the community's recovery (Röpke et al., 2017). This case demonstrates how changes in the flood pulse can lead to significant ecological shifts on a large spatial scale. A state change occurs when an ecosystem reaches a critical threshold or tipping point, leading to an alternative qualitative state for the ecosystem (Scheffer et al., 2001).

The hypothesis of oxygen- and capacity-limited thermal tolerance suggests that warmer water temperatures resulting from climate change can restrict individual fitness and population persistence (Pörtner & Knust, 2007). As water temperature rises, organisms' oxygen demand increases while the cardiovascular system's capacity to supply oxygen is constrained. Additionally, oxygen concentrations decrease with higher water temperatures. These interactions are expected to lead to declines in fish populations and potentially even extinctions as water temperatures continue to rise (Butzin & Pörtner, 2016).

Projections indicate that more intense and frequent flood pulses with greater variation between high- and low-water levels will most likely take place in the future (Barichivich et al., 2018; Costa et al., 2003; Junk, 2013). When combined with increased water temperature, these changes will result in lower oxygen levels (Barros et al., 2020) and higher concentrations of dissolved organic carbon (DOC) in Amazonian waters (Lauerwald et al., 2020). Elevated temperatures can exacerbate hypoxia or anoxia conditions, which are naturally observed in some Amazonian freshwater habitats. The inverse relationship between temperature and dissolved oxygen may lead to an expansion of hypoxic areas. While Amazonian fish exhibit various strategies for oxygen intake in response to hypoxia (Almeida-Val et al., 2000), there are limits to these adaptive strategies. The impacts of these changes on fish fauna and fisheries are currently unpredictable at both the species and ecosystem levels. The limited information on the biology of many Amazonian fish species makes it challenging to determine the current impacts of climate change on the Amazonian ecosystem, and even more difficult to predict future impacts (de Carvalho Freitas et al., 2012). Therefore, it is crucial to understand how these changes are affecting Amazonian food webs.

The effects of climate change also extend to the Amazonian aquatic mammals, which are considered to be keystone species in the ecosystem (Brum et al., 2021). During periods of extreme drought, large numbers of river dolphins and manatees often become trapped in pools, lakes, and river channels, leading to their deaths. Additionally, during low-water periods, dolphins, fishes, and fishers become concentrated in the remaining floodplain lakes, leading to conflicts as fishers see dolphins as competitors (Alves et al., 2012; Brum et al., 2021). The survival of manatees is reportedly affected by extreme droughts (Marsh et al., 2017) due to their limited morphological and physiological capacity to regulate body temperatures without significant metabolic costs (Elsner, 1999). Amazonian manatees have a narrow thermo-neutral zone, estimated to be between 23 and 32.5 °C (Gallivan et al., 1983).

Given that the present water temperatures in Amazonia vary from 28 to 32 °C and are forecasted to rise by 2.2–7.0 °C in the coming century (IPCC, 2014), the survival of Amazonian manatees is at risk. Additionally, the giant otter, a species with specialized diet and habitat requirements, is also of high concern (Loy & Cianfrani, 2018), as further climate warming may render their current core habitats unsuitable (Cianfrani et al., 2018). While there is currently no data available on the impacts of extreme weather events on semi-aquatic mustelids, it is likely that severe droughts could reduce food availability, leading to increased conflicts with fishers, while extreme flooding could destroy dens and have detrimental effects on reproduction.

5 Food Web Functioning in a Changing Environment

5.1 Carbon Sources for the Food Webs in Amazonian Floodplains

The central Amazon floodplains are highly productive ecosystems with a wide variety of organic carbon sources available for aquatic organisms. Aquatic macrophytes are a major source of organic carbon in these floodplains, accounting for over half of the ecosystem's net primary production (Melack & Forsberg, 2001). They contribute to both dissolved and particulate organic matter as well as nutrients that support phytoplankton production in the floodplain (Mortillaro et al., 2016). The different types of macrophytes, such as C_3 and C_4, have been extensively studied, including their biomass and production rates (e.g. Engle et al., 2008; Piedade et al., 2010; Silva et al., 2013). While C_4-macrophytes generally produce larger biomasses than C_3-macrophytes, the food webs in the central Amazon floodplains primarily rely on a C_3 carbon source (Araujo-Lima et al., 1986; Benedito-Cecilio et al., 2000; Lopes et al., 2007; Marchese et al., 2014). This is despite the fact that C_4-macrophytes, which form large floating meadows (dominated by *Echinochloa polystachya*, *Paspalum fasciculatum*, and *P. repens*; Piedade et al., 2010), are more rapidly mineralized by heterotrophic activity and released as CO_2 compared to C_3-plants (Mortillaro et al., 2016). The lower lability of C_3-macrophyte debris (e.g. *Eichhornia* sp., *Pistia stratiotes*, and *Salvinia auriculata*) makes them more available to be channelled through aquatic food webs, although C_4-macrophytes may provide a higher quality food source for specialized herbivorous fish species (Mortillaro et al., 2015).

C_3-macrophytes along with phytoplankton, attached algae and detritus derived from these, consist of the most important basal resources for the food webs on Amazonian flood plains (Arantes, Winemiller, Petrere, & Freitas, 2019; Araujo-Lima et al., 1986). This autochthonous production (produced within the aquatic system), especially that of phytoplankton and periphytic/benthic algae, is a valuable resource because of higher nutrient and energy content compared to that of material originating from the surrounding terrestrial areas (Brett et al., 2017). These resources support the energy to the various primary consumers in the food webs (zooplankton, benthic invertebrates, such as snails, mussels, and insect larvae) including the detritivorous fishes that form one of the most important functional groups for the subsistence economy of local Amazonian communities (Araujo-Lima et al., 1986; Campos-Silva et al., 2021; Marchese et al., 2014). Production of phytoplankton, including the differences in its taxonomic composition, biomass, and density are strongly impacted by the hydrological periods (Huszar & Reynolds, 1997; Kraus et al., 2019). During the high-water period, phytoplankton biomass and density are lower due to dilution processes and the large inputs of dissolved organic matter, which are impaired by nutrient and light availability (Huszar & Reynolds, 1997). In white-water systems, phytoplankton productivity is mainly limited to floodplain lakes, where lentic conditions allow for sediment deposition, leading to increased water clarity and enhanced light penetration. The organic matter and detritus produced by plants and phytoplankton in floodplains are transported via channels to the rivers (Moreira-Turcq et al., 2013; Mortillaro et al., 2011).

On the other hand, primary productivity is poor in Amazonian black-water lakes and rivers due to low nutrient levels, as these rivers originate from lowland forest areas with sandy soils and have minimal levels of suspended sediments and dissolved inorganic nutrients (Junk et al., 2015). In these aquatic systems, the aquatic macrophytes (except for *Orza* spp.) are generally rare or altogether absent (Lopes et al., 2019), but the concentrations of dissolved organic carbon (DOC) are high due to decomposing forest litter. This allochthonous DOC, resulting from a bacteria-mediated lignin degradation process, acidifies the entire Rio Negro watershed (pH typically 4.0–6; Morris et al., 2021), making it a challenging aquatic environment for the organisms inhabiting there. Additionally, the highly coloured molecules in allochthonous DOC limit light penetration and, therefore, the phytoplankton production in these black-water systems (Moreira-Turcq et al., 2003). Despite low in situ primary production, the black-water oligotrophic rivers maintain diverse fish communities and productive fisheries. Moreover, they are hotspots of endemic species of various organisms (Andrade et al., 2023). In these conditions, riparian vegetation largely supports the biomass of aquatic organisms, showing that the connection between land and water environment is of prime importance (Andrade et al., 2023).

During the annual hydrological cycle, as water levels increase, lakes expand in the area, and terrestrial organic matter from flooded forests, including higher inputs from C_4-macrophytes covering large extensions of lake and river banks, is carried to Amazonian floodplain lakes. This organic matter supports resources for consumers (Araujo-Lima et al., 1986). In the decomposition process, DOC derived from autochthonous (algae and macrophytes) or terrestrial allochthonous sources (e.g. tree leaves, fruits, and flowers) serves as the main substrate and energy source for the growth of planktonic bacteria. Bacteria play a crucial role in returning nutrients to the food web (microbial-loop concept, Azam et al., 1983), as they convert the DOC into their own biomass and are later consumed by various protists and multicellular consumers in higher trophic levels. Furthermore, the composition and concentration of DOC, as well as related bacterial metabolism, change throughout the hydrological cycle (Amaral et al., 2013). In clear-water systems, bacterial growth and DOC bioavailability are higher during the low-water period when phytoplankton serves as a significant extra source of DOC (Farjalla et al., 2006). Conversely, in black-water systems, the highest DOC concentrations and bacterial activity occur during the high water or falling water period (Amaral et al., 2013).

5.2 Primary Research Areas in Food Web Ecology in Amazonian Floodplains

Much of the knowledge about aquatic food web functioning in the Amazon region has been derived from research focused on fish biology and ecology. Stable isotopes, along with fish gut content analyses and community metrics, are commonly used methods for gaining a comprehensive understanding of trophic structure and ecosystem functioning in the variable conditions of Amazonian waters. Analysis of carbon and nitrogen stable isotopic ratios ($\delta^{13}C$ and $\delta^{15}N$, respectively) can reveal trophic pathways and food web structures, integrating temporal and spatial variation in feeding and reflecting the materials incorporated into animal tissue (Fry, 2006). A review by Santos et al. (2017) offers an overview of stable isotope ecological studies on fish in the Brazilian Amazon. Since the first works in the mid-1980s (Araujo-Lima et al., 1986), the use of stable isotopes in ecological studies of different fish taxonomic groups in the Amazon has increased (e.g. Caraballo et al., 2014, 2016; Lara et al., 2012; Magnusson et al., 1999; Schwamborn & Giarrizzo, 2015; Waichman, 1996), particularly focusing on trophic ecology, life history, migration, and the origin of carbon in fish and other aquatic organisms (e.g. manatees; Crema et al., 2019). The application of stable isotopes in food web ecology has provided insights into questions that are difficult to address with other methods, particularly in systems characterized by high species diversity and trophic complexity, such as the Amazonian floodplains (e.g. Rejas, 2018). This research has assessed the lengths of food chains and provided valuable information about trophic interactions in these complex ecosystems.

The majority of the research in trophic ecology has been centred on carbon sources assimilated by different species and fish-feeding guilds (e.g. Lopes et al., 2007; Marchese et al., 2014; Oliveira et al., 2006); larval fish from species forming the core of the fishing resources of the basins (*Semaprochilodus* spp., *Prochilodus nigricans*, *Mylossoma* spp., *Colossoma macropomum*, *Brycon cephalus*, and *Triportheus elongatus*; Leite et al., 2002), as well as portioning of trophic niches of various fish species (for example, herbivorous serrasalmids; Andrade et al., 2019). An important aspect has been revealing the diet shifts during the hydrological cycle as a response to fluctuating resources (e.g. Correa & Winemiller, 2014). Stable isotope research in the Amazon region has expanded our understanding of the interconnectedness of food webs, particularly through the migratory patterns of fish. For example, the carbon produced in black-water systems contributes to the fish stocks harvested from white-water systems, such as *Semaprochilodus insignis* (Benedito-Cecilio & Araujo-Lima, 2002). Moreover, beyond analyses of single species, functional groups, or multiple trophic levels from a single system (e.g. Marchese et al., 2014), the stable isotope method has been a useful tool towards more wider assessments of temporal and spatial variations of food webs across different watersheds as well as across the river-floodplain system (e.g. Arantes, Winemiller, Petrere, & Freitas, 2019).

The annual hydrological cycle and fish migratory patterns significantly influence the dynamics and quality of food sources, leading to spatial and temporal variability in the food webs of Amazonian river floodplains. Currently, the empirical research and knowledge of how food webs vary across broader regional landscapes in Amazon river-floodplain ecosystems are rather scarce. In a survey of 109 habitats across 19 floodplain lake systems distributed along ca. 250 km of the lower Amazon River, Arantes and co-workers (Arantes, Winemiller, Petrere, & Freitas, 2019) estimated the energy and matter flow to variable fish trophic guilds. This study showed that while the C_3-aquatic macrophytes and suspended particulate organic matter contributed the principal basal resource to the biomass of most of the fish species over the lake systems (regional scale), there was also a considerable variation within species and within lake systems (local scale). Some less important sources in terms of regional scale can form essential resources in certain species or in local lake systems. For example, in several lake systems, some fish species (e.g. *Pygocentrus natteri*, *P. squamosissimus*, and *Acestrorhynchus abbreviates*) had assimilated substantial amounts of material originated from C_4-grasses (Arantes, Winemiller, Petrere, & Freitas, 2019). This study highlights that for maintaining the local fish resources, research and knowledge on local resource–consumer interactions is highly important.

More recently, research has also focused on establishing baselines for trophic ecology to assess the potential impacts of environmental stressors, particularly in anticipation of reservoir construction or the operation of hydropower plants (e.g. Andrade et al., 2023). This research aims to provide valuable insights into the potential ecological impacts of such developments.

5.3 Sensitivity of Food Webs to Climate and Land-Use Changes

Modelling, in conjunction with other assessment methods, has provided valuable insights for a broader understanding of the processes and potential consequences of future changes (e.g. Capitani et al., 2021; Correa & Winemiller, 2018). Deforestation poses one of the most significant threats to Amazon ecosystems, leading to reduced inputs from forests and impacting fish production (Arantes, Winemiller, Asher, et al., 2019; Capitani et al., 2021). The food sources for Amazonian fish originate from the synthesis of two primary sources: autochthonous sources, which depend on the productivity of the aquatic ecosystem (such as phytoplankton, zooplankton, aquatic macro-invertebrates, insect larvae, etc.), and allochthonous sources, mainly from seasonally flooded forests (including fallen fruits and seeds). A growing body of research supports the idea that many fishes and other aquatic organisms in Amazonian rivers exploit allochthonous resources from floodplains, particularly during the high-water period (Correa & Winemiller, 2018; González-Bergonzoni et al., 2019; Manetta et al., 2003; Weiss et al., 2016, 2024). Even during the low-water period, the aquatic productivity seems to be mainly upheld by allochthonous sources, as the accumulated terrestrial organic matter in river channels serves as a nutrition source for the food webs (Andrade et al., 2023). In oligotrophic systems, these allochthonous inputs are very important in supporting the aquatic food web.

Research on Amazonian aquatic environments increasingly indicates disruptions in terrestrial–aquatic energy pathways. A modelling study in the lower Tapajós River in the Brazilian Amazon showed that deforestation can be a major driver of decreased fish and mammal biomass. Cutting down floodplain forests will most likely lead to an overall decline in fish biomass, with greater losses presumably threatening fish species that directly depend on floodplain forest resources (e.g. seeds and fruits) (Capitani et al., 2021). A survey of 462 aquatic habitats covering various land-cover conditions in the Amazon floodplain further confirms that not only is fish biomass affected, but also systems functional diversity at both local and regional scales. Certain functional groups, particularly those related to forest cover, are particularly vulnerable, while habitat generalists and those directly consuming autochthonous resources are unrelated to canopy coverage (Arantes, Winemiller, Asher, et al., 2019). An analysis of the diet of medium-sized omnivorous fish (*Bryconidae, Serrasalmidae*, and *Triportheidae*) from a floodplain of a black-water river (Apaporis River, western Amazonia) suggests that riparian vegetation, particularly fruits, supports metabolism by providing much of the energy, while arthropods (i.e. insects, spiders, centipedes, and millipedes) support fish biomass contributing proteins for their somatic growth (Correa & Winemiller, 2018). Food chains that originate from allochthonous sources support, especially, populations of large predatory fishes which are, therefore, most vulnerable in regard to deteriorating lateral connectivity as numerous prey species, including shrimp, crabs, and small fish, search for food in the shallow areas of the river channel and in the periodically flooded regions of the floodplain (Röpke et al., 2017).

The terrestrial input into aquatic food webs is heavily influenced by the duration of connections between these ecosystems (Roach, 2013). Changes in flood pulse patterns and the availability of fruit can lead to behavioural shifts in vertebrate frugivores. During a normal flood pulse regime, the seasonal abundance of fruit drives changes in frugivores' diets from being fruit-dominated during the flood season to consuming alternative, less preferred food types during the low-water period (Correa & Winemiller, 2014). If fruits become scarce or unavailable, permanent shifts in diet are probable, which can impact the health of frugivores by reducing body condition and altering their reproductive phenology. Floodplain forest specialists such as *Colossoma macropomum*, *Piaractus brachypomus*, *Myloplus* spp., and *Myleus* spp. are suggested to serve as indicators for floodplain ecosystems due to their reliance on floodplain habitat and food (Correa et al., 2022). Monitoring the feeding habits and health of floodplain frugivores could help document permanent diet shifts or reduced body condition, serving as ecosystem state indicators (Correa et al., 2022).

Similarly, the construction of large hydropower dams, especially in the Andean region, is expected to reduce aquatic productivity and fisheries yield in the Brazilian

Amazon (Forsberg et al., 2017; Latrubesse et al., 2021). Damming rivers affects sediment transport, floodplain dynamics, organism migration, aquatic population connectivity, habitat availability, and resource supply from seasonally flooded environments. Studies in South American rivers, particularly in the Upper Paraná River basin, have demonstrated that mean fish size and food web length decreased after the hydroelectric dams were constructed, with large migratory fish species being most influenced (Agostinho et al., 2016; Arantes, Fitzgerald, Hoeinghaus, & Winemiller, 2019; Santos et al., 2017). Arantes and colleagues (Arantes et al., 2022) analysed fish data from pre- and post-dam construction periods in the Madeira River basin and showed that dam operation impacted the functional composition of fish yields and reduced catches. Such a reduction would significantly affect indigenous and traditional populations in the floodplains, particularly those in densely populated white-water floodplains, who rely on fish as their main origin of animal protein (Begossi et al., 2018; Forsberg et al., 2017).

Loss of function in fish communities often occurs as a result of changes in feeding habitats or altered energy pathways that push certain species beyond their adaptive limits (e.g. Tribuzy-Neto et al., 2017). This has been observed following the loss of floodplain forests (Arantes et al., 2018; Arantes, Winemiller, Asher, et al., 2019), forest fires (Lugo-Carvajal et al., 2023), dam construction (Arantes et al., 2022), and extreme alterations in normal flood pulses in hydrological cycles (Freitas et al., 2013; Röpke et al., 2017). The extreme drought in 2005 dried out large areas of aquatic habitats and led to changes in fish assemblages by triggering functional shifts in community structure and decreasing overall diversity (Freitas et al., 2013; Röpke et al., 2017). Some research attests that large migratory omnivorous fish, particularly those species that depend on flooded habitats (e.g. *Colossoma macropomum*; *Prochilodus nigricans*, and *Semaprochilodus* spp.), could experience particularly adverse impacts (Freitas et al., 2013; Röpke et al., 2017; Tribuzy-Neto et al., 2017). During the extreme drought of 2005 (Marengo et al., 2008), a significant reduction in the body condition factor was observed for omnivorous fish species (e.g. *Colossoma macropomum* and *Mylossoma duriventre*) in several floodplain lakes (Baixio, Preto, Iauara, Ananá, Maracá, Araçá, Poraquê, and Aruã) close to the Solimões River, seemingly as a result of reduced access to habitat and food resources (Tribuzy-Neto et al., 2017). Alterations in fish functional groups could also adversely affect the regeneration of floodplain forests, as a decrease in omnivorous and frugivorous fish populations may reduce seed dispersal for numerous tree species (Lugo-Carvajal et al., 2023). Seeds of flooded forests consumed by fish highlight an important role of this group, which disperses the diaspores also upstream and between systems of floodplains and contributes to the maintenance and gene flow of floodplain forests (Correa et al., 2015; Gottsberger, 1978; Weiss et al., 2016, 2024).

Besides supporting resources to the food webs, annual flooding creating habitat heterogeneities is also the most important factor in the Amazon basin maintaining the various fish species' coexistence (van der Sleen & Rams, 2023). Recent analyses simulating the annual flood changes (van der Sleen & Rams, 2023) predict that the interruptions in flood pulse could result in species extinctions and lower fish diversity across tropical areas in the world. Hydrological changes caused by climate change that are accelerated further by deforestation and the construction of large hydropower dams are already taking place in the Amazon region (Almeida et al., 2020; Latrubesse et al., 2021; Marengo & Espinoza, 2016; Stropp et al., 2020; Timpe & Kaplan, 2017). Therefore, the research ongoing on trophic ecology in Amazonian floodplains is essentially important to develop more effective management and conservation strategies aimed at maintaining both the environment and fish stocks at sustainable levels of exploitation.

6 Impact of Changing Biogeochemistry on Food Web Processes

6.1 Methane Pathway in Aquatic Food Webs

River-floodplains systems produce and receive from surroundings substantial amounts of organic matter. Microbial metabolism plays a significant role in degrading the organic matter by transforming "the unused" or "less favourable" sources into mineralised nutrients (N, P) and into biochemical fluxes (CH_4, CO_2) that can be assimilated by autotrophs or incorporated into bacterial biomass that are further available for consumers in the food web. Here, we emphasize the role of biogenic methane (CH_4) in aquatic food webs, its relation to changing DOC concentrations, and its potential significance for Amazonian floodplains.

A recent estimation using different remote sensing techniques confirmed that 26% of the Amazon River basin is seasonally or permanently inundated (Fleischmann et al., 2022). The dominant wetland types in the Amazon basin are those with seasonal flooding shaped by flood pulses of varying frequency, predictability, and amplitude (Junk et al., 2011). These areas together with lakes act as organic matter producers, while the sediments store large amounts of carbon. The sediment accumulation rates in Amazonian floodplains are higher than those reported for lakes in arctic, boreal, temperate, and tropical zones (Sanders et al., 2017). This makes Amazon basin floodplains a significant source of greenhouse gases (CH_4 and CO_2). In a large-scale measurement, based on 297 lakes across different climate zones (Sanches et al., 2019), it was demonstrated that climate-related variables, primarily air temperature and precipitations, are correlated with CH_4 emissions with elevated emissions in regions where air temperature and precipitation are high—in the tropics and the southern temperate areas.

Amazonian wetlands have received increasing recognition for the importance of the global dynamics of CH_4 emission (e.g. Dalmagro, Zanella de Arruda, et al., 2019; Melack et al., 2022; Pangala et al., 2017; Ringeval et al., 2014). South American floodplain lakes and wetlands, particularly, are among the most high-emitting environments producing the highest rates of CH_4 emissions per area unit (e.g. fluxes m^{-2}) (Bastviken et al., 2010). Besides floodplains, vast flooded forested areas caused by the construction of reservoirs produce large quantities of methane as the organic material decays (Kemenes et al., 2007). Methane is an end product of organic matter degradation that is produced by methanogenic archaea in anaerobic conditions that develop in lake sediments and overflooded forest soil. Approximately, half of the Amazonian wetland CH_4 emissions are mediated by the stems of the seasonally flooded floodplain trees (Pangala et al., 2017). Still, a large part of the CH_4 produced in floodplain sediments is not outgassed to the atmosphere but is readily consumed by methane-oxidising bacteria (MOB) when it reaches the oxic sediment or water column (Barbosa et al., 2018; Bastviken et al., 2008). Barbosa et al. (2018) estimated that, on average, 80% of CH_4 produced was oxidized in an Amazonian floodplain lake (Janauacá Lake). These methanotrophic bacteria can in some instances form considerable biomasses (up to > 50% or comparable to primary production rates of phytoplankton; Taipale et al., 2011; Morana et al., 2015) and serve as alternative carbon and energy sources for the consumers in aquatic food webs (Grey, 2016; Jones & Grey, 2011).

MOB are found in a wide range of freshwater systems and in different habitats within these systems, i.e. sediment, water column, macrophyte root systems, and epiphyton of submersed parts of aquatic vascular plants (Hanson & Hanson, 1996; Trotsenko & Khmelenina, 2005; Yoshida et al., 2014). Through the consumption of biomass of MOB, and through the trophic linkages, the methane-derived carbon is potentially available for different types of organisms inhabiting aquatic biotopes (protozoans, zooplankton, macroinvertebrates, and fish) (e.g. Agasild et al., 2014, 2018; Murase & Frenzel, 2007; Ravinet et al., 2010; Taipale et al., 2008) (Fig. 6). Measurement of carbon stable isotopic compositions ($\delta^{13}C$) along with fatty acid and molecular methods have been the forefront tools revealing the methane-derived carbon contributions to aquatic consumers (Grey, 2016). Because biogenic methane is exceptionally depleted in ^{13}C, low $\delta^{13}C$ values, particularly measured for some

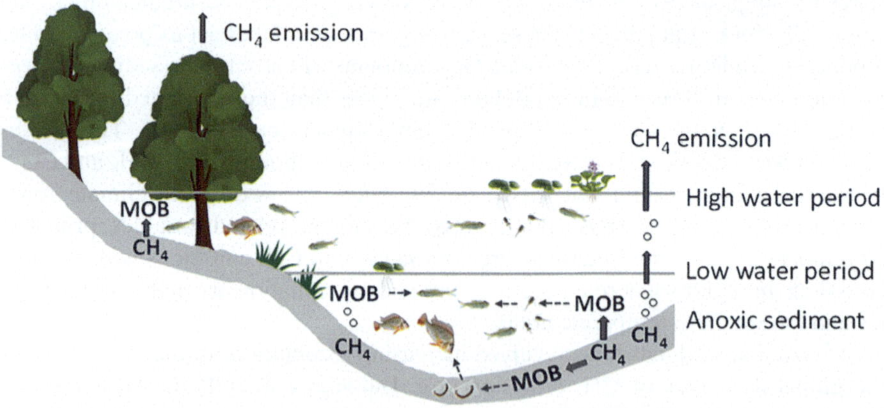

Fig. 6 Schematic presentation of methane (CH_4) production in anoxic floodplain sediments, emissions to the atmosphere via ebullition, and mediation by trees and aquatic macrophytes; oxidation by methane-oxidizing bacteria (MOB) in sediment and in water column. Via feeding interactions, MOB in sediment, water, and among periphyton on macrophytes contribute methane-derived carbon to benthic macroinvertebrates, zooplankton, and fish. During the inundation period, the anoxic conditions in the soils of floodplain forests favour methane production and subsequent oxidation by MOB

sediment-dwelling midge (Chironomidae) larvae ($\delta^{13}C$ down to −72‰; Jones et al., 2008), are indicative of assimilation from the methane-derived source (Grey, 2016). Benthic organisms particularly can benefit from the methane-derived carbon contributions (up to 70% of larval Chironomidae biomass; Jones et al., 2008). Similarly, pelagic organisms like crustacean zooplankton can periodically receive appreciable parts (up to 50%) of their diet from MOB, especially in lakes with high DOC loading (Kankaala et al., 2010). Current information on methane-derived carbon in food webs, however, is derived mostly from north-temperate lakes (Grey, 2016; Jones & Grey, 2011).

6.2 Recognition of Methane Pathway in Amazonian Aquatic Food Webs

The role of methane-derived carbon in freshwater food webs has received yet little attention and is largely unexplored in tropical floodplains of the Amazonian region despite the fact that Amazon lakes and flooded areas are characterized by high rates of methane production and oxidation (Barbosa et al., 2018; Bastviken et al., 2010; Conrad et al., 2011). Nevertheless, the increasing application of the stable isotope method for assessing the food web structures and fish diets in Amazonian rivers and floodplains has also revealed food web connections with the carbon from MOB (e.g. Jepsen & Winemiller, 2007; Rejas, 2018).

Depleted $\delta^{13}C$ signatures of detritivorous fishes seem to occur widely in Amazonian floodplains (Azevedo-Silva et al., 2016; Jepsen & Winemiller, 2007; Mortillaro et al., 2015; Pouilly et al., 2004; Rejas, 2018; Sanseverino et al., 2012; Wantzen et al., 2002). Similarly to detritivore fish, more depleted $\delta^{13}C$ values than photosynthetic sources have also been found for zooplankton in the Amazonian region with an explanation of consumption of a highly negative fraction of the microbial component of the periphyton and detritus (Caraballo et al., 2016; Torres-Bejarano et al., 2014). It is thought that fishes may encounter greater amounts of more isotopically depleted detritus, including MOB when feeding in the deeper zones (Wantzen et al., 2002). Piscivorous fishes are less numerous in deeper hypoxic conditions in habitats in the Amazonian floodplain lakes (Anjos et al., 2008). Oxygen-poor habitats provide a refuge to hypoxia-tolerant prey species relative to their not-so-tolerant predators (Anjos et al., 2008; Chapman et al., 2002; Robb & Abrahams, 2002), providing also an access to feeding for isotopically light methane-derived carbon sources. Similar conditions with low dissolved oxygen and high methane concentrations can occur under the floating macrophytes (Barbosa et al., 2018). So far, the only comprehensive research on the inclusion of carbon from biogenic methane in the lake food web from tropical South America originates from the Pantanal area, where Sanseverino and co-workers (Sanseverino et al., 2012) assessed the stable isotope and fatty acid tracers among zooplankton, macroinvertebrates, and fish from different feeding guilds. This research provides evidence that methane-derived carbon can contribute significantly not only to the microbial food web and invertebrates (16% and 22% for insects and zooplankton, respectively), but also to higher consumers (13% for fishes with benthivorous, omnivorous, and detritivorous feeding), thus supporting the whole food web.

Considering the vast areas of wetlands and floodplains rich in CH_4 production, the methane-derived carbon support to food webs could be a common phenomenon in these areas. An intensive research conducted in the Brazilian Pantanal has confirmed that methane production and emission are tightly connected to the hydrological dynamics. During periods of rising water, a large amount of organic material is received from the surroundings into the lakes including the organic matter produced by macrophytes which are rapidly converted to greenhouse gases, including CH_4 (Bastviken et al., 2010). The process occurs similarly in the seasonally inundated forest. Measurements from northern Pantanal revealed that during the flooded phase, CH_4 emission increased rapidly as soon as anaerobic conditions in the forest soil were established (Dalmagro, Zanella de Arruda, et al., 2019). With receding water, the forest soil became rapidly aerobic and methane emission decreased substantially. The methane fluxes from the forest soil were 50 times higher during the flooding period (Dalmagro, Zanella de Arruda, et al., 2019). The diffusive flux, which is the source of MOB assimilation, followed also the seasonal variation of the flooding (Marani & Alvalá, 2007). This research indirectly indicates, that also the processes of methane oxidation, MOB production and hence the MOB availability as a food source for the consumers (Deines et al., 2007) are impacted by the hydrological cycle. So far, however, only a few studies have indicated seasonal inundation effect on methane-derived carbon role in the food web (Wantzen et al., 2002).

6.3 Impacts from Altered Dissolved Organic Carbon Loading on Food Webs

Current projections indicate that greenhouse gas emissions from floodplains will increase as global average temperatures continue to rise, and this is particularly important for temperate and tropical lake ecosystems (Sanches et al., 2019). The diffusive methane flux, used by MOB (Sawakuchi et al., 2021), has been found to be positively related to dissolved organic carbon (DOC) concentration in the water (Sanches et al., 2019). DOC originating from both terrestrial and allochthonous sources (algae exudation and microbial degradation products) forms the substrate for methanogenesis (Casper, 1992; West et al., 2012). As the terrestrial DOC supply can increase the methane production rates (West et al., 2012), the lakes with higher DOC levels can apparently support higher methane-derived carbon inclusion in their food webs. This can especially be expected for black-water systems (Fig. 7). The deforestation in Amazonia will increasingly impact riverine carbon dynamics. According to future projections, the DOC leaching from soil organic carbon and litter remains the dominant source of DOC to riverine water; however, there will be an increase in DOC export in rivers in the near future due to deforestation and more intense floods (Lauerwald et al., 2020). It is demonstrated both in temperate and tropical areas, that in a forested watershed, the major part of DOC and particulate

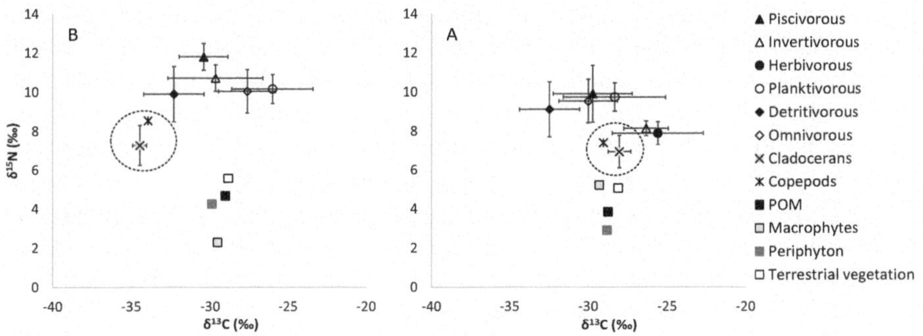

Fig. 7 Stable isotopic composition ($\delta^{13}C$, $\delta^{15}N$; ±SD) of fish feeding guilds, zooplankton (cladocerans and copepods) and potential food sources (particulate organic matter (POM)), periphyton, C_3-macrophytes, and terrestrial vegetation in white-water floodplain lake Janauacá connected to River Solimões River (A) and black-water Lake Prato connected to Negro River (B) sampled during high-water season (2016). Depleted $\delta^{13}C$ values of zooplankton (circled) in black-water lakes indicate a possible assimilation from methane-oxidizing bacteria; in white-water lakes, the zooplankton rely more on photosynthetic sources during the high-water season. Unpublished data from Zingel, Agasild, and Tuvikene

organic matter export from the land to the water can occur during short but extreme flow events (Clark et al., 2022; Raymond & Saiers, 2010).

Some predicting models suggest that the Amazon areas will experience not only more often rainfall but also dry period extremes in the future (see Sect. 4; Marengo et al., 2010, 2012), promoting also increasing drying and exposure of sediments. This is proposing the question of methanogenic activity in these conditions. The research with various types of Amazonian lake sediments has shown that the sediments both of black-water and clear-water lake types were all quite active in CH_4 production, mainly by the hydrogenotrophic pathway (Conrad et al., 2014; Ji et al., 2016). Methanogens seem to be resistant to the environmental changes between rising and decreasing water conditions (Gontijo et al., 2021), and their potential activity for methane production may even increase substantially after the sediment has been dried and rewetted (Conrad et al., 2014). High resistance of the methanogenic community along with predictions of increased, more labile DOC concentrations due to deforestation may also refer to the enhanced role of methane-derived carbon in food webs in the Amazonian region, especially that of in black-water and clear-water systems (Conrad et al., 2014; Gontijo et al., 2021; Ji et al., 2016).

The projected increase in allochthonous DOC concentration will affect biochemical processes in aquatic systems as well as response in bacterial production and community compositions (de Melo et al., 2020; Farjalla et al., 2006). The Amazon basin, especially the black-water systems, has very low rates of phytoplankton productivity (Wissmar et al., 1981), suggesting that DOC originating from terrestrial areas is an important source of carbon for bacterial growth (Júnior et al., 2020; Mayorga et al., 2005). It is generally shown based on research from the three Amazonian lake types that the overall quality of DOC, and not quantity, is the major driver factor for bacterial growth and composition (de Melo et al., 2020; Farjalla et al., 2006; Sylvain et al., 2021). The future changes in terrestrial areas will also change the DOC characteristics. The DOCs mobilized from deforested areas will probably consist of older, smaller molecules that are more labile to microbial degradation than DOC transferred from forested areas (Dalmagro, Lathuillière, et al., 2019; Drake et al., 2019). These future projections may indicate increasing food web reliance on chemosynthetically produced and microbially mobilized energy. This is also referring to an increasing importance of microbial pathways in Amazonian aquatic food webs. These processes overall may change the current balance between the microbial versus classical food chains in the food web which is, however, little addressed in Amazonian waters.

7 Integration of Microbial and Plankton Communities to Amazonian Food Web Analyses

The Amazon region's microbial diversity and its interaction with food webs have received relatively little attention compared to other tropical (e.g. Havens et al., 2007) or temperate areas (e.g. Weisse et al., 2016). Understanding these processes will be essential for comprehending changes in nutrient and carbon biochemical cycles under alterations in terrestrial–aquatic connectivity. Food webs are intricate systems of interconnected food chains, encompassing microbial components, through which the transfer of autochthonous (internal) and allochthonous (external) sources occurs via diverse microbial and multicellular consumers to higher trophic levels. Since the introduction of the microbial loop concept (Azam et al., 1983), there has been growing research attention on the complex interactions between bacteria and heterotrophic protists (such as flagellates and ciliates) and their role in the functioning of aquatic food webs, compared to the classical phytoplankton–zooplankton–fish food chain (e.g. Sommaruga, 1995; Zingel et al., 2023). Bacteria decompose DOC and plant material that cannot be directly consumed by herbivores—this decomposition process is essential to the food web and has also gained recognition in the floodplains of the Amazon (Marchese et al., 2014; Mortillaro et al., 2011; Rejas, 2018; Yossa & Araujo-Lima, 1998). Bacteria can be directly consumed by metazooplankton or passed through the microbial loop, which includes protozoa as an intermediate step (Jansson et al., 2000). As protozoa, especially ciliates, are often a major food source for metazooplankton (Agasild et al., 2012) and larval fish (Karus et al., 2014; Zingel et al., 2012), the microbial loop may be the primary pathway for mediating organic matter (e.g. Callieri et al., 1999). Research has shown that ciliates are crucial for the survival of fish larvae (e.g. Montagnes et al., 2010), as without them, the larvae would not meet their daily food requirements and would not survive in many circumstances (Zingel et al., 2012). Additionally, metazooplankton has been demonstrated to cause rapid shifts in ciliate community structure and abundances, indirectly affecting the food base and survivability of fish larvae (Agasild et al., 2013). Despite the importance of microbial loop functioning and its connection to fish development, these aspects remain understudied in Amazonian ecosystems.

In floodplains of large rivers, the lakes are important nursery areas for larval and juvenile fish. In these areas, the abundant food source and shelter from dense aquatic macrophytes, especially underneath the floating grass meadows, favour the development of fish larvae. Out of the very limited research on larval fish diet in Amazonian floodplains, the gut microscopic analyses of larvae from Itapu Reservoir, Brazil (Makrakis et al., 2005) indicated that zooplankton constituted the predominant food with insect and other fish larvae, and algae as less frequent items. Analysed larval species (*Iheringichthys labrosus*, *Hypophthalmus edentatus*, and *Plagioscion squamosissimus*) had intraspecific differences in their diet contribution from cladocerans, copepods, and rotifers, as well as ontogenetic shifts in diet spectrum. Collectively, the larval diet consisted of an incredibly high variability of zooplankton species (Makrakis et al., 2005) showing, indirectly, the nursery habitat richness in terms of food base and complex planktonic food web in fish feeding areas.

Current research on microbial communities (Feitosa et al., 2019; Meira et al., 2021; Rejas et al., 2005) indicates that bacterial, heterotrophic nanoflagellates (HNF), and ciliate abundances are relatively low in neotropical floodplains, comparable to those of oligotrophic temperate lakes (e.g. Tirok & Gaedke, 2007). However, these microbial communities can support significant carbon biomasses relative to phytoplankton (Feitosa et al., 2019). In one of the first comprehensive analyses from the Amazon region, Feitosa and colleagues (Feitosa et al., 2019) studied carbon partitioning in the microbial and classical food webs from Madeira River floodplains. They discovered that microbial compartments were the main component, representing about 90% of the total plankton biomasses. The significant role of the microbial food web was linked to the limitation of phosphorus for primary production (Feitosa et al., 2019). This study also demonstrated that the microbial food web exhibited greater complexity during the low-water period when nutrients and organisms were more concentrated. At this time, planktonic ciliates made significant contributions, while heterotrophic bacteria increased during the high-water period (Feitosa et al., 2019). An examination of the use of resources by crustacean zooplankton in the floodplain lakes of the Colombian Amazon also suggests a significant dependence on microbial resources, primarily supported by terrestrial dissolved organic carbon (DOC) and particulate organic matter (Aranguren-Riaño et al., 2018). The authors inferred that the increased detrital inputs through microbial pathways in these lakes could potentially account for the high functional richness of Amazonian zooplankton. Additional research is required to comprehend the carbon fluxes within the microbial compartment and its importance for the overall operation of the food web in Amazonian aquatic environments.

In tropical regions with higher temperatures (Havens et al., 2015) and year-round predation pressure from juvenile fishes, it is anticipated that the zooplankton community will be predominantly comprised of small-sized species. This is evident in the Amazonian region, where the zooplankton community is mostly dominated by small-sized species (de Azevedo & Costa Bonecker, 2003). In recent years, it has been widely observed that also global warming has led to a reduction in body sizes at both the species and community levels, which is considered a universal ecological response (Daufresne et al., 2009; Forster et al., 2013; Sheridan & Bickford, 2011). Previous research has shown that an increase in water temperature results in decreased cell size of protozoans (Zingel et al., 2018). It has also been found that the growth rate and division rate of protozoans are influenced by different factors, with many protozoans decreasing their cell sizes at higher temperatures (Laybourn-Parry, 1984). This aligns with the temperature-size rule, which suggests that individuals growing at lower temperatures tend to grow more slowly but reach a larger size at maturity (Atkinson, 1995). This rule has been observed in bacteria and many protozoans as well (Atkinson et al., 2003). Research by Forster et al. (2013) on *Cyclidium glaucoma* demonstrated that the cell size of this ciliate decreases with increasing temperature, indicating that maximizing reproductive rate at the expense of individual cell size enhances fitness, as faster-reproducing individuals make up a higher proportion of the final group of the species. Additionally, it was shown that the cell volume of protozoans is highly adaptable and can quickly respond to changes in environmental conditions and population abundances. However, a mesocosm study by Zingel et al. (2018) did not find a clear relationship between bacterial cell size and temperature, possibly due to elevated protozoan grazing masking the potential effects of temperature.

One important consideration is the critical role of protozoans as a vital food source for the larvae of various fish species (Zingel et al., 2012). Planktivorous fish determine prey density based on the "reactive distance," which refers to the distance at which a foraging fish can spot the prey item. Smaller prey items have a shorter reactive distance compared to larger items. As a result, fish may not perceive any difference in densities when larger prey items are significantly less abundant than smaller ones. For instance, if the reactive distance for one prey item is twice as high as for another, the volume of water in which a fish can spot the prey could be up to ten times larger for the prey species with the greater reactive distance (Gliwicz, 2001). The water volume in which the prey item can be seen by an individual fish is termed as the "reactive field volume," visualized as a sphere with the reactive distance as its radius (Wetterer & Bishop, 1985) (Fig. 8). Hypothetically, if a heatwave occurs and the cell size of planktonic protozoans diminishes to a level that reduces the larval fish's reactive distance by half (2: 1), the larval fish would perceive a relative tenfold drop in their prey densities (10: 1). In extreme cases, this could lead to larval malnutrition and should be thoroughly investigated in the future. These factors have the potential to significantly impact food web functioning but are not well understood in Amazonian communities and food webs.

8 Conclusions

A substantial amount of research has been conducted in studying the food chains of Amazonian water bodies. Many important aspects have been addressed. However, there are still many areas in which our knowledge just starts to arise. Therefore, it is important to be aware of and dedicate even more resources to study these areas in the future. This is especially crucial in the conditions of accelerating climate change.

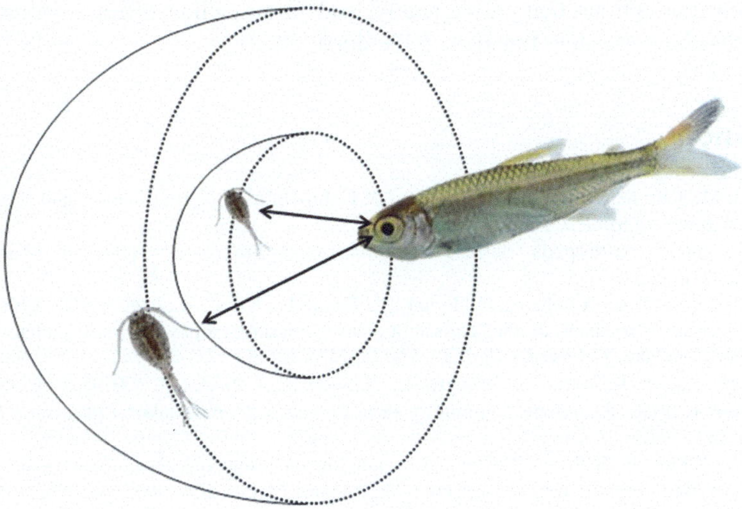

Fig. 8 Reactive distance and the reactive field volume of a planktivorous fish for two different-sized zooplankters. The fish perceives a relative tenfold difference in these zooplankter densities

Without understanding the fundamental processes, it will be very difficult to combat undesirable changes around us. Clearly, there is a need for a wider understanding of the entire food web dynamics in the face of ongoing environmental changes. Although our understanding is incomplete in several areas, a discernible pattern suggests that if immediate action is not taken to address climate and land-use change, the food chains in Amazonian water bodies will undergo inevitable changes, potentially leading to unfavourable outcomes for humans. This is particularly evident in the projected decline in future fish catches affecting the food security of indigenous and traditional populations who rely on fish as their main income and main origin of animal protein (Begossi et al., 2018; Forsberg et al., 2017).

A significant portion of the literature on freshwater ecosystems tends to focus on individual species or limited groups, resulting in a lack of comprehensive studies on entire communities encompassing more than two trophic levels. This is also the case in Amazonian food web research. Studies on fish and fish stocks in freshwater habitats often occur in isolation, as does research on DOC degradation by bacteria. The current focus is often on specific chains or segments of the food web. Environmental changes due to climate and land use are impacting aquatic food webs in various ways (bottom-up and top-down control), highlighting the necessity for a broader understanding of the dynamics of the entire food web in the Amazon. By solely examining the function of a section of the food web, there is a risk of biasing the comprehension of the entire system's response to ongoing environmental stressors. Conducting wider food web analyses is a significant challenge, yet it is essential for evaluating the ecosystem's response to current climate and land-use changes, which often act synergistically. Evidence of altered species ranges and abundances, along with the reshuffling of biological communities (Röpke et al., 2017), indicates ecosystem disruptions that subsequently lead to cascading effects, restructuring food webs and ecosystem functions (Oberdorff, 2022).

References

Adeney, J. M., Christensen, N. L., Vicentini, A., & Cohn-Haft, M. (2016). White-sand ecosystems in Amazonia. *Biotropica, 48*(1), 7–23.

Adis, J. (2007). Arthropods (terrestrial), Amazonian. In *Encyclopedia of biodiversity*. Academic Press.

Agasild, H., Kisand, A., Ainelo, E., Feldmann, T., Timm, H., Karus, K., et al. (2018). Chironomid incorporation of methane-derived carbon in plankton-and macrophyte-dominated habitats in a large shallow lake. *Freshwater Biology, 63*(11), 1433–1445.

Agasild, H., Zingel, P., Karus, K., Kangro, K., Salujoe, J., & Nõges, T. (2013). Does metazooplankton regulate the ciliate community in a shallow eutrophic lake? *Freshwater Biology, 58*(1), 183–191.

Agasild, H., Zingel, P., & Nõges, T. (2012). Live labeling technique reveals contrasting role of crustacean predation on microbial loop in two large shallow lakes. *Hydrobiologia, 684*, 177–187.

Agasild, H., Zingel, P., Tuvikene, L., Tuvikene, A., Timm, H., Feldmann, T., et al. (2014). Biogenic methane contributes to the food web of a large, shallow lake. *Freshwater Biology, 59*(2), 272–285.

Agostinho, A. A., Gomes, L. C., Santos, N. C., Ortega, J. C., & Pelicice, F. M. (2016). Fish assemblages in Neotropical reservoirs: Colonization patterns, impacts and management. *Fisheries Research, 173*, 26–36.

Alcantara-Rodriguez, M., Françozo, M., & Andel, T. v. (2019). Plant knowledge in the Historia Naturalis Brasiliae (1648): Retentions of seventeenth-century plant use in Brazil. *Economic Botany, 73*, 390–404.

Almeida, R. M., Hamilton, S. K., Rosi, E. J., Barros, N., Doria, C. R., Flecker, A. S., et al. (2020). Hydropeaking operations of two run-of-river mega-dams alter downstream hydrology of the largest Amazon tributary. *Frontiers in Environmental Science, 8*, 120.

Almeida-Val, V. M. F., Val, A. L., Duncan, W. P., Souza, F. C., Paula-Silva, M. N., & Land, S. (2000). Scaling effects on hypoxia tolerance in the Amazon fish *Astronotus ocellatus* (Perciformes: Cichlidae): Contribution of tissue enzyme levels. *Comparative Biochemistry and Physiology Part B: Biochemistry and Molecular Biology, 125*(2), 219–226.

Alves, L. C. P. D. S., Zappes, C. A., & Andriolo, A. (2012). Conflicts between river dolphins (Cetacea: Odontoceti) and fisheries in the Central Amazon: A path toward tragedy? *Zoologia (Curitiba), 29*, 420–429.

Amaral, J. H. F., Suhett, A. L., Melo, S., & Farjalla, V. F. (2013). Seasonal variation and interaction of photodegradation and microbial metabolism of DOC in black water Amazonian ecosystems. *Aquatic Microbial Ecology, 70*(2), 157–168.

AmphibiaWeb. (2023). AmphibiaWeb. https://amphibiaweb.org

Andrade, M. C., Fitzgerald, D. B., Winemiller, K. O., Barbosa, P. S., & Giarrizzo, T. (2019). Trophic niche segregation among herbivorous serrasalmids from rapids of the lower Xingu River, Brazilian Amazon. *Hydrobiologia, 829*, 265–280.

Andrade, M. C., Keppeler, F. W., Zuluaga-Gómez, M. A., Conceição, J. W., Lisboa, O. P., Andrades, R., et al. (2023). Food web structure in the Xingu River rapids prior to operation of the Amazon's largest hydropower plant. *Aquatic Sciences, 85*(3), 74.

Angelini, R. (2019). Aquatic food webs will need to stand on giants' shoulders. *Acta Limnologica Brasiliensia, 31*, e108.

Anjos, M. B. D., De Oliveira, R. R., & Zuanon, J. (2008). Hypoxic environments as refuge against predatory fish in the Amazonian floodplains. *Brazilian Journal of Biology, 68*, 45–50.

Antonelli, A., Smith, R. J., & Simmonds, M. S. J. (2019). Unlocking the properties of plants and fungi for sustainable development. *Nature Plants, 5*, 1100–1102.

Aranguren-Riaño, N. J., Guisande, C., Shurin, J. B., Jones, N. T., Barreiro, A., & Duque, S. R. (2018). Amino acid composition reveals functional diversity of zooplankton in tropical lakes related to geography, taxonomy and productivity. *Oecologia, 187*, 719–730.

Arantes, C. C., Fitzgerald, D. B., Hoeinghaus, D. J., & Winemiller, K. O. (2019). Impacts of hydroelectric dams on fishes and fisheries in tropical rivers through the lens of functional traits. *Current Opinion in Environmental Sustainability, 37*, 28–40.

Arantes, C. C., Laufer, J., Pinto, M. D. D. S., Moran, E. F., Lopez, M. C., Dutka-Gianelli, J., et al. (2022). Functional responses of fisheries to hydropower dams in the Amazonian Floodplain of the Madeira River. *Journal of Applied Ecology, 59*(3), 680–692.

Arantes, C. C., Winemiller, K. O., Asher, A., Castello, L., Hess, L. L., Petrere, M., Jr., & Freitas, C. E. (2019). Floodplain land cover affects biomass distribution of fish functional diversity in the Amazon River. *Scientific Reports, 9*(1), 16684.

Arantes, C. C., Winemiller, K. O., Petrere, M., Castello, L., Hess, L. L., & Freitas, C. E. (2018). Relationships between forest cover and fish diversity in the Amazon River floodplain. *Journal of Applied Ecology, 55*(1), 386–395.

Arantes, C. C., Winemiller, K. O., Petrere, M., & Freitas, C. E. (2019). Spatial variation in aquatic food webs in the Amazon River floodplain. *Freshwater Science, 38*(1), 213–228.

Araujo-Lima, C. A., Forsberg, B. R., Victoria, R., & Martinelli, L. (1986). Energy sources for detritivorous fishes in the Amazon. *Science, 234*(4781), 1256–1258.

Arrington, D. A., Winemiller, K. O., & Layman, C. A. (2005). Community assembly at the patch scale in a species rich tropical river. *Oecologia, 144*, 157–167.

Atkinson, D. (1995). Effects of temperature on the size of aquatic ectotherms: Exceptions to the general rule. *Journal of Thermal Biology, 20*(1–2), 61–74.

Atkinson, D., Ciotti, B. J., & Montagnes, D. J. (2003). Protists decrease in size linearly with temperature: Ca. 2.5% C− 1. Proceedings of the Royal Society of London. *Series B: Biological Sciences, 270*(1533), 2605–2611.

Avila-Pires, T. C. & Ramalho, W. P. Censo da Biodiversidade da Amazônia Brasileira—MPEG: Lagartos http://www.museugoeldi.br/censo/. Viewed 28 Mar 2019.

Azam, F., Fenchel, T., Field, J. G., Gray, J. S., Meyer-Reil, L. A., & Thingstad, F. J. M. E. P. S. (1983). The ecological role of water-column microbes in the sea. Marine ecology progress series. *Oldendorf, 10*(3), 257–263.

Azevedo-Silva, C. E., Almeida, R., Carvalho, D. P., Ometto, J. P., de Camargo, P. B., Dorneles, P. R., et al. (2016). Mercury biomagnification and the trophic structure of the ichthyofauna from a remote lake in the Brazilian Amazon. *Environmental Research, 151*, 286–296.

Bagley, J. E., Desai, A. R., Harding, K. J., Snyder, P. K., & Foley, J. A. (2014). Drought and deforestation: Has land cover change influenced recent precipitation extremes in the Amazon? *Journal of Climate, 27*(1), 345–361.

Bao, F., Assis, M. A., Arruda, R., & Pott, A. (2015). Effects of *Urochloa humidicola* on plant diversity in native grasslands in a Neotropical wetland. *Wetlands, 35*, 841–850.

Baranyi, C., Hein, T., Holarek, C., Keckeis, S., & Schiemer, F. (2002). Zooplankton biomass and community structure in a Danube River floodplain system: Effects of hydrology. *Freshwater Biology, 47*(3), 473–482.

Barbosa, P. M., Farjalla, V. F., Melack, J. M., Amaral, J. H. F., da Silva, J. S., & Forsberg, B. R. (2018). High rates of methane oxidation in an Amazon floodplain lake. *Biogeochemistry, 137*, 351–365.

Barichivich, J., Gloor, E., Peylin, P., Brienen, R. J., Schöngart, J., Espinoza, J. C., & Pattnayak, K. C. (2018). Recent intensification of Amazon flooding extremes driven by strengthened Walker circulation. *Science Advances, 4*(9), eaat8785.

Barneche, D. R., & Allen, A. P. (2018). The energetics of fish growth and how it constrains food-web trophic structure. *Ecology Letters, 21*(6), 836–844.

Barneche, D. R., Jahn, M., & Seebacher, F. (2019). Warming increases the cost of growth in a model vertebrate. *Functional Ecology, 33*(7), 1256–1266.

Barros, M. L. C., da Silva, T. D., da Cruz, A. G. B., & Rosman, P. C. C. (2020). Numerical simulation of wetland hydrodynamics and water quality. *Journal of the Brazilian Society of Mechanical Sciences and Engineering, 42*, 1–15.

Barros, G. G., Mascarenhas, B., Zuanon, J., & de Deus, C. P. (2024). Impacts of climate change and local disturbance on stream fish assemblages in the Amazon. *The Future of Amazonian Aquatic Biota*. (in press).

Bartley, T. J., McCann, K. S., Bieg, C., Cazelles, K., Granados, M., Guzzo, M. M., et al. (2019). Food web rewiring in a changing world. *Nature Ecology & Evolution, 3*(3), 345–354.

Bass, M. S., Finer, M., Jenkins, C. N., Kreft, H., Cisneros-Heredia, D. F., McCracken, S. F., et al. (2010). Global conservation significance of Ecuador's Yasuní National Park. *PLoS One, 5*(1), e8767.

Bastviken, D., Cole, J. J., Pace, M. L., & Van de Bogert, M. C. (2008). Fates of methane from different lake habitats: Connecting whole-lake budgets and CH_4 emissions. *Journal of Geophysical Research: Biogeosciences, 113*(G2).

Bastviken, D., Santoro, A. L., Marotta, H., Pinho, L. Q., Calheiros, D. F., Crill, P., & Enrich-Prast, A. (2010). Methane emissions from Pantanal, South America, during the low water season: Toward more comprehensive sampling. *Environmental Science & Technology, 44*(14), 5450–5455.

Begossi, A., Salivonchyk, S. V., Hallwass, G., Hanazaki, N., Lopes, P. F. M., Silvano, R. A. M., et al. (2018). Fish consumption on the Amazon: A review of biodiversity, hydropower and food security issues. *Brazilian Journal of Biology, 79*, 345–357.

Benedito-Cecilio, E., & Araujo-Lima, C. A. R. M. (2002). Variation in the carbon isotope composition of *Semaprochilodus insignis*, a detritivorous fish associated with oligotrophic and eutrophic Amazonian rivers. *Journal of Fish Biology, 60*(6), 1603–1607.

Benedito-Cecilio, E., Araujo-lima, C. M., Forsberg, B. R., Bittencourt, M. M., & Martinelli, L. C. (2000). Carbon sources of Amazonian fisheries. *Fisheries Management and Ecology, 7*(4), 305–314.

Berenguer, E., Lennox, G. D., Ferreira, J., Malhi, Y., Aragão, L. E., Barreto, J. R., et al. (2021). Tracking the impacts of El Niño drought and fire in human-modified Amazonian forests. *Proceedings of the National Academy of Sciences, 118*(30), e2019377118.

Bogotá-Gregory, J. D., Lima, F. C., Correa, S. B., Silva-Oliveira, C., Jenkins, D. G., Ribeiro, F. R., et al. (2020). Biogeochemical water type influences community composition, species richness, and biomass in megadiverse Amazonian fish assemblages. *Scientific Reports, 10*(1), 15349.

Boschilia, S. M., Oliveira, E. F., & Thomaz, S. M. (2008). Do aquatic macrophytes co-occur randomly? An analysis of null models in a tropical floodplain. *Oecologia, 156*, 203–214.

Bouvy, M., Nascimento, S. M., Molica, R. J. R., Ferreira, A., Huszar, V., & Azevedo, S. M. F. O. (2003). Limnological features in Tapacuá Reservoir (Northeast Brazil) during a severe drought. *Hydrobiologia, 493*, 115–130.

Brett, M. T., Bunn, S. E., Chandra, S., Galloway, A. W., Guo, F., Kainz, M. J., et al. (2017). How important are terrestrial organic carbon inputs for secondary production in freshwater ecosystems? *Freshwater Biology, 62*(5), 833–853.

Brum, S., Rosas-Ribeiro, P., Amaral, R. D. S., de Souza, D. A., Castello, L., & da Silva, V. M. F. (2021). Conservation of Amazonian aquatic mammals. *Aquatic Conservation: Marine and Freshwater Ecosystems, 31*(5), 1068–1086.

Burks, R. L., Mulderij, G., Gross, E., Jones, I., Jacobsen, L., Jeppesen, E., & Van Donk, E. (2006). *Center stage: The crucial role of macrophytes in regulating trophic interactions in shallow lake wetlands* (pp. 37–59). Springer.

Butzin, M., & Pörtner, H. O. (2016). Thermal growth potential of Atlantic cod by the end of the 21st century. *Global Change Biology, 22*(12), 4162–4168.

Callieri, C., Pugnetti, A., & Manca, M. (1999). Carbon partitioning in the food web of a high mountain lake: From bacteria to zooplankton. *Journal of Limnology, 58*(2), 144–151.

Campos-Silva, J. V., Peres, C. A., Amaral, J. H. F., Sarmento, H., Forsberg, B., & Fonseca, C. R. (2021). Fisheries management influences phytoplankton biomass of Amazonian floodplain lakes. *Journal of Applied Ecology, 58*(4), 731–743.

Capitani, L., Angelini, R., Keppeler, F. W., Hallwass, G., & Silvano, R. A. M. (2021). Food web modeling indicates the potential impacts of increasing deforestation and fishing pressure in the Tapajós River, Brazilian Amazon. *Regional Environmental Change, 21*(2), 42.

Caraballo, P., Forsberg, B. R., & Leite, R. G. (2014). Seasonal variation in the distribution and isotopic composition of phytoplankton in an Amazon floodplain lake, Brazil. *Acta Biológica Colombiana, 19*(2), 291–303.

Caraballo, P., Forsberg, B., & Leite, R. (2016). Abundance and isotopic composition of planktonic microcrustaceans in a Central Amazon floodplain lake: Implications for the trophic dynamics of the plankton community. *Caldasia, 38*(1), 149–164.

Cardoso, D., Särkinen, T., Alexander, S., et al. (2017). Amazon plant diversity revealed by a taxonomically verified species list. *Proceedings of the National Academy of Sciences, 114*, 10695–10700.

Casatti, L., Mendes, H. F., & Ferreira, K. M. (2003). Aquatic macrophytes as feeding site for small fishes in the Rosana reservoir, Paranapanema River, Southeastern Brazil. *Brazilian Journal of Biology, 63*, 213–222.

Casper, P. (1992). Methane production in lakes of different trophic state. *Arch Hydrobiol Beih Ergebn Limnol, 37*, 149–154.

Castello, L., McGrath, D. G., Hess, L. L., Coe, M. T., Lefebvre, P. A., Petry, P., Macedo, M. N., Renó, V. F., & Arantes, C. C. (2013). The vulnerability of Amazon freshwater ecosystems. *Conservation Letters, 6*, 217–229.

Castillo, M. M. (2000). Influence of hydrological seasonality on bacterioplankton in two neotropical floodplain lakes. *Hydrobiologia, 437*, 57–69.

Chaparro, G., Fontanarrosa, M. S., Schiaffino, M. R., de Tezanos Pinto, P., & O'Farrell, I. (2014). Seasonal-dependence in the responses of biological communities to flood pulses in warm temperate floodplain lakes: Implications for the "alternative stable states" model. *Aquatic Sciences, 76*, 579–594.

Chaparro, G., Marinone, M. C., Lombardo, R. J., Schiaffino, M. R., de Souza Guimarães, A., & O'Farrell, I. (2011). Zooplankton succession during extraordinary drought–flood cycles: A case study in a South American floodplain lake. *Limnologica, 41*(4), 371–381.

Chapman, L. J., Chapman, C. A., Nordlie, F. G., & Rosenberger, A. E. (2002). Physiological refugia: Swamps, hypoxia tolerance and maintenance of fish diversity in the Lake Victoria region. *Comparative Biochemistry and Physiology Part A: Molecular & Integrative Physiology, 133*(3), 421–437.

Chaudhari, S., Pokhrel, Y., Moran, E., & Miguez-Macho, G. (2019). Multi-decadal hydrologic change and variability in the Amazon River basin: Understanding terrestrial water storage variations and drought characteristics. *Hydrology and Earth System Sciences, 23*(7), 2841–2862.

Cianfrani, C., Broennimann, O., Loy, A., & Guisan, A. (2018). More than range exposure: Global otter vulnerability to climate change. *Biological Conservation, 221*, 103–113.

Clark, K. E., Stallard, R. F., Murphy, S. F., Scholl, M. A., González, G., Plante, A. F., & McDowell, W. H. (2022). Extreme rainstorms drive exceptional organic carbon export from forested humid-tropical rivers in Puerto Rico. *Nature Communications, 13*(1), 2058.

Comte, L., Buisson, L., Daufresne, M., & Grenouillet, G. (2013). Climate-induced changes in the distribution of freshwater fish: Observed and predicted trends. *Freshwater Biology, 58*(4), 625–639.

Conrad, R., Ji, Y., Noll, M., Klose, M., Claus, P., & Enrich-Prast, A. (2014). Response of the methanogenic microbial communities in Amazonian oxbow lake sediments to desiccation stress. *Environmental Microbiology, 16*(6), 1682–1694.

Conrad, R., Noll, M., Claus, P., Klose, M., Bastos, W. R., & Enrich-Prast, A. (2011). Stable carbon isotope discrimination and microbiology of methane formation in tropical anoxic lake sediments. *Biogeosciences, 8*(3), 795–814.

Constantine, J. A., Dunne, T., Ahmed, J., Legleiter, C., & Lazarus, E. D. (2014). Sediment supply as a driver of river meandering and floodplain evolution in the Amazon Basin. *Nature Geoscience, 7*(12), 899–903.

Correa, S. B., Costa-Pereira, R., Fleming, T., Goulding, M., & Anderson, J. T. (2015). Neotropical fish–fruit interactions: Eco-evolutionary dynamics and conservation. *Biological Reviews, 90*(4), 1263–1278.

Correa, S. B., Van Der Sleen, P., Siddiqui, S. F., Bogotá-Gregory, J. D., Arantes, C. C., Barnett, A. A., et al. (2022). Biotic indicators for ecological state change in Amazonian floodplains. *Bioscience, 72*(8), 753–768.

Correa, S. B., & Winemiller, K. O. (2014). Niche partitioning among frugivorous fishes in response to fluctuating resources in the Amazonian floodplain forest. *Ecology, 95*(1), 210–224.

Correa, S. B., & Winemiller, K. (2018). Terrestrial–aquatic trophic linkages support fish production in a tropical oligotrophic river. *Oecologia, 186*(4), 1069–1078.

Costa Bonecker, C., Da Costa, C. L., Velho, L. F. M., & Lansac-To^ha, F. A. (2005). Diversity and abundance of the planktonic rotifers in different environments of the Upper Paraná River floodplain (Paraná State—Mato Grosso do Sul State, Brazil). *Hydrobiologia, 546*, 405–414.

Costa, M. H., Botta, A., & Cardille, J. A. (2003). Effects of large-scale changes in land cover on the discharge of the Tocantins River, Southeastern Amazonia. *Journal of Hydrology, 283*(1–4), 206–217.

Crema, L. C., da Silva, V. M. F., Botta, S., Trumbore, S., & Piedade, M. T. F. (2019). Does water type influence diet composition in Amazonian manatee (Trichechus inunguis)? A case study comparing black and clearwater rivers. *Hydrobiologia, 835*, 1–19.

Dala-Corte, R. B., Melo, A. S., Siqueira, T., Bini, L. M., Martins, R. T., Cunico, A. M., et al. (2020). Thresholds of freshwater biodiversity in response to riparian vegetation loss in the Neotropical region. *Journal of Applied Ecology, 57*(7), 1391–1402.

Dalmagro, H. J., Lathuillière, M. J., Sallo, F. D. S., Guerreiro, M. F., Pinto, O. B., Jr., de Arruda, P. H., et al. (2019). Streams with riparian forest buffers versus impoundments differ in discharge and DOM characteristics for pasture catchments in Southern Amazonia. *Water, 11*(2), 390.

Dalmagro, H. J., Zanella de Arruda, P. H., Vourlitis, G. L., Lathuilliere, M. J. d. S., Nogueira, J., Couto, E. G., & Johnson, M. S. (2019). Radiative forcing of methane fluxes offsets net carbon dioxide uptake for a tropical flooded forest. *Global Change Biology, 25*(6), 1967–1981.

Daufresne, M., Lengfellner, K., & Sommer, U. (2009). Global warming benefits the small in aquatic ecosystems. *Proceedings of the National Academy of Sciences, 106*(31), 12788–12793.

de Azevedo, F., & Costa Bonecker, C. (2003). Community size structure of zooplanktonic assemblages in three lakes on the upper River Paraná floodplain, PR-MS, Brazil. *Hydrobiologia, 505*, 147–158.

de Carvalho Freitas, C. E., Rivas, A. A., Campos, C. P., Sant'Ana, I., Kahn, J. R., de Almeida Correa, M. A., & Catarino, M. F. (2012). The potential impacts of global climatic changes and dams on Amazonian fish and their fisheries. In *New advances and contributions to fish biology*. IntechOpen.

de Melo, M. L., Kothawala, D. N., Bertilsson, S., Amaral, J. H., Forsberg, B., & Sarmento, H. (2020). Linking dissolved organic matter composition and bacterioplankton communities in an Amazon floodplain system. *Limnology and Oceanography, 65*(1), 63–76.

Deines, P., Grey, J., Richnow, H. H., & Eller, G. (2007). Linking larval chironomids to methane: Seasonal variation of the microbial methane cycle and chironomid δ13C. *Aquatic Microbial Ecology, 46*(3), 273–282.

Deng, D., Zhang, S., Li, Y., Meng, X., Yang, W., Li, Y., & Li, X. (2010). Effects of Microcystis aeruginosa on population dynamics and sexual reproduction in two Daphnia species. *Journal of Plankton Research, 32*(10), 1385–1392.

Dias-Silva, K., Brasil, L. S., Veloso, G. K. O., Cabette, H. S. R., & Juen, L. (2020). Land use change causes environmental homogeneity and low beta-diversity in Heteroptera of streams. In *Annales de Limnologie-international journal of limnology* (Vol. 56, p. 9). EDP Sciences.

Domenici, P., Allan, B. J., Lefrançois, C., & McCormick, M. I. (2019). The effect of climate change on the escape kinematics and performance of fishes: Implications for future predator–prey interactions. *Conservation Physiology, 7*(1), coz078.

Drake, T. W., Van Oost, K., Barthel, M., Bauters, M., Hoyt, A. M., Podgorski, D. C., et al. (2019). Mobilization of aged and biolabile soil carbon by tropical deforestation. *Nature Geoscience, 12*(7), 541–546.
Elsner, R. (1999). Living in water: Solutions to physiological problems. In J. E. Reynolds & S. A. Rommel (Eds.), *Biology of marine mammals* (pp. 73–117). Smithsonian Institution Press.
Encalada, A. C., Flecker, A. S., Poff, N. L., Suárez, E., Herrera-R, G. A., Ríos-Touma, B., et al. (2019). A global perspective on tropical montane rivers. *Science, 365*(6458), 1124–1129.
Engle, D. L., Melack, J. M., Doyle, R. D., & Fisher, T. R. (2008). High rates of net primary production and turnover of floating grasses on the Amazon floodplain: Implications for aquatic respiration and regional CO2 flux. *Global Change Biology, 14*(2), 369–381.
Espinoza, J. C., Ronchail, J., Marengo, J. A., & Segura, H. (2019). Contrasting North–South changes in Amazon wet-day and dry-day frequency and related atmospheric features (1981–2017). *Climate Dynamics, 52*(9–10), 5413–5430.
Espinoza Villar, J. C., Ronchail, J., Guyot, J. L., Cochonneau, G., Naziano, F., Lavado, W., et al. (2009). Spatio-temporal rainfall variability in the Amazon basin countries (Brazil, Peru, Bolivia, Colombia, and Ecuador). *International Journal of Climatology: A Journal of the Royal Meteorological Society, 29*(11), 1574–1594.
Fares, A. L. B., Nonato, F. A. D. S., & Michelan, T. S. (2020). New records of the invasive macrophyte, *Urochloa arrecta* extend its range to eastern Brazilian Amazon altered freshwater ecosystems. *Acta Amazonica, 50*, 133–137.
Farjalla, V. F., Azevedo, D. A., Esteves, F. A., Bozelli, R. L., Roland, F., & Enrich-Prast, A. (2006). Influence of hydrological pulse on bacterial growth and DOC uptake in a clear-water Amazonian lake. *Microbial Ecology, 52*, 334–344.
Feitosa, I. B., Huszar, V. L. M., Domingues, C. D., Appel, E., Paranhos, R., Almeida, R. M., et al. (2019). Plankton community interactions in an Amazonian floodplain lake, from bacteria to zooplankton. *Hydrobiologia, 831*, 55–70.
Fernandes, R., Gomes, L. C., Pelicice, F. M., & Agostinho, A. A. (2009). Temporal organization of fish assemblages in floodplain lagoons: The role of hydrological connectivity. *Environmental Biology of Fishes, 85*(2), 99–108.
Ferreira, L. V., Parolin, P., Matos, D. C., Cunha, D. A., Chaves, P. P., & Neckel, S. O. (2016). The effect of exotic grass *Urochloa decumbens* (Stapf) RD Webster (Poaceae) in the reduction of species richness and change of floristic composition of natural regeneration in the Floresta Nacional de Carajás, Brazil. *Anais da Academia Brasileira de Ciências, 88*, 589–597.
Ficke, A. D., Myrick, C. A., & Hansen, L. J. (2007). Potential impacts of global climate change on freshwater fisheries. *Reviews in Fish Biology and Fisheries, 17*, 581–613.
Filizola, N., & Guyot, J. L. (2009). Suspended sediment yields in the Amazon basin: An assessment using the Brazilian national data set. *Hydrological Processes: An International Journal, 23*(22), 3207–3215.
Flecker, A. S., Shi, Q., Almeida, R. M., Angarita, H., Gomes-Selman, J. M., García-Villacorta, R., et al. (2022). Reducing adverse impacts of Amazon hydropower expansion. *Science, 375*(6582), 753–760.
Fleischmann, A. S., Papa, F., Fassoni-Andrade, A., Melack, J. M., Wongchuig, S., Paiva, R. C. D., et al. (2022). How much inundation occurs in the Amazon River basin? *Remote Sensing of Environment, 278*, 113099.
Fleischmann, A. S., Papa, F., Hamilton, S. K., et al. (2023). Increased floodplain inundation in the Amazon since 1980. *Environmental Research Letters, 18*, 034024.
Forsberg, B. R., Melack, J. M., Dunne, T., Barthem, R. B., Goulding, M., Paiva, R. C., et al. (2017). The potential impact of new Andean dams on Amazon fluvial ecosystems. *PLoS One, 12*(8), e0182254.
Forster, J., Hirst, A. G., & Esteban, G. F. (2013). Achieving temperature-size changes in a unicellular organism. *The ISME Journal, 7*(1), 28–36.
Franzke, C. L. E., Lee, J. Y., O'Kane, T., et al. (2023). Extreme weather and climate events: Dynamics, predictability and ensemble simulations. *Asia-Pacific Journal of Atmospheric Sciences, 59*, 1–2.

Freitas, C., Hurd, L., Andrade, B., Meireles, M., & Siqueira-Souza, F. (2024). Anthropogenic climate change is a present and future stressor for fish communities in the Amazon basin. *The Future of Amazonian Aquatic Biota*. (in press).

Freitas, C. E., Siqueira-Souza, F. K., Humston, R., & Hurd, L. E. (2013). An initial assessment of drought sensitivity in Amazonian fish communities. *Hydrobiologia, 705*, 159–171.

Fry, B. (2006). *Stable isotope ecology* (Vol. 521, p. 318). Springer.

Gallivan, G. J., Best, R. C., & Kanwisher, J. W. (1983). Temperature regulation in the Amazonian manatee Trichechus inunguis. *Physiological Zoology, 56*(2), 255–262.

Gardner, J. L., Peters, A., Kearney, M. R., Joseph, L., & Heinsohn, R. (2011). Declining body size: A third universal response to warming? *Trends in Ecology & Evolution, 26*(6), 285–291.

Gentry, A. (1997). Regional overview: South America. In S. D. Davis, V. H. Heywood, O. Herrera-Macbryde, J. Villa-Lobos, & A. C. Hamilton (Eds.), *Centers of plant diversity. A guide and strategy for their conservation*.

Gliwicz, Z. (2001). Species-specific population-density thresholds in cladocerans? *Hydrobiologia, 442*, 291–300.

Gloor, M. R. J. W., Brienen, R. J., Galbraith, D., Feldpausch, T. R., Schöngart, J., Guyot, J. L., et al. (2013). Intensification of the Amazon hydrological cycle over the last two decades. *Geophysical Research Letters, 40*(9), 1729–1733.

Godinho, M. B. D. C., & Da Silva, F. R. (2018). The influence of riverine barriers, climate, and topography on the biogeographic regionalization of Amazonian anurans. *Scientific Reports, 8*(1), 3427.

Gomes, L. F., Vieira, L. C. G., de Souza, C. A., Bonnet, M. P., & de Almeida, A. N. (2020). Environmental controls on zooplankton during hydrological periods of flooding and flushing in an Amazonian floodplain lake. *Limnetica, 39*(1), 35–48.

Gontijo, J. B., Paula, F. S., Venturini, A. M., Yoshiura, C. A., Borges, C. D., Moura, J. M. S., et al. (2021). Not just a methane source: Amazonian floodplain sediments harbour a high diversity of methanotrophs with different metabolic capabilities. *Molecular Ecology, 30*(11), 2560–2572.

González-Bergonzoni, I., D'Anatro, A., Vidal, N., Stebniki, S., Tesitore, G., Silva, I., & Teixeira de Mello, F. (2019). Origin of fish biomass in a diverse subtropical river: An allochthonic-supported biomass increase following flood pulses. *Ecosystems, 22*, 1736–1753.

Gottsberger, G. (1978). Seed dispersal by fish in the inundated regions of Humaitá, Amazonia. *Biotropica*, 170–183.

Gouveia, N. D. A., Gherardi, D. F. M., & Aragão, L. E. O. C. (2019). The role of the Amazon River plume on the intensification of the hydrological cycle. *Geophysical Research Letters, 46*(21), 12221–12229.

Grey, J. (2016). The incredible lightness of being methane-fuelled: Stable isotopes reveal alternative energy pathways in aquatic ecosystems and beyond. *Frontiers in Ecology and Evolution, 4*, 8.

Guayasamin, J. M., Ribas, C. C., Carnaval, A. C., Carrillo, J. D., Hoorn, C., Lohmann, L. G., et al. (2021). *Chapter 2: Evolution of Amazonian biodiversity*. Amazon Assessment Report.

Guimarães Silva, R., Zenni, R. D., Rosse, V. P., Bastos, L. S., & Berg, E. v. d. (2020). Landscape-level determinants of the spread and impact of invasive grasses in protected areas. *Biological Invasions, 22*, 3083–3099.

Hanson, P. (2016). *Insects and other arthropods of tropical America*. Cornell University Press.

Hanson, R. S., & Hanson, T. E. (1996). Methanotrophic bacteria. *Microbiological Reviews, 60*(2), 439–471.

Havens, K. E., Beaver, J. R., & East, T. L. (2007). Plankton biomass partitioning in a eutrophic subtropical lake: Comparison with results from temperate lake ecosystems. *Journal of Plankton Research, 29*(12), 1087–1097.

Havens, K. E., Pinto-Coelho, R. M., Beklioğlu, M., Christoffersen, K. S., Jeppesen, E., Lauridsen, T. L., et al. (2015). Temperature effects on body size of freshwater crustacean zooplankton from Greenland to the tropics. *Hydrobiologia, 743*, 27–35.

Heerspink, B. P., Kendall, A. D., Coe, M. T., & Hyndman, D. W. (2020). Trends in streamflow, evapotranspiration, and groundwater storage across the Amazon Basin linked to changing precipitation and land cover. *Journal of Hydrology: Regional Studies, 32*, 100755.

Hess, L. L., Melack, J. M., Affonso, A. G., Barbosa, C., Gastil-Buhl, M., & Novo, E. M. (2015). Wetlands of the lowland Amazon basin: Extent, vegetative cover, and dual-season inundated area as mapped with JERS-1 synthetic aperture radar. *Wetlands, 35*, 745–756.

Hibbert, K., Glenn, E., Smith, T. M., & González-Cruz, J. E. (2023). Changes to sea surface temperatures and vertical wind shear and their influence on tropical cyclone activity in the Caribbean and the main developing region. *Atmosphere, 14*(6), 999.

Hubbell, S. P., He, F., Condit, R., Borda-de-Água, L., Kellner, J., & Ter Steege, H. (2008). How many tree species are there in the Amazon and how many of them will go extinct? *Proceedings of the National Academy of Sciences, 105*(supplement_1), 11498–11504.

Humphreys, A. M., Govaerts, R., Ficinski, S. Z., Lughadha, E. N., & Vorontsova, M. S. (2019). Global dataset shows geography and life form predict modern plant extinction and rediscovery. *Nature Ecology & Evolution, 3*, 1043–1047.

Humphries, P., & Baldwin, D. S. (2003). Drought and aquatic ecosystems: An introduction. *Freshwater Biology, 48*(7), 1141–1146.

Huszar, V. L., & Reynolds, C. S. (1997). Phytoplankton periodicity and sequences of dominance in an Amazonian flood-plain lake (Lago Batata, Pará, Brasil): Responses to gradual environmental change. *Hydrobiologia, 346*, 169–181.

Illing, B., Downie, A. T., Beghin, M., & Rummer, J. L. (2020). Critical thermal maxima of early life stages of three tropical fishes: Effects of rearing temperature and experimental heating rate. *Journal of Thermal Biology, 90*, 102582.

IPCC. (2014). Climate change 2014: Synthesis report. In Core Writing Team, R. K. Pachauri, & L. A. Meyer (Eds.), *Contribution of Working Groups I, II and III to the fifth assessment report of the intergovernmental panel on climate change*. Intergovernmental Panel on Climate Change.

IPCC. (2022). Climate change 2022: Impacts, adaptation, and vulnerability. In H.-O. Pörtner, D. C. Roberts, M. Tignor, E. S. Poloczanska, K. Mintenbeck, A. Alegría, M. Craig, S. Langsdorf, S. Löschke, V. Möller, A. Okem, & B. Rama (Eds.), *Contribution of Working Group II to the sixth assessment report of the intergovernmental panel on climate change* (3056 pp). Cambridge University Press. https://doi.org/10.1017/9781009325844

Izaguirre, I., Pizarro, H., de Tezanos Pinto, P., Rodríguez, P., O'Farrell, I., Unrein, F., & Gasol, J. M. (2010). Macrophyte influence on the structure and productivity of photosynthetic picoplankton in wetlands. *Journal of Plankton Research, 32*(2), 221–238.

Jansson, M., Bergström, A. K., Blomqvist, P., & Drakare, S. (2000). Allochthonous organic carbon and phytoplankton/bacterioplankton production relationships in lakes. *Ecology, 81*(11), 3250–3255.

Jenkins, C. N., Pimm, S. L., & Joppa, L. N. (2013). Global patterns of terrestrial vertebrate diversity and conservation. *Proceedings of the National Academy of Sciences, 110*(28), E2602–E2610.

Jepsen, D. B., & Winemiller, K. O. (2007). Basin geochemistry and isotopic ratios of fishes and basal production sources in four neotropical rivers. *Ecology of Freshwater Fish, 16*(3), 267–281.

Jetz, W., Thomas, G. H., Joy, J. B., Hartmann, K., & Mooerset, A. O. (2012). The global diversity of birds in space and time. *Nature, 491*, 444–448.

Jézéquel, C., Tedesco, P. A., Bigorne, R., Maldonado-Ocampo, J. A., Ortega, H., Hidalgo, M., et al. (2020). A database of freshwater fish species of the Amazon Basin. *Scientific Data, 7*(1), 96.

Ji, Y., Angel, R., Klose, M., Claus, P., Marotta, H., Pinho, L., et al. (2016). Structure and function of methanogenic microbial communities in sediments of Amazonian lakes with different water types. *Environmental Microbiology, 18*(12), 5082–5100.

Jiménez-Muñoz, J. C., Mattar, C., Barichivich, J., Santamaría-Artigas, A., Takahashi, K., Malhi, Y., et al. (2016). Record-breaking warming and extreme drought in the Amazon rainforest during the course of El Niño 2015–2016. *Scientific Reports, 6*(1), 33130.

Joetzjer, E., Douville, H., Delire, C., et al. (2013). Present-day and future Amazonian precipitation in global climate models: CMIP5 versus CMIP3. *Climate Dynamics, 41*, 2921–2936.

Jones, R. I., Carter, C. E., Kelly, A., Ward, S., Kelly, D. J., & Grey, J. (2008). Widespread contribution of methane-cycle bacteria to the diets of lake profundal chironomid larvae. *Ecology, 89*(3), 857–864.

Jones, R. I., & Grey, J. (2011). Biogenic methane in freshwater food webs. *Freshwater Biology, 56*(2), 213–229.

Jose, S. (1993). Composition and seasonality of planktonic rotifers in limnetic and littoral region of a floodplain lake (Parana River System). *Revue Hydro Biologietropicale, 26*, 53–64.

Júnior, C. D. S., Sarmento, H., de Miranda, F. P., Henrique-Silva, F., & Logares, R. (2020). Uncovering the genomic potential of the Amazon River microbiome to degrade rainforest organic matter. *Microbiome, 8*(1), 151.

Junk, W. J. (1984). Ecology of the várzea, floodplain of Amazonian whitewater rivers. In *The Amazon: Limnology and landscape ecology of a mighty tropical river and its basin* (pp. 215–243). Springer.

Junk, W. J. (2013). Current state of knowledge regarding South America wetlands and their future under global climate change. *Aquatic Sciences, 75*(1), 113–131.

Junk, W. J., Bayley, P. B., & Sparks, R. E. (1989). The flood pulse concept in river-floodplain systems. *Canadian Special Publication of Fisheries and Aquatic Sciences, 106*(1), 110–127.

Junk, W. J., & Piedade, M. T. F. (1993). Herbaceous plants of the Amazon floodplain near Manaus: Species diversity and adaptations to the flood pulse. *Amaz Limnol Oecologia Reg Syst Fluminis Amaz, 12*, 467–484.

Junk, W. J., Piedade, M. T. F., Schöngart, J., Cohn-Haft, M., Adeney, J. M., & Wittmann, F. (2011). A classification of major naturally-occurring Amazonian lowland wetlands. *Wetlands, 31*, 623–640.

Junk, W. J., Piedade, M. T., Wittmann, F., Schöngart, J., Parolin, P., & (Eds.). (2010). *Amazonian floodplain forests: Ecophysiology, biodiversity and sustainable management* (Vol. 210). Springer Science & Business Media.

Junk, W. J., Wittmann, F., Schöngart, J., & Piedade, M. T. (2015). A classification of the major habitats of Amazonian black-water river floodplains and a comparison with their white-water counterparts. *Wetlands Ecology and Management, 23*, 677–693.

Kankaala, P., Taipale, S., Li, L., & Jones, R. I. (2010). Diets of crustacean zooplankton, inferred from stable carbon and nitrogen isotope analyses, in lakes with varying allochthonous dissolved organic carbon content. *Aquatic Ecology, 44*, 781–795.

Karus, K., Paaver, T., Agasild, H., & Zingel, P. (2014). The effects of predation by planktivorous juvenile fish on the microbial food web. *European Journal of Protistology, 50*(2), 109–121.

Kemenes, A., Forsberg, B. R., & Melack, J. M. (2007). Methane release below a tropical hydroelectric dam. *Geophysical Research Letters, 34*(12).

Kemenes, A., Forsberg, B. R., & Melack, J. M. (2011). CO2 emissions from a tropical hydroelectric reservoir (Balbina, Brazil). *Journal of Geophysical Research, 116*, G03004.

Kraus, C. N., Bonnet, M. P., de Souza Nogueira, I., Morais Pereira Souza Lobo, M. T., da Motta Marques, D., Garnier, J., & Cardoso Galli Vieira, L. (2019). Unraveling flooding dynamics and nutrients' controls upon phytoplankton functional dynamics in Amazonian floodplain lakes. *Water, 11*(1), 154.

Lara, N. R. F., Marques, T. S., Montelo, K. M., De Ataídes, Á. G., Verdade, L. M., Malvásio, A., & De Camargo, P. B. (2012). A trophic study of the sympatric Amazonian freshwater turtles *Podocnemis unifilis* and *Podocnemis expansa* (Testudines, Podocnemidae) using carbon and nitrogen stable isotope analyses. *Canadian Journal of Zoology, 90*(12), 1394–1401.

Latrubesse, E. M., Arima, E. Y., Dunne, T., Park, E., Baker, V. R., d'Horta, F. M., et al. (2017). Damming the rivers of the Amazon basin. *Nature, 546*(7658), 363–369.

Latrubesse, E. M., d'Horta, F. M., Ribas, C. C., Wittmann, F., Zuanon, J., Park, E., et al. (2021). Vulnerability of the biota in riverine and seasonally flooded habitats to damming of Amazonian rivers. *Aquatic Conservation: Marine and Freshwater Ecosystems, 31*(5), 1136–1149.

Lauerwald, R., Regnier, P., Guenet, B., Friedlingstein, P., & Ciais, P. (2020). How simulations of the land carbon sink are biased by ignoring fluvial carbon transfers: A case study for the Amazon Basin. *One Earth, 3*(2), 226–236.

Laybourn-Parry, J. (1984). *A functional biology of free-living protozoa*. University of California Press.

Leite, R. G., Araújo-Lima, C. A. R. M., Victoria, R. L., & Martinelli, L. A. (2002). Stable isotope analysis of energy sources for larvae of eight fish species from the Amazon floodplain. *Ecology of Freshwater Fish, 11*(1), 56–63.

Lewinsohn, T. M., & Prado, P. I. (2005). How many species are there in Brazil? *Conservation Biology, 19*, 619–624.

Lopes, C. A., Benedito-Cecilio, E., & Martinelli, L. A. (2007). Variability in the carbon isotope signature of *Prochilodus lineatus* (Prochilodontidae, Characiformes) a bottom-feeding fish of the Neotropical region. *Journal of Fish Biology, 70*(6), 1649–1659.

Lopes, A., Crema, L. C., Demarchi, L. O., Ferreira, A. B., Santiago, I. N., Ríos-Villamizar, E. A., & Piedade, M. T. F. (2019). Herbáceas aquáticas em igapós de água preta dentro e fora de Unidades de Conservação no Estado do Amazonas. *Biodiversidade Brasileira, 2*, 45–62.

Lopes, A., Schöngart, J., Mori, G. B., Quaresma, A., Demarchi, L. O., Paes, S.G., Mano, G.B., Guterres-Pazin, M.G., Klein, V.P., & Piedade, M.T.F. (2024). Aquatic herbaceous plants of the Amazonian wetlands: Key species under pressure. *The Future of Amazonian Aquatic Biota*. (in press).

Loy, A., & Cianfrani, C. (2018). Otters' vulnerability to climate change. In N. Duplaix & M. Savage (Eds.), *The global otter conservation strategy* (pp. 125–136). IUCN/SSC Otter Specialist Group.

Lugo-Carvajal, A., Holmgren, M., Zuanon, J., & van der Sleen, P. (2023). Fish on fire: Shifts in Amazonian fish communities after floodplain forest fires. *Journal of Applied Ecology, 60*, 1637.

Magnusson, W. E., Carmozina de Araújo, M., Cintra, R., Lima, A. P., Martinelli, L. A., Sanaiotti, T. M., et al. (1999). Contributions of C 3 and C 4 plants to higher trophic levels in an Amazonian savanna. *Oecologia, 119*, 91–96.

Makrakis, M. C., Nakatani, K., Bialetzki, A., Sanches, P. V., Baumgartner, G., & Gomes, L. C. (2005). Ontogenetic shifts in digestive tract morphology and diet of fish larvae of the Itaipu Reservoir, Brazil. *Environmental Biology of Fishes, 72*, 99–107.

Manetta, G. I., Benedito-Cecilio, E., & Martinelli, M. (2003). Carbon sources and trophic position of the main species of fishes of Baía River, Paraná River floodplain, Brazil. *Brazilian Journal of Biology, 63*, 283–290.

Mano, G. B., Lopes, A., & Piedade, M. T. F. (2023). Will climate change favor exotic grasses over native ecosystem engineer species in the Amazon Basin? *Ecological Informatics, 75*, 102102.

Marani, L., & Alvalá, P. C. (2007). Methane emissions from lakes and floodplains in Pantanal, Brazil. *Atmospheric Environment, 41*(8), 1627–1633.

Marchese, M. R., Saigo, M., Zilli, F. L., Capello, S., Devercelli, M., Montalto, L., et al. (2014). Food webs of the Paraná River floodplain: Assessing basal sources using stable carbon and nitrogen isotopes. *Limnologica, 46*, 22–30.

Marengo, J. A., Ambrizzi, T., Rocha, R. P., Alves, L. M., Cuadra, S. V., Valverde, M. C., et al. (2010). Future change of climate in South America in the late twenty-first century: Intercomparison of scenarios from three regional climate models. *Climate Dynamics, 35*, 1073–1111.

Marengo, J. A., Cunha, A. P., Cuartas, L. A., Deusdará Leal, K. R., Broedel, E., Seluchi, M. E., et al. (2021). Extreme drought in the Brazilian Pantanal in 2019–2020: Characterization, causes, and impacts. *Frontiers in Water, 3*, 639204.

Marengo, J. A., & Espinoza, J. C. (2016). Extreme seasonal droughts and floods in Amazonia: Causes, trends and impacts. *International Journal of Climatology, 36*(3), 1033–1050.

Marengo, J. A., Nobre, C. A., Tomasella, J., Oyama, M. D., Sampaio de Oliveira, G., De Oliveira, R., et al. (2008). The drought of Amazonia in 2005. *Journal of Climate, 21*(3), 495–516.

Marengo, J. A., Souza, C. M., Jr., Thonicke, K., Burton, C., Halladay, K., Betts, R. A., et al. (2018). Changes in climate and land use over the Amazon region: Current and future variability and trends. *Frontiers in Earth Science, 6*, 228.

Marengo, J. A., Tomasella, J., Soares, W. R., Alves, L. M., & Nobre, C. A. (2012). Extreme climatic events in the Amazon basin. *Theoretical and Applied Climatology, 107*, 73–85.

Marsh, H., Arraut, E. M., Diagne, L. K., Edwards, H., & Marmontel, M. (2017). Impact of climate change and loss of habitat on sirenians. In *Marine mammal welfare: Human induced change in the marine environment and its impacts on marine mammal welfare* (pp. 333–357).

Martins, R. T., Firmino, V. C. & Hamada, N. (2024). Amazonian aquatic insects in a changing landscape. *The Future of Amazonian Aquatic Biota*. (in press).

Mayer, M., Da Fonte, L. F., & Lötters, S. (2019). Mind the gap! A review of Amazonian anurans in GenBank. *Salamandra, 55*(2).

Mayorga, E., Aufdenkampe, A. K., Masiello, C. A., Krusche, A. V., Hedges, J. I., Quay, P. D., et al. (2005). Young organic matter as a source of carbon dioxide outgassing from Amazonian rivers. *Nature, 436*(7050), 538–541.

McClain, M. E., & Naiman, R. J. (2008). Andean influences on the biogeochemistry and ecology of the Amazon River. *Bioscience, 58*(4), 325–338.

McConnell, R., & Lowe-McConnell, R. H. (1987). *Ecological studies in tropical fish communities*. Cambridge University Press.

McKenzie, D. J. (2011). Energetics of fish swimming. *Encyclopedia of Fish Physiology, 3*, 1636–1644.

Meira, B. R., Progênio, M., Leite, E. C., Lansac-Tôha, F. M., Durán, C. L. G., Jati, S., et al. (2021). Functional feeding groups of Protist ciliates (Protist: Ciliophora) on a neotropical flood plain. In *Annales de Limnologie-international journal of limnology* (Vol. 57, p. 13). EDP Sciences.

Melack, J. M., Basso, L. S., Fleischmann, A. S., Botía, S., Guo, M., Zhou, W., et al. (2022). Challenges regionalizing methane emissions using aquatic environments in the Amazon Basin as examples. *Frontiers in Environmental Science, 10*, 866082.

Melack, J. M., & Forsberg, B. R. (2001). *Biogeochemistry of Amazon floodplain. The biogeochemistry of the Amazon Basin* (Vol. 1, pp. 235–274). Oxford University Press.

Menezes, J. A., Confalonieri, U., Madureira, A. P., Duval, I. D. B., Santos, R. B. D., & Margonari, C. (2018). Mapping human vulnerability to climate change in the Brazilian Amazon: The construction of a municipal vulnerability index. *PLoS One, 13*(2), e0190808.

Milá, B., Tavares, E. S., Muñoz Saldaña, A., Karubian, J., Smith, T. B., & Baker, A. J. (2012). A trans-Amazonian screening of mtDNA reveals deep intraspecific divergence in forest birds and suggests a vast underestimation of species diversity. *PLoS One, 7*, e40541.

Mittermeier, R. A., Mittermeier, C. G., Brooks, T. M., Pilgrim, J. D., Konstant, W. R., Da Fonseca, G. A., & Kormos, C. (2003). Wilderness and biodiversity conservation. *Proceedings of the National Academy of Sciences, 100*(18), 10309–10313.

Mittermeier, R. A., Mittermeier, C. G., Pilgrim, J., et al. (2002). *Wilderness: Earth's last wild places*. CEMEX.

Monaco, C. J., Wethey, D. S., & Helmuth, B. (2016). Thermal sensitivity and the role of behavior in driving an intertidal predator–prey interaction. *Ecological Monographs, 86*(4), 429–447.

Montagnes, D. J., Dower, J. F., & Figueiredo, G. M. (2010). The protozooplankton–ichthyoplankton trophic link: An overlooked aspect of aquatic food webs 1. *Journal of Eukaryotic Microbiology, 57*(3), 223–228.

Moraes, R. M., Bibiana Correa, S., Rodrigues da Costa Doria, C., Duponchelle, F., Miranda, G., Montoya, M., Phillips, O. L., Salinas, N., Silman, M., Ulloa, C., Zapata-Ríos, G., Arieira, J., & ter Steege, H. (2021). Chapter 4: Biodiversity and ecological functioning in the Amazon. In *Amazon assessment report 2021*. Science Panel for the Amazon.

Morana, C., Borges, A. V., Roland, F. A. E., Darchambeau, F., Descy, J. P., & Bouillon, S. (2015). Methanotrophy within the water column of a large meromictic tropical lake (Lake Kivu, East Africa). *Biogeosciences, 12*(7), 2077–2088.

Moreira-Turcq, P., Bonnet, M. P., Amorim, M., Bernardes, M., Lagane, C., Maurice, L., et al. (2013). Seasonal variability in concentration, composition, age, and fluxes of particulate organic carbon exchanged between the floodplain and Amazon River. *Global Biogeochemical Cycles, 27*(1), 119–130.

Moreira-Turcq, P., Seyler, P., Guyot, J. L., & Etcheber, H. (2003). Exportation of organic carbon from the Amazon River and its main tributaries. *Hydrological Processes, 17*(7), 1329–1344.

Morris, C., Val, A. L., Brauner, C. J., & Wood, C. M. (2021). The physiology of fish in acidic waters rich in dissolved organic carbon, with specific reference to the Amazon basin: Ionoregulation, acid–base regulation, ammonia excretion, and metal toxicity. *Journal of Experimental Zoology Part A: Ecological and Integrative Physiology, 335*(9–10), 843–863.

Mortillaro, J. M., Abril, G., Moreira-Turcq, P., Sobrinho, R. L., Perez, M., & Meziane, T. (2011). Fatty acid and stable isotope ($\delta^{13}C$, $\delta^{15}N$) signatures of particulate organic matter in the lower Amazon River: Seasonal contrasts and connectivity between floodplain lakes and the mainstem. *Organic Geochemistry, 42*(10), 1159–1168.

Mortillaro, J. M., Passarelli, C., Abril, G., Hubas, C., Albéric, P., Artigas, L. F., et al. (2016). The fate of C_4 and C_3 macrophyte carbon in Central Amazon floodplain waters: Insights from a batch experiment. *Limnologica, 59*, 90–98.

Mortillaro, J. M., Pouilly, M., Wach, M., Freitas, C. E. C., Abril, G., & Meziane, T. (2015). Trophic opportunism of Central Amazon floodplain fish. *Freshwater Biology, 60*(8), 1659–1670.

Muneepeerakul, R., Bertuzzo, E., Lynch, H. J., Fagan, W. F., Rinaldo, A., & Rodriguez-Iturbe, I. (2008). Neutral metacommunity models predict fish diversity patterns in Mississippi–Missouri basin. *Nature, 453*(7192), 220–222.

Murase, J., & Frenzel, P. (2007). A methane-driven microbial food web in a wetland rice soil. *Environmental Microbiology, 9*(12), 3025–3034.

Myers, B. J., Lynch, A. J., Bunnell, D. B., Chu, C., Falke, J. A., Kovach, R. P., et al. (2017). Global synthesis of the documented and projected effects of climate change on inland fishes. *Reviews in Fish Biology and Fisheries, 27*, 339–361.

Nobre, C. A., Obregón, G. O., Marengo, J. A., Fu, R., & Poveda, G. (2009). Characteristics of Amazonian climate: Main features. *Amazonia and Global Change, 186*, 149–162.

Nores, M. (2000). Species richness in the Amazonian bird fauna from an evolutionary perspective. *Emu, 100*, 419–430.

O'Farrell, I., Izaguirre, I., Chaparro, G., Unrein, F., Sinistro, R., Pizarro, H., et al. (2011). Water level as the main driver of the alternation between a free-floating plant and a phytoplankton dominated state: A long-term study in a floodplain lake. *Aquatic Sciences, 73*, 275–287.

Oberdorff, T. (2022). Time for decisive actions to protect freshwater ecosystems from global changes. *Knowledge & Management of Aquatic Ecosystems, 423*, 19.

Oberdorff, T., Dias, M. S., Jézéquel, C., Albert, J. S., Arantes, C. C., Bigorne, R., et al. (2019). Unexpected fish diversity gradients in the Amazon basin. *Science Advances, 5*(9), eaav8681.

Oliveira, A. C. B., Soares, M. G. M., Martinelli, L. A., & Moreira, M. Z. (2006). Carbon sources of fish in an Amazonian floodplain lake. *Aquatic Sciences, 68*, 229–238.

Paerl, H. W., & Huisman, J. (2009). Climate change: A catalyst for global expansion of harmful cyanobacterial blooms. *Environmental Microbiology Reports, 1*(1), 27–37.

Paidere, J. (2009). Influence of Flooding Frequency on Zooplankton in the Floodplains of the Daugava River (Latvia). *Acta Zoologica Lituanica, 19*(4), 306–313.

Pangala, S. R., Enrich-Prast, A., Basso, L. S., Peixoto, R. B., Bastviken, D., Hornibrook, E. R., et al. (2017). Large emissions from floodplain trees close the Amazon methane budget. *Nature, 552*(7684), 230–234.

Patterson, B. D. (2001). Fathoming tropical biodiversity: The continuing discovery of Neotropical mammals. *Diversity and Distributions, 7*, 191–196.

Peer, A. C., & Miller, T. J. (2014). Climate change, migration phenology, and fisheries management interact with unanticipated consequences. *North American Journal of Fisheries Management, 34*(1), 94–110.

Perry, A. L., Low, P. J., Ellis, J. R., & Reynolds, J. D. (2005). Climate change and distribution shifts in marine fishes. *Science, 308*(5730), 1912–1915.

Piedade, M. T. F., Junk, W., D'Ângelo, S. A., Wittmann, F., Schöngart, J., Barbosa, K. M. D. N., & Lopes, A. (2010). Aquatic herbaceous plants of the Amazon floodplains: State of the art and research needed. *Acta Limnologica Brasiliensia, 22*, 165–178.

Pivello, V. R. (2011). Invasões biológicas no cerrado brasileiro: efeitos da introdução de espécies exóticas sobre a biodiversidade. *Ecologia Info, 33*.

Pörtner, H. O., & Knust, R. (2007). Climate change affects marine fishes through the oxygen limitation of thermal tolerance. *Science, 315*(5808), 95–97.

Pouilly, M., Yunoki, T., Rosales, C., & Torres, L. (2004). Trophic structure of fish assemblages from Mamoré River floodplain lakes (Bolivia). *Ecology of Freshwater Fish, 13*(4), 245–257.

Rafael, J. A., Aguiar, A. P., & Amorim, D. D. S. (2009). Knowledge of insect diversity in Brazil: Challenges and advances. *Neotropical Entomology, 38*, 565–570.
Raven, P. H., Gereau, R. E., Phillipson, P. B., Chatelain, C., Jenkins, C. N., & Ulloa, C. (2020). The distribution of biodiversity richness in the tropics. *Science Advances, 6*(37), eabc6228.
Ravinet, M., Syväranta, J., Jones, R. I., & Grey, J. (2010). A trophic pathway from biogenic methane supports fish biomass in a temperate lake ecosystem. *Oikos, 119*(2), 409–416.
Raymond, P. A., & Saiers, J. E. (2010). Event controlled DOC export from forested watersheds. *Biogeochemistry, 100*, 197–209.
Reis, R. E., Albert, J. S., Di Dario, F., Mincarone, M. M., Petry, P., & Rocha, L. A. (2016). Fish biodiversity and conservation in South America. *Journal of Fish Biology, 89*(1), 12–47.
Rejas, D. (2018). Trophic structure of a floodplain fish assemblage in the upper Amazon basin, Bolivia. *Revista de biologia tropical, 66*(3), 1258–1271.
Rejas, D., Muylaert, K., & De Meester, L. (2005). Trophic interactions within the microbial food web in a tropical floodplain lake (Laguna Bufeos, Bolivia). *Revista de Biología Tropical, 53*(1–2), 85–96.
Resende, A. F., Piedade, M. T., Feitosa, Y. O., Andrade, V. H. F., Trumbore, S. E., Durgante, F. M., et al. (2020). Flood-pulse disturbances as a threat for long-living Amazonian trees. *New Phytologist, 227*(6), 1790–1803.
Ribas, C. C., & Aleixo, A. (2019). Diversity and evolution of Amazonian birds: Implications for conservation and biogeography. *Anais da Academia Brasileira de Ciências, 91*, e20190218.
Ringeval, B., Houweling, S., Van Bodegom, P. M., Spahni, R., Van Beek, R., Joos, F., & Röckmann, T. (2014). Methane emissions from floodplains in the Amazon Basin: Challenges in developing a process-based model for global applications. *Biogeosciences, 11*(6), 1519–1558.
Roach, K. A. (2013). Environmental factors affecting incorporation of terrestrial material into large river food webs. *Freshwater Science, 32*(1), 283–298.
Robb, T., & Abrahams, M. V. (2002). The influence of hypoxia on risk of predation and habitat choice by the fathead minnow, Pimephales promelas. *Behavioral Ecology and Sociobiology, 52*, 25–30.
Röpke, C. P., Amadio, S., Zuanon, J., Ferreira, E. J., Deus, C. P. D., Pires, T. H., & Winemiller, K. O. (2017). Simultaneous abrupt shifts in hydrology and fish assemblage structure in a floodplain lake in the central Amazon. *Scientific Reports, 7*(1), 40170.
Sampaio, G., et al. (2019). Assessing the possible impacts of a 4 °C or higher warming in Amazonia. In C. Nobre, J. Marengo, & W. Soares (Eds.), *Climate change risks in Brazil*. Springer.
Sanches, L. F., Guenet, B., Marinho, C. C., Barros, N., & de Assis Esteves, F. (2019). Global regulation of methane emission from natural lakes. *Scientific Reports, 9*(1), 255.
Sanders, L. M., Taffs, K. H., Stokes, D. J., Sanders, C. J., Smoak, J. M., Enrich-Prast, A., et al. (2017). Carbon accumulation in Amazonian floodplain lakes: A significant component of Amazon budgets? *Limnology and Oceanography Letters, 2*(1), 29–35.
Sanseverino, A. M., Bastviken, D., Sundh, I., Pickova, J., & Enrich-Prast, A. (2012). Methane carbon supports aquatic food webs to the fish level. *PLoS One*, e42723.
Santos, J. A., Barroco, L. S. A., de Souza, F. K. S., & de Carvalho Freitas, C. E. (2017). Isótopos estáveis em estudos ecológicos com peixes na Amazônia Brasileira. *Scientia Amazonia, 6*(3), 119–127.
Santos, R. E., Pinto-Coelho, R. M., Drumond, M. A., Fonseca, R., & Zanchi, F. B. (2020). Damming Amazon Rivers: Environmental impacts of hydroelectric dams on Brazil's Madeira River according to local fishers' perception. *Ambio, 49*, 1612–1628.
Sawakuchi, H. O., Bastviken, D., Enrich-Prast, A., Ward, N. D., Camargo, P. B., & Richey, J. E. (2021). Low diffusive methane emissions from the main channel of a large Amazonian run-of-the-river reservoir attributed to high methane oxidation. *Frontiers in Environmental Science, 9*, 655455.
Scheele, B. C., Pasmans, F., Skerratt, L. F., Berger, L., Martel, A. N., Beukema, W., et al. (2019). Amphibian fungal panzootic causes catastrophic and ongoing loss of biodiversity. *Science, 363*(6434), 1459–1463.

Scheffer, M., Carpenter, S., Foley, J. A., Folke, C., & Walker, B. (2001). Catastrophic shifts in ecosystems. *Nature, 413*(6856), 591–596.

Scheffer, M., Szabo, S., Gragnani, A., Van Nes, E. H., Rinaldi, S., Kautsky, N., et al. (2003). Floating plant dominance as a stable state. *Proceedings of the National Academy of Sciences, 100*(7), 4040–4045.

Schöngart, J., & Junk, W. J. (2020). Clima e hidrologia nas várzeas da Amazônia Central. In *Várzeas Amazônicas: Desafios para um Manejo Sustentável* (pp. 44–65).

Schöngart, J., Lopes, A., Wittmann, F., Parolin, P., Klein, V., Reboucas, E., Junk, W., Ferreira, C., Alves, Y., da Silva Costa, G., Wiess, B., Resende, A. F., da Rocha, M., & Piedade, M. T. F. (2024). Fate of floodplain forests in a changing Amazon. *The Future of Amazonian Aquatic Biota*. (in press).

Schöngart, J., Wittmann, F., Faria de Resende, A., Assahira, C., de Sousa Lobo, G., Rocha Duarte Neves, J., et al. (2021). The shadow of the Balbina dam: A synthesis of over 35 years of downstream impacts on floodplain forests in Central Amazonia. *Aquatic Conservation: Marine and Freshwater Ecosystems, 31*(5), 1117–1135.

Schwamborn, R., & Giarrizzo, T. (2015). Stable isotope discrimination by consumers in a tropical mangrove food web: How important are variations in C/N ratio? *Estuaries and Coasts, 38*, 813–825.

Sheridan, J. A., & Bickford, D. (2011). Shrinking body size as an ecological response to climate change. *Nature Climate Change, 1*(8), 401–406.

Silva, T. S., Melack, J. M., & Novo, E. M. (2013). Responses of aquatic macrophyte cover and productivity to flooding variability on the Amazon floodplain. *Global Change Biology, 19*(11), 3379–3389.

Simões, N. R., Lansac-Tôha, F. A., Velho, L. F., & Bonecker, C. C. (2012). Intra and inter-annual structure of zooplankton communities in floodplain lakes: A long-term ecological research study. *Revista de Biología Tropical, 60*(4), 1819–1836.

Sioli, H. (1954). Beiträge zur regionalen Limnologie des Amazonasgebietes. *Archiv für Hydrobiologie, 45*, 267–283.

Sioli, H. (1984). The Amazon and its main affluents: Hydrography, morphology of the river courses, and river types. In *The Amazon: Limnology and landscape ecology of a mighty tropical river and its basin* (pp. 127–165). Springer.

Solokas, M. A., Feiner, Z. S., Al-Chokachy, R., Budy, P., DeWeber, J. T., Sarvala, J., et al. (2023). Shrinking body size and climate warming: Many freshwater salmonids do not follow the rule. *Global Change Biology, 29*(9), 2478–2492.

Sommaruga, R. (1995). Microbial and classical food webs: A visit to a hypertrophic lake. *FEMS Microbiology Ecology, 17*(4), 257–270.

Sousa, R. G. C., de Freitas, H. C. P., Zacardi, D. M., & Faria-Junior, C. H. (2021). Effects of river dams on the fish guilds in the northwest region of the Brazilian Amazon. *Fisheries Research, 243*, 106091.

Stork, N. E. (2018). How many species of insects and other terrestrial arthropods are there on earth? *Annual Review of Entomology, 63*, 31–45.

Stropp, J., Umbelino, B., Correia, R. A., Campos-Silva, J. V., Ladle, R. J., & Malhado, A. C. M. (2020). The ghosts of forests past and future: Deforestation and botanical sampling in the Brazilian Amazon. *Ecography, 43*(7), 979–989.

Sylvain, F. É., Bouslama, S., Holland, A., Leroux, N., Mercier, P. L., Val, A. L., & Derome, N. (2021). The Amazon River microbiome, a story of humic carbon. *bioRxiv, 2021-07*. https://doi.org/10.1101/2021.07.21.453257

Taipale, S., Kankaala, P., Hahn, M. W., Jones, R. I., & Tiirola, M. (2011). Methane-oxidizing and photoautotrophic bacteria are major producers in a humic lake with a large anoxic hypolimnion. *Aquatic Microbial Ecology, 64*(1), 81–95.

Taipale, S., Kankaala, P., Tiirola, M., & Jones, R. I. (2008). Whole-lake dissolved inorganic 13C additions reveal seasonal shifts in zooplankton diet. *Ecology, 89*(2), 463–474.

Tedesco, P. A., Beauchard, O., Bigorne, R., Blanchet, S., Buisson, L., Conti, L., et al. (2017). A global database on freshwater fish species occurrence in drainage basins. *Scientific Data, 4*(1), 1–6.

ter Steege, H., Prado, P. I., Lima, R. A. D., Pos, E., de Souza Coelho, L., de Andrade Lima Filho, D., et al. (2020). Biased-corrected richness estimates for the Amazonian tree flora. *Scientific Reports, 10*(1), 10130.

Thomaz, S. M., Bini, L. M., & Bozelli, R. L. (2007). Floods increase similarity among aquatic habitats in river-floodplain systems. *Hydrobiologia, 579*, 1–13.

Timpe, K., & Kaplan, D. (2017). The changing hydrology of a dammed Amazon. *Science Advances, 3*(11), e1700611.

Tirok, K., & Gaedke, U. (2007). Regulation of planktonic ciliate dynamics and functional composition during spring in Lake Constance. *Aquatic Microbial Ecology, 49*(1), 87–100.

Torres-Bejarano, A. M., Duque, S. R., & Caraballo, P. (2014). The trophic role of zooplankton in a floodplain lake of Colombian amazon, through stable isotopes analysis. *Caldasia, 36*(2), 331–344.

Tregidgo, D., Piperata, B. A., de Lima, J. J. S., et al. (2023). Variation in food and nutritional stability among Amazonian populations living in a context of dramatic seasonal flooding. *Human Ecology, 51*, 907–922.

Tribuzy-Neto, I. A., Conceição, K. G., Siqueira-Souza, F. K., Hurd, L. E., & Freitas, C. E. C. (2017). Condition factor variations over time and trophic position among four species of Characidae from Amazonian floodplain lakes: Effects of an anomalous drought. *Brazilian Journal of Biology, 78*, 337–344.

Trivinho-Strixino, S. (2019). Ordem Diptera. Família Chironomidae. Guia de identificação de larvas. In Hamada et al. (Eds.), *Taxonomia, Biologia e Ecologia*. Editora INPA.

Trotsenko, Y. A., & Khmelenina, V. N. (2005). Aerobic methanotrophic bacteria of cold ecosystems. *FEMS Microbiology Ecology, 53*(1), 15–26.

Unrein, F., O'Farrell, I., Izaguirre, I., Sinistro, R., dos Santos Afonso, M., & Tell, G. (2010). Phytoplankton response to pH rise in a N-limited floodplain lake: Relevance of N2-fixing heterocystous cyanobacteria. *Aquatic Sciences, 72*, 179–190.

Vacher, J. P., Chave, J., Ficetola, F. G., Sommeria-Klein, G., Tao, S., Thébaud, C., et al. (2020). Large-scale DNA-based survey of frogs in Amazonia suggests a vast underestimation of species richness and endemism. *Journal of Biogeography, 47*(8), 1781–1791.

Val, L. A., & Wood, C. M. (2022). Global change and physiological challenges for fish of the Amazon today and in the near future. *Journal of Experimental Biology, 225*(10), jeb216440.

van der Sleen, P., & Rams, M. (2023). Flood pulses and fish species coexistence in tropical rivers-a theoretical food web model. *Environmental Biology of Fishes, 106*(8), 1785–1796.

Vinagre, C., Leal, I., Mendonca, V., Madeira, D., Narciso, L., Diniz, M. S., & Flores, A. A. (2016). Vulnerability to climate warming and acclimation capacity of tropical and temperate coastal organisms. *Ecological Indicators, 62*, 317–327.

Vismara-Rizzato, M., Nogueira, C. A., Carvalho, T. B., & Val, A. (2024). Effect of eggs incubation of *Podocnemis expansa* in different climatic scenarios on the morphology of neonates. *The Future of Amazonian Aquatic Biota*. (in press).

Waichman, A. V. (1996). Autotrophic carbon sources for heterotrophic bacterioplankton in a floodplain lake of Central Amazon. *Hydrobiologia, 341*, 27–36.

Wantzen, K. M., de Arruda Machado, F., Voss, M., Boriss, H., & Junk, W. J. (2002). Seasonal isotopic shifts in fish of the Pantanal wetland, Brazil. *Aquatic Sciences, 64*, 239–251.

Weiss, B., Santana, F. D., Petene Calvi, G., Costa, G., Zuanon, J., & Piedade, M. T. F. (2024). Effectiveness of fish assemblage as seed dispersers in Amazon oligotrophic flooded forests. *Austral Ecology, 49*(1), e13330.

Weiss, B., Zuanon, J. A., & Piedade, M. T. (2016). Viability of seeds consumed by fishes in a lowland forest in the Brazilian Central Amazon. *Tropical Conservation Science, 9*(4), 1940082916676129.

Weisse, T., Anderson, R., Arndt, H., Calbet, A., Hansen, P. J., & Montagnes, D. J. (2016). Functional ecology of aquatic phagotrophic protists–concepts, limitations, and perspectives. *European Journal of Protistology, 55*, 50–74.

West, W. E., Coloso, J. J., & Jones, S. E. (2012). Effects of algal and terrestrial carbon on methane production rates and methanogen community structure in a temperate lake sediment. *Freshwater Biology, 57*(5), 949–955.

Wetterer, J. K., & Bishop, C. J. (1985). Planktivore prey selection: The reactive field volume model vs. the apparent size model. *Ecology, 66*(2), 457–464.

Winemiller, K. O., & Rose, K. A. (1992). Patterns of life-history diversification in North American fishes: Implications for population regulation. *Canadian Journal of Fisheries and Aquatic Sciences, 49*(10), 2196–2218.

Wissmar, R. C., Richey, J. E., Stallard, R. F., & Edmond, J. M. (1981). Plankton metabolism and carbon processes in the Amazon river, its tributaries, and floodplain waters, Peru-Brazil, May-June 1977. *Ecology, 62*(6), 1622–1633.

Wittmann, H., von Blanckenburg, F., Maurice, L., Guyot, J. L., Filizola, N., & Kubik, P. W. (2011). Sediment production and delivery in the Amazon River basin quantified by in situ–produced cosmogenic nuclides and recent river loads. *Bulletin, 123*(5–6), 934–950.

Wuebbles, D. J. (2018). Climate change in the 21st century: Looking beyond the Paris Agreement. In C. Murphy, P. Gardoni, & R. McKim (Eds.), *Climate change and its impacts. Climate change management*. Springer.

Yoshida, N., Iguchi, H., Yurimoto, H., Murakami, A., & Sakai, Y. (2014). Aquatic plant surface as a niche for methanotrophs. *Frontiers in Microbiology, 5*, 30.

Yossa, M. I., & Araujo-Lima, C. A. R. M. (1998). Detritivory in two Amazonian fish species. *Journal of Fish Biology, 52*(6), 1141–1153.

Zapata-Ríos, G., Andreazzi, C. S., Carnaval, A. C., Doria, C. R. C., Duponchelle, F., Flecker, A., Guayasamín, J. M., Heilpern, S., Jenkins, C. N., Maldonado, C., Meneghelli, D., Miranda, G., Moraes, R. M., Silman, M., Silveira, M. A. P. A., Tabet, G., Trujillo, F., Ulloa, C., & Arieira, J. (2021). Chapter 3: Biological diversity and ecological networks in the Amazon. In *Amazon assessment report 2021*. United Nations Sustainable Development Solutions Network.

Zingel, P., Cremona, F., Nõges, T., Cao, Y., Neif, É. M., Coppens, J., et al. (2018). Effects of warming and nutrients on the microbial food web in shallow lake mesocosms. *European Journal of Protistology, 64*, 1–12.

Zingel, P., Jeppesen, E., Nõges, T., Hejzlar, J., Tavşanoğlu, Ü. N., Papastergiadou, E., et al. (2023). Changes in nutrient concentration and water level affect the microbial loop: A 6-month mesocosm experiment. *Aquatic Ecology, 57*(2), 369–381.

Zingel, P., Paaver, T., Karus, K., Agasild, H., & Nõges, T. (2012). Ciliates as the crucial food source of larval fish in a shallow eutrophic lake. *Limnology and Oceanography, 57*(4), 1049–1056.

Part II

Ecotoxicology: Fish and Amphibian Sensitivity

A Double Threat to Amazonian Amphibian Larvae: A Review of Toxic Contaminants and Their Interaction with Global Climate Change

Guilherme Azambuja, Igor Kaefer, Adalberto Luis Val, and Daiani Kochhann

Abstract

Amazonia holds global significance due to its essential role in climate regulation and its high biological diversity. However, in recent decades, the region has faced serious threats to its forest structure and rich biodiversity. Human activities, including habitat alterations, pollution, introduction of invasive species, pathogens, and accelerated climate change, have been identified as primary factors contributing to the decline of various populations in these environments. Approximately 76% of amphibian species are estimated to occur in tropical forests, highlighting their significance and the potential for discovering new species. Conversely, knowledge about the effects of environmental contamination in the region remains scarce, which is particularly concerning given the current scenario of climate change. Contaminants such as pesticides and microplastics can induce a range of morphological, physiological, and behavioral alterations in frogs and toads, especially during larval stages. When coupled with climate change, where rising temperatures play a pivotal role, these factors pose intricate challenges for amphibians, intensifying adverse effects and contributing to population declines. Here, we provide a review of the impacts of environmental pollution on various aspects of anuran amphibians, providing insight into

G. Azambuja (✉) · I. Kaefer
Department of Biology, Federal University of Amazonas, Manaus, AM, Brazil

A. L. Val
Laboratory of Ecophysiology and Molecular Evolution, Brazilian National Institute for Research of the Amazon (INPA), Manaus, Amazonas, Brazil

D. Kochhann
Laboratory of Behavioural Ecophysiology, Center of Agrarian and Biological Sciences, Acaraú Valley State University, Sobral, Ceará, Brazil

S. S. de Souza et al. (eds.), *The Future of Amazonian Aquatic Biota*,
https://doi.org/10.1007/978-3-031-66822-7_9

the current state of knowledge regarding Amazonia and highlighting knowledge gaps.

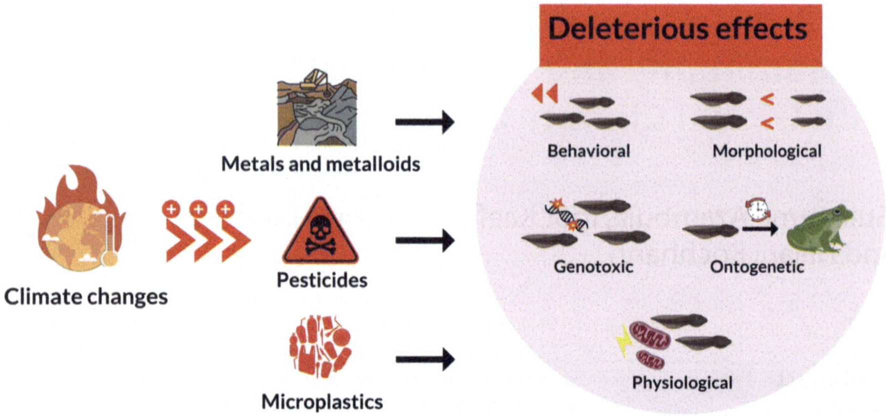

Major categories of negative effects resulting from aquatic contamination in amphibians, with a particular focus on how climate change interacts with contaminants, amplifying their deleterious effects.

The Neotropics harbor the highest amphibian species richness among biogeographical regions (Frost, 2023), and Brazil is the world's leading country in amphibian diversity (Guerra et al., 2020). The decline of anuran (frogs and toads) populations has been reported in urban, rural, and pristine areas at least since the 1940s (Bragg, 1952; Carnaval et al., 2006; Gardner, 2001) throughout the globe and; nowadays, Amphibia is the tetrapod group with the highest proportion of species under extinction risk. Presently, over 40% of the amphibian species are listed as globally threatened in the red list of threatened species (Luedtke et al., 2023). The scientific community gained significant awareness of this issue in the early 1990s during the First World Congress of Herpetology. Researchers from around the globe presented findings on population declines occurring on multiple continents, leading to heightened attention (Green et al., 2020). There is an emerging consensus that the decline of amphibians is intricately linked to the onset of the Anthropocene era and intensified human activity. The factors contributing to the decline of these populations are identified as follows: (1) habitat modification, (2) invasive species, (3) pollution, (4) exploitation, and (5) climate change (Balvanera et al., 2020; Bishop et al., 2012; Grant et al., 2016; McRae et al., 2020). The synergistic effects of these impacts exacerbate the already alarming situation for amphibians. Climate change impacts are heightened by alterations in land use, such as deforestation for agricultural production and urbanization. These changes can synergistically interact with pollutants, significantly amplifying their detrimental effects on amphibians. The projected global temperature for the period 2081–2100 is anticipated to be around 1–1.8 °C higher than the baseline period of 1850–1900 under the low greenhouse gas emission (GHG) scenario. Under the intermediate GHG emission scenario, the projected temperature increase ranges from 2.1 to 3.5 °C, while under the high GHG emission scenario, it is projected to be between 3.3 and 5.7 °C (Masson-Delmotte et al., 2021). However, without urgent mitigation measures, in all scenarios, the 1.5 °C limit proposed by the Paris Agreement is expected to be surpassed between 2021 and 2040 (Masson-Delmotte et al., 2021). In 2023, it has been confirmed as the warmest calendar year on record since 1850. For the first time in a given year, every single day exceeded 1 °C above the preindustrial level average. Moreover, approximately 50% of the days experienced temperatures 1.5 °C higher than the preindustrial period. The global mean temperature for 2023 stands as the highest on record, with an increase of 0.60 °C above the 1991–2020 average and 1.48 °C above the preindustrial average of 1850–1900 (Copernicus Climate Change Service, 2023). The heightened vulnerability of amphibians to anthropogenic changes can be attributed to intrinsic morphophysiological characteristics of the group that include pronounced water or moisture dependence due to their unshelled eggs, aquatic larval phase (tadpoles), and highly permeable skin in most species (Slaby et al., 2019). In comparison to other vertebrate groups, the life cycle of anurans is notably complex. It encompasses a wide range of reproductive modes, in which tadpoles can develop in either terrestrial or aquatic environments, or even be absent altogether in certain direct-developing species (Nunes-de-Almeida et al., 2021). The larval phase is considered a critical period in the development of anurans, as tadpoles undergo significant morphological and physiological changes while they transform into their adult form (Yaoita, 2019). During the larval stage, tadpoles develop morphological structures suited to the aquatic environment, including fins and external gills (McDiarmid & Altig, 1999). The transformation into an adult tetrapod capable of walking, jumping, and feeding on prey demands extensive external and internal morphological and physiological rearrangements, such as loss of gills and development of lungs, the formation of a urogenital apparatus, the rearrangement and ossification of skeletal elements, and modifications to the digestive tract (Tata, 2006; Yaoita, 2019). Concomitant with these changes, the highly permeable skin of amphibians, which enables rapid passage of oxygen and water, also facilitates the absorption of xenobiotics.

These foreign substances can diffuse into amphibians at a remarkably faster rate, ranging from one to two orders of magnitude (depending on the species' hydrophobicity), compared to mammals. This increased permeability makes amphibians highly sensitive organisms to anthropogenic activities (Quaranta et al., 2009). While they serve as excellent bioindicators for assessing environmental quality, and recent studies have highlighted the effects of freshwater contamination on physiological, morphological, and behavioral changes in amphibians, there is still a knowledge gap when compared to other freshwater organisms. Deleterious effects of different classes of contaminants have been demonstrated in recent years, with notable variations in their impacts observed throughout different stages of ontogenetic development, geographic regions, and species (Table 1). While scientific literature frequently focuses on examining the effects of individual contaminants, it is crucial to understand how species in their natural environments, influenced by multiple stressors simultaneously, will respond. In the aquatic environment, anurans are exposed to a mixture of different contaminants, as well as fluctuations in temperature, pH, oxygen levels, and UV radiation. Recent studies have documented both lethal and sublethal effects of contaminants combined with abiotic factors on anurans, including the following: (1) changes in tadpole hatching rates (Adams et al., 2021); (2) increased mortality (Vanzetto et al., 2019); (3) behavioral changes, such as alterations in swimming speed, activity patterns, and habitat choice (Sievers et al., 2018); (4) physiological changes, including shifts in tadpole development time (Adelizzi et al., 2019), which can influence the size of the juvenile individual, as well as metabolic changes; (5) morphological alterations, such as reduction in body size (Dulumoni et al., 2020), and structural alterations, both externally (deformities) and internally (changes in internal organs) (Rathanayaka & Rajakaruna, 2018); and (6) cellular-level changes, such as an increase in micronuclei in cells (Herek et al., 2021), and modifications in the antioxidant and defense profiles of tadpoles (Rutkoski et al., 2020). These studies highlight the wide range of effects that contaminants combined with abiotic factors can have on anurans, affecting their development, behavior, physiology, morphology, and overall health.

1 Amazonia and Its Frog Diversity

The global importance of the Amazonian biome is widely acknowledged due to its critical role in climate regulation and its biological diversity (Fearnside, 2022). Amazonia plays a significant part in the hydrological cycle of South America by recycling wind-borne moisture from the evaporation of the tropical Atlantic Ocean. Approximately, 35% of this moisture is recycled within the region, favoring local precipitation and contributing to the maintenance of the forest ecosystem (Dirmeyer et al., 2009; Zemp et al., 2014). Indeed, Amazonia has a significant contribution to the hydrological cycle beyond its local impact. During the wet season, wind patterns create a flow of moisture from the Andes eastward toward the subtropical regions, providing water vapor to the Midwest, Southeast, and South regions of Brazil (Arraut et al., 2012). This highlights the role of the biome in providing water resources even to geographically distant areas. Another crucial function of Amazonia is its capacity to store atmospheric carbon, which contributes to the mitigation of greenhouse gasses. It is estimated that the region holds between 350 and 600 Pg (petagrams) of carbon and experiences an annual loss of 1.1 Pg of carbon per year due to deforestation and forest degradation (Baccini et al., 2012; Silva Junior et al., 2018). This emphasizes the importance of the Amazon forest for climate regulation.

The importance of Amazonian forests and associated ecosystems goes beyond their recognized significance. Unfortunately, intense land-use changes over the past few decades have posed a serious threat to the forest's structure, rich biodiversity, and ecosystemic services. Within just 35 years (1985–2021), approximately, 44

Table 1 Summary of reported literature results on the main effects of contaminants on amphibian larvae. Abbreviations: GS, Gosner Stage (Gosner, 1960); a.i./L, active ingredient per liter

Contaminant	Species	Family	Larval stage	Tested concentrations	LC50	Effects	Region	Reference
Cadmium (Cd) Lead (Pb) Cd + Pb	*Bufo gargarizans*	Bufonidae	GS 2–3	Control and 108 μg/L Control and 204 μg/L Control and 106 + 201 μg/L	Not evaluated	Morphological: reduced and increased body mass, reduction total length and snout-vent length, reduction bone length of the ilium, decreased maximum jump distance Histological: toward colloid depletion in follicular cells of thyroid gland (TG), reduction and increased TG area, increase in phagophores per TG, decrease in thickness of the epithelial cells, reduction in the colloid area of TG Genotoxic: dysregulation of genes involved in TG action and endochondral ossification	Shaanxi Province, China, East Asia	Chen et al., 2022
Copper (Cu)	*Rhinella granulosa*	Bufonidae	GS 25	Control, 15, 20, 25, 33, 43, 56, 72, and 94 ug/L	23.4 ug/L (stream water)	Mortality	Central Amazonia, Brazil, South America	Franco-de-Sá & Val, 2014)
GLY 2,4-D Piroclam Piroclam +2,4-D	*Rhinella marina*	Bufonidae	GS 25	Control, 4.9, 9.8, and 14.7 Control, 128.9, 257.8, and 386.7 mg/L Control, 0.12, 0.25, and 0.37 mg/L Control, 6.9, 13.9, and 20.9 mg/L	32 mg/L 282.8 mg/L 0.3 mg/L 42.4 mg/L	Mortality Morphological: increased and decreased body mass	Southern Amazonia, Brazil, South America	Figueiredo & Rodrigues, 2014

(continued)

Table 1 (continued)

Contaminant	Species	Family	Larval stage	Tested concentrations	LC50	Effects	Region	Reference
Glyphosate + predator cue	*Anaxyrus americanus*	Bufonidae	GS 25	Control, 0.1, 1, 5, 10, and 20 mg a.i./L + present/absent predator cue	2.52 mg a.i./L	Mortality	Pittsburgh, USA, North America	Relyea, 2005
Glyphosate + temperature	*Bufo bufo*	Bufonidae	GS 8 GS 21–24	15 °C and 20 °C: control, 0.5, 1.0, and 1.5 mg a.e./L and a pulse of 0.5 a.e./mg/L	Not evaluated	GS 8: Mortality Morphological: body length, body width, tail length, and total length of tadpoles at 20 °C were increased. Ontogenetic: accelerated development Glyphosate + temperature Morphological: increased body length of herbicide-treated tadpoles, body width, tail length, and total length at 15 °C. At 20 °C body length, tail length, and total length increased, and body width was decreased by 2% compared to no-herbicide control GS 21–24 Mortality Morphological: body length, body width, tail length, and total length of tadpoles were increased at 20 °C Glyphosate + temperature Ontogenetic: accelerated development	Vienna, Austria, Central Europe	Baier et al., 2016

Mercury (Hg)	*Bufo gargarizans*	Bufonidae	GS 26	Control, 6, 12, 18, 24, and 30 μg/L	Not evaluated	Mortality Morphological: reduction in total length, tail length, hindlimb, and body mass Ontogenetic: reduction in metamorphic rates Histopathological: enlarged thyroid follicular space, severe deformation in follicular cells in thyroid glands (TG) Follicular cell hyperplasia and deformation in TG Alteration in frequency of melanomacrophage centers in the liver, unclear cell boundaries, and degeneration of hepatocytes Enlarged hepatic cells space Increased number of lipid droplets, nuclear deformation, and degenerated mitochondria Reduction in the number of mitochondria and lipid droplets, degeneration and disintegration of hepatocytes, mitochondrial vacuolation, and endoplasmic reticulum breakdown Physiological: alterations in thyroid hormone regulation, antioxidant response, and lipid metabolism	Qinling Mountains, Shaanxi Province, China	Shi et al., 2018
Methomyl + temperature	*Duttaphrynus melanostictus*	Bufonidae	GS 25	15 °C: control, 500, 700, 900, 1100, and 1250 mg/L 20 °C: control, 500, 750, 1000, 1250, and 1500 mg/L 25 °C: control, 500, 750, 1000, 1250, and 1500 mg/L 30 °C: control, 100, 200, 350, 500, and 600 mg/L 35 °C: control, 200, 350, 500, 600, and 700 mg/L	847.8 mg/L 1109 mg/L 940.7 mg/L 378.3 mg/L 265.5 mg/L	Mortality	Hong Kong, China, East Asia	Lau et al., 2015

(continued)

Table 1 (continued)

Contaminant	Species	Family	Larval stage	Tested concentrations	LC50	Effects	Region	Reference
MPs mixture: expanded polystyrene + high-density polyethylene + polyester fibers + polyvinyl chloride	*Bufotes viridis*	Bufonidae	GS 26–28	Control, 1, 7, and 50 mg/L	2.5 mg/L	Mortality Morphological: decreased weight Behavior: activity level decrease	Lombardy, Italy, North Europe	Balestrieri et al., 2022
α- and β-endosulfan	*Sclerophrys maculata*	Bufonidae	GS 25	Control (acetone), 1, 10, and 100 μg a.i./L	Not evaluated	Mortality Behavioral: uncoordinated, conspicuous, and unnatural movements Morphological: shorter tail	Tanguiéta, Benin, West Africa	Stechert et al., 2015
Cadmium (Cd)	*Fejervarya limnocharis*	Dicroglossidae	GS 26–28	0.1, 0.2, 0.3, 0.4, 0.5, 1, 2, and 3 mg/L	10-day: 0.32 mg/L 24 h: 1.85 mg/L	Mortality Morphological: decreased body mass and higher body length Ontogenetic: accelerated metamorphosis Genotoxicity: increased micronuclei in erythrocytes, DNA damage evidenced by migration of DNA fragments in gels forming a comet-like structure Decreased DNA content in the comet head region with concurrent increase in the tail region Increase in the olive tail moment and mean distance of migration in the tail	Assam, India	Patar et al., 2016
Tamoxifen	*Euphlyctis cyanophlyctis*	Dicroglossidae	GS 25–42	Control, 25, 50, 100, and 200 ug/L	Not evaluated	Morphological: testis maturation in males and increased oocytes size in females Ontogenetic: delayed in metamorphosis and heavier and larger body	Maharashtra, India, South Asia	Phuge, 2020
Zinc chloride	*Fejervarya limnocharis*	Dicroglossidae	GS 22–25	LC50: control, 3, 4, 5, and 6 mg/L Sublethal: control, 0.5, 1, 1.5, and 2 mg/L	3.61 mg/L	Mortality Ontogenetic: delayed development Genotoxic: decreased amount DNA in comet head, increased tail region and olive tail moment and increased micronuclei	Assam, India, South Asia	Patar et al., 2021

Atrazine	*Dendropsophus minutus*	Hylidae	GS 25–33 GS 36–39 GS 42–43	Control, positive control (cuclophosphamide 40 mg/L), 2.25, 4.5, 9, 18, and positive control	36.4 mg/L	Mortality Genotoxic: increase % of DNA tail damage and increased micronuclei frequency	Goiás, Brazil, South America	Gonçalves et al., 2017
Chlorpyrifos	*Acris crepitans*	Hylidae	GS 25–27	Control, 1, 10, 100, and 200 μg/l with/without sediment	Not evaluated	Morphological: reduced snout-vent length Physiological: cholinesterase activity decreased	Oklahoma, USA, North America	Widder & Bidwell, 2008
Chlorpyrifos	*Dryophytes chrysoscelis*	Hylidae	GS 25–27	Control, 1, 10, 100, and 200 μg/l with/without sediment	Not evaluated	Morphological: reduced mass Physiological: decreased cholinesterase activity Behavioral: slower swimming	Oklahoma, USA, North America	Widder & Bidwell, 2008
Copper (Cu)	*Scinax ruber*	Hylidae	Eggs GS 25	Control, 15, 20, 25, 33, 43, 56, 72, and 94 ug/L	36.3 ug/L 50 ug/L (stream water) 15.9ug/L (water from well)	Mortality	Central Amazonia, Brazil, South America	Franco-de-Sá & Val, 2014
Water from a contaminated urban stream—Iron (Fe), zinc (Zn), nickel (Ni), manganese (Mn), copper (Cu), chromium (Cr), Lead (Pb), and cobalt (Co)	*Osteocephalus taurinus*	Hylidae	Eggs GS 25	Control, 16, 20, 27, 36, 48, 60, 75, 90, and 117 mL/L	Eggs: 25.19 mL/L Tadpoles: 47.63 mL/L	Mortality Morphological: edema in the abdominal region and increased body size Ontogenetic: accelerated development	Central Amazonia, Brazil, South America	Franco-de-Sá et al., 2012
Cypermethrin (95% purity) Cypermethrin (commercial)	*Boana pulchella*	Hylidae	GS 25	Control, 0.1, 0.5, 1, 5, 10, 50, 100, 250, 500, 750, 1.000, 2.500, 5.000, 7.500, and 10.000 ug a.i./L	479.7 mg/L 175.2 mg/L	Mortality Behavioral: irregular swimming, delayed response, slow swimming, and immobility Morphological: abnormalities in head and axial, eye, gut, and face Growth inhibition	Buenos Aires, Argentina, South America	Agostini et al., 2010

(continued)

Table 1 (continued)

Contaminant	Species	Family	Larval stage	Tested concentrations	LC50	Effects	Region	Reference
Glyphosate Polyoxyethylene amine (POEA)	*Dendropsophus minutus*	Hylidae	GS 25	Control, 65, 130, 260, and 520 μg/L Control, 1.25, 2.5, 5, and 10 μg/L	Not evaluated	Genotoxic: all types of genotoxic damage (%DNA in the tail, tail length, and olive tail moment, and increased micronuclei) were observed Histopathological: vacuolization of the hepatocytes, hyperchromatic nuclei, and presence of spots with inflammatory cells. Presence of edematous degeneration and melanomacrophages were more extensive and more pigmented	Goiás, Brazil, South America	Lopes et al., 2021
Glyphosate	*Dendropsophus minutus*	Hylidae	GS 26–31	Control, positive control (10 mg/L of mitomycin C), 0.28, 1.0, 2.0, and 4.0 mg a.i./L	2.4 mg a.i./L	Mortality Genotoxic: increase in DNA damage for all parameters of the comet: tail length, % DNA in the tail and olive tail moment	Goiás, Brazil, South America	Carvalho et al., 2018
Glyphosate + predator cue	*Dryophytes versicolor*	Hylidae	GS 25	Control, 0.1, 1, 5, 10, and 20 mg a.i./L + present/absent predator cue	1.35 mg a.i./L	Mortality	Pittsburgh, USA, North America	Relyea, 2005
2,4-D	*Leptodactylus fuscus*	Leptodactylidae	GS 25	Control, 5, 10, 20, 40, and 80 mg/L	28.81 mg/L	Mortality	São Paulo, Brazil, South America	Freitas et al., 2019
2,4-D	*Physalaemus nattereri*	Leptodactylidae	GS 25	Control, 20, 40, 80, 160, and 320 mg/L	143.08 mg/L	Mortality	São Paulo, Brazil, South America	Freitas et al., 2019
Cypermethrin Deltamethrin	*Physalaemus gracilis*	Leptodactylidae	GS 20 GS 25	Control, 0.05, 0.1, 0.5, 1.0, 1.5, and 2.0 mg/L of cypermethrin control, 0.01, 0.02, 0.03, 0.05, 0.07, and 0.1 mg/L of cypermethrin and 0.001, 0.0015, 0.003, 0.004, 0.006, and 0.009 mg/L of deltaethrin	Not evaluated	Behavioral: reduction swimming activity, stopped swimming and lethargy, and spasms Morphological: malformation in oral morphology	Rio Grande do Sul, Brazil, South America	Vanzetto et al., 2019

Glyphosate 2,4-D Piroclam Piroclam +2,4-D	*Physalaemus centralis*	Leptodactylidae	GS 25	Control, 4.9, 9.8, and 14.7 Control, 128.9, 257.8, and 386.7 mg/L Control, 0.12, 0.25, and 0.37 mg/L Control, 6.9, 13.9, and 20.9 mg/L	19.7 mg/L 515.7 mg/L 0.5 mg/L 27.9 mg/L	Mortality Ontogenetic: accelerated time to metamorphosis Morphological: increased and decreased body mass	Southern Amazonia, Brazil, South America	Figueiredo & Rodrigues, 2014
Polyethylene microplastics	*Physalaemus cuvieri*	Leptodactylidae	GS 28–29	Control and 60 mg/L	Not evaluated	Morphological: increased body mass Behavioral: decreased locomotor activity	Goiás, Brazil, South America	da Costa Araújo & Malafaia, 2020
Chlorpyrifos	*Gastrophryne olivacea*	Microhylidae	GS 25–27	Control, 1, 10, 100, and 200 μg/l with/without sediment	Not evaluated	Morphological: reduced mass and snout-vent length Physiological: decreased cholinesterase activity	Oklahoma, USA, North America	Widder & Bidwell, 2008
Methomyl + temperature	*Microhyla pulchra*	Microhylidae	GS25	15 °C: control, 25, 50, 100, 175, and 250 mg/L 20 °C: control, 50, 100, 150, 200, and 250 mg/L 25 °C: control, 20, 40, 60, 80, and 100 mg/L 30 °C: control, 20, 40, 60, 80, and 100 mg/L 35 °C: control, 2.5, 5, 10, 15, and 20 mg/L	97.7 mg/L 88.8 mg/L 84.5 mg/L 46.5 mg/L 15.4 mg/L	Mortality	Hong Kong, China, East Asia	Lau et al., 2015
Pyriproxyfen	*Odontophrynus americanus*	Odontophrynidae	GS 26–30	Control, positive control (DTW), 0.01, 0.1, 1, 10 mg/L	2.51 mg/L	Mortality Physiological: oxidative stress, increased acetylcholinesterase, and carboxylesterase activity, increased T4 levels and heart rate Behavioral: decreased distance moved, mean speed, and global activity	Santa Fe Province, Argentina, South America	Lajmanovich et al., 2019

(continued)

Table 1 (continued)

Contaminant	Species	Family	Larval stage	Tested concentrations	LC50	Effects	Region	Reference
Carbaryl	*Xenopus laevis*	Pipidae	GS 8	Control, 1, 2, 4, 8, 16, and 24 mg/L	20.28 mg/L	Mortality Morphological: abnormal gut coiling, generalized edema, cardiac edema, abdominal edema, abnormal tail flexure, and microphthalmia Histopathological: skeletal muscle lesions consisted of myotomes reduced in size, showing myocytes with disorganized contractile systems and irregular myosepta, coupled with disarranged myocyte apexes, necrotic tissue with vacuolated cytoplasms, pyknotic nuclei and degenerating contractile systems, abnormally developed notochords Physiological: cholinesterase inhibition	Italy, Southern Europe	Bacchetta et al., 2008
Cadmium (Cd) Copper (Cu) Lead (Pb) Zinc (Zn)	*Xenopus laevis*	Pipidae	Eggs	Control (modified Barth solution) for all treatments Cd: 0.5, 0.6, 0.7, 0.8, and 0.9 ppm Cu: 0.1, 0.2, 0.3, 0.4, and 0.5 ppm Pb: 0.5, 0.6, 0.7, 0.8 and 0.9 ppm Zn: 5, 10, 15 and 20 ppm	Not evaluated	Decreased embryo hatching and tadpole survival Morphological: body length decreased and malformations (kinking and curling of the tail and edema)	Witwatersrand, South Africa	Haywood et al., 2004
Metamifop	*Xenopus laevis*	Pipidae	GS 45–46	LC50: control, 0.4, 0.5, 0.6, 0.7, 0.8, and 0.9 mg/L Sublethal: control and 0.063 mg/L	0.634 mg/L	Mortality Behavioral: increased swirling, side-lying, air-breathing, activity, and use of peripheral area. Inhibited feeding Morphological: decreased body weight Physiological: decreased triglycerides and total cholesterol and decreased neurotransmitters contents. Oxidative stress. Inhibition of T3 and T4 contents and expression genes	Beijing, China, East Asia	Liu et al., 2021

Polyethylene microplastics	*Xenopus laevis*	Pipidae	GS 25	Control, 60 mg/L cellulose, and 60 mg/L PE MPs	Not evaluated	Ontogenetic: accelerated development Morphology: decreased snout-vent length, longer and heavier intestine	Hamburg, Germany, Central Europe	Ruthsatz et al., 2022
2,4-D	*Aquarana catesbeiana*	Ranidae	GS 25	LC50: control, 450, 500, 550, 600, and 650 mg/L Sublethal: control, 125, 250, and 500 µg/L	574.52 mg/L	Mortality Morphological: decreased growth Ontogenetic: accelerated metamorphosis Physiological: Changes in total lipid contents, total carbohydrates and proteins, and decreased respiration rates Behavioral: reduced speed response to the stimulus	São Paulo, Brazil, South America	Freitas et al., 2019
2-hydroxyatrazine	*Aquarana catesbeiana*	Ranidae	GS 26–32	25 °C and 32 °C: control, 10, 50, and 200 ng/L	Not evaluated	Physiological: oxidative stress and increased acetylcholinesterase activity Histopathological: in the liver, increase in cells with vacuolization, mild steatosis, number of melanomacrophages, inflammatory infiltration, and cell congestion were observed. In the kidney, glomerular shrinkage, Bowman's space enlargement, inflammatory infiltrates, and fibrosis were observed	Santa Catarina, Brazil, South America	Carneiro et al., 2023
Arsenic (As)	*Lithobates pipiens*	Ranidae	GS 19	Control, 10, 20, 150, 500, and 1.000 µg/L	Not evaluated	Behavioral: slower swimming	Madison, Wisconsin, USA	Chen et al., 2009
Chlorpyrifos	*Lithobates sphenocephalus*	Ranidae	GS 25–27	Control, 1, 10, 100, and 200 µg/l + absent/present sediment	Not evaluated	Morphological: reduced body mass Physiological: decreased cholinesterase activity Behavioral: slower swimming	Oklahoma, USA, North America	Widder & Bidwell, 2008

(continued)

Table 1 (continued)

Contaminant	Species	Family	Larval stage	Tested concentrations	LC50	Effects	Region	Reference
Chromium (Cr) Lead (Pb) Cr + Pb	*Rana chensinensis*	Ranidae	GS 38	Control, 104 ug/L (Cr), 50 ug/L (Pb), and 104 ug/L (Cr) + 50 ug/L (Pb)	Not evaluated	Morphological: reduction in SVL and body mass and decrease in relative intestine length Histopathological: irregular cell nuclei, indistinct cells, and brush-like borders were observed in histological structures Physiological: elevated amylase activity, decrease in pepsin activity, and inhibition of lipase activity Perturbation of intestinal homeostasis: changes in bacterial richness and diversity, alterations in antioxidant enzymes, and pro-apoptotic gene modifications	Qinling Mountains, Shaanxi Province, China,	Zhang et al., 2023
Glyphosate + predator cue	*Aquarana catesbeiana*	Ranidae	GS 25	Control, 0.1, 1, 5, 10, and 20 mg a.i./L + present/absent predator cue	2.07 mg a.i./L	Mortality	Pittsburgh, USA, North America	Relyea, 2005
Glyphosate + predator cue	*Aquarana clamitans*	Ranidae	GS 25	Control, 0.1, 1, 5, 10, and 20 mg a.i./L + present/absent predator cue	2.17 mg a.i./L	Mortality	Pittsburgh, USA, North America	Relyea, 2005
Glyphosate + predator cue	*Boreorana sylvatica*	Ranidae	GS 25	Control, 0.1, 1, 5, 10, and 20 mg a.i./L + present/absent predator cue	2.17 mg a.i./L	Mortality	Pittsburgh, USA, North America	Relyea, 2005
Glyphosate + predator cue	*Lithobates pipiens*	Ranidae	GS 25	Control, 0.1, 1, 5, 10, and 20 mg a.i./L + present/absent predator cue	2.46 mg a.i./L	Mortality	Pittsburgh, USA, North America	Relyea, 2005

Methylmercury chloride	*Lithobates pipiens*	Ranidae	GS not specified	Control, positive control (saline solution), 0.001, 0.01, 0.05, 0.1, 0.5, and 1.0 ppm	Not evaluated	Mortality Behavioral: irritative movements, abnormal swimming, difficulties in breathing and lethargy Morphological: edema in leg and renal tubules were replaced by an infiltration of large hyperchromatic cells resembling bone marrow hematopoietic cells. Reduced cellularity and necrotic changes were observed in many renal glomeruli	Madison, Wisconsin, USA	Chang et al., 1974
MPs mixture: expanded polystyrene (PS), high-density polyethylene (HDPE), polyester fibers (PES) and polyvinyl chloride (PVC)	*Rana latastei*	Ranidae	GS 26–28	Control, 1, 7, and 50 mg/L	2.5 mg/L	Mortality Morphological: decreased weight Behavioral: decreased activity	Lombardy, Italy, North Europe	Balestrieri et al., 2022
Propiconazole + Cyproconazole	*Pelophylax saharicus*	Ranidae	GS 26–36	Control, positive control (MMS 1.56 mg/L), 50, 75, 100, and 150 µl/L	Not evaluated	Hematotoxicity: red blood cells' miotic indices higher Genotoxicity: increased micronuclei and binucleated cells	Algeria, North Africa	Bouhafs et al., 2009
Imidacloprid	*Polypedates teraiensis*	Rhacophoridae	GS 26–28	LC50: control, 18.5, 20, 21.5, 23, 24.5, 26, and 27.5 mg/L Sublethal: control, 0.69, 1.04, and 2.09 mg/L	20.91 mg/L	Morphological: decreased weight and snout-vent length Behavioral: hyperactivity followed by loss of balance, erratic movements, becoming motionless, lying down on the bottom, and leading to death	Assam, India, South Asia	Dulumoni et al., 2020
Methomyl + temperature	*Polypedates megacephalus*	Rhacophoridae	GS 25	15 °C: control, 200, 300, 400, 500, and 600 mg/L 20 °C: control, 200, 350, 500, 650, and 800 mg/L 25 °C: control, 100, 200, 300, 400, and 500 mg/L 30 °C: control, 200, 350, 500, 600, and 700 mg/L 35 °C: control, 25, 50, 100, 150, and 200 mg/L	>600 mg/L 466.9 mg/L 380 mg/L 449.3 mg/L 101.1 mg/L	Mortality Morphological: increased malformations	Hong Kong, China, East Asia	Lau et al., 2015

million hectares (13%) of the native area have been converted for agricultural activities (Projeto MapBiomas, 2022). Deforestation in Amazonia has surged to historic levels in recent years, escalating from 4545 km^2 in 2018 to 9696 km^2 in 2022, representing a more than 143% increase since 2012 (Leite-Filho et al., 2021; PRODES, 2023). Anthropogenic landscape changes associated with urbanization, agriculture, and cattle ranching have profoundly transformed ecosystems, leading to pervasive landscape and biotic homogenization that negatively impacts the abundance and diversity of amphibian populations and assemblages (Cordier et al., 2021; Fulgence et al., 2022).

Additionally, the region faces a persistent problem of wildfires. The year 2022 has seen a record number of active fires compared to 2021, with only February–April showing lower values due to high precipitation levels. As of September 2022, there have been 41,282 fires, a 147% increase compared to the same period in 2021. The cumulative number of fires in 2022 (87,304) has already been surpassed, reaching a total of 98,646 fires recorded thus far (INPE, 2023). These fires, primarily ignited by human activities such as land clearance for agriculture, are facilitated by landscape modifications that elevate temperatures and reduce humidity, creating favorable conditions for the rapid spread of fire.

The characteristics of tropical forests such as Amazonia favor a magnificent diversity of anurans. Approximately, 76% of the world's total amphibian species are estimated to inhabit tropical forests (Pillay et al., 2022). In a recent update, the biodiversity census of the Brazilian Amazon reported 383 frog species in the region (Hoogmoed, 2023)—a likely significant underestimate due to the vast area, challenging accessibility, and the potential for sampling bias (Vacher et al., 2020; Verdade et al., 2011). Furthermore, tropical and subtropical moist broadleaf forests are projected to have the highest percentage of potential future amphibian species discoveries worldwide, particularly due to cryptic taxonomic diversity (Moura & Jetz, 2021).

2 Aquatic Contamination and Effects on Anurans

2.1 Agriculture and the Use of Pesticides

The rapid growth of the global human population in recent decades has led to extensive land-use changes and increased demand for food and resources, which have had detrimental effects on amphibians (Balvanera et al., 2020). Agricultural activities play a significant role in these changes, with approximately 38% of the global land surface (5 billion hectares) being dedicated to agriculture, of which 1.66 billion hectares are used for food production and 3.32 billion hectares are used for livestock grazing (FAO, 2022). Contamination of the aquatic environment can arise from the agricultural sector, which annually consumes substantial quantities of fertilizers and pesticides (Brovini et al., 2021). Between 2000 and 2008, Brazil ranked third in pesticide use, following the United States and China, respectively, claiming first and second place (IBAMA, 2022). However, Brazil surpassed China in 2009, occupying the second spot (FAO, 2022). The sale of pesticides in Brazil has witnessed an exponential growth year after year, with sales increasing by nearly 100% in just 11 years (2010–2021). Sales figures escalated from 361.66 tons in 2010 to 719.50 tons in 2021 (Fig. 1). In the same period, Brazilian Amazonia alone experienced an increase from 67.42 tons in 2010 to 179.36 tons in 2020 (IBAMA, 2022). Despite the existence of several recommendations on pesticide use, such as specifying suitable crops, ideal plant stages, climatic conditions, dosage per hectare, and spacing between applications, these guidelines are often disregarded (Kotz et al., 2021).

Pesticides are frequently used indiscriminately, leading to the development of resistance in target organisms. Moreover, the application of pesticides in croplands does not confine their impact to this specific environment; instead, they can disperse widely (Kjellstrom et al., 2006; Manisalidis et al., 2020). Soil serves as a significant reservoir for pesticides, as it accumulates a significant portion of these compounds during crop application. Soil characteristics, including texture, temperature, and pH, influence its capacity to absorb pesticides, thereby determining their persistence within the environment and their potential mobility into water bodies and other areas (Vryzas, 2018). Various processes contribute to the dissipation of pesticides through the soil, such as leaching, runoff, and drainage. Additionally, these chemicals find their way into the aquatic environment through precipitation subsequent to their application in agricultural areas and via air currents during aerial spraying (Casara et al., 2012). The persistence of pesticides in the aquatic environment is further influenced by factors such as water temperature, pH, radiation, and microbial community (Vryzas, 2018).

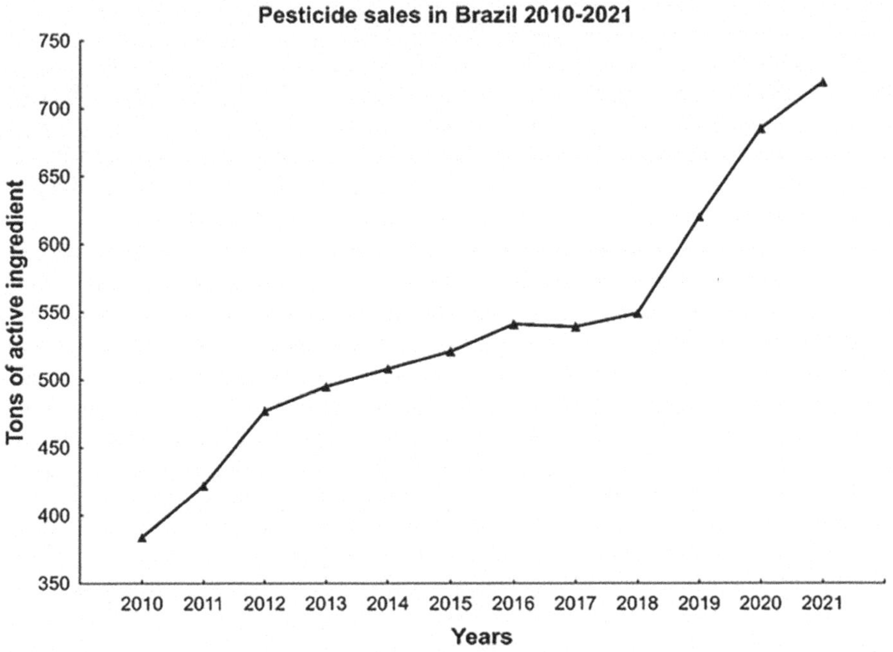

Fig. 1 Annual pesticide sales in Brazil from 2010 to 2021

The adverse impacts of contaminants on amphibians have been extensively documented (Trudeau et al., 2020), but there is a noticeable bias toward studies conducted in temperate regions (Ghose et al., 2014; Schiesari et al., 2007), particularly in North America (Carrasco et al., 2021; D'Avila et al., 2020). Conversely, there is a scarcity of information on this subject in tropical regions, which is concerning considering that amphibians are a dominant vertebrate group in these areas (Duellman, 1999; Ghose et al., 2014; Yang et al., 2022). Research on pesticide effects also tends to focus more on particular families such as Bufonidae, Hylidae, and Leptodactylidae, which alone do not represent the tropical anuran diversity (D'Avila et al., 2020). In Brazilian Amazonia, specifically, there is a significant knowledge gap in terms of ecotoxicological studies involving anurans, which mirrors the overall scarcity of such research in Brazil. Despite the region's high species richness, there is a wide array of pesticides used, totaling approximately 300 different compounds. Notably, the most commonly sold pesticides in Brazilian Amazonia include glyphosate, 2-4,D, mancozeb, and atrazine (IBAMA, 2022).

Glyphosate (GLY)

Glyphosate [N-(phosphonomethyl) glycine (C3H8NO5P)] is a synthetic herbicide widely used for its broad-spectrum, nonselective properties, effectively targeting various types of weeds, perennials, monocotyledons, and dicotyledons (Agostini et al., 2020; Peillex & Pelletier, 2020). Classified as a class III product hazardous to the environment, glyphosate (GLY) has gained popularity due to its efficacy. It is a highly polar molecule which contributes to its high-water solubility and facilitates its uptake by aquatic organisms (Leite-Filho et al., 2021; Williams et al., 2000). The global consumption of glyphosate has significantly increased in recent decades, with usage escalating from approximately 56,000 tons in 1994 to over 825,000 tons in 2014 (Benbrook, 2016). Forecasts indicate that glyphosate consumption will surpass the milestone of 1,000,000 tons worldwide in the coming years (Agostini et al., 2020; Landrigan & Belpoggi, 2018; Soares et al., 2021). Presently, glyphosate stands as the most extensively employed herbicide globally, constituting 60% of herbicides within its class (de Maria et al., 2021). Moreover, it holds the distinction of being the most widely sold herbicide in Amazonia (IBAMA, 2022).

The presence of elevated concentrations of glyphosate (GLY) in aquatic environments is a global concern. In Brazil, regulatory agencies allow maximum GLY levels of 65 μg/L in class I waters and 280 μg/L in class III waters, which are significantly higher than the limits set by the European Union (0.1 μg/L) (Brasil, 2005; Pavan et al., 2021). Class I waters are designated for various purposes, including human consumption after simplified treatment, protection of aquatic communities, primary contact recreation, irrigation of raw-eaten vegetables and fruits, and protection of aquatic communities in indigenous territories. Class II waters can be used for human consumption after conventional or advanced treatment, irrigation of crops, amateur fishing, secondary contact recreation, and animal watering (Brasil, 2005). The World Health Organization does not recognize any safe concentration of GLY for human consumption (WHO, 2022). However, studies on GLY contamination in Brazil are scarce, with only one study conducted in the Amazon region. In various Brazilian rivers, GLY concentrations exceeding 100 μg/L have been detected in the Passo do Pilão River in south Brazil, a maximum concentration of 6.80 μg/L in the Pirapó River in the Paraná River basin in the south, and values up to 1.06 μg/L in the Samambaia River sub-basin in the Midwest Brazil (Correia et al., 2020; da Silva et al., 2003; Delmonico et al., 2014). In the Amazonian biome, water samples collected from the Santarém plateau in Pará state revealed GLY concentrations between 1.5 μg/L and 9.7 μg/L (Pires et al., 2020). The contamination of aquatic environments by GLY has negative effects on tadpoles of various species worldwide. For example, *Microhyla fissipes* exhibited a 10-day median lethal concentration (LC50) of 77.5 mg/L, and exposure to GLY at concentrations of 60 and 90 mg/L suppressed tadpole growth, with more pronounced effects at higher concentrations. GLY also led to transcriptional downregulation of carbohydrate and lipid catabolism and upregulation of amino acid catabolism (Wang et al., 2019). *Xenopus laevis* embryos and tadpoles at Gosner stage 8 and 46 exhibited 96-h LC50 values of 32.1 mg active ingredient (a.i.)/L and 35.1 mg a.i./L, respectively, with a reduction in embryo length at concentrations of 31–37.6 mg a.i./L (Turhan et al., 2020). Histopathological alterations were observed in *Dendropsophus molitor* after 1-month exposure to GLY in concentrations ranging from 350 to 750 μg glyphosate acid equivalent (a.e.)/L (Riaño et al., 2020). Studies on Brazilian anurans have shown varied effects. Exposure of *Scinax nasicus* to GLY concentrations between 3.07 and 7.5 mg/L resulted in increased mortality and malformations, with LC50 values of 4.78 mg/L at 24 h and 2.64 mg/L at 96 h (Lajmanovich et al., 2003). *Physalaemus cuvieri* and *Boana pardalis* tadpoles exhibited 96-h LC50 values of 115 and 106 mg a.i./L, respectively (Daam et al., 2019). Malformations and developmental changes were observed in *Rhinella arenarum* tadpoles, including abnormal eye shape, visceral alterations, lack of organization in the gill chamber, bilateral asymmetry, alterations in mouth structure, and a reduction in thyroid hormone (T4), resulting in developmental delays (Boccioni et al., 2021). GLY exposure in *Melanophryniscus admirabilis* tadpoles induces oxidative stress, characterized by a decrease in catalase and superoxide dismutase enzyme activity, along with an increase in glutathione *S*-transferase activity (da Silva et al., 2021). In Brazilian Amazonia, studies have provided valuable data on the effects of GLY on tadpoles of three anuran species (Figueiredo & Rodrigues, 2014). *Rhinella marina* tadpoles exhibited a 96-h LC50 value of approximately 32 mg/L, while *Physalaemus centralis* tadpoles were found to be more sensitive to acute contamination, with a 96-h LC50 value of 19.70 mg/L. Chronic exposure to sublethal concentrations, ranging from 25% to 75% of the LC50, mainly resulted in reduced survival for both species. Interestingly, the development and body size of *R. marina* tadpoles were unaffected, whereas *P. centralis* tadpoles exhibited accelerated development and a larger body size. For *Dendropsophus minutus* tadpoles, the LC50 value was determined to be 2.49 mg active ingredient a.i./L. Exposure to a sublethal concentration of 0.28 mg a.i./L for 96 h resulted in observed DNA damage based on the comet assay (Carvalho et al., 2018). Furthermore, exposure to 520 μg/L of GLY increased the presence of micronuclei in cells and induced histopathological hepatic damage in *D. minutus* tadpoles (Lopes et al., 2021).

2.2 2,4-D

2,4-D is a widely used systemic organochlorine herbicide in pre and postemergent applications on various crops including soybean, rice, wheat, and corn (Ruiz de Arcaute et al., 2016). It is the second most sold pesticide in Brazil and is primarily employed for weed control (Barbosa et al., 2015; IBAMA, 2022). In Amazonia, it ranks second in herbicide sales (IBAMA, 2022). Due to its high solubility in water and low adsorption coefficient, 2,4-D can contaminate surface water and be readily taken up by aquatic organisms (Freitas et al., 2019; Gervais et al., 2008; Silva et al., 2007). The Brazilian legislation allows a maximum concentration of 4 μg/L in class I waters and 30 μg/L in class II waters (Brasil, 2005), which aligns with the World Health Organization (WHO) guidelines (WHO, 2022). Although 2,4-D contamination has been reported worldwide, there is limited information on its presence in the aquatic environment in Brazil. Studies conducted in the Vacacaí and Vacacaí-Mirim River basins in southern Brazil have detected concentrations ranging from 0.4 to 2.7 μg/L (Marchesan et al., 2010). In Amazonia, concentrations of 3–5 ng/L have been found in the city of Manaus, with a frequency of occurrence of 35% in the collected samples (Rico et al., 2022).

Regarding the effects on amphibians, studies have focused on a few species. Tadpoles of *Leptodactylus fuscus* exhibited a 96-h LC50 value of 28.8 mg/L when exposed to 2,4-D (Freitas et al., 2019). In the case of *Rhinella marina* tadpoles, exposure to 2,4-D resulted in reduced survival, while tadpoles of *Physalaemus centralis* showed higher mortality when exposed to 75% of the LC50. Furthermore, *P. centralis* tadpoles exhibited faster development and larger body size at concentrations of 25% and 50% of the LC50 (Figueiredo & Rodrigues, 2014). The LC50 values were also determined for another South American anuran species, *Physalaemus albonotatus*, which varied with exposure time. The 48-h LC50 was 1040.2 mg/L, decreasing to 754.2 mg/L at 72 h and 350 mg/L at 96 h. Chronic exposure of *P. albonotatus* tadpoles to 2,4-D concentrations of 87.5, 175, and 262.5 mg/L resulted in a significant reduction in survival after 21 days of exposure. Tadpoles exposed to 2,4-D exhibited a shorter larval development time compared to the control group, as well as an increase in malformations affecting the oral disc, intestine, abdomen, and body shape (Curi et al., 2019). Recent studies have also investigated the effects of pesticide mixtures on tadpoles, such as the combination of 2,4-D and glyphosate. Exposure of *Boana faber* and *Leptodactylus latrans* tadpoles to this mixture led to irregular swimming activity, lethargy, convulsions, and morphological damage, particularly in the mouths of tadpoles. Cellular damage, including increased micronuclei and erythrocyte nuclear abnormalities (the most frequent being lobed nucleus), was also reported (Pavan et al., 2021).

2.3 Mancozeb (MZ)

Mancozeb is a widely used fungicide introduced in 1962, belonging to the dithiocarbamate class. It is commonly applied to crops such as soybeans, wheat, beans, and coffee. In Amazonia, mancozeb is the most popular fungicide and ranks third among the top-selling pesticides (IBAMA, 2022). However, its misuse is prevalent among farmers in the Manaus region, as they often exceed the recommended dose and number of applications (Römbke et al., 2008). Mancozeb (MZ) exhibits low solubility and persistence in soil. In the aquatic environment, it can be hydrolyzed within 24 h to 7 days. Nevertheless, its degradation product, ethylene thiourea (ETU), is more water-soluble and mobile in soils and is considered more toxic than MZ itself (Xu, 2000). Brazilian regulations allow a maximum concentration of 8 μg/L of MZ in drinking water for human consumption (Brasil, 2011). The specific effects of mancozeb on Amazonian species are not extensively documented. However, there are two reports concerning Brazilian tadpoles. Tadpoles of *Rhinella arenarum* exposed to 5 mg/L of MZ for 24 h did not show mortality, but a 100% mortality rate was observed after 48 h at the same concentration. Prolonged exposure for 336 days resulted in mortality at lower concentrations ranging from 0.01 to 1 mg/L. Chronic exposure also induced malformations in tadpoles (Asparch et al., 2020). Similarly, *Physalemus henselii* tadpoles exposed to a concentration of 2 mg/L of MZ for 24 h showed increased expression of the heat shock protein HSP70 compared to the control group. HSP70 is a cellular stress response protein commonly used as a marker to assess temperature and contamination effects on organisms (Azambuja et al., 2021).

2.4 Atrazine (ATZ)

Atrazine is a triazine herbicide that was introduced in the 1950s and is widely used for controlling grasses and broad-leaved weeds, particularly in sugarcane plantations. It exhibits high persistence and mobility in soil, which contributes to its contamination of the aquatic environment. Atrazine has been detected in public water supplies, groundwater, lakes, and rivers (de Albuquerque et al., 2020). Although it is not permitted in the European Union due to its potential for groundwater contamination, in Brazil, it is allowed in freshwater at concentrations up to 2 μg/L (Brasil, 2005). In Amazonia, atrazine has been detected with a frequency of 70% in water samples, ranging from 3 to 25 ng/L (Rico et al., 2022). As of 2021, atrazine ranks sixth among the top pesticides used in the biome (IBAMA, 2022). The effects of atrazine on anuran species are diverse and have been extensively studied. These effects include changes in biochemical parameters, genotoxic and mutagenic effects, behavioral changes, and abnormalities (de Albuquerque et al., 2020). For example, the species *Dendropsophus minutus*, which is also found in the Amazon basin, has a 96-h median lethal concentration (LC50) of 36.41 mg/L for atrazine. Although atrazine did not alter cell viability, the comet assay showed that tadpoles exhibited increased DNA damage compared to the control group (Gonçalves et al., 2017). The extent of DNA damage varied depending on the developmental stage, with tadpoles in the early and late stages showing a higher occurrence of DNA damage, while those in the intermediate stages showed no significant differences. Additionally, there was an increased frequency of micronuclei in tadpoles of

D. minutus in the early stages (Gonçalves et al., 2017). Similar effects were observed in tadpoles of *Rhinella diptycha* (at developmental stage 30) exposed to 19 mg/L of atrazine, resulting in an increased frequency of micronuclei and other nuclear abnormalities. Atrazine also induced histopathological lesions in the liver such as sinusoidal enlargement after 48 and 96 h of exposure to 12.5 mg/L atrazine and erythrocyte accumulation in vessels causing hypervascularization after 96 h of exposure to 1.5 mg/L atrazine. Morphological abnormalities were observed, including abnormalities of the caudal axis, mouth (losses or absence of keratodonts), lateral compression of the body, and invaginations in the dermis (Pérez-Iglesias et al., 2019). Other species, such as *Xenopus laevis*, have shown morphological abnormalities such as intersex gonads, heart malformation, visceral hemorrhage, intestine coiling malformations, curved body axes, and edema (Carr et al., 2003; Hayes et al., 2002; Lenkowski et al., 2008). Tadpoles of *Lithobates pipiens* exposed to atrazine also exhibited smaller size at metamorphosis, delayed gonadal development, and thymus damage (Hayes et al., 2006). Behavioral alterations have been reported for *Physalaemus cuvieri* at concentrations above 0.24 mg/L, and concentrations above 0.1 mg/L have been found to alter larval development time in tadpoles of *Rhinella arenarum* (Brodeur et al., 2009). The sensitivity of species to atrazine varies, with different responses observed among species. When comparing lethal endpoints in a sensitivity-distribution curve, *R. diptycha* exhibited intermediate sensitivity to atrazine compared to other globally distributed species. It was found to be more sensitive to atrazine than *Dendropsophus minutus* and *Physalaemus gracilis* (Rutkoski et al., 2018).

2.5 Metals and Metalloids

The expansion of illegal mining in the Amazon poses a significant environmental challenge for amphibians due to aquatic contamination by metals and metalloids, coupled with deforestation associated with these areas (Sonter et al., 2017). The extraction of various minerals in Amazonia contributes to contamination, notably in the extraction of cassiterite and gold, which constitute major sources of sediment in watercourses (Fearnside, 2019). During the cassiterite extraction process, substantial quantities of metals and metalloids are released, such as arsenic (As), mercury (Hg), copper (Cu), lead (Pb), and cadmium (Cd). Similarly, in gold extraction, mercury is employed to amalgamate fine gold particles, resulting in the release of tons of this metal into rivers (Moulatlet et al., 2023). Additionally, agricultural systems and solid waste disposal contribute to the introduction of metals into Amazonian aquatic environments, as fertilizers may contain elevated concentrations of these metals (Gupta et al., 2014). Pesticides may also contain metals associated with their compounds, such as manganese and zinc in association with the fungicide carbamate mancozeb.

Similar to pesticide contamination, the impact of mining extends beyond areas directly associated with mining activities, as metals can be transported and deposited through soil and rivers. In the case of the Amazon, during the rainy season, rising water levels transport this contaminated water to marginal regions such as puddles and shallows associated with streams and to environments frequently utilized by various amphibian species for reproduction. Elevated levels of these metals are currently identified in various regions of the Amazon basin, both in water and sediment samples, surpassing the limits established by legislation and deemed toxic to aquatic fauna (Ribeiro et al., 2017).

Analysis of samples from the Xingu and Fresco rivers in the state of Pará, Eastern Amazonia, revealed contamination by As, Cd, Cu, Pb, Hg, Chromium (Cr), and Nickel (Ni) in water, soil, and sediment samples (Ribeiro et al., 2017). Particularly noteworthy are the maximum values found for As (0.43 μg/L) and Hg (0.22 μg/L) in water samples, Cr (61.91 mg/kg) and Ni (26.62 mg/kg) in sediment samples, and Cr (339.89 mg/kg), Pb (106 mg/kg), and Ni (40.36 mg/kg) in soil samples, all of which exceeded the values permitted by CONAMA (Brasil, 2005). Samples from the Araguari River in the state of Amapá, Eastern Amazon also exhibited water samples with levels above those allowed by CONAMA for metals such as Cd (9 μg/L), Pb (160 μg/L), Fe (4530 μg/L), Hg (9 μg/L), Cu (11 μg/L), and Zn (4800 μg/L) (Viana et al., 2020). Many anuran species utilize small water formations such as holes in fallen trees, holes in trunks and roots of living trees, and fallen palm bracts, in addition to puddles formed on rocky substrates and artificial plastic or metallic containers. Contamination data in these ephemeral environments are scarce, but they can be polluted through airborne metal pollution during ore extraction, as exemplified by Hg in gold mining. Water and sediment samples from these environments, used by a species in French Guiana, *Dendrobates tinctorius*, also revealed high concentrations of Hg ranging from 0.007 to 13.040 ppm (Schlippe-Justicia et al., 2024).

While awareness of contamination in the Amazon exists, there is still limited information regarding the impacts of metals and metalloids on amphibian species in the region. Tadpoles of *Scinax ruber* and *Rhinella granulosa*, exposed to Cu concentrations, exhibited distinct sensitivities, as evidenced by 96 h LC50 values of 23.48 and 36.37 μg/L, respectively. In the case of *Scinax ruber* eggs, the LC50 was 50.02 μg/L (Franco-de-Sá & Val, 2014). Similarly, eggs and tadpoles of *Osteocephalus taurinus*, subjected to water contaminated with heavy metals from polluted streams, showed 96 h LC50 values of 25.19 and 47.53 mL/L, respectively. Additionally, exposure to the water from the polluted streams induced an increase in larval development time and body size during metamorphosis. Among the identified metals in the streams, notable ones include Iron (Fe) (3.60 mg/L), Zinc (Zn) (0.13 mg/L), Ni (0.08 mg/L), Manganese (Mn) (0.17 mg/L), Cu (<0.02 mg/L), Cr (0.20 mg/L), Pb (0.19 mg/L), and Cobalt (Co) (< 0.05 mg/L). Except for Cu and Co, all exceeded the limits permitted by legislation (Franco-de-Sá et al., 2012). Commonly observed effects in other species facing contamination by metals and metalloids include mortality and varying levels of sensitivity, as well as morphological, physiological, and behavioral alterations (Chen et al., 2009; Thambirajah et al., 2019). For instance, tadpoles of *Bufo gargarizans* and *Lithobates pipiens* exposed to mercury and methylmercury, derived from mercury methylation in water, experienced mortality, thyroid disruptions, liver damage, oxidative stress, and lipid metabolism disorders (Chang et al., 1974; Shi et al., 2018). Tadpoles of *Rana chensinensis* exposed to Cr and Pb, alone or in combination, exhibited malformations, intestinal histological injuries, and reduced activities of digestive enzymes (Zhang et al., 2023). Moreover, metals such as Pb, Cu, Cd, and Zn affected the survival of *Xenopus laevis* tadpoles, declining as the concentration of these metals increased, while for eggs, only Pb, Cu, and Cd reduced hatching success (Haywood et al., 2004). In *Fejervarya limnocharis* tadpoles, exposure to Cd led to concentration-dependent mortality, along with accelerated metamorphosis, reduced body weight, and genotoxicity (Patar et al., 2016).

2.6 Microplastics (MP)

Plastics and their associated waste have become a significant environmental concern due to their widespread use and improper disposal. Plastic-based products offer various advantages such as flexibility, corrosion resistance, versatility, durability, and low production cost (Shen et al., 2020). However, the exponential growth in plastic production has led to the accumulation of plastic waste, with approximately 5800 million tons (Mt) of plastic waste generated between 1950 and 2015. Out of this total, around 800 Mt. was incinerated, 100 Mt. was reclaimed, and 4900 Mt. was landfilled (Geyer et al., 2017; Yang et al., 2021). Recent years have also shown an increasing trend in plastic production, with 335 Mt. produced in 2016 and 367 Mt. in 2020 (PlasticsEurope, 2021). Microplastics are particles smaller than 5 mm, and nanoplastics are smaller than 1u g, and nowadays are widely distributed in aquatic aerial and terrestrial ecosystems (Ringwood, 2021). They can be classified as primary or secondary microplastics: primary microplastics are intentionally produced for specific purposes, such as microbeads in personal care products, while secondary microplastics result from the degradation of larger plastic items, including synthetic fibers (Balestrieri et al., 2022; Castro-Castellon et al., 2022). Microplastics exhibit various morphologies, including pellets, fragments, lines, and fibers (Massarelli et al., 2021). While the presence of microplastics has been extensively documented globally, research on microplastic pollution in Amazonian rivers is relatively limited and new. The first record of microplastic pollution in the region was based on sediment samples from the Solimões, Negro, and Amazonas rivers, revealing concentrations ranging from 417 to 8178 particles/kg of dry sediment (Gerolin et al., 2020). Water samples from various rivers in the Amazon region revealed distinct levels of microplastic contamination. Specifically, the Amazon River displayed concentrations ranging from 0.003 to 0.67 mg/m^3. In the Tapajós River, the values were measured between 0.53 and 1.9 mg/m^3. The Negro River exhibited concentrations from 0.007 to 0.29 mg/m^3, with higher values near Manaus. The Tocantins River showed concentrations ranging from 0.31 to 0.33 mg/m^3. These findings highlight the variability in microplastic pollution across different water bodies in the Amazon region (Rico et al., 2023).

Once microplastics enter the aquatic environment, they become suspended in the water column and can adhere to algae and substrates. Tadpoles, due to their feeding habits, are prone to ingesting microplastics as they filter phytoplankton from the water and scrape algae from substrates. Microplastics can enter tadpoles' bodies primarily through the mouth and gills. After ingestion, microplastics can accumulate in various organs, including the heart, liver, and intestine. In the intestine, microplastics may accumulate and create a false sense of satiety, leading to reduced food intake and absorption. This can result in deleterious effects such as changes in growth rate, behavior, histopathological damage, erythrocyte changes, and malformations. Furthermore, microplastics can interact with pesticides in the aquatic environment, increasing the toxicity of these chemicals to tadpoles (Lajmanovich et al., 2022). The greatest contributions to the knowledge of the effects of microplastics on Brazilian tadpoles are the studies conducted with *Physalaemus cuvieri*, a widely distributed species in the country. Exposure to environmentally relevant concentrations was able to induce different hepatotoxic and cytotoxic effects (da Costa Araújo et al., 2020). The accumulation of microplastics in the liver indicates the ability to absorb microplastics from water, and this accumulation induces blood vessel dilation, infiltration, congestion, hydropic degeneration, hypertrophy, and hyperplasia (da Costa Araújo et al., 2020). Microplastics also affected erythrocyte nuclear morphometry (da Costa Araújo et al., 2020) and induced behavioral alterations such as reduced locomotor activity, increased anxiety index, and the absence of the common antipredator response of tadpole aggregation (da Costa Araújo & Malafaia, 2020). Studies on *Scinax squalirostris* tadpoles have shown that the activity of metabolic enzymes, specifically carboxylesterases (CbEs) and alkaline phosphatase (ALP), is negatively affected by microplastics (MPs). This indicates significant alterations in physiological and immunological processes, which can lead to a loss of fitness for the tadpoles (Attademo et al., 2022). Furthermore, microplastics can be transferred through the trophic chain. Tadpoles of *Physalaemus cuvieri*, contaminated with microplastics, were preyed upon by juvenile tambatinga fish (*Colossoma macropomum* x *Piaractus brachypomus*). These fish, in turn, were preyed upon by healthy Swiss mice (*Mus musculus*). As a result of this transfer, Swiss mice individuals were found to have an accumulation of microplastics in their livers (da Costa Araújo & Malafaia, 2021). This highlights the potential for microplastics to enter and persist in higher levels of the food chain, posing risks to various omnivore and carnivore consumers.

3 Interaction Between Temperature and Contaminants

Anurans, being sensitive organisms, are greatly influenced by the environmental characteristics of their habitat. The most significant abiotic factor impacting amphibians is temperature, which has been increasing since the end of the industrial revolution, with a more pronounced rise in recent decades. A comparison between the preindustrial period (1850–1900) and the years 2011–2020 reveals a global temperature increase of 1.09 °C, with 1.07 °C attributed to intense human activities (Masson-Delmotte et al., 2021). Human activities, such as the burning of fossil fuels and deforestation, contribute to the emission of greenhouse gases (GHGs) into the atmosphere, further exacerbating the global temperature increase (Manabe, 2019). Temperature plays a crucial regulatory role in the physiological and behavioral processes of anurans. Fluctuations in temperature on a daily and seasonal basis elicit adaptive responses that influence various aspects of anuran biology, including locomotor performance, oxygen consumption, digestion, and gas exchange routes (Navas et al., 2008). The thermal characteristics of anurans are defined by thermal performance curves, which include the thermal optimum temperature (Topt), thermal breadth, and critical thermal limits. The Gaussian-shaped curve is bounded by the maximum (CTMax) and minimum (CTMin) thermal limits. Anurans typically have an optimum temperature for development (Topt) that falls within the CTMax and CTMin, with temperatures outside this range potentially causing physiological damage or death (Duarte et al., 2012; Katzenberger et al., 2012). The thermal tolerance range of anurans shows a positive correlation with the variation in ambient temperature, with higher latitudes experiencing greater seasonal temperature variations and consequently broader thermal amplitudes. Conversely, amphibians in tropical regions, where seasonal temperature variation is lower, tend to have narrower thermal amplitudes (Deutsch et al., 2008). Anurans employ strategies to mitigate the negative effects of temperature, including phenotypic plasticity and local adaptation.

Phenotypic plasticity allows individuals to express different phenotypes from a single genotype in response to environmental challenges. Plasticity in thermal tolerance is associated with the magnitude of temperature variation, as proposed by the climate variability hypothesis. Consequently, tropical amphibians may be more susceptible to climate change due to their proximity to their optimal temperature range (Janzen, 1967; Levins, 1968). Although robust data specific to amphibians are currently lacking, an alternative hypothesis regarding the susceptibility of tropical ectotherms suggests that these species already inhabit environments where ambient temperatures are near their upper thermal limits (Nati et al., 2021; Vinagre et al., 2019). These findings suggest that the susceptibility of ectothermic organisms, including amphibians, to climate change, may indeed be influenced by the proximity of current temperatures to their upper thermal limits. However, it is important to note that further research specifically focusing on amphibians is needed to validate these trends and understand the unique factors contributing to their vulnerability. Local adaptation, on the other hand, refers to changes in genotype and phenotype that occur after exposure to a stressor, even if these changes do not confer greater fitness in other environments. Individuals with locally adapted traits may have a competitive advantage and higher fitness within their specific habitat compared to genotypes from other habitats (Kawecki & Ebert, 2004).

Thermal thresholds in organisms can be influenced by exposure to contaminants, which is a matter of concern, particularly for thermosensitive organisms. Notably, studies on tadpoles have revealed the impact of specific contaminants on their thermal tolerance. For instance, research conducted on *Rhinella arenarum* tadpoles exposed to chlorpyrifos demonstrated a reduction in their CTmax, the maximum critical thermal limit, by approximately 4 °C (Quiroga et al., 2019). Similarly, tadpoles of *Physalaemus henselii* exposed to mancozeb exhibited a decrease in their thermal tolerance by 2.41 °C (Azambuja et al., 2021). Another commonly observed phenomenon is the synergistic effect of contaminants and elevated temperatures, resulting in increased toxicity. In the case of *Ranoidea citropa* tadpoles, exposure to endosulfan at fluctuating temperatures proved to be more lethal compared to exposure at a constant temperature (Broomhall, 2002). This indicates that temperature fluctuations enhance the toxicity of endosulfan in this species. Temperature can also interact with glyphosate, a widely used herbicide. Research on *Boana faber* tadpoles demonstrated that exposure to glyphosate at higher temperatures (32 °C) led to increased mortality compared to lower temperatures (25 °C) (Alves-Ferreira et al., 2023). This interaction suggests that elevated temperatures intensify the toxic effects of glyphosate on tadpoles. Interestingly, the effects of contaminants on tadpole development and growth can be temperature-dependent. Exposure to Tebuthiuron at 25 °C accelerated the development of *Aquarana catesbeianus* larvae without affecting their mass gain. However, at 32 °C, tadpole development was delayed, mass gain reduced, and total length unaffected (Grott et al., 2022). This illustrates the intricate relationship between contaminants, temperature, and their combined effects on tadpole physiology. These findings underscore the significance of considering the impact of contaminants and temperature on thermosensitive organisms. By exploring these interactions, we can gain insights into the vulnerability and responses of organisms to environmental stressors, facilitating informed conservation and management efforts.

4 Implications of Deleterious Effects for Tadpoles

As previously discussed, contaminants interact in intricate ways with temperature. The combination of contaminants and elevated temperature can induce a spectrum of adverse effects on tadpoles, with documented alterations in morphological, behavioral, and physiological parameters throughout the larval developmental phase. These effects not only lead to immediate lethal consequences but also result in adults with lower fitness, as there is a correlation between tadpole performance and adult fitness. During the larval stage, tadpoles acquire energy and build up sufficient energetic reserves, crucial for the metamorphic climax, a period characterized by profound morphological and physiological changes (Orlofske & Hopkins, 2009; Shi, 2000). Malformations in tadpoles are frequently observed, particularly in the mouth apparatus. These malformations can involve the total or partial absence of keratodont rows and the loss of the upper jaw sheath (Herek et al., 2020; Samojeden et al., 2022). Changes in eye structure, such as the absence of eyes or lens abnormalities, and abnormally small eyes have also been reported (Sánchez-Domene et al., 2018). Tadpoles may also experience damage to the caudal structure (da Silva et al., 2020), the gut, and increased edema in the body (Herek et al., 2020). These alterations in mouthparts and intestines, essential for feeding and nutrient absorption, can result in nutrient deficiencies (Tolledo et al., 2014; Venesky et al.,

2010). In the larval phase, the degeneration of tadpole mouthparts occurs as adult mandibles develop, leading to a temporary interruption of feeding and requiring the use of energy accumulated during the larval period (Beck & Congdon, 2003). Consequently, increased mortality and reduced body size of tadpoles can be observed. Smaller body size is also associated with higher vulnerability to predation, and since tadpole size influences the size of adult individuals, smaller tadpoles are likely to become smaller adults with potentially reduced reproductive success and predator avoidance efficiency (Adams et al., 2021; Beasley, 2020; Samojeden et al., 2022; Smith, 1987). These effects not only compromise healthy tadpole development but also result in lower energy accumulation during the larval phase (Beck & Congdon, 2003). Furthermore, damage to the caudal structure and eye components can alter tadpole behavioral patterns, including reduced effectiveness in predator avoidance and foraging (Bach et al., 2016; Lajmanovich et al., 2003). Changes in tadpole growth and development rates are also frequently observed, reflecting the overall disruption caused by contaminants and increased temperature.

Tadpoles' metamorphosis exhibits high plasticity in their response to adverse environments, enabling them to adjust their developmental rates. However, there is a trade-off between accelerated developmental rates and body growth in tadpoles, with early metamorphosing individuals typically displaying smaller body sizes. As mentioned earlier, smaller tadpole body size translates into smaller adults with lower reproductive success. Contaminants and temperature can induce changes in development and growth by disrupting endocrine function (Grott et al., 2022; Liu et al., 2021). The thyroid gland plays a primary role in initiating metamorphosis in amphibians (Vieira, 2010). Under stressful conditions, the levels of stress hormones, such as corticotropin-releasing factor, corticotropin, and glucocorticoids, increase. These hormones act on the hypothalamus-pituitary-thyroid axis, which regulates the production of thyroid hormones (THs) (Denver, 2009; Ruthsatz et al., 2019). The main THs are tetraiodothyronine (T4) and triiodothyronine (T3), and they serve as important mediators of tadpole metamorphosis. It is through these hormones that the rate of larval development is established (Vieira, 2010). Elevated levels of THs accelerate tadpole development and increase metabolic rate, leading to smaller-bodied metamorphosed tadpoles (Denver, 2021). In the long term, these changes in amphibian populations can alter trophic dynamics in many communities. The smaller body sizes of metamorphosed tadpoles can impact their roles as prey or predators within their respective ecosystems, influencing the overall balance of food webs and trophic interactions. In addition to the well-known effects of accelerated development and reduced size of tadpoles in response to pesticides, it is noteworthy that exposure to low concentrations of these chemicals can, conversely, result in an increase in tadpole size (Figueiredo & Rodrigues, 2014). This outcome may be attributed to the phenomenon known as hormesis, which is a biphasic response characterized by stimulation at low doses and inhibition at high doses (Calabrese, 2008). The ability of tadpoles to exhibit an increase in body size at lower doses underscores the importance of understanding the underlying mechanisms of hormesis for a comprehensive assessment of the environmental risks associated with pesticide exposure.

Within Neotropical ecosystems, amphibians play a vital role in energy flow and nutrient cycling (Valencia-Aguilar et al., 2013). As adult organisms, they act as important carnivores, primarily preying on insects, worms, and other invertebrates, thus contributing to the regulation of pest populations (Halliday, 2008). Similarly, during their larval stage, herbivorous tadpoles play a crucial role in controlling algae populations. Both life stages serve as prey for a diverse array of predators, and any reduction in their biomass can lead to significant changes in ecosystem dynamics (Whiles et al., 2006).

5 Conclusion

Tadpoles face a myriad of environmental stressors that significantly impact their morphology, physiology, and behavior. Among these stressors, temperature plays a crucial role, as it not only influences fundamental processes for amphibians but also interacts with other stressors, potentially magnifying their effects. It is noteworthy that there is a notable lack of ecotoxicological studies on amphibians worldwide, a gap that becomes even more pronounced when considering research conducted in Brazil and the Amazonian region. Existing information examining the effects of contaminants and temperature increase in isolation indicates a wide range of important consequences, including increased mortality rates, the occurrence of malformations and structural damage, as well as biochemical, genotoxic, and neurotoxic effects. However, the limited amount of studies that have investigated the interaction between contaminants and temperature suggest that temperature can enhance the toxicity of contaminants, leading to adverse effects even at lower concentrations. It is crucial to acknowledge that the specific temperature and contamination thresholds required to induce these effects can vary among species. Additionally, in natural environments, other stressors such as oxygen levels, UV radiation exposure, predator presence, food availability, and hydroperiod (the duration and timing of wet and dry periods) also interact with temperature and pollutants.

To gain a comprehensive understanding of the relationship between amphibians, contamination, and temperature, conducting baseline research is imperative. However, this task is challenging due to the vast number of amphibian species, their idiosyncratic responses to environmental stressors, the variety of contaminants present, and the complex projections related to climate change. Amphibians, highly sensitive to environmental changes, serve as crucial ecological indicators, highlighting the potential impacts of contamination and climate change on biodiversity as a whole.

Therefore, prioritizing research and conservation efforts for amphibians is vital to mitigate the threats they face. The future prospects for amphibians and their habitats are alarming, as the next three decades will be critical in achieving the long-term mitigation goals outlined in the Paris Agreement. In effect since 2016, the agreement aims to limit global temperature increase to well below 2 °C above preindustrial levels and pursue efforts to limit it to 1.5 °C. Addressing these challenges and ensuring biological conservation worldwide requires international collaboration and concerted efforts. This necessitates implementing efficient public policies, scientific monitoring, and a global collaborative effort to adopt climate change mitigation strategies that consider the unique vulnerabilities of amphibians. Investments in technologies aimed at reducing CO2 emissions, such as renewable energies, along with habitat restoration initiatives, are important components of this collective task. By protecting amphibians and their habitats, we can contribute to the overall health and resilience of ecosystems and safeguard the biodiversity upon which we all depend.

References

Adams, E., Leeb, C., & Brühl, C. A. (2021). Pesticide exposure affects reproductive capacity of Common toads (Bufo bufo) in a viticultural landscape. *Ecotoxicology, 30*, 213–223. https://doi.org/10.1007/s10646-020-02335-9

Adelizzi, R., Portmann, J., & Van Meter, R. (2019). Effect of individual and combined treatments of pesticide, fertilizer, and salt on growth and corticosterone levels of larval southern Leopard frogs (Lithobates sphenocephala). *Archives of Environmental Contamination and Toxicology, 77*, 29–39. https://doi.org/10.1007/s00244-019-00629-6

Agostini, M. G., Natale, G. S., & Ronco, A. E. (2010). Lethal and sublethal effects of cypermethrin to Hypsiboas pulchellus tadpoles. *Ecotoxicology, 19*, 1545–1550. https://doi.org/10.1007/s10646-010-0539-3

Agostini, M. G., Roesler, I., Bonetto, C., Ronco, A. E., & Bilenca, D. (2020). Pesticides in the real world: The consequences of GMO-based intensive agriculture on native amphibians. *Biological Conservation, 241*, 108355. https://doi.org/10.1016/j.biocon.2019.108355

Alves-Ferreira, G., Katzenberger, M., Fava, F. G., Costa, R. N., Carilo Filho, L. M., & Solé, M. (2023). Roundup Original DI® and thermal stress affect survival, morphology and thermal tolerance in tadpoles of Boana faber (Hylidae, Anura). *Ecotoxicology, 32*, 93–101. https://doi.org/10.1007/s10646-023-02622-1

Arraut, J. M., Nobre, C., Barbosa, H. M. J., Obregon, G., & Marengo, J. (2012). Aerial rivers and lakes: Looking at large-scale moisture transport and its relation to Amazonia and to subtropical rainfall in South America. *Journal of Climate, 25*, 543–556. https://doi.org/10.1175/2011JCLI4189.1

Asparch, Y., Svartz, G., & Pérez Coll, C. (2020). Toxicity characterization and environmental risk assessment of Mancozeb on the South American common toad Rhinella arenarum. *Environmental Science and Pollution Research, 27*, 3034–3042. https://doi.org/10.1007/s11356-019-07248-7

Attademo, A. M., Cuzziol Boccioni, A. P., Peltzer, P. M., Franco, V. G., Simoniello, M. F., Passeggi, M. C. G., & Lajmanovich, R. C. (2022). Effect of microplastics on the activity of carboxylesterase and phosphatase enzymes in Scinax squalirostris tadpoles. *Environmental Monitoring and Assessment, 194*, 718. https://doi.org/10.1007/s10661-022-10322-2

Azambuja, G., Martins, I. K., Franco, J. L., & dos Santos, T. G. (2021). Effects of mancozeb on heat shock protein 70 (HSP70) and its relationship with the thermal physiology of Physalaemus henselii (Peters, 1872) tadpoles (Anura: Leptodactylidae). *Journal of Thermal Biology, 98*, 102911. https://doi.org/10.1016/j.jtherbio.2021.102911

Bacchetta, R., Mantecca, A. M., Vismara, C., & Vailati, G. (2008). Axial–skeletal defects caused by Carbaryl in Xenopus laevis embryos. *Science of the Total Environment, 392*(1), 110–118. https://doi.org/10.1016/j.scitotenv.2007.11.031

Baccini, A., Goetz, S. J., Walker, W. S., Laporte, N. T., et al. (2012). Estimated carbon dioxide emissions from tropical deforestation improved by carbon-density maps. *Nature Climate Change, 2*, 182–185. https://doi.org/10.1038/nclimate1354

Bach, N. C., Natale, G. S., Somoza, G. M., & Ronco, E. R. (2016). Effect on the growth and development and induction of abnormalities by a glyphosate commercial formulation and its active ingredient during two developmental stages of the South-American Creole frog, Leptodactylus latrans. *Environmental Science and Pollution Research, 23*, 23959–23971. https://doi.org/10.1007/s11356-016-7631-z

Baier, F., Jedinger, M., Gruber, E., & Zaller, J. G. (2016). Temperature-dependence of glyphosate-based herbicide's effects on egg and tadpole growth of Common Toads. *Frontiers in Environmental Sciences, 4*(28), 51. https://doi.org/10.3389/fenvs.2016.00051

Balestrieri, A., Winkler, A., Scribano, G., Gazzola, A., Lastrico, G., Grioni, A., Pellitteri-Rosa, D., & Tremolada, P. (2022). Differential effects of microplastic exposure on anuran tadpoles: A still underrated threat to amphibian conservation? *Environmental Pollution, 303*, 119137. https://doi.org/10.1016/j.envpol.2022.119137

Balvanera, P., Pfaff, A., Merino, L., Molnár, Z., & Purvis, A. (2020). Our world in 2020. In R. E. A. Almond, M. Grooten, & T. Petersen (Eds.), *Living planet report 2020 - Bending the curve of biodiversity loss* (pp. 52–55). WWF.

Barbosa, A. M. C., Solano, M. d. L. M., & Umbuzeiro, G. d. A. (2015). Pesticides in drinking water—The Brazilian monitoring program. *Frontiers in Public Health, 3*, 246. https://doi.org/10.3389/fpubh.2015.00246

Beasley, V. R. (2020). Direct and indirect effects of environmental contaminants on amphibians. In *Reference module in earth systems and environmental sciences*. Elsevier. https://doi.org/10.1016/B978-0-12-409548-9.11274-6

Beck, C. W., & Congdon, J. D. (2003). Energetics of metamorphic climax in the Southern toad (Bufo terrestris). *Oecologia, 137*, 344–351. https://doi.org/10.1007/s00442-003-1374-5

Benbrook, C. M. (2016). Trends in glyphosate herbicide use in the United States and globally. *Environmental Sciences Europe, 28*, 3. https://doi.org/10.1186/s12302-016-0070-0

Bishop, P. J., Angulo, A., Lewis, J. P., Moore, R. D., Rabb, G. B., & Moreno, J. G. (2012). The amphibian extinction crisis - What will it take to put the action into the amphibian conservation action plan? *Sapiens, 5*(2), 97–111.

Boccioni, A. P. C., Lajmanovich, R. C., Peltzer, P. M., Attademo, A. M., & Martinuzzi, C. S. (2021). Toxicity assessment at different experimental scenarios with glyphosate, chlorpyrifos and antibiotics in Rhinella arenarum (Anura: Bufonidae) tadpoles. *Chemosphere, 273*, 128475. https://doi.org/10.1016/j.chemosphere.2020.128475

Bouhafs, N., Berrebbah, H., Devaux, A., Rouabhi, R., & Djebar, M. R. (2009). Micronucleus induction in erythrocytes of tadpole Rana saharica (Green frog of North Africa) exposed to Artea 330EC. *American-Eurasian Journal of Toxicological Sciences, 1*(1), 07–12.

Bragg, A. N. (1952). Decline in toad populations in Central Oklahoma. *Proceedings of the Oklahoma Academy of Science, 33*, 1.

Brasil. (2005). *Resolução CONAMA n° 357, de 17 de março de 2005*. Conselho Nacional de Meio Ambiente. Available: https://www.icmbio.gov.br/cepsul/images/stories/legislacao/Resolucao/2005/res_conama_357_2005_classificacao_corpos_agua_rtfcda_altrd_res_393_2007_397_2008_410_2009_430_2011.pdf. Accessed 25 May 2023

Brasil. (2011). *Portaria n° 2.914, de 12 de dezembro de 2011*. Ministério da Saúde. Available: https://bvsms.saude.gov.br/bvs/saudelegis/gm/2011/prt2914_12_12_2011.html. Accessed 20 Jan 2023

Brodeur, J. C., Svartz, G., Perez-Coll, C. S., Marino, D. J. G., & Herkovits, J. (2009). Comparative susceptibility to atrazine of three developmental stages of Rhinella arenarum and influence on metamorphosis: Non-monotonous acceleration of the time to climax and delayed tail resorption. *Aquatic Toxicology, 91*, 161–170. https://doi.org/10.1016/j.aquatox.2008.07.003

Broomhall, S. (2002). The effects of endosulfan and variable water temperature on survivorship and subsequent vulnerability to predation in Litoria citropa tadpoles. *Aquatic Toxicology, 61*, 243–250. https://doi.org/10.1016/S0166-445X(02)00061-9

Brovini, E. M., de Deus, B. C. T., Vilas-Boas, J. A., Quadra, G. R., Carvalho, L., Mendonça, R. F., Pereira, R. d. O., & Cardoso, S. J. (2021). Three-bestseller pesticides in Brazil: Freshwater concentrations and potential environmental risks. *Science of the Total Environment, 771*, 144754. https://doi.org/10.1016/j.scitotenv.2020.144754

Calabrese, E. J. (2008). Hormesis: Why it is important to toxicology and toxicologists. *Environmental Toxicology and Chemistry, 27*, 1451–1474. https://doi.org/10.1897/07-541

Carnaval, A. C. O. D. Q., Puschendorf, R., Peixoto, O. L., Verdade, V. K., & Rodrigues, M. T. (2006). Amphibian chytrid fungus broadly distributed in the Brazilian Atlantic Rain Forest. *EcoHealth, 3*, 41–48. https://doi.org/10.1007/s10393-005-0008-2

Carneiro, F. E., Grott, S. C., Israel, N. G., Bitschinski, D., Abel, G., Alves, T. C., de Albuquerque, C. A. C., da Silva, E. B., & de Almeida, E. A. (2023). Influence of temperature on the biomarker responses of bullfrog tadpoles (Lithobates catesbeianus) to 2-hydroxyatrazine exposure. *Aquatic Toxicology, 257*, 106468. https://doi.org/10.1016/j.aquatox.2023.106468

Carr, J. A., Gentles, A., Smith, E. E., Goleman, W. L., Urquidi, L. J., Thuett, K., Kendall, R. J., Giesy, J. P., Gross, T. S., Solomon, K. R., & Van Der Kraak, G. (2003). Response of larval Xenopus laevis to atrazine: Assessment of growth, metamorphosis, and gonadal and laryngeal morphology. *Environmental Toxicology and Chemistry, 22*, 396–405. https://doi.org/10.1002/etc.5620220222

Carrasco, G. H., de Souza, M. B., & de Souza Santos, L. R. (2021). Effect of multiple stressors and population decline of frogs. *Environmental Science and Pollution Research, 28*, 59519–59527. https://doi.org/10.1007/s11356-021-16247-6

Carvalho, W. F., Franco, F. C., Godoy, F. R., Folador, D., Avelar, J. B., Nomura, F., da Cruz, A. D., de Sabóia-Morais, S. M. T., Bastos, R. P., & Silva, D. d. M. (2018). Evaluation of genotoxic and mutagenic effects of glyphosate Roundup Original ® in Dendropsophus minutus Peters, 1872 tadpoles. *South American Journal of Herpetology, 13*, 220–229. https://doi.org/10.2994/SAJH-D-17-00016.1

Casara, K. P., Vecchiato, A. B., Lourencetti, C., Pinto, A. A., & Dores, E. F. G. C. (2012). Environmental dynamics of pesticides in the drainage area of the São Lourenço River headwaters, Mato Grosso State, Brazil. *Journal of the Brazilian Chemical Society, 23*, 1719–1731. https://doi.org/10.1590/S0103-50532012005000037

Castro-Castellon, A. T., Horton, A. A., Hughes, J. M. R., Rampley, C., Jeffers, E. S., Bussi, G., & Whitehead, P. (2022). Ecotoxicity of microplastics to freshwater biota: Considering exposure and hazard across trophic levels. *Science of the Total Environment, 816*, 151638. https://doi.org/10.1016/j.scitotenv.2021.151638

Chang, L. W., Reuhi, K. R., & Dudley, A. W., Jr. (1974). Effects of methylmercury chloride on Rana pipiens tadpoles. *Environmental Research, 8*, 82–91. https://doi.org/10.1016/0013-9351(74)90065-6

Chen, A., Deng, H., Song, X., Liu, X., & Chai, L. (2022). Effects of separate and combined exposure of cadmium and lead on the endochondral ossification in Bufo gargarizans. *Environmental Toxicology and Chemistry, 41*(5), 1228–1245. https://doi.org/10.1002/etc.52967

Chen, T., Gross, J. A., & Karasov, W. (2009). Chronic exposure to pentavalent arsenic of larval leopard frogs (Rana pipiens): Bioaccumulation and reduced swimming performance. *Ecotoxicology, 18*, 587–593. https://doi.org/10.1007/s10646-009-0316-3

Copernicus Climate Change Service (C3S) - Global Climate Highlights 2023. https://climate.copernicus.eu/copernicus-2023-hottest-year-record Accessed 10 Jan 2024.

Cordier, J. M., Aguilar, R., Lescano, J. N., Leynaud, G. C., Bonino, A., Miloch, D., Loyola, R., & Nori, J. (2021). A global assessment of amphibian and reptile responses to land-use changes. *Biological Conservation, 253*, 108863. https://doi.org/10.1016/j.biocon.2020.108863

Correia, N. M., Carbonari, C. A., & Velini, E. D. (2020). Detection of herbicides in water bodies of the Samambaia River sub-basin in the Federal District and eastern Goiás. *Journal of Environmental Science and Health Part B, 55*, 574–582. https://doi.org/10.1080/03601234.2020.1742000

Curi, L. M., Peltzer, P. M., Sandoval, M. T., & Lajmanovich, R. C. (2019). Acute toxicity and sublethal effects caused by a commercial herbicide formulated with 2,4-D on Physalaemus albonotatus tadpoles. *Water, Air, and Soil Pollution, 230*, 22. https://doi.org/10.1007/s11270-018-4073-x

D'Avila, R. d. S., Brum, B. R., Hurtado, T. C., & Ignácio, Á. R. A. (2020). Análise quantitativa temporal sobre os efeitos do uso de agrotóxicos em anfíbios—anuros. *Research, Society and Development, 9*, e383985682. https://doi.org/10.33448/rsd-v9i8.5682

da Costa Araújo, A. P., Gomes, A. R., & Malafaia, G. (2020). Hepatotoxicity of pristine polyethylene microplastics in neotropical physalaemus cuvieri tadpoles (Fitzinger, 1826). *Journal of Hazardous Materials, 386*, 121992. https://doi.org/10.1016/j.jhazmat.2019.121992

da Costa Araújo, A. P., & Malafaia, G. (2020). Can short exposure to polyethylene microplastics change tadpoles' behavior? A study conducted with neotropical tadpole species belonging to order anura (Physalaemus cuvieri). *Journal of Hazardous Materials, 391*, 122214. https://doi.org/10.1016/j.jhazmat.2020.122214

da Costa Araújo, A. P., & Malafaia, G. (2021). Microplastic ingestion induces behavioral disorders in mice: A preliminary study on the trophic transfer effects via tadpoles and fish. *Journal of Hazardous Materials, 401*, 123263. https://doi.org/10.1016/j.jhazmat.2020.123263

da Silva, P. R., Borges-Martins, M., & Oliveira, G. T. (2021). Melanophryniscus admirabilis tadpoles' responses to sulfentrazone and glyphosate-based herbicides: An approach on metabolism and antioxidant defenses. *Environmental Science and Pollution Research, 28*, 4156–4172. https://doi.org/10.1007/s11356-020-10654-x

da Silva, M. B., Fraga, R. E., Silva, F. L., De Oliveira, L. A. A., De Queiroz, T. S., Rocha, M. A., & Juncá, F. A. (2020). Effects of acute exposure of chlorpyrifos on the survival, morphology, and swimming ability of Odontophrynus carvalhoi tadpoles. *Ecotoxicology and Environmental Contamination, 15*, 37–42. https://doi.org/10.5132/eec.2020.01.05

da Silva, M. D., Peralba, M. D. C. R., & Mattos, M. L. T. (2003). determinação de glifosato e ácido aminometilfosfônico em águas superficiais do arroio passo do pilão. *Pestic Rev Ecotoxicol e Meio Ambien, 13*, 19–28. https://doi.org/10.5380/pes.v13i0.3161

Daam, M. A., Moutinho, M. F., Espíndola, E. L. G., & Schiesari, L. (2019). Lethal toxicity of the herbicides acetochlor, ametryn, glyphosate and metribuzin to tropical frog larvae. *Ecotoxicology, 28*, 707–715. https://doi.org/10.1007/s10646-019-02067-5

de Albuquerque, F. P., de Oliveira, J. L., Moschini-Carlos, V., & Fraceto, L. F. (2020). An overview of the potential impacts of atrazine in aquatic environments: Perspectives for tailored solutions based on nanotechnology. *Science of the Total Environment, 700*, 134868. https://doi.org/10.1016/j.scitotenv.2019.134868

de Maria, S. F., Parizi, J. L. S., Odorizzi, G. A. S. d. M., Sato, G. M. R. H., Patrão, I. B., Chagas, P. H. N., de Azevedo Mello, F., & Nai, G. A. (2021). Subchronic exposure to a glyphosate-based herbicide causes dysplasia in the digestive tract of Wistar rats. *Environmental Science and Pollution Research, 28*, 61477–61496. https://doi.org/10.1007/s11356-021-15051-6

Delmonico, E. L., Bertozzi, J., Souza, N. E. d., & Oliveira, C. C. (2014). Determination of glyphosate and aminomethylphosphonic acid for assessing the quality tap water using SPE and HPLC. *Acta Scientiarum. Technology, 36*, 513. https://doi.org/10.4025/actascitechnol.v36i3.22406

Denver, R. J. (2009). Stress hormones mediate environment-genotype interactions during amphibian development. *General and Comparative Endocrinology, 164*, 20–31. https://doi.org/10.1016/j.ygcen.2009.04.016

Denver, R. J. (2021). Stress hormones mediate developmental plasticity in vertebrates with complex life cycles. *Neurobiology of Stress, 14*, 100301. https://doi.org/10.1016/j.ynstr.2021.100301

Deutsch, C. A., Tewksbury, J. J., Huey, R. B., Sheldon, K. S., Ghalambor, C. K., Haak, D. C., & Martin, P. R. (2008). Impacts of climate warming on terrestrial ectotherms across latitude. *Proceedings of the National Academy of Sciences, 105*, 6668–6672. https://doi.org/10.1073/pnas.0709472105

Dirmeyer, P. A., Brubaker, K. L., & DelSole, T. (2009). Import and export of atmospheric water vapor between nations. *Journal of Hydrology, 365*, 11–22. https://doi.org/10.1016/j.jhydrol.2008.11.016

Duarte, H., Tejedo, M., Katzenberger, M., Marangoni, F., Baldo, D., Beltrán, J. F., Martí, D. A., Richter-Boix, A., & Gonzalez-Voyer, A. (2012). Can amphibians take the heat? Vulnerability to climate warming in subtropical and temperate larval amphibian communities. *Global Change Biology, 18*, 412–421. https://doi.org/10.1111/j.1365-2486.2011.02518.x

Duellman, W. E. (1999). *Patterns of distribution of amphibians: A global perspective*. Johns Hopkins University Press.

Dulumoni, T., Caroline, B., Ratul, N., & Mithra, D. (2020). Acute and chronic effects of imidacloprid to anuran tadpoles (Polypedates teraiensis). *Research Journal of Chemistry and Environment, 24*(10), 25–30.

FAO. Food and Agriculture Organization of the United Nations. (2022). FAOSTAT. Food and Agriculture Statistics database. https://www.fao.org/faostat/en/#data. Accessed 22 Aug 2022.

Fearnside, P. M. (2019). Exploração mineral na Amazônia brasileira: o custo ambiental. In E. Castro & E. do Carmo (Eds.), *Dossiê desastres e crimes da mineração em Barcarena, Mariana e Brumadinho* (pp. 35–42). NAEA.

Fearnside, P. M. (2022). Amazon environmental services: Why Brazil's highway BR-319 is so damaging. *Ambio, 51*, 1367–1370. https://doi.org/10.1007/s13280-022-01718-y

Figueiredo, J., & Rodrigues, D. d. J. (2014). Size to metamorphosis of two species of tadpoles (Rhinella marina and Physalaemus centralis) from the southern Amazon, Brazil. *Herpetological Journal, 24*, 7–15.

Franco-de-Sá, J. F. O., Keller, C., Santana, G. P., & Menin, M. (2012). Efeito da poluição do igarapé do Educandos (Manaus, Amazonas, Brasil) sobre ovos e girinos de Osteocephalus taurinus (Anura, Hylidae). In J. L. Marcon, M. Menin, M. G. P. Araújo, & T. Hrbek (Eds.), *Biodiversidade Amazônica: caracterização, ecologia e conservação* (pp. 301–323). Edua.

Franco-de-Sá, J. F. O., & Val, A. L. (2014). Copper toxicity for Scinax ruber and Rhinella granulosa (Amphibia: Anura) of the Amazon: Potential of biotic ligand model to predict toxicity in urban streams. *Acta Amazonica, 44*(4), 491–498. https://doi.org/10.1590/1809-4392201400383

Freitas, J. S., Girotto, L., Goulart, B. V., Alho, L. d. O. G., Gebara, R. C., Montagner, C. C., Schiesari, L., & Espíndola, E. L. G. (2019). Effects of 2,4-D-based herbicide (DMA® 806) on sensitivity, respiration rates, energy reserves and behavior of tadpoles. *Ecotoxicology and Environmental Safety, 182*, 109446. https://doi.org/10.1016/j.ecoenv.2019.109446

Frost, D. R. (2023). *Amphibian species of the world: An online reference. Version 6.2*. American Museum of Natural History database. https://amphibiansoftheworld.amnh.org/index.php. Accessed 22 Jun 2023. https://doi.org/10.5531/db.vz.0001

Fulgence, T. R., Martin, D. A., Randriamanantena, R., Botra, R., Befidimanana, E., Osen, K., Wurz, A., Kreft, H., Andrianarimisa, A., & Ratsoavina, F. M. (2022). Differential responses of amphibians and reptiles to land-use change in the biodiversity hotspot of North-Eastern Madagascar. *Animal Conservation, 25*, 492–507. https://doi.org/10.1111/acv.12760

Gardner, T. (2001). Declining amphibian populations: a global phenomenon in conservation biology. *Animal Biodiversity and Conservation, 24*, 25–44.

Gerolin, C. R., Pupim, F. N., Sawakuchi, A. O., Grohmann, C. H., Labuto, G., & Semensatto, D. (2020). Microplastics in sediments from Amazon rivers, Brazil. *Science of the Total Environment, 749*, 141604. https://doi.org/10.1016/j.scitotenv.2020.141604

Gervais, J., Luukinen, B., Buhl, K., & Stone, D. (2008). *2,4-D technical fact sheet; national pesticide information center*. Oregon State University Extension Services. http://npic.orst.edu/factsheets/archive/2,4-DTech.html. Accessed 15 Jun 2023

Geyer, R., Jambeck, J. R., & Law, K. L. (2017). Production, use, and fate of all plastics ever made. *Science Advances, 3*, e1700782. https://doi.org/10.1126/sciadv.1700782

Ghose, S. L., Donnelly, M. A., Kerby, J., & Whitfield, S. M. (2014). Acute toxicity tests and meta-analysis identify gaps in tropical ecotoxicology for amphibians: Toxicity of pesticides used in Costa Rica to amphibians. *Environmental Toxicology and Chemistry, 33*, 2114–2119. https://doi.org/10.1002/etc.2665

Gonçalves, M. W., Marins de Campos, C. B., Batista, V. G., da Cruz, A. D., de Marco, J. P., Bastos, R. P., & de Melo e Silva, D. (2017). Genotoxic and mutagenic effects of Atrazine Atanor 50

SC on Dendropsophus minutus Peters, 1872 (Anura: Hylidae) developmental larval stages. *Chemosphere, 182*, 730–737. https://doi.org/10.1016/j.chemosphere.2017.05.078

Gosner, K. (1960). A simplified table for staging anuran embryos and larvae with notes on identification. *Herpetologica, 16*, 183–190.

Grant, E. H. C., Miller, D. A. W., Schmidt, B. R., Adams, M. J., et al. (2016). Quantitative evidence for the effects of multiple drivers on continental-scale amphibian declines. *Scientific Reports, 6*, 25625. https://doi.org/10.1038/srep25625

Green, D. M., Lannoo, M. J., Lesbarrères, D., & Muths, E. (2020). Amphibian population declines: 30 years of progress in confronting a complex problem. *Herpetologica, 76*, 97. https://doi.org/10.1655/0018-0831-76.2.97

Grott, S. C., Israel, N., Lima, D., Bitschinski, D., Abel, G., Alves, T. C., Da Silva, E. B., De Albuquerque, C. A. C., Mattos, J. J., Bainy, A. C. D., & De Almeida, E. A. (2022). Influence of temperature on growth, development and thyroid metabolism of American bullfrog tadpoles (Lithobates catesbeianus) exposed to the herbicide tebuthiuron. *Environmental Toxicology and Pharmacology, 94*, 103910. https://doi.org/10.1016/j.etap.2022.103910

Guerra, V., Jardim, L., Llusia, D., Márquez, R., & Bastos, R. P. (2020). Knowledge status and trends in description of amphibian species in Brazil. *Ecological Indicators, 118*, 106754. https://doi.org/10.1016/j.ecolind.2020.106754

Gupta, D. K., Chatterjee, S., Datta, S., Veer, V., & Walther, C. (2014). Role of phosphate fertilizers in heavy metal uptake and detoxification of toxic metals. *Chemosphere, 108*, 134–144. https://doi.org/10.1016/j.chemosphere.2014.01.030

Halliday, T. R. (2008). Why amphibians are important: Why amphibians are important. *International Zoo Yearbook, 42*, 7–14. https://doi.org/10.1111/j.1748-1090.2007.00037.x

Hayes, T. B., Case, P., Chui, S., et al. (2006). Pesticide mixtures, endocrine disruption, and amphibian declines: Are we underestimating the impact? *Environmental Health Perspectives, 114*(1), 40–50. https://doi.org/10.1289/ehp.8051

Hayes, T. B., Collins, A., Lee, M., et al. (2002). Hermaphroditic, demasculinized frogs after exposure to the herbicide atrazine at low ecologically relevant doses. *Proceedings of the National Academy of Sciences of the United States of America, 99*(8), 5476–5480. https://doi.org/10.1073/pnas.082121499

Haywood, L. K., Alexander, G. J., Byrne, M. J., & Cukrowska, E. (2004). Xenopus laevis embryos and tadpoles as models for testing for pollution by zinc, copper, lead and cadmium. *African Zoology, 39*(2), 163–174. https://doi.org/10.1080/15627020.2004.11657213

Herek, J. S., Vargas, L., Trindade, S. A. R., Rutkoski, C. F., Macagnan, N., Hartmann, P. A., & Hartmann, M. T. (2020). Can environmental concentrations of glyphosate affect survival and cause malformation in amphibians? Effects from a glyphosate-based herbicide on Physalaemus cuvieri and P. gracilis (Anura: Leptodactylidae). *Environmental Science and Pollution Research, 27*, 22619–22630. https://doi.org/10.1007/s11356-020-08869-z

Herek, J. S., Vargas, L., Trindade, S. A. R., Rutkoski, C. F., Macagnan, N., Hartmann, P. A., & Hartmann, M. T. (2021). Genotoxic effects of glyphosate on Physalaemus tadpoles. *Environmental Toxicology and Pharmacology, 81*, 103516. https://doi.org/10.1016/j.etap.2020.103516

Hoogmoed, M. (2023). Censo da biodiversidade da Amazônia brasileira. Grupo: Anura database http://censo.museu-goeldi.br:8080/museugoeldi-web-1.2.0/paginas/index.xhtml. Accessed 10 Jul 2023.

IBAMA. Instituto Brasileiro do Meio Ambiente e dos Recursos Naturais Renováveis. (2022) Relatórios de comercialização de agrotóxicos. http://www.ibama.gov.br/agrotoxicos/relatorios-de-comercializacao-de-agrotoxicos. Accessed 22 Aug 2022.

INPE—Instituto Nacional de Pesquisas Espaciais. Banco de Dados de queimadas. Available: https://queimadas.dgi.inpe.br/queimadas/portal-static/estatisticas_estados/. Accessed 25 May 2023.

Janzen, D. H. (1967). Why mountain passes are higher in the tropics. *The American Naturalist, 101*, 233–249.

Katzenberger, M., Tejedo, M., Duarte, H., Marangoni, F., & Beltrán, J. F. (2012). Tolerância e sensibilidade térmica em anfíbios. *Revista da Biologia, 8*, 25–32. https://doi.org/10.7594/revbio.08.05

Kawecki, T. J., & Ebert, D. (2004). Conceptual issues in local adaptation. *Ecology Letters, 7*, 1225–1241. https://doi.org/10.1111/j.1461-0248.2004.00684.x

Kjellstrom, T., Lodh, M., McMichael, T., Ranmuthugala, G., et al. (2006). Air and water pollution: Burden and strategies for control. In D. T. Jamison, J. G. Breman, A. R. Measham, G. Alleyne, et al. (Eds.), *Disease control priorities in developing countries* (pp. 817–832). Oxford University Press.

Kotz, E. J., Cabral, F. B., Trezzi, I., Dias, G. L., Spanevello, R. M., Hildebrandt, L. M., Centenaro, A. P. F. C., & Silveira, A. (2021). Notions and pesticide use: A case study with family farmers. *Research, Society and Development, 10*(7), e50510716898. https://doi.org/10.33448/rsd-v10i7.16898

Lajmanovich, R. C., Attademo, A. M., Lener, G., Cuzziol Boccioni, A. P., Peltzer, P. M., Martinuzzi, C. S., Demonte, L. D., & Repetti, M. R. (2022). Glyphosate and glufosinate ammonium, herbicides commonly used on genetically modified crops, and their interaction with microplastics: Ecotoxicity in anuran tadpoles. *Science of the Total Environment, 804*, 150177. https://doi.org/10.1016/j.scitotenv.2021.150177

Lajmanovich, R. C., Peltzer, P. M., Martinuzzi, C. S., Attademo, A. M., Bassó, A., & Colussi, C. L. (2019). Insecticide pyriproxyfen (Dragón®) damage biotransformation, thyroid hormones, heart rate, and swimming performance of Odontophrynus americanus tadpoles. *Chemosphere, 220*, 714–722. https://doi.org/10.1016/j.chemosphere.2018.12.181

Lajmanovich, R. C., Sandoval, M. T., & Peltzer, P. M. (2003). Induction of mortality and malformation in Scinax nasicus tadpoles exposed to glyphosate formulations. *Bulletin of Environmental Contamination and Toxicology, 70*, 612–618. https://doi.org/10.1007/s00128-003-0029-x

Landrigan, P. J., & Belpoggi, F. (2018). The need for independent research on the health effects of glyphosate-based herbicides. *Environmental Health, 17*, 51. https://doi.org/10.1186/s12940-018-0392-z

Lau, E. T. C., Karraker, N. E., & Leung, K. M. Y. (2015). Temperature-dependent acute toxicity of methomyl pesticide on larvae of 3 Asian amphibian species. *Environmental Toxicology and Chemistry, 34*(10), 2322–2327. https://doi.org/10.1002/etc.3061

Leite-Filho, A. T., Soares-Filho, B. S., Davis, J. L., Abrahão, G. M., & Börner, J. (2021). Deforestation reduces rainfall and agricultural revenues in the Brazilian Amazon. *Nature Communications, 12*, 2591. https://doi.org/10.1038/s41467-021-22840-7

Lenkowski, J. R., Reed, J. M., Deininger, L., & McLaughlin, K. A. (2008). Perturbation of organogenesis by the herbicide atrazine in the amphibian Xenopus laevis. *Environmental Health Perspectives, 116*, 223–230. https://doi.org/10.1289/ehp.10742

Levins, R. (1968). *Evolution in changing environments: Some theoretical explorations*. Princeton University Press.

Liu, R., Qin, Y., Diao, J., & Zhang, H. (2021). Xenopus laevis tadpoles exposed to metamifop: Changes in growth, behavioral endpoints, neurotransmitters, antioxidant system and thyroid development. *Ecotoxicology and Environmental Safety, 220*, 112417. https://doi.org/10.1016/j.ecoenv.2021.112417

Lopes, A., Benvindo-Souza, M., Carvalho, W. F., Nunes, H. F., de Lima, P. N., et al. (2021). Evaluation of the genotoxic, mutagenic, and histopathological hepatic effects of polyoxyethylene amine (POEA) and glyphosate on Dendropsophus minutus tadpoles. *Environmental Pollution, 289*, 117911. https://doi.org/10.1016/j.envpol.2021.117911

Luedtke, J. A., Chanson, J., Neam, K., et al. (2023). Ongoing declines for the world's amphibians in the face of emerging threats. *Nature, 622*, 308–314. https://doi.org/10.1038/s41586-023-06578-4

Manabe, S. (2019). Role of greenhouse gas in climate change. *Tellus A: Dynamic Meteorology and Oceanography, 71*, 1620078. https://doi.org/10.1080/16000870.2019.1620078

Manisalidis, I., Stavropoulou, E., Stravropoulos, A., & Bezirtzoglou, E. (2020). Environmental and health impacts of air pollution: A review. *Frontiers in Public Health, 8*, 14. https://doi.org/10.3389/fpubh.2020.00014

Marchesan, E., Sartori, G. M. S., Avila, L. A. d., Machado, S. L. d. O., Zanella, R., Primel, E. G., Macedo, V. R. M., & Marchezan, M. G. (2010). Resíduos de agrotóxicos na água de rios da Depressão Central do Estado do Rio Grande do Sul, Brasil. *Ciência Rural, 40*, 1053–1059. https://doi.org/10.1590/S0103-84782010005000078

Massarelli, C., Campanale, C., & Uricchio, V. F. (2021). A handy open-source application based on computer vision and machine learning algorithms to count and classify microplastics. *Water, 13*, 2104. https://doi.org/10.3390/w13152104

Masson-Delmotte, V., Zhai, P., Pirani, A., et al. (2021). Summary for policymakers - Climate change 2021: The physical science basis. In *Contribution of Working Group I to the sixth assessment report of the intergovernmental panel on climate change* (pp. 3–32). Cambridge University Press. https://doi.org/10.1017/9781009157896.001

McDiarmid, R. W., & Altig, R. (1999). *Tadpoles: The biology of anuran larvae*. University of Chicago Press.

McRae, L., Deinet, S., Marconi, V., Scott-Gatty, K., & Freeman, R. (2020). The living planet index: An early warming indicator the health of nature. In R. E. A. Almond, M. Grooten, & T. Petersen (Eds.), *Living planet report 2020 - Bending the curve of biodiversity loss* (pp. 14–21). WWF.

Moulatlet, G. M., Yacelga, N., Rico, A., Mora, A., Hauser-Davis, R. A., Cabrera, M., & Capparelli, M. V. (2023). A systematic review on metal contamination due to mining activities in the Amazon basin and associated environmental hazards. *Chemosphere, 339*, 139700. https://doi.org/10.1016/j.chemosphere.2023.139700

Moura, M. R., & Jetz, W. (2021). Shortfalls and opportunities in terrestrial vertebrate species discovery. *Nature Ecology & Evolution, 5*, 631–639. https://doi.org/10.1038/s41559-021-01411-5

Nati, J. J. H., Svendsen, M. B. S., Marras, S., et al. (2021). Intraspecific variation in thermal tolerance differs between tropical and temperate fishes. *Scientific Reports, 11*, 21272. https://doi.org/10.1038/s41598-021-00695-8

Navas, C. A., Gomes, F. R., & Carvalho, J. E. (2008). Thermal relationships and exercise physiology in anuran amphibians: Integration and evolutionary implications. *Comparative Biochemistry and Physiology. Part A, Molecular & Integrative Physiology, 151*, 344–362. https://doi.org/10.1016/j.cbpa.2007.07.003

Nunes-de-Almeida, C. H. L., Haddad, C. F. B., & Toledo, L. F. (2021). A revised classification of the amphibian reproductive modes. *Salamandra, 57*, 413–427.

Orlofske, S. A., & Hopkins, W. A. (2009). Energetics of metamorphic climax in the pickerel frog (Lithobates palustris). *Comparative Biochemistry and Physiology. Part A, Molecular & Integrative Physiology, 154*, 191–196. https://doi.org/10.1016/j.cbpa.2009.06.001

Patar, A., Das, I., Giri, S., & Giri, A. (2021). Zinc contamination is an underestimated risk to amphibians: Toxicity evaluation in tadpoles of Fejervarya limnocharis. *Journal of Environmental Engineering and Landscape Management, 29*, 489–498. https://doi.org/10.3846/jeelm.2021.15814

Patar, A., Giri, A., Boro, F., et al. (2016). Cadmium pollution and amphibians - Studies in tadpoles of Rana limnocharis. *Chemosphere, 144*, 1043–1049. https://doi.org/10.1016/j.chemosphere.2015.09.088

Pavan, F. A., Samojeden, C. G., Rutkoski, C. F., Folador, A., Da Fré, S. P., Müller, C., Hartmann, P. A., & Hartmann, M. T. (2021). Morphological, behavioral, and genotoxic effects of glyphosate and 2,4-D mixture in tadpoles of two native species of South American amphibians. *Environmental Toxicology and Pharmacology, 85*, 103637. https://doi.org/10.1016/j.etap.2021.103637

Peillex, C., & Pelletier, M. (2020). The impact and toxicity of glyphosate and glyphosate-based herbicides on health and immunity. *Journal of Immunotoxicology, 17*, 163–174. https://doi.org/10.1080/1547691X.2020.1804492

Pérez-Iglesias, J. M., Franco-Belussi, L., Natale, G. S., & de Oliveira, C. (2019). Biomarkers at different levels of organization after atrazine formulation (SIPTRAN 500SC®) exposure in

Rhinella schineideri (Anura: Bufonidae) Neotropical tadpoles. *Environmental Pollution, 244*, 733–746. https://doi.org/10.1016/j.envpol.2018.10.073
Phuge, S. (2020). Tamoxifen stimulates gonad development and somatic growth in the tadpoles of Indian skipper frog, Euphlyctis cyanophlyctis. *Journal of Basic and Applied Zoology, 81*, 19. https://doi.org/10.1186/s41936-020-00161-3
Pillay, R., Venter, M., Aragon-Osejo, J., González-del-Pliego, P., Hansen, A. J., Watson, J. E., & Venter, O. (2022). Tropical forests are home to over half of the world's vertebrate species. *Frontiers in Ecology and the Environment, 20*, 10–15. https://doi.org/10.1002/fee.2420
Pires, N. L., Passos, C. J. S., Morgado, M. G. A., Mello, D. C., Infante, C. M. C., & Caldas, E. D. (2020). Determination of glyphosate, AMPA and glufosinate by high performance liquid chromatography with fluorescence detection in waters of the Santarém Plateau, Brazilian Amazon. *Journal of Environmental Science and Health, Part B, 55*, 794–802. https://doi.org/10.1080/03601234.2020.1784668
PlasticsEurope - Association of Plastic Manufacturers. (2021). Plastics - The facts 2021. Available: https://plasticseurope.org/wp-content/uploads/2021/12/Plastics-the-Facts-2021-web-final.pdf. Accessed 25 May 2023.
PRODES. Monitoramento da Floresta Amazônica Brasileira por Satélite. (2023). INPE database. http://terrabrasilis.dpi.inpe.br/queimadas/bdqueimadas/#graficos. Accessed 7 Jan 2023.
Projeto MapBiomas—Mapeamento Anual de Cobertura e Uso da Terra no Brasil - Coleção 7. (2022). https://mapbiomas-br-site.s3.amazonaws.com/Fact-Sheet-Colecao7.pdf. Accessed 20 May 2023.
Quaranta, A., Bellantuono, V., Cassano, G., & Lippe, C. (2009). Why amphibians are more sensitive than mammals to xenobiotics. *PLoS One, 4*, e7699. https://doi.org/10.1371/journal.pone.0007699
Quiroga, L. B., Sanabria, E. A., Fornés, M. W., Bustos, D. A., & Tejedo, M. (2019). Sublethal concentrations of chlorpyrifos induce changes in the thermal sensitivity and tolerance of anuran tadpoles in the toad Rhinella arenarum? *Chemosphere, 219*, 671–677. https://doi.org/10.1016/j.chemosphere.2018.12.059
Rathanayaka, R. R. P. Y. K., & Rajakaruna, R. S. (2018). Cocktail effect of Profenophos and Abamectin on tadpoles of Asian common toad (Duttaphrynus melanostictus). *Ceylon J Sci, 47*, 185. https://doi.org/10.4038/cjs.v47i2.7515
Relyea, R. A. (2005). The lethal impacts of roundup and predatory stress on six species of North American tadpoles. *Archives of Environmental Contamination and Toxicology, 48*, 351–357. https://doi.org/10.1007/s00244-004-0086-0
Riaño, C., Ortiz-Ruiz, M., Pinto-Sánchez, N. R., & Gómez-Ramírez, E. (2020). Effect of glyphosate (roundup active®) on liver of tadpoles of the Colombian endemic frog Dendropsophus molitor (amphibia: Anura). *Chemosphere, 250*, 126287. https://doi.org/10.1016/j.chemosphere.2020.126287
Ribeiro, D. R. G., Faccin, H., Molin, T. R. D., de Carvalho, L. M., & Amado, L. L. (2017). Metal and metalloid distribution in different environmental compartments of the middle Xingu River in the Amazon, Brazil. *Science of the Total Environment, 605-606*, 66–74. https://doi.org/10.1016/j.scitotenv.2017.06.143
Rico, A., de Oliveira, R., de Souza, S., Nunes, G., Rizzi, C., Villa, S., De Caroli, V. B., Montagner, C. C., & Waichman, A. V. (2022). Ecological risk assessment of pesticides in urban streams of the Brazilian Amazon. *Chemosphere, 291*, 132821. https://doi.org/10.1016/j.chemosphere.2021.132821
Rico, A., Redondo-Hasselerharm, P. E., & Vighi, M. (2023). Large-scale monitoring and risk assessment of microplastics in the Amazon River. *Water Research, 232*, 119707. https://doi.org/10.1016/j.watres.2023.119707
Ringwood, A. H. (2021). Bivalves as biological sieves: Bioreactivity pathways of microplastics and nanoplastics. *The Biological Bulletin, 241*, 185–195. https://doi.org/10.1086/716259
Römbke, J., Waichman, A. V., & Garcia, M. V. B. (2008). Risk assessment of pesticides for soils of the Central Amazon, Brazil: Comparing outcomes with temperate and tropical data. *Integrated Environmental Assessment and Management, 4*, 94. https://doi.org/10.1897/IEAM_2007-052.1

Ruiz de Arcaute, C., Soloneski, S., & Larramendy, M. L. (2016). Toxic and genotoxic effects of the 2,4-dichlorophenoxyacetic acid (2,4-D)-based herbicide on the Neotropical fish Cnesterodon decemmaculatus. *Ecotoxicology and Environmental Safety, 128*, 222–229. https://doi.org/10.1016/j.ecoenv.2016.02.027

Ruthsatz, K., Dausmann, K. H., Reinhardt, S., Robinson, T., Sabatino, N. M., Peck, M. A., & Glos, J. (2019). Endocrine disruption alters developmental energy allocation and performance in Rana temporaria. *Integrative and Comparative Biology, 59*, 70–88. https://doi.org/10.1093/icb/icz041

Ruthsatz, K., Domscheit, M., Engelkes, K., & Vences, M. (2022). Microplastics ingestion induces plasticity in digestive morphology in larvae of Xenopus laevis. *Comparative Biochemistry and Physiology Part A: Molecular & Integrative Physiology, 269*, 111210. https://doi.org/10.1016/j.cbpa.2022.111210

Rutkoski, C. F., Macagnan, N., Folador, A., Skovronski, V. J., do Amaral, A. M. B., Leitemperger, J., Costa, M. D., Hartmann, P. A., Müller, C., Loro, V. L., & Hartmann, M. T. (2020). Morphological and biochemical traits and mortality in Physalaemus gracilis (Anura: Leptodactylidae) tadpoles exposed to the insecticide chlorpyrifos. *Chemosphere, 250*, 126162. https://doi.org/10.1016/j.chemosphere.2020.126162

Rutkoski, C. F., Macagnan, N., Kolcenti, C., Vanzetto, G. V., Sturza, P. F., Hartmann, P. A., & Hartmann, M. T. (2018). Lethal and sublethal effects of the herbicide atrazine in the early stages of development of Physalaemus gracilis (Anura: Leptodactylidae). *Archives of Environmental Contamination and Toxicology, 74*, 587–593. https://doi.org/10.1007/s00244-017-0501-y

Samojeden, C. G., Pavan, F. A., Rutkoski, C. F., Folador, A., da Fré, S. P., Müller, C., Hartmann, P. A., & Hartmann, M. (2022). Toxicity and genotoxicity of imidacloprid in the tadpoles of Leptodactylus luctator and Physalaemus cuvieri (Anura: Leptodactylidae). *Scientific Reports, 12*, 11926. https://doi.org/10.1038/s41598-022-16039-z

Sánchez-Domene, D., Navarro-Lozano, A., Acayaba, R., Picheli, K., Montagner, C., de Cerqueira, R.-F. D., Rodrigues da Silva, F., & Alves de Almeida, E. (2018). Eye malformation baseline in Scinax fuscovarius larvae populations that inhabit agroecosystem ponds in southern Brazil. *Amphibia-Reptilia, 39*, 325–334. https://doi.org/10.1163/15685381-20181038

Schiesari, L., Grillitsch, B., & Grillitsch, H. (2007). Biogeographic biases in research and their consequences for linking amphibian declines to pollution. *Conservation Biology, 21*(836), 465–471. https://doi.org/10.1111/j.1523-1739.2006.00616.x

Schlippe-Justicia, L., Lemaire, J., Dittrich, C., et al. (2024). Poison in the nursery: Mercury contamination in the tadpole-rearing sites of an Amazonian frog. *Science of the Total Environment, 912*, 169450. https://doi.org/10.1016/j.scitotenv.2023.169450

Shen, M., Song, B., Zhu, Y., Zeng, G., Zhang, Y., Yang, Y., Wen, X., Chen, M., & Yi, H. (2020). Removal of microplastics via drinking water treatment: Current knowledge and future directions. *Chemosphere, 251*, 126612. https://doi.org/10.1016/j.chemosphere.2020.126612

Shi, Y. B. (2000). *Amphibian metamorphosis*. Wiley-Liss.

Shi, Q., Sun, N., Kou, H., Wang, H., & Zhao, H. (2018). Chronic effects of mercury on Bufo gargarizans larvae: Thyroid disruption, liver damage, oxidative stress and lipid metabolism disorder. *Ecotoxicology and Environmental Safety, 164*, 500–509. https://doi.org/10.1016/j.ecoenv.2018.08.058

Sievers, M., Hale, R., Swearer, S. E., & Parris, K. M. (2018). Contaminant mixtures interact to impair predator-avoidance behaviours and survival in a larval amphibian. *Ecotoxicology and Environmental Safety, 161*, 482–488. https://doi.org/10.1016/j.ecoenv.2018.06.028

Silva Junior, C., Aragão, L., Fonseca, M., Almeida, C., Vedovato, L., & Anderson, L. (2018). Deforestation-induced fragmentation increases forest fire occurrence in Central Brazilian Amazonia. *Forests, 9*, 305. https://doi.org/10.3390/f9060305

Silva, T. M., Stets, M. I., Mazzetto, A. M., Andrade, F. D., Pileggi, S. A. V., Fávero, P. R., Cantú, M. D., Carrilho, E., Carneiro, P. I. B., & Pileggi, M. (2007). Degradation of 2,4-D herbicide by microorganisms isolated from Brazilian contaminated soil. *Brazilian Journal of Microbiology, 38*, 522–525. https://doi.org/10.1590/S1517-83822007000300026

Slaby, S., Marin, M., Marchand, G., & Lemiere, S. (2019). Exposures to chemical contaminants: What can we learn from reproduction and development endpoints in the amphibian toxicology literature? *Environmental Pollution, 248*, 478–495. https://doi.org/10.1016/j.envpol.2019.02.014

Smith, D. C. (1987). Adult recruitment in chorus frogs: Effects of size and date at metamorphosis. *Ecology, 68*, 344–350. https://doi.org/10.2307/1939265

Soares, D., Silva, L., Duarte, S., Pena, A., & Pereira, A. (2021). Glyphosate use, toxicity and occurrence in food. *Food, 10*, 2785. https://doi.org/10.3390/foods10112785

Sonter, L. J., Herrera, D., Barrett, D. J., et al. (2017). Mining drives extensive deforestation in the Brazilian Amazon. *Nature Communications, 8*, 1013. https://doi.org/10.1038/s41467-017-00557-w

Stechert, C., Kolb, M., Rödel, M., & Bahadir, M. (2015). Effects of insecticide formulations used in cotton cultivation in West Africa on the development of flat-backed toad tadpoles (Amietophrynus maculatus). *Environmental Science and Pollution Research, 22*, 2574–2583. https://doi.org/10.1007/s11356-014-3462-y

Tata, J. R. (2006). Amphibian metamorphosis as a model for the developmental actions of thyroid hormone. *Molecular and Cellular Endocrinology, 246*, 10–20. https://doi.org/10.1016/j.mce.2005.11.024

Thambirajah, A. A., Koide, E. M., Imbery, J. J., & Helbing, C. C. (2019). Contaminant and environmental influences on thyroid hormone action in amphibian metamorphosis. *Frontiers in Endocrinology, 10*, 276. https://doi.org/10.3389/fendo.2019.00276

Tolledo, J., Silva, E. T., Nunes-de-Almeida, C. H. L., & Toledo, L. F. (2014). Anomalous tadpoles in a Brazilian oceanic archipelago: Implications of oral anomalies on foraging behaviour, food intake and metamorphosis. *Herpetological Journal, 24*, 7.

Trudeau, V. L., Thomson, P., Zhang, W. S., Reynaud, S., et al. (2020). Agrochemicals disrupt multiple endocrine axes in amphibians. *Molecular and Cellular Endocrinology, 513*, 110861. https://doi.org/10.1016/j.mce.2020.110861

Turhan, D. Ö., Güngördü, A., & Ozmen, M. (2020). Developmental and lethal effects of glyphosate and a glyphosate-based product on Xenopus laevis embryos and tadpoles. *Bulletin of Environmental Contamination and Toxicology, 104*, 173–179. https://doi.org/10.1007/s00128-019-02774-z

Vacher, J. P., Chave, J., Ficetola, F. G., et al. (2020). Large-scale DNA-based survey of frogs in Amazonia suggests a vast underestimation of species richness and endemism. *Journal of Biogeography, 47*, 1781–1791. https://doi.org/10.1111/jbi.13847

Valencia-Aguilar, A., Cortés-Gómez, A. M., & Ruiz-Agudelo, C. A. (2013). Ecosystem services provided by amphibians and reptiles in Neotropical ecosystems. *International Journal of Biodiversity Science, Ecosystem Services & Management, 9*, 257–272. https://doi.org/10.1080/21513732.2013.821168

Vanzetto, G. V., Slaviero, J. G., Sturza, P. F., Rutkoski, C. F., Macagnan, N., Kolcenti, C., Hartmann, P. A., Ferreira, C. M., & Hartmann, M. T. (2019). Toxic effects of pyrethroids in tadpoles of Physalaemus gracilis (Anura: Leptodactylidae). *Ecotoxicology, 28*, 1105–1114. https://doi.org/10.1007/s10646-019-02115-0

Venesky, M. D., Wassersug, R. J., & Parris, M. J. (2010). The impact of variation in labial tooth number on the feeding kinematics of tadpoles of southern leopard frog (Lithobates sphenocephalus). *Copeia, 2010*, 481–486. https://doi.org/10.1643/CG-09-093

Verdade, V. K., Carnaval, A. C., Rodrigues, M. T., et al. (2011). Decline of amphibians in Brazil. In H. Heatwole, C. L. Barrio-Amorós, & J. W. Wilkinson (Eds.), *Amphibian biology, Volume 9, Part 2: Status of decline of amphibians: Western hemisphere: Uruguay, Brazil, Colombia, and Ecuador* (pp. 85–127). Australia.

Viana, L. F., Cardoso, C. A. L., Lima-Junior, S. E., et al. (2020). Bioaccumulation of metal in liver tissue of fish in response to water toxicity of the Araguari-Amazon River, Brazil. *Environmental Monitoring and Assessment, 192*, 781. https://doi.org/10.1007/s10661-020-08696-2

Vieira, V. A. R. O. (2010). Regulação gênica dos receptores dos hormônios tireoideanos durante a metamorfose de anfíbios anuros. *Review Biology, 4*, 16–21. https://doi.org/10.7594/revbio.04.04

Vinagre, C., Dias, M., Cereja, R., et al. (2019). Upper thermal limits and warming safety margins of coastal marine species – Indicator baseline for future reference. *Ecological Indicators, 102*, 644–649. https://doi.org/10.1016/j.ecolind.2019.03.030

Vryzas, Z. (2018). Pesticide fate in soil-sediment-water environment in relation to contamination preventing actions. *Current Opinion in Environmental Science & Health, 4*, 5–9. https://doi.org/10.1016/j.coesh.2018.03.001

Wang, X., Chang, L., Zhao, T., Liu, L., Zhang, M., Li, C., Xie, F., Jiang, J., & Zhu, W. (2019). Metabolic switch in energy metabolism mediates the sublethal effects induced by glyphosate-based herbicide on tadpoles of a farmland frog Microhyla fissipes. *Ecotoxicology and Environmental Safety, 186*, 109794. https://doi.org/10.1016/j.ecoenv.2019.109794

Whiles, M. R., Lips, K. R., Pringle, C. M., Kilham, S. S., Bixby, R. J., Brenes, R., Connelly, S., Colon-Gaud, J. C., Hunte-Brown, M., Huryn, A. D., Montgomery, C., & Peterson, S. (2006). The effects of amphibian population declines on the structure and function of Neotropical stream ecosystems. *Frontiers in Ecology and the Environment, 4*, 27–34. https://doi.org/10.1890/1540-9295(2006)004[0027:TEOAPD]2.0.CO;2

WHO. (2022). Guidelines for drinking-water quality: Fourth edition incorporating the first and second addenda. Available: https://www.who.int/publications/i/item/9789240045064. Accessed 17 Feb 2023.

Widder, P. D., & Bidwell, J. R. (2008). Tadpole size, cholinesterase activity, and swim speed in four frog species after exposure to sub-lethal concentrations of chlorpyrifos. *Aquatic Toxicology, 88*(1), 9–18. https://doi.org/10.1016/j.aquatox.2008.02.008

Williams, G. M., Kroes, R., & Munro, I. C. (2000). Safety evaluation and risk assessment of the herbicide roundup and its active ingredient, glyphosate, for humans. *Regulatory Toxicology and Pharmacology, 31*, 117–165. https://doi.org/10.1006/rtph.1999.1371

Xu, S. (2000). *Environmental fate of mancozeb*. Environmental Monitoring and Pest Management.

Yang, Z., Lü, F., Zhang, H., Wang, W., Shao, L., Ye, J., & He, P. (2021). Is incineration the terminator of plastics and microplastics? *Journal of Hazardous Materials, 401*, 123429. https://doi.org/10.1016/j.jhazmat.2020.123429

Yang, L., Zhao, S., & Liu, S. (2022). A global analysis of urbanization effects on amphibian richness: Patterns and drivers. *Global Environmental Change, 73*, 102476. https://doi.org/10.1016/j.gloenvcha.2022.102476

Yaoita, Y. (2019). Tail resorption during metamorphosis in Xenopus tadpoles. *Frontiers in Endocrinology, 10*, 143. https://doi.org/10.3389/fendo.2019.00143

Zemp, D. C., Schleussner, C.-F., Barbosa, H. M. J., van der Ent, R. J., Donges, J. F., Heinke, J., Sampaio, G., & Rammig, A. (2014). On the importance of cascading moisture recycling in South America. *Atmospheric Chemistry and Physics, 14*, 13337–13359. https://doi.org/10.5194/acp-14-13337-2014

Zhang, S., Chen, A., Deng, H., Jiang, L., & Liu, X. (2023). Intestinal response of Rana chensinensis larvae exposed to Cr and Pb, alone and in combination. *Ecotoxicology and Environmental Safety, 255*, 114774. https://doi.org/10.1016/j.ecoenv.2023.114774

Toxicological Impacts of Pesticides and Heavy Metals on Amazon Fish: Trends and Future Perspectives

Susana Braz-Mota, Adalberto Luis Val, and Rafael Mendonça Duarte

Abstract

The Amazon region has faced substantial anthropogenic pressures, resulting in an increase in the presence of pesticides and heavy metals in aquatic ecosystems. This situation is primarily attributed to population growth, uncontrolled urban expansion, high deforestation rates, and the expansion of mining activities. This review assessed more than 100 papers and reports published between 1999 and 2024 describing different effects of metals (Cu, Hg, Mn, Cd, Pb, and Ni) and pesticides (herbicides, insecticides, fungicides, and parasiticides) in the water of the Amazon basin. This chapter delves into the critical issue of pesticide and heavy metal pollution in Amazon's aquatic environments, focusing on the toxicological effects and their implications for Amazonian fish populations. We also examine the varying degrees of toxicological vulnerability among different Amazon fish species based on recent studies. Furthermore, we offer insights into the prospects of the Amazonian aquatic biota within the context of this pollution and present emerging trends and prospective developments.

S. Braz-Mota (✉) · A. L. Val
Laboratory of Ecophysiology and Molecular Evolution, Brazilian National Institute for Research of the Amazon (INPA), Manaus, Amazonas, Brazil

R. M. Duarte
Laboratory of Ecophysiology and Aquatic Toxicology, São Paulo State University (UNESP), Institute of Biosciences, São Vicente, São Paulo, Brazil

S. S. de Souza et al. (eds.), *The Future of Amazonian Aquatic Biota*,
https://doi.org/10.1007/978-3-031-66822-7_10

1 Context of Contamination of Amazonian Aquatic Environments by Pesticides and Heavy Metals

The Amazon aquatic ecosystem has undeniable global importance, with a vast biodiversity of endemic species distributed in exuberant hydrographic systems, surpassing world records in extension and water volume, such as the underground aquifer Alter do Chão and the Amazon River. Maintaining the integrity of these aquatic systems is strictly related to the preservation of forests since both play a fundamental role in ecosystem services such as carbon sequestration, water cycling, and regulating regional climate balance (Nobre et al., 2016). However, the Amazon region has suffered intense anthropic pressures, mainly due to population growth, disorderly urbanization, and high deforestation rates (Gomes et al., 2023). The increased vulnerability of the Amazonian environment was strengthened in the late 1960s and 1970s, when policies motivated by the Brazilian army stimulated exploratory development in the region, including incentives for mining, industrial action, agricultural expansion, rural exploration, construction of highways, and hydroelectric power stations, in addition to several other significant developments in the region. Consequently, there was a significant increase in demographic rates, whether driven by the displacement of the local population to large urban centers or even by the strong incentive of immigration and occupation of the region, which forced the disorderly development of large cities, such as Manaus, Belém, Porto Velho, and Santarém (Staevie, 2015). Furthermore, in the last 5 years, many projects of laws were approved in Brazil, weakening environmental regulating efforts, which increased the threat to the conservation of Amazon aquatic systems by reducing the requirement for licenses to new infrastructure plant projects—such as mining activities over protected areas and facilitating the registration of new pesticides that are prohibited in European Union (Abessa et al., 2019).

Indeed, there was a dramatic increase in demand for food in the last decades that stimulated small-scale agricultural activities around the large metropolitan areas of the Amazon, increasing the rampant use of pesticides to control agricultural pests (Parry et al., 2010; Römbke et al., 2008). Although agricultural activity is more intensive in the south and southeast of Brazil, the expansion of pesticide use has been boosted in the Legal Amazon region, as it constitutes one of the leading agricultural frontiers in the country, mainly for soy production (Laabs et al., 2002). Soybean cultivation occurs in approximately one-third of the Legal Amazon territory, concentrated in the State of Mato Grosso, followed by Tocantins, Maranhão, Pará, and Rondônia.

According to the Brazilian Institute of Environment and Renewable Natural Resources 800,652.17 tons of pesticides were sold in Brazil in 2021, in which almost 202 tons/y of active ingredients were sold in the nine states that comprise the Brazilian Amazon (where 74% were commercialized in Mato Grosso state with the predominance of herbicides, fungicides, and insecticides) (IBAMA, 2021). In 2022, more than 500 active ingredients were allowed for sale and use in Brazil, and in the Amazon region, the pesticides mostly used and commercialized are glyphosate (herbicides); copper oxychloride and nancozeb (fungicides); and deltamethrin, malathion (MAL), and methyl-parathion (insecticides) (Römbke et al., 2008; Waichman, 2008; Waichman et al., 2002). Although almost 60% of pesticides commercialized have been classified as moderately toxic (toxicological class III), there is a marked increase in the use of both extremely and highly toxic active ingredients (toxicological class I and II) in the Amazon region in the last decade (Schiesari et al., 2013).

Some pesticides have also been widely used in the Amazon around fish farming ponds to control aquatic weeds (Araújo et al., 2008; Braz-Mota et al., 2015). Most of these compounds have high solubility, facilitating their absorption and enhancing contamination. Even in urban and peri-urban areas of the Brazilian Amazon, it is possible to identify 11 compounds in surface waters, including insecticides, herbicides, and fungicides (Rico et al., 2022). Thus, the widespread and nonselective use of pesticides combined with the lack of training and education programs for safe use potentially increases the risk of Amazonian environmental contamination, especially the aquatic environment, when considering the high solubility of these substances, which can result in severe toxic effects on nontarget species such as invertebrates, amphibians, turtles, and fishes (Mendonça et al., 2023; Römbke et al., 2008; Schiesari et al., 2013). In addition, the indiscriminate handling of these compounds also promotes human contamination by inhalation, as well as through the ingestion of contaminated food, resulting in acute and chronic intoxication (Abessa et al., 2019; Waichman et al., 2002, 2007).

Amazonian aquatic ecosystems have also been contaminated by the presence of heavy metals from various anthropogenic sources, such as raw sewage from urban areas, effluents from industries, agricultural activities, and waste from mining areas and gold mines (Capparelli et al., 2020). With the population increase, the need for housing also became urgent, concentrating the population in the peripheral regions and close to the streams. Associated with this, rapid urbanization has increased the pollution of these places, mainly due to the lack of basic sanitation. On average, only 14% of all Amazonian households have a sewage collection and treatment service (Viana et al., 2016), with most domestic sewage discharged into aquatic environments. As a result, the pollution of urban streams is frequently observed and enhanced by the region's high industrial activity, promoting the release of heavy metals in the Amazonian water systems. Industrial action releases a wide variety of heavy metals as they are different sources of activities aimed at handling and transforming raw materials for the production of consumer goods; the effects of environmental pollution being much more complex and synergistic. In addition, these effects can also be cumulative and chronic, a fact that makes their remediation more complex. The lack of basic sanitation also impacts the prevalence of endemic vector diseases in urban areas of the Amazon (Castro et al., 2019), further increasing the use of insecticide to control disease vectors, such as mosquitoes and kissing bugs, in addition to urban pests and households, such as ants, termites, and cockroaches (Fernandes et al., 2020).

Agriculture is also an important nonpoint source of metal pollution in water bodies, not only due to the use of fertilizers that release Cd, Cr, Pb, and Zn, but also pesticides that contain metals such as Cu, Pb, Mn, and Zn in its composition and waste from intensive production of cattle, pigs, and poultry (Cu, As, and Zn) (Liu et al., 2020; Santos et al., 2023). In addition, both metals and pesticides released into the soil from this activity are carried to rivers by the runoff of surface water from rainfall, with metals persisting in the aquatic environment because of their free/ionic form, which facilitates their accumulation in the tissues, mainly of fish.

Additionally, mining activities have been considered a hot spot for metal pollution in Amazon aquatic environments, affecting both the surrounding areas of mining and distant places from the emission source. Mining activities are widespread in all geographical regions of Amazon and can be classified into three types: (1) underground mining, where metals and minerals are extracted by digging tunnels on rocks below the ground surface; (2) alluvial mining that consists of removing alluvial material on riverbeds and riverbanks; and (3) open-pit mining where metals and minerals (especially gold and silver) are found in surface deposits and require pumping water against the pit walls to collect rich-metals sediment, which is further separated by using mercury sulfide (HgS) through amalgamation process to concentrate and extract gold and silver (Asner & Tupayachi, 2016; Moulatlet et al., 2023). In this regard, recent studies have pointed out that mining activities are now responsible for about 10% of deforestation in the Brazilian Amazon (Soares-Filho & Rajão, 2018) and markedly contribute to metal enrichment in both sediment and water matrices (Moulatlet et al., 2023).

Concerning water contamination, a systematic review assessing metal background concentrations in the Amazon basin revealed that metals such as Hg, Fe, Mn, Zn, Cd, Cr, Cu, Ni, and Pb were in higher concentration (over 100-fold) in most samples from Northwestern Amazon (NWA). In addition, Hg, Cd, Ni, Pb, and Zn are present at levels above the allowed by the US Environmental Protection Agency (USEPA) for acute and chronic criteria for water quality in Northwestern and Southern Amazon (Moulatlet et al., 2023).

The increase of mining activities results in pronounced alterations in physicochemical properties of the Amazonian aquatic environments, where the metals enrichment has as consequence its bioaccumulation in tissues of several fish species (Costa et al., 2022; Lima et al., 2015; Santana, 2016). For example, levels of Hg, Cd, Cr, and Pb were found above the legal limits allowed by the Brazilian Health Regulatory Agency (ANVISA, 1998) in flesh edible muscle of several fish species in Cassiporé River (Northern Amazon, Amapá state), especially in carnivorous and omnivorous fishes, an area impacted by gold mining activities (Hacon et al., 2020; Lima et al., 2015). In addition, the bioaccumulation of Hg in white muscle was seen in several carnivorous and piscivorous fish species from both Madeira and Purus Rivers (Central Amazon) in levels higher than those recommended by the World Health Organization (WHO) (limit of > 0.50 mg/kg of total mercury), in comparison with detritivorous and herbivorous species (Bastos et al., 2015; Castro et al., 2016).

Other important sources of metal water pollution and bioaccumulation in fishes are related to the expansion of anthropogenic activities in Amazon, especially in the surrounding areas of the main cities, such as domestic sewage release, construction of hydroelectric dams, and industrial plants (Duarte & Val, 2020; Val et al., 2023). For example, in Manaus—the biggest city of the Amazon—the large stream networks crossing the urban area are facing substantial degradation of water quality and enrichment by metals, which has resulted in the loss of species accomplished by a great shift in aquatic community composition reducing local biodiversity (Anjos, 2007; Duarte & Val, 2020; Guarido, 2014). In streams crossing the industrial area in Manaus, the levels of Co, Ni, Fe, Zn, Cd, Cu, and Pb are above the limits allowed by Resolution No. 357 of the Brazilian Environmental National Council (CONAMA, 2005) in both water and sediment, resulting in bioaccumulation of Zn and Cu in the liver and white muscle of the facultative air-breathing armored catfish (*Hoplosternum littorale*) (Santana, 2016; Santana & Barroncas, 2005; Silva et al., 1999).

In the Amazon River mouth region, close to Macapá city, studies have demonstrated a strong correlation between increased levels of metals in water with liver bioconcentration of these compounds. In the Araguari River, hepatic bioaccumulation of metals was seen in both detritivorous fish, *Curimata vittata* (Fe, Pb, Hg, and Cu) and *Hemiodus unimaculatus* (Mn and Zn), and carnivorous *Plagioscion squamosissimus* (Cd, Ni, and Zn), which are strongly correlated to metal content in water and suggesting that tropic level of fishes is important to determine the levels of bioaccumulation (Viana et al., 2020). In addition, in the Amazon River estuary, females of the endemic pufferfish *Colomesus asellus* showed high bioaccumulation of Ni, Mn, Pb, Cd, and Hg in the liver, with some metals being also accumulated in skin and gonads (Viana et al., 2023). Furthermore, water contamination by metals in the Gulf of Maranhão, in the coastal area of Northeastern Amazon, resulted in the bioaccumulation of Fe, Mn, and Cu in the white muscle of fishes, especially in *Macrodon ancylodon* (carnivorous) and *Stellifer rastrifer* (omnivorous), in levels that possess a severe risk for human consumption (Silva et al., 2023).

2 Toxicological Vulnerability of Amazon Fish to Different Pesticides and Heavy Metals

In spite of the high environmental risk of both pesticides and metals contamination to the conservation of Amazon aquatic biota, there is a lack of systematic studies evaluating the toxicity of these xenobiotics in Amazon fish. This fact is highly concerning in the context of the high diversity of fish species in this ecosystem, and the increment in several substances released is detected in Amazonian water bodies. To our knowledge, the toxicity of only 12 pesticides: four herbicides (atrazine, duron, glyphosate, and paraquat), six insecticides (deltamethrin, fipronil, imidaclorid, lufenuron, methyl parathion, and malathion), one fungicide (carbendazim) and one parasiticide (trichlorfon) and six metals (Cu, Ni, Hg, Pb, Cd, and Mn) were previously evaluated in just over 21 fish species (Tables 1 and 2). Overall, Amazon fish species seem to be moderate to highly sensitive to pesticides and metals, making them extremely vulnerable to water contamination.

There is a crescent number of reports showing an increase in pesticide contamination in the Amazon, especially of different formulations of insecticides and fungicides that can reach environmental concentrations posing high ecological risk to the aquatic biota (Duarte & Val, 2020; Rico et al., 2021; Soumis et al., 2003). In

Table 1 Acute toxicity (LC_{50}) of pesticides with different chemical composition on Amazon fishes

Pesticide classification	Chemical group	Family	Species	96 h LC_{50}	Weight of fish (g)	Reference
Herbicides	Atrazine	Serrasalmidae	*Colossoma macropomum*	22.00 mg/L	–	Chapadense et al., 2009
	Diuron	Characidae	*Paracheirodon axelrodi*	17.61 mg/L	0.12 ± 0.03	Souza, 2014
		Serrasalmidae	*Colossoma macropomum*	21.44 mg/L	0.42 ± 0.19	Souza, 2014
		Gasteropelecidae	*Carnegiella strigata*	28.81 mg/L	0.37 ± 0.29	Souza, 2014
		Characidae	*Hemigrammus rhodostomus*	32.41 mg/L	0.42 ± 0.16	Souza, 2014
		Callichthyidae	*Corydoras schwartzi*	42.65 mg/L	0.41 ± 0.04	Souza, 2014
	Glyphosate	Serrasalmidae	*Colossoma macropomum*	3.607 mg/L	0.42 ± 0.19	Souza, 2014
		Gasteropelecidae	*Carnegiella strigata*	5.75 mg/L	0.37 ± 0.29	Souza, 2014
		Characidae	*Paracheirodon axelrodi*	6.991 mg/L	0.12 ± 0.03	Souza, 2014
		Characidae	*Hemigrammus rhodostomus*	7.53 mg/L	0.42 ± 0.16	Souza, 2014
		Callichthyidae	*Corydoras schwartzi*	13.835 mg/L	0.41 ± 0.04	Souza, 2014
		Serrasalmidae	*Colossoma macropomum*	20 mg/L	–	Miyazaki et al., 2004
	Paraquat	Serrasalmidae	*Colossoma macropomum*	48.05 mg/L	87.69 ± 34.23	Salazar-Lugo et al., 2009

Insecticides	Deltamethrin	Hypopomidae	*Microsternarchus* cf. *bilineatus*	2.15 μg/L	0.89 ± 0.33	Chaves et al., 2020
		Serrasalmidae	*Colossoma macropomum*	6.69 μg/L	0.42 ± 0.19	Souza et al., 2020
		Gasteropelecidae	*Carnegiella strigata*	7.83 μg/L	0.37 ± 0.29	Souza et al., 2020
		Characidae	*Paracheirodon axelrodi*	22.49 μg/L	0.42 ± 0.16	Souza et al., 2020
		Characidae	*Hemigrammus rhodostomus*	23.63 μg/L	0.42 ± 0.16	Souza et al., 2020
		Bryconidae	*Brycon amazonicus*	36 μg/L	10.5 ± 3	Moraes et al., 2013
		Serrasalmidae	*Colossoma macropomum*	56 μg/L	9.15 ± 1.49	Cunha et al., 2018
		Callichthyidae	*Corydoras schwartzi*	183.51 μg/L	0.41 ± 0.04	Souza et al., 2020
	Fipronil	Serrasalmidae	*Colossoma macropomum*	0.33 mg/L	0.08 ± 0.01	Saldaña-Serrano et al., 2021
	Imidaclorid	Gasteropelecidae	*Carnegiella strigata*	145.63 mg/L	0.37 ± 0.29	Souza, 2014
		Serrasalmidae	*Colossoma macropomum*	245.38 mg/L	0.42 ± 0.19	Souza, 2014
		Callichthyidae	*Corydoras schwartzi*	250.70 mg/L	0.41 ± 0.04	Souza, 2014
		Characidae	*Paracheirodon axelrodi*	261.03 mg/L	0.12 ± 0.03	Souza, 2014
		Characidae	*Hemigrammus rhodostomus*	302.13 mg/L	0.42 ± 0.16	Souza, 2014
	Lufenuron	Serrasalmidae	*Colossoma macropomum*	0.58 mg/L	–	Soares et al., 2016
	Methyl parathion	Cichlidae	*Dicrossus maculatus*	2.90 mg/L	0.40 ± 0.10	Rico et al., 2010
		Serrasalmidae	*Colossoma macropomum*	2.91 mg/L	45.23 ± 0.43	Silva et al., 2017
		Callichthyidae	*Corydoras pygmaeus*	4.09 mg/L	0.20 ± 0.10	Rico et al., 2010
		Serrasalmidae	*Colossoma macropomum*	4.98 mg/L	0.30 ± 0.10	Rico et al., 2010
		Lebiasinidae	*Nannostomus unifasciatus*	5.38 mg/L	0.40 ± 0.10	Rico et al., 2010
		Characidae	*Paracheirodon axelrodi*	6.09 mg/L	0.40 ± 0.30	Rico et al., 2010
		Loricariidae	*Otocinclus affinis*	6.83 mg/L	0.50 ± 0.20	Rico et al., 2010
		Characidae	*Hyphessobrycon erythrostigma*	7.27 mg/L	0.90 ± 0.30	Rico et al., 2010
	Malathion	Characidae	*Paracheirodon axelrodi*	0.25 mg/L (72 h LC_{50})	0.40 ± 0.30	Rico et al., 2011
		Characidae	*Hyphessobrycon erythrostigma*	0.25 mg/L	0.90 ± 0.30	Rico et al., 2011
		Loricariidae	*Otocinclus affinis*	1.07 mg/L	0.50 ± 0.20	Rico et al., 2011
		Serrasalmidae	*Colossoma macropomum*	1.51 mg/L	–	Rico et al., 2011
		Serrasalmidae	*Colossoma macropomum*	15.8 ± 3.30 mg/L	39.26 ± 3.40	Souza et al., 2020

(continued)

Table 1 (continued)

Pesticide classification	Chemical group	Family	Species	96 h LC_{50}	Weight of fish (g)	Reference
Fungicide	Carbendazim	Characidae	*Paracheirodon axelrodi*	1.65 mg/L	0.40 ± 0.30	Rico et al., 2011
		Characidae	*Hyphessobrycon erythrostigma*	3.69 mg/L	0.90 ± 0.30	Rico et al., 2011
		Serrasalmidae	*Colossoma macropomum*	4.16 mg/L	–	Rico et al., 2011
		Loricariidae	*Otocinclus affinis*	4.24 mg/L	0.50 ± 0.20	Rico et al., 2011
Parasiticide	Trichlorfon	Serrasalmidae	*Colossoma macropomum*	0.87 mg/L	13.6 ± 4.3	Da Silva et al., 2019
		Serrasalmidae	*Colossoma macropomum*	0.87 mg/L	–	Duncan et al., 2020

Table 2 Acute toxicity (LC_{50}) of metals to Amazon fishes determined in different types of water in toxicity tests

Metal	Family	Species	96 h LC_{50}	Weight of fish (g)	Type of water	Reference
Cadmium	Serrasalmidae	*Colossoma macropomum*	38.47 mg/L	87.69 ± 34.23	Dechlorinated tap water	Salazar-Lugo et al., 2013
Copper	Characidae	*Hemigrammus rhodostomus*	12.81 μg/L	0.338 ± 0.02	Well water	Duarte et al., 2009
	Loricariidae	*Otocinclus hasemani*	18.01 μg/L	0.389 ± 0.02	Well water	Duarte et al., 2009
	Gasteropelecidae	*Carnegiella strigata*	26.25 μg/L	0.28 ± 0.01	Well water	Duarte et al., 2009
	Characidae	*Hyphessobrycon socolofi*	26.85 μg/L	0.795 ± 0.02	Well water	Duarte et al., 2009
	Cichlidae	*Apistogramma hyppolytae*	32.95 μg/L	1.215 ± 0.08	Well water	Duarte et al., 2009
	Cichlidae	*Dicrossus maculatus*	39.32 μg/L	0.68 ± 0.01	Well water	Duarte et al., 2009
	Cichlidae	*Apistogramma agassizii*	40.15 μg/L	1.267 ± 0.02	Well water	Duarte et al., 2009
	Characidae	*Paracheirodon axelrodi*	45.87 μg/L	0.245 ± 0.01	Well water	Duarte et al., 2009
	Callichthyidae	*Corydoras schwartzi*	53.02 μg/L	0.940 ± 0.02	Well water	Duarte et al., 2009
	Characidae	*Paracheirodon axelrodi*	56.30 μg/L	0.07 ± 0.01	Well water	Crémazy et al., 2016
	Callichthyidae	*Dianema urostriatum*	74.14 μg/L	3.04 ± 0.10	Well water	Duarte et al., 2009
	Characidae	*Paracheirodon axelrodi*	194.0 μg/L	0.07 ± 0.01	Rio Solimões water (filtered)	Crémazy et al., 2016
	Characidae	*Hemmigramus bellote*	229.7 μg/L	0.34 ± 0.02	Rio Solimões water (dry season)	Belivacqua, 2009
	Characidae	*Hyphessobrycon* sp	238.5 μg/L	0.53 ± 0.02	Rio Amazonas water (wet season)	Belivacqua, 2009
	Gasteropelecidae	*Carnegiella strigata*	307.00 μg/L	0.280 ± 0.01	Rio Negro water	Duarte et al., 2024
	Cichlidae	*Dicrossus maculatus*	333.00 μg/L	0.680 ± 0.01	Rio Negro water	Duarte et al., 2024
	Characidae	*Hyphessobrycon socolofi*	414.00 μg/L	0.870 ± 0.03	Rio Negro water	Duarte et al., 2024
	Characidae	*Hemigrammus rhodostomus*	443.00 μg/L	0.340 ± 0.02	Rio Negro water	Duarte et al., 2024
	Loricariidae	*Otocinclus hasemani*	454.00 μg/L	0.390 ± 0.02	Rio Negro water	Duarte et al., 2024
	Characidae	*Hemmigramus bellote*	460.3 μg/L	0.34 ± 0.02	Rio Solimões water (wet season)	Belivacqua, 2009

(continued)

Table 2 (continued)

Metal	Family	Species	96 h LC_{50}	Weight of fish (g)	Type of water	Reference
	Characidae	*Hyphessobrycon* sp	464.7 μg/L	0.53 ± 0.02	Rio Amazonas water (dry season)	Belivacqua, 2009
	Characidae	*Paracheirodon axelrodi*	558.05 μg/L	0.24 ± 0.01	Rio Negro water (wet season)	Belivacqua, 2009
	Loricariidae	*Otocinclus vitatus*	572.1 μg/L	0.27 ± 0.04	Rio Solimões water	Dal Pont et al., 2017
	Characidae	*Paracheirodon axelrodi*	620.0 μg/L	0.07 ± 0.01	Rio Solimões water	Crémazy et al., 2019
	Characidae	*Paracheirodon axelrodi*	672.00 μg/L	0.240 ± 0.01	Rio Negro water	Duarte et al., 2024
	Serrasalmidae	*Colossoma macropomum*	740 μg/L	44.7 ± 1.0	Well water	Oliveira, 2003
	Cichlidae	*Apistogramma hyppolytae*	831.00 μg/L	1.210 ± 0.83	Rio Negro water	Duarte et al., 2024
	Callichthyidae	*Corydoras schwartzi*	915.00 μg/L	0.940 ± 0.02	Rio Negro water	Duarte et al., 2024
	Cichlidae	*Apistogramma agassizii*	1090.00 μg/L	1.270 ± 0.15	Rio Negro water	Duarte et al., 2024
	Characidae	*Paracheirodon axelrodi*	1090.0 μg/L	0.07 ± 0.01	Rio Negro water (filtered)	Crémazy et al., 2016
	Characidae	*Hyphessobrycon socolofi*	1396.02 μg/L	0.795 ± 0.02	USEPA soft water	Duarte et al., 2009
	Characidae	*Paracheirodon axelrodi*	1701.27 μg/L	0.24 ± 0.01	Rio Negro water (dry season)	Belivacqua, 2009
	Serrasalmidae	*Colossoma macropomum*	4300.00 μg/L	40.8 ± 7.1	Dechlorinated tap water	Tavares-Dias et al., 2011
Copper nanoparticle (nCuO)	Characidae	*Paracheirodon axelrodi*	139.2 μg/L	0.25 ± 0.01	Well water	Braz-Mota et al., 2018
	Cichlidae	*Apistogramma agassizii*	116.6 μg/L	0.89 ± 0.07	Well water	Braz-Mota et al., 2018
Lead	Serrasalmidae	*Colossoma macropomum*	28.71 mg/L	44.7 ± 1.0	Well water	Oliveira, 2003
Manganese	Serrasalmidae	*Colossoma macropomum*	4.03 mg/L	7–10	Well water	Gabriel et al., 2013
Mercury	Serrasalmidae	*Piaractus brachypomus*	0.56 mg/L	6.55 ± 1.3	Dechlorinated tap water	Naranjo-Gómez et al., 2013
	Bryconidae	*Brycon amazonicus*	0.71 mg/L	17.94 ± 3.13	Dechlorinated tap water	Monteiro et al., 2010

Nickel	Characidae	*Paracheirodon axelrodi*	28.9 mg/L	0.1–0.2	Rio Negro water (pH 7.0 wet season)	Holland et al., 2017
	Characidae	*Paracheirodon axelrodi*	17.8 mg/L	0.1–0.2	Rio Negro water (pH 7.0 dry season)	Holland et al., 2017
	Characidae	*Paracheirodon axelrodi*	13.8 mg/L	0.1–0.2	Rio Tapajós water (pH 7.0 wet season)	Holland et al., 2017
	Characidae	*Paracheirodon axelrodi*	24.1 mg/L	0.1–0.2	Rio Tapajós water (pH 7.0 dry season)	Holland et al., 2017
	Characidae	*Paracheirodon axelrodi*	22.2 mg/L	0.1–0.2	Rio Solimões water (pH 7.0 wet season)	Holland et al., 2017
	Characidae	*Paracheirodon axelrodi*	21.8 mg/L	0.1–0.2	Rio Solimões water (pH 7.0 dry season)	Holland et al., 2017
	Characidae	*Paracheirodon axelrodi*	9.72 mg/L	0.1–0.2	Rio Negro water (pH 4.0 wet season)	Holland et al., 2017
	Characidae	*Paracheirodon axelrodi*	41.5 mg/L	0.1–0.2	Rio Negro water (pH 4.0 dry season)	Holland et al., 2017
	Characidae	*Paracheirodon axelrodi*	27.8 mg/L	0.1–0.2	Rio Tapajós water (pH 4.0 wet season)	Holland et al., 2017
	Characidae	*Paracheirodon axelrodi*	71.3 mg/L	0.1–0.2	Rio Solimões water (pH 4.0 dry season)	Holland et al., 2017
	Characidae	*Paracheirodon axelrodi*	90.3 mg/L	0.1–0.2	Well water	Holland et al., 2017

these regards, these formulations have been classified as moderately to highly toxic to Amazon fish, where insecticides (e.g., deltamethrin, fipronil, lufenuron and trichlorfon) are by far more toxic than fungicides (carbendazim) and herbicides (e.g., glyphosate, atrazine, diuron, and paraquat). Table 1 shows the lethal concentration values (LC_{50}) of different pesticides for different species of Amazonian fish. The pyrethroid insecticide deltamethrin is extremely toxic to Amazon fishes, where the electric knife fish *Microsternarchus* cf. *bilineatus* (Gymnotiformes) was the most sensitive species—2.15 mg/L 96 h LC_{50}—followed by Characiformes—*Carnegiella strigatta* (7.83 mg/L), *Paracheirodon axelrodi* (22.49 mg/L; cardinal tetra), and *Hemigrammus rhodostomus* (23.63 mg/L; rummy-nose tetra)—and the Siluriformes species—*Corydoras schwartzi* (183.51 mg/L) (Chaves et al., 2020; Souza et al., 2020). In addition, the ontogenetic development stage of fish influences deltamethrin toxicity, where fry of *Colossoma macropomum* (0.42 g) were eight times more sensitive than juvenile fish (9.15 g) (96 h LC_{50} of 6.69 mg/L and 56 mg/L, respectively) (Cunha et al., 2018; Souza et al., 2020).

Acute toxicity comparison among other formulations reveals that the fish species model *C. macropomum* is moderately sensitive to insecticides, in which the most toxic pesticides were fipronil (0.33 mg/L 96 h LC_{50}), luferon (0.58 mg/L), and trichlorfon (0.87 mg/L), followed by malathion (1.51 mg/L), methyl parathion (2.91–4.98 mg/L), and imidaclorid (245.38 mg/L), respectively (Duncan et al., 2020; Rico et al., 2010, 2011; Saldaña-Serrano et al., 2021; Silva et al., 2017; Soares et al., 2016; Souza, 2014). Similarly, acute toxicity of malathion was substantially higher to *Nannostomus unifasciatus*—One-line pencil fish—(0.11 mg/L), *P. axelrodi* (0.25 mg/L), *Hyphessobrycon erythrostigma*—Bleeding-heart tetra—(0.25 mg/L), and *Otocinclus affinis* (1.07 mg/L) compared to methyl parathion (5.38, 6.09, 7.27, and 6.83 mg/L, respectively) (Rico et al., 2010; Rico et al., 2011). On this point, the insecticide imidaclorid is the less toxic pesticide to Amazon fish species with acute toxicity ranging from 145.63 (*C. strigatta*) to 302.13 mg/L (*H. rhodostomus*) (Souza, 2014).

Carbendazim is the only fungicide that had its toxicity studied at the moment and displays moderate toxicity to Amazon fish with 96 h LC_{50} ranging from 1.65 (*P. axelrodi*) to 4.24 mg/L (*O. affinis*) (Rico et al., 2011). Already in the class of herbicides, acute toxicity to *C. macropomum* was higher in the following order: glyphosate (3.6 mg/L) > diuron (21.4 mg/L) > atrazine (22.0 mg/L) > paraquat (48.05 mg/L) (Chapadense et al., 2009; Salazar-Lugo et al., 2009; Souza, 2014), where larger juveniles were more tolerant to glyphosate (20 mg/L) than fish in fry stage (Miyazaki et al., 2004; Souza, 2014). In fact, acute toxicity of diuron was also markedly lower to *P. axelrodi* (2.5x), *C. schwartzi* (3.1x), *H. rhodostomus* (4.3x), and *C. strigatta* (5.1x), compared to glyphosate, with 96 h LC_{50} ranging from 17.61 (*P. axelrodi*) to 42.65 mg/L (*C. schwartzi*) (Souza, 2014).

Regarding metal toxicity in Amazon, the most hazards found in sediment are the Cu, Hg Cd, Pb, and Zn, while for the water environment is Cu, Ni, Hg, Fe, Cd, Pb, and Zn in water (Moulatlet et al., 2023). Only a few studies addressed the toxicological vulnerability of Amazon fish to Cd (Salazar-Lugo et al., 2013) and Pb (Oliveira, 2003), while there are no studies regarding the lethal toxicity of Fe and Zn for Amazon fish. The Cu is by far the most toxic metal to Amazon fish—and the most studied metal—with Ni and Pb displaying moderate toxicity. Table 2 shows the lethal concentration values (LC_{50}) of different heavy metals for different species of Amazonian fish. Interestingly, there is robust evidence that the physicochemical properties of Amazon waters, especially those found in both black and white waters, strongly modulate metal toxicity to Amazon fish. Under controlled laboratory conditions, using ion-poor groundwater as a test solution, Cu is extremely toxic to Characidae species (tetras) where 96 h LC_{50} was lower than 29 μgCu/L to *C. strigatta*, *H. rhodostomus*, and *H. socolofi*, whereas to the Callichthyidae *C. schwartzi* (armored catfish) and *D. urostriatum* (flagtail catfish) 96 h LC_{50} was 53.0 and 74.1 μgCu/L, respectively (Duarte et al., 2009). Among Characidae species, *P. axelrodi* displays a remarkable tolerance to Cu presenting 96 h LC_{50} almost two times higher (45.0–56.30 μgCu/L) than the other species (Crémazy et al., 2016; Duarte et al., 2009). In addition, the toxicity of Cu nanoparticle (nCuO) was seen to be lower for both *P. axelrodi* (139.2 μg nCuO/L) and *A. agassizii* (116.6 μg nCuO/L) compared to the ionic form (Cu^{2+}) (approximately 40 μgCu/L) (Braz-Mota et al., 2018; Duarte et al., 2009). The studies have shown a moderate toxicity of Hg to Amazon fish with 96 h LC_{50} ranging from 0.56 mg/L in *Piaractus brachypomus* to 0.71 mg/L in *Brycon amazonicus* (Monteiro et al., 2010; Naranjo-Gómez et al., 2013). Furthermore, to the fish species model *C. macropomum* kept in well water Cu (96 h LC_{50} ranging from 0.74 to 4.3 mg/L) and Mn (4.03 mg/L) were significantly more toxic than Pb (28.71 mg/L) and Cd (38.47 mg/L) (Oliveira, 2003; Salazar-Lugo et al., 2013; Tavares-Dias et al., 2011). Finally, Ni toxicity was determined only to cardinal tetra (*P. axelrodi*) in groundwater that displays a high tolerance to acute metal exposure (96 h LC_{50} of 90.3 mg/L) (Holland et al., 2017).

On the other hand, increased competition and complexation of Cu^{2+} by enhancing ionic content, suspended solids, and dissolved organic carbon (DOC) in water promote a remarkable reduction of Cu toxicity to Amazon fish. For example, Cu toxicity determined in standard USEPA soft water was 52 times lower to *H. socolofi* (96 h LC_{50} of 1396,02 μg/L) compared to toxicity test run in groundwater (26.85 μg/L) (Duarte et al., 2009). In studies conducted with Amazon white waters—rich in suspended solids and higher ionic content—Cu toxicity to *O. vittatus* was more than seven times reduced with the increment of water from Rio Solimões (RSW) in test solution, ranging from 97.91 μg/L at 20% of RSW to 697.60 μg/L in full strength of RSW (Dal Pont et al., 2017). The amount of suspended solids in RSW also influences copper toxicity, as seen by Crémazy et al. (2016, 2019), where a three times lower 96 h LC_{50} was determined to *P. axelrodi* in filtered RSW (194.0 μg/L) compared to unfiltered RSW (620.0 μg/L). In addition, natural variation in suspended solids and ionic content of white waters between dry and wet seasons modulate Cu toxicity. In water from Rio Solimões, Cu was two times more toxic to *Hemmigramus bellote* in the dry season (229.7 μg/L), whereas in water from Rio Amazonas, *Hyphessobrycon* sp. was more tolerant to Cu in dry (464.7 μg/L) than in wet season (238.5 μg/L) (Belivacqua, 2009). Similarly, the increment of DOC in water has a significant effect in reducing Cu toxicity in Amazon fish. In toxicity tests with Rio Negro water (RNW) reductions in Cu toxicity (i.e., increased 96 h LC_{50} values) were seen to *D. maculatus* (8.5-fold), *C. strigatta* (11.7-fold), *P. axelrodi* (14.7-fold), *H. socolofi* (15.4-fold), *C. schwartzi* (17.3-fold), *O. hasemani* (25.2-fold), *A. hippolytae* (25.2-fold), *A. agassizzi* (27.1-fold), and *H. rhodostomus* (34.6-fold) (Duarte et al., 2024), while Cu toxicity was three times lower to *P. axelrodi* at RNW in wet season (higher DOC content) compared to dry season (1701.27 and 558.05 μg/L, respectively) (Belivacqua, 2009).

Although Ni LC_{50} has been determined only to cardinal tetra (*P. axelrodi*), there is strong evidence of Ni toxicity modulation by natural variation in physicochemical properties of Amazon waters (Holland et al., 2017). In general, the toxicity of Ni was significantly higher at pH 7.0 than at pH 4.0, as seen by lower 96 h LC_{50} values and determined in tests with natural water from Rio Tapajós (clearwater), Rio Solimões (whitewater), and Rio Negro (blackwater). However, at neutral pH, the influence of seasonal changes in water chemistry on Ni toxicity differs among the types of waters, with a higher toxicity seen in the dry season at blackwater (17.8–28.9 mgNi/L), contrasting with observed in clearwater where Ni is more toxic at wet season (13.8–24.1 mg/L). In contrast, at toxicity tests run at low pH (4.0), Ni was markedly more toxic in the wet than in the dry season. It was especially seen in Rio Negro waters where the 96 LC_{50} values rose from 9.7 to 41.5 mg/L compared to Ni toxicity reported at both Tapajós and Solimões waters (96 h LC_{50} around 28 and 71 mg/L, respectively). Interestingly, seasonal changes in some properties of DOC molecules seem to drive Ni speciation in natural waters controlling the concentration of Ni toxic form (Ni^{2+}), with Ni toxicity being positively related to the amount of humic-like and fulvic-like compounds, as well as to DOC concentration (Holland et al., 2017). Therefore, these effects on metal toxicity suggest that variations in natural conditions of Amazon aquatic environments might exert a protective action on fishes. It seems especially true for the amount and composition (e.g., humic-like fraction) of DOC and the suspended solids, which control metal speciation by enhancing the ionic competition and metal complexation resulting in reduced metal bioavailability.

While Cu and Ni are the most harmful to Amazon fish, and the DOC can attenuate their toxicity, there is no information about DOC's impact on metals like Hg. Due to its prolonged persistence in the environment, high toxicity even at low concentrations, and increased contamination due to mining activities, Hg toxicity raises serious concerns (Meneses et al., 2022). For the fish, the carnivorous species are the most affected at the top of the food chain, which accumulates significantly higher levels of Hg, exceeding the levels recommended by the WHO, including for the humans who consume these fish and also show levels of Hg in their hair that indicate toxicity (Albuquerque et al., 2020). Currently, we have only two values of the LC_{50} for Hg in the Amazon species *Piaractus brachypomus* (0.56 mg/L) and *Brycon amazonicus* (0.71 mg/L), demonstrating the sensitivity of Amazonian fish to this metal. The environmental concentrations are a concern once, even in regions far from gold mining sites, the Hg level can reach 0.25 mg/L (Meneses et al., 2022), showing that exposure to Hg is not restricted to the gold mining site areas. Interestingly, the Hg can make associations with essential trace elements, such as Se and also Fe, Co, and Mn, which also would make it interesting to understand the influence of DOC and the suspended solids on Hg toxicity once these compounds could play a role in Hg cycling and methylation, reducing Hg toxicity in aquatic environments (Albuquerque et al., 2020).

3 Physiological Effects Based on Exposure and Biomarkers

3.1 Toxic Effects of Pesticides on Amazonian Fish

The adverse effects of pesticides on Amazonian fish are multifaceted and influenced by factors such as exposure time, the specific fish species involved, the type of pesticide employed, and the contamination route (whether the contaminant is in the water or trophic pathway). The contamination initiates a series of harmful changes that disrupt the regular function of the systems of these animals, resulting in a wide range of consequences, from minor impairments to severe, even fatal outcomes for the animals (Fig. 1).

Enzymatic changes can hinder the operation of vital enzymes in Amazon fish and are usually one of the first alterations after pesticide exposure. This interference disrupts fundamental metabolic functions essential for the fish's overall well-being. Herbicides prompt the excessive production of reactive oxygen within fish cells. Herbicides such as Roundup® (RD) and insecticides such as malathion (MAL) and deltamethrin (DM) commonly increase the activation of antioxidant enzymes and biotransformation mechanisms in Amazon fish (Braz-Mota et al., 2015; Souza et al., 2020, 2021). As a result of higher production of reactive oxygen species (ROS), as seen by imbalances in biotransformation and antioxidant systems,

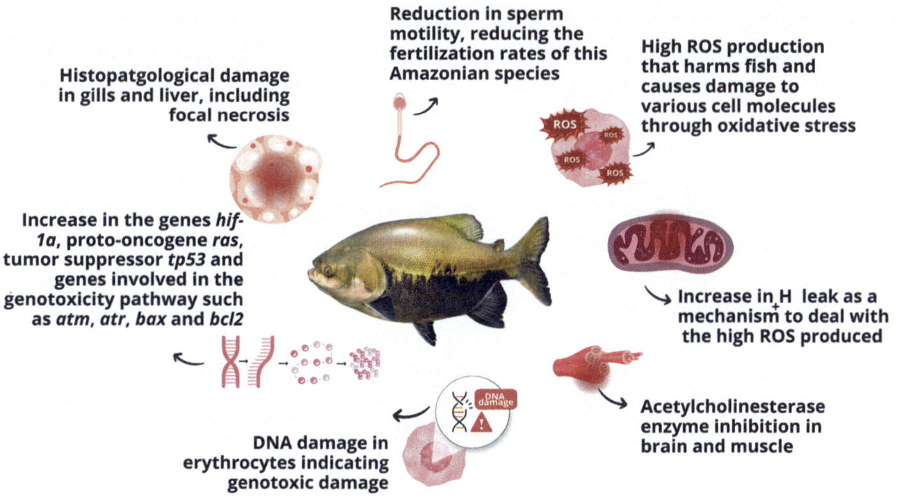

Fig. 1 General diagram showing the main toxic effects of pesticides on Amazonian fish

oxidative stress occurs, promoting the oxidation of lipids and proteins and resulting in oxidative damage in DNA. The Amazon fish surubim (*Pseudoplatystoma* sp) displays changes in oxidative stress parameters after exposure to RD. This species is very sensitive to this pesticide, presenting alterations in important antioxidant enzymes, such as superoxide dismutase (SOD) and catalase (CAT) in tissues such as the liver and brain and also in biomarkers of oxidative stress, such as thiobarbituric acid reactive substances (TBARS) and carbonyl protein (Sinhorin et al., 2014). All these changes show how toxic these different pesticides are to Amazonian fish and promote high ROS production that harms fish and causes damage to various cell molecules through oxidative stress.

Mitochondrial energy demand is also affected by exposure to pesticides and contributes to imbalances of ATP generation and acts as one of the main sources of ROS for cells (Lesnefsky et al., 2001). These disruptions impair fish's ability to efficiently generate energy through cellular respiration, as has already been demonstrated experimentally in Amazonian fish for *Colossoma macropomum* (tambaqui) exposed to MAL (Souza et al., 2020, 2021). For this fish species, the MAL effects on mitochondria respiration are time-dependent and affect the adaptation mechanism of this fish to maintain homeostasis and ATP demand. The increase in H^+ leak is observed until 48 h after exposure to MAL, suggesting a disturbance of the mechanisms to deal with the high ROS produced (Brookes, 2005). Furthermore, MAL exposure also increases the capacity of the electron transport system (ETS), increases respiration stimulated by ADP, and consequently increases ATP synthesis (Souza et al., 2021). Such an increase in ETS capacity must be related to the amount of ATP needed for activating ROS neutralization mechanisms.

Failure in biotransformation processes, alongside the decline in antioxidant enzyme activity, is frequently employed with other enzymes and biomarkers to assess the extent of damage in fish exposed to pesticides, such as the neurotoxicity assessment (van der Oost et al., 2003). The acetylcholinesterase enzyme (AChE) regulates several nervous functions in animals. Typically, it exhibits high sensitivity to organophosphate compounds. This inhibition disrupts its normal function, accumulating acetylcholine in synaptic junctions and elevating AChE levels (Bretaud et al., 2000).

Exposure to organophosphate herbicide RD promotes inhibition of brain cholinergic activity in the Amazon fish tambaqui (Braz-Mota et al., 2015). In contrast, MAL, an insecticide also based on organophosphate, showed no AChE inhibition, suggesting an absence of neurotoxicity. Such results show that the neurotoxic effects can vary significantly depending on the type of pesticide, the organism's size, environmental conditions, species, individual variations, and pesticide concentration (Sturm et al., 1999). For the Amazon fish species, *Microsternarchus* cf. *bilineatus*, there were no changes in AChE after exposure to the pyrethroid insecticide DM (Chaves et al., 2020), though the effects of this pesticide have previously been associated with a compensatory mechanism of stimulation of this enzyme to perform an adaptive up-regulation of acetylcholine hydrolysis (Hossain et al., 2004).

As a result of the oxidative stress, genotoxic effects are also observed in Amazon fish exposure to pesticides. For the tambaqui, the herbicide RD promotes DNA damage in erythrocytes indicating genetic damage in these animals (Braz-Mota et al., 2015; Silva et al., 2019). These results show that the damage caused by this herbicide overrides the activity of DNA repair systems, such as base excision repair (BER) and/or nucleotide excision repair (NER). They indicate that the repair system does not compensate for the constant production of free radicals and generates oxidative stress. In the Amazon, RD is widely used in fish farms to control floating aquatic plants (Araújo et al., 2008). The effect of this herbicide associated with the hypoxic waters of the Amazon could enhance RD's toxic effect, which was not observed under laboratory conditions since the synergistic effect of genotoxic damage was not observed between hypoxia and RD (Silva et al., 2019).

Pesticide exposure promotes modifications in the way genes are expressed in fish. Such genetic changes can have far-reaching effects on the fish's physiology, potentially disrupting normal cellular processes. Once again, the studies of fish of the Amazon are evaluating RD and MAL's effects on tambaqui (Silva et al., 2019; Souza et al., 2020, 2023). The gene hypoxia-inducible factor-1α (*hif-1α*) is a master regulator of many physiological responses to hypoxia, controlling the transcription of more than 100 genes regulating diverse functions, such as angiogenesis, erythropoiesis, glucose metabolism, vasodilatation, apoptosis, cell growth, and cell proliferation (Wu et al., 2010). The *hif-1α* usually decreases in hypoxia, and some recent studies also show that during exposure to RD plus hypoxia, a synergistic effect is observed, evidencing that its suppression is potentialized due to the impairment of molecular machinery in response to oxidative stress and DNA damage.

For the tambaqui, the MAL induced an overexpression of proto-oncogene *ras* (Souza et al., 2020, 2023), while for RD, the increase in this gene expression was potentialized by hypoxic events, demonstrating high toxicity of RD and hypoxia together (Silva et al., 2019). The MAL also increased the gene tumor suppressor *tp53* expression and genes involved in the central response pathway to genotoxicity such as *atm*, *atr*, *atm*, *bax*, and *bcl2*. The positive regulation of *atm* and *tp53* can explain the increased DNA damage and the occurrence of apoptosis in tissues such as the liver. The alterations in all these genes after exposure to MAL show that apoptosis pathways were activated since the *tp53* gene regulates both pro and antiapoptotic target genes. Such findings indicate that pesticide exposure poses a significant threat, even to resilient fish species like tambaqui. This underscores the severity of stress-induced conditions resulting from exposure, which, in turn, prompt notable genetic responses related to both tumor suppression and apoptotic reactions.

The pesticides also disrupt the delicate balance between cell proliferation and cell death (apoptosis) in fish tissues, resulting in tissue damage and hindrance to the regeneration of damaged cells. The parasiticide trichlorfon, widely used by pisciculture to mitigate the effects of parasitic infections in fish, plays a substantial role in the cell division process due to the increase in the metaphases number in tambaqui exposed to 50% of the LC_{50}–96 h (0.435 mg/L) (dos Santos Costa et al., 2022). In addition, this same insecticide has a considerable effect on the mobilization of repetitive sequences in the tambaqui genome, observed by the increased dispersed distribution profile in the karyotypes of the retroelements *Rex1* and *Rex3* (dos Santos Costa et al., 2022). Thus, the trichlorfon can transpose Rex retrotransposable elements and change their distribution in the heterochromatin of the mitotic chromosomes of tambaqui.

The tambaqui also show cellular alterations when exposed to insecticide, such as the MAL, which frequently leads to adverse changes in their gill structure, including hyperplasia, lamellar fusion, and necrosis. These alterations inflict irreparable harm on the gill tissue, compromising the organ's functionality (Souza et al., 2020). Furthermore, when these same fish are exposed to the RD herbicide, their gills exhibit similar alterations such as hypertrophy and hyperplasia in filament and lamellar epithelium, epithelial lifting, aneurysm formation, rupture of lamellae epithelium, and an increased proliferation of chloride and mucous cells (Braz-Mota et al., 2015). These histopathologies disrupt osmorespiratory functions, promoting detrimental effects on the respiratory system and ionic regulation.

Severe alterations in the liver have been observed in tambaqui exposed to RD. These alterations include granular and vacuolar-drop dystrophy, which leads to the death of hepatocytes and necrotic changes (Silva et al., 2019), and, consequently, liver failure. Interestingly, the same study revealed again that the histopathological effects on tambaqui are even more severe when RD exposure is combined with hypoxia, leading to focal necrosis and further tissue damage. In the Amazon region, hypoxia is a natural phenomenon in flooded areas, leading to periods of severe hypoxia with oxygen levels dropping below 1 mg.L^{-1} O_2 (Braz-Mota & Almeida-Val, 2021). These findings raise significant concerns for the Amazon environment, as the synergistic effects of pesticides in conjunction with the region's hypoxic conditions can exacerbate gene expression disturbances and damage to fish tissues.

In addition, hormonal disruptions can be observed after pesticide exposure in fish and can have widespread consequences for various physiological processes, including reproduction and growth. Recently, a noteworthy observation has been made regarding a reduction in sperm motility associated with two widely used agricultural pesticides: glyphosate and fenitrothion (Santos et al., 2023). For both pesticides, sperm motility rates were significantly reduced after 30 s of exposure across all tested concentrations (6, 12, 24, and 120 mg L^{-1}). Exposure to 240 mg L^{-1} results in no activation at all. These tests have provided compelling evidence that the semen of tambaqui fish is remarkably sensitive to the effects of exposure to these two pesticides. It is worth noting that the entire process of fertilization, from sperm motility to embryonic development, can be significantly impacted by the presence of excessive contaminants (Kime & Nash, 1999). Consequently, it becomes a worrying condition since these pesticides can reduce the fertilization rates of this Amazonian species. These contaminants yield a substantial influence on the reproductive capacity and replenishment of fish species, potentially leading to environmental imbalances and a decline in species diversity.

It is crucial to acknowledge that significant challenges are associated with physiological alterations of herbicides on Amazon fish. Addressing these challenges is imperative to protect aquatic ecosystems and sustain fish populations, particularly those fish species used as food by local human populations. Furthermore, despite the large number of studies evaluating the toxicity of pesticides (Table 1), few studies focused on the effects of these compounds on fish. In summary, pesticides promote a complex and severe impact on the physiology of Amazon fish. These findings underscore the importance of investigating and addressing these issues to preserve the health of aquatic ecosystems and ensure the sustainability of fish populations.

3.2 Toxic Effects of Heavy Metals on Amazonian Fish

Metal pollution in the Amazon is also a significant concern for the fish that inhabit this region since these metals can bioaccumulate in the animal's body or even biomagnify (Braz-Mota et al., 2018; Crémazy et al., 2016, 2019; Viana et al., 2023). Bioaccumulation refers to the process by which fish absorb elevated levels of metals dissolved in their contaminated aquatic environment or food sources over time, subsequently accumulating these metals within their tissues. On the other hand, biomagnification refers to the increase in the concentration of a contaminant along a food chain that occurs along successive trophic levels in a food chain (Ali & Khan, 2019). The bioaccumulation and biomagnification of toxic heavy metals have a significant public health concern, as fish are a common part of the human diet, and elevated levels of heavy metals in these animals can harm human health (Ali et al., 2017; Viana et al., 2023). The contamination by heavy metals through food can result in several toxic effects, including the interruption of many biological and biochemical processes, not only of fish but also of the animals that consume them (Braz-Mota et al., 2018; Crémazy et al., 2016, 2019).

In this context, it is essential to mention that Amazonian rivers present ecosystem patterns in floodplain areas that are determined by annual fluctuations in water level and by the alternation between dry (low water) and flood (high water) phases, where the variation of water level is an important factor for the bioavailability of metals (França, 2017). These fluctuations can either enhance or reduce the accessibility of these metals depending on the season. For example, in the middle Xingu River, in the Amazon, high concentrations of some metals such as mercury (Hg), chromium (Cr), lead (Pb), and nickel (Ni) were found due to the proximity to the mining area, and the rainy season increased the bioavailability of these metals (Ribeiro et al., 2017), probably due to the high leaching of these metals into aquatic bodies, putting the local aquatic biota at risk.

During the flood period, high methylmercury concentrations were detected in both the water and sediment of Lake Janauacá, located in the Amazonas region (França, 2017). This phenomenon can be attributed to an increased rate of methylation, primarily influenced by several variables such as water column depth, dissolved oxygen levels, dissolved organic carbon content, pH levels, and electrical conductivity. These findings indicate that during this specific time of the year, methylmercury has a higher propensity for bioaccumulation and biomagnification, particularly in fish at the top of the food chain. This same work also showed that fish such as *Pterygoplichthys pardalis* and *Hoplias malabaricus* collected in this environment with a high concentration of this metal showed erythrocyte nuclear abnormalities, such as reniform, vacuolated, lobed, segmented nuclei, and binucleate cells. Furthermore, metallothionein concentration and liver injury indices indicate that fish from Lake Janauacá may be under severe environmental stress. Thus, the toxic effects of metals can be diverse and alter the physiology, biochemistry, and morphology of Amazon fish.

Various metals, including Cu, have been found to cause metabolic and ionoregulatory disruptions in Amazonian fish. When the fish *Colossoma macropomum* (tambaqui), adapted to very soft water conditions (10 μM Ca^{2+}), is exposed to acute Cu concentrations ranging from 100 to 400 μgCu/L, there is an inhibition of Na^+ uptake and an increase in the passive loss of Na^+, resulting in a significant net Na^+ deficiency (Matsuo et al., 2005). However, a mild mitigating effect occurs when commercial humic acid (HA) is introduced at 5–40 mgC/L concentrations. Humic acid reduces Na^+ efflux while promoting Na^+ uptake, helping the fish to maintain a more stable Na^+ balance during Cu exposure (Matsuo et al., 2005). Similarly, Cu exposure has been observed to significantly increase the loss of Na^+ (and K^+ and Cl^-) in *P. axelrodi*, both in ion-poor water conditions and in filtered water from the Rio Solimões and Rio Negro (Crémazy et al., 2016, 2019). For the Cd toxicity, when tambaqui are acclimated to extremely soft water conditions, they exhibit a significant reduction in the rate of Ca^{2+} influx throughout their entire body when exposed to varying concentrations of Cd, ranging from 10 to 80 μg Cd/L, for durations spanning from 3 to 24 h (Matsuo et al., 2005).

In another study with Cu and Cd, Giacomin et al. (2018) fed tambaqui chronically (36 days) with diets containing 500 μg Cu/g and 500 μg Cd/g food. They did not observe ionoregulatory changes in the gills, intestine, and plasma, nor plasma cortisol, and lactate, regardless of diet. However, the Cd-contaminated diet seemed to suppress feeding. Interestingly, in this study, the fish fed Cd-contaminated diet appeared to have increased tolerance to hypoxia as evidenced by a decreased Pcrit (critical O_2 tension—partial pressure of O_2 below which the animal can no longer maintain a stable MO_2), increased time to loss of equilibrium, but blood O_2 transport characteristics (P_{50}, Bohr coefficient, hemoglobin, and hematocrit) were unaffected compared to control in the same experiment. Additionally, cardinal tetra fish (*P. axelrodi*) exhibited a notable increase in oxygen consumption (MO_2) when exposed to 50% of the 96-h LC_{50} Cu; in contrast, dwarf cichlid fish (*Apistogramma agassizii*) showed no significant change in MO_2 (Braz-Mota et al., 2018), demonstrating the effects of this metal on metabolism. Copper also promotes mitochondrial respiration alteration, such as uncoupling mechanisms, as a response to increased ETS and ROS generation, which also promotes the activation of the antioxidant enzymes and reduced effects in gill lamellar epithelium (Braz-Mota et al., 2018).

Physical factors, notably temperature, can exacerbate Cu effects. This concern arises due to the gravity of the current global climate change, which is closely associated with the rapid increase of temperature and CO_2 levels. As water temperature climbs, the solubility and bioavailability of metals, including Cu, tend to rise as well, as highlighted by Schiedek et al. (2007). For the Amazonian species of *Hoplosternum littorale* acclimatized to 34 °C in the natural waters of the Amazon, the survival time was significantly reduced under exposure to two different Cu concentrations (50 and 500 μg Cu/L), indicating a synergistic effect between Cu and temperature, mainly at 50 μg Cu/L (Braz-Mota et al., 2017). In periods of extreme drought in the Amazon, such as the historic event of 2010, which is currently being repeated in 2023, the consequence is a high warming of the water level, which increases the synergistic effect of these metals on fish.

The spermatozoids of tambaqui are notably sensitive to the effects of metals exposure, such as Cu and Cd. These transition metals have been observed to significantly impact the motility rate of spermatozoids, leading to a noticeable decrease in their mobility. Elevated concentrations of Cd and Cu have been found to have detrimental effects on the fertilization rate in tambaqui populations and have been associated with a marked reduction in the successful fertilization of tambaqui (Pinto et al., 2021). The adverse influence of Cd and Cu extends to the hatching rate of tambaqui eggs, causing a noticeable reduction in the successful hatching of these eggs. Intriguingly, the impact of Cu and Cd is not limited to reproductive outcomes alone (Pinto et al., 2021). Also, these metals can promote an increase in the antioxidant enzymes found in the spermatozoids, indicating a higher production of cellular stress due to the increase in reactive oxygen species (ROS), demonstrating the toxic effects of exposure to Cu and Cd.

Lemus et al. (2013) analyzed Cd, Cu, Ni, Fe, and Pb concentrations, as well as their associations with metallothionein levels and RNA/DNA ratios in tambaqui collected from a lake located in the lower Orinoco River. The study findings demonstrated that the primary accumulation of these metals occurred predominantly in the liver. Notably, the increased metallothionein levels in the liver were attributed to Cu. It was linked to Cd in the gills, and in the muscle tissue, it was associated with both Pb and Cd. While the authors did not identify a direct correlation between the physiological condition of the fish and the concentration of metals, they did observe a connection between the RNA/DNA ratio and the levels of metallothionein in both muscle and liver tissues.

Exposure to individual metals affected tambaqui physiology, but it has been necessary to consider mixtures of metals as they occur in the environment. Thus, Pavione (Pryscilla Pavione, personal unpublished data) decided to evaluate the acute effects of dissolved Cu, Ni (25% of respective LC_{50}), and their interaction (mixture, 12.5% of their respective LC_{50}) on juvenile tambaqui at pH 4 and pH 7. Overall, she observed a more pronounced effect of pH compared to metal, and exposure to pH 7 exerted the most significant impacts in tambaqui, considering the increase found for hematological parameters ([Hb]; Ht; MCH and MCHC) and for the activity of gills Na^{+}/ K^{+}-ATPAse, and liver SOD, GPX, GST, LPO, and AChe.

Summarizing, the toxic effects of metals are highly harmful to Amazonian fish and can be exacerbated by regular environmental variations occurring in the region (Fig. 2). River flooding leaches metals into water bodies, increasing their bioavailability. On the other hand, extreme drought events, such as the current one, raise water temperatures, increasing metal solubility, and bioavailability. These factors increase metal bioaccumulation and biomagnification, causing physiological changes such as metabolic and ionoregulatory dysfunctions, oxidative stress, and neurotoxic damage, including reduced fish population fertilization rates and

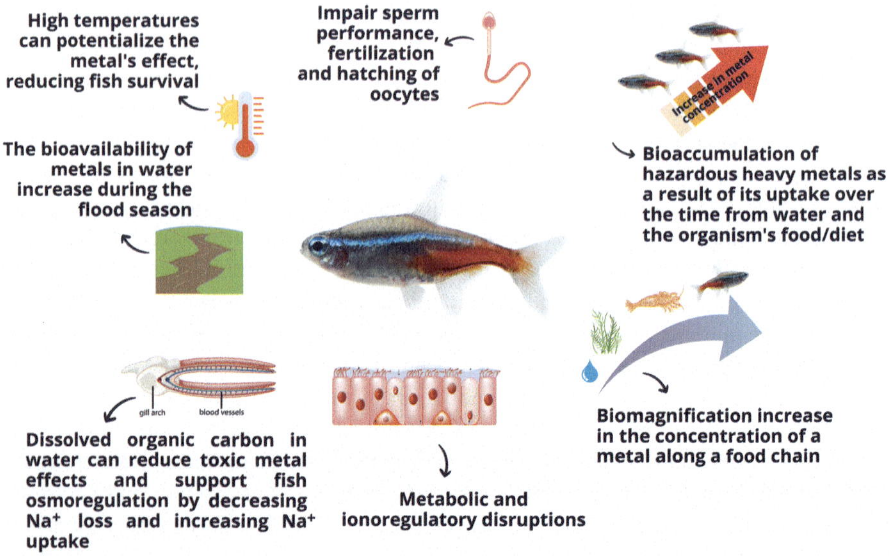

Fig. 2 General diagram showing the main toxic effects of transition metals on fish of the Amazon

hatching success. Recent studies highlight the vital role of HA in waters of the Amazon in attenuating metal toxicity and benefiting ionic regulation. Further research is needed because the high metal concentrations in the Amazon require urgent attention due to its potential unprecedented damage to fish populations.

4 Perspectives for the Future of the Amazonian Aquatic Biota

Although there has been an increasing number of studies on the contamination, detection, exposure, and toxicological effects of pesticides and metals in the Brazilian Amazon, there remain a large knowledge gaps. The unprecedented fish diversity of the Amazon displays unique adaptations that allow them to survive the natural environmental conditions. Still, the indiscriminate release and enrichment of pesticides and metals driven by anthropogenic action in the Amazon basin are critically threatening the conservation of local fauna, including ichthyofauna, meaning that if pollution levels are not controlled in aquatic biota, the Amazon fish species will not be able to survive, and there will be a loss of biodiversity.

Climate change, also due to anthropogenic action, offers an additional concern about the toxicity of the heavy metals and pesticides. The high temperatures are used to decrease the membrane permeabilities, thus facilitating greater absorption of these contaminants both through the gills, as well as through the body and through ingestion. This can promote greater absorption of these compounds by fish, causing consequences resulting from the combined effect between temperature and the metabolic and reproductive disruptions mentioned in this text. However, there are very few studies analyzing the effects of temperature on the toxicity of heavy metals and pesticides (Braz-Mota et al., 2017), which makes it even more worrying, given that climate change is already a reality today. Higher temperatures decrease oxygen solubility in water, but metabolic activity is elevated, increasing the oxygen demand of tissues. An increase in metabolic rate is connected to elevated oxidative stress which is already known to be increased by heavy metals and contaminants, as shown in this chapter. The contamination by heavy metals and pesticides can be an additional challenge for Amazonian fish, as living at high temperatures near their upper thermal limit (as in equatorial regions such as the Amazon basin) appear to have a limited capacity to tolerate elevated temperatures (Campos et al., 2019; Jung et al., 2020). Empirically, we can assume that the toxicity of these substances is increased due to the increase in temperature, making not only fish but several other species of living beings vulnerable.

It is important to note that climate change is not just limited to rising temperatures but also a significant increase in the concentration of CO_2 in the atmosphere, with forecasts of values exceeding 900 ppm by the end of the century (IPCC, 2023). There are no studies regarding the combined effects of CO_2 and heavy metals or pesticides on Amazon fish. Especially in the aquatic environment, the interaction of contaminants with CO_2 is poorly understood. It should be given more attention because the CO_2 reacts when in the water, leading to an increase in H^+ and HCO_{3-} concentrations and lower CO_3^{2-} levels. This results in water acidification, and due to climate change, surface water acidification is predicted to increase by 0.14–0.42 units by the end of the century (Pörtner, 2021). Thus, combined with contaminants, it represents an additional challenge for the Amazon since the waters are naturally more acidic. Therefore, global warming and deforestation will significantly impact the balance of ecosystems and the services they provide, especially in the Amazon, where the thermal vulnerability of fish, which already live close to their thermal limits, is closely related to other stressors, such as hypoxia and high levels of CO_2. Thus, anthropogenic actions such as the use and dispersion of pesticides and heavy metals, in addition to ongoing climate change, represent a potential risk for the future of the Amazonian aquatic biota, and they need to be a warning for the urgent implementation of public policies to preserve and guarantee the maintenance and survival of these species. In summary, some key areas for future investigations deserve greater attention by researchers and policy-makers:

- Establishment of a permanent station for monitoring the environmental levels of pesticides and metals throughout the Amazon basin, especially in the areas surrounding mining and industrial facilities where studies have shown significant levels of contamination
- Determination of the acute and chronic toxicity of new formulations of pesticides and metals in a greater number of fish species, both under natural and controlled laboratory conditions

- Use of fish models with different trophic levels to allow a better understanding of the main routes of exposure, modes of action, and biological effects of pesticides and metals
- Establishment of specific quality standards for the protection of aquatic life against pesticides and metals, taking into account natural background levels, the rate of entry into surface waters, and the effects of local physicochemical characteristics on speciation and bioavailability
- Evaluation of the interaction of environmental parameters (e.g., pH, ionic strength, DOC concentration and composition, suspended solids, and levels of ammonia and other contaminants) on toxicity and biological effects—especially for pesticides—in order to determine possible additive or synergistic effects in a more realistic contamination scenario
- Define clearly the effects that climate change represents for Amazon fish, considering they are living at high temperatures near their upper thermal limit which appear to have a limited capacity to tolerate elevated temperatures, and there is no information about interactions of CO_2 with these compounds

References

Abessa, D., Famá, A., & Buruaem, L. (2019). The systematic dismantling of Brazilian environmental laws risks losses on all fronts. *Nature Ecology & Evolution, 3*, 510–511. https://doi.org/10.1038/s41559-019-0855-9

Albuquerque, F. E. A., Minervino, A. H. H., Miranda, M., Herrero-Latorre, C., Júnior, R. A. B., Oliveira, F. L. C., Sucupira, M. C. A., Ortolani, E. L., & López-Alonso, M. (2020). Toxic and essential trace element concentrations in fish species in the Lower Amazon, Brazil. *Science of the Total Environment, 732*, 138983. https://doi.org/10.1016/j.scitotenv.2020.138983

Ali, H., Ali, W., Ullah, K., Akbar, F., Ahrar, S., Ullah, I., Ahmad, I., Ahmad, A., Ilahi, I., & Sajad, M. A. (2017). Bioaccumulation of Cu and Zn in *Schizothorax plagiostomus* and *Mastacembelus armatus* from River Swat, River Panjkora and River Barandu in Malakand Division, Pakistan. *Pakistan Journal of Zoology, 49*(5), 1555–1561. https://doi.org/10.17582/journal.pjz/2017.49.5.1555.1561

Ali, H., & Khan, E. (2019). Trophic transfer, bioaccumulation, and biomagnification of non-essential hazardous heavy metals and metalloids in food chains/webs—Concepts and implications for wildlife and human health. *Human and Ecological Risk Assessment, 25*, 6. https://doi.org/10.1080/10807039.2018.1469398

Anjos, H. D. B. (2007). *Efeitos da Fragmentação Florestal Sobre as Assembléias de Peixes de Igarapés da Zona Urbana de Manaus, Amazonas* (p. 114). Dissertação de Mestrado, Instituto Nacional de Pesquisas da Amazônia (INPA).

ANVISA. (1998). National Health Surveillance Agency-ANVISA ORDINANCE No. 29, OF JANUARY 13, 1998.

Araújo, R. S., Garcia, M. V. B., & Garcia, T. B. (2008). Toxicidade aguda do herbicida Glifosato (Roundup®) para *Danio rerio* (Teleostei, Cyprinidae), em condições tropicais. In *Anais da III Jornada de Iniciação Científica da Embrapa Amazônia Ocidental* (p. 57) ISSN 1517–3135.

Asner, G. P., & Tupayachi, R. (2016). Accelerated losses of protected forests from gold mining in the Peruvian Amazon. *Environmental Research Letters, 12*, 094004. https://doi.org/10.1088/1748-9326/aa7dab

Bastos, W. R., Dórea, J. G., Bernardi, J. V. E., Lauthartte, L. C., Mussy, M. H., Lacerda, L. D., & Malm, O. (2015). Mercury in fish of the Madeira river (temporal and spatial assessment), Brazilian Amazon. *Environmental Research, 140*, 191–197. https://doi.org/10.1016/j.envres.2015.03.029

Belivacqua, A. H. V. (2009). *O uso do Modelo do Ligante Biótico na avaliação da contaminação por cobre em águas da Amazônia*. Dissertação de Mestrado, Instituto Nacional de Pesquisas da Amazônia (INPA).

Braz-Mota, S., & Almeida-Val, V. M. F. (2021). Ecological adaptations of Amazonian fishes acquired during evolution under environmental variations in dissolved oxygen: A review of responses to hypoxia in fishes, featuring the hypoxia tolerant spp. *Journal of Experimental Zoology Part A: Ecological and Integrative Physiology, 335*, 771–786. https://doi.org/10.1002/jez.2531

Braz-Mota, S., Campos, D. F., Maccormack, T. J., Duarte, R. M., Val, A. L., & Almeida-Val, V. M. F. (2018). Mechanisms of toxic action of copper and copper nanoparticles in two Amazon fish species: Dwarf cichlid (*Apistogramma agassizii*) and cardinal tetra (*Paracheirodon axelrodi*). *Science of the Total Environment, 630*, 1168–1180. https://doi.org/10.1016/j.scitotenv.2018.02.216

Braz-Mota, S., Fé, L. M. L., Delunardo, F. A. C., Sadauskas-Henrique, H., Almeida-Val, V. M. F., & Val, A. L. (2017). Exposure to waterborne copper and high temperature induces the formation of reactive oxygen species and causes mortality in the Amazonian fish *Hoplosternum littorale*. *Hydrobiologia, 789*, 157–166. https://doi.org/10.1007/s10750-016-2847-y

Braz-Mota, S., Sadauskas-Henrique, H., Duarte, R. M., Val, A. L., & Almeida-Val, V. M. F. (2015). Roundup® exposure promotes gills and liver impairments, DNA damage and inhibition of brain cholinergic activity in the Amazon teleost fish *Colossoma macropomum*. *Chemosphere, 135*, 53–60.

Bretaud, S., Toutant, J. P., & Saglio, P. (2000). Effects of carbofuran, diuron, and nicosulfuron on acetylcholinesterase activity in goldfish (*Carassius auratus*). *Ecotoxicology and Environmental Safety, 47*(2), 117–124. https://doi.org/10.1006/eesa.2000.1954

Brookes, P. S. (2005). Mitochondrial H^+ leak and ROS generation: An odd couple. *Free Radical Biology & Medicine, 38*(1), 12–23. https://doi.org/10.1016/j.freeradbiomed.2004.10.016

Campos, D. F., Braz-Mota, S., Val, A. L., & Almeida-Val, V. M. F. (2019). Predicting thermal sensitivity of three Amazon fishes exposed to climate change scenarios. *Ecological Indicators, 101*, 533–540. https://doi.org/10.1016/j.ecolind.2019.01.051

Capparelli, M. V., Moulatlet, G. M., Abessa, D. M. S., Lucas-Solis, O., Rosero, B., Galarza, E., Tuba, D., Carpintero, N., Ochoa-Herrera, V., & Cipriani-Avila, I. (2020). An integrative approach to identify the impacts of multiple metal contamination sources on the Eastern Andean foothills of the Ecuadorian Amazonia. *Science of the Total Environment, 709*, 136088. https://doi.org/10.1016/j.scitotenv.2019.136088

Castro, M. C., Baeza, A., Codeço, C. T., Cucunuba, Z. M., Dal'Asta, A. P., De Leo, G. A., Dobson, A. P., Carrasco-Escobar, G., Lana, R. M., Lowe, R., Monteiro, A. M. V., Pascual, M., & Santos-Vega, M. (2019). Development, environmental degradation, and disease spread in the Brazilian Amazon. *PLoS Biology, 17*, 4–11. https://doi.org/10.1371/journal.pbio.3000526

Castro, N. S. S., Braga, C. M., Trindade, P. A. A., Giarrizzo, T., & Lima, M. O. (2016). Mercury in fish and sediment of Purus River, Acre State, Amazon. *Cadernos Saúde Coletiva, 24*(3), 294–300. https://doi.org/10.1590/1414-462X201600030142

Chapadense, P. F. G., Castro, F. J., Almeida, J. A., & Moron, S. E. (2009). Toxicity of atrazine herbicide in *Colossoma macropomum*. *Revista Brasileira de Saúde e Produção Animal, 10*, 398–405.

Chaves, V. S., Marcon, J. L., Duncan, W. P., & Alves-Gomes, J. A. (2020). Acute toxicity of a deltamethrin based pesticide (DBP) to the Neotropical electric fish *Microsternarchus* cf. *bilineatus* (Gymnotiformes). *Acta Amazonica, 50*, 355–362. https://doi.org/10.1590/1809-4392201904001

CONAMA. (2005). *Resolução no 357, de 17 de março de 2005. Dispõe Sobre a Classificação dos Corpos de Água e Diretrizes Ambientais para seu Enquadramento, bem como Estabelece as Condições e Padrões de Lançamento de Efluentes, e dá Outras Providências*. Diário Oficial da República Federativa do Brasil.

Costa, M. S., Viana, L. F., Cardoso, C. A. L., Isacksson, E. D. G. S., Silva, J. C., & Florentino, C. (2022). Landscape composition and inorganic contaminants in water and muscle tissue of *Plagioscion squamosissimus* in the Araguari River (Amazon, Brazil). *Environmental Research, 208*, 112691. https://doi.org/10.1016/j.envres.2022.112691

Crémazy, A., Wood, C. M., Scott Smith, D., Ferreira, M. S., Johannsson, O. E., Giacomin, M., & Val, A. L. (2016). Investigating copper toxicity in the tropical fish cardinal tetra (*Paracheirodon axelrodi*) in natural Amazonian waters: Measurements, modeling, and reality. *Aquatic Toxicology, 180*, 353–363. https://doi.org/10.1016/j.aquatox.2016.10.011

Crémazy, A., Wood, C. M., Scott Smith, D., & Val, A. L. (2019). The effects of natural suspended solids on copper toxicity to the cardinal tetra in Amazonian river waters. *Environmental Toxicology and Chemistry, 38*, 2708–2718. https://doi.org/10.1002/etc.4586

Cunha, F. S., Sousa, N. C., Santos, R. F. B., Meneses, J. O., Couto, M. V. S., Almeida, F. T. C., Filho, J. G. S., Carneiro, P. C. F., Maria, A. N., & Fujimoto, R. Y. (2018). Deltamethrin-induced nuclear erythrocyte alteration and damage to the gills and liver of *Colossoma macropomum*. *Environmental Science and Pollution Research, 25*, 15102–15110. https://doi.org/10.1007/s11356-018-1622-1

da Silva, H. C. M., Silva, A. G., Idalino, J. J. S., Sousa, F. B., Gomes, A. L., Duncan, W. P., & Matoso, D. A. (2019). Trichlorfon acute lethal toxicity to juvenile tambaqui (*Colossoma macropomum*). *Aquaculture Research, 00*, 1–4. https://doi.org/10.1111/are.14412

Dal Pont, G., Valdez-Domingos, F. X., Fernandes-Castilho, M., & Val, A. L. (2017). Potential of the biotic ligand model (BLM) to predict copper toxicity in the white-water of the Solimões-Amazon River. *Bulletin of Environmental Contamination and Toxicology, 98*, 27–32. https://doi.org/10.1007/s00128-016-1986-1

dos Santos Costa, M., da Silva, H. C. M., Soares, S. C., Favarato, R. M., Feldberg, E., Gomes, A. L. S., Artoni, R. F., & Matoso, D. A. (2022). A perspective of molecular cytogenomics, toxicology, and epigenetics for the increase of heterochromatic regions and retrotransposable elements in Tambaqui (*Colossoma macropomum*) exposed to the parasiticide trichlorfon. *Animals, 12*, 1945. https://doi.org/10.3390/ani12151945

Duarte, R. M., Crémazy, A., Wood, C. M., Almeida-Val, V. M. F., & Val, A. L. (2024). The biotic ligand model as a promising tool to predict Cu toxicity in amazon blackwaters. *Environmental Pollution, 341*, 122988. https://doi.org/10.1016/j.envpol.2023.122988

Duarte, R. M., Menezes, A. C. L., Rodrigues, L. S., Almeida-Val, V. M. F., & Val, A. L. (2009). Copper sensitivity of wild ornamental fish of the Amazon. *Ecotoxicology and Environmental Safety, 72*, 693–698. https://doi.org/10.1016/j.ecoenv.2008.10.003

Duarte, R. M., & Val, A. L. (2020). Water-related problem with special reference to global climate change in Brazil. In O. Singh, Y. Milshina, K. Tian, D. Gusain, & J. P. Bassin (Eds.), *Water conservation and wastewater treatment in BRICS Nations: Technologies challenges, strategies and policies* (pp. 3–21). Elsevier.

Duncan, W. P., Idalino, J. J. S., Silva, A. G., Modal, R. F., Silva, H. C. M., Matoso, D. A., & Gomes, A. L. S. (2020). Acute toxicity of the pesticide trichlorfon and inhibition of acetylcholinesterase in *Colossoma macropomum* (Characiformes: Serrasalmidae). *Aquaculture International, 28*, 815–830. https://doi.org/10.1007/s10499-019-00497-w

Fernandes, C. L. F., Volcão, L. M., Ramires, P. F., De Moura, R. R., & Júnior, F. M. R. D. S. (2020). Distribution of pesticides in agricultural and urban soils of Brazil: A critical review. *Environmental Sciences: Processes & Impacts, 22*(2), 256–270. https://doi.org/10.1039/C9EM00433E

França, A. J. (2017). *Dinâmica espaço temporal e influência do metilmercúrio em peixes do Lago Janauacá, AM*. Dissertação de Mestrado, Instituto Nacional de Pesquisas da Amazônia (INPA). https://repositorio.inpa.gov.br/handle/1/11365

Gabriel, D., Riffel, A. P. K., Finamor, I. A., Saccol, E. M. H., Ourique, G. M., Goulart, L. O., Kochhann, D., Cunha, M. A., Garcia, L. O., Pavanato, M. A., Val, A. L., Baldisserotto, B., & Llesuy, S. F. (2013). Effects of subchronic manganese chloride exposure on Tambaqui (*Colossoma macropomum*) tissues: Oxidative stress and antioxidant defenses. *Archives of Environmental Contamination and Toxicology, 64*, 659–667. https://doi.org/10.1007/s00244-012-9854-4

Giacomin, M., Vilarinho, G. C., Castro, K. F., Ferreira, M., Duarte, R. M., Wood, C. M., & Val, A. L. (2018). Physiological impacts and bioaccumulation of dietary Cu and Cd in a model teleost: The Amazonian tambaqui (*Colossoma macropomum*). *Aquatic Toxicology, 199*, 30–45. https://doi.org/10.1016/j.aquatox.2018.03.021

Gomes, D. F., Pinto, T. J. S., Raymundo, L. B., Sperandei, V. F., Michiel Daam, M., Moreira, R. A., & Rocha, O. (2023). Ecological risk assessment for metals in sediment and waters from the Brazilian Amazon region. *Chemosphere, 345*, 140413. https://doi.org/10.1016/j.chemosphere.2023.140413

Guarido, P. C. P. (2014). *Degradação Ambiental e Presença de Espécies de Peixes Não Nativas em Pequenos Igarapés de Terra Firma de Manaus, Amazonas* (p. 58). Dissertação de Mestrado, Instituto Nacional de Pesquisas da Amazônia (INPA).

Hacon, S. S., Oliveira-da-Costa, M., Gama, C. S., Ferreira, R., Basta, P. C., Schramm, A., & Yokota, D. (2020). Mercury exposure through fish consumption in traditional communities in the Brazilian northern Amazon. *International Journal of Environmental Research and Public Health, 17*, 5269. https://doi.org/10.3390/ijerph17155269

Holland, A., Wood, C. M., Scott Smith, D., Correia, T. G., & Val, A. L. (2017). Nickel toxicity to cardinal tetra (*Paracheirodon axelrodi*) differs seasonally and among the black, white and clear river waters of the Amazon basin. *Water Research, 126*, 21–29. https://doi.org/10.1016/j.watres.2017.06.044

Hossain, M. M., Suzuki, T., Sato, I., Takewaki, T., Suzuki, K., & Kobayashi, H. (2004). The modulatory effect of pyrethroids on acetylcholine release in the hippocampus of freely moving rats. *Neurotoxicology, 25*, 825–833. https://doi.org/10.1016/j.neuro.2004.01.002

IBAMA - Brazilian Institute of Environment and Renewable Natural Resources. (2021). Relatório de comercialização de agrotóxicos. Available at: https://www.gov.br/ibama/pt-br/assuntos/quimicos-e-biologicos/agrotoxicos/relatorios-de-comercializacao-de-agrotoxicos/relatorios-de-comercializacao-de-agrotoxicos. Accessed 15/08/2023.

IPCC. (2023). Climate change 2023: The physical science basis. In *Contribution of Working Group I to the sixth assessment report of the intergovernmental panel on climate change*. Cambridge University Press. Disponível em. https://report.ipcc.ch/ar6wg1/pdf/IPCC_AR6_WG1_SPM.pdf

Jung, E. H., Brix, K. V., Richards, J. G., Val, A. L., & Brauner, C. J. (2020). Reduced hypoxia tolerance and survival at elevated temperatures may limit the ability of Amazonian fishes to survive in a warming world. *Science of the Total Environment, 748*(15), 141349. https://doi.org/10.1016/j.scitotenv.2020.141349

Kime, D. E., & Nash, J. P. (1999). Gamete viability as an indicator of reproductive endocrine disruption in fish. *Science of the Total Environment, 233*(1–3, 123), –129. https://doi.org/10.1016/S0048-9697(99)00219-3

Laabs, V., Amelung, W., Pinto, A. A., Wantzen, M., da Silva, C. J., & Zech, W. (2002). Pesticides in surface water, sediment, and rainfall of the northeastern Pantanal basin, Brazil. *Journal of Environmental Quality, 31*, 1636–1648.

Lemus, M., Blanco, I., Hernández, M., León, A., Centeno, L., Chung, K., & Salazar-Lugo, R. (2013). Relation between metallothionein, RNA/DNA and heavy metals in juveniles of *Colossoma macropomun* (Cuvier, 1818) in natural conditions. *E3S Web of Conferences, 1*, 12001. https://doi.org/10.1051/e3sconf/20171302003

Lesnefsky, E. J., Slabe, T. J., Stoll, M. S., Minkler, P. E., & Hoppel, C. L. (2001). Myocardial ischemia selectively depletes cardiolipin in rabbit heart subsarcolemmal mitochondria. *American Journal of Physiology. Heart and Circulatory Physiology, 280*(6), H2770–H2778. https://doi.org/10.1152/ajpheart.2001.280.6.H2770

Lima, D. P., Santos, C., Silva, R. D. S., & Yoshioka, E. T. O. (2015). Heavy metal contamination in fish and water from Cassiporé River basin, state of Amapá, Amazonia, Brazil. *Acta Amazonica, 45*, 405–414. https://doi.org/10.1590/1809-4392201403995

Liu, W. R., Zeng, D., She, L., Su, W. X., He, D. C., Wu, G. Y., Ma, X. R., Jiang, S., Jiang, C. H., & Ying, G. G. (2020). Comparisons of pollution characteristics, emission situations, and mass loads for heavy metals in the manures of different livestock and poultry in China. *Science of the Total Environment, 734*, 139023. https://doi.org/10.1016/j.scitotenv.2020.139023

Matsuo, A. Y. O., Wood, C. M., & Val, A. L. (2005). Effects of copper and cadmium on ion transport and gill metal binding in the Amazonian teleost tambaqui (*Colossoma macropomum*) in extremely soft water. *Aquatic Toxicology, 74*, 351–364. https://doi.org/10.1016/j.aquatox.2005.06.008

Mendonça, J. S., Vieira, L. G., Hirano, L. Q. L., Qualhato, G., Almeida, J. C. N., André Santos, L. Q., Fernandes, E. A., Malafaia, G., & Andrade, D. V. (2023). Eggshell composition of Amazon turtle (*Podocnemis expansa*) is altered after incubation in substrates containing glyphosate and fipronil formulations. *Science of the Total Environment, 893*, 164901. https://doi.org/10.1016/j.scitotenv.2023.164901

Meneses, H. N. M., Oliveira-da-Costa, M., Basta, P. C., Morais, C. G., Pereira, R. J. B., de Souza, S. M. S., & Hacon, S. S. (2022). Mercury contamination: A growing threat to riverine and urban communities in the Brazilian Amazon. *International Journal of Environmental Research and Public Health, 19*, 2816. https://doi.org/10.3390/ijerph19052816

Miyazaki, D. M. Y., Machado Neto, J. G., & Castagnolli, N. (2004). Toxicidade aguda de triclorfon, partion metflico e glifosato para alevinos de tambaqui (*Colossoma macropormum*) e tambacu (*C. macropromum* X *Piaractus mesopotamicus*). In *Anais Encontro Brasileiro de Patologistas de Organismos Aquáticos (ENBRAPOA), Laguna* (Vol. 8, p. 203). Biblioteca Setorial da UEM.

Monteiro, D. A., Rantin, F. T., & Kalinin, A. L. (2010). Inorganic mercury exposure: Toxicological effects, oxidative stress biomarkers and bioaccumulation in the tropical freshwater fish matrinxã, *Brycon amazonicus* (Spix and Agassiz, 1829). *Ecotoxicology, 19*, 105–123. https://doi.org/10.1007/s10646-009-0395-1

Moraes, F. D., Venturini, F. P., Cortella, L. R. X., Rossi, P. A., & Moraes, G. (2013). Acute toxicity of pyrethroid-based insecticides in the Neotropical freshwater fish *Brycon amazonicus*. *Ecotoxicology and Environmental Contamination, 8*(2), 59–64. https://doi.org/10.5132/eec.2013.02.009

Moulatlet, G. M., Yacelga, N., Rico, A., Mora, A., Hauser-Davis, R. A., Cabrera, M., & Capparelli, M. V. (2023). A systematic review on metal contamination due to mining activities in the Amazon basin and associated environmental hazards. *Chemosphere, 339*, 139700. https://doi.org/10.1016/j.chemosphere.2023.139700

Naranjo-Gómez, J., Vargas-Rojas, L. F., & Rondón-Barragán, I. (2013). Acute toxicity of mercury chloride ($HgCl_2$) to Cachama Blanca, *Piaractus brachypomus* (Cuvier, 1818). *Actualités Biologiques, 35*(98), 85–93.

Nobre, C. A., Sampaio, G., Borma, L. S., Castilla-Rubio, J. C., Silva, J. C., & Cardoso, M. (2016). Land-use and climate change risks in the Amazon and the need of a novel sustainable development paradigm. *Proceedings of the National Academy of Sciences of United States of America, 13*(39), 10759–10768. https://doi.org/10.1073/pnas.1605516113

Oliveira, C. P. F. (2003). *Efeito do cobre e chumbo, metais pesados provenientes na água de formação derivada da extração de petróleo da província petroleira do Urucu—AM, sobre o tambaqui, Colossoma macropomum (Cuvier, 1818)* (86p). Dissertação de Mestrado, Dissertação de Mestrado, Instituto Nacional de Pesquisas da Amazônia (INPA).

Parry, L., Peres, C. A., Day, B., & Amaral, S. (2010). Rural-urban migration brings conservation threats and opportunities to Amazonian watersheds. *Conservation Letters, 3*, 251–259. https://doi.org/10.1111/j.1755-263X.2010.00106.x

Pinto, G. L., Castro, J. S., & Val, A. L. (2021). Copper and cadmium impair sperm performance, fertilization and hatching of oocytes from Amazonian fish *Colossoma macropomum Chemosphere*. *Chemosphere, 266*, 128957. https://doi.org/10.1016/j.chemosphere.2020.128957

Pörtner, H. O. (2021). Climate impacts on organisms, ecosystems and human societies: integrating OCLTT into a wider context. *The Journal of Experimental Biology*. Feb 24; 224 (Pt Suppl 1): jeb238360. https://doi.org/10.1242/jeb.238360. PMID: 33627467.

Ribeiro, D. R. G., Faccin, H., Molin, T. R. D., Carvalho, L. M., & Amado, L. L. (2017). Metal and metalloid distribution in different environmental compartments of the middle Xingu River in the Amazon, Brazil. *Science of the Total Environment, 605–606*, 66–74. https://doi.org/10.1016/j.scitotenv.2017.06.143

Rico, A., de Oliveira, R., Nunes, G. S. S., Rizzi, C., Villa, S., Vizioli, B. C., Montagner, C. C., & Waichman, A. V. (2021). Ecological risk assessment of pesticides in urban streams of the Brazilian Amazon. *Chemosphere, 291*(1), 132821. https://doi.org/10.1016/j.chemosphere.2021.132821

Rico, A., Geber-Corrêa, R., Campos, P. S., Garcia, M. V., Waichman, A. V., & van den Brink, P. J. (2010). Effect of parathion-methyl on Amazonian fish and freshwater invertebrates: A comparison of sensitivity with temperate data. *Archives of Environmental Contamination and Toxicology, 58*, 765–771. https://doi.org/10.1007/s00244-009-9409-5

Rico, A., Oliveira, R., Nunes, G. S. S. N., Rizzi, C., Villa, S., Vizioli, B. C., Montagner, C. C., & Waichman, A. V. (2022). Ecological risk assessment of pesticides in urban streams of the Brazilian Amazon. *Chemosphere, 291*, 132821. https://doi.org/10.1016/j.chemosphere.2021.132821

Rico, A., Waichman, A. V., Geber-Corrêa, R., & van den Brink, P. J. (2011). Effects of malathion and carbendazim on Amazonian freshwater organisms: Comparison of tropical and temperate species sensitivity distributions. *Ecotoxicology, 20*, 625–634. https://doi.org/10.1007/s10646-011-0601-9

Römbke, J., Waichman, A. V., & Garcia, M. V. (2008). Risk assessment of pesticides for soils of the central amazon, Brazil: Comparing outcomes with temperate and tropical data. *Integrated Environmental Assessment and Management, 4*, 94–104. https://doi.org/10.1897/IEAM_2007-052.1

Salazar-Lugo, R., Estrella, A., Oliveros, A., Rojas-Villarroel, E., Villalobos, L., & Lemus, M. (2009). Paraquat and temperature affect nonspecific immune response of *Colossoma macropomum*. *Environmental Toxicology and Pharmacology, 27*, 321–326. https://doi.org/10.1016/j.etap.2008.11.010

Salazar-Lugo, R., Vargas, A., Rojas, L., & Lemus, M. (2013). Histopathological changes in the head kidney induced by cadmium in a neotropical fish *Colossoma macropomum*. *Open Veterinary Journal, 3*(2), 145–150.

Saldaña-Serrano, M., Espinoza-Morriberon, D., De-la-Torre, G. E., Álvarez-Béjar, G. O., & López, A. (2021). Toxicity of fipronil insecticide on the early life stages of *Colossoma macropomum* (gamitana). *International Aquatic Research, 13*(4), 301–307. https://doi.org/10.22034/IAR.2021.1943467.1207

Santana, G. P. (2016). Heavy metal distribution in the sediment and *Hoplosternum littorale* from manaus industrial. *The Journal of Engineering and Exact Sciences, 2*(2), 070–081. https://doi.org/10.18540/jcecvl2iss2pp070-081

Santana, G. P., & Barroncas, P. S. R. (2005). Heavy metals (Co, Cu, Fe, Cr, Ni, Mn, Pb and Zn) study in the Tarumã-Açu Basin—Manaus (AM). *Acta Amazonica, 37*, 111–118. https://doi.org/10.1590/S0044-59672007000100013

Santos, J. P., Almeida, S. J. M., Costa, C. C., Ferreira, A. N. S., Teixeira, E. G., Guimarães, E. C., Brito, P. S., Ottoni, F. P., & Carvalho-Neta, R. N. F. (2023). Changes in sperm motility of amazonian fish Tambaqui *Colossoma macropomum* (Cuvier 1816) (Characiformes: Serrasalmidae) exposed to two pesticides. *Biota Neotropica, 23*(2), e20231471. https://doi.org/10.1590/1676-0611-BN-2023-1471

Schiedek, D., Sundelin, B., Readman, J. W., & Macdonald, R. W. (2007). Interactions between climate change and contaminants. *Marine Pollution Bulletin, 54*, 1845–1856. https://doi.org/10.1016/j.marpolbul.2007.09.020

Schiesari, L., Waichman, A., Brock, T., Adams, C., & Grillitsch, B. (2013). Pesticide use and biodiversity conservation in the Amazonian agricultural frontier. *Philosophical Transactions of the Royal Society B, 368*, 20120378. https://doi.org/10.1098/rstb.2012.0378

Silva, A. L. F., Araújo, L. D., Gomes, L. C., & Chagas, E. C. (2017). Acute toxicity and sublethal effects of methyl parathion on tambaqui (*Colossoma macropomum*). *Journal of Chemical and Pharmaceutical Research, 9*, 93–97.

Silva, M. H. L., Castro, A. C. L., Silva, I. S., Cabral, P. F. P., Azevedo, J. W. J., Silva Soares, L., Bandeira, A. M., Marianna Basso, M. J., & Nunes, L. S. (2023). Determination of metals in estuarine fishes in a metropolitan region of the coastal zone of the Brazilian Amazon. *Marine Pollution Bulletin, 186*, 114477. https://doi.org/10.1016/j.marpolbul.2022.114477

Silva, G. S., Matos, L. V., Freitas, J. O. S., Campos, D. F., & Val, V. M. F. (2019). Gene expression, genotoxicity, and physiological responses in an Amazonian fish, *Colossoma macropomum* (CUVIER 1818), exposed to Roundup® and subsequent acute hypoxia. *Comparative Biochemistry and physiology. Toxicology & Pharmacology, 222*, 49–58. https://doi.org/10.1016/j.cbpc.2019.04.010

Silva, M. S. R., Ramos, J. F., & Pinto, A. G. N. (1999). Transition metals in the sediments of the Manaus Igarapés. *Acta Limnologica Brasiliensia, 11*, 89–100.

Sinhorin, V. D. G., Sinhorin, A. P., Teixeira, J. M. S., Miléski, K. M. L., Hansen, P. C., Moreira, P. S. A., Kawashita, N. H., Baviera, A. M., & Loro, V. L. (2014). Effects of the acute exposition to glyphosate-based herbicide on oxidative stress parameters and antioxidant responses in a hybrid Amazon fish Surubim (*Pseudoplatystoma sp*). *Ecotoxicology and Environmental Safety, 106*, 181–187. https://doi.org/10.1016/j.ecoenv.2014.04.040

Soares, P. R. L., Andrade, A. L. C., Santos, T. P. S., Silva, S. C. B. L., Silva, J. F., Santos, A. R., Souza, E. H. L. S., Cunha, F. M., Teixeira, V. W., Cadena, M. R. S., Sá, F. B., Júnior, L. B. C., & Cadena, P. G. (2016). Acute and chronic toxicity of the benzoylurea pesticide, lufenuron, in the fish, *Colossoma macropomum*. *Chemosphere, 161*, 412–421. https://doi.org/10.1016/j.chemosphere.2016.07.033

Soares-Filho, B., & Rajão, R. (2018). Traditional conservation strategies still the best option. *Nature Sustainability, 1*, 608–610. https://doi.org/10.1038/s41893-018-0179-9

Soumis, N., Lucotte, M., Sampaio, D., Almeida, D. C., Giroux, D., Morais, S., & Pichet, P. (2003). Presence of organophosphate insecticides in fish of the Amazon River. *Acta Amazonica, 33*, 325–338. https://doi.org/10.1590/1809-4392200332338

Souza, T. C. (2014). *Toxicidade aguda de agrotóxicos e curva de sensibilidade de espécies para peixes Amazônicos* (p. 65). Dissertação de Mestrado, Universidade Federal do Amazonas (UFAM).

Souza, S. S., Castro, J. S., Campos, D. F., Pereira, R. S., Bataglion, G. A., Da Silva, G. S., & Almeida-Val, V. M. F. (2021). Temporal exposure to malathion: Biochemical changes in the Amazonian fish tambaqui, *Colossoma macropomum*. *Aquatic Toxicology, 241*, 105997. https://doi.org/10.1016/j.aquatox.2021.105997

Souza, S. S., da Costa, J. C., da Silva, G. S., & Alemida-Val, V. M. F. (2023). Malathion alters the transcription of target genes of the tumour suppressor *tp53* and cancerous processes in *Colossoma macropomum*: Mechanisms of DNA damage response, oxidative stress and apoptosis. *Chemico-Biological Interactions, 374*, 110405. https://doi.org/10.1016/j.cbi.2023.110405

Souza, T. C., Silva, S. L. R., Marcon, J. L., & Waichman, A. V. (2020). Acute toxicity of deltamethrin to Amazonian freshwater fish. *Toxicology and Environmental Health Sciences, 12*, 149–155. https://doi.org/10.1007/s13530-020-00048-4

Staevie, P. M. (2015). Crescimento demográfico e exclusão social nas capitais periféricas da Amazônia. *Revista Textos e Debates, 28*, 185–204.

Sturm, A., Wogram, J., Hansen, P. D., & Liess, M. (1999). Potential use of cholinesterase in monitoring low levels of organophosphates in small streams: Natural variability in three-spined stickleback (*Gasterosteus aculeatus*) and relation to pollution. *Environmental Toxicology and Chemistry, 18*(02), 194–2000. https://doi.org/10.1002/etc.5620180214

Tavares-Dias, M., Ferreira, J. S., Affonso, E. G., Ono, E. A., & Martins, M. L. (2011). Toxicity and effects of copper sulfate on parasitic control and hematological response of tambaqui *Colossoma macropomum*. *Boletim do Instituto de Pesca, 37*(4), 355–365.

Val, A. L., Duarte, R. M., Campos, D., & Almeida-Val, V. M. F. (2023). Environmental stressors in Amazonian riverine systems. In N. A. Fangue, S. J. Cooke, A. P. Farrell, C. J. Brauner, &

E. J. Eliason (Eds.), *Fish physiology* (Vol. 39., Part B. ISBN 9780128242681). https://doi.org/10.1016/bs.fp.2022.05.001

van der Oost, R., Beyer, J., & Vermeulen, N. P. (2003). Fish bioaccumulation and biomarkers in environmental risk assessment: A review. *Environmental Toxicology and Pharmacology, 13*(2), 57–149. https://doi.org/10.1016/s1382-6689(02)00126-6

Viana, L. F., Cardoso, C. A. L., Lima-Junior, S. E., Súarez, Y. R., & Florentino, A. C. (2020). Bioaccumulation of metal in liver tissue of fish in response to water toxicity of the Araguari Amazon River, Brazil. *Environmental Monitoring and Assessment, 192*, 781. https://doi.org/10.1007/s10661-020-08696-2

Viana, R. L., de Freitas, C. M., & Giatti, L. L. (2016). Saúde ambiental e desenvolvimento na Amazônia legal: Indicadores socioeconômicos, ambientais e sanitários, desafios e perspectivas. *Saúde e Sociedade, 25*, 233–246. https://doi.org/10.1590/S0104-12902016140843

Viana, L. F., Kummrow, F., Cardoso, C. A. L., Lima, N. A., Crispim, B. A., Barufatti, A., & Florentino, A. C. (2023). Metal bioaccumulation in fish from the Araguari River (Amazon biome) and human health risks from fish consumption. *Environmental Science and Pollution Research, 30*, 4111–4122. https://doi.org/10.1007/s11356-022-22457-3

Waichman, A. V. (2008). Uma proposta de avaliação integrada de risco do uso de agrotóxicos no estado do Amazonas, Brasil. *Acta Amazonica, 38*(1), 45–50. https://doi.org/10.1590/S0044-5967200800

Waichman, A. V., Eve, E., & Nina, N. C. S. (2007). Do farmers understand the information displayed on pesticide product labels? A key question to reduce pesticides exposure and risk of poisoning in the Brazilian Amazon. *Crop Protection, 26*, 576–583. https://doi.org/10.1016/j.cropro.2006.05.011

Waichman, A. V., Römbke, J., Ribeiro, M. O. A., & Nina, N. C. S. (2002). Use and fate of pesticides in the Amazon State, Brazil. Risk to human health and the environment. *Environmental Science and Pollution Research, 9*, 423–428. https://doi.org/10.1007/BF02987596

Wu, W., Chen, X., Hu, C., Li, J., Yu, Z., & Cai, W. (2010). Transplantation of neural stem cells expressing hypoxia-inducible factor-1α (HIF-1α) improves behavioral recovery in a rat stroke model. *Journal of Clinical Neuroscience, 17*(1), 92–95. https://doi.org/10.1016/j.jocn.2009.03.039

Impact of Organic Pollution on the Genome of Native Amazonian Fish Species: Highlights for the Cytogenomics and Gene Expression

Daniele Aparecida Matoso, Hallana Cristina Menezes da Silva, Francijara Araújo Silva, Roberto Ferreira Artoni, and Eliana Feldberg

D. A. Matoso
Federal University of Amazonas, Department of Genetics, Laboratory of Biotechnology and Animal Cytogenomics, Manaus, AM, Brazil

State University of Ponta Grossa, Laboratory of Genetics and Evolution, Ponta Grossa, PR, Brazil

Graduate Program in Genetics, Conservation and Evolutionary Biology, Brazilian National Institute for Research in the Amazon, Manaus, AM, Brazil

H. C. M. da Silva (✉)
State University of Ponta Grossa, Laboratory of Genetics and Evolution, Ponta Grossa, PR, Brazil

Graduate Program in Genetics, Conservation and Evolutionary Biology, Brazilian National Institute for Research in the Amazon, Manaus, AM, Brazil

F. A. Silva
Graduate Program in Genetics, Conservation and Evolutionary Biology, Brazilian National Institute for Research in the Amazon, Manaus, AM, Brazil

Northern University Center, School of Health Sciences and Fametro University Center, Central Campus and Nova Cidade Unit, Manaus, AM, Brazil

R. F. Artoni
State University of Ponta Grossa, Laboratory of Genetics and Evolution, Ponta Grossa, PR, Brazil

State University Ponta Grossa, Department of Structural, Molecular Biology and Genetics, Laboratory of Genetics and Evolution, Ponta Grossa, PR, Brazil

E. Feldberg
Graduate Program in Genetics, Conservation and Evolutionary Biology, Brazilian National Institute for Research in the Amazon, Manaus, AM, Brazil

Brazilian National Institute for Research in the Amazon, Animal Genetics Laboratory, Manaus, AM, Brazil

S. S. de Souza et al. (eds.), *The Future of Amazonian Aquatic Biota*,
https://doi.org/10.1007/978-3-031-66822-7_11

Abstract

Organophosphate pollution severely harms aquatic ecosystems in the Amazon. The urban area of Manaus is crossed by several rivers, the majority of which are contaminated by some pollutant. Using fish as model organisms, this work discussed the major investigations that have been conducted in the fields of cytogenomics and gene expression. Regarding cytogenomics, the retrotransposable elements *Rex*1, *Rex*3, and *Rex*6 have already been mapped in several species and their activation has been observed after environmental stress by temperature, toxic metal, and organic pollution. A study was conducted using the organophosphate Trichlorfon in populations of grown tambaqui, and it showed that genes involved in cellular apoptosis, the immune system, and oxidative stress were significantly expressed.

1 Introduction

One of the animal species that most damages the ecosystem is man, and among the many negative effects he causes, we can mention habitat destruction, an increase in greenhouse gas emissions, over-exploitation of non-renewable resources, pollution, and even the introduction of exotic species. Pollution can occur in a variety of ways, from improper waste disposal to severe contamination caused by industrial effluents or agroforestry pesticides.

In aquatic environments, contamination happens when unwanted materials are released into the water and/or discarded there, changing the water's quality (Alrumman et al., 2016), and harming both the environment and human health. In the past 50 years, there has been an increase in industrial and agricultural production that can be characterized by the intensive production and use of synthetic chemical insecticides, herbicides, and plasticizers, among other things. This increase is due to the high rate of population growth that is linked to the search for a better quality of life. Whether naturally or anthropically, the wastes produced by these compounds have shown a tendency to conflict with the maintenance of the environment's balance (Sunjog et al., 2013). In this sense, mitigation measures that aim to reduce or eliminate the pollution from hydrocarbons are important and necessary to ensure the sustainability and long-term viability of the management of animal, plant, and human populations.

According to Arias et al. (2007), the aquatic biota is continuously exposed to a large number of toxic substances that are released daily into untreated, open environments from various sources. Due to this fact, contamination may occur directly or indirectly: directly, through sewage overflow and industrial and hospital contaminants, or indirectly, through the accumulation of substances in the air, which may reach hydric resources (Desai & Smt, 2014). Repeatedly, pollutants released into aquatic ecosystems are not removed by natural purification processes, leaving these sources useless, causing major problems to water bodies, and seriously affecting the health of the human population that utilizes these waters (Akhtar et al., 2021).

As a result, discharges of waste from domestic, medical, and industrial effluents are the main causes of water pollution. Toxins in industrial and hospital waste are the main contributor to immune suppression, reproductive failure, and severe poisoning. Infectious diseases like cholera, typhoid fever (Juneja & Chaudhary, 2013), and other illnesses like gastroenteritis, diarrhea, vomiting, skin problems, and rhinitis are spreading due to contaminated water (Khan & Ghouri, 2011). In this way, water pollutants are removing algae, mollusks, birds, fish, crustaceans, and other aquatic organisms that are food for humans (FAO, 2020) and balance the biota.

A huge amount of domestic organic waste is discharged into the water, and the majority of it is not treated. These compositions include solid waste, plastic trash, bacterial contaminants, and toxic materials that, if left untreated, pollute the water. The greatest biodiversity in the world, distributed across many biomes, is concentrated in Brazil (Lewinsohn & Prado, 2005). The aquatic biota is significantly present in the Amazonian ecosystems, which have an extensive water network made up of several rivers and numerous watercourses (Sioli, 1984). However, in recent years, the Amazonian region has made significant progress in urban development, with the population of its metropolitan areas rising significantly. As a result, cities turned to technological advancement, causing extra issues related to disorderly population growth (Nelson & Lucivaldo, 2011; Vianna, 2015), among them the loss of natural characteristics of urban aquatic ecosystems and the dynamics, and organizational structure of their biotic communities (Bruno et al., 2021; Omar et al., 2015).

In this research, it was applied an analytical approach toward the contamination of Amazonian aquatic systems with a particular emphasis on fish. In topic 1, the aim was to discuss research on organophosphate pollution, with an emphasis on Trichlorfon and Malathion; in topic 2, it was discussed the effects of environmental stressors on repetitive DNA sequences—transposable elements and retroelements, focusing on SINES *Rex1*, *Rex3*, and *Rex6*, the most currently studied in fish.

2 Contamination by Organophosphate Pesticides Trichlorfon and Malathion

Organophosphate pesticides (OPs) are used in agriculture to combat pests. Once their potential as pesticides was established, OPs began to be used as antiparasitics in agriculture and pisciculture. These pesticides continue to be under observation because numerous studies indicate their potential for long- and short-term toxicity (Silva et al., 2022). Due to its potential for bioaccumulation, its toxicology is particularly concerning for human health (Dzudzevic Cancar et al., 2016). The first organophosphate compound approved for commercial use was Malathion®, which faced numerous chemical tests before being declared safe in the 1950s (Chambers et al., 2010a). These compounds' potential toxicological effects result from their ability to inhibit the action of the enzyme acetylcholinesterase, which can happen not just in the target organism but also in a widespread way.

Characterized as cholinesterase, acetylcholinesterase (AChE) is an enzyme whose main function in all living organisms is to stop the transmission of nervous impulses. This occurs as a result of the neurotransmitter acetylcholine (ACh) being broken down by hydrolysis into acetic acid (Ac) and choline (Ch), which prevents the transmission of nerve impulses once they are no longer necessary. When an organophosphate compound is available in the environment, it binds to AChE via phosphorylation in an irreversible manner, blocking the enzyme and preventing the interruption of the nervous impulse (Bates & Campbell, 2008; Kubitza et al., 2007) (Fig. 1). Blocking of AChE leads to an accumulation of ACh in the central nervous system, which in humans can result in diseases like Alzheimer's disease and in fish can lead to already described physiologic changes that intensify the nervous impulse (De Aguiar et al., 2004). In the OP's structure, the group responsible for connecting and interrupting the AChE operation is the leaving group, which is more reactive and more variable because of the change that occurs during this binding (Chambers et al., 2010a).

The two main chemical groups that organophosphates are derived from are phosphoric acid (H3PO4) and phosphorous acid (H3PO3). Their nomenclature and definition are based primarily on the atoms that are linked to the central phosphor (P) in

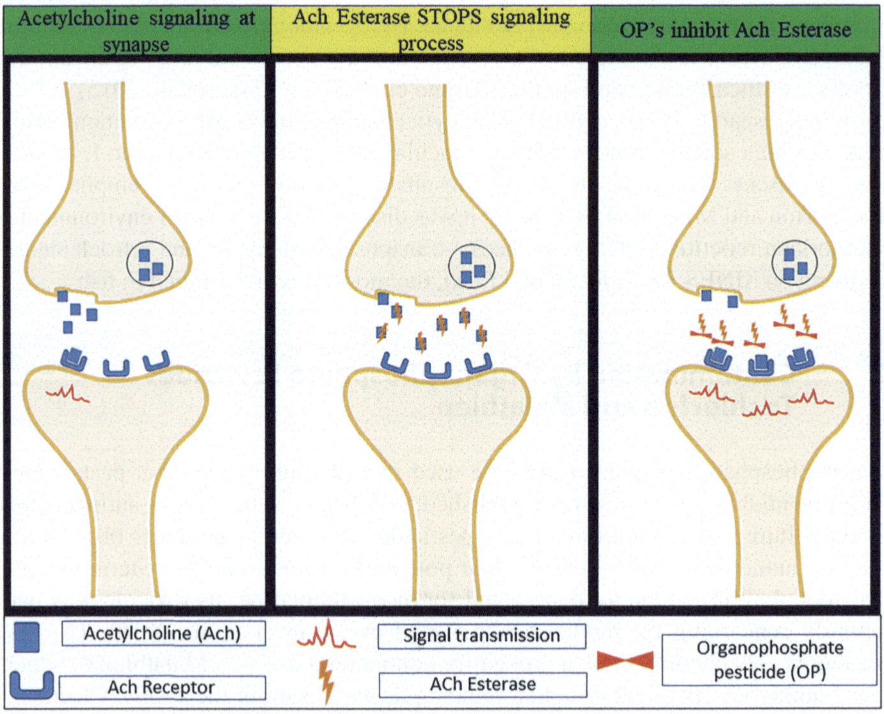

Fig. 1 Mechanism of action of OPs on the AChE enzyme. Acetylcholine signaling at the synapse: normal functioning of synaptic transmission through the binding of ChE to the postsynaptic neuron receptor. ACh Esterase STOPS signaling process: the action of AChE on ChE, stopping the signal from the synaptic transmission. OPs inhibit ACh Esterase: the action of OPs on AChE preventing the interruption of synaptic signal transmission (Silva et al., 2022)

their chemical structure, along with the binding of additional three atoms (H, O, and C); this structure is supported by a covalent bond (Chambers et al., 2010b). The main pathway for organophosphate metabolization involves the enzymes responsible for xenobiotics metabolism, which break down the OP into secondary metabolites that later also need to be metabolized. For this reason, it is known that the metabolism of OPs is split into two phases: Phase one involves the oxidation, reduction, and hydrolysis of the compound, while phase two involves conjugation reactions (Chambers et al., 2010b).

Due to their widespread use in both large-scale productions and domestic settings for treating parasitosis in domestic animals, certain OPs were banned from being commercialized in the last decade due to their toxicological effects (Bates & Campbell, 2008). Despite this, the list of authorized compounds is still very long, especially in Brazil, and requires continued attention. These compounds may be absorbed by various organisms by ingestion, cutaneous penetration, and inhalation when in direct contact with the substance (Bates & Campbell, 2008).

When OP effects on humans are evaluated, the toxicity caused by them may result in the following symptoms: abdominal pain, convulsions, difficulty breathing, confusion, depression, irritability, disorientation, impaired memory and concentration, difficulty speaking, eye irritability, and severe neurological disorders (Dhull et al., 2013). Most often, these damages are brought on by indirect contamination through the use of this compound as pesticides and parasite medications in agriculture and pisciculture. The primary purpose of using these compounds in pisciculture is to control ectoparasites, which act as acaricidal, insecticidal, and anti-helminthic agents and may result in production losses (Arauco et al., 2005). According to ANVISA (Agência Nacional de Vigilância Sanitária), the most widely used OPs in pisciculture are made from compounds such as Trichlorfon and Dichlorvos, which are classified under toxicity class II (highly toxic), and Malathion, with use classified according to each product (art. 38 of Resolução da Diretoria Colegiada—RDC nº 294, of July 29th 2019). Several studies have documented the effects of these compounds on organisms besides the parasitic ones they are supposed to treat, as well as their capacity to bioaccumulate in sediments and their long-lasting effects.

In their review of the literature, Silva et al. (2022) presented the effects of using this organophosphate as reported in studies conducted worldwide and in Brazil, demonstrating that the effects are not restricted to the studied organisms. The studies that were presented mostly demonstrated the AChE enzyme's vulnerability in organisms exposed to Trichlorfon-based antiparasitic. When it comes to studies carried out in order to investigate the effects of organophosphates on Amazonian fish, some recent work has been done at the Instituto Nacional de Pesquisas da Amazônia, in partnership with the Universidade Federal do Amazonas. In the Amazonian region, pisciculture frequently uses the organophosphate Trichlorfon to treat parasites that interfere with production, although certain deleterious effects have been noted in these products (Gomes 2018; personal communication), such as loss of swimming balance, physiological changes, hemorrhages (for more details, see: Silva, Guimarães, et al., 2020, supplementary material).

Silva, Guimarães, et al. (2020), and Duncan et al. (2020) evaluated the LC50-96 h values in juvenile tambaqui (*Colossoma macropomum*), one of the most popular and widely sold fish in the Amazonian region, as well as the enzymatic activities of AChE and GST (glutathione S-transferase) in juvenile fish exposed to Trichlorfon-based antiparasitic. Other morphological effects were noticed in fish exposed to a concentration of 3.2 mg/L of Trichlorfon. The amount of LC50-96 h discovered was 0.870 mg/L of Trichlorfon (Fig. 2). In terms of the enzymatic activities of AChE and GST, Duncan et al. (2020) noted a decline in AChE activity; however, the levels of GST activity remained consistent in both control and experimental samples, demonstrating that GST activity is not compromised when fish are exposed to Trichlorfon. The data obtained are consistent with numerous studies that have been previously described for other species of fish around the world.

Silva de Souza et al. (2020) evaluated the biological, histopathological, and genetic effects of the organophosphate-based antiparasitic Malathion used in tambaqui (*Colossoma macropomum*). The animals were exposed for 96 h to a sub-lethal concentration of 7.30 mg/L. In contrast to what was shown in animals exposed to Trichlorfon, levels of acetylcholinesterase remained steady after the exposure period, and DNA and mitochondrial respiration were unaffected. On the other hand, the activity of antioxidant and biotransformation enzymes increased, suggesting that they are involved in the Malathion metabolism. Additionally, *ras*

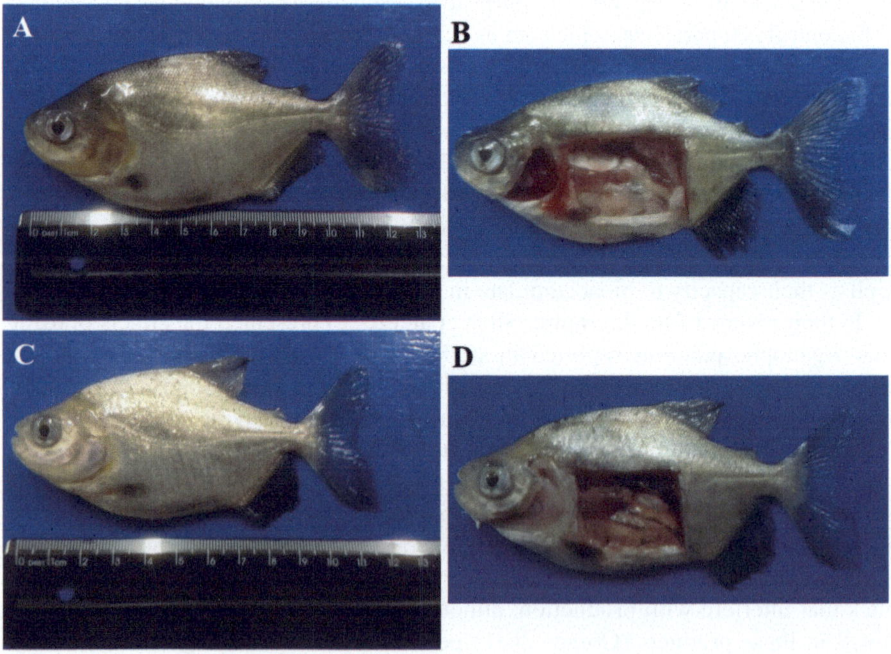

Fig. 2 In A and B, fish in the control condition. In C and D, fish exposed to a concentration of 3.2 mg/L of Trichlorfon for 11 h. In the opened fish, it is possible to observe, mainly, the difference in the color of the organs (Silva, da Silva, et al., 2020)

proto-oncogenes showed an increase in their expression. The findings suggest that 96 h is the best window of time for activating this organophosphate response, and the observation of an increase in proto-oncogene expression suggests that exposure may be linked to tumorigenic events.

Silva et al. (in preparation) also evaluated the effects of the Trichlorfon-based antiparasitic drug on the transcriptome of tambaqui's liver (*Colossoma macropomum*) exposed to two sub-lethal concentrations: 30% and 50% of LC50-96 h (0.261 mg/L and 0.435 mg/L, respectively). The results from the transcriptomic analysis revealed 176 genes with significantly differential expression levels when compared to the control group. Among these genes, it was found that the ones involved in the metabolism of xenobiotics, such as the enzyme *Cytochrome P450 2 J6-like*, antigens of histocompatibility class I, such as *F10 alpha chain-like and Q9 alpha chain-like*, genes of tumor response, such as the *proto-oncogene Pim-2 serine/threonine kinase* and tumor *protein p53*, and genes involved in cell death induction, such as *DFFA-like effector C* and *BCL2*.

The main mechanism of action of the organophosphates is acetylcholinesterase inhibition; however, studies with exposure to two of these compounds suggest that the responses vary depending on the drug used. On the other hand, by using genetic analysis, it is possible to estimate the tumor potential of these mixtures given that the proto-oncogenes *Pim-2* and *p53* are highly expressed in both of these cases. Some studies are still being conducted in order to better understand the mechanisms that are activated by the organophosphates found in Amazonian fish. Additionally, efforts are being made to develop biomarkers that will support the monitoring of their use in the cultivation ponds of production in the Amazon state.

Regarding the development of markers, in the epigenetic study developed by Nonato et al. (2023), using trichlorfon as a stressor agent for the tambaqui genome, it was verified the methylation pattern of the 18S ribosomal gene (18S rDNA). The smaller subunit of a cell's ribosomes is made up of a ribosomal RNA that is transcript by this gene. In this study, it was discovered that there was a rise in methylation at the experiment's concentration 1 (C1 = 0.261 mg/L of Trichlorfon diluted in water), at 24 h and 96 h, pointing to an epigenetic mechanism of action for the inactivation of 18S copies. These 18S copies' inactivation has not yet been studied for their effects on cellular function. It is known that at least half of the hundreds of copies in the genome are naturally inactivated by methylation (Baccalini et al., 2014). However, it seems clear that chemical stress-induced inactivation can have a direct impact on protein synthesis in these organisms. Future approaches on DNA methylation patterns and other epigenetic mechanisms will be able to identify the metabolic pathways that organophosphates' action on the genomes of fish species native to the Amazon are activating and inactivating. Toxic metals, high temperatures, and organophosphate pesticides are just a few of the environmental stressors that have already been proven to be potent inducers of genomic alterations. All of these stressors have an immediate impact on the expression and dispersion of mobile genetic elements, which is the subject of the following section in this chapter.

3 Action of Environmental Stressors on Repetitive DNA Sequences—Transposable Elements and Retroelements

Many questions remain unanswered regarding the origin, evolution, and purpose of the repeating sequences in species' genomes. The transposable and retroelements are repeating regions dispersed throughout the genome with sequences that are located in relatively narrow portions, such as centromeres and subtelomeric regions (Fig. 3) (Biscotti et al., 2015). These sequences have the ability to travel around the genome, which may disrupt the gene transcription machine or result in the generation of variability. They constitute a substantial portion of the DNA content in the cell's nucleus (Marsano & Dimitri, 2022).

Numerous studies conducted by the research group of Instituto Nacional de Pesquisas da Amazônia, in partnership with the Universidade Federal do Amazonas, have mapped repetitive DNA sequences in aquatic organisms, specifically fish. Highlighted among these studies are those by Costa et al. (2022), in which the authors mapped the retroelements *Rex1*, *Rex3*, and *Rex6* in *Colossoma macropomum* exposed to 30% and 50% of the LC50-96 h of Trichlorfon diluted in water and found an increase in *Rex3* marks at the highest concentration and for the longest period of the experiment. Furthermore, the authors correlated the placement of these retroelements with the heterochromatin distribution patterns, which was coincident. In another study, using tambaqui, Ferreira et al. (2019), after exposing the fish to higher temperatures in accordance with the Intergovernmental Panel on Climate Change (IPCC), observed a gradual increase in retroelement *Rex3* markings, which was also coincident with an increase in the distribution of heterochromatin in the experiment's most extreme conditions (Scenario A2, which simulates a temperature increase of 4.5 °C and 850 ppm of CO_2).

In a similar study, also using tambaqui, Ribeiro (2013) investigated variations in the pattern of heterochromatin distribution among individuals kept at different temperatures (22, 28, and 30 °C). In this experiment, a noticeable increase in the number of heterochromatic blocks was seen in fish that had been maintained at a

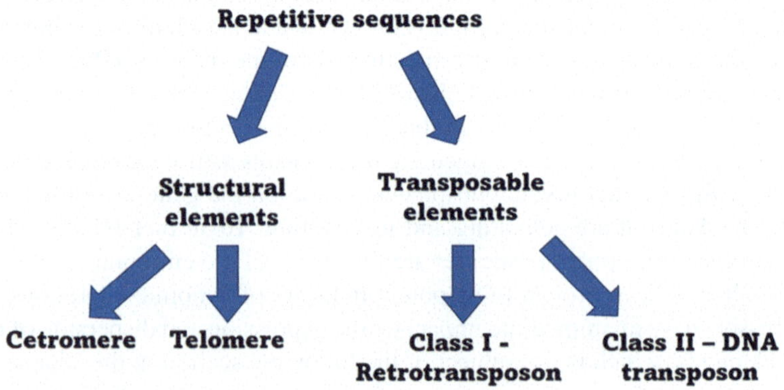

Fig. 3 Diagram showing different types of repetitive sequences in the genome

temperature of 22 °C, indicating a correlation between the increase in heterochromatin and the presence of the retroelements *Rex* (1, 3, and 6) in chromosomes. The retrotransposons *Rex1* and *Rex6* were dispersedly distributed, occurring in both heterochromatic and euchromatic portions. However, in animals maintained at a temperature of 22 °C, there was an increase in the markings, which extended nearly the whole length of some of the complement's chromosomes. According to the author, the increase in heterochromatic DNA blocks is an epigenetic response to a temperature decrease that provides a quick adaptation to a stressful situation. Additionally, given that this is one of the mechanisms that can prevent the transposition of mobile elements, the increase in the transposable elements *Rex1* and *Rex6* may be related to the heterochromatinization process. As a result, it can be shown that the tambaqui genome also responds to stressful conditions like cold temperatures. This is because, even though this species can easily adapt to lower temperatures, it reaches its greatest at temperatures of about 28 °C.

In a related work but investigating the deleterious action of contamination by organic material, Silva, da Silva, et al. (2020) analyzed the genome of the species *Pterygoplichthys pardalis*, *Semaprochilodus insignis*, and *Cichlasoma amazonarum* and mapped retroelements *Rex1*, *Rex3*, and *Rex6* in the three species. The authors noticed an increase in markings, especially *Rex3* and *Rex6*, in the individuals from polluted environments compared to those from clean environments. They concluded that the studied species' genomes are dynamic and that the pollution of urban rivers, caused by domestic and industrial effluents, is influencing the modulation of the phenotypic plasticity displayed by Amazonian species. When TEs are incorporated into genetic regions, many types of genomic rearrangements, including gene insertions, deletions, inversions, and duplications, are produced by transpositions (Alzohairy et al., 2013). TEs may have a significant impact on the phenotype of host organisms by maintaining gene regulatory activity (Lisch, 2013). According to numerous studies, adaptation occurs after invasion events in response to selective pressures in the new environment (Brawand et al., 2014). These organisms can respond by changing their biological characteristics (Frederico et al., 2016), but primarily through genetic plasticity (Heinrichs-Caldas et al., 2019; Silva et al., 2016). The cellular synthesis machinery is directly affected by these chromosomal and genomic alterations. Because it impairs an organism's ability to respond metabolically and physiologically to various sorts of environmental changes, this phenomenon becomes hazardous for fish that must constantly deal with environmental obstacles, as is the case with Amazonian fish. As a result, it is possible to experience a decline in the biochemical and physiological phenotypic plasticity that is essential for reducing stressful events.

In this way, TEs that are influenced by environmental stress may produce the raw diversity that species need over the course of their evolution to survive in these conditions (Brawand et al., 2014; Casacuberta & Gonzalez, 2013; Silva, da Silva, et al., 2020).

This becomes even more crucial when we take into account how actions by domestic, commercial, and industrial effluents are causing modifications to the urban Amazon environments that also help establish the complex Amazon aquatic ecosystem. Additionally, taking into account that fish health reflects the environmental conditions in which these ecosystems exist (Barbieri & Bondioli, 2013; Ramsdorf, 2007), this type of assessment has emerged as a key tool for environmental monitoring, and on this premise, an evaluation of the quality of these environments, which are frequently used by human populations downstream, is absolutely necessary.

Thus, it is clear that stress has significantly increased the expression and dispersion of repetitive sequences, such as the retroelements *Rex*. Stress factors include changes in temperature, aquatic contamination by toxic substances such as organophosphates and toxic metals, as well as domestic and industrial wastes. Furthermore, this has modulated the phenotypic plasticity of Amazonian fish in such a way that environmental changes can bring serious problems to the ecological balance and to the maintenance of fish species in their natural environments. Given the numerous polluting residues that are currently affecting both natural and artificial aquatic environments in the Amazon, their monitoring is vital and necessary. For educating society and promoting the sustainable use of such aquatic environments, mitigation actions against pollution are urgently needed. These actions are related to waste management education programs aimed at high school and college students as well as the general public and environmental organizations.

4 Conclusions

Although it is still necessary to expand basic studies of cytogenetic and genomic characterization on the megadiversity of Amazonian fish, several studies have shown harmful effects on fish genetic material resulting from anthropic effects that end up polluting water courses with sources ranging from organic to agricultural and hospital waste. DNA structural variations and its functional plasticity highlight the uneven battle that aquatic organisms have fought for survival. These same organisms are listed as bioindicators that are guardians of life and health of the environment. Monitoring contamination levels and their effects on animal life is consistent with the most recent One Health concepts. One of humanity's greatest challenges is realizing that we are an integral component of the balanced biota; as a result, human health depends on the well-being of animals and the environment.

Acknowledgements The authors are grateful to the anonymous reviewers for their suggestions. This study was financed in part by the Coordenação de Aperfeiçoamento de Pessoal de Nível Superior—Brasil (CAPES), FAPEAM-POSGRAD, INCT/ADAPTA-CNPq 465540/2014-7.

References

Akhtar, N., Syakir-Ishak, M. I., Bhawani, S. A., & Umar, K. (2021). Various natural and anthropogenic factors responsible for water quality degradation: A review. *Water, 13*, 2660. https://doi.org/10.3390/w13192660

Alrumman, S. A., El-kott, A. F., & Kehsk, M. A. (2016). Water pollution: Source and treatment. *American Journal of Environmental Engineering, 6*(3), 88–98.

Alzohairy, A. M., Gyulai, G., Jansen, R. K., & Bahieldin, A. (2013). Transposable elements domesticated and neo functionalized by eukaryotic genomes. *Plasmid, 69*, 1–15.

Arauco, L. R. R., Cruz, C., & Machado Neto, J. G. (2005). Efeito da Presença de Sedimento na Toxicidade aguda do Sulfato de Cobre e do Triclorfon para três espécies de Daphnia. *Pesticidas: Revista de Ecotoxicologia e Meio Ambiente, 15*, 55–64.

Arias, A. R. L., Inacio, A. F., Novo, L. A., Albuquerque, C., & Moreira, J. C. (2007). Multibiomarker approach in fish to assess the impact of pollution in a large Brazilian river, Paraiba do Sul. *Environmental Pollution, 156*(3), 974–979.

Baccalini, M. G., Pacilli, A., Giuliani, C., Penzo, M., & Treré, D. (2014). The nucleolar size is associated to the methylation status of ribosomal DNA in breast carcinomas. *BMC Cancer, 14*, 1–11.

Barbieri, E., & Bondioli, A. C. V. (2013). Acute toxicity of ammonia in Pacu fish (*Piaractus mesopotamicus*, Holmberg, 1887) at different temperatures levels. *Aquaculture Research, 46*, 565. https://doi.org/10.1111/are.12203

Bates, N., & Campbell, A. (2008). Organophosphate insecticides. In *Handbook of poisoning in dogs and cats* (pp. 199–204).

Biscotti, M. A., Ettore Olmo, E., & Heslop-Harrison, J. S. (2015). Repetitive DNA in eukaryotic genomes. *Chromosome Research, 23*, 415. https://doi.org/10.1007/s10577-015-9499-z

Brawand, D., Wagner, C. E., Li, Y. I., Malinsky, M., Keller, I., Fan, S., Simakov, O., Ng, A. Y., Lim, Z. W., & Bezault, E. (2014). The genomic substrate for adaptive radiation in African cichlid fish. *Nature, 513*, 375–381.

Bruno, D., Willmer, I., Pereira, L., Rocha, R., Saint'Pierre, T., Baldassin, P., & Hauser-Davis, R. (2021). Metal and metalloid contamination in Green Sea Turtles (*Chelonia mydas*) found stranded in Southeastern Brazil. *Frontiers in Marine Science, 8*, 1–11.

Casacuberta, E., & Gonzalez, J. (2013). The impact of transposable elements in environmental adaptation. *Molecular Ecology, 22*, 1503–1517.

Chambers, H. W., Meek, E. C., & Chambers, J. E. (2010a). *Chemistry of organophosphorus insecticides* (Vol. 2, pp. 1395–1398). Elsevier.

Chambers, J. E., Meek, E. C., & Chambers, H. W. (2010b). *The metabolism of organophosphorus insecticides* (Vol. 2, pp. 1399–1407). Elsevier.

Costa, M., Silva, H. C. M., Soares, S. C., Favarato, R. M., Feldberg, E., Silva-Gomes, A. L., Artoni, R. F., & Matoso, D. A. (2022). A perspective of molecular cytogenomics, toxicology, and epigenetics for the increase of heterochromatic regions and retrotransposable elements in Tambaqui (*Colossoma macropomum*) exposed to the parasiticide Trichlorfon. *Animals, 12*, 1945–1911.

De Aguiar, L. H., Moraes, G., Avilez, I. M., Altran, A. E., & Corrêa, C. F. (2004). Metabolical effects of Folidol 600 on the neotropical freshwater fish matrinxã, *Brycon cephalus*. *Environmental Research, 95*, 224–230.

Desai, N., & Smt, V. (2014). A study on the water pollution based on the environmental problem. *Indian Journal of Research, 3*(12), 95–96.

Dhull, V., Gahlaut, A., Dilbaghi, N., & Hooda, V. (2013). Acetylcholinesterase biosensors for electrochemical detection of organophosphorus compounds: A review. *Biochemistry Research International, 2013*, 731501. https://doi.org/10.1155/2013/731501

Duncan, W. P., Idalino, J. J. S., da Silva, A. G., Moda, R. F., Silva, H. C. M., Matoso, D. A., & Gomes, A. L. S. (2020). Acute toxicity of the pesticide trichlorfon and inhibition of acetylcholinesterase in *Colossoma macropomum* (Characiformes: Serrasalmidae). *Aquaculture International, 28*, 815–830.

Dzudzevic Cancar, H., Soylemez, S., Akpinar, Y., Kesik, M., Göker, S., & Gunbas, G. (2016). A novel acetylcholinesterase biosensor: Core-shell magnetic nanoparticles incorporating a conjugated polymer for the detection of organophosphorus pesticides. *ACS Applied Materials & Interfaces, 8*, 8058–8067.

FAO. (2020). *The state of world fisheries and aquaculture*. Sustainability in action. https://doi.org/10.4060/ca9229en

Ferreira, A. M. V., Marajó, L., Matoso, D. A., Ribeiro, L. B., & Feldberg, E. (2019). Chromosomal mapping of *Rex* retrotransposons in Tambaqui (*Colossoma macropomum* Cuvier, 1818) exposed to three climate change scenarios. *Cytogenetic and Genome Research, 159*(1), 39–47. https://doi.org/10.1159/000502926

Frederico, R. G., Olden, J. D., & Zuanon, J. (2016). Climate change sensitivity of threatened, and largely unprotected, Amazonian fishes. *Aquatic Conservation, 26*, 91–102.

Heinrichs-Caldas, W., Campos, D. F., Paula-Silva, M. N., & Almeida-Val, V. M. F. (2019). Oxygen dependent distinct expression of hif-1α gene in aerobic and anaerobic tissues of the amazon Oscar, *Astronotus crassipinnis*. *Comparative Biochemistry and Physiology. Part B, Biochemistry & Molecular Biology, 227*, 31–38. https://doi.org/10.1016/j.cbpb.2018.08.011

Juneja, T., & Chaudhary, A. (2013). Assessment of water quality and its effects on the health of residents of Jhunjhunu District, Rajasthan: A cross sectional study. *Journal of Public Health and Epidemiology, 5*, 186–191.

Khan, A., & Ghouri, A. (2011). Environmental pollution: Its effects on life and its remedies. *Researcher World: Journal of Arts, Science & Commerce, 2*, 276–285.

Kubitza, F., Ono, E. A., & Campos, J. L. (2007). Os caminhos da produção de peixes nativos no Brasil : Uma análise da produção e obstáculos da piscicultura. *Panorama da Aquicultura, 17*, 1–10.

Lewinsohn, T. M., & Prado, P. I. (2005). How many species are there in Brazil? *Conservation Biology, 19*, 619–624.

Lisch, D. R. (2013). How important are transposons for plant evolution? *Nature Reviews Genetics, 14*, 49–61.

Marsano, R. M., & Dimitri, P. (2022). Constitutive heterochromatin in Eukaryotic genomes: A mine of transposable elements. *Cells, 11*, 761. https://doi.org/10.3390/cells11050761

Nelson, P., & Lucivaldo, J. (2011). *Dados estatísticos—Igarapés Manaus*. Equipe Vencer.

Nonato, L. S., da Silva, H. C. M., Silva-Gomes, A. L., Traldi, J. B., Artoni, R. F., & Matoso, D. A. (2023). Methylation profile of 18S rDNA gene in brain and muscle of tambaqui exposed to parasiticide trichlorfon. *International Journal of Zoology and Animal Biology, 6*(1), 000445. https://doi.org/10.23880/izab-16000445

Omar, W. A., Mikhail, W. Z. A., Abdo, H. M., Abou El Defan, T. A., & Poraas, M. M. (2015). Avaliação do Risco Ecológico da Poluição Metálicos ao longo do Grande Setor do Cairo no Rio Nilo, no Egito, Usando Tilapia do Nilo, *Oreochromis niloticus*, como Bioindicador. *Journal of Toxicology, 167*, 319.

Ramsdorf, W. A. (2007). *Utilização de duas species de Astyanax (Astyanax sp. B e A. altiparanae) como bioindicadores de região contaminada por agrotóxico (Fazenda Canguiri—UFPR)*. Dissertação de Mestrado.

Ribeiro, L. B. (2013). *Mapeamento físico cromossômico de elementos repetitivos em Colossoma macropomum (Characiformes, Serrasalmidae) associado à piscicultura*. Instituto Nacional de Pesquisas da Amazônia. https://bdtd.inpa.gov.br/handle/tede/726

Silva, H. C., Carvalho, A. P. C., Silva-Gomes, A. L., Artoni, R. F., & Matoso, D. A. (2022). Impact of trichlorfon organophosphate use in pisciculture: A review. *Studies in Environmental and Animal Sciences (SEAS), 3*, 1399–1423.

Silva, H. C. M., da Silva, A. G., Idalino, J. J. S., de Sousa, F. B., Gomes, A. L. S., Duncan, W. P., & Matoso, D. A. (2020). Trichlorfon acute lethal toxicity to juvenile tambaqui (*Colossoma macropomum*). *Aquaculture Research, 51*, 863–866. https://doi.org/10.1111/are.14412

Silva de Souza, S., Machado, R. N., Custódio da Costa, J., Campos, D. F., Sebrenski da Silva, G., & Almeida-Val, V. M. F. (2020). Severe damages caused by Malathion exposure in Colossoma macropomum. *Ecotoxicology and Environmental Safety, 205*, 111340. https://doi.org/10.1016/j.ecoenv.2020.111340

Silva, F. A., Guimarães, E. M. C., Carvalho, N. D. M., Ferreira, A. M. V., Schneider, C. H., Carvalho-Zilse, G. A., Feldberg, E., & Gross, M. C. (2020). Transposable DNA elements in

Amazonian fish: From genome enlargement to genetic adaptation to stressful environments. *Cytogenetic and Genome Research, 160*, 148–155.
Silva, F. A., Schneider, C. H., Feldberg, E., Baccaro, F. B., Carvalho, N. D. M., & Gross, M. C. (2016). Genomic organization under different environmental conditions: *Hoplosternum littorale* as a model. *Zebrafish, 13*(3), 197–208.
Sioli, H. (1984). The Amazon and its main effluents: Hydrography, morphology of the river courses, and river types. In H. Sioli (Ed.), *The Amazon: Limnology landscape ecology of a mighty tropical river and its basin* (pp. 127–166). W. Junk Publications.
Sunjog, K., Kolarević, S., Héberger, K., Gačić, Z., Knežević-Vukčević, J., & Vuković-Gačić, B. (2013). Comparison of comet assay parameters for estimation of genotoxicity by sum of ranking differences. *Analytical and Bioanalytical Chemistry, 14*, 4879–4885.
Vianna, A. M. (2015). Poluição ambiental, um problema de urbanização e crescimento desordenado das cidades. *Revista Sustinere, 3*(1), 22. https://doi.org/10.12957/sustinere.2015.17325

Part III

Phenotypic Plasticity and Adaptive Mechanisms: Change to Survive

Amazonian Fish and Their Bacterial Symbionts: An Intricate Co-Dependence Relationship in a Changing Environment

François-Étienne Sylvain, Alizée Thomas, and Nicolas Derome

Abstract

Fish are hosts to symbiotic bacteria that form diverse communities, which play key functions for their host immunity, nutrition, development, and physiology. Numerous studies documented the existence of these bacterial communities in the guts, skin mucus, and gills of Amazonian fish. Dominated by Betaproteobacteria, Amazonian fish microbiotas play critical roles for their host, notably in the modulation of ionoregulatory processes, which enable fish to thrive in physiologically challenging acidic and ion-poor blackwater found throughout the Amazonian basin. In this report, we detail the composition and roles of Amazonian fish microbial symbionts, and we discuss their sensitivity to various host- and environment-specific factors such as host genotype, parasitism level, diet, and environmental physicochemistry and bacterioplankton composition. All these factors will be affected by future anthropogenic perturbations such as deforestation, dam construction, riverine agriculture, mining, oil exploitation, and climate change. Downstream effects on fish microbiotas are suspected to include a dysbiosis, an increase in the abundance of pathogenic bacteria, a shift in community diversity, and a perturbation of the equilibrium in fish-microbiota-parasite interactions. We conclude that Amazonian fish can instantly acquire adaptive functions via the recruitment of new microbial symbionts, which could benefit their ability to cope with present and future threats.

F.-É. Sylvain (✉)
Fisheries and Oceans Canada, Moncton, NB, Canada
e-mail: francois-etienne.sylvain@dfo-mpo.gc.ca

A. Thomas · N. Derome
Institute of integrative biology and systems, Laval University, Québec, QC, Canada

S. S. de Souza et al. (eds.), *The Future of Amazonian Aquatic Biota*,
https://doi.org/10.1007/978-3-031-66822-7_12

1 Fish-Microbe Systems

Symbioses are a type of biological interaction involving an association between organisms of different species during a significant portion of their life (Margulis & Fester, 1991). The microbiota of an organism can be defined by "*all the microorganisms, cultivable or not, associated with an individual or an ecosystem*" (Bertand et al., 2011), thus including bacteria, archaea, viruses, and unicellular eukaryotes. Symbiotic interactions between eukaryotic hosts and their microbiota contribute significantly to the fitness of hosts since they allow the production of different phenotypic variants, and they can contribute to the reproductive isolation of hosts and can facilitate the adaptation of hosts to variable environmental conditions (Bang et al., 2018; Rowe et al., 2020).

Just like humans and all Metazoa, fish harbor symbiotic microbial communities. Our knowledge of fish microbiota is in its infancy, the first studies on fish microbiota dating from the 1970s (Horsley, 1977). Since that time, numerous studies have been conducted, thanks to the explosion of next-generation sequencing technologies in the last decade, as well as the development and widespread use of gnotobiotic models such as the zebrafish (*Danio rerio*) and the three-spined stickleback (*Gasterosteus aculeatus*) (Lescak & Milligan-Myhre, 2017).

Fish microbial communities are mostly dominated by bacteria with the most abundant phylum being Proteobacteria (Ghanbari et al., 2015). The following phyla are also commonly listed: Fusobacteria, Bacteroidetes, Firmicutes, and Actinobacteria (Egerton et al., 2018; Llewellyn et al., 2014). Although fish microbiota samples from the literature often harbor a set of common characteristics (high diversity, high abundance of Proteobacteria), several studies have identified inter-community differences, i.e., bacterial biomarkers, making it possible to differentiate samples from (1) different species (Kuang et al., 2020), (2) host clades with different evolutionary history (Doane et al., 2020), (3) diseased or healthy hosts (Boutin et al., 2013; Llewellyn et al., 2017), (4) hosts with contrasting diets (Parata et al., 2020), (5) fish with different genotypes (Boutin et al., 2014), and (6) hosts exploiting different environments (freshwater versus saltwater, healthy versus polluted ecosystems, etc.) (Kim et al., 2021; Sylvain et al., 2020). Thus, the existence of bacterial biomarkers enabling the differentiation of individuals according to these criteria testifies to the great sensitivity of fish microbial communities to a multitude of environmental and host-specific factors.

The Fish Microbiota Plays Essential Roles for Its Host Fitness Several studies have investigated the general roles of microbiota in fish immunity, nutrition, development, and physiology (Burns et al., 2016; Nie et al., 2017; Smith et al., 2017; Yan et al., 2016).

Immunity Numerous reviews have focused on the interactions between gut microbiota and innate immunity in fish (Gómez & Balcázar, 2008; Nie et al., 2017). In healthy fish, the microbiota constitutes the first barrier against pathogens (Boutin et al., 2014). It antagonizes the colonization or growth of pathogens through niche exclusion, competition for resources, and suppression of virulence factors (Nie et al., 2017). Studies on gnotobiotic zebrafish have also shown that the gut microbiota promotes intestinal cell proliferation over normal ontogenesis (Cheesman et al., 2011; Rawls et al., 2006). Compared with germ-free zebrafish, conventionally raised zebrafish exhibit a higher expression level of epithelial proliferation genes, but also of innate immune response genes (Rawls et al., 2004). For instance, the microbiota colonization in germ-free zebrafish has been shown to induce the expression of the innate immune response genes amyloid A1 (*Saa1*), complement component 3 (*C3*), and myeloperoxidase (*MPO*), indicating the role of the microbiota in the priming of fish innate immunity (Rawls et al., 2004). Other studies have also shown the implication of the microbiota in fish adaptive immune responses. For instance, microbiota colonization was found to activate the NF-κB pathway, which plays important roles in innate and adaptive immune responses, in the intestine of germ-free zebrafish, further supporting the role of the microbiota in immune regulation (Kanther et al., 2011).

Nutrition Commensal bacteria can also regulate many key aspects of fish nutrition, including feeding behavior, energy balance, and nutrient metabolism (Butt & Volkoff, 2019). For instance, studies have found that the gut shows differential metabolic capacity in fish with different microbiotas, which suggest that the gut microbiota may contribute to the nutritional metabolism of the host (Liu et al., 2016, 2021). Furthermore, the gut microbiota composition is closely related to the activity of the digestive enzymes of fish, such as amylase, cellulase, trypsin, and lipase (Liu et al., 2021). The gut microbiota also promotes energy absorption and regulates the expression of genes associated with energy and lipid metabolism in zebrafish (Semova et al., 2012; Sheng et al., 2018). For example, Semova et al. (2012) showed that the microbiota can stimulate fatty acid uptake and lipid droplet formation in the intestinal epithelium and liver of gnotobiotic zebrafish. Overall, this suggests that the gut microbiota plays a key role in regulating the energy metabolism of fish hosts.

Development and Physiology Finally, the fish microbiota also affects its host development, nervous system, and endocrinology (de La Torre Canny et al., 2021). For instance, a study on germ-free zebrafish found that early colonization of the gut microbiota by bacterial species *Aeromonas veronii* or *Vibrio cholerae* was required for normal neurobehavioral development of host fish (Phelps et al., 2017). Another study on zebrafish showed that commensal microbiota colonization resulted in a widespread transcriptional response in the olfactory organ, which was associated with the overexpression of RE1 silencing transcription factor, that is related to neuronal development and differentiation (Casadei et al., 2019). In addition, the control of the fish endocrine system by commensal microbiota has also been reported. For instance, Gioacchini et al. (2018) revealed that the use of *Lactobacillus* probiotics in tilapia triggered gut microbiota changes that were associated with the activation of the endocrine system, including the increased expression of genes from the insulin-like growth factors (*igfs*) system.

2 The Microbiota of Amazonian Fish

Although microbial communities usually vary according to numerous host-associated or environment-associated factors (reviewed by Lazado & Caipang, 2014), core microbiomes, i.e., groups of microorganisms that are consistently present, are often detectable between phylogenetically related hosts or hosts sharing the same environment (Hamady et al., 2010). Recent studies have begun to shed light on the core microbial communities inhabiting the gut, skin, and gill of various Amazonian fish species (Di Maiuta et al., 2013; Sylvain et al. 2016, 2019, 2020, 2022; Ramírez et al., 2018; do Vale Pereira et al., 2017; McCauley et al., 2020; McDonald et al., 2019). Those communities could provide various benefits to their host fish, including implications in disease resistance, food digestion, and adaptation to challenging environments, such as black water.

2.1 Gut Communities

Studies conducted on the pirarucu (*Arapaima gigas*), tambaqui (*Colossoma macropomum*), flag cichlid (*Mesonauta festivus*), black piranha (*Serrasalmus rhombeus*), maculatus piranha (*Serrasalmus maculatus*), red-belly piranha (*Pygocentrus nattereri*), pacu peva (*Mylossoma duriventre*), and *Panaque* sp. have shown collectively that the main bacterial phyla characterizing the gut microbiota of Amazonian fish were Proteobacteria (especially Gammaproteobacteria, Alphaproteobacteria, Betaproteobacteria), Fusobacteria (particularly the genus *Cetobacterium* sp.), Bacteroidetes (Bacteroidia), Firmicutes (Clostridia), and Actinobacteria (Di Maiuta et al., 2013; Sylvain et al. 2016, 2019, 2020, 2022; Ramírez et al., 2018; do Vale Pereira et al., 2017; McCauley et al., 2020; McDonald et al., 2019).

2.2 Skin Mucus Communities

Similarly, studies on the flag cichlid, black piranha, maculatus piranha, red-belly piranha, pacu peva, tambaqui, and discus (*Symphysodon aequifasciata*), have found greater abundance of Proteobacteria (Gammaproteobacteria, Alphaproteobacteria, Proteobacteria), Mollicutes, Actinobacteria, Fusobacteria, Flavobacteria, and Clostridia in the cutaneous mucus microbiota (Sylvain et al., 2016, 2019, 2020; Sylvain & Derome, 2017; Sylvain, Leroux, et al., 2022).

2.3 Gill Communities

Research focused on the gill microbial communities of the flag cichlid, black piranha, freshwater sardine (*Triportheus albus*), and peacock bass (*Cichla* spp.), has demonstrated collectively a high abundance of Proteobacteria (mostly Betaproteobacteria, Alphaproteobacteria, Gammaproteobacteria), Actinobacteria, and Bacteroidetes (*Bacteroides*), Firmicutes (Clostridia), Flavobacteria (Sylvain et al., 2023; Sylvain, Normandeau, et al., 2022).

Interestingly, the aforementioned bacterial phyla are found on freshwater fish globally, across various ecosystems, thus highlighting their potential importance in global fish biology. For instance, similarly to what was reported on Amazonian fish, studies on the gut microbiota from different host fish species have revealed the most common phyla to be Proteobacteria, Actinobacteria, Bacteroidetes, Firmicutes, and Fusobacteria (Al-Hisnawi et al., 2015; Eichmiller et al., 2016; Sullam et al., 2012). Regarding the skin, Proteobacteria, Firmicutes, Fusobacteria, and Bacteroidetes have generally been observed in the literature (Larsen et al., 2013; Reinhart et al., 2019; Rosado et al., 2019; Scheifler et al., 2022). On the gills, a predominance of Proteobacteria, Firmicutes, Fusobacteria, and Bacteroidetes was reported (Pratte et al., 2018; Rosado et al., 2019). However, although the phyla of bacteria found in the microbiota of Amazonian fish are similar to those found in fish from other regions, even temperate ones, our understanding at the species level remains limited due to the fact that most studies focused on Amazonian fish microbial communities used a 16S rRNA gene metabarcoding approach to characterize community composition. This approach, although powerful and cost-effective, does not offer the high taxonomic resolution to the species or to the strain level that other contemporary methods do (e.g., Oxford Nanopore or PacBio long-read sequencing). Furthermore, the 16S rRNA metabarcoding approach does not provide any indication on the functional repertory of the taxa identified (although functions can be predicted with a high level of uncertainty, from 16S rRNA amplicons, using tools such as PICRUSt2). There is a possibility that the bacterial communities present in the Amazonian fish microbiota are distinct from temperate fish, showcasing unique functions specific to the Amazonian ecosystem. One way to move forward would be to use a shotgun metagenome approach, which would generate information on the functional repertory of these communities. More research is needed to fully understand the diversity and roles of these bacteria in the Amazonian fish microbiota.

2.4 Potential Functions

The bacterial taxa documented on Amazonian fish have also been the subject of other investigations, which detailed the functions of such taxa on other fish models.

For instance, bacteria from the genus *Cetobacterium*, often reported in Amazonian fish guts, are from the Fusobacteria phylum and showcase a diverse array of potentially key functions for their host organisms. Notably, *Cetobacterium* bacteria possess the ability to synthesize vitamin B12 (Finegold et al., 2003; Sugita et al., 1991;

Tsuchiya et al., 2008), which plays a vital role in fish development (Eichmiller et al., 2016) and has been suggested to potentially contribute to the inhibition of bacterial pathogen growth (Sugita et al., 1996). In addition, *Cetobacterium* could facilitate peptide fermentation (Romero et al., 2014; Tsuchiya et al., 2008) and enhance carbohydrate metabolism (Finegold et al., 2003; Sugita et al., 1991). Additionally, some species within this genus contribute to the decomposition of organic debris, phytoplankton, and zooplankton consumed by fish (Borsodi et al., 2017). Moreover, *Cetobacterium* has shown relationship with parasitic infections which indirectly affect host metabolism (Gaulke et al., 2019; Scheifler et al., 2022). Also, certain bacteria belonging to the Fusobacteria phylum can enhance the host's external protective barrier against pathogens, by promoting increased mucus production and exhibiting anti-inflammatory effects (Andoh et al., 1999; von Engelhardt et al., 1998).

Clostridia is another clade commonly encountered in Amazonian fish guts and is known for synthesizing short-chain fatty acids such as butyrate, propionate, and acetate, which provide energy to intestinal cells, maintain gut integrity, and regulate immune responses. Furthermore, some species from the class Clostridia possess enzymes that enhance nutrient absorption, improve digestion, help prevent pathogenic infections by producing antimicrobial substances, and interact with the immune system by promoting anti-inflammatory responses and strengthening the gut barrier (Boutard et al., 2014; Eichmiller et al., 2016; reviewed by Guo et al., 2020).

Among the Proteobacteria phylum, members of the class Betaproteobacteria could contribute to the prevention of toxic ammonia build-up trough the cycling of nitrogenous wastes. For instance, studies on common carp (*Cyprinus carpio*) and zebrafish (*Danio rerio*) have shown that some Betaproteobacteria, like *Nitrosomonas* species, could potentially convert ammonia to dinitrogen gas in the gill, protecting their host fish from the toxicity of ammonia (van Kessel et al., 2016). In addition, recent research by Sylvain et al. (2023) has revealed that *Aquitela* and *Vogesella*, two other members of the Betaproteobacteria class identified as potential gill endosymbionts, have the potential to influence gill epithelial permeability. These bacteria could play key roles in the regulation of ionoregulatory processes in the gills of Amazonian fish found in ion-poor blackwater environments. Indeed, the study from Sylvain et al. (2023) showed that the expression of the ATP-sensitive K^+ transporter of black piranha (*S. rhombeus)* was associated with the expression of ArgT (an ABC lysine/arginine/ornithine transporter) by *Aquitalea* bacteria, and with the expression of the porin porB by *Aquitalea* and *Vogesella* bacteria. Another study conducted on the skin mucus of black piranha (*S. rhombeus*) and flag cichlid (*M. festivus*), two species that are phylogenetically divergent and exhibit contrasting ecology, showed that the *Reyranella massiliensis* and *Roseiarcus fermentans* (Alphaproteobacteria) bacteria were consistently found in their skin mucus, exhibiting varying abundance depending on the water type (higher abundance in black—than in whitewater), while being absent in the bacterioplankton community. These niche specificities suggest a potential role in adapting to water types in the Amazon River, although their exact function remains to be discovered (Sylvain et al., 2019).

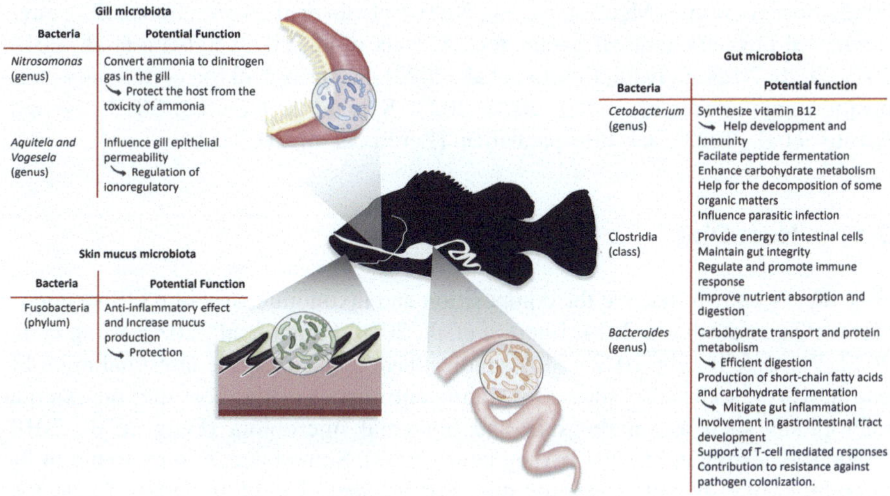

Fig. 1 Diagram representing Amazonian fish microbiomes from different tissues, the main bacterial taxa detected, and the potential functions of these taxa reported from the literature. Original drawing by Alizée Thomas

The Bacteroidetes phylum, specifically the Bacteroides genus, commonly found in Amazonian fish gills and guts, plays a vital role in various functions essential to the host metabolism. These include carbohydrate transport and protein metabolism, which are crucial for efficient diet digestion (Karlsson et al., 2011). Bacteroides are also known for their production of short-chain fatty acids, involvement in gastrointestinal tract development, support of T-cell-mediated responses, and contribution to resistance against pathogen colonization. They also play a significant role in mitigating gut inflammation through short-chain fatty acid production and carbohydrate fermentation (Den Besten et al., 2013).

The phylogeny and functions of bacterial taxa reported in Amazonian fish microbiomes from different tissues are summarized in Fig. 1.

3 Factors Driving the Structure of Fish Microbiotas and Reports on Amazonian Fish

The general factors that interfere in the assembly of fish microbiotas have been the subject of numerous studies and have been reviewed in Llewellyn et al. (2014), Wang et al. (2018), and Luan et al. (2023). However, on Amazonian fish, only a handful of studies investigated the effect of host-specific and environmental factors on fish microbiotas. Up to now, the impact of five main factors (two host-specific, two environment-specific, and one host X environment-specific factor) on Amazonian fish microbiota has been studied: host diet (Di Maiuta et al., 2013; McDonald et al., 2019; Sylvain & Derome, 2017; Sylvain, Normandeau, et al.,

2022), host genotype (McCauley et al., 2020; Sylvain et al. 2019, 2020, 2022), environmental physicochemical parameters (Leroux et al., 2023; Sylvain et al., 2016, 2019, 2020, 2023; Sylvain, Leroux, et al., 2022), bacterioplankton community composition (Sylvain et al., 2019, 2020, 2023; Sylvain & Derome, 2017; Sylvain, Leroux, et al., 2022), and host parasitism (Leroux et al., 2023).

3.1 Host Diet

Host diet is known to drive the composition and taxonomic structure of fish microbiotas (Clements et al., 2014; Egerton et al., 2018; Parata et al., 2020; Wang et al., 2022; Wong & Rawls, 2012). Most studies have focused on the intestinal microbiota, however, some also showed a significant impact of the host diet on external microbiotas such as on the skin and branchial microbiota (Feng et al., 2010; Landeira-Dabarca et al., 2013; Yang et al., 2007). Some taxa are only found in the microbiota of fish with a specific diet. For instance, Liu et al. (2016) found that bacteria from the *Cetobacterium* genus were only detected in carnivorous fish, while those from the *Clostridium* genus were only present in herbivorous fish. Additionally, Sullam et al. (2012) documented that omnivorous fish generally have gut microbial communities that more closely resemble those of carnivorous fish than those of herbivorous species; for instance, Desulfovibrionales and Aeromonadales were abundant in omnivorous and carnivorous species, while Clostridiales, Bacteroidales, and Verrucomicrobiales dominated in herbivorous species. Moreover, studies have shown that not all fish microbial communities are equally affected by the host trophic level: Autochthonous (resident) and allochthonous (transient) microbiota do not show the same sensitivity. Autochthonous communities can adhere to fish mucosal surfaces and can sometimes colonize epithelial tissues where they are intimately bound to host cells (Lowrey et al., 2015; Ringo et al., 2016). In fish guts, autochthonous communities are resistant to the acidic environment created by bile and stomach acids (Nayak, 2010). Autochthonous communities are strongly driven by the host fish genetics and can play key roles in the metabolism of fish-microbe systems (Banerjee & Ray, 2017; Kim et al., 2007; Wang et al., 2010). Allochthonous communities differ, as they are usually non-adherent and colonize the host only temporarily, especially in times of stress (Banerjee & Ray, 2017; Olsen et al., 2005). Unlike autochthonous communities, the allochthonous microbiota is mostly driven by environmental factors (Sylvain et al., 2016, 2020). For instance, allochthonous communities from the gut are mainly found in the alimentary bolus and their composition varies according to the diet and the trophic level of the host fish (Ringo et al., 2016).

Reports on Amazonian Fish The effect of host diet on Amazonian fish microbiotas has been studied on fish from the Cichlidae, Serrasalmidae, and Loriicaridae families, and all investigations focused on the gut microbiota (McDonald et al., 2019; Sylvain & Derome, 2017; Sylvain, Normandeau, et al., 2022). A study on discus fish (*Symphysodon aequifasciata*), a Cichlidae kept in an aquarium setting, has shown that the fry diet plays a major role in structuring the assembly of the early stages of the gut microbiota ontogeny. Indeed, the main source of bacteria for larval fish coincided with the diet of the host fish; in very early life stages, the bacterioplankton taxa were the main taxa found in discus fry. Then, the main intestinal taxa shifted to symbionts recruited from their parents' skin mucus, which provided fry with a unique food source for the first 3 weeks. After the parental feeding stage, the main intestinal taxa were recruited from standard fish food (Sylvain & Derome, 2017).

A recent investigation of four phylogenetically close species of the Serrasalmidae clade (*Serrasalmus rhombeus, Serrasalmus maculatus, Mylossoma duriventre, Pygocentrus nattereri*) has shown that gut microbiotas vary according to diet at the inter-specific level (Sylvain et al., 2023). However, this diet X gut microbiota correlation was not detected at the intra-specific level, potentially due to the stochasticity of the diet of these opportunistic predators.

Another study was focused on the Amazonian catfish *Panaque nigrolineatus*, a Loriicaridae that can switch between xylivorous and detritivorous dietary strategies depending on environmental conditions (McDonald et al., 2019). After exposing *P. nigrolineatus* to both diets, the authors found that the gut microbiota from this species selects enteric community members based on function (e.g., to sustain a xylivorous metabolism), rather than through vertically transferred symbiotic relationships. This functional selection strategy suggests that fish-microbe systems can instantly adapt their metabolic potential, through microbiota restructuration, when dietary switches are needed to survive in variable environments. This phenomenon (discussed further later in this chapter) is termed as metagenomic plasticity (Alberdi et al., 2016) where recruitment of alternative microbial symbionts allows the host to gain instantly new microbial functions.

3.2 Host Genotype

Host-specific factors exercise some control over the selection of members of their microbiota before hatching: Receptors on the surface of the chorion, which interact with environmental microbes, are specific to the genotype of the host fish (Hansen & Olafsen, 1989; Llewellyn et al., 2014). The fish phylogeny significantly influences the structure of the skin, gut, and gill microbiotas (Chiarello et al., 2018; Larsen et al., 2013; Sylvain et al., 2020; Wang et al., 2010). Fish show strong inter-specific differences in their microbiota. For instance, Pratte et al. (2018) found a strong influence of host phylogeny in the structuring of the gill bacterial communities from 53 reef fish species. However, the attribution of intra-specific differences in microbiota to genotype variations between hosts from the same species has shown divergent results in fish. Studies on lake whitefish (*Coregonus clupeaformis*), brook trout (*Salvelinus fontinalis*), and Atlantic salmon (*Salmo salar*) microbiotas showed a significant influence of the fish genotype (Boutin et al., 2014; Sevellec et al., 2018; Webster et al., 2018). However, other studies showed a non-significant influence of the latter (Bledsoe et al., 2018; Chiarello et al., 2019; Riiser et al., 2020).

Reports on Amazonian Fish Several studies suggest that the microbiota of Amazonian fish might be significantly affected by the host genotype, as they show that different fish species sharing the same environment have contrasting microbiotas. This has been shown for flag cichlids (*Mesonauta festivus*) and black piranhas (*Serrasalmus rhombeus*) (Sylvain et al., 2019, 2020), and more recently in a study that included the two aforementioned species in addition to the freshwater sardine (*Triportheus albus*) and the peacock bass (*Cichla* sp.) (Sylvain et al., 2023). However, in these studies, the effect of the host genotype is confounded with other host-specific effects such as the fish diet.

A study focused on four Serrasalmidae with similar diets has shown a significant influence of the host genotype on fish microbiotas at the inter-specific and at the intra-specific level (Sylvain, Leroux, et al., 2022; Sylvain, Normandeau, et al., 2022). This influence was stronger for skin bacterial communities than for those from the gut.

The influence of intra-specific genotype variations on fish microbiotas has been studied further on the flag cichlid model (Sylvain, Leroux, et al., 2022; Sylvain, Normandeau, et al., 2022). This recent study documented a significant correlation between flag cichlid genotypes and gill microbiotas; however, it also revealed that the influence of genotypic factors was weak in comparison with environmental factors in shaping gill communities.

3.3 Environmental Physicochemistry

Numerous studies have investigated the association between environmental physicochemical conditions and the structure of fish microbiotas (Boutin et al., 2013; Dehler et al., 2017; Legrand et al., 2018; Lokesh & Kiron, 2016; Zhang et al., 2016). For instance, salinity is known to be an important driver of fish microbiotas: Saltwater fish generally show an enrichment in Vibrionales, while freshwater fish have an increased abundance of Aeromonadales and Enterobacteriales (Legrand et al., 2018). In the euryhaline Nile tilapia (*Oreochromis niloticus*), the proportion of pathogens increases when fish are exposed to hypo- or hyper-saline stress (Zhang et al., 2016). The environmental dissolved oxygen level is also a factor commonly associated with the structure of fish microbiota: A study on brook trout (*Salvelinus fontinalis*) showed that hypoxia led to a loss of beneficial and commensal bacterial symbionts (e.g., *Methylobacterium, Sphingomonas*, and *Propionibacterium*) (Boutin et al., 2013).

Reports on Amazonian Fish Recent studies on Amazonian fish skin and gut microbiomes (from *Mesonauta festivus* and *Serrasalmus rhombeus*) revealed the existence of significant correlations between the relative abundance of specific bacterial taxa and the values of several physicochemical parameters, including the hardness, pH, and the concentrations of dissolved organic carbon (DOC), sodium, magnesium, potassium, chlorine, and calcium (Sylvain et al., 2019, 2020, 2023).

Among the effects of all those factors on fish skin and gut microbiotas, the effect of environmental pH is one of the most studied. For instance, a study on the tambaqui fish (*Colossoma macropomum*), an Amazonian fish raised in aquaculture in Brazil, showed that environmental acidification led to a restructuring of skin and gut microbiotas (Sylvain et al., 2016). In this study, the microbial community resistance and resilience to acidification were tissue-specific.

The microbial community from another tissue, the fish gills, was also previously studied. Two recent studies (Sylvain et al., 2023; Sylvain, Leroux, et al., 2022; Sylvain, Normandeau, et al., 2022) documented the physicochemical parameters driving the gill microbiota structure of four Amazonian fish species (*M. festivus, T. albus, S. rhombeus,* and *Cichla* sp.). They showed that conductivity, concentration of silicate, chlorophyll a, aluminum, and DOC were significantly associated with gill microbiota structure. DOC was an especially important component, since its overall concentration and the relative abundances of its different fluorescent fractions (humic-, fluvic-, or protein-like) were strongly associated with the gill microbiota for these four species. DOC has been shown to interact directly with (and to even attach to) fish gills (Campbell et al., 1997), and Amazon river bacteria found on fish gills, such as *Polynucleobacter* (Sylvain et al., 2023), are also known to transform/degrade DOC; thus, the interaction between the gill microbiota and DOC will merit further investigation.

3.4 Bacterioplankton

Just like fish skin, gut, and gill microbiotas, bacterioplanktonic communities are also highly sensitive to environmental physicochemistry (Niño-García et al., 2016), and due to their close proximity with external mucous membranes of fish, they could play critical roles in the structuring of fish microbiota. Bacterioplankton plays key roles in fish microbiota assembly, especially in the initial recruitment of symbionts during early life stages of the larvae (Lavoie et al., 2021; Sylvain & Derome, 2017). Studies conducted on several species of fish consistently showed that fish egg surface is mostly colonized by the microbial taxa from the surrounding water, which also quickly colonize the fish skin and gut soon after hatching (Romero & Navarrete, 2006; McIntosh et al., 2008; Giatsis et al., 2015; Sylvain & Derome, 2017). Studies on Atlantic cod larvae (*Gadus morhua*) (Bakke et al., 2013) and tilapia larvae (Giatsis et al., 2015) have shown that fish microbiota is similar in composition to bacterioplankton during the early life stages, although some vertical intergenerational transmission may also affect these communities (Beemelmanns et al., 2019; Hansen & Olafsen, 1999; Long et al., 2014; Sylvain & Derome, 2017). Other reports indicate that the bacterioplankton can also affect the microbiota of adult fish, for example, through the secretion of cyanotoxins (Duperron et al., 2019; Gallet et al., 2021).

Reports on Amazonian Fish Studies on two species of Amazonian Cichlidae are especially interesting on that matter. A first investigation, conducted on the discus fish (*S. aequifasciata*), showed that fry microbiota was almost entirely composed of bacterioplanktonic taxa during the first few days after hatching (Sylvain & Derome, 2017). A second study, focused on adult specimens of flag cichlids (*M. festivus*), revealed that, among several host- and environmental-specific factors (e.g., bacterioplankton, host genotype, and environmental physicochemistry), the most important factor driving fish gill microbiotas was the bacterioplankton structure.

3.5 Host Parasitism

The impressive diversity of fishes and the various habitats in the Amazon River come along with a great diversity of parasites (Baia et al., 2018). Parasites are organisms that depend on a host organism for their survival and reproductive success (Thatcher, 1991). Parasites obtain nutrients and resources from the host and can manipulate the host's behavior, physiology, and immune responses (Gonçalves et al., 2016; Tavares-Dias et al., 2017; Thatcher, 1991). Some parasites require different intermediate hosts to complete their life cycle, while others can remain within a single host species (Baia et al., 2018; Thatcher, 1991). The major parasite groups found in the neotropical zone include Protozoa, Myxozoa, Ciliophora, Platyhelminthes, Nematoda, Acanthocephala, Arthropoda, Annelida, and Pentastomida (Baia et al., 2018; Thatcher, 1991).

The establishment and progression of parasitic infections rely not only on the virulence or pathogenicity of the parasite but also on environmental conditions and characteristics of the host (Alcântara & Tavares-Dias, 2015). The host's immune system plays a crucial role in this process, and recent research highlighted the significant influence of the host's microbiota on immune responses (Hahn et al., 2022; Kaplan et al., 2011) and, therefore, the outcome of parasite infections (Gaulke et al., 2019; Leung et al., 2018).

Fish microbiota has the potential to interact with parasites, and certain bacteria have been identified as capable of either preventing or enhancing parasite infections, notably by influencing their growth and development (Gaulke et al., 2019; Hayes et al., 2010; Hennersdorf et al., 2016; Scheifler et al., 2022). For instance, an investigation conducted on the Amazonian pirarucu (*A. gigas*) demonstrated that supplementing the diet with *Enterococcus faecium* resulted in a reduced infection of the ectoparasite *Trichodina* sp. in the fish's cutaneous mucus. Similarly, in a study conducted on bogue (*Boops boops*), it was observed that uninfected fish by the parasite *Lamellodiscus* had significantly higher abundance of *Shewanella*, *Cetobacterium*, and *Vibrio* bacterial species (Scheifler et al., 2022), suggesting a potential relationship between the fish's microbiota composition and parasite infection status. In addition to their immunological functions, certain bacteria have the potential to facilitate the attachment of ectoparasites to fish by secreting specific molecules (Ohashi et al., 2007).

While the microbiota can play a crucial role in preventing parasite infections, it is important to note that parasites themselves can have a significant impact on the composition and functioning of the host's microbiota. Infections caused by parasites can disrupt the balance within the microbial community, resulting in changes in both the diversity and abundance of microorganisms (Gaulke et al., 2019). Indeed, Gaulke et al. (2019) conducted a study which demonstrated that exposure to *Pseudocapillaria tomentosa* leads to alterations in microbial community diversity in zebrafish (*Danio rerio*) gut. The study revealed a negative correlation between the exposure and certain genera such as *Plesiomonas*, *Shewanella*, and *Cetobacterium*. In addition, Hahn et al. (2022) have also highlighted a disturbance of three-spined stickleback (*Gasterosteus aculeatus*) microbiota by *Schistocephalus solidus*, suggesting that this parasite impacts the host microbiome by influencing the host immune system. In infected fish, the study observed a correlation between the expression of genes involved in immunity and the abundance of specific bacteria, namely Synechococcophycideae, Lactobacillaceae, and Pseudanabaenacea (which showed a negative relationship), as well as Vibrionaceae (which showed a positive relationship). Furthermore, they discovered a correlation between the parasite microbiome and host immune gene expression, revealing an additional layer of complexity in understanding the interactions between hosts and parasites. Finally, a study conducted on the grass carp (*Ctenopharyngodon idellus*) demonstrated that the effects of parasites extend beyond the local microbiota in direct contact with the parasite and can also disrupt other microbial communities. For instance, infection of the grass carp (*C. idellus*) gills by *Dactylogyrus lamellatus* resulted as well in gut microbiota disturbances (Wang et al., 2023).

The presence of parasites can induce changes in the composition of the microbiota, leading to dysbiosis and cascading effects on host health and susceptibility to other parasites infections, weakening their immunity (Colin et al., 2022; Gaulke et al., 2019). Furthermore, these infections can promote the presence of pathogenic bacteria. Gaulke et al. (2019) demonstrated that the presence of *P. tomentosa* in zebrafish microbiota was correlated with the presence of *Mycoplasma* sp., a bacteria associated with neoplasms. In fact, some infected fish showed the presence of tumors, prompting the authors to speculate on the implication of the parasite in these tumors. Furthermore, an investigation examining the impact of the salmon lice (*Lepeophtheirus salmonis*) on the skin surface of Atlantic salmon (*Salmo salar*) revealed significant correlations between lice abundance and the prevalence of various potentially pathogenic genera, including *Vibrio*, *Flavobacterium*, *Tenacibaculum*, and *Pseudomonas* (Llewellyn et al., 2017).

However, the disturbance signals induced by the presence of parasites can vary from a case study to another. For instance, in contrast to the findings of Gaulke et al. (2019), which found a negative relationship between parasite infection and *Cetobacterium* abundance, another study revealed a positive relationship between the infection of *Dactylogyrus lamellatus* on *C. idellus* and the presence of *Cetobacterium* (Wang et al., 2023). Additionally, while some studies have reported an increase in diversity following infection (Gaulke et al., 2019; Llewellyn et al., 2017; Rosa et al., 2018; Vasemägi et al., 2017), others have observed a diversity decrease (Colin et al., 2022; Hahn et al., 2022; Wang et al., 2023). These observations are reminders that the interaction between microbiota and parasites is influenced by various factors such as parasite and host phylogenies (Hahn et al., 2022) in addition to environmental factors, as observed in the Amazon (Leroux et al., 2023). To this respect, seasonality of water level was observed to impact parasite load and diversity in the Amazonian Cichlidae *Astronotus ocellatus* (Neves et al., 2013). In addition, fasting and/or diet change in omnivores during the dry season has a stronger impact on fish microbiota composition and significantly stimulates the immune response (Xia et al., 2014), which in turn will impacts the colonization success, replication, and virulence of parasites (Leung et al., 2018), as discussed above. Taken together, the influence of both host-specific and environmental factors on parasitism emphasizes the importance of conducting further studies on host-microbiota-parasite interactions in the Amazon region, which harbors its own specificities. Considering the significant influence of parasites on the structure, size, behavior, feeding, growth rate, and reproduction of fish populations (Hagmayer et al., 2020; Moreira et al., 2009; Tavares-Dias et al., 2014), understanding these interactions is critical for the effective management and conservation of fish populations in the Amazon.

4 The Critical Roles of Amazonian Fish Microbiome in Hostile Blackwater Habitats

On the planet, life is found in a multitude of different conditions, including extreme environments such as hot springs (Roy et al., 2020), the deep sea (Van Dover, 2000), the cold waters of Antarctica (Anesio et al., 2017), streams contaminated with heavy metals (Cheaib et al., 2018; Custodio et al., 2022), and several others. Although the metabolic capacities encoded in metazoan genomes are generally limited to a narrow range of environmental conditions (Nursall, 1959; Portner, 2001), metazoan life is found in the most extreme environments (Weber et al., 2007). Can microorganisms help their eukaryotic hosts survive in such environments? In fact, it turns out that the co-dependency between hosts and their microbiota is particularly emphasized in extreme environments (reviewed in Bang et al., 2018).

In the Amazon, we find such extreme environments, which should exceed the metazoans' metabolic limitations: the blackwater habitats. Amazonian blackwater habitats contain some of the most naturally acidic, ion-poor, and dissolved organic carbon (DOC)-rich waters globally (Furch, 1984; reviewed by Val & Almeida-Val, 1995, and by Wood et al., 1998). Their pH can be as acidic as ~3.0, due to the high input of decaying plant material from the dense surrounding forests (Walker & Henderson, 1996), resulting in high humic DOC concentrations, typically 5–15 mg L^{-1} (Morris et al., 2021). The chemistry of these habitats appears so extreme that one would expect them to be toxic to most freshwater fish, yet the ichthyofauna found in these habitats is one of the most biodiverse on Earth, encompassing approximately 8% of all freshwater fish species (Morris et al., 2021). To adapt to these challenging habitats, fish must rely on ionoregulatory strategies, which are still not well understood and do not fit current models from non-native fish (Morris et al., 2021).

A recent study has made a breakthrough in our understanding of the fish adaptation to blackwater physicochemistry (Sylvain et al., 2023). This study tested a novel hypothesis concerning the role of the gill microbiome in the regulation of the iono-regulatory processes involved in the adaptation to blackwater. To do so, four species of Amazonian teleosts (*M. festivus, T. albus, S. rhombeus,* and *Cichla* sp.) were sampled at 15 sampling sites from a hydrochemical gradient encompassing five blackwater habitats. Gill samples were collected, and a metatranscriptomic analysis was conducted to document the transcriptomes of the hosts and of their gill bacterial symbionts. The results showed that the host physiological responses to blackwaters were host-species-specific, but occasionally included the overexpression of integrins and Toll-receptors, which are cell surface proteins associated with interkingdom communication. Furthermore, gill microbiotas from blackwater fish were characterized by the presence of a transcriptionally active betaproteobacterial cluster, which had the genomic potential to interfere with the host's epithelial permeability (on which rely the integrity of ionoregulatory processes). One of these betaproteobacteria was from the genus *Aquitalea*, which is known to be involved in host-microbe interactions via its production of indole (Williams et al., 2020). Indole can modulate numerous host metabolic and immunological processes (Chimerel et al., 2013). An interesting possibility is that bacterial indole could modulate the transcription of host ion channels (Chimerel et al., 2010) and increase the host epithelial-cell tight-junction resistance (Bansal et al., 2010), which could be beneficial in ion-poor environments such as blackwater habitats.

The study from Sylvain et al. (2023) further explored blackwater fish-microbe interactions in the laboratory, through the analysis of the metatranscriptomes from 320 sterile zebrafish larvae exposed to sterile, non-sterile, and inverted (non-native bacterioplankton) blackwater during 18 h. The results from that experiment showed that sterile zebrafish larvae do not survive well (19% mortality after 18 h) when exposed to sterile/inverted blackwater, but do survive well in non-sterile blackwater (4% mortality). The response of zebrafish larvae to blackwater depended on 31 host genes involved in ionoregulatory processes, but 22 of those genes were not overexpressed on zebrafish in sterile/inverted blackwater, and thus, the host expression of those genes was dependent on the presence of blackwater bacterioplankton. These results suggest an essential role of autochthonous blackwater bacterioplankton taxa in blackwater fish physiology.

Could We Refer to Blackwater Fish as *Holobionts*? The study from Sylvain et al. (2023) revealed the co-dependency relationship between Amazonian blackwater fish and their gill bacterial symbionts. In the last decades, the discovery of the co-dependency between communities of symbiotic microorganisms and a vast number of eukaryotic hosts changed the definition of the biological unit subject to the action of natural selection (from the eukaryotic organism only, to the host and its symbiotic microbial partners): "*Biology has entered a new era with the ability to understand that an organism's genetics and fitness include that of its microbiome*"

(Brucker & Bordenstein, 2014). The redefinition of the unit of natural selection led to the concept of "*meta-organism*" or "*holobiont*" (Bosch & McFall-Ngai, 2011; McFall-Ngai et al., 2013; Rosenberg et al., 2007; Theis et al., 2016; Zilber-Rosenberg & Rosenberg, 2008), which considers that symbiotic communities of microbes are an integral part of the hosts' functional and metabolic potential. Since symbiotic microorganisms are responsible for an essential part of the phenotype subject to the force of natural selection, these microbes influence the fitness of their host organism (McFall-Ngai, 2002; O'Hara & Shanahan, 2006; Fraune & Bosch, 2010; Bordenstein & Theisn, 2015; Deines et al., 2017). Although the concept of "*holobiont*" still remains criticized within the field of microbial ecology, especially regarding the degree of host-microbe co-dependency and the fidelity of the microorganisms involved (Douglas & Werren, 2016), the recent results from Sylvain et al. (2023) support the holobiont theory. Indeed, the results suggest that Amazonian blackwater fish are dependent on the presence of specific bacterial gill symbionts for the host survival in blackwater. The bacteria involved also seem to be dependent on the presence of fish, as they are found only in extremely low abundances in blackwater bacterioplankton, suggesting a strong niche preference for blackwater fish gills. Furthermore, the results from that study showed that blackwater symbionts might confer a rapid adaptation to the blackwater habitat, as shown from the aforementioned zebrafish experiment (Sylvain et al., 2023). This result suggests that these bacterial symbionts might have facilitated the colonization of blackwater habitats. Overall, these evidences point to an important role of bacterial symbionts in supporting the fitness of their hosts in blackwater habitats.

5 Impact of Current and Future Threats on Amazonian Fish-Microbe Systems

Predicting the impact of climate change on Amazonian fish and on their interactions with their microbial partners is undoubtedly a complex question. To begin, it is necessary to review the ongoing direct threats to the various fish habitats in the Amazon basin. The Brazilian Amazon houses some of the world's biggest and most productive aquatic ecosystems, which are currently facing tremendous changes, potentially outpacing their resilience capacities. These already fragile aquatic ecosystems are synergistically threatened by numerous anthropogenic activities including deforestation, dam construction, pollution via agricultural runoff, oil extraction, and mining, in addition to climate change. The negative impacts of these activities may also be worsened by climate fluctuations such as El Niño and La Niña (Moura et al., 2019). Furthermore, the high hydrological connectivity during the rainy season in Brazil (7 months) has exacerbated the impacts of riverine metropolitan area, which harbors more than 60% of the total population (e.g., Manaus, Belém, Santarém).

5.1 Deforestation

Accelerated industry development has tremendous negative impacts on aquatic ecosystem productivity, thus inducing further the migration and/or conversion of Indigenous fishermen from doing highly sustainable economic activities such as traditional fishing to short-term activities like *slash-and-burn* agriculture (e.g., as observed extensively in the Brazilian state of Para). Such economic shift is thus amplifying deforestation, which will further impact the aquatic ecosystem integrity and productivity. To that respect, deforestation impacts water physiochemistry by decreasing evapotranspiration which in turn increases water runoff and stream discharge (Hayhoe et al., 2011). In addition, the loss of tree roots promotes soil erosion, which combined with an increased release of terrestrial sediments into the rivers (Neill et al., 2001), will exert a tremendous negative impact on fish habitats at regional scales. Sediment release can be expected to reduce water quality, and loss of canopy will increase water temperature, and as a primary driver of microbial assemblages (Freixa et al., 2017), will alter higher trophic level biotic assemblages and thus ecosystem productivity. It should be emphasized here that the overall metabolism of the aquatic ecosystem and the potential transfer of energy to higher trophic levels are tightly linked to the balance between primary production of the forest and respiration of aquatic organisms, as river systems are net heterotrophic (Gagne-Maynard et al., 2017). Therefore, deforestation will inevitably lead to an increase in CO_2 in the rivers, which is already particularly high in acidic black waters, potentially driving fish communities beyond the limit of their physiological capacity with regard to hypoxia and/or ionoregulation (Sylvain et al., 2023). In terms of habitat structure, fish assemblages will be strongly affected by the removal of trees that no longer provide shelter for aquatic biota during the rainy season (Neill et al., 2001; Williams et al., 1997). In addition, the highly productive plant communities from floodplains that sustain abundant fish populations (e.g., species such as the omnipresent flag cichlid, which feeds on and hides in floodplain plants (Pires et al., 2015)) during the wet season will be altered in terms of abundance and diversity following deforestation (Melack & Forsberg, 2001). Furthermore, previous reports have shown that in some area (e.g., Tocantins and Araguaia basins), deforestation rates were strong enough to move the flood pulse by 1 month in local rivers (Coe et al., 2009; Costa, 2004), thus severely perturbating the annual synchronicity with the reproduction season of hundreds of fish species (Jiménez-Segura et al., 2016). For instance the tambaqui, one of the most important species for Amazonian traditional fishery, performs an annual migration tightly coordinated with the seasonal flood pulse (Goulding & Carvalho, 1982), which is threatened by increasing deforestation. The magnitude of these threats can be worsened or mitigated by the climatic phenomenon of El Niño and La Niña, respectively. For instance, evapotranspiration is severely affected, either decreased by La Niña phenomenon, or increased by El Niño (Moura et al., 2019).

Initiatives to control deforestation have been implemented in the Brazilian Amazon, notably through the Native Vegetation Protection Law (Forest Code) which regulates the clearing of forests on private lands, or the Action Plan for the

Prevention and Control of Deforestation in the Legal Amazon [*Plano de Ação para a Prevenção e Controle do Desmatamento na Amazônia* (PPCDAm)] which accounted for a 37% decrease in deforestation between 2005 and 2007 (reviewed in Garret et al., 2021).

5.2 Dam Construction

Like deforestation, the construction and operation of dams and waterways impacts fish habitat structure and the water physiochemical parameters essential to aquatic life, such as temperature and oxygen. Dams disrupt longitudinal river connectivity with tributaries and floodplains, stream hydrodynamics, especially upstream, forest submergence, and/or active deforestation, the combined action of which leads to the alteration of sediment transport dynamics and longitudinal and lateral fish migrations (Agostinho et al., 2008; Consoli et al., 2022; Poff & Hart, 2002). The lifecycle of several Amazonian fish depends on their capacity to migrate over long distances, such as the Goliath catfish which migrates over an impressive stretch of >11,000 km over 2 years to spawn at the foothill of the Andes (Barthem et al., 2017). Fish passages are usually the principal means for minimizing impacts on migratory stocks; however, in the neotropics they have unfortunately proven to be unsuccessful (Pompeu et al., 2012) and even harmful (Pelicice & Agostinho, 2008).

Furthermore, the establishment of waterways during dam construction involves the removal of natural barriers (e.g., boulders, submerged trees, attenuation of river meanderings). Such habitat transformation will necessarily exert greater disturbances in small streams and associated floodplains, where hydrodynamic shifts destroy many fragile breeding sites, thereby significantly decreasing the taxonomic diversity of fish and other aquatic organism assemblages both upstream and downstream of dams (Agostinho et al., 2008). Upstream, the building of a reservoir establishes a stratification of water layers leading to the formation of hypoxic area and increases temperature and sediment deposition. Downstream, fish habitat is degraded from reduced disturbance in terms of flow variability and sediment supply (Consoli et al., 2022).

5.3 Agriculture

The practice of *slash-and-burn* agriculture involves the burning of forests to clear land and fertilize the nutrient-poor Amazonian soil. In addition to the deforestation and loss of habitats associated with this activity, modern riverine agriculture releases into freshwater ecosystems large amounts of nitrogen and phosphorus waste from fertilizers and many toxic molecules derived from phytosanitary agents (Williams et al., 1997). Once again, these threats exert greater disturbances in small fish habitats: on one hand, nitrogen and phosphorus waste loading increases primary production, creating algal blooms, which in turn establish hypoxic conditions (Neill et al., 2001); on the other hand, pesticides bioaccumulate in food webs. Both of these stresses strongly alter food web structures and, as endocrine disruptors, alter fish and aquatic invertebrates' reproduction (Yang et al., 2021) and survival. For instance, a study on the surubim (*Pseudoplatystoma* sp., a large Amazonian catfish) exposed juvenile specimens to 15 mg L^{-1} of glyphosate-based herbicide and showed a 50% mortality rate after 96 h of exposition (Sinhorin et al., 2014).

Different, more sustainable, practices than *slash-and-burn* agriculture were once the norm in the region: Evidence from pollen records has shown that ancient Amazonian populations utilized raised-bed farming (crops planted in mounds to protect against flooding and improve soil fertility) (Iriarte et al., 2012). Common sustainable agriculture practices that would reduce the need for chemical fertilizers and herbicides include polyculture, crop rotation, and the use of animal manures.

5.4 Mining

Another pollutant is of particular concern because of the harm it can cause to riparian populations: mercury (Hg). This metal is particularly abundant in the upper soil layer of the Rio Negro (blackwater tributary of the upper Amazon) and is released into streams as a result of leaching from deforestation, or directly when used to extract gold (Pfeiffer & Lacerda, 1988) and, to less extent, from atmospheric transport and deposition (Gerson et al., 2022). These anthropogenic activities result in a massive release of mercury into fish habitats. Mercury is converted by microorganisms to methylmercury (MeHg), which is then bioaccumulated throughout the aquatic food chain, including fish. Factors such as high bacterial activity, low pH, low conductivity, and high concentrations of dissolved organic matter promote mercury methylation (Lázaro et al., 2018). Because these factors are combined in the Rio Negro river, the highest levels of MeHg have been measured in this basin.

A study on Amazonian fish have found that Hg exposure caused modifications in the activity of superoxide dismutase (SOD), catalase (CAT), glutathione S-transferase (GST), and GPx in trahira (*Hoplias malabaricus*) (Mela et al., 2014). Another study has shown that Hg exposure induced significant alterations in the expression of the antioxidant enzyme genes sod, cat, gst, gpx, and gr in matrinxã (*Brycon amazonicus*), which lead to oxidation of lipids and proteins (Monteiro et al., 2010).

A study from Eissa et al. (2012) investigated fish-parasite-metals interactions and suggested that parasites can help their host fish to survive in heavy metal-polluted environments by collecting larger quantities of heavy metals and therefore acting as metal sinks. The role of fish parasites in the mitigation of the pervasive effects of MeHg in blackwater fish merits further investigation.

5.5 Oil Exploitation

Another major source of pollution is oil exploration, which has developed significantly in the last decades in the northwestern Amazon. In the Ecuador, Peru, and

Colombia, oil and gas exploration and exploitation resulted in spills of crude oil, drilling mud, and other toxic components that have polluted large areas of rainforest, without sparing Amazon waterways. For instance, the Ecuadorian Amazon has become one of the world's major sites of recurrent oil pollution, involving substantial amounts of toxic waste and crude oil (Finer et al., 2008; Jernelöv, 2010), contaminating nearby waters with concentrations of hydrocarbon-related toxins well above the toxicity thresholds allowed by North American or European regulations, leading to human health problems (Sebastian & Hurtig, 2004). Finally, ships that transport crude oil and derivatives are potentially prone to accidental spills in the lower Amazon (Araújo et al., 2023).

Studies on Amazonian fish have shown that exposition to crude oil affected the gas exchange and ion regulation capacity of Urucu (*Hoplosternum littorale*) (Brauner et al., 1999), and exposition of the tambaqui (*Colossoma macropomum*) to phenanthrene, a polycyclic aromatic hydrocarbon, caused significant disturbances of physiological functions (decrease growth rate, blood hemoglobin, in addition to increased lactate and plasma glucose) (Chávez-Veintemilla & Val, 2019).

5.6 Global Effect of Anthropogenic Threats on Amazonian Fish-Microbe Systems

Although many additional studies are needed to anticipate the medium- and long-term effect of currently observed threats to the Amazon freshwater ecosystem, ecological theory points to the alteration of natural hydrology as the primary threat (Junk, 1989; Vannote et al., 1980). We argue that hydrological alterations in the Amazon basin come primarily from two sources: large-scale deforestation, which significantly alters river flow and flood magnitude (Coe et al., 2009), and dam construction, which reduces flood magnitude (Poff & Hart, 2002). Alterations caused by these two sources can be amplified or mitigated by the El Niño and La Niña climate events, respectively (Moura et al., 2019). El Niño reduces regional precipitation and river flow while increasing the frequency of extreme droughts (Malhi et al., 2009). Contrastingly, La Niña would increase precipitation; however, its impact on temperature is not clear (Hua et al., 2017). Taken together, these hydrological alterations could reduce the magnitude of floods and increase the frequency and severity of low-flow events (Costa, 2004). By disrupting the lateral migration of commercially exploited fish, these hydrological alterations could threaten the food security of riparian populations (Isaac et al., 2015; Jackson & Marmulla, 2001).

Therefore, it is expected that higher occurrence of El Niño phenomenon combined with current threats will affect fish physiology, both in terms of oxidative stress and ionoregulation, due to increases of temperature, DOC, and MeHg, for example. These environmental shifts would increase Amazonian fish cardiac rhythm (van der Walt et al., 2021), modulate their growth speed (Lindmark et al., 2022), and alter their digestive processes and energy storage (Volkoff & Rønnestad, 2020). The host physiology is known to have an impact on the structure of host-associated microbial communities (Uren Webster et al., 2020), and vice-versa (Butt & Volkoff,

2019). However, these microbial communities respond not only to the host, but also to the environmental conditions (e.g., water physicochemistry) which will also vary according to anthropogenic perturbations. Thus, due to the complex nature of fish-microbe-environment interactions, one cannot predict precisely the future impact of the increasing anthropogenic perturbations on Amazonian fish-microbe systems. Experiments in controlled environments (e.g., microcosms that can create the predicted environmental conditions according to specific physicochemical simulations), where fish-microbe metatranscriptomes are studied in control and test environments, would be one way to move forward.

Furthermore, as previously detailed in the Sects. 3.1–3.5 of this chapter, Amazonian fish microbiotas are sensitive to numerous host- and environment-specific factors (Fig. 2). The aforementioned anthropogenic perturbations can have an impact on several factors known to drive fish microbiotas, such as the host genotype, the host diet, the bacterioplankton community composition, the environmental physiochemical parameters, and the host parasitism. The frequency of different host genotypes could change in a population due to temperature change (Rank &

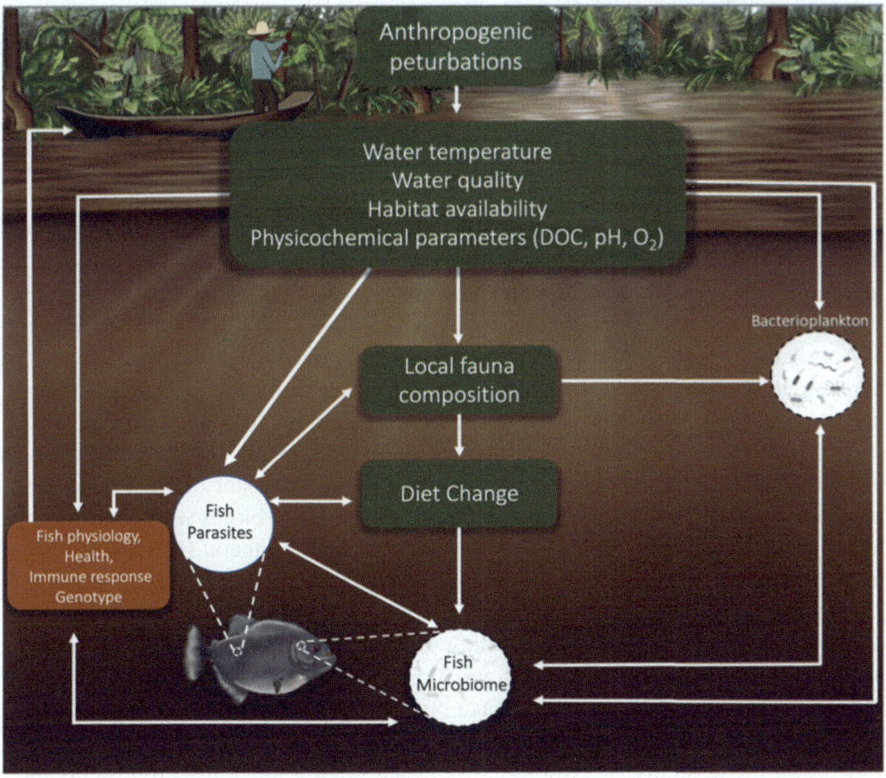

Fig. 2 Summary of the effect of anthropogenic perturbations on several processes associated with the composition or functional potential of fish-microbe systems in the Amazon. Original drawing by Alizée Thomas

Dahlhoff, 2002), e.g., due to a spatial redistribution of the different fish populations in the Amazonian basin, or due to a population being favored compared to others with a decreased fitness in the new environmental conditions. The construction of dams is also expected to have a significant impact on fish migration capacity/genetic flow, which could drive new population structures in the Amazon. The host diet could also vary according to anthropogenic disturbances. Changes in environmental physicochemistry could lead to a spatial redistribution of the food sources (animal preys or plant) or to a change in metabolic/energetic needs. For instance, Hardison et al. (2021) studied the opaleye (*Girella nigricans*) and showed that the magnitude of thermal acclimation responses in fish is associated with the fish diet. Finally, the composition of Amazonian bacterioplanktonic communities is also highly sensitive to environmental physicochemistry (Sylvain et al., 2023). The results of Apple et al. (2006), collected from another bacterioplanktonic community, reveal that a temperature increase can also have an impact on bacterioplankton production, respiration, and growth efficiency. The study from Sylvain et al. (2023) also showed that the main factor driving Amazonian bacterioplanktonic communities is the abundance and type of environmental DOC, which is expected to be affected by anthropogenic perturbations such as dam construction, deforestation, and increases in temperature associated with global climate change. Changes in water physicochemistry associated with these changes will probably have major effects on fish microbiotas. A recent study of Ghosh et al. (2022) on the microbiota of the Chum Salmon (*Oncorhynchus keta*) exposed to increased temperatures showed that temperature changes triggered a significant dysbiosis (disequilibrium) in the fecal and skin mucus microbiota. The genus *Vibrio*, which is typically pathogenic, showed an increased abundance at higher temperatures, and the authors suspected the opportunistic growth of this pathogenic species to impede host immunity. Whether an increase in temperature translates in an increase in the relative abundance of pathogenic taxa in Amazonian fish microbiotas would need to be investigated further.

Another potential effect of environmental physicochemistry changes on Amazonian fish microbiota would be a change in alpha diversity. Indeed, environmental perturbations such as temperature increases or pH decreases are expected to cause physiological stress to Amazonian fish, and stress is associated with a dysbiosis of the fish microbiota (Sylvain et al., 2016; Uren Webster et al., 2021). Fish-microbe systems in a dysbiosis state typically harbor microbiotas with a lower alpha diversity than healthy individuals, as shown on bronze gudgeon (*Coreius guichenoti*) (Li et al., 2016), crucian carp (*Carassius auratus*) (Li et al., 2017), and ayu (*Plecoglossus altivelis*) (Nie et al., 2017). A possible explanation for this pattern is that under the state of dysbiosis, invading opportunistic pathogens out-compete the gut commensals, thereby reducing diversity. Dysbiosis can also translate in the rise of previously rare and opportunistic taxa both in water and fish (Cheaib et al. 2020, 2021), which could also translate into a rise in alpha diversity. Several studies have shown that the diversity of symbiotic microbial communities directly affects their protective effects on the host (De Schryver & Vadstein, 2014; Johnson, 2008), thus a change in the alpha diversity of fish microbiotas due to climate change could weaken the host's first barrier of defense against pathogens and affect the overall host immunity.

Finally, the nature and occurrence of fish parasitism is another important aspect associated with the health of fish-microbe systems that will also be affected by the aforementioned anthropogenic threats, in addition to climate change. As temperatures increase, parasites are believed to exhibit greater virulence and higher population densities due to accelerated growth and maturation rates (reviewed by Budria & Candolin, 2013). In addition, temperature changes can modify the expression of immune genes and overall decrease host immune resistance in fish. Indeed, Da Costa and Val (2020) observed that exposure to higher temperatures led to an increased parasite load and inflammation in the tambaqui (*Colossoma macropomum*), a commercially important fish species in the Amazon basin and the second most produced in Brazil (IBGE, 2022). Other conditions, such as eutrophication, can favor parasites, resulting in a general trend of increased parasite virulence and densities, particularly among generalist parasites (reviewed by Brunner & Eizaguirre, 2016). Furthermore, the composition of local fauna can undergo changes due to variations in the resilience of fish species (reviewed by Val & Wood, 2022) under environmental changes. These changes in local composition and host density can impact the diet of individuals and availability of intermediate hosts, thereby altering the dynamics of parasite infections (Baia et al., 2018). Furthermore, climate change can induce changes in host behaviors, such as alterations in foraging strategies or habitat preferences, which can subsequently impact the encounters between hosts and parasites (reviewed by Budria & Candolin, 2014 and by Brunner & Eizaguirre, 2016).

Specifically, the Amazon River water is anticipated to undergo various changes due to anthropogenic perturbations and climate change, including higher temperatures, increased ion levels, elevated CO_2 levels, and lower oxygen levels (reviewed by Val & Wood, 2022). These changes are expected to have direct effects on fish physiology and influence the microbiota composition (Sylvain et al., 2016, 2019; Vasemägi et al., 2017). As microbiota plays a vital role in fish resistance against parasitic infections, alterations in its composition, resulting from changes in environmental microbial populations, environmental physicochemistry, and the host's physiological response to these stressors, could increase the susceptibility of Amazonian host-microbe systems to parasite infections.

Higher parasite load can have harmful consequences for both fish populations and the human communities that depend on them. Understanding the complex interplay between fish microbiota and parasites in the Amazon basin is crucial because of the great ecological and economic importance that fish communities have as the primary source of high-quality protein for both super predators (e.g., predatory fishes, dolphin, alligators, and osprey) and riverine fishermen communities (Dufour et al., 2016).

5.7 The Microbiota, a Tool That Will Facilitate Fish Adaptation to Climate Change?

The genetic adaptation of diploid or multiploid multicellular eukaryotes is relatively slow, since following the emergence of new genetic variants, it takes several generations before an advantageous allele reaches high frequencies in a population (Urban et al., 2014). A faster response can be obtained by phenotypic plasticity (the ability of a single genotype to alter its expression in order to express different phenotypes in response to environmental pressures (Pigliucci, 2005; Scheiner, 2013), or by epigenetic modifications (alterations of DNA methylation profiles) (Stankiewicz et al., 2013). However, following environmental changes, the fastest response of host-microbiota systems comes from the symbiotic microbes (Bang et al., 2018). The speed of this particular response is directly linked to the very short generation time of bacteria, which can be as short as 20–60 min for several species (Kaiser, 2021). By altering their metabolism through microbiota restructuring, rather than depending on genetic mutations and recombination events on host DNA, host-microbiota systems can respond quickly and efficiently to changing environmental conditions (Rosenberg & Zilber-Rosenberg, 2016). Alberdi et al. (2016) described this concept as the *metagenomic plasticity*, and defined it as the ability of a microbial community to alter its composition or gene expression according to its host physiological and environmental variations.

Although plasticity has its own limits and certain costs (Murren et al., 2015), variable environments often lead to a positive selection of phenomes with high plasticity (Scheiner, 2013), since these are more likely to develop a phenotype with a high fitness corresponding to the new environmental conditions, in comparison with less plastic phenomes. The concept also applies to metagenomic plasticity; plastic microbiota are flexible and thus allow microbial community rearrangements that can provide greater adaptive capacity in new environmental conditions.

Productive freshwater ecosystems are bacteria rich; the bacterioplankton of several lakes shows bacterial densities varying between $1.3 \pm 0.7 \times 10^9$ and $22.0 \pm 4.7 \times 10^9$ cells per liter (Teixeira et al., 2011) as well as a great diversity of taxa (e.g., 24 different phyla detected in some Chinese lakes (Li et al., 2017)). Thus, the bacterioplankton constitutes an important and diversified genetic pool that can potentially allow rapid adaptation of fish-microbiota systems, if they have the capacity to recruit beneficial strains from the environment. Thus, a high level of microbiota plasticity, varying in tandem with bacterioplankton, could potentially constitute an evolutionary advantage allowing fish-microbiota systems to adapt rapidly to changing environmental conditions (Alberdi et al., 2016). The involvement of bacterial symbionts in Amazonian fish acclimation to physiologically challenging conditions characteristic of perturbed ecosystems merits further investigation in controlled environment experiments.

6 Conclusion

In this chapter, we highlighted the crucial role of microbiota in fish immunity, digestion, development, physiology, and their potential to help in survival in extreme environments like black water in the Amazon River. Although the microbiota of fish from the Amazon region shares taxonomic similarities at the phyla level with fish from other biomes, there is still limited information at the species-level and associated functions; undoubtedly, the Amazon fish microbiota harbors its own distinct specificities. Various factors including diet, genotype, environment, and parasites influence the composition of fish microbiota. However, anthropogenic perturbations such as deforestation, dam construction, pollution, and climate change can disrupt these factors, impacting the host microbiota in addition to the fish host physiology. While it is challenging to precisely predict the effect of anthropogenic perturbations on complex fish-microbe holobionts, the rapid evolutionary rate of bacteria could potentially facilitate the host's adaptation to these environmental shifts. Research on the Amazon fish microbiome is still in its early stages, with much left to uncover about its composition and function. Further studies are needed to better understand the interactions and functions of Amazon fish microbiota, which are not only ecologically important but also holds economic significance due to the vital role of Amazon fish species in supporting local livelihoods and fisheries.

References

Agostinho, A. A., Pelicice, F. M., & Gomes, L. C. (2008). Dams and the fish fauna of the Neotropical region: Impacts and management related to diversity and fisheries. *Brazilian Journal of Biology, 68*(4), 1119–1132.

Alberdi, A., Aizpurua, O., Bohmann, K., et al. (2016). Do vertebrate gut metagenomes confer rapid ecological adaptation? *Trends in Ecology & Evolution, 31*(9), 689–699.

Alcântara, N. M., & Tavares-Dias, M. (2015). Structure of the parasites communities in two Erythrinidae fish from Amazon river system (Brazil). *Revista Brasileira de Parasitologia Veterinária, 24*, 183–190.

Al-Hisnawi, A., Ringø, E., Davies, S. J., et al. (2015). First report on the autochthonous gut microbiota of brown trout (*Salmo trutta* Linnaeus). *Aquaculture Research, 46*(12), 2962–2971.

Andoh, A., Bamba, T., & Sasaki, M. (1999). Physiological and anti-inflammatory roles of dietary fiber and butyrate in intestinal functions. *Journal of Parenteral and Enteral Nutrition, 23*(5 Suppl), S70–S73.

Anesio, A. M., Lutz, S., Chrismas, N. A. M., et al. (2017). The microbiome of glaciers and ice sheets. *NPJ Biofilms and Microbiomes, 3*, 10.

Apple, J. K., del Giorgio, P. A., & Kemp, W. M. (2006). Temperature regulation of bacterial production, respiration, and growth efficiency in a temperate salt-marsh estuary. *Aquatic Microbial Ecology, 43*, 243.

Araújo, E. P., de Abreu, C. H. M., Cunha, H. F. A., et al. (2023). Vulnerability of biological resources to potential oil spills in the Lower Amazon River, Amapá, Brazil. *Environmental Science and Pollution Research, 30*(12), 35430–35449.

Baia, R. R. J., Florentino, A. C., Silva, L. M. A., et al. (2018). Patterns of the parasite communities in a fish assemblage of a river in the Brazilian Amazon region. *Acta Parasitologica, 63*(2), 304–316.

Bakke, I., Skjermo, J., Vo, T. A., & Vadstein, O. (2013). Live feed is not a major determinant of the microbiota associated with cod larvae (Gadus morhua). *Environmental Microbiology Reports, 5*(4), 537–548.

Banerjee, G., & Ray, A. K. (2017). Bacterial symbiosis in the fish gut and its role in health and metabolism. *Symbiosis, 72*(1), 1–11.

Bang, C., Dagan, T., Deines, P., et al. (2018). Metaorganisms in extreme environments: Do microbes play a role in organismal adaptation? *Zoology, 127*, 1–19.

Bansal, T., Alaniz, R. C., Wood, T. K., et al. (2010). The bacterial signal indole increases epithelial-cell tight-junction resistance and attenuates indicators of inflammation. *PNAS, 107*(1), 228–233.

Barthem, R. B., Goulding, M., Leite, R. G., et al. (2017). Goliath catfish spawning in the far western Amazon confirmed by the distribution of mature adults, drifting larvae and migrating juveniles. *Scientific Reports, 7*, 41784.

Beemelmanns, A., Poirier, M., Bayer, T., et al. (2019). Microbial embryonal colonization during pipefish male pregnancy. *Scientific Reports, 9*, 3.

Bertand, J. C., Caumette, P., Lebaron, P., et al. (2011). *Écologie microbienne: Microbiologie des milieux naturels et anthropisés* (p. 1002). Presses universitaires de Pau et des Pays de l'Adour.

Bledsoe, J. W., Waldbieser, G. C., Swanson, K. S., et al. (2018). Comparison of channel catfish and blue catfish gut microbiota assemblages shows minimal effects of host genetics on microbial structure and inferred function. *Frontiers in Microbiology, 9*, 1073.

Bordenstein, S. R., & Theisn, K. R. (2015). Host biology in light of the microbiome: Ten principles of holobionts and hologenomes. *PLoS Biology, 13*(8), e1002226.

Borsodi, A. K., Szabo, A., Krett, G., et al. (2017). Gut content microbiota of introduced big-headed carps (*Hypophthalmichthys* spp.) inhabiting the largest shallow lake in Central Europe. *Microbiological Research, 195*, 40–50.

Bosch, T. C. G., & McFall-Ngai, M. J. (2011). Metaorganisms as the new frontier. *Zoology, 114*(4), 185–190.

Boutard, M., Cerisy, T., Nogue, P. Y., et al. (2014). Functional diversity of carbohydrate-active enzymes enabling a bacterium to ferment plant biomass. *PLoS Genetics, 10*, e1004773.

Boutin, S., Bernatchez, L., Audet, C., et al. (2013). Network analysis highlights complex interactions between pathogen, host and commensal microbiota. *PLoS One, 8*(12), e84772.

Boutin, S., Sauvage, C., Bernatchez, L., Audet, C., et al. (2014). Inter individual variations of the fish skin microbiota: Host genetics basis of mutualism? *PLoS One, 9*(7), e102649.

Brauner, C. J., Ballantyne, C. L., Vijayan, M. M., et al. (1999). Crude oil exposure affects air-breathing frequency, blood phosphate levels and ion regulation in an air-breathing teleost fish, *Hoplosternum littorale. Comparative Biochemistry and Physiology. Part C, Pharmacology, Toxicology & Endocrinology, 123*(2), 127–134.

Brucker, R. M., & Bordenstein, S. R. (2014). Response to comment on "The hologenomic basis of speciation: Gut bacteria cause hybrid lethality in the genus Nasonia". *Science, 345*(6200), 667.

Brunner, F. S., & Eizaguirre, C. (2016). Can environmental change affect host/parasite-mediated speciation? *Zoology, 119*(4), 384–394.

Budria, A., & Candolin, U. (2013). How does human-induced environmental change influence host-parasite interactions? *Parasitology, 141*, 462–474.

Budria, A., & Candolin, U. (2014). How does human-induced environmental change influence host-parasite interactions? *Parasitology, 141*(4), 462–474.

Burns, A. R., Stephens, W. Z., Stagaman, K., et al. (2016). Contribution of neutral processes to the assembly of gut microbial communities in the zebrafish over host development. *The ISME Journal, 10*(3), 655–664.

Butt, R. L., & Volkoff, H. (2019). Gut microbiota and energy homeostasis in fish. *Frontiers in Endocrinology, 10*, 9.

Campbell, P. G. C., Twiss, M. R., & Wilkinson, K. J. (1997). Accumulation of natural organic matter on the surfaces of living cells: Implications for the interaction of toxic solutes with aquatic biota. *Canadian Journal of Fisheries and Aquatic Sciences, 54*(11), 2543–2554.

Casadei, E., Tacchi, L., Lickwar, C. R., et al. (2019). Commensal bacteria regulate gene expression and differentiation in vertebrate olfactory systems through transcription factor rest. *Chemical Senses, 44*(8), 615–630.

Chávez-Veintemilla, C., & Val, A. L. (2019). Effects of phenanthrene on the Amazonian fish tambaqui *Colossoma macropomum*: LC50, growth and haematology. *Environment and Ecology Research, 7*(5), 293–302.

Cheaib, B., Le Boulch, M., Mercier, P. L., et al. (2018). Taxon-function decoupling as an adaptive signature of lake microbial metacommunities under a chronic polymetallic pollution gradient. *Frontiers in Microbiology, 9*, 869.

Cheaib, B., Seghouani, H., Ijaz, U. Z., & Derome, N. (2020). Community recovery dynamics in yellow perch microbiome after gradual and constant metallic perturbations. *Microbiome, 8*(1), 14.

Cheaib, B., Seghouani, H., Llewellyn, M., Vandal-Lenghan, K., Mercier, P. L., & Derome, N. (2021). The yellow perch (Perca flavescens) microbiome revealed resistance to colonisation mostly associated with neutralism driven by rare taxa under cadmium disturbance. *Animal Microbiome, 3*(1), 3.

Cheesman, S. E., Neal, J. T., Mittge, E., et al. (2011). Epithelial cell proliferation in the developing zebrafish intestine is regulated by the wnt pathway and microbial signaling via myd88. *PNAS, 108*(Suppl 1), 4570–4577.

Chiarello, M., Auguet, J. C., Bettarel, Y., et al. (2018). Skin microbiome of coral reef fish is highly variable and driven by host phylogeny and diet. *Microbiome, 6*, 1–14.

Chiarello, M., Paz-Vinas, I., Veyssiere, C., et al. (2019). Environmental conditions and neutral processes shape the skin microbiome of European catfish (*Silurus glanis*) populations of Southwestern France. *Environmental Microbiology Reports, 11*(4), 605–614.

Chimerel, C., Emery, E., Summers, D. K., et al. (2010). Bacterial metabolite indole modulates incretin secretion from intestinal enteroendocrine l cells. *Cell Reports, 9*(4), 1202–1208.

Chimerel, C., Murray, A. J., Oldewurtel, E. R., et al. (2013). The effect of bacterial signal indole on the electrical properties of lipid membranes. *ChemPhysChem, 14*(2), 417–423.

Clements, K. D., Angert, E. R., Montgomery, W. L., et al. (2014). Intestinal microbiota in fishes: What's known and what's not. *Molecular Ecology, 23*, 1891–1898.

Coe, M. T., Costa, M. H., & Soares-Filho, B. S. (2009). The influence of historical and potential future deforestation on the stream flow of the Amazon river-land surface processes and atmospheric feedbacks. *Journal of Hydrology, 369*, 165–174.

Colin, Y., Molbert, N., Berthe, T., et al. (2022). Dysbiosis of fish gut microbiota is associated with helminths parasitism rather than exposure to PAHs at environmentally relevant concentrations. *Scientific Reports, 12*(1), 11084.

Consoli, G., Haller, R. M., Doering, M., et al. (2022). Tributary effects on the ecological responses of a regulated river to experimental floods. *Journal of Environmental Management, 303*, 114122.

Costa, M. H. (2004). Large-scale hydrological impacts of tropical forest conversion. In M. Bonell & L. A. Bruijnzeel (Eds.), *Forests, water and people in the humid tropics* (pp. 590–597). Cambridge University Press.

Custodio, M., Espinoza, C., Penaloza, R., et al. (2022). Microbial diversity in intensively farmed lake sediment contaminated by heavy metals and identification of microbial taxa bioindicators of environmental quality. *Scientific Reports, 12*(1), 80.

da Costa, J. C., & Val, A. L. (2020). Extreme climate scenario and parasitism affect the Amazonian fish *Colossoma macropomum*. *Science of the Total Environment, 726*, 138628.

de La Torre Canny, S. G., Mueller, O., Craciunescu, C. V., et al. (2021). Tributyltin exposure leads to increased adiposity and reduced abundance of leptogenic bacteria in the zebrafish intestine. *bioRxiv*.

De Schryver, P., & Vadstein, O. (2014). Ecological theory as a foundation to control pathogenic invasion in aquaculture. *The ISME Journal, 8*(12), 2360–2368.

Dehler, C. E., Secombes, C. J., & Martin, S. A. M. (2017). Seawater transfer alters the intestinal microbiota profiles of Atlantic salmon (*Salmo salar* L.). *Scientific Reports, 7*, 13877.

Deines, P., Lachnit, T., & Bosch, T. C. G. (2017). Competing forces maintain the hydra metaorganism. *Immunological Reviews, 279*(1), 123–136.
Den Besten, G., Van Eunen, K., Groen, A. K., et al. (2013). The role of short-chain fatty acids in the interplay between diet, gut microbiota, and host energy metabolism. *Journal of Lipid Research, 54*(9), 2325–2340.
Di Maiuta, N., Schwarzentruber, P., Schenker, M., et al. (2013). Microbial population dynamics in the faeces of wood-eating loricariid catfishes. *Letters in Applied Microbiology, 56*(6), 401–407.
do Vale Pereira, G., da Cunha, D. G., Pedreira Mourino, J. L., et al. (2017). Characterization of microbiota in Arapaima gigas intestine and isolation of potential probiotic bacteria. *Journal of Applied Microbiology, 123*(5), 1298–1311.
Doane, M. P., Morris, M. M., Papudeshi, B., et al. (2020). The skin microbiome of elasmobranchs follows phylosymbiosis, but in teleost fishes, the microbiomes converge. *Microbiome, 8*(1), 1–15.
Douglas, A. E., & Werren, J. H. (2016). Holes in the hologenome: Why host-microbe symbiosis are not holobionts. *mBio, 7*(2), e02099.
Dufour, D. L., Piperata, B. A., Murrieta, R. S., et al. (2016). Amazonian foods and implications for human biology. *Annals of Human Biology, 43*(4), 330–348.
Duperron, S., Halary, S., Habiballah, M., et al. (2019). Response of fish gut microbiota to toxin-containing cyanobacterial extracts: A microcosm study on the medaka (*Oryzias latipes*). *Environmental Science & Technology Letters, 6*(6), 341–347.
Egerton, S., Culloty, S., Whooley, J., et al. (2018). The gut microbiota of marine fish. *Frontiers in Microbiology, 9*, 873.
Eichmiller, J. J., Hamilton, M. J., Stale, C., et al. (2016). Environment shapes the fecal microbiome of invasive carp species. *Microbiome, 4*(1), 44.
Eissa, I. A. M., Gehan, I. S., Wafeek, M., et al. (2012). Bioremediation for heavy metals in some Red Sea fishes in Suez, Egypt. *SCVMJ, 17*(2), 341–356.
Feng, J. B., Hu, C. Q., Luo, P., et al. (2010). Microbiota of yellow grouper (*Epinephelus awoora* Temminck Schlegel, 1842) fed two different diets. *Aquaculture Research, 41*(12), 1778–1790.
Finegold, S. M., Vaisanen, M. L., Molitoris, D. R., et al. (2003). *Cetobacterium somerae sp* nov from human feces and emended description of the genus *Cetobacterium. Systematic and Applied Microbiology, 26*, 177–181.
Finer, M., Jenkins, C. N., Pimm, S. L., et al. (2008). Oil and gas projects in the western Amazon: Threats to wilderness, biodiversity, and indigenous peoples. *PLoS One, 8*, e2932.
Fraune, S., & Bosch, T. C. G. (2010). Why bacteria matter in animal development and evolution. *BioEssays, 32*(7), 571–580.
Freixa, A., Acuña, V., Casellas, M., et al. (2017). Warmer night-time temperature promotes microbial heterotrophic activity and modifies stream sediment community. *Global Change Biology, 23*(9), 3825–3837.
Furch, K. (1984). Water chemistry of the Amazon basin: The distribution of chemical elements among freshwaters. In H. Sioli (Ed.), *The Amazon. Monographiae biologicae* (Vol. 56, pp. 167–199). Springer.
Gallet, A., Halary, S., Duval, C., et al. (2021). Disruption of fish gut microbiota composition and holobiont's metabolome by cyanobacterial blooms. *bioRxiv*.
Gagne-Maynard, W. C., Ward, N. D., Keil, R. G., et al. (2017). Evaluation of primary production in the lower amazon river based on a dissolved oxygen stable isotopic mass balance. *Frontiers in Marine Science, 4*, 26.
Garret, R. D., Cammelli, F., Ferreira, J., et al. (2021). Forests and sustainable development in the Brazilian Amazon: History, trends, and future prospects. *Annual Review of Environment and Resources, 46*, 625–652.
Gaulke, C. A., Martins, M. L., Watral, V. G., et al. (2019). A longitudinal assessment of host-microbe-parasite interactions resolves the zebrafish gut microbiome's link to *Pseudocapillaria tomentosa* infection and pathology. *Microbiome, 7*(1), 10.

Gerson, J. R., Szponar, N., Zambrano, A. A., et al. (2022). Amazon forests capture high levels of atmospheric mercury pollution from artisanal gold mining. *Nature Communications, 13*(1), 559.
Ghanbari, M., Kneifel, W., & Domig, K. J. (2015). A new view of the fish gut microbiome: Advances from next-generation sequencing. *Aquaculture, 448*, 464–475.
Ghosh, S. K., Wong, M. K.-S., Hyodo, S., et al. (2022). Temperature modulation alters the gut and skin microbial profiles of chum salmon (*Oncorhynchus keta*). *Frontiers in Marine Science, 9*, 1027621.
Giatsis, C., Sipkema, D., Smidt, H., et al. (2015). The impact of rearing environment on the development of gut microbiota in tilapia larvae. *Scientific Reports, 5*, 18206.
Gioacchini, G., Ciani, E., Pessina, A., et al. (2018). Effects of lactogen 13, a new probiotic preparation, on gut microbiota and endocrine signals controlling growth and appetite of *Oreochromis niloticus* juveniles. *Microbial Ecology, 76*(4), 1063–1074.
Gómez, G. D., & Balcázar, J. L. (2008). A review on the interactions between gut microbiota and innate immunity of fish. *FEMS Immunology and Medical Microbiology, 52*(2), 145–154.
Gonçalves, R. A., Oliveira, M. S. B., Neves, L. R., et al. (2016). Seasonal pattern in parasite infracommunities of *Hoplerythrinus unitaeniatus* and *Hoplias malabaricus* (Actinopterygii: Erythrinidae) from the Brazilian Amazon. *Acta Parasitologica, 61*(1), 119–129.
Goulding, M., & Carvalho, M. L. (1982). Life history and management of the tambaqui (*Colossoma macropomum*, Characidae): An important Amazonian food fish. *Revista Brasileira de Zoologia, 1*(2), 107–133.
Guo, P., Zhang, K., Ma, X., et al. (2020). *Clostridium* species as probiotics: Potentials and challenges. *Journal of Animal Science and Biotechnology, 11*(1), 24.
Hagmayer, A., Furness, A. I., & Pollux, B. J. (2020). Parasite infestation influences life history but not boldness behavior in placental live-bearing fish. *Oecologia, 194*(4), 635–648.
Hahn, M. A., Piecyk, A., Jorge, F., et al. (2022). Host phenotype and microbiome vary with infection status, parasite genotype, and parasite microbiome composition. *Molecular Ecology, 31*(5), 1577–1594.
Hamady, M., Lozupone, C., & Knight, R. (2010). Fast UniFrac: Facilitating high-throughput phylogenetic analyses of microbial communities including analysis of pyrosequencing and PhyloChip data. *The ISME Journal, 4*, 17–27.
Hansen, G. H., & Olafsen, J. A. (1989). Bacterial-colonization of cod (*Gadus morhua* l) and halibut (*Hippoglossus hippoglossus*) eggs in marine aquaculture. *Applied and Environmental Microbiology, 55*(6), 1435–1446.
Hansen, G. H., & Olafsen, J. A. (1999). Bacterial interactions in early life stages of marine cold water fish. *Microbial Ecology, 38*(1), 1–26.
Hardison, E. A., Kraskura, K., Van Wert, J., et al. (2021). Diet mediates thermal performance traits: Implications for marine ectotherms. *The Journal of Experimental Biology, 224*(21), jeb242846.
Hayes, K. S., Bancroft, A. J., Goldrick, M., et al. (2010). Exploitation of the intestinal microflora by the parasitic nematode *Trichuris muris*. *Science, 328*, 1391–1394.
Hayhoe, S., Neill, C., McHorney, R., et al. (2011). Conversion to soy on the Amazonian agricultural frontier increases streamflow without affecting stormflow dynamics. *Global Change Biology, 17*, 1821–1833.
Hennersdorf, P., Kleinertz, S., Theisen, S., et al. (2016). Microbial diversity and parasitic load in tropical fish of different environmental conditions. *PLoS One, 11*(3), e0151594.
Horsley, R. W. (1977). Review of bacterial-flora of Teleosts and elasmobranchs, including methods for its analysis. *Journal of Fish Biology, 10*(6), 529–553.
Hua, L., Lu, Z., Yuan, N., et al. (2017). Percolation phase transition of surface air temperature networks: A new test bed for El Niño/La Niña simulations. *Scientific Reports, 7*(1), 8324.
IBGE. (2022). *Produção da pecuária municipal* (Vol. 50, pp. 1–12). IBGE, Rio de Janeiro.
Iriarte, J., Power, M. J., Rostain, S., et al. (2012). Fire-free land use in pre-1492 Amazonian savannas. *Proceedings of the National Academy of Sciences, 109*(17), 6473–6478.
Isaac, V. J., Almeida, M. C., Cruz, R. E., et al. (2015). Artisanal fisheries of the Xingu River basin in Brazilian Amazon. *Brazilian Journal of Biology, 75*(3 Suppl 1), 125–137.

Jackson, D. C., & Marmulla, G. (2001). The influence of dams on river fisheries. In G. Marmulla (Ed.), *Dams, fish and fisheries: Opportunities, challenges and conflict resolution* (pp. 1–44). Food Agriculture Organization, Fisheries Technical Paper 419.

Jernelöv, A. (2010). The threats from oil spills: Now, then, and in the future. *Ambio, 39*(5–6), 353–366.

Jiménez-Segura, L. F., Galvis-Vergara, G., Cala-Cala, P., et al. (2016). Freshwater fish faunas, habitats and conservation challenges in the Caribbean river basins of North-Western South America. *Journal of Fish Biology, 89*(1), 65–101.

Johnson, D. B. (2008). Biodiversity and interactions of acidophiles: Key to understanding and optimizing microbial processing of ores and concentrates. *Transactions of Nonferrous Metals Society of China, 18*, 1367–1373.

Junk, W. J. (1989). Flood tolerance and tree distribution in central Amazonian floodplains. In L. B. Holm-Nielsen, I. C. Nielsen, & H. Balslev (Eds.), *Tropical forests: Botanical dynamics, speciation and diversity* (pp. 47–64). Academic Press.

Kaiser, G. (2021). 2.2: Introduction to bacterial growth and aseptic techniques. The Community College of Baltimore County (Cantonsville). https://bio.libretexts.org/@go/page/3385. Accessed 12/07/2023.

Kanther, M., Sun, X., Mühlbauer, M., et al. (2011). Microbial colonization induces dynamic temporal and spatial patterns of NF-κB activation in the zebrafish digestive tract. *Gastroenterology, 141*(1), 197–207.

Kaplan, J. L., Shi, H. N., & Walker, W. (2011). The role of microbes in developmental immunologic programming. *Pediatric Research, 69*(6), 465–447.

Karlsson, F. H., Ussery, D. W., Nielsen, J., et al. (2011). Closer look at Bacteroides: Phylogenetic relationship and genomic implications of a life in the human gut. *Microbial Ecology, 61*, 473–485.

Kim, D. H., Brunt, J., & Austin, B. (2007). Microbial diversity of intestinal contents and mucus in rainbow trout (*Oncorhynchus mykiss*). *Journal of Applied Microbiology, 102*(6), 1654–1664.

Kim, P. S., Shin, N. R., Lee, J. B., et al. (2021). Host habitat is the major determinant of the gut microbiome of fish. *Microbiome, 9*(1), 166.

Kuang, T. X., He, A. Y., Lin, Y. F., et al. (2020). Comparative analysis of microbial communities associated with the gill, gut, and habitat of two filter-feeding fish. *Aquaculture Reports, 18*, 100501.

Landeira-Dabarca, A., Sieiro, C., & Alvarez, M. (2013). Change in food ingestion induces rapid shifts in the diversity of microbiota associated with cutaneous mucus of Atlantic salmon *Salmo salar*. *Journal of Fish Biology, 82*(3), 893–906.

Larsen, A., Tao, Z., Bullard, S., et al. (2013). Diversity of the skin microbiota of fishes: Evidence for host species specificity. *FEMS Microbiology Ecology, 85*(3), 483–494.

Lavoie, C., Wellband, K., Perreault, A., et al. (2021). Artificial rearing of Atlantic salmon juveniles for supportive breeding programs induces long-term effects on gut microbiota after stocking. *Microorganisms, 9*(9), 1932.

Lazado, C. C., & Caipang, C. M. (2014). Mucosal immunity and probiotics in fish. *Fish and Shellfish Immunology, 39*(1), 78–89.

Lázaro, W. L., Díez, S., da Silva, C. J., et al. (2018). Seasonal changes in peryphytic microbial metabolism determining mercury methylation in a tropical wetland. *Science of the Total Environment, 15*(627), 1345–1352.

Legrand, T., Catalano, S. R., Wos-Oxley, M. L., et al. (2018). The inner workings of the outer surface: Skin and gill microbiota as indicators of changing gut health in yellowtail kingfish. *Frontiers in Microbiology, 8*, 2664.

Leroux, N., Sylvain, F. E., Holland, A., et al. (2023). Gut microbiota of an Amazonian fish in a heterogeneous riverscape: Integrating genotype, environment, and parasitic infections. *Microbiology Spectrum, 11*(5), e02755-22.

Lescak, E. A., & Milligan-Myhre, K. C. (2017). Teleosts as model organisms to understand host-microbe interactions. *Journal of Bacteriology, 199*(15), 10–1128.

Leung, J. M., Graham, A. L., & Knowles, S. C. (2018). Parasite-microbiota interactions with the vertebrate gut: Synthesis through an ecological lens. *Frontiers in Microbiology, 9*, 843.

Li, X., Yan, Q., Ringø, E., et al. (2016). The influence of weight and gender on intestinal bacterial community of wild largemouth bronze gudgeon (*Coreius guichenoti*, 1874). *BMC Microbiology, 16*(1), 191.

Li, H., Zeng, J., Ren, L., et al. (2017). Contrasting patterns of diversity of abundant and rare bacterioplankton in freshwater lakes along an elevation gradient. *Limnology and Oceanography, 62*, 1570.

Lindmark, M., Audzijonyte, A., Blanchard, J. L., et al. (2022). Temperature impacts on fish physiology and resource abundance lead to faster growth but smaller fish sizes and yields under warming. *Global Change Biology, 28*(21), 6239–6253.

Liu, H., Guo, X., Gooneratne, R., et al. (2016). The gut microbiome and degradation enzyme activity of wild freshwater fishes influenced by their trophic levels. *Scientific Reports, 6*, 24340.

Liu, Y., Li, X., Li, J., et al. (2021). The gut microbiome composition and degradation enzymes activity of black Amur bream (*Megalobrama terminalis*) in response to breeding migratory behavior. *Ecology and Evolution, 11*(10), 5150–5163.

Llewellyn, M. S., Boutin, S., Hoseinifar, S. H., et al. (2014). Teleost microbiomes: The state of the art in their characterization, manipulation and importance in aquaculture and fisheries. *Frontiers in Microbiology, 5*, 207.

Llewellyn, M. S., Leadbeater, S., Garcia, C., et al. (2017). Parasitism perturbs the mucosal microbiome of Atlantic Salmon. *Scientific Reports, 7*, 43465.

Lokesh, J., & Kiron, V. (2016). Transition from freshwater to seawater reshapes the skin-associated microbiota of Atlantic salmon. *Scientific Reports, 6*, 19707.

Long, A., Call, D. R., & Cain, K. D. (2014). Investigation of the link between broodstock infection, vertical transmission, and prevalence of *Flavobacterium psychrophilum* in eggs and progeny of rainbow trout and coho salmon. *Journal of Aquatic Animal Health, 26*(2), 66–77.

Lowrey, L., Woodhams, D. C., Tacchi, L., et al. (2015). Topographical mapping of the rainbow trout (*Oncorhynchus mykiss*) microbiome reveals a diverse bacterial community with antifungal properties in the skin. *Applied and Environmental Microbiology, 81*(19), 6915–6925.

Luan, Y., Li, M., Zhou, W., et al. (2023). The fish microbiota: Research progress and potential applications. *Engineering, 29*, 137.

Malhi, Y., Aragao, L. E. O. C., Galbraith, D., et al. (2009). Exploring the likelihood and mechanism of a climate-change-induced dieback of the Amazon rainforest. *PNAS, 106*, 20610.

Margulis, L., & Fester, R. (1991). *Symbiosis as a source of evolutionary innovation: Speciation and morphogenesis*. MIT Press.

McCauley, M., German, D. P., Lujan, N. K., et al. (2020). Gut microbiomes of sympatric Amazonian wood-eating catfishes (Loricariidae) reflect host identity and little role in wood digestion. *Ecology and Evolution, 10*(14), 7117–7128.

McDonald, R. C., Watts, J. E. M., & Schreier, H. J. (2019). Effect of diet on the enteric microbiome of the wood-eating catfish *Panaque nigrolineatus*. *Frontiers in Microbiology, 10*, 2687.

McFall-Ngai, M. J. (2002). Unseen forces: The influence of bacteria on animal development. *Developmental Biology, 242*(1), 1–14.

McFall-Ngai, M., Hadfield, M. G., Bosch, T. C. G., et al. (2013). Animals in a bacterial world, a new imperative for the life sciences. *PNAS, 110*(9), 3229–3236.

McIntosh, D., Ji, B., Forward, B. S., et al. (2008). Culture-independent characterization of the bacterial populations associated with cod (*Gadus morhua* L.) and live feed at an experimental hatchery facility using denaturing gradient gel electrophoresis. *Aquaculture, 275*(1–4), 42–50.

Mela, M., Neto, F. F., Yamamoto, F. Y., et al. (2014). Mercury distribution in target organs and biochemical responses after subchronic and trophic exposure to Neotropical fish *Hoplias malabaricus*. *Fish Physiology and Biochemistry, 40*, 245–256.

Melack, J. M., & Forsberg, B. R. (2001). Biogechemistry of Amazon floodplain lakes and associated wetlands. In M. E. McClain, R. L. Victoria, & J. E. Richey (Eds.), *The biogeochemistry of the Amazon basin* (pp. 235–274). Oxford University Press.

Monteiro, D. A., Rantin, F. T., & Kalinin, A. L. (2010). Inorganic mercury exposure: Toxicological effects, oxidative stress biomarkers and bioaccumulation in the tropical freshwater fish matrinxã, *Brycon amazonicus* (Spix and Agassiz, 1829). *Ecotoxicology, 19*, 105–123.

Moreira, L. H. A., Takemoto, R. M., Yamada, F. H., et al. (2009). Ecological aspects of metazoan endoparasites of *Metynnis lippincottianus* (Cope, 1870) (Characidae) from upper Paraná River floodplain, Brazil. *Helminthologia, 46*(4), 214–219.

Morris, C., Val, A. L., Brauner, C. J., et al. (2021). The physiology of fish in acidic waters rich in dissolved organic carbon, with specific reference to the Amazon basin: Ionoregulation, acid-base regulation, ammonia excretion, and metal toxicity. *Journal of Experimental Zoology Part A: Ecological and Integrative Physiology, 335*(9–10), 843–863.

Moura, M. M., Dos Santos, A. R., Pezzopane, J. E. M., et al. (2019). Relation of El Niño and La Niña phenomena to precipitation, evapotranspiration and temperature in the Amazon basin. *Science of the Total Environment, 651*, 1639–1651.

Murren, C. J., Auld, J. R., Callahan, H., et al. (2015). Constraints on the evolution of phenotypic plasticity: Limits and costs of phenotype and plasticity. *Heredity, 115*(4), 293–301.

Nayak, S. K. (2010). Role of gastrointestinal microbiota in fish. *Aquaculture Research, 41*(11), 1553–1573.

Neill, C., Deegan, L., Thomas, S., et al. (2001). Deforestation for pasture alters nitrogen and phosphorus in small Amazonian streams. *Ecological Applications, 11*, 1817–1828.

Neves, L. R., Pereira, F. B., Tavares-Dias, M., et al. (2013). Seasonal influence on the parasite fauna of a wild population of *Astronotus* ocellatus (Perciformes: Cichlidae) from the Brazilian Amazon. *The Journal of Parasitology, 99*, 718–721.

Nie, L., Zhou, Q. J., Qiao, Y., et al. (2017). Interplay between the gut microbiota and immune responses of ayu (*Plecoglossus altivelis*) during *Vibrio anguillarum* infection. *Fish & Shellfish Immunology, 68*, 479–487.

Niño-García, J. P., Ruiz-González, C., & del Giorgio, P. A. (2016). Interactions between hydrology and water chemistry shape bacterioplankton biogeography across boreal freshwater networks. *The ISME Journal, 10*, 1755–1766.

Nursall, J. R. (1959). Oxygen as a prerequisite to the origin of the metazoa. *Nature, 183*(4669), 1170–1172.

O'Hara, A. M., & Shanahan, F. (2006). The gut flora as a forgotten organ. *EMBO Reports, 7*(7), 688–693.

Ohashi, H., Umeda, N., Hirazawa, N., et al. (2007). Purification and identification of a glycoprotein that induces the attachment of oncomiracidia of *Neobenedenia girellae* (Monogenea, Capsalidae). *International Journal for Parasitology, 37*(13), 1483–1490.

Olsen, R. E., Sundell, K., Mayhew, T. M., et al. (2005). Acute stress alters intestinal function of rainbow trout, *Oncorhynchus mykiss* (Walbaum). *Aquaculture, 250*(1–2), 480–495.

Parata, L., Mazumder, D., Sammut, J., et al. (2020). Diet type influences the gut microbiome and nutrient assimilation of genetically improved farmed tilapia (*Oreochromis niloticus*). *PLoS One, 15*(8), e0237775.

Pelicice, F. M., & Agostinho, A. A. (2008). Fish-passage facilities as ecological traps in large neotropical rivers. *Conservation Biology, 22*(1), 180–188.

Pfeiffer, W. C., & Lacerda, L. D. (1988). Mercury inputs into the Amazon Region, Brazil. *Environmental Technology, 9*, 325–330.

Phelps, D., Brinkman, N. E., Keely, S. P., et al. (2017). Microbial colonization is required for normal neurobehavioral development in zebrafish. *Scientific Reports, 7*(1), 11244.

Pigliucci, M. (2005). Evolution of phenotypic plasticity: Where are we going now? *Trends in Ecology & Evolution, 20*(9), 481–486.

Pires, T. H. S., Campos, D. F., Röpke, C. P., et al. (2015). Ecology and life-history of *Mesonauta festivus*: Biological traits of a broad ranged and abundant Neotropical cichlid. *Environmental Biology of Fishes, 98*, 789–799.

Poff, N., & Hart, D. (2002). How dams vary and why it matters for the emerging science of dam removal. *Bioscience, 8*, 659–668.

Pompeu, P., Agostinho, A. A., & Pelicice, F. (2012). Existing and future challenges: The concept of successful fish passage in South America. *River Research and Applications, 28*(4), 504–512.

Portner, H. O. (2001). Climate change and temperature-dependent biogeography: Oxygen limitation of thermal tolerance in animals. *Die Naturwissenschaften, 88*(4), 137–146.

Pratte, Z. A., Besson, M., Hollman, R. D., et al. (2018). The gills of reef fish support a distinct microbiome influenced by host-specific factors. *Applied and Environmental Microbiology, 84*(9), e00063–e00018.

Ramírez, C., Coronado, J., Silva, A., et al. (2018). *Cetobacterium* is a major component of the microbiome of giant amazonian fish (*Arapaima gigas*) in Ecuador. *Animals, 8*(11), 189.

Rank, N. E., & Dahlhoff, E. P. (2002). Allele frequency shifts in response to climate change and physiological consequences of allozyme variation in a montane insect. *Evolution: International Journal of Organic Evolution, 56*(11), 2278–2289.

Rawls, J. F., Mahowald, M. A., Ley, R. E., et al. (2006). Reciprocal gut microbiota transplants from zebrafish and mice to germ-free recipients reveal host habitat selection. *Cell, 127*(2), 423–433.

Rawls, J. F., Samuel, B. S., & Gordon, J. I. (2004). Gnotobiotic zebrafish reveal evolutionarily conserved responses to the gut microbiota. *PNAS, 101*(13), 4596–4601.

Reinhart, E., Korry, B., Rowan-Nash, A., et al. (2019). Defining the distinct skin and gut microbiomes of the Northern Pike (*Esox lucius*). *Frontiers in Microbiology, 10*, 2118.

Riiser, E. S., Haverkamp, T. H. A., Varadharajan, S., et al. (2020). Metagenomic shotgun analyses reveal complex patterns of intra- and interspecific variation in the intestinal microbiomes of codfishes. *Applied and Environmental Microbiology, 86*(6), e02788-19.

Ringo, E., Zhou, Z., Vecino, J. L. G., et al. (2016). Effect of dietary components on the gut microbiota of aquatic animals. A never-ending story? *Aquaculture Nutrition, 22*(2), 219–282.

Romero, J., & Navarrete, P. (2006). 16S rDNA-based analysis of dominant bacterial populations associated with early life stages of coho salmon (*Oncorhynchus kisutch*). *Microbial Ecology, 51*(4), 422–430.

Romero, J., Ringø, E., & Merrifield, D. L. (2014). The gut microbiota of fish. *Aquaculture Nutrition: Gut Health, Probiotics and Prebiotics*, 75–100.

Rosa, B. A., Supali, T., Gankpala, L., et al. (2018). Differential human gut microbiome assemblages during soil-transmitted helminth infections in Indonesia and Liberia. *Microbiome, 6*(1), 1–19.

Rosado, D., Pérez-Losada, M., Severino, R., et al. (2019). Characterization of the skin and gill microbiomes of the farmed seabass (*Dicentrarchus labrax*) and seabream (*Sparus aurata*). *Aquaculture, 500*, 57–64.

Rosenberg, E., Koren, O., Reshef, L., et al. (2007). The role of microorganisms in coral health, disease and evolution. *Nature Reviews Microbiology, 5*(5), 355–362.

Rosenberg, E., & Zilber-Rosenberg, I. (2016). Microbes drive evolution of animals and plants: The hologenome concept. *Mbio, 7*(2), 10.

Rowe, M., Veerus, L., Trosvik, P., et al. (2020). The reproductive microbiome: An emerging driver of sexual selection, sexual conflict, mating systems, and reproductive isolation. *Trends in Ecology & Evolution, 35*(3), 220–234.

Roy, C., Rameez, M. J., Haldar, P. K., et al. (2020). Microbiome and ecology of a hot spring-microbialite system on the Trans-Himalayan Plateau. *Scientific Reports, 10*(1), 5917.

Scheifler, M., Sanchez-Brosseau, A., Magnanou, E., et al. (2022). Diversity and structure of sparids external microbiota (Teleostei) and its link with monogenean ectoparasites. *Animal Microbiome, 4*, 27.

Scheiner, S. M. (2013). The genetics of phenotypic plasticity: Temporal and spatial heterogeneity. *Ecology and Evolution, 3*(13), 4596–4609.

Sebastian, S. M., & Hurtig, A. K. (2004). Oil exploitation in the Amazon basin of Ecuador: A public health emergency. *Pan American Journal of Public Health, 3*, 205–211.

Semova, I., Carten, J. D., Stombaugh, J., et al. (2012). Microbiota regulate intestinal absorption and metabolism of fatty acids in the zebrafish. *Cell Host & Microbe, 12*(3), 277–288.

Sevellec, M., Derome, N., & Bernatchez, L. (2018). Holobionts and ecological speciation: The intestinal microbiota of lake whitefish species pairs. *Microbiome, 6*, 1–15.

Sheng, Y., Ren, H., Limbu, S. M., et al. (2018). The presence or absence of intestinal microbiota affects lipid deposition and related genes expression in zebrafish (*Danio rerio*). *Frontiers in Microbiology, 9*, 1124.

Sinhorin, V. D., Sinhorin, A. P., Teixeira, J. M., et al. (2014). Effects of the acute exposition to glyphosate-based herbicide on oxidative stress parameters and antioxidant responses in a hybrid Amazon fish Surubim (*Pseudoplatystoma* sp). *Ecotoxicology and Environmental Safety, 106*, 181–187.

Smith, P., Willemsen, D., Popkes, M., et al. (2017). Regulation of life span by the gut microbiota in the short-lived african turquoise killifish. *eLife, 6*, e27014.

Stankiewicz, A. M., Swiergiel, A. H., & Lisowski, P. (2013). Epigenetics of stress adaptations in the brain. *Brain Research Bulletin, 98*, 76–92.

Sugita, H., Miyajima, C., & Deguchi, Y. (1991). The vitamin B12-producing ability of the intestinal microflora of freshwater fish. *Aquaculture, 92*, 267–276.

Sugita, H., Shibuya, K., Shimooka, H., et al. (1996). Antibacterial abilities of intestinal bacteria in freshwater cultured fish. *Aquaculture, 145*(1–4), 195–203.

Sullam, K. E., Essinger, S. D., Lozupone, C. A., et al. (2012). Environmental and ecological factors that shape the gut bacterial communities of fish: A meta-analysis. *Molecular Ecology, 21*, 3363–3378.

Sylvain, F. É., Cheaib, B., Llewellyn, M., et al. (2016). PH drop impacts differentially skin and gut microbiota of the Amazonian fish tambaqui (*Colossoma macropomum*). *Scientific Reports, 6*(1), 32032.

Sylvain, F. É., & Derome, N. (2017). Vertically and horizontally transmitted microbial symbionts shape the gut microbiota ontogenesis of a skin-mucus feeding discus fish progeny. *Scientific Reports, 7*(1), 5263.

Sylvain, F. É., Holland, A., Audet-Gilbert, É., et al. (2019). Amazon fish bacterial communities show structural convergence along widespread hydrochemical gradients. *Molecular Ecology, 28*(15), 3612–3626.

Sylvain, F. É., Holland, A., Bouslama, S., et al. (2020). Fish skin and gut microbiomes show contrasting signatures of host species and habitat. *Applied and Environmental Microbiology, 86*(16), e00789–e00720.

Sylvain, F. É., Leroux, N., Normandeau, E., et al. (2022). Genomic and environmental factors shape the active gill bacterial community of an Amazonian teleost holobiont. *Microbiology Spectrum, 10*(6), e02064–e02022.

Sylvain, F. E., Leroux, N., Normandeau, E., et al. (2023). Important role of endogenous microbial symbionts of fish gills in the challenging but highly biodiverse Amazonian blackwaters. *Nature Communications, 14*, 3903.

Sylvain, F. É., Normandeau, E., Holland, A., et al. (2022). Genomics of Serrasalmidae teleosts through the lens of microbiome fingerprinting. *Molecular Ecology, 31*(18), 4656–4671.

Tavares-Dias, M., Gonçalves, R. A., Oliveira, M. S. B., et al. (2017). Ecological aspects of the parasites in *Cichlasoma bimaculatum* (Cichlidae), ornamental fish from the Brazilian Amazon. *Acta Biológica Colombiana, 22*, 175–180.

Tavares-Dias, M., Oliveira, M. S. B., Gonçalves, R. A., et al. (2014). Ecology and seasonal variation of parasites in wild *Aequidens tetramerus*, a Cichlidae from the Amazon. *Acta Parasitologica, 59*(1), 158–164.

Teixeira, M. C., Santana, N. F., de Azevedo, J. C. R., et al. (2011). Bacterioplankton features and its relations with doc characteristics and other limnological variables in Parana river floodplain environments (PR/MS-Brazil). *Brazilian Journal of Microbiology, 42*(3), 897–908.

Thatcher, V. E. (1991). Amazon fish parasites. *Amazoniana, 11*(3/4), 263–572.

Theis, K. R., Dheilly, N. M., Klassen, J. L., et al. (2016). Getting the hologenome concept right: An eco-evolutionary framework for hosts and their microbiomes. *Msystems, 1*(2), e00068.

Tsuchiya, C., Sakata, T., & Sugita, H. (2008). Novel ecological niche of *Cetobacterium somerae*, an anaerobic bacterium in the intestinal tracts of freshwater fish. *Letters in Applied Microbiology, 46*, 43–48.

Urban, M. C., Richardson, J. L., & Freidenfelds, N. A. (2014). Plasticity and genetic adaptation mediate amphibian and reptile responses to climate change. *Evolutionary Applications, 7*(1), 88–103.

Uren Webster, T. M., Consuegra, S., & Garcia de Leaniz, C. (2021). Early life stress causes persistent impacts on the microbiome of Atlantic salmon. *Comparative Biochemistry and Physiology Part D: Genomics and Proteomics, 40*, 100888.

Uren Webster, T. M., Rodriguez-Barreto, D., Consuegra, S., et al. (2020). Cortisol-related signatures of stress in the fish microbiome. *Frontiers in Microbiology, 11*, 1621.

Val, A. L., & Almeida-Val, V. M. F. (1995). *Fishes of the Amazon and their environment*. Springer-Verlag.

Val, A. L., & Wood, C. M. (2022). Global change and physiological challenges for fish of the Amazon today and in the near future. *Journal of Experimental Biology, 225*(10), jeb216440.

van der Walt, K.-A., Potts, W. M., Porri, F., et al. (2021). Marine heatwaves exceed cardiac thermal limits of adult sparid fish (*Diplodus capensis*, Smith 1884). *Frontiers in Marine Science, 8*, 702463.

Van Dover, C. L. (2000). *The ecology of deep-sea hydrothermal vents* (p. 448). Princeton University Press.

van Kessel, M. A. H. J., Mesman, R. J., Arshad, A., et al. (2016). Branchial nitrogen cycle symbionts can remove ammonia in fish gills. *Environmental Microbiology Reports, 8*(5), 590–594.

Vannote, R. L., Minshall, G. W., Cummins, K. W., et al. (1980). The river continuum concept. *Canadian Journal of Fisheries and Aquatic Sciences, 37*, 130–137.

Vasemägi, A., Visse, M., & Kisand, V. (2017). Effect of environmental factors and an emerging parasitic disease on gut microbiome of wild salmonid fish. *MSphere, 2*(6), e00418–e00417.

Volkoff, H., & Rønnestad, I. (2020). Effects of temperature on feeding and digestive processes in fish. *Temperature, 7*(4), 307–320.

von Engelhardt, W., Bartels, J., Kirschberger, S., et al. (1998). Role of short-chain fatty acids in the hind gut. *Veterinary Quarterly, 20*(Suppl 3), S52–S59.

Walker, I., & Henderson, P. A. (1996). Ecophysiological aspects of Amazonian Blackwater litterbank fish communities. In A. L. Val, V. M. F. Almeida-Val, & D. J. Randall (Eds.), *Physiology and biochemistry of fishes of the Amazon*. Instituto Nacional de Pesquisas da Amazonia.

Wang, M., Fan, Z., Zhang, Z., et al. (2022). Effects of diet on the gut microbial communities of Nile Tilapia (*Oreochromis niloticus*) across their different life stages. *Frontiers in Marine Science, 9*, 926132.

Wang, A. R., Ran, C., Ringø, E., et al. (2018). Progress in fish gastrointestinal microbiota research. *Reviews in Aquaculture, 10*(3), 626–640.

Wang, L., Zhang, D., Xie, J., et al. (2023). Do ectoparasites on fish gills "talk" with gut microbiota far away? *Aquaculture, 562*, 738880.

Wang, W. W., Zhou, Z. G., He, S., et al. (2010). Identification of the adherent microbiota on the gills and skin of poly-cultured gibel carp (*Carassius auratus gibelio*) and bluntnose black bream (*Megalobrama amblycephala*). *Aquaculture Research, 41*(9), e72–e83.

Weber, A. P. M., Horst, R. J., Barbier, G. G., et al. (2007). Metabolism and metabolomics of eukaryotes living under extreme conditions. *International Review of Cytology - A Survey of Cell Biology, 256*, 1–34.

Webster, T. M. U., Consuegra, S., Hitchings, M., et al. (2018). Interpopulation variation in the Atlantic salmon microbiome reflects environmental and genetic diversity. *Applied and Environmental Microbiology, 84*(16).

Williams, K. B., Bischof, J., Lee, F. J., et al. (2020). Regulation of axial and head patterning during planarian regeneration by a commensal bacterium. *Mechanisms of Development, 163*, 103614.

Williams, M. R., Fisher, T. R., & Melack, J. M. (1997). Solute dynamics in soil water and groundwater in a Central Amazon catchment undergoing deforestation. *Biogeochemistry, 3*, 303–335.

Wong, S. D., & Rawls, J. F. (2012). Intestinal microbiota composition in fishes is influenced by host ecology and environment. *Molecular Ecology, 21*(13), 3100–3102.

Wood, C. M., Wilson, R. W., Gonzalez, R. J., et al. (1998). Responses of an Amazonian teleost, the tambaqui (*Colossoma macropomum*), to low pH in extremely soft water. *Physiological Zoology, 71*(6), 658–670.

Xia, J. H., Lin, G., Fu, G. H., et al. (2014). The intestinal microbiome of fish under starvation. *BMC Genomics, 15*, 266.

Yan, Q., Li, J., Yu, Y., et al. (2016). Environmental filtering decreases with fish development for the assembly of gut microbiota. *Environmental Microbiology, 18*(12), 4739–4754.

Yang, G. M., Bao, B. L., Peatman, E., et al. (2007). Analysis of the composition of the bacterial community in puffer fish *Takifugu obscurus*. *Aquaculture, 262*(2–4), 183–191.

Yang, C., Lim, W., & Song, G. (2021). Reproductive toxicity due to herbicide exposure in freshwater organisms. *Comparative Biochemistry and Physiology Part C: Toxicology & Pharmacology, 248*, 109103.

Zhang, M. L., Sun, Y. H., Liu, Y. K., et al. (2016). Response of gut microbiota to salinity change in two euryhaline aquatic animals with reverse salinity preference. *Aquaculture, 454*, 72–80.

Zilber-Rosenberg, I., & Rosenberg, E. (2008). Role of microorganisms in the evolution of animals and plants: The hologenome theory of evolution. *FEMS Microbiology Reviews, 32*(5), 723–735.

Environmental Nucleic Acids: A Toolkit for Monitoring Diversity, Conservation, and Environmental/Population Health

Waldir Heinrichs-Caldas, Samara Silva de Souza, Alexandre Vergueiro Júnior, Jefferson Almeida Silva, Carolina Sousa de Sá-Leitão, Anderson Andrade de Souza, Maria de Nazaré Paula, and Vera Maria Fonseca de Almeida e Val

Abstract

The Amazon basin comprehends several aquatic environments and a great diversity of ichthyofauna, with a high rate of endemism. The number of species found in the region, combined with its extension, makes the identification and monitoring of biodiversity a problem, which might be solved with the application of the environmental nucleic acid technique, such as eDNA and eRNA. The eDNA technique uses aquatic or terrestrial environmental samples, where there are genetic traces of the organisms that frequent the local. This method can be used for the identification of target species, for the study of diversity of several *taxa*, and for the identification of invasive species, minimizing the stress of the organisms once it is not an invasive technique. In addition, environmental genetic analyses can still use other tools, such as the analysis of environmental mRNA and microRNA, where, through the expression of specific genes, it is possible to identify genes already known to respond to different stresses, such as pollution, thermal stress, and natural variation in dissolved oxygen levels, generally determining the possible pressures that organisms are exposed in their environment. As it is a technique under development, its application in the Amazon water basin still faces problems, such as the physical-chemical conditions of the aquatic environments and the lack of a genetic database for the species of such a diverse environment. Despite this, there are prospects of development applications in

W. Heinrichs-Caldas (✉) · A. V. Júnior · J. A. Silva · C. S. de Sá-Leitão · A. A. de Souza ·M. de Nazaré Paula · V. M. F. de Almeida e Val
Laboratory of Ecophysiology and Molecular Evolution (LEEM), Brazilian National Institute for Research of the Amazon, INPA, Manaus, Amazonas, Brazil

S. S. de Souza
Laboratory of Ecophysiology and Molecular Evolution, Brazilian National Institute for Research of the Amazon (INPA), Manaus, Amazonas, Brazil

S. S. de Souza et al. (eds.), *The Future of Amazonian Aquatic Biota*,
https://doi.org/10.1007/978-3-031-66822-7_13

biodiversity monitoring, in addition to the identification of environmental genetic parameters related to human and animal health. In this chapter, we will discuss the applications of the environmental DNA technique for the Amazon region, the main results found with the technique, so far, and a comparison with studies that used traditional approaches to assess diversity. Finally, we will list the several perspectives for its future applications, both those concerning the monitoring of environmental and those concerning human health.

1 Environmental Nucleic Acids: A Toolkit in a Biodiversity Hotspot Environment

The Amazon basin is home to an astounding array of fish species, with estimates ranging from 2500 to over 6000 species, making it a globally significant hotspot of freshwater biodiversity (Val, 2019). Despite the great known diversity, the number of Amazonian aquatic species remains underestimated, since vast areas, especially those with difficult access, remain unexplored (Alho et al., 2015; Santana et al., 2021). This is particularly interesting if we consider that the Amazon basin is distributed over an area of more than seven million km^2 (Sioli, 1990). The Amazon rainforest and its intricate network of rivers and tributaries are teeming with an astonishing diversity of fish species. However, traditional methods of fish biodiversity assessment, such as netting or visual surveys, can be logistically challenging due to the vastness and inaccessibility of the region. Additionally, the sheer number of fish species, many of which are elusive or occur in remote and hard-to-reach habitats, poses significant challenges for comprehensive biodiversity surveys (Tsuji et al., 2022).

New tools that use environmental nucleic acids (eDNA and eRNA) have proven to be efficient in the assessment and study of biodiversity (Santana et al., 2021). By using eDNA analysis, changes in fish communities can be monitored, the presence of invasive species can be detected, and the overall health and integrity of aquatic ecosystems in the Amazon can be assessed. On the other hand, environmental RNA analysis involves the study of RNA molecules present in the environment. This approach provides unique insights into the functional activity and gene expression of organisms in their natural habitats. Environmental DNA analysis was shown as an effective measure to determine diversity and biomass of specific species or communities (Rourke et al., 2022), and for the Amazon, it has been more explored than the environmental RNA technique, whose application remains unexplored in the region. By collecting water samples from different parts of the Amazon rivers and tributaries, researchers can extract and analyze the DNA and RNA present in the water. As fish move through the water, they leave behind genetic material such as scales, skin cells, mucus, and feces, collectively referred to as eDNA. Both eDNA and eRNA can be extracted from water samples and subjected to genetic analysis, enabling the detection and identification of fish species present in the sampled area. In general, the study of eRNA is still taking its first steps, and this is no different for the Amazon, where most studies made use of the eDNA technique. Therefore, in this chapter, we will explore the importance of environmental nucleic acid in studying Amazon fish biodiversity and its implications for conservation and ecological understanding.

2 Uncovering Amazonian Biodiversity with eDNA

The use of eDNA in studying Amazon biodiversity can offer several significant advantages: (1) provides a non-invasive method of sampling that does not require capturing or harming fish specimens. This approach is particularly crucial in delicate and sensitive ecosystems like the Amazon, where minimizing disturbance is essential for conservation efforts; (2) eDNA analysis can detect rare and elusive fish species that may be difficult to be captured or directly observed. Many Amazon fish species are cryptic, occurring in remote or hard-to-access habitats, and exhibit elusive behavior, making them challenging to study using traditional methods. However, by detecting their DNA signatures in sediment or water samples, researchers can obtain insights into the presence and distribution of these species, expanding our knowledge of Amazon fish biodiversity. The environmental DNA technique also provides a snapshot of the environment, which can be analyzed and reanalyzed in the future as a source of biological information of the given environment.

In summary, the eDNA analysis involves a series of steps, starting with the collection of environmental samples such as water or soil. Once collected, the nucleic acid is extracted from these samples. This extracted DNA is then subjected to polymerase chain reaction (PCR) to amplify specific target regions, allowing for easier detection. Subsequently, the amplified DNA is sequenced using high-throughput sequencing technologies. The resulting sequence data are then processed and analyzed bioinformatically, often by comparing them to reference databases to identify the species present in the sampled environment. Quantitative PCR or other quantitative methods may be employed to estimate the abundance of each species. Each step has its own peculiarities, which needs to be determined depending on the sample (Shokralla et al., 2012). For example, distinct water types, as found in the Amazon basin, might need a different treatment when extracting the DNA.

To our knowledge, the first study using eDNA-based technique in the Amazon tested seven DNA extraction methods from water samples collected in the Cacau floodplain lake in the Central Amazonia. The DNeasy Blood & Tissue kit (Qiagen), with modifications developed during the study, showed the best overall result. The study shows that this method, among all analyzed DNA extraction protocols, can be used to obtain a higher eDNA concentration, which can be used to research the ichthyofauna of the Amazon basin, in order to evaluate fish diversity or detect the presence of specific fish species. These first findings contribute to the standardization of this method in the region, considering the peculiarities found in Amazon aquatic ecosystems (Bevilaqua et al., 2020).

In the study carried out by Sales et al. (2020), whose objective was to evaluate the biodiversity of neotropical mammals in two biomes, the Amazon and the Atlantic Forest, the eDNA metabarcoding technique was applied. The authors observed the co-occurrence of species identified by eDNA metabarcoding with those identified by alternative camera trap methods. Furthermore, mammal species listed in different categories of the IUNC Red List were identified. Santana et al. (2021), while analyzing the relationship between the fish diversity information achieved through traditional catch and by eDNA from water samples, found that eDNA metabarcoding is a powerful tool for detecting fish diversity in the Amazon basin, with comparable results to traditional capture-based sampling methods. The use of MiFish primers (Miya et al., 2015), whose amplicon is a fragment of 12S mitochondrial gene, and existing public reference libraries allowed for the identification of a diverse range of fish species, including rare and elusive species. The study also highlighted the importance of natural history museums in contributing to the development and validation of eDNA technology, as well as the need for standardized protocols and reference databases to ensure accurate and reliable results. Overall, the study demonstrated the potential of eDNA metabarcoding for advancing our understanding of biodiversity in the Amazon basin and beyond. Studying fish diversity in the Tarumã-Mirim River, a tributary of Negro River, located in the Brazilian Amazon, Batista et al. (2021) conducted an assessment of ichthyofaunistic diversity along the river, also analyzing the MiFish fragment of eDNA sampled at 15 equidistant sites to identify fish species present in the area. They found a total of 54 fish species, including several that were not previously known to inhabit the river. The study also found that eDNA analysis was more effective at detecting rare and elusive species than traditional methods of fish sampling.

The first study of fish diversity in igarapés basin was conducted in the Reserva Florestal Adolpho Ducke (RFAD), a preservation area located on the outskirts of Manaus, State of Amazonas, Brazil. Vergueiro et al. (2024, in prep.), using the 12S fragment (MiFish) amplification to analyze the genetic information on eDNA, assess the fish diversity present in the Acará Basin at RFAD during rainy season. Igarapés are small forest streams, with mostly stable physicochemical conditions, and can be divided in orders, based in its size. Vergueiro et al. (2024, in prep.) found 28 molecular operational taxonomic units (MOTUs) related to fishes species and that the first-order igarapés harbor a richness of species greater than that found in second- and third-order igarapés, respectively, finding 24, 8, and 13 MOTUs (Fig. 1a), and the fish diversity during rainy season appears to have most similarity between second- and third-order igarapés than the diversity found in first order (Fig. 1c).

Half (50%) of the fish species (MOTUS-related sequences) were only detected at the first-order igarapé harbor. Second- and third-order igarapés harbor one and three MOTUs, respectively (Fig. 1a, b). The richness of fish species found in first-order igarapés highlights the importance of these ecosystems to the Amazon basin as a whole. This is mainly due to their elevated fragility and susceptibility to anthropogenic disturbance. eDNA-based assessments are a promising new way to conduct ecological studies in these systems.

Lastly, eDNA analysis allows for a more comprehensive assessment of fish diversity in the Amazon basin. Traditional methods, such as netting or visual

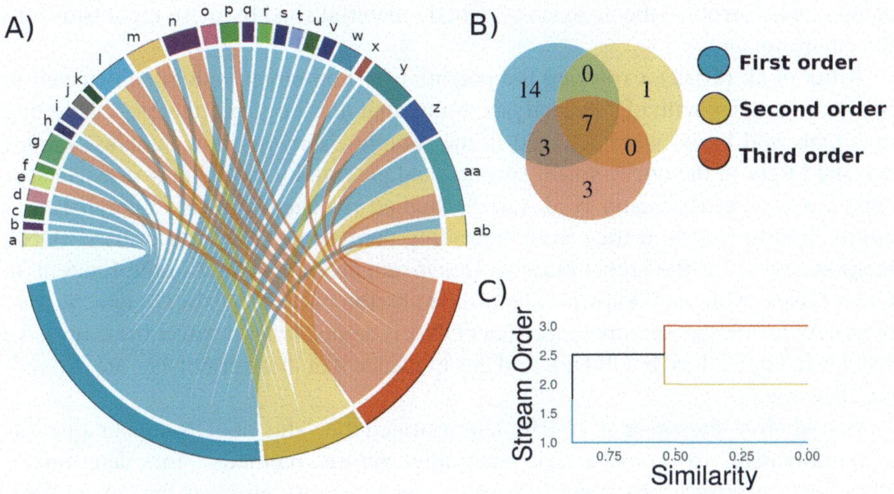

Fig. 1 Fish diversity found in Igarapés orders at Acará Basin. (**a**) The species richness found in the igarapés orders, letters (above) represent each MOTU assigned to fish species in the igarapés orders (below); (**b**) occurrence and overlap of detected MOTUs in the three igarapés order analyzed, and (**c**) fish diversity similarity dendrogram, determined by Jaccard method, represents the diversity similarity between the different igarapés orders

surveys, have limitations in terms of time, effort, and sampling coverage (Packer et al., 2009).

In contrast, eDNA analysis enables broader spatial coverage, facilitating the detection of both common and rare fish species across different habitats and river systems (Shaw et al., 2016). This comprehensive approach may contribute to a more accurate understanding of fish biodiversity patterns in the Amazon region. By harnessing genetic information present in environmental samples, research can uncover hidden aspects of ecosystems and inform conservation efforts. As technology advances and our understanding of eDNA improves, this technique is expected to play an increasingly significant role in environmental research and management, even though it still has problems and limitations.

3 Detection and Monitoring of Alien, Invasive, or Rare Species

Environmental DNA (eDNA) analysis has proven to be a valuable tool for detecting and monitoring alien and exotic species in various ecosystems (Fonseca et al., 2023). The concept of exotic and alien species refers to organisms that have been introduced to regions outside their native range, often as a result of human activities (Rotherham, 2009). These species can have significant ecological, economic, and social impacts on the recipient ecosystems. The detection of alien and exotic species using eDNA involves the analysis of genetic material shed by these organisms into the environment.

Ritter et al. (2022), exploring the potential for non-target taxa to be detected in by-catch samples with eDNA analysis, found that non-target vertebrates, including mammals and birds, were detected in the samples, showing how aquatic samples can shed light in the environment species' and potentially record new invasive species in a region. Robinson et al. (2018) studied two major invasive crayfishes in North America to see if they were expanding their range. The authors used eDNA method to test for the signal crayfish (*Pacifastacus leniusculus*) in the Laurentian Great Lakes, which are known to be invaded by the rusty crayfish (*Orconectes rusticus*). As result, the authors found that eDNA is a good way to detect these invasive species in large lakes, but did not find any evidence that these crayfishes are expanding their range.

A study by Davison et al. (2019) demonstrated the value of eDNA techniques, as a complement to conventional sampling/capture methods, for determining *Pseudorasbora parva* distribution within a given area to inform management decisions on eradication procedures. This invasive species is a small cyprinid fish that first arrived in Europe as a contaminant of Asian carp consignments to Romania in the 1960s and is now widespread throughout the continent.

The Nile Tilapia (*Oreochromis niloticus*) is an introduced species in Amazonian environments. However, few studies have been conducted to assess the occurrence of this alien species in Amazonian environments. Souza et al. (2021, in prep.) developed specific primers to a 12S mitochondrial DNA fragment that can be used to monitor the distribution of Nile Tilapia in Amazon aquatic environments. In this research, realized on an igarapé located in the urban zone of Manaus, Amazonas, Brazil, a comparative analysis was conducted between the caught Nile Tilapia specimens along the water-body and the positive amplification from the 12S eDNA fragment using specific pair primers. This work shows the possibilities of this technique in identifying invasive species in the Amazon basin.

By detecting the presence of alien and exotic species through eDNA analysis, we can obtain valuable information for conservation and management efforts before it turns into a bigger problem. Early detection of these species is crucial for implementing effective control and eradication measures in the region, as well as preventing their further spread.

4 Environmental RNA

Environmental RNA (eRNA) is an emerging field of research that complements the use of environmental nucleic acids in ecological studies. Fish eRNA is abundant in aquatic environment and can be used to accurately detect a wide range of fish species. It can be more sensitive than eDNA for detecting fish in low abundance and having a higher positive predictivity than eDNA, meaning that it is less likely to produce false positives (Miyata et al., 2021).

One of the key advantages of eRNA analysis is its ability to provide real-time information about the biological processes occurring in ecosystems. RNA molecules are transient and reflect the immediate state of gene expression, offering a dynamic snapshot of the activity of organisms in response to environmental conditions (Li et al., 2015). By studying eRNA, researchers can gain a deeper understanding of ecological processes, such as species interactions, metabolic activity, and responses to environmental changes.

eRNA analysis has a wide range of applications in environmental research. Along with eDNA (Horodesky et al., 2020), eRNA can be used to study microbial communities, including bacteria, archaea, and fungi, and their functional roles in nutrient cycling, decomposition, and symbiotic relationships. By examining the eRNA profiles, key microbial players and their metabolic pathways in specific ecosystems can be identified (McGrath et al., 2008; Poretsky et al., 2009).

In addition to microbial communities, eRNA analysis can also shed light on the functional dynamics of others organisms. For example, it can be used to investigate gene expression patterns in plants, providing insights into stress responses, developmental processes, and interactions with other organisms (Whangbo & Hunter, 2008). Furthermore, eRNA analysis can aid in the study of animal behavior, including mating calls, communication signals, and responses to environmental stimuli (Cristescu, 2019).

eRNA is a more comprehensive tool for studying the community structure of living organisms in a given environment than eDNA, as eRNA only detects living organisms, and eDNA can detect both living and dead organisms. Additionally, eRNA metabarcoding can detect more unique organisms and greater alpha diversity than eDNA metabarcoding because eRNA from larger organisms does not interfere with the detection of eRNA from smaller organisms. This higher detection of microfauna using eRNA can lead to greater statistical power in studies of community differences between treatments. Utilizing eRNA for environmental monitoring can improve the detection of benthic organisms that are rarely seen through morphological identification (Giroux et al., 2022).

For the Amazon, the field of eRNA remains unexplored. However, it has the potential to be used to assess the environmental health of the region. Several studies with Amazonian fishes can help to determine key RNAs as target of responses to environmental and anthropogenic stressors, such as *hif-1α* and hypoxia tolerance (Heinrichs-Caldas et al., 2019), *hif-1α* in response to benzo[a]pyrene (Silva et al., 2017), *tp53* and exposition to pesticides (Souza et al., 2023), and genes related to immune response in relation to temperature and parasitism (Costa et al., 2021). The studies with fish and eRNA have a number of implications for fish conservation and management. For example, eRNA could be used to monitor fish populations over time to track changes in abundance and distribution and could also be used to identify areas that are important for fish conservation. Additionally, eRNA could be used to assess the impact of human activities on fish populations (Miyata et al., 2021).

eRNA analysis offers a novel and powerful approach to investigate ecological processes and biodiversity in their natural context. Its integration with eDNA analysis can provide a comprehensive understanding of ecosystems, bridging the gap between genetic potential and functional activity. Continued advancements in technology and bioinformatics will further enhance the applications and reliability of eRNA analysis in environmental research.

5 Troubles with Environmental Nucleic Acid in the Amazon

Despite its tremendous potential, environmental nucleic acid analysis also has limitations. The physicochemical properties of water, such as temperature, pH, salinity, and organic matter content, can significantly impact the analysis (Barnes et al., 2014). These factors can affect DNA degradation rates, DNA extraction efficiency, and the ability to detect and amplify target DNA sequences. For example, high temperatures and low pH levels can accelerate DNA degradation, making it more challenging to detect and analyze accurately (Strickler et al., 2015). Considering the greater instability of RNA, rapid degradation is also expected (Tsuri et al., 2020); thus, the development of fixation/conservation of samples is a crucial process to ensure the quality of the data.

Considering that the physicochemical characteristics of the Amazon rivers vary depending on the location, and the season, some general trends can be observed; the water temperature in the Amazon rivers ranges from 20 to 28 °C. The temperature is generally higher in the shallower, more exposed areas of the river, and lower in the deeper, more shaded areas. Jo et al. (2022) found that warm temperature increases the degradation of eRNA, which is related to its instability when compared to eDNA, making it more susceptible to degradation by enzymes and other factors. These results suggest that eRNA may not be as reliable as eDNA for monitoring aquatic ecosystems in warm environments.

Other relevant environmental variable is the pH. Amazon rivers is typically slightly acidic, with a pH of around 5.0. Its known that pH influences in the degradation and stability of nucleic acids. Strickler et al. (2015) found that neutral pH, a condition favorable to microbial growth, increases the eDNA degradation rate, which indicates that these factors can be biologically mediated. On the other hand, Jo et al. (2022) found that an alkaline environment speeds up the degradation of eRNA. The results observed with eRNA suggest that eRNA may be useful for detecting recent or ongoing biological activity, as it degrades more rapidly than eDNA, and that the environmental conditions should be carefully considered when designing eRNA monitoring studies.

Similarly, high levels of organic matter in water samples can inhibit DNA extraction or introduce PCR inhibitors, leading to false-negative results. Dissolved organic carbon (DOC) is a major component of the total carbon in Amazonian rivers. The relation between DOC and environmental nucleic acids showed a strong relation to other physicochemical parameters (Barnes et al., 2014). Overall, the research on the effects of DOC on eDNA persistence is still limited. However, the available evidence suggests that DOC can play a significant role in influencing eDNA persistence in aquatic systems. Using eDNA metabarcoding, Sales et al. (2020) detected a difference in tropical mammal families between the Atlantic Forest and Amazon biomes and associated this variation with the difference in pH between the biomes. The black waters of the Amazon have an acidic pH, and according to Sales et al. (2020), the eDNA detected in this environment may be derived from recent contact of the individuals.

Additionally, the presence of inhibitors and contaminants in environmental samples can hinder the accurate detection of eDNA. Water samples may contain substances that can interfere with DNA extraction, amplification, and sequencing. These inhibitors can arise from various sources, such as humic acids, heavy metals, and other organic and inorganic compounds present in the environment (Barnes et al., 2014; Gomes et al., 2017; Tsuji et al., 2017). In order to ensure PCR success, adjuvant chemical substances can be used to enhance the success of PCR in eDNA analysis. Some common adjuvants are bovine serum albumin, dimethyl sulfoxide, and Triton X-100 during the DNA extraction and formamide (Kawato et al., 2021). Managing and mitigating the effects of these inhibitors is crucial to ensure reliable and accurate eDNA analysis.

Another significant challenge in eDNA analysis is the lack of comprehensive and standardized databases for reference sequences. DNA sequences obtained from environmental samples need to be compared to existing reference databases to identify the species or taxonomic group accurately. However, many organisms, especially those from less-studied or newly discovered species, may not have their DNA sequences included in existing databases (Ruppert et al., 2019). This limitation can make it difficult to confidently assign taxonomic identities to the detected eDNA sequences, which feels harder when working with a high biodiverse biome such as the Amazon rivers.

Furthermore, the establishment of reference databases requires considerable effort and resources, as it involves the collection and sequencing of DNA samples from a wide range of organisms across various ecosystems. Building a comprehensive database with representative genetic information from different Amazonian taxonomic groups is an ongoing endeavor, and it may take time to reach a level of completeness that ensures accurate species identification through eDNA analysis. Jackman et al. (2021) using eDNA metabarcoding in Amazonian aquatic environments indicated the limitation of the results obtained precisely due to a lack of a reference database for neotropical fish species.

6 Relation Between Environmental Nucleic Acids and One Health

The One Health concept is an integrative approach that recognizes the interconnectedness of human health, animal health, and the health of the environment in which they coexist (WHO, 2017). It emphasizes the collaborative efforts of multiple disciplines, including human medicine, veterinary medicine, environmental science, and public health, to address complex health challenges (Zinsstag et al., 2011).

At its core, the One Health concept acknowledges that the health of humans, animals, and the environment are intimately linked. It recognizes that diseases and health issues can transcend species boundaries and can arise from the complex interactions between humans, animals, and their shared ecosystems (Keune et al., 2022). By considering the interdependencies between these domains, the One Health approach aims to improve health outcomes for all.

One Health recognizes that the health of humans is interconnected with the health of animals. Many infectious diseases are zoonotic, meaning they can be transmitted between animals and humans. Examples include Ebola, avian influenza, and rabies. By understanding the dynamics of disease transmission between animals and humans, and by promoting collaboration between human and veterinary medicine, early detection and prevention of zoonotic diseases can be enhanced (Cunningham et al., 2017).

Furthermore, the One Health concept highlights the importance of the environment in supporting health. Environmental factors, such as climate change, pollution, and habitat loss, can have profound impacts on the health of both humans and animals (Zinsstag et al., 2018). For instance, changes in temperature and rainfall patterns can influence the spread of vector-borne diseases, while pollution can lead to adverse health effects in both humans and wildlife. By addressing environmental challenges and promoting sustainable practices, the One Health approach aims to protect and improve the well-being of all living beings.

The applications of the One Health concept are diverse and far-reaching. It involves collaboration and communication across disciplines, fostering partnerships between human health professionals, veterinarians, ecologists, environmental scientists, policymakers, and other stakeholders. One Health recognizes the importance of preserving biodiversity and maintaining healthy ecosystems for the benefit of all species. By promoting sustainable practices, protecting wildlife habitats, and addressing environmental degradation, the One Health approach supports the long-term health and resilience of ecosystems and the species within them, and this where enter the environmental nucleic acids.

Environmental nucleic acid analysis has emerged as a valuable tool for studying rivers and their ecosystems, offering insights into biodiversity, ecological health, and the One Health concept. By extracting and analyzing genetic material shed by organisms into the water, eDNA analysis provides a non-invasive and efficient approach to monitor riverine ecosystems and understand the interconnectedness of human, animal, and environmental health (Ogden, 2022).

The use of environmental nucleic acids in river ecosystems is closely linked to the One Health concept, which recognizes the interconnectedness of human, animal, and environmental health (Rock et al., 2009). Rivers serve as vital resources for human populations, providing drinking water, irrigation, and recreational opportunities. Additionally, they support diverse aquatic species that contribute to ecological balance and ecosystem services. By monitoring environmental nucleic acids in rivers, we can assess the health and functioning of these ecosystems and understand the potential impacts on both human and animal health.

In the context of One Health, the use of environmental nucleic acids has proven effective, especially in monitoring human and veterinary parasitic diseases. The use of eDNA can be useful in disease surveillance; however, of all publications, the application of the technique in parasitological studies represents a small proportion (Sengupta et al., 2022). Furthermore, there is a large gap in studies in developing countries, where parasitic diseases are prevalent, the review carried out by Sengupta et al. (2022) clearly demonstrates that monitoring of parasitic diseases using the eDNA tool is mostly carried out in European countries and the United States.

eDNA can also be used to detect insect vectors of diseases that are public health problems such as malaria and dengue and can be used as early detection and identification strategies, assisting in vector control (Gutiérrez-López et al., 2023; Odero et al., 2018; Sakata et al., 2022). The application of eDNA to identify insect vectors has proven successful (Gutiérrez-López et al., 2023; Odero et al., 2018). However, as far as we know, no study has applied eDNA to detect insect vectors in the Amazon, where populations are highly vulnerable to these endemics. eRNA has also been used in an attempt to better understand the occurrence and transmission of pathogens that harm the conservation of wild species and, more recently, to monitor the virus that drove the current global COVID-19 pandemic, as well as other human pathogens (Farrel et al., 2021).As stated in this text before, the application of environmental nucleic acids analysis in rivers has several benefits, which can be implicated for the One Health concept:

- Biodiversity Monitoring: eDNA analysis enables comprehensive monitoring of riverine biodiversity, including fish, amphibians, and invertebrates. This information is crucial for assessing ecosystem health, identifying species at risk, and informing conservation efforts (Harrison et al., 2019).
- Early Detection of Pathogens: eDNA analysis can be used to detect the presence of pathogens and disease-causing organisms in rivers. By monitoring for the genetic signatures of specific pathogens, such as bacteria or viruses, it is possible to identify potential health risks to both human and animal populations (Amarasiri et al., 2021).
- Assessing Anthropogenic Impacts: The analysis of eDNA can provide insights into the impact of human activities on river ecosystems. This includes the detection of pollutants, invasive species, and changes in biodiversity patterns, which can have direct and indirect effects on human and animal health (Liu et al., 2020).
- Integrated Management Approaches: The use of eDNA in river monitoring supports an integrated and holistic approach to riverine ecosystem management. By considering the interactions between human, animal, and environmental health, policymakers and stakeholders can make informed decisions to promote sustainable resource use and protect the health of both ecosystems and communities.

In conclusion, the use of environmental nucleic acid analysis in studying Amazon fish biodiversity and health can revolutionize our understanding of this biodiverse region. By providing a non-invasive, comprehensive, and efficient method of species detection and monitoring, this analysis can contribute to our knowledge of fish

Fig. 2 Figure shows how the environmental nucleic acids in the aquatic environment. (**a**) Environmental DNA and RNA spread through the aquatic system. (**b**) Local fish species, in this case *Colossoma macropomum*, shed eDNA and eRNA in the environment. (**c**) The technique can also be used to identify invasive species, such as *Oreochromis niloticus*. (**d**) The relation between the environment health and the population living nearby, applying the environmental nucleic acid in the One Health concept

diversity, health, and adaptations, assists in conservation efforts, aids in the sustainable management of Amazonian freshwater ecosystems, and provides a direct talk between environment state and its relation to One Health. We know that the application of eDNA and eRNA has limitations and that they have not solved the problems in the Amazon. However, what is evident here are tools that can be applied to improve knowledge of the Amazon megadiversity and even in public health issues. The potential is undeniable, but there is still a long way to go in terms of applying the techniques. We believe that the first step is certainly the establishment of a robust reference database for neotropical fish species (Fig. 2).

References

Alho, C. J. R., Reis, R. E., & Aquino, P. U. (2015). Amazonian freshwater habitats experiencing environmental and socioeconomic threats affecting subsistence fisheries. *Ambio, 44*, 412–425.

Amarasiri, M., Furukawa, T., Nakajima, F., & Sei, K. (2021). Pathogens and disease vectors/hosts monitoring in aquatic environments: Potential of using eDNA/eRNA based approach. *Science of the Total Environment, 796*, 148810.

Barnes, M. A., Turner, C. R., Jerde, C. L., Renshaw, M. A., Chadderton, W. L., & Lodge, D. M. (2014). Environmental conditions influence eDNA persistence in aquatic systems. *Environmental Science & Technology, 48*(3), 1819–1827.

Batista, L. M., Sá-Leitão, C. S., Souza, É. M. S., Anjos-Santos, C. H., & Almeida-Val, V. M. F. (2021). Addressing Amazonian fish diversity using environmental DNA (eDNA): A first glance.

Bevilaqua, D. R., Melo, S. A., Carvalho Freitas, C. E., Silva, A. C. V., & Silva Batista, J. (2020). First environmental DNA (eDNA) record of Central Amazon in a floodplain lake: Extraction method selection and validation. *Brazilian Journal of Development, 6*(11), 87606–87621.

Costa, J. C., Souza, S. S., Castro, J. D. S., Amanajás, R. D., & Val, A. L. (2021). Climate change affects the parasitism rate and impairs the regulation of genes related to oxidative stress and ionoregulation of *Colossoma macropomum*. *Scientific Reports, 11*(1), 22350.

Cristescu, M. E. (2019). Can environmental RNA revolutionize biodiversity science? *Trends in Ecology & Evolution, 34*(8), 694–697.

Cunningham, A. A., Daszak, P., & Wood, J. L. (2017). One health, emerging infectious diseases and wildlife: Two decades of progress. *Philosophical Transactions of the Royal Society B: Biological Sciences, 372*(1725), 20160167.

Davison, P. I., Falcou-Préfol, M., Copp, G. H., Davies, G. D., Vilizzi, L., & Créach, V. (2019). Is it absent or is it present? Detection of a non-native fish to inform management decisions using a new highly-sensitive eDNA protocol. *Biological Invasions, 21*, 2549–2560.

Farrel, J. A., Whitmore, L., & Duffy, D. J. (2021). The promise and pitfalls of environmental DNA and RNA approaches for the monitoring of human and animal pathogens from aquatic sources. *BioScience, 71*, 609–625.

Fonseca, V. G., Davison, P. I., Creach, V., Stone, D., Bass, D., & Tidbury, H. J. (2023). The application of eDNA for monitoring aquatic non-indigenous species: Practical and policy considerations. *Diversity, 15*(5), 631.

Giroux, M. S., Reichman, J. R., Langknecht, T., Burgess, R. M., & Ho, K. T. (2022). Environmental RNA as a tool for marine community biodiversity assessments. *Scientific Reports, 12*(1), 17782.

Gomes, G. B., Hutson, K. S., Domingos, J. A., Chung, C., Hayward, S., Miller, T. L., & Jerry, D. R. (2017). Use of environmental DNA (eDNA) and water quality data to predict protozoan parasites outbreaks in fish farms. *Aquaculture, 479*, 467–473.

Gutiérrez-López, R., Egetar, B., Paupy, C., Rahola, N., Makanga, B., Jiolle, D., Bourret, V., Melo, M., & Loiseau, C. (2023). Monitoring mosquito richness in an understudied area: Can environmental DNA metabarcoding be a complementary approach to adult trapping? *Bulletin of Entomological Research, 113*(4), 456–468.

Harrison, J. B., Sunday, J. M., & Rogers, S. M. (2019). Predicting the fate of eDNA in the environment and implications for studying biodiversity. *Proceedings of the Royal Society B, 286*(1915), 20191409.

Heinrichs-Caldas, W., Campos, D. F., Paula-Silva, M. N., & Almeida-Val, V. M. F. (2019). Oxygen-dependent distinct expression of hif-1α gene in aerobic and anaerobic tissues of the Amazon Oscar, *Astronotus crassipinnis*. *Comparative Biochemistry and Physiology Part B: Biochemistry and Molecular Biology, 227*, 31–38.

Horodesky, A., Castilho-Westphal, G. G., Pont, G. D., Faoro, H., Balsanelli, E., Tadra-Sfeir, M. Z., Cozer, N., Pie, M. R., & Ostrensky, A. (2020). A. Metagenomic analysis of the bacterial microbiota associated with cultured oysters (*Crassostrea sp.*) in estuarine environments. *Anais da Academia Brasileira de Ciências, 92*(suppl 1), e20180432. https://doi.org/10.1590/0001-3765202020180432

Jackman, J. M., Benevenuto, C., Coscia, I., Carvalho, C. O., Ready, J. S., Boubli, J. P., Magnusson, W. E., McDevitt, A. D., & Sales, N. G. (2021). eDNA in a bottleneck: Obstacles to fish metabarcoding studies in megadiverse freshwater systems. *Environmental DNA, 3*, 837–849.

Jo, T., Tsuri, K., Hirohara, T., & Yamanaka, H. (2022). Warm temperature and alkaline conditions accelerate environmental RNA degradation. *Environmental DNA, 5*, 836.

Kawato, M., Yoshida, T., Miya, M., Tsuchida, S., Nagano, Y., Nomura, M., Yabuki, A., Fujiwara, Y., & Fujikura, K. (2021). Optimization of environmental DNA extraction and amplification methods for metabarcoding of deep-sea fish. *MethodsX, 8*, 101238. https://doi.org/10.1016/j.mex.2021.101238

Keune, H., Payyappallimana, U., Morand, S., Rüegg, S. R., Visseren-Hamakers, I., & Kok, M. T. (2022). One health and biodiversity. In *Transforming biodiversity governance* (pp. 93–114).

Li, Y., Varala, K., & Coruzzi, G. M. (2015). From milliseconds to lifetimes: Tracking the dynamic behavior of transcription factors in gene networks. *Trends in Genetics, 31*(9), 509–515. https://doi.org/10.1016/j.tig.2015.05.005

Liu, Q., Zhang, Y., Wu, H., Liu, F., Peng, W., Zhang, X., & Zhang, H. (2020). A review and perspective of eDNA application to eutrophication and HAB control in freshwater and marine ecosystems. *Microorganisms, 8*(3), 417.

McGrath, K. C., Thomas-Hall, S. R., Cheng, C. T., Leo, L., Alexa, A., Schmidt, S., & Schenk, P. M. (2008). Isolation and analysis of mRNA from environmental microbial communities. *Journal of Microbiological Methods, 75*(2), 172–176.

Miya, M., Sato, Y., Fukunaga, T., Sado, T., Poulsen, J. Y., Sato, K., & Iwasaki, W. (2015). MiFish, a set of universal PCR primers for metabarcoding environmental DNA from fishes: Detection of more than 230 subtropical marine species. *Royal Society Open Science, 2*(7), 150088.

Miyata, K., Inoue, Y., Amano, Y., Nishioka, T., Yamane, M., Kawaguchi, T., et al. (2021). Fish environmental RNA enables precise ecological surveys with high positive predictivity. *Ecological Indicators, 128*, 107796.

Odero, J., Gomes, B., Fillinger, U., & Weetman, D. (2018). Detection and quantification of *Anopheles gambiae* sensu lato mosquito larvae in experimental aquatic habitats using environmental DNA (eDNA). *Wellcome Open Research, 3*, 26.

Ogden, L. E. (2022). The emergence of eDNA: An interdisciplinary tool helps monitor biodiversity and health. *Bioscience, 72*(1), 5–12.

Packer, L., Gibbs, J., Sheffield, C., & Hanner, R. (2009). DNA barcoding and the mediocrity of morphology. *Molecular Ecology Resources, 9*, 42–50.

Poretsky, R. S., Gifford, S., Rinta-Kanto, J., Vila-Costa, M., & Moran, M. A. (2009). Analyzing gene expression from marine microbial communities using environmental transcriptomics. *JoVE (Journal of Visualized Experiments), 24*, e1086.

Ritter, C. D., Dal Pont, G., Stica, P. V., Horodesky, A., Cozer, N., Netto, O. S. M., et al. (2022). Wanted not, wasted not: Searching for non-target taxa in environmental DNA metabarcoding by-catch. *Environmental Advances, 7*, 100169.

Robinson, C. V., Webster, T. M. U., Cable, J., James, J., & Consuegra, S. (2018). Simultaneous detection of invasive signal crayfish, endangered white-clawed crayfish and the crayfish plague pathogen using environmental DNA. *Biological Conservation, 222*, 241–252.

Rock, M., Buntain, B. J., Hatfield, J. M., & Hallgrímsson, B. (2009). Animal–human connections,"one health," and the syndemic approach to prevention. *Social Science & Medicine, 68*(6), 991–995.

Rotherham, I. (2009). Exotic and alien species in a changing world. *Ecos-British Association of Nature Conservationists, 30*(2), 42–49.

Rourke, M. L., Fowler, A. M., Hughes, J. M., Broadhurst, M. K., DiBattista, J. D., Fielder, S., & Furlan, E. M. (2022). Environmental DNA (eDNA) as a tool for assessing fish biomass: A review of approaches and future considerations for resource surveys. *Environmental DNA, 4*(1), 9–33.

Ruppert, K. M., Kline, R. J., & Rahman, M. S. (2019). Past, present, and future perspectives of environmental DNA (eDNA) metabarcoding: A systematic review in methods, monitoring, and applications of global eDNA. *Global Ecology and Conservation, 17*, e00547.

Sakata, M. K., Sato, M., Sato, M. O., Watanabe, T., Mitsuishi, H., Hikitsuchi, T., Kobayashi, J., & Minamoto, T. (2022). Detection and persistence of environmental DNA (eDNA) of the different developmental stages of a vector mosquito, *Culex pipiens pallens*. *PLoS One, 17*(8), e0272653.

Sales, N. G., Kaizer, M. C., Coscia, I., Perkins, J. C., Highlands, A., Boubli, J. P., Magnusson, W. E., Silva, M. N. F., Benevenuto, C., & Mcdevitt, A. (2020). Assessing the potential of environmental DNA metabarcoding for monitoring Neotropical mammals: A case study in the Amazon and Atlantic Forest, Brazil. *Mammal Review, 50*(3), 221–225.

Santana, C. D., Parenti, L. R., Dillman, C. B., Coddington, J. A., Bastos, D. A., Baldwin, C. C., Zuanon, J., Torrente-Vilara, G., Covain, R., Menezes, N. A., Datovo, A., Sado, T., & Miya, M. (2021). The critical role of natural history museums in advancing eDNA for biodiversity studies: A case study with Amazonian fishes. *Scientific Reports, 11*(1), 18159. https://doi.org/10.1038/s41598-021-97128-3

Sengupta, M. E., Lynggaard, C., Mukaratirwa, S., Vennervald, B. J., & Stensgaard, A. S. (2022). Environmental DNA in human and veterinary parasitology - Current applications and future prospects for monitoring and control. *Food and Waterborne Parasitology, 29*, e00183.

Shaw, J. L., Clarke, L. J., Wedderburn, S. D., Barnes, T. C., Weyrich, L. S., & Cooper, A. (2016). Comparison of environmental DNA metabarcoding and conventional fish survey methods in a river system. *Biological Conservation, 197*, 131–138.

Shokralla, S., Spall, J. L., Gibson, J. F., & Hajibabaei, M. (2012). Next-generation sequencing technologies for environmental DNA research. *Molecular Ecology, 21*(8), 1794–1805.

Silva, G. S. D., Fé, L. M. L., Silva, M. D. N. P. D., & Almeida-Val, V. M. F. (2017). Ras oncogene and hypoxia-inducible factor-1 alpha (hif-1α) expression in the Amazon fish *Colossoma macropomum* (Cuvier, 1818) exposed to benzo [a] pyrene. *Genetics and Molecular Biology, 40*, 491–501.

Sioli, H. (1990). *Amazônia: fundamentos da ecologia da maior região de florestas tropicais* (p. 73). Editora Vozes.

Souza, S. S., da Silva Castro, J., Campos, D. F., Pereira, R. S., Bataglion, G. A., da Silva, G. S., & de Almeida-Val, V. M. F. (2021). Temporal exposure to malathion: Biochemical changes in the Amazonian fish tambaqui, colossoma macropomum. *Aquatic Toxicology, 241*, 105997.

Souza, S. S., Costa, J. C., Silva, G. S., & Almeida-Val, V. M. F. (2023). Malathion alters the transcription of target genes of the tumour suppressor tp53 and cancerous processes in *Colossoma macropomum*: Mechanisms of DNA damage response, oxidative stress and apoptosis. *Chemico-Biological Interactions, 374*, 110405.

Strickler, K. M., Fremier, A. K., & Goldberg, C. S. (2015). Quantifying effects of UV-B, temperature, and pH on eDNA degradation in aquatic microcosms. *Biological Conservation, 183*, 85–92.

Tsuji, S., Inui, R., Nakao, R., Miyazono, S., Saito, M., Kono, T., & Akamatsu, Y. (2022). Quantitative environmental DNA metabarcoding shows high potential as a novel approach to quantitatively assess fish community. *Scientific Reports, 12*(1), 21524.

Tsuji, S., Yamanaka, H., & Minamoto, T. (2017). Effects of water pH and proteinase K treatment on the yield of environmental DNA from water samples. *Limnology, 18*, 1–7.

Tsuri, K., Ikeda, S., Hirohara, T., Shimada, Y., Minamoto, T., & Yamanaka, H. (2020). Messenger RNA typing of environmental RNA (eRNA): A case study on zebrafish tank water with perspectives for the future development of eRNA analysis on aquatic vertebrates. *Environmental DNA, 3*, 14–12.

Val, A. L. (2019). Fishes of the Amazon: Diversity and beyond. *Anais da Academia Brasileira de Ciências, 91*, e20190260.

Vergueiro, A. M. K. J., Almeida-Val, V. M. F., & Sá-Leitão, C. S. (2024). Influência da sazonalidade pluviométrica e de variáveis ambientais na diversidade de peixes em igarapés da Reserva Florestal Adolpho Ducke: Abordagem via dna ambiental. *Manaus: [s.n]*. x, 80 f.

Whangbo, J. S., & Hunter, C. P. (2008). Environmental RNA interference. *Trends in Genetics, 24*(6), 297–305.

WHO, O. (2017). *One health* (p. 736). World Health Organization.

Zinsstag, J., Crump, L., Schelling, E., Hattendorf, J., Maidane, Y. O., Ali, K. O., & Cissé, G. (2018). Climate change and one health. *FEMS Microbiology Letters, 365*(11), fny085.

Zinsstag, J., Schelling, E., Waltner-Toews, D., & Tanner, M. (2011). From "one medicine" to "one health" and systemic approaches to health and well-being. *Preventive Veterinary Medicine, 101*(3–4), 148–156.

Increasing Temperatures Enhance Hypoxic Encounters in the Amazon: Consequences for Air-Breathing Fish

Bernd Pelster and Chris M. Wood

Abstract

To cope with frequently hypoxic conditions in the Amazon, many fish refer to breathing air. In order to avoid a possible loss of oxygen taken up in the air-breathing organ to hypoxic water through the gills in many of these fish, gill surface area is reduced. Gills, however, are multifunctional organs, so that breathing air affects not only gas exchange, but also other physiological functions typically located at the gills, for example, ion regulation, acid-base regulation, or nitrogen excretion. Elevated temperatures cause an increase in metabolic activity, reduce oxygen solubility in water, and therefore are expected to stimulate air-breathing. This may potentially increase the risk of predation at the water surface. An increased reliance on aerial respiration may also stimulate progressive reduction of the gill surface area, which would affect other functions located at the gills.

1 Introduction

The Amazon basin is well known for hosting a rich fauna of freshwater fish. More than 2700 species have been described so far, representing about 20% of all freshwater fish (Jézéquel et al., 2020; Jung et al., 2020; Tencatt et al., 2020; Val & Wood, 2022), although Amazon waters represent a challenging environment. Amazon

B. Pelster (✉)
Institute for Zoology, Universität Innsbruck, Innsbruck, Austria

Center for Molecular Biosciences, University Innsbruck, Innsbruck, Austria
e-mail: bernd.pelster@uibk.ac.at

C. M. Wood
Department of Zoology, University of British Columbia, Vancouver, BC, Canada

S. S. de Souza et al. (eds.), *The Future of Amazonian Aquatic Biota*,
https://doi.org/10.1007/978-3-031-66822-7_14

waters are ion-poor, and so-called black water is rich in humic substances (humic and fulvic acids) and other organic compounds (collectively termed "dissolved organic carbon = DOC") which lower the pH (Thurman, 1985), thereby complicating the necessary ion uptake from freshwater (Gonzalez et al., 2005). Seasonal variations with rainy season and dry season are connected to frequently hypoxic conditions, impeding oxygen uptake (Val & Wood, 2022). Similarly, circadian cycles with light-dependent photosynthetic oxygen production often result in hyperoxic conditions during daytime, and hypoxic conditions at night. Hypoxic events therefore are regularly encountered in stagnant water and during darkness (Val et al., 2006; Val & Almeida-Val, 1995).

Considering the current conditions of climate change, the overall forecast for Amazonia is an increase in extreme environmental conditions, more frequent and prolonged hypoxic events, and desiccation (Ficke et al., 2007; IPCC, 2014; Val & Wood, 2022). The frequency of unusually high-temperature events has already increased over the last decade, and Tmax anomalies in the order of 3–3.5 °C have been recorded for the Amazon (Geirinhas et al., 2018). A temperature increase of up to 3.5 °C or even 5 °C has been predicted for Manaus and the Amazon (Val & Wood, 2022). Over the last 25 years, the amplitude between the high and the low water level has increased (Schöngart & Junk, 2020; Val & Wood, 2022). This results in an increase in flooded forest areas (Barichivich et al., 2018), and more organic material is eluted, increasing the DOC content in water. Intensified and prolonged flood events will create prolonged periods of stagnant water and intensify hypoxic events. Intensified and prolonged flood events causing hypoxia will also result in greater amounts of organic matter in the water. In vegetation-rich environments, hypoxia is often paired with hypercarbia, and the degradation of this organic matter may also intensify hypoxic events and increase PCO_2 (Furch & Junk, 1997; Ultsch, 1996; Val & Wood, 2022). Elevated concentrations of humic substances and an elevated PCO_2 will acidify the poorly buffered water.

Higher temperatures decrease oxygen solubility in water, while oxygen diffusivity increases slightly. The Q_{10} of Krogh's constant of diffusion therefore is close to 1.1 (Dejours, 1981), and applying mass-transfer modeling to assess the oxygen bioavailability resulted in a Q_{10} value of ~1.24 (Atkinson et al., 2022). The Q_{10} of metabolism, however, is close to 2, resulting in an increasing oxygen demand of tissues with increasing temperatures. An increase in metabolic rate connected to elevated oxidative stress has already been reported for Amazon fishes when exposed to higher temperatures and elevated PCO_2 (Campos et al., 2019). Fishes living at high temperatures near their upper thermal limit (as in equatorial regions such as the Amazon basin) appear to have a limited capacity to tolerate elevated temperatures (Campos et al., 2019; Jung et al., 2020). General predictions, however, are difficult, and Campos et al. pointed out that individual responses to thermal stress depend on metabolic activity, and a lower thermal sensitivity of metabolic rate was associated with greater thermal tolerance in two Characiform fish (Campos et al., 2017). Based on this and other studies, Val and Wood (2022) recently concluded that climate change over the longer term will likely favor the long-term survival of predominantly water-breathing species with low routine metabolic rates, low temperature sensitivity of routine metabolic rates, high anaerobic capacity, high hypoxia tolerance, and high thermal tolerance.

Frequently hypoxic conditions often connected to hypercarbia apparently stimulated the transition to air-breathing, and indeed, a large number of air-breathing fishes are found in the Amazon region (Bayley et al., 2019; Damsgaard et al., 2020; Graham, 1997, 2011a; Lefevre et al., 2014). Air-breathing behavior is seen as a short visit at the water surface taking an air-breath, but the fish remains in the water. Alternatively, the fish may leave the water for more or less extended periods of time, and these fish are considered amphibious fishes (Damsgaard et al., 2020; Sayer, 2005; Turko et al., 2021; Wright, 2021). Aerial surface respiration may also be induced by aquatic hypoxia, but in this case, fish actually rely on the higher oxygen partial pressure of the top water layers in contact to air, they do not breath air (Kramer & Mehegan, 1981). In this review, the main focus is on air-breathing fish, which are frequently encountered in the Amazon basin, and on the consequences of an elevated air-breathing activity, expected to occur under elevated temperatures.

2 More Frequent and Long-Lasting Hypoxic Events Are Expected to Intensify Air-Breathing Activity

In tropical areas and in particular in the Amazon basin, a large number of fishes have developed strategies to use air as an additional source for oxygen (Graham, 1997, 2011b; Val & Almeida-Val, 1995). Gills and skin may be modified to allow for the uptake of oxygen from the air, as observed in *Monopterus* (Asian swamp eel), blennies, and mudskippers, for example (Graham, 2011b; Ishimatsu, 2012; Pelster et al., 1988). Cutaneous respiration is observed in a large number of amphibious fishes, which leave the water for more or less prolonged periods of time (Turko et al., 2021; Wright, 2021). Members of several families of fishes, such as Loricariidae, Trichomycteridae, Cobitidae, or Cannichthyidae, have been reported to use either the stomach or gut segments for aerial respiration (Nelson, 2014; Scott et al., 2017). The tambaqui *Colossoma macropomum* significantly extends its lower lip to pick up oxygen from the water surface or the air, and the buccal cavity or extensions of the gill cavity are used for aerial respiration. Various modifications of the gill, buccal, pharyngeal, or opercular chambers are used for aerial respiration (Graham, 1997, 2011b). The swim bladder may be used as an air-breathing organ (Graham, 1997, 2011b; Pelster, 2021), and in *Arapaima gigas*, a special air-breathing organ (ABO) or respiratory gas bladder is located underneath the vertebral column (Scadeng et al., 2020). With *Lepidosiren paradoxa*, a true lungfish is encountered in the Amazon basin. Some of these fishes are obligatory air-breathing fish, i.e., these fish are severely stressed and often drown when access to air is denied. Others are facultative air-breathers, i.e., they breathe air only under certain conditions, typically under hypoxic conditions.

Considering the changing environmental conditions predicted for the Amazon, it is expected that the more frequent hypoxic and possibly hypercapnic conditions will stimulate air-breathing activity in order to meet the increasing oxygen demand of tissues. Many facultative air-breathing fishes do so only under hypoxic conditions. It has repeatedly been shown that hypoxia increases air-breathing frequency. In *Clarias lazera*, for example, the number of air-breaths increases with decreasing PO_2, and in *Arapaima gigas*, the time elapsing between two air-breaths significantly decreases under hypoxic conditions (Ar & Zacks, 1989; Pelster et al., 2020) (Fig. 1). Thus, fishes already referring to air-breathing may do so more often. In obligatory air-breathing fish, the frequency of taking an air-breath may increase, and facultative air-breathing fish will be forced to breath air more often due to frequently encountered hypoxic events. Water-breathing species that are not very tolerant toward hypoxia may have trouble to take up sufficient oxygen from the hypoxic water, which may cause problems to survive or perhaps, through natural selection, stimulate a rapid evolutionary transition to air-breathing. Furthermore, in the spotted gar *Lepisosteus oculatus* an increase in temperature alone stimulated air-breathing activity (Smatresk & Cameron, 1982). We therefore would expect that with increasing temperatures, air-breathing becomes a more common phenomenon in the short term, but air-breathing is connected to a large number of necessary

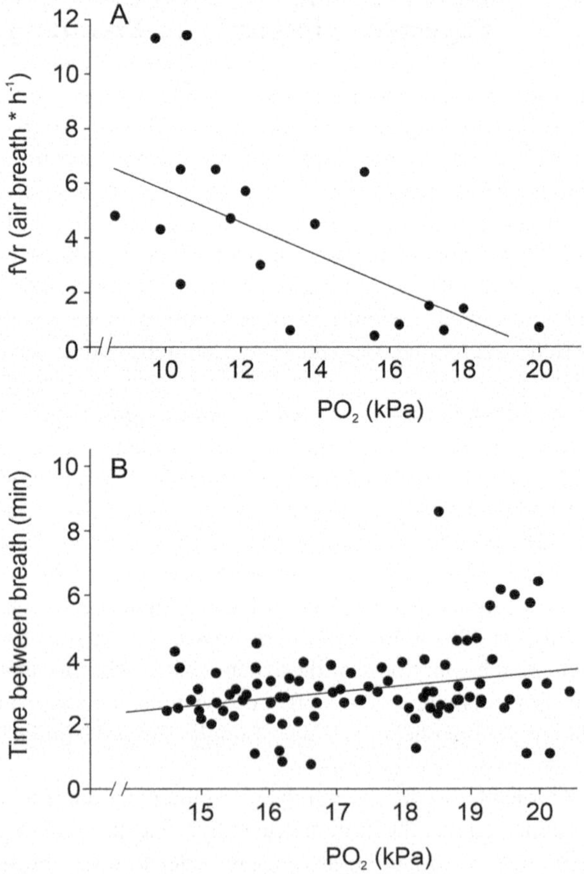

Fig. 1 (**a**) In *Clarias lazera*, the number of air-breaths taken increases with decreasing PO_2 in the environmental water. Data taken from Ar and Zacks (1989). (**b**) In small *Arapaima gigas* (body mass about 5 g), the time elapsing between subsequent air-breaths decreases with decreasing PO_2 in the water. Data taken from Pelster et al. (2020)

physiological adaptations in a fish, as reviewed by Pelster and Wood (Pelster & Wood, 2018). An enhanced reliance on air-breathing therefore is expected to affect homeostasis of a number of physiological phenomena ranging from aquatic gas exchange through ion and acid-base homeostasis to nitrogen excretion and predator avoidance, for example.

3 Enhanced Reliance on Air-Breathing May Result in a Reduction of Gill Surface Area

In fishes with a single atrium and ventricle, oxygenated blood returning from an ABO to the heart is passing the gills prior to supplying other organs. In consequence, oxygen taken up in the ABO may be lost to the water at the site of the gills. This has been demonstrated for the bowfin *Amia calva* (Johansen et al., 1970; Randall et al., 1981), and also for the armored catfish *Hypostomus aff. pyreneusi* (Scott et al., 2017). Loss of oxygen to hypoxic water at the gills has also been reported for spotted gar *Lepisosteus osseus* (Smatresk, 1986; Smatresk & Cameron, 1982). However, a recent study indicates that it is negligible in adult specimens of the obligate air-breathing pirarucu *Arapaima gigas* (Aaskov et al., 2022).

To reduce the danger of losing oxygen to the hypoxic environmental water, the gas exchange capacity of the gills may be reduced, and this can be achieved in different ways, like a reduction of gill surface area or an increase in the interlamellar cell mass (ILCM), which covers the surface of the lamellar or causes fusion of lamellar (Nilsson et al., 2012; Sollid et al., 2003). This is seen to its greatest extent in adult *Arapaima gigas* where the lamellae are completely buried and the filaments become finger-like projections (Brauner, Matey, et al., 2004).

Many amphibious or air-breathing fishes have a reduced gill surface area (Sayer, 2005). Among mudskippers, there are species relying completely on water-breathing, while others frequently refer to air-breathing. Comparing the surface area of the gills revealed that the reduction in gill surface area is correlated to the air-breathing activity; mudskippers with high air-breathing frequency have the lowest gill surface area (Ishimatsu, 2012). Similarly, in two closely related erythrinids, the gill area in the facultatively air-breathing jeju (*Hoplerythrinus unitaeniatus*) was reduced by 50% relative to the exclusively water-breathing traira (*Hoplias malabaricus)* (Cameron & Wood, 1978). The pirarucu *Arapaima gigas* is expected to undergo a rapid transition from a water-breathing hatchling to an obligatory air-breathing adult fish, and this transition is accompanied by a remarkable reduction in relative gill surface area, while the gas exchange capacity of the air-breathing organ is significantly improved (Brauner, Matey, et al., 2004; Frommel et al., 2021; Gonzalez et al., 2010). However, even early developmental stages 2–3 weeks after hatching acquire a significant fraction of oxygen from the air under resting conditions (Pelster et al., 2020). In larger fish, the fraction of oxygen taken up from the air increases, gill lamellar size and surface area decrease, and the gas exchange capacity of the air-breathing organ increases (Frommel et al., 2021; Pelster et al., 2020).

In the amphibious *Kryptolebias marmoratus* (formerly *Rivulus marmoratus*), 1 week of air exposure resulted in a significant reduction in lamellar area as a result of an extension of the interlamellar cell mass (ILCM), not as a result of a decrease in lamellar size (Ong et al., 2007).

Air exposure does not mean hypoxia for these amphibious fishes, but gills are no longer functional for oxygen uptake, and in amphibious fish, the skin appears to be a multifunctional organ, taking over functions typically located at the site of the gills (Wright, 2021).

Taken together, these data clearly suggest that more frequent hypoxic conditions concomitant with elevated temperatures increasing the oxygen demand of fishes will favor additional reductions in gill surface area, either by reducing the size of the lamellae or by increasing the ILCM, covering the lamellar surface. This will be advantageous to prevent the loss of oxygen (taken up at the ABO) to hypoxic water at the gills, but it will also reduce the surface area available for CO_2 excretion, because in air-breathing fish, a large fraction of CO_2 is typically excreted into the water at the site of the gills, while oxygen is taken up in the ABO. Moreover, a reduction in gill surface area reduces the surface area available for ion exchange, nitrogenous waste excretion, or acid-base regulation (Evans et al., 2005; Laurent & Perry, 1991). Thus, additional compensation to retain homeostasis in these crucial physiological parameters will be required.

4 Enhanced Reliance on Air-Breathing Affects Acid-Base Balance

Due to a high solubility of CO_2 in water, water-breathing fishes are characterized by low PCO_2 and low HCO_3^- concentrations in their blood. With the high ventilatory rate required to take up sufficient oxygen, CO_2 is easily excreted, so that a low venous PCO_2 in water-breathers is sufficient for an adequate excretion of CO_2 (Shartau & Brauner, 2014; Ultsch, 1996). Air-breathers in turn have higher PCO_2 values in order to maintain an outward diffusion gradient, combined with elevated $[HCO_3^-]$ so as to maintain about the same pH in their body fluids as found in water-breathers (Dejours, 1981).

Using the air as an additional oxygen source would reduce the necessity for a high rate of gill ventilation, and given the high CO_2 capacitance coefficient of water a lower rate of ventilation and even a reduced gill surface area observed in many air-breathing fishes (see above) might be sufficient to excrete CO_2 into the water, and thus to separate the main site of oxygen uptake from the site of CO_2 elimination. Although the Krogh's diffusion constant for CO_2 is much larger than the constant for oxygen, CO_2 excretion at the gills appears to be diffusion limited (Perry & Gilmour, 2002). Moreover, bimodal-breathing fish typically refer to intensified air-breathing and concurrently reduce gill ventilation in hypoxic water. With reduced contact to water, especially in amphibious fishes, higher PCO_2 values in blood are required to assure a proper diffusion gradient for CO_2 excretion, and, compared to water-breathers, in air-breathing fishes, higher PCO_2 values have been reported

(Bayley et al., 2019; Shartau & Brauner, 2014). While arterial PCO_2 in water-breathing fishes typically is around 2–3 mmHg, in air-breathing *Arapaima gigas* values of up to 30 mmHg and even higher have been recorded in arterial blood and in the air-breathing organ (Gonzalez et al., 2010; Pelster et al., 2020; Randall et al., 1978). Thus, the transition to breathing air was expected to affect acid-base homeostasis in fishes (Dejours, 1981).

A comparison of three water-breathing and three air-breathing fishes including the Amazonian electric eel *Electrophorus electricus* revealed similar pH values in blood, but as expected PCO_2 and [HCO_3^-] were significantly elevated in the air-breathing species (Ultsch, 1987). In air-breathing *Amia calva* and *Hypostomus spec.*, a significant elevation in PCO_2 was detected, but there was no change in pH (McKenzie & Randall, 1990; Wood et al., 1979). A high capacity for extracellular pH regulation has also been reported for *Pangasianodon hypophthalmus* (Damsgaard et al., 2015). In *Arapaima gigas* concomitant with elevated PCO_2 values, elevated [HCO_3^-] values have been recorded, so that arterial pH remained at 7.6–7.7 (Gonzalez et al., 2010).

Alternatively, PCO_2 may be elevated during air-breathing without concomitant elevation of [HCO_3^-], which would result in a blood acidification as observed in the marbled swamp eel *Synbranchus marmoratus*, another air-breathing species found in the Amazon (Heisler, 1982). Similarly, in the spotted gar *Lepisosteus oculatus,* air-breathing induced by an increase in temperature caused an elevation of PCO_2 at constant CCO_2, meaning that blood [HCO_3^-] was not elevated to compensate the acidosis (Smatresk & Cameron, 1982). A thorough literature survey revealed that an uncompensated acidosis characterized by a decrease in blood pH without appropriate increase in [HCO_3^-] appears to be the common response of many air-breathing fish (Brauner & Baker, 2009). Concomitant measurements of intracellular pH in various tissues revealed constant pHi values in spite of the pronounced extracellular acidosis (Brauner, Wang, et al., 2004; Heisler, 1982). Brauner and his team therefore proposed that preferential intracellular pH regulation would be a common strategy of air-breathing fish, and of fish exposed to hypercarbia (Shartau et al., 2016, 2020; Shartau & Brauner, 2014).

In air-breathing fish staying in contact with water, CO_2 excretion thus does not appear to be a problem, but an increase in PCO_2 and thus a modification of the acid-base status are encountered. The acidosis may be compensated by accumulation of HCO_3^- in the plasma, or it may be uncompensated, while intracellular pH is stabilized.

Increasing temperatures result in a slight acidification of blood, possibly in the order of −0.015 to −0.020 pH units $°C^{-1}$ (Wang & Jackson, 2016). Although not pronounced, with the expected enhanced reliance on air-breathing due to the increased metabolic oxygen demand at higher temperatures, this decline in pH may contribute to a prolongation of the periods of uncompensated severe extracellular acidosis in those air-breathing fishes that do not compensate the elevated PCO_2 by increasing HCO_3^- in plasma.

5 Possible Influence on Ion Homeostasis

Gills are the main site of ion regulation in fish, and a reduction in gill surface area will interfere with ion homeostasis. Direct comparison of closely related water-breathing and air-breathing erythrinid fishes revealed that ion fluxes under aquatic and aerial normoxic conditions may not be significantly lower in air-breathing fish, in spite of a significantly reduced gill surface area (Pelster & Wood, 2018; Wood et al., 2016). The reduction in surface area may be compensated by an increase in the transport capacity of the remaining tissue (Hulbert et al., 1978; Wood et al., 2016), or by a repositioning of mitochondria-rich ionocytes to the remaining gill lamellae, as detected by analysis of a large number of air-breathing and water-breathing fishes (Lin & Sung, 2003). In *Arapaima gigas*, ionocytes are relocated to the water-exposed surface of the ILCM, covering the secondary lamellae, and thus reducing the area available for gas exchange (Brauner, Matey, et al., 2004). Air-breathing fish in fact use a number of different strategies to compensate for the reduction in gill surface area, including a transfer of ion regulation to other tissues or an improvement in kidney function (Pelster & Wood, 2018; Wood et al., 2020). In the facultative air-breathing jeju *Hoplerythrinus unitaeniatus*, relative to the obligate water-breathing traira *Hoplias malabaricus*, a greater fraction of ion uptake and ion regulation is transferred to the gut (Wood et al., 2016). Ion transport capacities and also glucose and ammonia transport capacities by far exceed the capacities measured in the closely related water-breathing traira. An interesting consequence of this strategy is that under hypoxic conditions, the traira loses ions to environment, probably at the gills due to an increase in ventilation or a decrease in diffusion distance to get enough oxygen from the hypoxic water (osmoregulatory compromise), while the jeju refers to breathing air with negligible disturbance of the ion homeostasis. During fasting, however, the jeju faces a net ion loss, while the traira is able to maintain ion balance so as to keep internal ion concentrations constant (Pelster et al., 2016; Wood et al., 2016).

A thorough analysis of gill ion flux rates in *Arapaima gigas* revealed about equal Na^+ influx and efflux rates, while Cl^- and K^+ net flux rates were negative (Wood et al., 2020). The high Na^+ uptake rates may be due to the relocation of mitochondria-rich cells to the surface of the interlamellar cell mass completely covering the second lamellar in pirarucu of this size (Brauner, Matey, et al., 2004; Ramos et al., 2013). Using water with lower Na^+ concentration, a much lower Na^+-uptake rate was recorded in *Arapaima* (Gonzalez et al., 2010). Together with the negative Cl^- and K^+ balance recorded at the gills (Wood et al., 2020), these data suggest that another tissue is involved in ion uptake from the water, and as observed in *Hoplerythrinus*, this could be the gut.

A number of air-breathing fishes use the gut as additional air-breathing organ (Nelson, 2014). The part of the gut used for air-breathing cannot be used for ion regulation, which means that there must be different sections responsible for one or the other function, or that air-breathing and ion uptake cannot occur at the same time. Most gut-breathing fishes appear to be herbivorous (Nelson, 2014), and herbivorous fish have a longer gut than carnivorous species (Pelster et al., 2015). This may enable the fish to use separate gut sections for ion regulation and breathing.

The kidney cannot be used to take up ions from the water, necessary for a freshwater fish to compensate ion loss to the hypoosmotic environment, but its function can be optimized to reduce ion loss connected to necessary waste excretion as much as possible. This is established in *Arapaima gigas*, which is able to produce a urine with extremely low Na^+ and Cl^- concentrations (Wood et al., 2020). With a high urine flow rate of 10–12 ml kg^{-1} h^{-1}, the inorganic ion concentrations of the urine (only about 200 μmol L^{-1}), and the urine-to-plasma concentration ratios (0.001–0.003) were the lowest ever measured in a teleost fish (Wood et al., 2020). This was likely explained by Na^+, K^+ ATPase activities that were more than 10-fold higher in kidney tissues than in gill tissues.

Aquatic hypoxia caused a transient decrease in urinary flow rate in *Arapaima* (Wood et al., 2020). This has been observed in several Amazonian hypoxia-tolerant fish species, and it is assumed that the decrease in urine flow rate is connected to a general reduction in gill permeability during hypoxia as a cost-saving measure (De Boeck et al., 2013; Giacomin et al., 2020; Pelster, 2009; Robertson et al., 2015; Wood et al., 2007, 2009, 2019). A decrease in urinary flow rate has also been observed in the air-exposed swamp eel *Monopterus albus* (Thinh et al., 2019). In this species under hypercapnic conditions, often encountered together with hypoxia, the increasing PCO_2 was progressively compensated by elevation of plasma HCO_3^-, and the kidney contributed by excreting acidic equivalents (Thinh et al., 2019). More frequent hypoxic events caused by elevated temperatures may thus cause an impairment of excretory activity.

6 Possible Influence on Nitrogen Excretion

Nucleotides and amino acids contain nitrogen, and the catabolism of these compounds results in the production of ammonia. Dietary protein not used for anabolic metabolism is degraded for energy metabolism, resulting in the production of ammonia (Ballantyne, 2001; Campbell, 1991). Higher concentrations of ammonia are toxic, however, and it must be continuously excreted or converted to less toxic compounds (Ip et al., 2001; Randall & Ip, 2006). Ammonia is highly soluble in water, and due to the alkaline pK' value, at physiological pH values established in most body fluids, it is mostly present as NH_4^+, with a small fraction as the dissolved gas NH_3 (Ip & Chew, 2018; Randall & Ip, 2006). Compared to CO_2, the membrane solubility of NH_3 is very low, but due to the large gill surface area and high blood perfusion and ventilation rates, NH_3 transfer appears to be the main route for ammonia excretion (Evans et al., 2005; Ip & Chew, 2018; Randall & Ip, 2006). The diffusion of NH_3 may be supported by a metabolon including Rhesus glycoproteins (Rh proteins) which serve as channels to accelerate the diffusion of NH_3 (Nawata et al., 2010) and acid excretion into the environmental water via V-H^+ ATPases and NHE proteins, trapping NH_3 on the apical membrane of gill cells by converting it to NH_4^+, thereby maintaining the gradient for NH_3 diffusion (Weihrauch et al., 2009; Wright

& Wood, 2009, 2012). This way, a small NH_3 partial pressure gradient would be sufficient to maintain continuous NH_3 outward diffusion. Some aquaporins have also been reported to facilitate NH_3 diffusion through membranes (Musa-Aziz et al., 2009; Nakhoul et al., 2001). Due to similar hydration radii, some K^+ channels or transport proteins appear to show a certain permeability to NH_4^+, and a small fraction of ammonia may be excreted via these K^+ transport proteins (Choe et al., 2000; Randall & Ip, 2006).

Some air-breathing fish actively enhance NH_4^+ excretion (Chew et al., 2005; Chew & Ip, 2014; Ip et al., 2010; Ip & Chew, 2018). The mudskipper, *Periophthalmodon schlosseri*, exposed to up to 30 mM NH_4^+ was still able to secrete NH_4^+, and it was concluded that this was achieved via activity of basolateral Na^+/K^+-ATPase and an apical Na^+/H^+ exchanger (Chew et al., 2014; Randall et al., 1999, 2004). Exposure to elevated NH_4^+ resulted in an increase in membrane cholesterol content and in a modified fatty acid composition to reduce or avoid back-diffusion of NH_3 (Ip, Randall, et al., 2004). Active NH_4^+ excretion has also been observed in the climbing perch *Anabas testudineus*, and it was proposed that the $Na^+/K^+/2Cl^-$ co-transporter (nkcc) is involved in basolateral uptake of NH_4^+ in mitochondria-rich gill cells (Loong et al., 2012; Tay et al., 2006). Activity of Na^+/K^+-ATPase and of cystic fibrosis transmembrane conductance regulator (CFTR) has been connected to NH_4^+ secretion in the climbing perch (Ip et al. 2012, b). It has been suggested that Cl^- secretion via the CFTR generates a favorable electrical potential supporting the secretion of NH_4^+ (Ip, Wilson, et al., 2012).

The reduction in gill ventilation observed in many bimodal-breathing fishes during aquatic hypoxia and a reduced gill surface area may impair ammonia excretion, and elevated plasma ammonia concentrations compared with those of typical water-breathing fish have been reported for *Arapaima gigas* (Gonzalez et al., 2010; Wood et al., 2020). In the loach *Misgurnus anguillicaudatus*, buildup of elevated ammonia concentrations has been reported during aerial exposure (Chew et al., 2001). A temperature-induced enhancement of aerial respiration may therefore be associated with a transient accumulation of ammonia in body fluids.

Temperature has already been shown to influence ammonia as well as urea excretion in the tambaqui *Colossoma macropomum* (Wood et al., 2017). The Q_{10} for urea excretion was with 2.6 and 2.1 in the temperature range of 28–33 °C and 33–38 °C slightly higher than the Q_{10} for oxygen consumption. In the higher temperature range, the Q_{10} for ammonia excretion, however, was with 3.8 almost twice as high as the Q_{10} for oxygen consumption, indicating a remarkable increase in ammonia production and excretion at higher temperatures (Wood et al., 2017). A higher temperature sensitivity of ammonia excretion has in fact been reported for other teleosts as well (Wood, 2001).

Several strategies have been adopted to reduce deleterious effects of elevated ammonia concentrations connected to air-breathing, or especially air exposure (Ip, Chew, & Randall, 2004). A reduction of proteolysis and amino acid catabolism results in a decreased rate of ammonia production, and this strategy has been adopted by fish emerging from water, but may also be used by air-breathing fish remaining in the water (see (Ip & Chew, 2018) for review).

In aestivating African lungfish *Protopterus annectens* and *Protopterus aethiopicus*, a decrease in ammonia production combined with an increase in urea synthesis has been observed when aestivating in air, but aestivation in mud resulted in a decrease in urea and ammonia production (Ip et al., 2005; Loong et al., 2005, 2008). Similar changes in ammonia and urea production have been reported for *Protopterus dolloi* during aerial aestivation (Chew, 2003; Chew et al., 2004). These results suggest that at least when aerial respiration is connected to limited access to water, ammonia production may be reduced, while urea production gains increasing importance. When the lungfish returned to water, there was only a modest burst of ammonia excretion but a massive increase in urea excretion which was accomplished by activation of a facilitated diffusion UT-A type urea transporter, mainly in the skin (Hung et al., 2009; Wood et al., 2005). Nevertheless, some air-breathing fish, like the swamp eel *Monopterus albus,* have developed a comparatively high brain ammonia tolerance (Ip et al., 2010, 2013; Tng et al., 2009). With elevated internal ammonia levels or during prolonged emersion even a volatilization of NH_3 has been detected in the rivulus killifish (Frick & Wright, 2002; Tsui et al., 2002).

In air-breathing fish mostly remaining aquatic, a decrease in ammonia production may not be required because ammonia can easily be excreted into the water using the gills or the skin, which may be more vascularized than in water-breathing fishes (Graham, 1997; Randall & Ip, 2006). Although urea synthesis is of minor importance in most teleost fish, as they lack a functional ornithine-urea cycle, there is nevertheless a general trend for increased urea excretion under conditions where ammonia excretion is impeded, such as high environmental pH and high environmental ammonia (Wood, 1996). Air-breathing is another circumstance that would potentially impede ammonia excretion. The generally small amounts of urea excreted by freshwater or marine fish are facilitated by diffusive urea transporters at the branchial epithelium (McDonald et al., 2006), similar to those in the skin of the lungfish (see above). As urea is much less toxic than ammonia, urea production would be a strategy to prevent ammonia toxicity, but in a number of fish no increase in urea production associated with air-breathing or emersion has been detected (see (Ip & Chew, 2018) for review), while multiple studies demonstrate urea production in the African lungfish (see above).

A comparison of two erythrinid teleosts, the water-breathing traira *Hoplias malabaricus* and the obligatory air-breathing jeju *Hoplerythrinus unitaeniatus*, revealed a more than twofold higher rate of urea excretion in the air-breathing jeju (Wood et al., 2016). In the traira, urea-N-excretion amounted to 7% of total nitrogen excretion, while in the jeju, urea-N-excretion contributed 16% of total nitrogen excretion, suggesting that even in some air-breathing fish retaining access to water, urea production may be of elevated importance for nitrogen excretion to reduce the danger of ammonia toxicity (Fig. 2A). In small *Arapaima gigas* with a body mass of about 5 g, fractional contribution of urea-N-excretion to total N-excretion was even higher than the value determined for the jeju (Fig. 2B). Hypoxic water combined with normoxic air supply resulted in a significant stimulation of aerial respiration, and both urea and ammonia excretion significantly decreased. Moreover, net Na^+ loss also decreased, suggesting that blood flow over the gills was reduced (Pelster et al.,

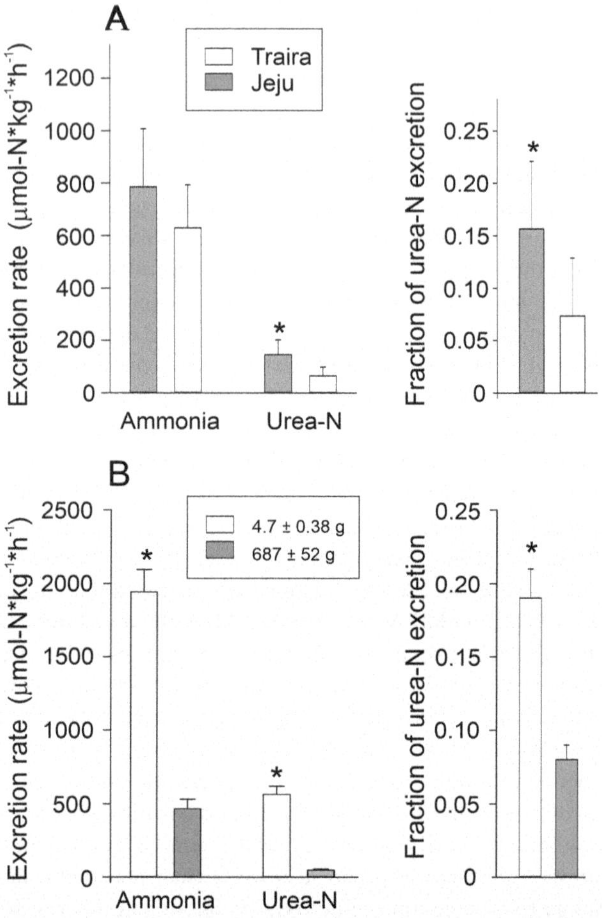

Fig. 2 (**a**) Ammonia, urea excretion, and the fractional contribution of urea to total N-excretion in two erythrinid fishes, the water-breathing *Hoplias malabaricus* (Traira) and the air-breathing *Hoplerythrinus unitaeniatus* (Jeju). Data taken from Wood et al. (2016). (**b**) Ammonia, urea excretion, and the fractional contribution of urea to total N-excretion in small- and medium-sized obligatory air-breathing *Arapaima gigas* (Pirarucu). Data taken from Wood et al. (2020)

2016; Wood et al., 2016), reducing the capacity to excrete nitrogenous wastes into the water.

In large obligatory air-breathing *Arapaima gigas*, urea also contributes a minor fraction (about 5%) to total nitrogen excretion. Interestingly, in the *Arapaima* the kidney contributes more than 20% to total ammonia and urea excretion (Wood et al., 2020). Compared to other teleosts, this is an unusually high value (Ip & Chew, 2018; Weihrauch et al., 2009; Wilkie, 2002; Wood, 1993), indicating that nitrogenous waste excretion is partially shifted from the gills to the kidney. High levels of ammonia (as NH_4^+) in the urine appeared to be coupled to approximately equal high levels of total CO_2 (as HCO_3^-) in the urine, thereby achieving acid-base neutrality, though the mechanisms involved remain to be investigated. In the *Arapaima*, the kidney is located within the ABO, so that the organ may have a direct supply with oxygen from the ABO, though Wood et al. (2020) found that manipulating the aerial O_2 concentrations did not affect the performance of the kidney.

So far, there is no clear answer to the question whether air-breathing fish show an elevated rate of urea excretion as compared to water-breathing species. It is quite possible that obligatory air-breathing fish and facultative air-breathing fish differ in their strategy of nitrogen waste excretion. Obligatory air-breathing may result in an elevated reliance on urea excretion.

7 Visits at the Water Surface May Be Costly and Associated with the Risk of Predation

Taking an air-breath at the water surface requires muscular activity and thus energy. Quantification of this energy spent in addition to the basal energy requirements, i.e., quantification of the cost of surfacing, is a difficult task, and the results are controversial. In the air-breathing catfish *Pangasianodon hypophthalmus*, oxygen consumption rates with and without surfacing and air-breathing were compared suggesting that the net cost of transport was 25% lower in fish that did not breathe air (Lefevre et al., 2013). This would suggest that air-breathing is associated with a significantly elevated oxygen demand. In *Arapaima gigas*, however, increased surfacing activity induced by aquatic hypoxia did not result in an elevated oxygen consumption, so in this species the cost of surfacing appeared to be very low (Aaskov et al., 2022).

Visiting the water surface, either to perform aquatic surface respiration (ASR) and/or to take an air-breath, may place the fish at risk of predation from above or from below the water surface, during daytime as well as at nighttime (Pratt & Franklin, 2009; Sloman et al., 2009). Large oscars (*Astronotus ocellatus*), which are at low risk of predation, began ASR at the PO_2 where O_2 uptake from the water first started to decline ("Pcrit"). Smaller oscars, living closer to the water surface and at much greater risk of predation, postponed ASR to a PO_2 well below their Pcrit (Sloman et al., 2006). That predation is an important parameter can be concluded from the observation that taking an air-breath is an extremely short event taking only seconds. Moreover, air-breaths are preferably taken under the cover of vegetation in nature (Shingles et al., 2005) and simultaneously with other fish (Shelton et al., 1986). Building groups and staying close together is a typical behavior of many species to reduce the risk of predation (Godin, 1986; Partridge, 1982). For the Siamese fighting fish (*Betta splendens*), it was even reported that they stop fighting to take an air-breath together so as to reduce the risk of predation, and then resume fighting afterward (Alton et al., 2013)! The elevated oxygen demand resulting from display behavior for courtship or territory defense is met by more frequent air-breathing. Synchronous air-breathing has in fact been described for a number of teleosts including Amazonian species and appears to be a common phenomenon

(Chapman & Chapman, 1994; Kramer & Graham, 1976; Lima Filho et al., 2012; Sloman et al., 2009).

Fish apparently have adopted different strategies to reduce the danger of predation near or at the surface. Nevertheless, with an increased necessity to visit the surface during more frequently and perhaps prolonged encountered hypoxic events, the danger of predation will increase.

8 Perspectives

The current climate change is expected to cause more frequent and possibly more intense hypoxic events in tropical areas like the Amazon region. Concomitant increases in temperature cause an elevation of metabolic rates, and the oxygen required to fuel aerobic metabolism is more easily acquired from the air than from the water. Accordingly, bimodal-breathing fish may intensify air-breathing activity, and some water-breathing species may attempt to switch to air-breathing. Air-breathing, however, is accompanied by complex and sophisticated physiological adjustments to retain homeostasis in acid-base regulation, ion regulation, and nitrogen excretion. As discussed above, all these physiological adaptations connected to air-breathing therefore must be adjusted to an elevated reliance on air-breathing. Given the rapid climate changes currently encountered, it is difficult to predict whether all these necessary adjustments will be possible in the different species in order to retain internal homeostasis of all parameters. In a recent study, it was predicted that climate change over the longer term may favor survival of predominantly water-breathing species with low routine metabolic rates, low temperature sensitivity of routine metabolic rates, high hypoxia tolerance, and high thermal tolerance (Val & Wood, 2022). Given the advantage of air as a source for oxygen with a much higher oxygen capacity and much lower density compared to water, however, it cannot be excluded that some fish will be able to successfully explore the transition to air-breathing or will be able to intensify their air-breathing activity. Several studies demonstrated that phenotypic plasticity varies significantly between species (Burggren, 2018; Pelster & Burggren, 2018; Westneat et al., 2019), and given the rapid climatic changes, currently encountered phenotypic plasticity may become a crucial characteristic for survival of a species.

References

Aaskov, M. L., Jensen, R. J., Skov, P. V., et al. (2022). *Arapaima gigas* maintains gas exchange separation in severe aquatic hypoxia but does not suffer branchial oxygen loss. *The Journal of Experimental Biology, 225*, jeb243672. https://doi.org/10.1242/jeb.243672

Alton, L. A., Portugal, S. J., & White, C. R. (2013). Balancing the competing requirements of air-breathing and display behaviour during male–male interactions in Siamese fighting fish *Betta splendens*. *Comparative Biochemistry and Physiology Part A: Molecular & Integrative Physiology, 164*, 363–367. https://doi.org/10.1016/j.cbpa.2012.11.012

Ar, A., & Zacks, D. (1989). Alterations in the bimodal gas exchange of the African catfish *Clarias lazera*. In *Physiological function in special environments* (pp. 172–190). Springer.
Atkinson, D., Leighton, G., & Berenbrink, M. (2022). Controversial roles of oxygen in organismal responses to climate warming. *The Biological Bulletin, 243*, 207–219.
Ballantyne, J. S. (2001). Amino acid metabolism. In P. A. Wright & P. Anderson (Eds.), *Fish physiology* (Vol. 20, pp. 77–107). Elsevier.
Barichivich, J., Gloor, E., Peylin, P., et al. (2018). Recent intensification of Amazon flooding extremes driven by strengthened Walker circulation. *Science Advances, 4*, eaat8785. https://doi.org/10.1126/sciadv.aat8785
Bayley, M., Damsgaard, C., Thomsen, M., et al. (2019). Learning to air-breathe: The first steps. *Physiology, 34*, 14–29. https://doi.org/10.1152/physiol.00028.2018
Brauner, C. J., & Baker, D. W. (2009). Patterns of acid–base regulation during exposure to hypercarbia in fishes. In *Cardio-respiratory control in vertebrates* (pp. 43–63). Springer.
Brauner, C. J., Matey, V., Wilson, J. M., et al. (2004). Transition in organ function during the evolution of air-breathing; insights from *Arapaima gigas*, an obligate air-breathing teleost from the Amazon. *The Journal of Experimental Biology, 207*, 1433–1438.
Brauner, C. J., Wang, T., Wang, Y., et al. (2004). Limited extracellular but complete intracellular acid-base regulation during short-term environmental hypercapnia in the armoured catfish, *Liposarcus pardalis*. *The Journal of Experimental Biology, 207*, 3381–3390. https://doi.org/10.1242/jeb.01144
Burggren, W. (2018). Developmental phenotypic plasticity helps bridge stochastic weather events associated with climate change. *The Journal of Experimental Biology, 221*, jeb161984. https://doi.org/10.1242/jeb.161984
Cameron, J. N., & Wood, C. M. (1978). Renal function and acid-base regulation in two Amazon erythrinid fishes: *Hoplias malabaricus*, a water breather, and *Hoplerythrinus unitaeniatus*, a facultative air breather. *Canadian Journal of Zoology, 56*, 917–930.
Campbell, J. W. (1991). Excretory nitrogen metabolism. In C. L. Prosser (Ed.), *Comparative animal physiology* (Environmental and metabolic animal physiology) (pp. 277–324). Wiley.
Campos, D. F., Braz-Mota, S., Val, A. L., & Almeida-Val, V. M. F. (2019). Predicting thermal sensitivity of three Amazon fishes exposed to climate change scenarios. *Ecological Indicators, 101*, 533–540. https://doi.org/10.1016/j.ecolind.2019.01.051
Campos, D. F., Jesus, T. F., Kochhann, D., et al. (2017). Metabolic rate and thermal tolerance in two congeneric Amazon fishes: *Paracheirodon axelrodi* Schultz, 1956 and *Paracheirodon simulans* Géry, 1963 (Characidae). *Hydrobiologia, 789*, 133–142. https://doi.org/10.1007/s10750-016-2649-2
Chapman, L. J., & Chapman, C. A. (1994). Observations on synchronous air breathing in *Clarias liocephalus*. *Copeia, 1994*, 246. https://doi.org/10.2307/1446696
Chew, S. F. (2003). Urea synthesis in the African lungfish *Protopterus dolloi* - Hepatic carbamoyl phosphate synthetase III and glutamine synthetase are upregulated by 6 days of aerial exposure. *The Journal of Experimental Biology, 206*, 3615–3624. https://doi.org/10.1242/jeb.00619
Chew, S. F., Chan, N. K. Y., Loong, A. M., et al. (2004). Nitrogen metabolism in the African lungfish (*Protopterus dolloi*) aestivating in a mucus cocoon on land. *The Journal of Experimental Biology, 207*, 777–786. https://doi.org/10.1242/jeb.00813
Chew, S. F., Hiong, K. C., Lam, S. P., et al. (2014). Functional roles of Na^+/K^+-ATPase in active ammonia excretion and seawater acclimation in the giant mudskipper, *Periophthalmodon schlosseri*. *Frontiers in Physiology, 5*, 158. https://doi.org/10.3389/fphys.2014.00158
Chew, S. F., & Ip, Y. K. (2014). Excretory nitrogen metabolism and defence against ammonia toxicity in air-breathing fishes. *Journal of Fish Biology, 84*, 603–638. https://doi.org/10.1111/jfb.12279
Chew, S. F., Jin, Y., & Ip, Y. K. (2001). The Loach *Misgurnus anguillicaudatus* reduces amino acid catabolism and accumulates alanine and glutamine during aerial exposure. *Physiological and Biochemical Zoology, 74*, 226–237. https://doi.org/10.1086/319663
Chew, S. F., Wilson, J. M., Ip, Y. K., & Randall, D. J. (2005). Nitrogen excretion and defense against ammonia toxicity. *Fish Physiology, 21*, 307–395.

Choe, H., Sackin, H., & Palmer, L. G. (2000). Permeation properties of inward-rectifier potassium channels and their molecular determinants. *The Journal of General Physiology, 115*, 391–404. https://doi.org/10.1085/jgp.115.4.391

Damsgaard, C., Baliga, V. B., Bates, E., et al. (2020). Evolutionary and cardio-respiratory physiology of air-breathing and amphibious fishes. *Acta Physiologica, 228*, e13406. https://doi.org/10.1111/apha.13406

Damsgaard, C., Gam, L. T. H., Dang, D. T., et al. (2015). High capacity for extracellular acid-base regulation in the air-breathing fish *Pangasianodon hypophthalmus*. *The Journal of Experimental Biology, 218*, 1290. https://doi.org/10.1242/jeb.117671

De Boeck, G., Wood, C. M., Iftikar, F. I., et al. (2013). Interactions between hypoxia tolerance and food deprivation in Amazonian oscars, *Astronotus ocellatus* (Agassiz). *The Journal of Experimental Biology, 216*(24), 4590–4600. https://doi.org/10.1242/jeb.082891

Dejours, P. (1981). *Principles of comparative respiratory physiology*. Elsevier.

Evans, D. H., Piermarini, P. M., & Choe, K. P. (2005). The multifunctional fish gill: Dominant site of gas exchange, osmoregulation, acid-base regulation, and excretion of nitrogenous waste? *Physiological Reviews, 85*, 97–177. https://doi.org/10.1152/physrev.00050.2003

Ficke, A. D., Myrick, C. A., & Hansen, L. J. (2007). Potential impacts of global climate change on freshwater fisheries. *Reviews in Fish Biology and Fisheries, 17*, 581–613. https://doi.org/10.1007/s11160-007-9059-5

Frick, N. T., & Wright, P. A. (2002). Nitrogen metabolism and excretion in the mangrove killifish *Rivulus marmoratus* II. Significant ammonia volatilization in a teleost during air-exposure. *The Journal of Experimental Biology, 205*, 91–100. https://doi.org/10.1242/jeb.205.1.91

Frommel, A. Y., Kwan, G. T., Prime, K. J., et al. (2021). Changes in gill and air-breathing organ characteristics during the transition from water- to air-breathing in juvenile *Arapaima gigas*. *Journal of Experimental Zoology Part A: Ecological and Integrative Physiology, 335*, 801–813. https://doi.org/10.1002/jez.2456

Furch, K., & Junk, W. J. (1997). Physicochemical conditions in the floodplains. *Ecology Studies, 12*(6), 69–108.

Geirinhas, J. L., Trigo, R. M., Libonati, R., et al. (2018). Climatic and synoptic characterization of heat waves in Brazil. *International Journal of Climatology, 38*, 1760–1776. https://doi.org/10.1002/joc.5294

Giacomin, M., Onukwufor, J. O., Schulte, P. M., & Wood, C. M. (2020). Ionoregulatory aspects of the hypoxia-induced osmorespiratory compromise in the euryhaline Atlantic killifish (*Fundulus heteroclitus*): The effects of salinity. *The Journal of Experimental Biology, 223*, jeb216309. https://doi.org/10.1242/jeb.216309

Godin, J. G. J. (1986). Antipredator function of shoaling in teleost fishes: A selective review. *Naturalist canadien (Review of Ecology, and Systematics), 113*, 241–250.

Gonzalez, R. J., Brauner, C. J., Wang, Y. X., et al. (2010). Impact of ontogenetic changes in branchial morphology on gill function in *Arapaima gigas*. *Physiological and Biochemical Zoology, 83*, 322–332. https://doi.org/10.1086/648568

Gonzalez, R. J., Wilson, R. W., & Wood, C. M. (2005). Ionoregulation in tropical fishes from ion-poor, acidic blackwaters. *Fish Physiology, 21*, 397–442.

Graham, J. B. (1997). *Air-breathing fishes. Evolutions, diversity, and adaptation*. Academic Press.

Graham, J. B. (2011a). Air-breathing fishes | The biology, diversity, and natural history of air-breathing fishes: An introduction. In *Encyclopedia of fish physiology* (pp. 1850–1860). Elsevier.

Graham, J. B. (2011b). Air-breathing fishes | Respiratory adaptations for air-breathing fishes. In *Encyclopedia of fish physiology* (pp. 1861–1874). Elsevier.

Heisler, N. (1982). Intracellular and extracellular acid-base regulation in the tropical fresh-water teleost fish *Synbranchus marmoratus* in response to the transition from water breathing to air breathing. *The Journal of Experimental Biology, 99*, 9–28.

Hulbert, W. C., Moon, T. W., & Hochachka, P. W. (1978). The erythrinid gill: Correlations of structure, function, and metabolism. *Canadian Journal of Zoology, 56*, 814–819.

Hung, C. Y. C., Galvez, F., Ip, Y. K., & Wood, C. M. (2009). Increased gene expression of a facilitated diffusion urea transporter in the skin of the African lungfish (*Protopterus annectens*) during massively elevated post-terrestrialization urea excretion. *The Journal of Experimental Biology, 212*, 1202–1211. https://doi.org/10.1242/jeb.025239

Ip, Y. K., & Chew, S. F. (2018). Air-breathing and excretory nitrogen metabolism in fishes. *Acta Histochemica, 120*, 680–690. https://doi.org/10.1016/j.acthis.2018.08.013

Ip, A., Chew, S., & Ip, Y. (2010). Ammonia production, excretion, toxicity, and defense in fish: A review. *Frontiers in Physiology, 1*, 134. https://doi.org/10.3389/fphys.2010.00134

Ip, Y. K., Chew, S. F., & Randall, D. J. (2001). Ammonia toxicity, tolerance, and excretion. *Fish Physiology, 20*, 109–148.

Ip, Y. K., Chew, S. F., & Randall, D. J. (2004). Five tropical air-breathing fishes, six different strategies to defend against ammonia toxicity on land. *Physiological and Biochemical Zoology, 77*, 768–782. https://doi.org/10.1086/422057

Ip, Y. K., Hou, Z., Chen, X. L., et al. (2013). High brain ammonia tolerance and down-regulation of Na^+:K^+:$2Cl^-$ cotransporter 1b mRNA and protein expression in the brain of the swamp eel, *Monopterus albus*, exposed to environmental ammonia or terrestrial conditions. *PLoS One, 8*, e69512. https://doi.org/10.1371/journal.pone.0069512

Ip, Y. K., Loong, A. M., Kuah, J. S., et al. (2012). Roles of three branchial Na^+-K^+-ATPase α-subunit isoforms in freshwater adaptation, seawater acclimation, and active ammonia excretion in *Anabas testudineus*. *American Journal of Physiology-Regulatory, Integrative and Comparative Physiology, 303*, R112–R125. https://doi.org/10.1152/ajpregu.00618.2011

Ip, Y. K., Randall, D. J., Kok, T. K. T., et al. (2004). The giant mudskipper *Periophthalmodon schlosseri* facilitates active NH_4^+ excretion by increasing acid excretion and decreasing NH_3 permeability in the skin. *The Journal of Experimental Biology, 207*, 787–801. https://doi.org/10.1242/jeb.00788

Ip, Y. K., Wilson, J. M., Loong, A. M., et al. (2012). Cystic fibrosis transmembrane conductance regulator in the gills of the climbing perch, *Anabas testudineus*, is involved in both hypoosmotic regulation during seawater acclimation and active ammonia excretion during ammonia exposure. *Journal of Comparative Physiology B, 182*, 793–812. https://doi.org/10.1007/s00360-012-0664-9

Ip, Y. K., Yeo, P. J., Loong, A. M., et al. (2005). The interplay of increased urea synthesis and reduced ammonia production in the African lungfish *Protopterus aethiopicus* during 46 days of aestivation in a mucus cocoon. *Journal of Experimental Zoology Part A: Comparative Experimental Biology, 303A*, 1054–1065. https://doi.org/10.1002/jez.a.237

IPCC. (2014). *Impacts, adaptation, and vulnerability, Part. B: Regional aspects, climate change 2014*. Cambridge University Press.

Ishimatsu, A. (2012). Evolution of the cardiorespiratory system in air-breathing fishes. *Aqua-BioScience Monographs, 5*, 1–28.

Jézéquel, C., Tedesco, P. A., Darwall, W., et al. (2020). Freshwater fish diversity hotspots for conservation priorities in the Amazon Basin. *Conservation Biology, 34*, 956–965. https://doi.org/10.1111/cobi.13466

Johansen, K., Hanson, D., & Lenfant, C. (1970). Respiration in a primitive air breather, *Amia calva*. *Respiration Physiology, 9*, 162–174.

Jung, E. H., Brix, K. V., Richards, J. G., et al. (2020). Reduced hypoxia tolerance and survival at elevated temperatures may limit the ability of Amazonian fishes to survive in a warming world. *Science of the Total Environment, 748*, 141349. https://doi.org/10.1016/j.scitotenv.2020.141349

Kramer, D. L., & Graham, J. B. (1976). Synchronous air breathing, a social component of respiration in fishes. *Copeia, 1976*, 689–697.

Kramer, D. L., & Mehegan, J. P. (1981). Aquatic surface respiration, an adaptive response to hypoxia in the guppy, *Poecilia reticulata* (Pisces, Poecilidae). *Environmental Biology of Fishes, 6*, 299–313.

Laurent, P., & Perry, S. F. (1991). Environmental effects on fish gill morphology. *Physiological Zoology, 64*, 4–25.

Lefevre, S., Bayley, M., Mckenzie, D. J., & Craig, J. F. (2014). Air-breathing fishes. *Journal of Fish Biology, 84*, 547–553. https://doi.org/10.1111/jfb.12349

Lefevre, S., Wang, T., Huong, D. T. T., et al. (2013). Partitioning of oxygen uptake and cost of surfacing during swimming in the air-breathing catfish *Pangasianodon hypophthalmus*. *Journal of Comparative Physiology B, 183*, 215–221. https://doi.org/10.1007/s00360-012-0701-8

Lima Filho, J. A., Martins, J., Arruda, R., & Carvalho, L. N. (2012). Air-breathing behavior of the jeju fish *Hoplerythrinus unitaeniatus* in amazonian streams. *Biotropica, 44*, 512–520. https://doi.org/10.1111/j.1744-7429.2011.00839.x

Lin, H. C., & Sung, W. T. (2003). The distribution of mitochondria-rich cells in the gills of air-breathing fishes. *Physiological and Biochemical Zoology, 76*, 215–228.

Loong, A. M., Chew, S. F., Wong, W. P., et al. (2012). Both seawater acclimation and environmental ammonia exposure lead to increases in mRNA expression and protein abundance of Na^+:K^+:$2Cl^-$ cotransporter in the gills of the climbing perch, *Anabas testudineus*. *Journal of Comparative Physiology. B, 182*, 491–506. https://doi.org/10.1007/s00360-011-0634-7

Loong, A. M., Hiong, K. C., Lee, S. M. L., et al. (2005). Ornithine-urea cycle and urea synthesis in African lungfishes, *Protopterus aethiopicus* and *Protopterus annectens*, exposed to terrestrial conditions for six days. *Journal of Experimental Zoology Part A: Comparative Experimental Biology, 303A*, 354–365. https://doi.org/10.1002/jez.a.147

Loong, A. M., Pang, C. Y. M., Hiong, K. C., et al. (2008). Increased urea synthesis and/or suppressed ammonia production in the African lungfish, *Protopterus annectens*, during aestivation in air or mud. *Journal of Comparative Physiology B, 178*, 351–363. https://doi.org/10.1007/s00360-007-0228-6

McDonald, M. D., Smith, C. P., & Walsh, P. J. (2006). The physiology and evolution of urea transport in fishes. *The Journal of Membrane Biology, 212*, 93–107. https://doi.org/10.1007/s00232-006-0869-5

McKenzie, D. J., & Randall, D. J. (1990). Does *Amia calva* aestivate? *Fish Physiology and Biochemistry, 8*, 147–158. https://doi.org/10.1007/BF00004442

Musa-Aziz, R., Chen, L.-M., Pelletier, M. F., & Boron, W. F. (2009). Relative CO_2/NH_3 selectivities of AQP1, AQP4, AQP5, AmtB, and RhAG. *Proceedings of the National Academy of Sciences, 106*, 5406–5411. https://doi.org/10.1073/pnas.0813231106

Nakhoul, N. L., Hering-Smith, K. S., Abdulnour-Nakhoul, S. M., & Hamm, L. L. (2001). Transport of NH_3 /NH_4^+ in oocytes expressing aquaporin-1. *American Journal of Physiology-Renal Physiology, 281*, F255–F263. https://doi.org/10.1152/ajprenal.2001.281.2.F255

Nawata, C. M., Wood, C. M., & O'Donnell, M. J. (2010). Functional characterization of Rhesus glycoproteins from an ammoniotelic teleost, the rainbow trout, using oocyte expression and SIET analysis. *The Journal of Experimental Biology, 213*, 1049–1059. https://doi.org/10.1242/jeb.038752

Nelson, J. A. (2014). Breaking wind to survive: Fishes that breathe air with their gut. *Journal of Fish Biology, 84*, 554–576. https://doi.org/10.1111/jfb.12323

Nilsson, G. E., Dymowska, A., & Stecyk, J. A. W. (2012). New insights into the plasticity of gill structure. *Respiratory Physiology & Neurobiology, 184*, 214–222. https://doi.org/10.1016/j.resp.2012.07.012

Ong, K. J., Stevens, E. D., & Wright, P. A. (2007). Gill morphology of the mangrove killifish (*Kryptolebias marmoratus*) is plastic and changes in response to terrestrial air exposure. *The Journal of Experimental Biology, 210*, 1109–1115. https://doi.org/10.1242/jeb.002238

Partridge, B. L. (1982). The structure and function of fish schools. *Scientific American, 246*, 114–123. https://doi.org/10.1038/scientificamerican0682-114

Pelster, B. (2009). Buoyancy control in aquatic vertebrates. In M. L. Glass & S. C. Wood (Eds.), *Cardio-respiratory control in vertebrates* (pp. 65–98). Springer Verlag.

Pelster, B. (2021). Using the swimbladder as a respiratory organ and/or a buoyancy structure—Benefits and consequences. *Journal of Experimental Zoology Part A: Ecological and Integrative Physiology, 335*, 831–842. https://doi.org/10.1002/jez.2460

Pelster, B., Bridges, C. R., & Grieshaber, M. K. (1988). Physiological adaptations of the intertidal rockpool teleost *Blennius pholis* L., to aerial exposure. *Respiration Physiology, 71*, 355–373. https://doi.org/10.1016/0034-5687(88)90028-X

Pelster, B., & Burggren, W. W. (2018). Responses to environmental stressors in developing animals: Costs and benefits of phenotypic plasticity. In *Development and environment* (pp. 97–113). Springer.

Pelster, B., Giacomin, M., Wood, C. M., & Val, A. L. (2016). Improved ROS defense in the swimbladder of a facultative air-breathing erythrinid fish, jeju, compared to a non-air-breathing close relative, traira. *Journal of Comparative Physiology. B, 186*, 615–624.

Pelster, B., & Wood, C. M. (2018). Ionoregulatory and oxidative stress issues associated with the evolution of air-breathing. *Acta Histochemica, 120*, 667–679. https://doi.org/10.1016/j.acthis.2018.08.012

Pelster, B., Wood, C. M., Braz-Mota, S., & Val, A. L. (2020). Gills and air-breathing organ in O_2 uptake, CO_2 excretion, N-waste excretion, and ionoregulation in small and large pirarucu (*Arapaima gigas*). *Journal of Comparative Physiology B, 190*, 569–583. https://doi.org/10.1007/s00360-020-01286-1

Pelster, B., Wood, C. M., Speers-Roesch, B., et al. (2015). Gut transport characteristics in herbivorous and carnivorous serrasalmid fish from ion-poor Rio Negro water. *Journal of Comparative Physiology B, 185*, 225–241. https://doi.org/10.1007/s00360-014-0879-z

Perry, S. F., & Gilmour, K. M. (2002). Sensing and transfer of respiratory gases at the fish gill. *The Journal of Experimental Zoology, 293*, 249–263. https://doi.org/10.1002/jez.10129

Pratt, K. L., & Franklin, C. E. (2009). Predator or prey? The dive response to aerial and aquatic predators of Arafura filesnakes. *Australian Journal of Zoology, 57*, 423. https://doi.org/10.1071/ZO09080

Ramos, C. A., Fernandes, M. N., da Costa, O. T. F., & Duncan, W. P. (2013). Implications for osmorespiratory compromise by anatomical remodeling in the gills of *Arapaima gigas*. *The Anatomical Record, 296*, 1664–1675. https://doi.org/10.1002/ar.22758

Randall, D. J., Cameron, J. N., Daxboeck, C., & Smatresk, N. (1981). Aspects of bimodal gas exchange in the bowfin. *Respiration Physiology, 43*, 339–348. https://doi.org/10.1016/0034-5687(81)90114-6

Randall, D. J., Farrell, A. P., & Haswell, M. S. (1978). Carbon dioxide excretion in the pirarucu (*Arapaima gigas*), an obligate air-breathing fish. *Canadian Journal of Zoology, 56*, 977–982. https://doi.org/10.1139/z78-136

Randall, D. J., & Ip, Y. K. (2006). Ammonia as a respiratory gas in water and air-breathing fishes. *Respiratory Physiology & Neurobiology, 154*, 216–225.

Randall, D. J., Ip, Y. K., Chew, S. F., & Wilson, J. M. (2004). Air breathing and ammonia excretion in the giant mudskipper, *Periophthalmodon schlosseri*. *Physiological and Biochemical Zoology, 77*, 783–788. https://doi.org/10.1086/423745

Randall, D. J., Wilson, J. M., Peng, K. W., et al. (1999). The mudskipper, *Periophthalmodon schlosseri*, actively transports NH_4^+ against a concentration gradient. *American Journal of Physiology-Regulatory, Integrative and Comparative Physiology, 277*, R1562–R1567. https://doi.org/10.1152/ajpregu.1999.277.6.R1562

Robertson, L. M., Val, A. L., Almeida-Val, V. F., & Wood, C. M. (2015). Ionoregulatory aspects of the osmorespiratory compromise during acute environmental hypoxia in 12 tropical and temperate Teleosts. *Physiological and Biochemical Zoology, 88*, 357–370. https://doi.org/10.1086/681265

Sayer, M. D. J. (2005). Adaptations of amphibious fish for surviving life out of water. *Fish and Fisheries, 6*, 186–211. https://doi.org/10.1111/j.1467-2979.2005.00193.x

Scadeng, M., McKenzie, C., He, W., et al. (2020). Morphology of the Amazonian Teleost genus Arapaima using advanced 3D imaging. *Frontiers in Physiology, 11*, 260. https://doi.org/10.3389/fphys.2020.00260

Schöngart, J., & Junk, W. J. (2020). Clime e hidrologia nas várzeas da Amazonia Central. In W. J. Junkj, M. Piedale, F. Wittmann, & J. Schöngart (Eds.), *Desafios para um Manejo Sustentavel* (pp. 44–65).

Scott, G. R., Matey, V., Mendoza, J. A., et al. (2017). Air breathing and aquatic gas exchange during hypoxia in armoured catfish. *Journal of Comparative Physiology. B, 187*, 117–133.

Shartau, R. B., Baker, D. W., Crossley, D. A., & Brauner, C. J. (2016). Preferential intracellular pH regulation: Hypotheses and perspectives. *The Journal of Experimental Biology, 219*, 2235–2244. https://doi.org/10.1242/jeb.126631

Shartau, R. B., Baker, D. W., Harter, T. S., et al. (2020). Preferential intracellular pH regulation is a common trait amongst fishes exposed to high environmental CO_2. *The Journal of Experimental Biology, 223*, jeb.208868. https://doi.org/10.1242/jeb.208868

Shartau, R. B., & Brauner, C. J. (2014). Acid-base and ion balance in fishes with bimodal respiration. *Journal of Fish Biology, 84*, 682–704.

Shelton, G., Jones, D. R., & Milsom, W. K. (1986). Control of breathing in ectothermic vertebrates. In *Comprehensive physiology* (pp. 857–909). Wiley.

Shingles, A., McKenzie, D. J., Claireaux, G., & Domenici, P. (2005). Reflex cardioventilatory responses to hypoxia in the flathead gray mullet (*Mugil cephalus*) and their behavioral modulation by perceived threat of predation and water turbidity. *Physiological and Biochemical Zoology, 78*, 744–755. https://doi.org/10.1086/432143

Sloman, K. A., Sloman, R. D., De Boeck, G., et al. (2009). The role of size in synchronous air breathing of *Hoplosternum littorale*. *Physiological and Biochemical Zoology, 82*, 625–634. https://doi.org/10.1086/605936

Sloman, K. A., Wood, C. M., Scott, G. R., et al. (2006). Tribute to R. G. Boutilier: The effect of size on the physiological and behavioural responses of oscar, *Astronotus ocellatus*, to hypoxia. *The Journal of Experimental Biology, 209*, 1197–1205. https://doi.org/10.1242/jeb.02090

Smatresk, N. J. (1986). Ventilatory and cardiac reflex responses to hypoxia and NaCN in *Lepisosteus osseus*, an air-breathing fish. *Physiological Zoology, 59*, 385–397. https://doi.org/10.1086/physzool.59.4.30158592

Smatresk, N. J., & Cameron, J. N. (1982). Respiration and acid-base physiology of the spotted gar, a bimodal breather: II. Responses to temperature change and hypercapnia. *The Journal of Experimental Biology, 96*, 281–293. https://doi.org/10.1242/jeb.96.1.281

Sollid, J., De Angelis, P., Gundersen, K., & Nilsson, G. E. (2003). Hypoxia induces adaptive and reversible gross morphological changes in crucian carp gills. *The Journal of Experimental Biology, 206*, 3667–3673. https://doi.org/10.1242/jeb.00594

Tay, Y. L., Loong, A. M., Hiong, K. C., et al. (2006). Active ammonia transport and excretory nitrogen metabolism in the climbing perch, *Anabas testudineus*, during 4 days of emersion or 10 minutes of forced exercise on land. *The Journal of Experimental Biology, 209*, 4475–4489. https://doi.org/10.1242/jeb.02557

Tencatt, L. F. C., Muriel-Cunha, J., Zuanon, J., et al. (2020). A journey through the Amazon Middle Earth reveals *Aspidoras azaghal* (Siluriformes: Callichthyidae), a new species of armoured catfish from the rio Xingu basin, Brazil. *Journal of Fish Biology, 97*, 1072–1086. https://doi.org/10.1111/jfb.14467

Thinh, P. V., Huong, D. T. T., Gam, L. T. H., et al. (2019). Renal acid excretion contributes to acid-base regulation during hypercapnia in air-exposed swamp eel (*Monopterus albus*). *The Journal of Experimental Biology, 222*, jeb.198259. https://doi.org/10.1242/jeb.198259

Thurman, E. (1985). *Geochemistry of natural waters*. Martinus Nijhof/Dr. W Junk Publishers.

Tng, Y. Y. M., Chew, S. F., Wee, N. L. J., et al. (2009). Acute ammonia toxicity and the protective effects of methionine sulfoximine on the swamp eel, *Monopterus albus*. *Journal of Experimental Zoology Part A: Ecological Genetics and Physiology, 311A*, 676–688. https://doi.org/10.1002/jez.555

Tsui, T. K. N., Randall, D. J., Chew, S. F., et al. (2002). Accumulation of ammonia in the body and NH_3 volatilization from alkaline regions of the body surface during ammonia loading and exposure to air in the weather loach *Misgurnus anguillicaudatus*. *The Journal of Experimental Biology, 205*, 651–659. https://doi.org/10.1242/jeb.205.5.651

Turko, A. J., Rossi, G. S., & Wright, P. A. (2021). More than breathing air: Evolutionary drivers and physiological implications of an amphibious lifestyle in fishes. *Physiology, 36*, 307–314. https://doi.org/10.1152/physiol.00012.2021

Ultsch, G. R. (1987). The potential role of hypercarbia in the transition from water-breathing to air-breathing in vertebrates. *Evolution, 41*, 442–445. https://doi.org/10.1111/j.1558-5646.1987.tb05811.x

Ultsch, G. R. (1996). Gas exchange, hypercarbia and acid-base balance, paleoecology, and the evolutionary transition from water-breathing to air-breathing among vertebrates. *Palaeogeography Palaeoclimatology Palaeoecology, 123*, 1–27. https://doi.org/10.1016/0031-0182(96)00121-6

Val, A. L., & Almeida-Val, V. M. F. (1995). *Fishes of the Amazon and their environment zoophysiology* (Vol. 32). Springer verlag.

Val, A. L., Almeida-Val, V. M. F., & Randall, D. J. (2006). Tropical environment. In A. L. Val, V. M. F. Almeida-Val, & D. J. Randall (Eds.), *Fish physiology* (Vol. 21, pp. 1–45). Elsevier B.V.

Val, A., & Wood, C. M. (2022). Global change and physiological challenges for fish of the Amazon today and in the near future. *The Journal of Experimental Biology, 225*, jeb.216440. https://doi.org/10.1242/jeb.216440

Wang, T., & Jackson, D. C. (2016). How and why pH changes with body temperature: The α-stat hypothesis. *The Journal of Experimental Biology, 219*, 1090–1092.

Weihrauch, D., Wilkie, M. P., & Walsh, P. J. (2009). Ammonia and urea transporters in gills of fish and aquatic crustaceans. *The Journal of Experimental Biology, 212*, 1716–1730. https://doi.org/10.1242/jeb.024851

Westneat, D. F., Potts, L. J., Sasser, K. L., & Shaffer, J. D. (2019). Causes and consequences of phenotypic plasticity in complex environments. *Trends in Ecology & Evolution, 34*, 555–568. https://doi.org/10.1016/j.tree.2019.02.010

Wilkie, M. P. (2002). Ammonia excretion and urea handling by fish gills: Present understanding and future research challenges. *The Journal of Experimental Zoology, 293*, 284–301.

Wood, C. M. (1993). Ammonia and urea metabolism and excretion. In D. H. Evans (Ed.), *The physiology of fishes* (pp. 379–425). CRC Press.

Wood, C. M. (1996). Is there a relationship between urea production and acid-base balance in fish? In A. L. Val & D. J. Randall (Eds.), *Physiology and biochemistry of the fishes of the Amazon* (pp. 339–357). INPA.

Wood, C. M. (2001). The influence of feeding, exercise, and temperature on nitrogen metabolism and excretion. In P. A. Anderson & P. A. Wright (Eds.), *Fish physiology* (Vol. 20, pp. 201–238). Academic Press.

Wood, C. M., de Souza Netto, J. G., Wilson, J. M., Duarte, R. M., & Val, A. L. (2017). Nitrogen metabolism in tambaqui (*Colossoma macropomum*), a neotropical model teleost: Hypoxia, temperature, exercise, feeding, fasting, and high environmental ammonia. *Journal of Comparative Physiology B, 187*, 135–151.

Wood, C. M., Iftikar, F. I., Scott, G. R., et al. (2009). Regulation of gill transcellular permeability and renal function during acute hypoxia in the Amazonian oscar (*Astronotus ocellatus*): New angles to the osmorespiratory compromise. *The Journal of Experimental Biology, 212*, 1949–1964. https://doi.org/10.1242/jeb.028464

Wood, C. M., Kajimura, M., Sloman, K. A., et al. (2007). Rapid regulation of Na^+ fluxes and ammonia excretion in response to acute environmental hypoxia in the Amazonian oscar, *Astronotus ocellatus*. *American Journal of Physiology-Regulatory, Integrative and Comparative Physiology, 292*, R2048–R2058. https://doi.org/10.1152/ajpregu.00640.2006

Wood, C. M., Pelster, B., Braz-Mota, S., & Val, A. L. (2020). Gills versus kidney for ionoregulation in the obligate air-breathing *Arapaima gigas*, a fish with a kidney in its air-breathing organ. *The Journal of Experimental Biology, 223*, jeb232694. https://doi.org/10.1242/jeb.232694

Wood, C. M., Pelster, B., Giacomin, M., et al. (2016). The transition from water-breathing to air-breathing is associated with a shift in ion uptake from gills to gut: A study of two closely related erythrinid teleosts, *Hoplerythrinus unitaeniatus* and *Hoplias malabaricus*. *Journal of Comparative Physiology. B, 186*, 431–445. https://doi.org/10.1007/s00360-016-0965-5

Wood, C. M., Ruhr, I. M., Schauer, K. L., et al. (2019). The osmorespiratory compromise in the euryhaline killifish: Water regulation during hypoxia. *The Journal of Experimental Biology, 222*, jeb204818. https://doi.org/10.1242/jeb.204818

Wood, C. M., Walsh, P. J., Chew, S. F., & Ip, Y. K. (2005). Greatly elevated urea excretion after air-exposure appears to be carrier-mediated in the slender lungfish (*Protopterus dolloi*). *Physiological and Biochemical Zoology, 78*, 893–907.

Wood, S. C., Weber, R. E., & Davis, B. J. (1979). Effects of air-breathing on acid-base balance in the catfish, *Hypostomus* sp. *Comparative Biochemistry and Physiology Part A: Physiology, 62*, 185–187. https://doi.org/10.1016/0300-9629(79)90753-9

Wright, P. A. (2021). Cutaneous respiration and osmoregulation in amphibious fishes. *Comparative Biochemistry and Physiology Part A: Molecular & Integrative Physiology, 253*, 110866. https://doi.org/10.1016/j.cbpa.2020.110866

Wright, P. A., & Wood, C. M. (2009). A new paradigm for ammonia excretion in aquatic animals: Role of Rhesus(Rh) glycoproteins. *The Journal of Experimental Biology, 212*, 2303–2312. https://doi.org/10.1242/jeb.023085

Wright, P. A., & Wood, C. M. (2012). Seven things fish know about ammonia and we don't. *Respiratory Physiology & Neurobiology, 184*, 231–240. https://doi.org/10.1016/j.resp.2012.07.003

Part IV

Physiological and Biochemical Responses

Amazonian Crustaceans: *Macrobrachium amazonicum* (Decapoda, Palaemonidae)

Francisco A. Leone, Daniela P. Garçon, Leonardo M. Fabri, Cintya M. Moraesl, Cristiana R. Maciel, Jhonatan W. Pimentel Costa, Fernando A. Abrunhosa, Mariana V. Capparelli, and John C. McNamara

Abstract

We discuss the distribution and reproduction, functional morphology, osmoregulatory capability, digestive enzyme expression, and metal ecotoxicology of *Macrobrachium amazonicum*, a palaemonid shrimp from tropical and subtropical, fresh or brackish waters in South America. Morphophysiological findings on the development of the larval digestive system and of digestive enzyme machinery as reflected in the functional genome are discussed. Mechanisms of isosmotic intracellular and anisosmotic extracellular regulation and osmoregulatory modulation through ion transporter gene expression are outlined. Kinetic characteristics of the gill (Na^+, K^+)- and V(H^+)-ATPases, including the effects of phosphorylation by protein kinases on enzyme activity in different salinities, are

F. A. Leone (✉)
Department of Chemistry, University of Sao Paulo, São Paulo, SP, Brazil
e-mail: fdaleone@ffclrp.usp.br

D. P. Garçon
Federal University of Triângulo Mineiro, Iturama, MG, Brazil

L. M. Fabri · C. M. Moraesl
Department of Biochemistry and Immunology, Faculty of Medicine of Ribeirão Preto, University of Sao Paulo, São Paulo, SP, Brazil

C. R. Maciel · J. W. Pimentel Costa · F. A. Abrunhosa
Institute of Coastal Studies, Federal University of Pará, Bragança, PA, Brazil

M. V. Capparelli
Faculty of Earth and Water Sciences, University Regional Amazónica, Tena, Ecuador

J. C. McNamara
Department of Biology, Faculty of Philosophy, Sciences and Letters of Ribeirão Preto, University of Sao Paulo, São Paulo, SP, Brazil

Marine Biology Center, University of Sao Paulo, São Paulo, SP, Brazil

S. S. de Souza et al. (eds.), *The Future of Amazonian Aquatic Biota*,
https://doi.org/10.1007/978-3-031-66822-7_15

considered. Given the effects of climate change on Amazonian biological diversity, we also consider threats to *M. amazonicum* and their consequences in future ecological and economic scenarios.

1 Geographical Distribution and Ontogenesis

Decapods are found throughout the most varied environments across the planet. The Infraorder Caridea is a successful and diverse taxon given its abundant species that inhabit marine, brackish, and fresh waters (De Grave & Fransen, 2011; WoRMS, 2023). Of the 36 caridean families, 18 occur in Brazil, the Palaemonidae (Crustacea, Decapoda) being the most representative in terms of species numbers. One of the genera, *Macrobrachium* Spence Bate 1868, includes more than 265 species worldwide that constitute an important contribution to the diversity of this palaemonid shrimp family (WoRMS, 2023).

Macrobrachium amazonicum (Heller, 1862) is a palaemonid shrimp encountered in freshwater rivers, lakes, and reservoirs and in brackish water estuaries and coastal lagoons. Restricted to South America, it is distributed widely throughout Venezuela, Colombia, Ecuador, Peru, Guyana, Suriname, French Guiana, Brazil, Bolivia, Paraguay, and Argentina (Maciel & Valenti, 2009; Melo, 2003). In Brazil, the shrimp occurs in the Amazon, Araguaia-Tocantins, São Francisco, and Paraná-Paraguay River basins (Coelho & Ramos-Porto, 1985; Melo, 2003; Valencia & Campos, 2007). It often has been considered a diadromous species that inhabits fresh water and migrates to brackish water to spawn (McNamara et al., 1983, 1986). However, many populations are landlocked and hololimnetic, lacking access to estuarine or coastal marine waters, and their larval development takes place entirely in fresh water (Boudour-Boucheker et al., 2016; Magalhães, 1985; Pantaleão et al., 2012). Thus, two reproductive strategies (Jalihal et al., 1993) are found in *M. amazonicum*. R-strategists are diadromous, produce many small eggs and spawn in brackish waters where larval development takes place. The metamorphic postlarval stage returns actively to fresh water. In contrast, the K-strategists are hololimnetic and produce fewer larger eggs, with larval development occurring entirely in fresh water.

Embryogenesis in *M. amazonicum* requires between 12 and 18 days. After hatching, the larvae of r-strategists are carried by currents to estuarine waters where salinities are around 10–12‰S. This ontogenetic phase covers a period of about 23 days and consists of 9–11 larval stages. After metamorphosis to the postlarval stage, the juvenile phase begins, and the immature shrimps migrate actively back to fresh water (Maciel & Valenti, 2009). Differentiation of the reproductive organs is not visible (Jalihal et al., 1993; Maciel & Valenti, 2009; Moraes-Riodades & Valenti, 2004). Information on the species' development and reproductive biology is important for aquacultural practices (Maciel & Valenti, 2009; New, 2005) and development differs between wild and cultivated shrimps. Heterogeneous growth is a major aquacultural challenge and concerns the repression of the development of smaller male shrimp in the presence of larger males (Preto et al., 2010).

Macrobrachium amazonicum is economically important and is widely used as a model species in morphological, biochemical, physiological, ecological, and environmental studies (Maciel & Valenti, 2009). In fact, the species' wide distribution throughout Brazil may have resulted in part from anthropogenic activities like aquaculture. The species thus also is of relevant biological and conservation value (Maciel & Valenti, 2009).

Based on molecular and morphological evidence (Vergamini et al., 2011), the various populations of *M. amazonicum* in Brazil are considered to form a clade, suggesting the existence of a single taxonomic entity divided into three groups: (1) populations from the Amazon River Basin; (2) populations in the Paraná-Paraguay River Basin; and (3) coastal populations in northern and northeastern Brazil. Further, populations from the Pantanal region of Mato Grosso do Sul State in central Brazil are geographically isolated from these other populations and show differences in their morphological, physiological, reproductive, and ecological characteristics (Boudour-Boucheker et al., 2016; Charmantier & Anger, 2011; Hayd & Anger, 2013; Meireles et al., 2013; Vercesi & Hayd, 2015; Weiss et al., 2015). A new species has been proposed for such populations, *Macrobrachium pantanalense* (Santos et al., 2013). Recent studies corroborate the speciation of *M. pantanalense*, which does not reproduce with *M. amazonicum* from the Tietê River (Nogueira et al., 2020). Other populations from the upper Parana River basin also have been identified as *M. pantanalense* (see Calixto-Cunha et al., 2021).

The first zoeae of diadromous and hololimnetic populations lack gills and the (Na^+, K^+)-ATPase is found in the inner epithelium lining the branchiostegite. Gill (Na^+, K^+)-ATPase expression also differs between the intermediate (zoea V) and the late (decapodid) larval stages in both populations and is found in both the gills and branchiostegites of juveniles, and in the antennal glands of all developmental stages (Boudour-Boucheker et al., 2013). There is no structural differentiation between the anterior-most and posterior-most gills of palaemonid shrimps (Freire & McNamara, 1995). Adult *M. amazonicum* has seven pairs of pleurobranchs located on either side of the cephalothoracic branchial chambers. Each gill consists of a series of two phyllobranchiate hemi-lamellae that project from a triangular base characterized by the presence of afferent and efferent vessels (Freire et al., 2008). The (Na^+, K^+)-ATPase is located in the membrane invaginations of the intralamellar septal cells (McNamara & Torres, 1999).

Ontogenetic variation in palaemonid osmoregulatory capability underlies salinity tolerance and involves morphological changes that enable the occupation of biotopes of different salinities (Anger, 2003; Boudour-Boucheker et al., 2013; Charmantier, 1998). During ontogenesis, three osmoregulatory patterns can be identified: (1) osmoregulation varies little with development and is unaltered by metamorphosis, adults are osmoconformers or weak osmoregulators; (2) adult osmoregulatory ability is established during the first postembryonic stage (zoea I); (3) adult osmoregulatory ability begins at metamorphosis (Charmantier, 1998).

Osmoregulatory ability in *M. amazonicum* depends on ontogenetic development (Anger, 2003; Boudour-Boucheker et al., 2013; Charmantier, 1998). In estuarine populations, all stages strongly hyper-osmoregulate in brackish water (1–17‰S) and hypo-osmoregulate at higher salinities. This pattern is consistent with a diadromous life cycle consisting of hatching in fresh water, rapid transport of the early zoeae to brackish estuarine waters, and the active countercurrent migration of postlarvae and juveniles to fresh water (Anger, 2013; Augusto et al., 2007; Magalhães & Walker, 1988; Moreira et al., 1986; Odinetz-Collart, 1991). The hololimnetic Pantanal population of *M. amazonicum* spends its entire life cycle in fresh water (Anger & Hayd, 2010; Hayd & Anger, 2013) and hyper-osmoregulates in fresh and brackish water (0.2–17‰S). Mortality increases with increasing experimental salinity and no ontogenetic stage survives in seawater; hypo-osmoregulatory capability is absent in this Pantanal population (Anger & Hayd, 2010; Charmantier & Anger, 2011; Hayd & Anger, 2013).

The hypo-osmoregulatory ability of the early developmental stages of the Amazonian populations is linked to ion transport by the internal epithelium lining the branchiostegite. (Na^+, K^+)-ATPase is incipient or even absent from the branchiostegite of the Pantanal population. Fully functional gills expressing the (Na^+, K^+)-ATPase are essential for hyper-osmoregulation in the later developmental stages of the Pantanal population (Boudour-Boucheker et al., 2013). In adult shrimps in fresh water and at high salinity (25‰S), the gill (Na^+, K^+)-ATPase is restricted to the invaginations of the intralamellar septal cells (Boudour-Boucheker et al., 2014) as seen in *M. olfersi* (see McNamara & Torres, 1999). In the branchiostegite, the (Na^+, K^+)-ATPase is found in the inner epithelium regardless of salinity (Boudour-Boucheker et al., 2014).

2 Osmoregulatory Capability

The South American, Amazon River shrimp, *Macrobrachium amazonicum*, has diversified into many distinct populations that inhabit a wide range of osmotic niches, from freshwater streams, rivers, lakes, and reservoirs (<0.5‰ salinity, <15 mOsm/kg H_2O) to brackish water coastal estuaries (28‰ salinity, 840 mOsm/kg H_2O). Some populations are hololimnetic, undergoing ontogenesis entirely in fresh water. Others are diadromous and depend on estuarine and coastal marine waters for development from the first zoea, newly hatched in fresh water, through to the early metamorphic estuarine postlarva that returns to fresh water from the estuary. Although modest in overall osmoregulatory ability, hyper-osmoregulatory ability in all these populations is well-conserved. However, hypo-osmoregulatory ability at salinities above isosmotic (<17‰S) is variable and the different populations show patterns ranging from isosmoticity to modest hypo-osmoregulatory ability (Augusto et al., 2007; Charmantier & Anger, 2011; Faleiros et al., 2010). *Macrobrachium amazonicum* exhibits low mortality when acclimated in increased salinity, hemolymph osmolality being hyper-regulated at salinities below 20‰S (Augusto et al., 2007; Charmantier & Anger, 2011; Faleiros et al., 2010).

Here we examine the species' isosmotic intracellular regulation, anisosmotic extracellular osmotic and ionic regulatory capability, the ultrastructure of the gill epithelial cells responsible for ionic regulation, and enzyme activity and gene expression of the main ion transporters found in the gill ionocyte cell membranes.

2.1 Isosmotic Intracellular Regulation

Isosmotic intracellular regulation includes homeostatic cellular processes that regulate cell swelling and shrinkage, consequently maintaining the intracellular fluid isosmotic with the extracellular fluid (Claybrook, 1983; Péqueux, 1995). These volume regulatory increases and decreases are underpinned by the active transport of two very distinct osmotically active components: (1) inorganic ions like Na^+, K^+, Ca^{2+}, and Cl^-; and (2) organic osmolytes such as nonessential free amino acids (FAA) like glycine, arginine, alanine, proline, and taurine, together with peptides, sugars, lactate, phosphate, and bicarbonate (Gilles & Péqueux, 1981; Pierce, 1982; Siebers et al., 1972). Ion transport across the cell membrane is effected mainly by the (Na^+, K^+)-ATPase, $Na^+/K^+/2Cl^-$ symporter, and $Na^+/H^+(NH_4^+)$ exchanger, but also depends on Na^+, K^+, Ca^{2+}, and Cl^- channels (Foster et al., 2010; Freire et al., 2013). The osmotically active FAA/peptide pool is regulated by alterations in the rates of FAA export to and import from the extracellular fluid across the cell membrane, in the rates of cellular FAA oxidation and synthesis, and in the rates of protein degradation and synthesis, and FAA deamination (McNamara et al., 2004). These two mechanisms create osmotic gradients down which water moves via the regulated expression of aquaporins in the plasma membrane (Foguesatto et al., 2017, 2019).

Isosmotic intracellular regulation appears to be an ancient ability, appearing early in evolutionary time, likely in prokaryotes or early eukaryotes. The later advent of anisosmotic extracellular regulation requires multicellularity and epithelial specialization (McNamara & Freire, 2022). Together, mechanisms of intra- and extracellular volume regulation preserve cytosolic conditions such as ion, water, enzyme, and substrate concentrations optimal for cellular metabolic reactions. Isosmotic intracellular regulation is especially important when anisosmotic extracellular mechanisms are insufficient to maintain hemolymph osmolality within species-specific limits, and the extracellular fluid becomes isosmotic with the external medium at salinities above the isosmotic point.

There is very little information available on isosmotic intracellular osmoregulation in *M. amazonicum* (Augusto et al., 2007). Total FAA concentrations in *M. amazonicum* in fresh water and dilute media depend on the ontogenetic stage. They are similar in the embryo, second zoea, and adult muscle tissue, reaching around 120–180 nmoles/mg dry weight. In the first zoea, the total FAA concentration is about 350 nmoles/mg dry weight (Augusto et al., 2007). Glutamic acid, glycine, alanine, and proline predominate in the embryonic and zoeal stages, while glycine, alanine, arginine, and leucine prevail in adult shrimp muscle, gill, and nerve tissues. Hemolymph FAA concentration in adult shrimps in fresh water is about 3 mmol L^{-1}.

On salinity challenge at 25‰S, total FAA increased 1.4-fold within 48 h in the first zoea, and 1.7-fold in muscle tissue of adult shrimp within 5 days; gill FAA increased 1.6-fold within 24 h. These increases are due mainly to proline in the first zoea; arginine, alanine, and proline in adult muscle, and glycine, alanine, and proline in gill tissue. Total FAA concentrations are unchanged in the second zoea, and in nerve tissue and hemolymph of adult shrimps (Augusto et al., 2007). The effective contribution of total FAA to intracellular osmolality is ≈16% in the first zoea and 6–8% in the second zoea and adult shrimp tissues. Since *M. amazonicum* is isosmotic at 22‰S and above (see below), clearly, isosmotic intracellular regulation compensates to some degree for the inability to hypo-regulate the extracellular fluid up to the critical salinity limit (30‰S), employing increased cellular FAA as osmolytes, which reduces cell water loss.

2.2 Anisosmotic Extracellular Regulation

Hemolymph osmolality in landlocked freshwater populations of *M. amazonicum* is held at around 400 mOsm/kg H_2O (Augusto et al., 2007; Faleiros et al., 2010; Maraschi et al., 2022; Zanders & Rodriguez, 1992), representing an osmotic gradient of about 40: 1. Hemolymph sodium concentration ranges from 120 to 170 mmol L^{-1} (Augusto et al., 2007; Zanders & Rodriguez, 1992) and chloride from 149 to 218 mmol L^{-1} (Augusto et al., 2007; Faleiros et al., 2010; Maraschi et al., 2022; Zanders & Rodriguez, 1992), both showing ionic gradients of around 20: 1. These high osmotic and ionic concentrations and gradients are characteristic of estuarine palaemonid species (McNamara & Faria, 2012; McNamara & Freire, 2022).

In these landlocked populations, on salinity exposure up to 30‰S, the shrimp's hemolymph becomes isosmotic at 684–720 mOsm/kg H_2O (22–24‰S) after about 48 h (Augusto et al., 2007; Faleiros et al., 2010; Zanders & Rodriguez, 1992), suggestive of a body surface permeable to ions, characteristic of estuarine palaemonid species. Hemolymph sodium is not well regulated on salinity challenge, becoming iso-natriuretic at 310 mmol L^{-1} (14‰S), and maintaining iso-natriuremicity at higher salinities (Augusto et al., 2007). Chloride is well-regulated and becomes isochloremic at 250–277 mmol L^{-1} (15.6–17‰S), hypo-conforming at higher salinities (Augusto et al., 2007; Faleiros et al., 2010; Maraschi et al., 2022; Zanders & Rodriguez, 1992). Osmotic and ionic regulatory capabilities (Δhemolymph/Δexternal medium) are 0.40–0.45 for osmolality (Augusto et al., 2007; Faleiros et al., 2010), 0.67 for sodium (Augusto et al., 2007), and 0.13–0.48 for chloride (Augusto et al., 2007; Faleiros et al., 2010), revealing modest regulatory ability (Augusto et al., 2007).

However, overall osmoregulatory ability differs distinctly among the different populations (Fig. 1). Estuarine populations from northern Brazil exhibit good hyper-osmoregulation and modest hypo-osmoregulation. Populations from central Brazilian freshwater rivers also show good hyper-osmotic regulation but become isosmotic at ≈600 mOsm kg^{-1} H_2O (30‰S) and above as do landlocked populations from southeastern Brazil (≈680 mOsm/kg^{-1} H_2O, 22‰S) that also are good hyper-osmoregulators.

Shrimps representing estuarine populations were obtained from the Aquaculture Center, UNESP, Jaboticabal, São Paulo State, southeastern Brazil. These specimens were broodstock, originally collected from a stretch of the Fura das Marinhas channel at Orla de Caiçauna, (1° 14′ 30" S, 48° 19′ 52" W), on the southern side of the Ilha do Mosqueiro in the Amazon River Delta, Pará State, northern Brazil. This

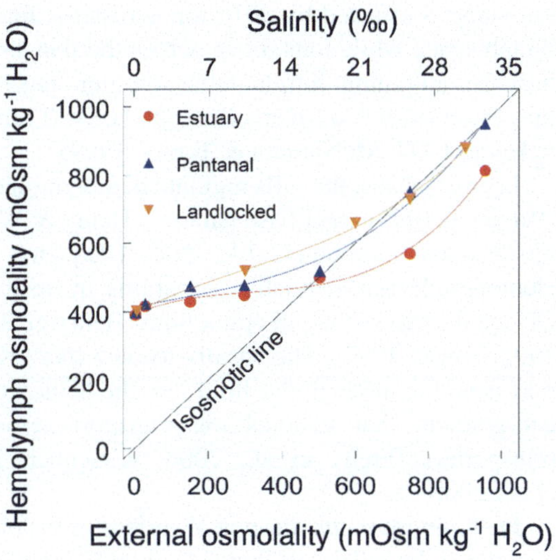

Fig. 1 Osmoregulatory ability in three geographically distinct populations of the Amazon River shrimp, *Macrobrachium amazonicum*

population (Estuary) exhibits good hyper-osmoregulation but modest hypo-osmoregulation. Data are redrawn from Fig. 1a in Charmantier and Anger (2011).

Shrimps representing the freshwater Pantanal populations were obtained from a hatchery at UEMS, Aquidauana, Mato Grosso do Sul State, Brazil. These specimens were collected in fresh water from the Miranda River, 6.5 km downstream from the Rio Salobra bridge over highway BR262 (20° 8.9' S, 56° 30.4' W) in Mato Grosso do Sul State, central Brazil. This population (Pantanal) shows good hyper-osmotic regulation but becomes isosmotic at ≈600 mOsm kg^{-1} H_2O (20‰S). Data are redrawn from Fig. 1b in Charmantier and Anger (2011).

Shrimps from a landlocked freshwater population were collected from the Santa Elisa Reservoir (21° 6.616' S, 48° 3.080' W) near Sertãozinho in São Paulo State, southeastern Brazil. This population (Landlocked) shows hyper-osmoregulatory ability and isosmoticity (≈680 mOsm/kg^{-1} H_2O, 22‰S) similar to the freshwater Pantanal population. Data are redrawn from Fig. 1 in Augusto et al. (2007).

3 Gill Epithelial Cell Ultrastructure and Ion Transport Pathways

Like all palaemonid shrimps, *M. amazonicum* possesses phyllobranchiate pleurobranchiae that are functionally undifferentiated in terms of their ion transport and gas exchange capabilities compared to brachyuran crabs (Freire & McNamara, 1995). The gill lamellae perform both functions indiscriminately owing to their particular microanatomy and epithelial cell ultrastructure. Hemolymph enters the gill through a single, central afferent vessel, percolates across each lamella to the bounding marginal canal, through the capillary spaces formed by the pillar cell bodies and the intralamellar septum and exits through the two lateral efferent vessels. Gas exchange is effected by diffusion across the fine pillar cell flanges into the hemolymph space while ions move across the flanges and through the pillar cell bodies into the adjoining intralamellar septum, reaching the hemolymph through the mitochondria-rich intralamellar septal cells (Freire & McNamara, 1995; McNamara & Lima, 1997; McNamara & Torres, 1999).

In crustaceans, the gills regulate hemolymph osmolality and effect gas exchange (Péqueux, 1995), acid-base balance (Henry & Wheatly, 1992), and the excretion of nitrogenous compounds like NH_3 (Péqueux, 1995; Weihrauch et al., 2004). Hemolymph osmoregulatory capability in fresh water depends mainly on Na^+ and Cl^- uptake across the gill epithelium (Henry et al., 2002; Kirschner, 2004; Péqueux, 1995; Towle, 1997). The basally located (Na^+, K^+)-ATPase drives active Na^+ transport from the intracellular fluid into the hemolymph, establishing an electrochemical gradient that sustains ion transport across a variety of other membrane transporters (Freire et al., 2008; Kirschner, 2004; Sáez et al., 2009; Wilder et al., 2000).

In *M. amazonicum*, the apical surface of the pillar cells, both in the flange regions and above the pillar cell bodies, is highly amplified by numerous evaginations that abut on the cuticle, forming a subapical space (Faleiros et al., 2010). This apical membrane contains a suite of ion transporters such as the V(H^+)-ATPase (Pinto et al., 2016; Boudour-Boucheker et al., 2014; Lucena et al., 2015), $Na^+/H^+(NH_4^+)$ and Cl^-/HCO_3^- exchangers, and Na^+ and K^+ channels (Freire et al., 2008; Maraschi et al., 2022; McNamara & Faria, 2012). Cytosolic protons, transported by the V(H^+)-ATPase into the subapical space, and cytosolic HCO_3^- exchanged for external Cl^-, are furnished from the hydration of CO_2 by carbonic anhydrase in the pillar cell cytosol. The membranes of the intralamellar septal cells are also highly amplified by deep, fine invaginations, each associated with numerous mitochondria that furnish ATP for the (Na^+, K^+)-ATPase located in the invagination membranes (Leone et al., 2014; McNamara & Torres, 1999). Ions like Na^+ present in the external medium (fresh water) are transported down the outside positive electrical gradient created by the apical V(H^+)-ATPase and/or exchanged for counter protons or NH_4^+, across the evaginations of the apical pillar cell membrane. They then diffuse through the pillar cell body cytosol, cross into the adjoining intralamellar septal cells, and are exchanged for hemolymph K^+ by the (Na^+, K^+)-ATPase. Cl^- exchanged apically for HCO_3^- follows a similar pathway, exiting into the hemolymph via Cl^- channels in the septal cell invaginations, although this ion may enter the hemolymph via Cl^- channels in the membranes of the lower pillar cell flange (Freire et al., 2008; McNamara & Faria, 2012).

4 Ion Transporter Activity and Expression in the Gill Epithelium

(Na^+, K^+)-ATPase-specific activity in gill homogenates from *M. amazonicum* is very salinity sensitive. Activity decreases 4-fold, from 260 U mg^{-1} protein in shrimps in fresh water to 68 U mg^{-1} protein in shrimps after 5 days acclimation at 25‰S

(Faleiros et al., 2010). Likewise, in gill microsomal preparations, activity decreases 2.3-fold, from 122 U mg^{-1} protein in shrimps in fresh water to 52 U mg^{-1} protein in shrimps after 10 days acclimation at 25‰S (Fabri et al., 2022). Relative expression of gill (Na^+, K^+)-ATPase α-subunit mRNA in shrimps acclimated to 25‰ S from fresh water increases markedly by 6.5-fold after 1 h but decreases to 1.7-fold after 10 days. Decreased (Na^+, K^+)-ATPase activity in full salinity-acclimated *M. amazonicum* is not accompanied by decreased (Na^+, K^+)-ATPase α-subunit mRNA expression. However, the transporter may play a role in salinity acclimation, particularly in Cl^- secretion through the $Na^+/K^+/2Cl^-$ symporter, which is partially dependent on (Na^+, K^+)-ATPase activity (Maraschi et al., 2022; McNamara & Faria, 2012). Although $Na^+/K^+/2Cl^-$ symporter mRNA expression is unaltered on salinity challenge, $Na^+/K^+/2Cl^-$ protein expression increases twofold (Maraschi et al., 2022).

Gill V(H^+)-ATPase-specific activity is 23 U mg^{-1} protein in shrimps in fresh water (Faleiros et al., 2010). In shrimps exposed to 25‰ S from fresh water, gill V(H^+)-ATPase B-subunit mRNA expression also increases 1.5-fold after 1 h, but decreases markedly to just 0.25-fold that of shrimps in fresh water after 10 days acclimation as does V(H^+)-ATPase activity (Faleiros et al., 2010). Clearly, gill V(H^+)-ATPase activity is sustained by V(H^+)-ATPase B-subunit mRNA expression in *M. amazonicum* when in fresh water but transporter mRNA expression and activity are strongly downregulated at high salinity.

These findings demonstrate that *M. amazonicum* employs strategies of isosmotic intracellular and anisosmotic extracellular regulation both during its diadromous life cycle and on salinity challenge. Such regulation is dependent on alterations in both the intracellular FAA pool and in the expression and activity of the main transporters like the (Na^+, K^+)-ATPase, V(H^+)-ATPase, and $Na^+/K^+/2Cl^-$ symporter that affect gill-based ionic regulation.

5 The Gill (Na^+, K^+)-ATPase

The (Na^+, K^+)-ATPase transports three Na^+ from the intracellular medium into the extracellular medium, and two K^+ in the opposite direction, using energy from ATP hydrolysis (Apell, 2017; Kaplan, 2002). In crustacean gills, the (Na^+, K^+)-ATPase predominates in specialized cells that mediate NaCl uptake from the external medium (Compere et al., 1989; Lucu & Towle, 2003). The basic functional unit of the (Na^+, K^+)-ATPase consists of a catalytic α subunit (Mr ≈ 110 kDa) responsible for ATP hydrolysis, and a noncatalytic subunit β subunit (Mr ≈ 35 kDa) associated with the γ subunit, a protein belonging to the FXYD family, expressed in a tissue-specific manner (Clausen et al., 2017; Kaplan, 2002; Morth et al., 2007). The α-subunit contains the substrate, ouabain, Mg^{2+}, K^+, and Na^+ binding sites, and is phosphorylated during the catalytic cycle (Palmgren & Nissen, 2011). The β-subunit modulates enzyme affinity for ions and acts as a chaperone (Geering, 2008). The γ-subunit regulates pump activity (Geering, 2008). The Alberts-Postmodel is the most accepted model explaining the (Na^+, K^+)-ATPase catalytic cycle (Clarke, 2009; Kaplan, 2002) and involves transitions between two conformational states,

E1 and E2, which are reversibly phosphorylated at the D369 aspartate residue (Post et al., 1972). The E1 conformational state shows a high affinity for intracellular Na^+ while the E2 shows a high affinity for extracellular K^+ (Clausen et al., 2017).

The kinetic characteristics of the gill (Na^+, K^+)-ATPase in the different *M. amazonicum* populations (Table 1) show considerable differences, likely attributable to their adaptation to different freshwater habitats (Fabri et al., 2019). The maximum rate of ATP hydrolysis is ≈130 nmol Pi min^{-1} mg^{-1} of protein in various populations (Fabri et al., 2019; Garçon et al., 2022; Leone et al., 2012); this rate is only half that seen in a landlocked population inhabiting the Santa Elisa reservoir in Sertãozinho, São Paulo State (Santos et al., 2007). In contrast, the maximum rate of hydrolytic activity for *M. pantanalense* is ≈80 nmol Pi min^{-1} mg^{-1} protein (Table 1). The gill (Na^+, K^+)-ATPase of the Santa Elisa reservoir population (Santos et al., 2007) and *M. pantanalense* from the Araguari River (Table 1) show modulation by ATP differently from other *M. amazonicum* populations (Fabri et al., 2019; Garçon et al., 2022; Leone et al., 2012).

When acclimated for 10 days in salinities close to its isosmotic point, the maximum rate of ATP hydrolysis by the gill (Na^+, K^+)-ATPase from *M. amazonicum* decreases by 60% (Belli et al., 2009; Fabri et al., 2022; Faleiros et al., 2010), accompanied by a small decrease in α subunit protein expression (Fabri et al., 2022) and a small increase in α subunit mRNA expression (Faleiros et al., 2010). Affinity for Mg^{2+} and Na^+ increases after acclimation while K^+ affinity decreases (Fabri et al., 2022).

The regulation of (Na^+, K^+)-ATPase activity involves a variable and complex set of factors (Pirkmajer & Chibalin, 2019; Therien & Blostein, 2000), including interaction with different FXYD proteins (Geering, 2006, 2008; Pirkmajer & Chibalin, 2019). In *M. amazonicum*, while exogenous pig FXYD2 has no effect on gill (Na^+, K^+)-ATPase activity in shrimps in fresh water, the peptide stimulates activity in 21‰ S-acclimated shrimps by ≈50%, increased further by NH_4^+ (Fabri et al., 2022). Phosphorylation of the (Na^+, K^+)-ATPase constitutes a link between hormonal regulation and enzymatic activity (Pirkmajer & Chibalin, 2019; Therien & Blostein, 2000). PKA-mediated phosphorylation inhibits gill (Na^+, K^+)-ATPase activity by 85% in 21‰S-acclimated shrimps while PKC phosphorylation markedly inhibits activity in both freshwater- and salinity-acclimated *M. amazonicum*. The alterations seen in activity and ion affinity after salinity acclimation may result from regulatory mechanisms mediated by phosphorylation via protein kinases A and C and the FXYD peptide (Fabri et al., 2022).

In crustaceans, most ammonia is excreted as NH_4^+ by the gills (Weihrauch et al., 1998, 1999); the (Na^+, K^+)-ATPase is involved in this process (Leone et al., 2017; Masui et al., 2005; Weihrauch et al., 1998, 1999). The synergistic stimulation by NH_4^+ seen in various crustaceans suggests an additional NH_4^+-stimulated component of (Na^+, K^+)-ATPase activity, inhibited by Mg^{2+}, in Brachyura (Leone et al., 2005, 2017; Masui et al., 2005). In *M. amazonicum*, exposure to NH_4^+ increases (Na^+, K^+)-ATPase activity (Garçon et al., 2022; Pinto et al., 2016) although mRNA expression and α-subunit distribution are unaltered (Pinto et al., 2016).

Table 1 Comparison of kinetic parameters calculated for the stimulation by ATP, Mg^{2+}, Na^+, K^+, and NH_4^+, and inhibition by ouabain, of (Na^+, K^+)-ATPase activity in microsomal fractions from the gills of different species and populations of freshwater shrimps *Macrobrachium*

Macrobrachium amazonicum									*Macrobrachium pantanalense*			
	Tietê River[a]		Cultivated[b]		Santa Elisa Reservoir[c]		Grande River[d]		Baiazinha Lagoon[e]		Araguari River[e]	
	V_M	$K_{0.5}$	V_M	$K_{0.5}$	V_M	$K_{0.5}$	V_M	$K_{0.5}$	V_M	$K_{0.5}$	V_M	$K_{0.5}$
Modulator	(U mg^{-1})	(mmol L^{-1})	(U mg^{-1})	(mmol L^{-1})	(U mg^{-1})	(mmol L^{-1})	(U mg^{-1})	(mmol L^{-1})	(U mg^{-1})	(mmol L^{-1})	(U mg^{-1})	(mmol L^{-1})
ATP	130.3 ± 4.8 -	0.065 ± 0.0016 -	133.3 ± 6.4 -	0.21 ± 0.01 -	33.9 ± 1.9[f] 232.9 ± 15.3[g]	0.004 ± 0.0002 0.144 ± 0.01	137.3 ± 8.1 -	0.083 ± 0.010 -	67.9 ± 3.4 -	0.11 ± 0.04 -	42.4 ± 2.0[f] 50.9 ± 2.1[g]	0.012 ± 0.003 0.102 ± 0.01
Mg^{2+}	130.6 ± 6.8	0.33 ± 0.042	139.4 ± 6.7	1.03 ± 0.05	261.9 ± 18.3	0.79 ± 0.06	135.6 ± 2.5	0.16 ± 0.011	71.4 ± 4.8	0.19 ± 0.06	90.1 ± 4.7	0.68 ± 0.06
Na^+	127.5 ± 5.3	5.30 ± 0.42	126.4 ± 6.3	3.00 ± 0.15	275.1 ± 15.1	5.50 ± 0.3	134.2 ± 4.3	5.5 ± 0.48	71.9 ± 2.7	7.89 ± 0.64	95.7 ± 8.8	22.74 ± 4.4
K^+	126.7 ± 7.7	0.65 ± 0.079	137.1 ± 7.0	2.02 ± 0.10	271.8 ± 10.9	0.88 ± 0.04	132.3 ± 5.6	0.75 ± 0.14	67.6 ± 3.8	2.81 ± 0.43	91.2 ± 4.6	2.33 ± 0.5
NH_4^+	134.5 ± 8.6	1.28 ± 0.44	194.2 ± 8.5	4.76 ± 0.23	385.9 ± 15.8	5.00 ± 0.2	149.2 ± 7.1	2.5 ± 0.27	–	–	–	–
		K_I (mmol L^{-1})		**K_I (mmol L^{-1})**		**K_I (mmol L^{-1})**		**K_I (mmol L^{-1})**		**K_I (mmol L^{-1})**		**K_I (mmol L^{-1})**
Ouabain		0.18 ± 0.058		0.13 ± 0.007		0.147 ± 0.009		0.14 ± 0.01		0.143 ± 0.004		0.140 ± 0.016

[a]Data from Fabri et al. (2019)
[b]Data from Leone et al. (2012)
[c]Data from Santos et al. (2007)
[d]Data from Garçon et al. (2022)
[e]Data from Fabri et al. (2024)
[f]High-affinity ATP-binding sites
[g]Low-affinity ATP-binding sites

The osmoregulatory capability of *M. amazonicum* changes as a function of gill development (Anger, 2003; Boudour-Boucheker et al., 2013), and is accompanied by differences in the expression and location of the (Na^+, K^+)-ATPase (Boudour-Boucheker et al., 2013; Leone et al., 2014). Distribution in juvenile and adult gill lamellae is similar, predominating in the intralamellar septum (Leone et al., 2014). Enzyme activity and affinity for ATP increase twofold in decapodid III compared to zoeae I, juveniles, and adults. Affinity for different metal ion modulators also changes during development. These marked kinetic differences underscore a role for the (Na^+, K^+)-ATPase in larval osmoregulatory ability (Leone et al., 2012).

Polyamines and/or biogenic amines are found in the hemolymph and gills of crustaceans (Péqueux et al., 2002). Their intracellular content is regulated by two general mechanisms: ATP and sodium gradient-dependent transport across the cell membrane, and de novo biosynthesis (Kobayashi et al., 1999; Seiler & Dezeure, 1990). Crustacean gill (Na^+, K^+)-ATPase activity is likely regulated by neuroendocrine factors (Morris & Edwards, 1995), and various biogenic amines are involved in the control of water and salt movements in euryhaline decapods (Garçon et al., 2011; Liu et al., 2008a; Lovett & Watts, 1995; Silva et al., 2008). Polyamine levels increase during adaptation to hyperosmotic media (Liu et al., 2008b; Péqueux et al., 2002) and inhibit (Na^+, K^+)-ATPase activity in several vertebrate tissues (Heinrich-Hirsch et al., 1977; Robinson et al., 1986). Spermidine and putrescine also inhibit gill (Na^+, K^+)-ATPase activity in *M. amazonicum* in an ontogenetic stage-specific manner; spermine has only a negligible inhibitory effect on kinetic characteristics (Lucena et al., 2019; Lucena, Garçon, et al., 2017; Lucena, McNamara, & Leone, 2017). (Na^+, K^+)-ATPase activity in the gills of juvenile and adult *M. amazonicum* responds differently to enzyme phosphorylation/dephosphorylation (Lucena, Garçon, et al., 2017). Dopamine stimulates activity by 40% in adults, binding directly to the enzyme molecule at a site different from ouabain binding (Lucena et al., 2019).

6 The Gill V(H^+)-ATPase

The V(H^+)-ATPase is a large protein complex arrayed as two functional structures acting in concert through a rotary mechanism that provides proton transport: the integral membrane domain V_0, responsible for H^+ transport, and the hydrophilic V_1 domain, facing the cytoplasm, that hydrolyzes ATP (Collins & Forgac, 2018; Toei et al., 2010). In the crustacean gill epithelium, the apical V(H^+)-ATPase represents an additional force together with the (Na^+, K^+)-ATPase in driving ion uptake in dilute media. It actively transports H^+ across the apical membrane into the subapical space below the cuticle, creating a positive extracellular potential that drives Na^+ influx through apical Na^+ channels (Lee et al., 2022; McNamara & Faria, 2012).

The V(H^+)-ATPase is located apically in the gill pillar cells of different *M. amazonicum* populations (Boudour-Boucheker et al., 2014; Fabri et al., 2023; Lucena et al., 2015; Pinto et al., 2016). V(H^+)-ATPase activity in cultivated *M. amazonicum* (Lucena et al., 2015) is 50% greater than in various wild populations (Table 2; Fabri et al., 2023; Faleiros et al., 2010). The apparent affinity of the V(H^+)-ATPase for ATP in these various populations ranges from 0.10 to 0.30 mmol L^{-1} (Table 2; Faleiros et al., 2010; Lucena et al., 2015; Fabri et al., 2023) and between 0.15 and 0.45 mmol L^{-1} for Mg^{2+} (Table 2; Faleiros et al., 2010; Lucena et al., 2015; Fabri et al., 2023). Modulation by ATP of V(H^+)-ATPase activity in populations from the Tietê and Grande Rivers follows Michaelis-Menten kinetics (Fabri et al., 2023) in contrast with the cooperative kinetics seen for the Santa Elisa Reservoir (Faleiros et al., 2010) and cultivated (Lucena et al., 2015) *M. amazonicum* populations.

V(H^+)-ATPase activity and protein expression disappear from the gill tissue of *M. amazonicum* acclimated to 21 or 25‰S for 10 days (Faleiros et al., 2010; Fabri et al., 2023), accompanied by a fourfold decrease in B subunit mRNA expression (Faleiros et al., 2010). In juvenile and adult *M. amazonicum*, B subunit mRNA expression and V(H^+)-ATPase activity are similar, while the initial larval stages, dependent on brackish water, do not show V(H^+)-ATPase activity (Lucena et al., 2015). These differences in V(H^+)-ATPase activity suggest the expression of distinct isoenzymes (Lucena et al., 2015). The exposure of *M. amazonicum* to ammonia also increases V(H^+)-ATPase activity (Garçon et al., 2022; Pinto et al., 2016). The increase of V(H^+)-ATPase activity, mRNA expression, and cytoplasmic distribution suggests that the V(H^+)-ATPase plays an important role in NH_4^+ excretion (Garçon et al., 2022; Pinto et al., 2016). In cultivated *M. amazonicum* populations, increased V(H^+)-ATPase activity reflects a 2.5-fold increase in B subunit mRNA

Table 2 Comparison of calculated kinetic parameters estimated for ATP and Mg^{2+} stimulation, and bafilomycin inhibition, of V(H^+)-ATPase activity in gill microsomal fractions from different populations of *Macrobrachium amazonicum*

Population	Parameter	ATP	Mg^{2+}	Bafilomycin
Tietê River[a]	V_M (U mg^{-1})	27.2 ± 1.1	28.0 ± 2.4	
	K_M (mmol L^{-1})	0.27 ± 0.04	0.28 ± 0.05	
	n_H	1.1	1.2	
	K_I (nmol L^{-1})	–	–	35.1 ± 1.1
Grande River[a]	V_M (U mg^{-1})	16.7 ± 1.7	17.4 ± 1.3	
	K_M (mmol L^{-1})	0.16 ± 0.03	0.14 ± 0.02	
	n_H	1.1	1.2	
	K_I (nmol L^{-1})	–	–	21.2 ± 1.3
Santa Elisa Reservoir[b]	V_M (U mg^{-1})	22.7 ± 1.3	23.3 ± 1.2	
	K_M (mmol L^{-1})	0.11 ± 0.03	0.44 ± 0.02	
	n_H	2.8	3.5	
	K_I (nmol L^{-1})	–	–	3.97 ± 0.32
Cultivated[c]	V_M (U mg^{-1})	41.6 ± 1.3	39.8 ± 1.1	
	K_M (mmol L^{-1})	0.21 ± 0.02	0.26 ± 0.02	
	n_H	1.3	1.6	
	K_I (nmol L^{-1})	–	–	24.2 ± 1.2

[a]Data from Fabri et al. (2023)
[b]Data from Faleiros et al. (2010)
[c]Data from Lucena et al. (2015)

expression; part of this newly synthesized enzyme is directed to the cytoplasm (Pinto et al., 2016).

7 Functionality of the Feeding Apparatus and Foregut During Ontogeny: Implications for Larviculture

The larvae of some populations of *M. amazonicum* hatch as zoeae I containing a large yolk reserve, suggesting obligatory lecithotrophy (Anger & Hayd, 2009; Araújo & Valenti, 2007) and nonfeeding behavior (Queiroz et al., 2011).

The larvae seek live food from stage II on, and inert food from stage III on. Larvae at stages IV, V, and VI accept both foods indiscriminately, while inert food is preferred by stages VII to IX. Feeding *M. amazonicum* larvae on *Artemia* before stage II or on an inert diet before stage IV is unnecessary and increases production costs while reducing water quality. From stages IV to VI, feeding on *Artemia* and an inert diet is necessary, while an inert diet should be offered from stage VII on. This schedule optimizes feeding management and production costs (Araújo & Valenti, 2007).

Nevertheless, the larval development of prawn species may be strongly related to the morphogenesis and functionality of the digestive system (Anger, 2001). Some species, for example, show lecithotrophy during a specific stage or throughout larval development and disregard external feeding. Therefore, their nutritional needs are supplied by internal lipid stores (endogenous reserves), which may be essential to the larvae overcome natural environmental barriers, such as the lack of food resources during their development (Abrunhosa et al., 2008; Anger, 2001; Anger & Hayd, 2009; Magalhães & Walker, 1988; Queiroz et al., 2011; Rocha et al., 2016, 2017).

Similar feeding behavior strategies are seen among some *Macrobrachium* species. *Macrobrachium rosenbergii* zoeae I do not feed (Araújo & Valenti, 2007) and are indifferent to the food offered (Moller, 1978) although they show tactile perception (Barros & Valenti, 1997). However, feeding habits do change during the larval development of *M. amazonicum*, which is key information for adequate culture management. Studies on the morphology of the buccal appendages and the internal foregut structures during larval development are required to better understand such feeding processes.

Morphofunctional development of the pereiopods is another important aspect of larval feeding behavior. In *M. amazonicum*, the pereiopods undergo significant morphological changes, especially in the fifth pair of pereiopods, that strongly indicate a direct relationship with food capture. These pereiopods are lacking in zoeae I but appear as a nonfunctional bud in zoeae II, remaining unchanged in zoeae III, and becoming functional and complex in zoeae IV. The larvae of *M. amazonicum* use the fifth pair of pereiopods to attach themselves firmly to food with which they can swim. They also ingest small food items using their buccal appendages, like mandibles (Fig. 2a and b).

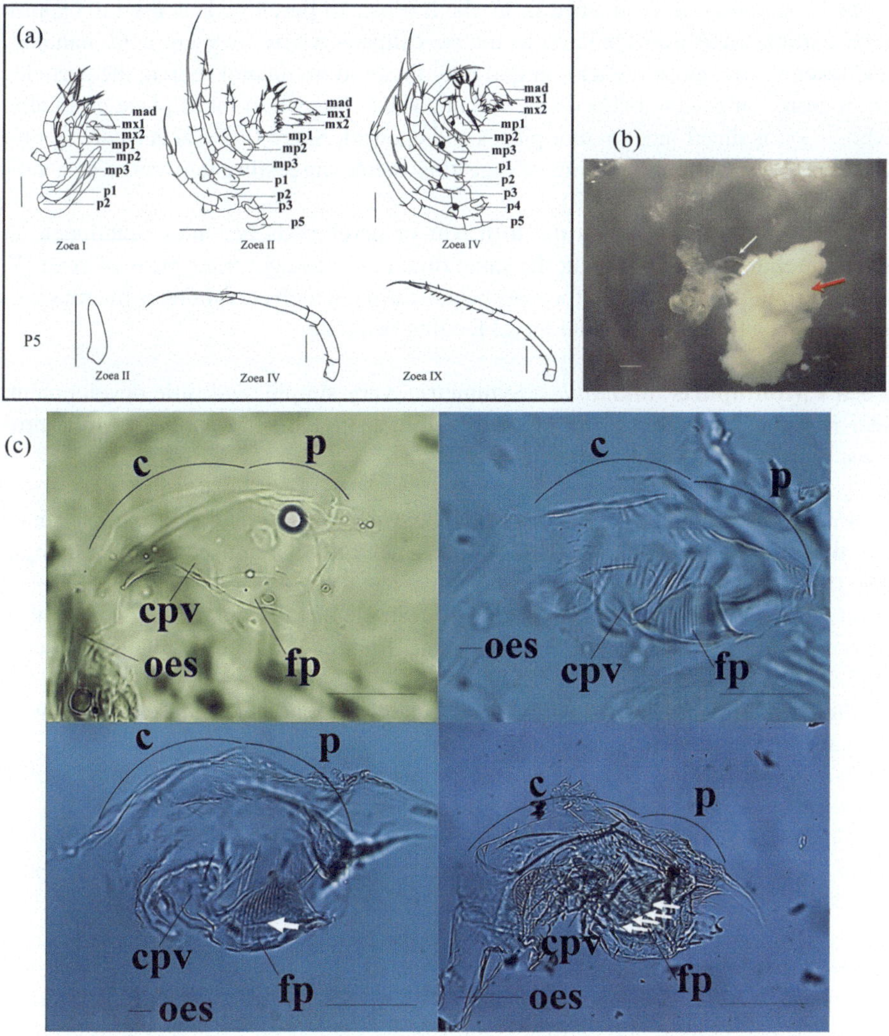

Fig. 2 (**a**) Ventral right-side view of feeding appendage and pereiopod development in zoea I, II, and IV of the Amazon River shrimp *Macrobrachium amazonicum*. Exopods of all appendages are not shown. Abbreviations: mad, mandible; mx1, maxillule; mx2, maxila; mp1, first maxilliped; mp2, second maxilliped; mp3, third maxilliped; p1, first pereiopod; p2, second pereiopod; p3, third pereiopod; p4, fourth pereiopod; p5, fifth pereiopod. Scale bar, 135 μm. P5, development of fifth pereiopods in *M. amazonicum*. Scale bar. Zoea I, II, and IV = 135 μm; Zoea IX = 300 μm. (X) = Absent. (**b**) Zoea IV larva of the Amazonian River shrimp *M. amazonicum* capturing inert food (red arrow) using the fifth pair of pereiopods (white arrows). Scale bar, 270 μm. (**c**) Foregut development in the Amazon River shrimp *M. amazonicum*: above, zoea I and zoea II; below, zoea IV and zoea IX. Abbreviations: oes, esophagus; cpv, cardiopyloric valve; fp, filter press; c, cardiac chamber; p, pyloric chamber. White arrow indicates ampullary ridges of the filter press. Scale bar, zoea I and II = 50 μm; zoea IV and zoea IX = 200 μm

In *M. rosenbergii*, from zoea II on, the last pair of pereiopods is used to capture food particles and transfer them to the maxillipeds where they are held, handled, and taken to the mouth. After manipulating the food in the oral region, the particles are ingested or rejected (Barros & Valenti, 1997). Ingestion takes place gradually, while food is diced into smaller pieces and chewed. Stage I to V zoeae can capture food particles without handling or ingesting them, suggesting they serve as a substrate (Barros & Valenti, 1997).

In *M. amazonicum*, when the fifth pair of pereiopods becomes functional, an interampullary crest appears at the same time in the foregut filter press of zoea IV, characterizing an increase in foregut specialization, which corroborates the relationship between this structure and larval feeding behavior.

Zoea I Mouthparts *Mandibles.* Symmetric, very simple, and little development. Incisive process: 2 short teeth. Median region: a single *lacinia mobilis.* Molar process: rounded, without protuberances (Fig. 2a).

Maxillule. Showing few setae on the lobe. Coxal endite: 3–4 short simple setae. Basal endite: 2 min cuspidate +2–3 simple setae. Endopod: unsegmented +1 single seta (Fig. 2a).

Maxila. Coxal endite: lobes fused with 2 simple distal setae. Basal endite: lobes incompletely fused, each with 3 simple setae. Endopod: bilobed with 3 small setae (Fig. 2a).

First Maxilliped. Basipod with 3 small setae. Endopod: short, unsegmented with 4 small, simple setae (Fig. 2a).

Second Maxilliped. Basipod: with a simple seta. Endopod: 4-segmented with 5 setae (0,0,1 denticulate, 3 simple +1 long serrulate on apex) (Fig. 2a).

Third Maxilliped. Basipod: with 1 simple seta. Endopod: 4-segmented with 6 setae (1 simple, 0, 1 simple +1 serrulate, 2 simple +1 long serrulate on apex) (Fig. 2a).

Gross Morphology of the First Zoea Foregut The foregut of zoea I is poorly developed, almost indiscernible, and entirely free of setae or hard structures (Fig. 2c). The pyloric and cardiac chambers are enveloped by a very thin layer of chitin. Both structures and the filter press lack setae and other structures like ossicles or teeth. Setae are also absent on the cardiopyloric valve, which strongly suggests that food processing is not present.

Zoea II Mouthparts *Mandibles.* A little more developed than at stage 1. Incisive process: with 3 teeth. Median region: with 1 *lacinia mobilis.* Molar process: with one short tooth (Fig. 2a).

Maxillule. Increased number of setae. Coxal endite: with 6 simple setae. Basal endite: with 7 setae (4 cuspidate +3 simple). Endopod: unsegmented with 2 simple setae (Fig. 2a).

Maxila. Coxal endite: showing an increase in seta size. Basal endite: with 3 median simple setae. Endopod: with 3 median setae (Fig. 2a).

First Maxilliped. Basipod: with 3 simple median setae. Endopod: with 2 + 3 simple setae. (Fig. 1a).

Second and Third Maxilliped. Basipod: unchanged. Endopod: unchanged (Fig. 2a).

Gross Morphology of Foregut of the Second, Third, and Fourth Zoeae The foregut of zoea I undergoes remarkable changes during the transition to zoea II. Newly molted zoea II exhibit numerous setae (Fig. 2c). The cardiac chamber is narrow and similar in size to the pyloric chamber, showing numerous fine short setae distributed over the cardiac floor. The cardiopyloric valve is rounded, bearing 7 long robust setae surrounding the posterolateral portion. The pyloric chamber is more enlarged compared to the cardiac chamber, showing medium length, elongated setae on the pyloric roof; numerous elongated setae are arranged in rows on the pyloric floor. The filter press is enlarged, lacking ampullary ridges, occupying the whole lower portion; the elongate ampullary setae arise from the lower edge of the filter press. The cardiopyloric valve and filter press are more developed compared to the previous stage. These features provide evidence of their function in food processing, suggesting that feeding behavior occurs at this stage.

The foregut of zoea III is like that of zoea II without significant morphological alterations. The filter press is altered after molting to zoea IV, showing an ampullary ridge, suggesting a gradually increased capacity for processing food processing (Fig. 2c).

Zoea IV Mouthparts *Mandibles.* Increased numbers of teeth and denticles. Incisive process: with 3 strong teeth. Median region: with eight short denticles. Molar process: with 3 strong teeth (Fig. 2a).

Maxillule. Increasing setae size. Coxal endite: with 6 simple setae. Basal endite: with 7 setae (4 cuspidate +3 simple). Endopod: unsegmented with 2 simple setae (Fig. 2a).

Maxila. No significant change (Fig. 1a).

First, Second, and Third Maxilliped. No significant change (Fig. 2a).

Gross Morphology of the Zoea IX Foregut The foregut is more developed and complex at this stage. Changes include an increased number of setae, although lacking ossicles, teeth, or other rigid structures (Fig. 2c). The cardiac chamber is narrow and much longer than the pyloric chamber, showing numerous medium setae, arranged in rows along the cardiac floor and roof; the cardiac floor has numerous fine, long medial setae distributed in 2 lateral rows. The cardiopyloric valve is rounded and stronger with around 9 long erect setae that surround the posterolateral portion of the valve; other small setae are distributed in the anterior, superior, and lateral valve portions. Pyloric chamber shows numerous short setae arranged in a row along the roof; the pyloric wall has long setae concentrated in the posterior portion with long rigid medial setae on the interampullary ridge. The large filter press is large and much more complex, showing 5 ampullary ridges and numerous lower ampullary setae.

Development of the Pereiopods *Pereiopod 1.* Zoea I: Present and bilobed; Zoea II: Functional. Segmented, with several setae and a significantly developed and elongated apical spine and curve on apex; Zoea IV: Similar to zoea II, without substantial alterations, except for increments and changes of setae; Zoea IX: Well developed, with increments of numerous setae (simple, serrated, and plumose). Chela developed, with the presence of a short apical spine on the movable finger.

Pereiopod 2. Zoea I, II, and IV: Similar to pereiopod 1, without substantial changes; Zoea IX: Well developed, with increments of numerous setae (simple and serrated). Chela developed, with the presence of short apical spines on the fixed and movable fingers. Longer than the first pereiopod.

Pereiopod 3. Absent in zoea I and appears in an undeveloped bilobed form in zoea II; Zoea IV: Functional. Segmented, showing several setae and a rigid, slightly curved apical spine; Zoea IX: Similar to previous zoeae, except for increments and modifications of some setae (simple and serrated).

Pereiopod 4. Absent in zoeae I and II, and bilobed in zoea IV; Zoea IX: No substantial modifications in relation to the third pereiopod.

Pereiopod 5. Absent in zoea I and appears as a single bud in zoea II; Zoea IV: Functional, presenting a significant change compared to zoea III. Segmented, with setae and an extremely long, terminally curved apical spine. Larger compared to the other pereiopods; Zoea IX: Similar to zoea IV, with a large increase in size and in the number of smooth setae, as well as a shortened apical spine (Fig. 2d).

8 Adult Digestive Machinery: An In Silico Approach Using Functional Genomics

Our next-generation sequencing has provided massive data from the functional genome of the hepatopancreas, refined using bioinformatic tools in silico, allowing the identification of digestive enzymes. The enzymatic apparatus is compatible with the species' omnivorous feeding behavior, suggesting that *M. amazonicum* can digest foods rich in proteins, lipids, and carbohydrates (Rocha et al., 2022).

Among the proteases, a serine protease (trypsin) and two cysteine proteases (cathepsins B and L) were found. Trypsin is one of the most abundant crustacean proteases, responsible for hydrolysis at the carboxyl region of lysine and arginine residues in protein substrates (Jiang et al., 2011). Cathepsins B and L belong to the papain family, known as lysosomal proteolytic enzymes. These enzymes are active in both intracellular and extracellular environments since they are found in the B and F cells and in the lumen of hepatopancreas tubules (Hu & Leung, 2007; Jung et al., 2013; Stephens et al., 2012). Cathepsin B is distinct from the other members of the papain superfamily as it functions as both endopeptidase and exopeptidase (McGrath, 1999).

Little is known of the enzymatic machinery related to the digestion and metabolism of lipids in *M. amazonicum*, except for small fragments of the pancreatic triacylglycerol lipase and lipid transporter (LDL) genes (Rocha et al., 2022). Information on the enzymes involved in carbohydrates digestion is limited to descriptions of partial genes encoding chitinase, amylase, glucanase, and mannanase (Rocha et al., 2022).

Carbohydrates are not considered essential for crustaceans since they rely on more complex molecules such as polysaccharides than on simple sugars (Saborowski, 2015). However, analyses of stomach content in *M. amazonicum* collected from different locations in the wild, suggest omnivorous behavior, with detritus and plant material consumed more frequently than items of animal origin (Cruz et al., 2021; Mulati, 2017; Odinetz-Collart, 1988). *Macrobrachium amazonicum* prefers to feed on babassu coconut than on other food sources, even when pieces of fish are offered (Bentes et al., 2014). Nonradioactive isotope studies in captive shrimps reveal the incorporation of carbon from plant tissues, which increases growth rate (Heldt et al., 2019), suggesting that the enzymatic apparatus for the digestion of plant-based foods may be more complex than previously inferred from the functional hepatopancreas genome, as seen in crabs like *Parasesarma pictum* and *Parasesarma affine* (see Wang et al., 2022).

8.1 Novel Findings

Knowledge of the digestive machinery of *M. amazonicum* is limited. This section incorporates novel in silico scans of the functional genome database for the hepatopancreas obtained using the Illumina platform. The sequenced database includes ~40,000 contigs; hepatopancreas genes differentially expressed (DE) or overexpressed compared to muscle tissue were selected, resulting in 3860 fragments. Of the annotated fragments, 600 were associated with proteases and 494 were assigned to general lipid metabolism. Considering the scarce records of genes associated with carbohydrate digestion in crustaceans and the evidence that cultivated *M. amazonicum* might benefit from consuming nutrients of plant origin (Heldt et al., 2019), the search was enriched by mining using the Carbohydrate-Active enZYmes Database (CAZY), resulting in 1116 paired sequences.

After excluding redundant sequences, the 77 resulting protease-related transcripts were classified as: 51 cysteine proteases, 24 serine proteases, and 2 aspartyl proteases. Enzymes belonging to the three protease classes were characterized, such as cathepsins B and D, chymotrypsin, and six trypsin isoforms. All showed a high degree of similarity with proteases from other crustaceans, including the complete amino acid coding region with active sites and conserved substrate binding sites. Besides adding trypsin isoforms, cathepsin D and chymotrypsin to the current list of enzymes found in *M. amazonicum*, these data corroborate findings by Rocha et al.

(2022), who have reported small fragments of these enzymes. This diversity in proteases suggests an evolutionary strategy to enable digestion of different protein substrates, as seen in other crustaceans (Aguiñaga-Cruz et al., 2019; Muhlia-Almazán et al., 2008; Perera et al., 2015). In *Litopenaeus vannamei*, protease diversity may be an adaptive strategy that avoids the effects of proteolysis inhibitors found in certain food items (Maytorena-Verdugo et al., 2017).

While the scan for functions related to lipid metabolism resulted in 494 differentially expressed fragments, only three genes could be linked to lipid digestion: pancreatic triacylglycerol lipase, low density lipoprotein receptor, and triacylglycerol lipase. All sequences encompassed the complete amino acid coding regions and conserved substrate binding sites and are similar to other crustacean genes. Triacylglycerol lipase was encountered for the first time in *M. amazonicum* compared to a previous study (Rocha et al., 2022). The limited number of records concerning genes related to lipid digestion as seen for *M. amazonicum* is typical of crustaceans (Dammannagoda et al., 2015; Wang et al., 2020; Wei et al., 2014). Although the crabs *Parasesarma pictum* and *Parasesarma affine* exhibit a low diversity of enzymes related to lipid digestion, they show many differentially expressed enzymes related to the digestion of plant substrates, attributable to their herbivorous feeding habits (Wang et al., 2022).

The carbohydrate-related enzymes from the present scans of the *M. amazonicum* genome comprised 1116 sequences. Only glycosyl hydrolases were selected as they represent the main enzymes involved in carbohydrate digestion, resulting in 178 fragments. The enzymes involved in the digestion of polysaccharides of plant origin then were mined, totaling 119 fragments, distributed as follows: 44 related to starch digestion, 28 to cellulose digestion, and 27 to hemicellulose. After removing the repeated sequences, 11 different enzymes were identified, including their entire coding region, most of the active sites and the conserved substrate binding sites; homologies with other crustaceans were identified. The category containing the greatest number of annotations (7) was related to the degradation of hemicellulose: alpha-mannosidase, beta-mannosidase, beta-galactosidase, mannan-endo-1,4-beta-mannosidase, beta-glucuronidase, alpha-L-fucosidase, and lysosomal acid glucosylceramidase. Regarding cellulose degradation, endo-beta1,4-glucanase was identified; genes related to starch degradation included alpha-amylase, sucrase-isomaltase, and maltase (oligo-1,6-glucosidase). We thus infer that the enzymatic machinery discernable in the functional genome of *M. amazonicum* is compatible with the digestion of polysaccharides of plant origin, even though further nutrigenomic studies are required. It is also necessary to establish whether these enzymes are synthesized by *M. amazonicum* or derive from symbiotic organisms inhabiting the shrimp's digestive tract.

These novel data on genes related to digestive enzymes corroborate the hypothesis that *M. amazonicum* is an omnivorous species that has evolved strategies for the digestion of various protein and energy substrates, including vegetal material as an energy source and the digestion of plant fibers like cellulose and hemicellulose. The carbohydrates contained in the ingredients used in formulating artificial diets can act as important dietary protein savers, reducing feed costs and potential environmental impacts consequent to cultivation practices.

9 Ecotoxicology of *Macrobrachium amazonicum*

Amazon forests face ongoing degradation of their ecosystems (Dias et al., 2021; Lapola et al., 2023; Pelicice & Castello, 2021) due to mining activities, urbanization, and agricultural expansion (Capparelli et al., 2021). Pollution from these activities, such as wastewater from fish farming, legal and illegal gold mining, urban contamination and leachates from nonfunctional landfills, accrete large quantities of inorganic (i.e., metals) and organic (i.e., pesticides) material, together with plastics and microplastics (Lucas-Solis et al., 2021; Rico et al., 2023) into the freshwater ecosystems (Capparelli et al., 2020; Hoang & Kanemoto, 2021). Consequently, Amazonian freshwater species are at risk.

Ecotoxicological studies using *M. amazonicum* are scarce (Galvão et al., 2018) although aquaculture experiments reveal how decreased water quality including alterations in pH, temperature, salinity, ammonia, and nitrate impairs ontogenetic development (Araujo & Valenti, 2010; Dutra et al., 2016, 2017; Hayd et al., 2014). Nitrite concentrations can affect the larval development and growth of *M. amazonicum* exposed to the pesticide Cypermethrin (Soares et al., 2020). Intersexuality resulting from hormonal disruption by certain contaminants is known in wild populations (Paschoal & Zara, 2017).

Mercury and other metals originating in mining areas of the Brazilian Amazon accumulate in *M. amazonicum* (see Costa et al., 2020). However, toxic elements like As, Cd, Hg, and Pb, and essential trace elements such as Co, Cu, Cr, Fe, Mn, Mo, Ni, Se, and Zn, evaluated in tissues of *M. amazonicum* from a mining region near Santarém on the lower Amazon River are not higher than in shrimps from nonmining areas (Albuquerque et al., 2020). While studies to understand the effects of metals on *M. amazonicum* are necessary (Chalkiadaki et al., 2014), the number of recent studies has increased exponentially due to concerns regarding potential risks, ecological effects, persistence, nondegradability, bioaccumulation, and public food safety (Suami et al., 2019). The tolerance of the marine Pacific Whiteleg shrimp *Litopenaeus vannamei* to cadmium and zinc increases progressively with increasing salinity. Exposure to sublethal cadmium and zinc reduces the hyper-osmoregulatory ability of shrimp exposed to 5 and 15‰S and increases hypo-osmoregulatory ability at 27‰S (Ardiansyah et al., 2012). Exposure of *L. vannamei* to sublethal concentrations of copper for 15 days increases gill metal accumulation causing hyperplasia, necrosis, and loss of intercellular hemolymph gaps (Soegianto et al., 2013). Exposure of the African sharptooth catfish *Clarias gariepinus* to lead and zinc simultaneously for 3 weeks decreases gill (Na^+, K^+)-ATPase activity (Apiamu et al., 2022). Exposure of the endemic South Africa freshwater crab *Potamonautes warreni* to elevated zinc and cadmium for 21 days reduces hemolymph sodium although gill (Na^+, K^+)- and the V(H^+)-ATPase activities are unaffected (Glendinning et al.,

2021). Gill (Na^+, K^+)-ATPase and carbonic anhydrase activities in the fiddler crab *Minuca rapax* are markedly inhibited depending on delivery regime, salinity, and copper concentration (Capparelli et al., 2017). In vitro studies show that cobalt can stimulate or inhibit Blue crab *Callinectes danae* gill (Na^+, K^+)-ATPase K^+-phosphatase activity depending on the presence of magnesium ions (Leone et al., 2023).

Studies on the effects of contaminants like hydrocarbons and pesticides on *M. amazonicum*, and of emerging xenobiotics such as pharmaceuticals and microplastics are scant, despite their increasing impact on aquatic ecosystems. *While* microplastic titers *in Amazonian rivers* are elevated (Lucas-Solis et al., 2021; Rico et al., 2023), only a single study has addressed microplastic contamination in *M. amazonicum* (dos Anjos Guimarães et al., 2023). On the Amazon River at Itacoatiara (Amazonas State, northern Brazil), 90% of shrimps sampled contained polypropylene and nylon microplastic particles, particularly those closest to the urban center. Most studies of microplastic bioaccumulation and effects focus on marine shrimps (Gurjar et al., 2021) where they can accumulate in the tissues and cause biological, physiological, and genetic effects (Hsieh et al., 2021; Suman et al., 2020). Similar effects are seen in the freshwater shrimp *Macrobrachium nipponense* (see Li et al., 2022).

Knowledge of bioaccumulation and the toxic effects of pollutants on aquatic organisms also contributes to the monitoring and conservation of aquatic ecosystems (Sá et al., 2019). *Macrobrachium amazonicum* could be used as a bioindicator of contamination, as proposed for mercury contamination in mining areas due to tissue mercury accumulation (Costa et al., 2020).

In addition to pollutants, Amazonian ecosystems are affected by multiple stressors related to global climate change. These include decreasing pH, increasing substrate and water temperatures, increasing frequency and intensity of storms and flooding cycles, and increasing periods of extreme drought and exacerbated precipitation. The adverse effects of climate change on freshwater species biodiversity are likely to be exacerbated by various synergic stressors, including land use changes, habitat loss, and contamination by pollutants, also increasing the risk to human health (Rainbow, 2002). Considering the *wide* distribution *of M. amazonicum*, its ecological and commercial importance, and that it is a component of ecosystems facing augmented anthropogenic impact, the species soon may be seriously threatened.

10 Future Perspectives

The degradation of the Amazon rainforest, a biome rich in hydrological and mineral resources and in biodiversity, harbors the potential for considerable regional and global climate change. The major challenges to this complex sytem of habitats are uncontrolled deforestation due to cattle raising and soybean production, the impact of pesticides, and especially illegal trade and activities in timber and gold mining. If unchecked, the forest will be transformed into a savannah ecosystem, with extensive loss of biodiversity. Temperature increases will affect the hydrological and other biogeochemical cycles and increase the production of greenhouse gases, affecting the stability of rivers, leading to large-scale species extinctions. Potential decrease in humidity within the Amazon basin will affect the climate of the entire South American continent. Understanding the effects of the degradation of the Amazon rainforest is imperative, including its causes and impacts on species. Future agricultural expansion, and logging and mining activities must be predicted accurately, and the interventions necessary to restrict them must be implemented to maintain evapotranspiration and to prevent loss of biodiversity. The current biochemical, physiological, and morphological findings provided here on *Macrobrachium amazonicum*, an important component of the Amazon biota, should contribute to the development of programs aimed at preserving this species and at reducing the environmental impact caused by Amazon rainforest destruction.

References

Abrunhosa, F. A., Arruda, D. C. B., Simith, D. J. B., et al. (2008). The importance of feeding in the larval development of the ghost shrimp *Callichirus major* (Decapoda: Callianassidae). *Anais da Academia Brasileira de Ciências, 80*, 445–453. https://doi.org/10.1590/S0001-37652008000300006

Aguiñaga-Cruz, J. A., Sainz-Hernández, J. C., García-Rodríguez, L., et al. (2019). Trypsin polymorphism and modulation in *Penaeus vannamei* (Boone, 1931): A review. *Latin American Journal of Aquatic Research, 47*, 723–732. https://doi.org/10.3856/vol47-issue5-fulltext-1

Albuquerque, F. E. A., Minervino, A. H. H., Miranda, M., et al. (2020). Toxic and essential trace element concentrations in the freshwater shrimp *Macrobrachium amazonicum* in the Lower Amazon, Brazil. *Journal of Food Composition and Analysis, 86*, 103361. https://doi.org/10.1016/j.jfca.2019.103361

Anger, K. (2001). *The biology of decapod crustacean larvae* (p. 429). AA Balkema.

Anger, K. (2003). Salinity as a key parameter in the larval biology of decapod crustaceans. *Invertebrate Reproduction and Development, 43*, 29–45. https://doi.org/10.1080/07924259.2003.9652520

Anger, K. (2013). Neotropical Macrobrachium (Caridea: Palaemonidae): On the biology origin and radiation of freshwater-invading shrimp. *Journal of Crustacean Biology, 33*, 151–183. https://doi.org/10.1163/1937240X-00002124

Anger, K., & Hayd, L. (2009). From lecithotrophy to planktotrophy: Ontogeny of larval feeding in the Amazon River prawn *Macrobrachium amazonicum*. *Aquatic Biology, 7*, 19–30. https://doi.org/10.3354/ab00180

Anger, K., & Hayd, L. (2010). Feeding and growth in early larval shrimp *Macrobrachium amazonicum* from the Pantanal southwestern Brazil. *Aquatic Biology, 9*, 251–261. https://doi.org/10.3354/ab00259

Apell, H. J. (2017). Mechanistic principles of ion transport in the Na,K-ATPase. *Russian Journal of Electrochemistry, 53*, 237–247. https://doi.org/10.1134/s1023193517030028

Apiamu, A., Osawaru, S. U., Asagba, S. O., et al. (2022). Exposure of African catfish (*Clarias gariepinus*) to lead and zinc modulates membrane-bound transport protein: A plausible effect on Na^+/K^+-ATPase activity. *Biological Trace Element Research, 200*, 4160–4170. https://doi.org/10.1007/s12011-021-03005-5

Araújo, M. C., & Valenti, W. C. (2007). Feeding habit of the Amazon river prawn *Macrobrachium amazonicum* larvae. *Aquaculture, 265*, 187–193. https://doi.org/10.1016/j.aquaculture.2007.01.016

Araujo, M. C., & Valenti, W. C. (2010). Effect of low salinities on *Macrobrachium amazonicum* (Crustacea, Palaemonidae) larvae subjected to starvation. *Magistra, Cruz das Almas, BA, 22*, 191–197.

Ardiansyah, S., Irawan, B., & Soegianto, A. (2012). Effect of cadmium and zinc in different salinity levels on survival and osmoregulation of white shrimp (*Litopenaeus vannamei* Boone). *Marine and Freshwater Behaviour and Physiology, 45*, 291–302. https://doi.org/10.1080/10236244.2012.734056

Augusto, A., Greene, L. J., Laure, H. J., et al. (2007). The ontogeny of isosmotic intracellular regulation in the diadromous freshwater palaemonid shrimps *Macrobrachium amazonicum* and *M. olfersii* (Crustacea Decapoda). *Journal of Crustacean Biology, 27*, 626–634. https://doi.org/10.1651/S-2796.1

Barros, H. P., & Valenti, W. C. (1997). Comportamento alimentar do camarão de água doce, *Macrobrachium rosenbergii* (De Man) (Crustacea, Palaemonidae) durante a fase larval: Análise qualitativa. *Revista Brasileira de Zoologia, 14*, 785–793. https://doi.org/10.1590/S0101-81751997000400003

Belli, N. M., Faleiros, R. O., Firmino, K. C. S., et al. (2009). Na,K-ATPase activity and epithelial interfaces in gills of the freshwater shrimp *Macrobrachium amazonicum* (Decapoda, Palaemonidae). *Comparative Biochemistry and Physiology Part A: Molecular & Integrative Physiology, 152*, 431–439. https://doi.org/10.1016/j.cbpa.2008.11.017

Bentes, B., Martinelli-Lemos, J., Paes, E., et al. (2014). Experimental study on the efficiency of different types of traps and baits for harvesting *Macrobrachium amazonicum* (Heller, 1862). *Acta Scientiarum. Biological Sciences, 36*, 383–391. https://doi.org/10.4025/actascibiolsci.v36i4.22852

Boudour-Boucheker, N., Boulo, V., Charmantier-Daures, M., et al. (2014). Differential distribution of V-type H^+-ATPase and Na^+/K^+-ATPase in the branchial chamber of the palaemonid shrimp *Macrobrachium amazonicum*. *Cell and Tissue Research, 357*, 195–206. https://doi.org/10.1007/s00441-014-1845-5

Boudour-Boucheker, N., Boulo, V., Charmantier-Daures, M., et al. (2016). Osmoregulation in larvae and juvenile of two recently separated Macrobrachium species: Expression patterns of ion transporter genes. *Comparative Biochemistry and Physiology, 195A*, 39–45. https://doi.org/10.1016/j.cbpa.2016.02.005

Boudour-Boucheker, N., Boulo, V., Lorin-Nebel, C., et al. (2013). Adaptation to freshwater in the palaemonid shrimp *Macrobrachium amazonicum*: Comparative ontogeny of osmoregulatory organs. *Cell and Tissue Research, 353*, 87–98. https://doi.org/10.1007/s00441-013-1622-x

Calixto-Cunha, M., Rodrigues, T. S., Ueira-Vieira, C., et al. (2021). Genetic and phenotypic variability in populations of the *Macrobrachium amazonicum* complex: New findings to the upper Parana Hydrographic Basin. *Zoologischer Anzeiger, 293*, 26–36. https://doi.org/10.1016/j.jcz.2021.05.006

Capparelli, M. V., Cabrera, M., Rico, A., et al. (2021). An integrative approach to assess the environmental impacts of gold mining contamination in the Amazon. *Toxics, 9*(7), 149. https://doi.org/10.3390/toxics9070149

Capparelli, M. V., McNamara, J. C., & Grosell, M. (2017). Effects of waterborne copper delivered under two different exposure and salinity regimes on osmotic and ionic regulation in the mudflat fiddler crab, *Minuca rapax* (Ocypodidae, Brachyura). *Ecotoxicol Environ Safety, 143*, 201–209. https://doi.org/10.1016/j.ecoenv.2017.05.042

Capparelli, M. V., Moulatlet, G. M., Abessa, D. M. S., et al. (2020). An integrative approach to identify the impacts of multiple metal contamination sources on the Eastern Andean foothills of the Ecuadorian Amazonia. *Science of the Total Environment, 709*, 136088. https://doi.org/10.1016/j.scitotenv.2019.136088

Chalkiadaki, O., Dassenakis, M., & Lydakis-Simantiris, N. (2014). Bioconcentration of Cd and Ni in various tissues of two marine bivalves living in different habitats and exposed to heavily polluted seawater. *Chemistry and Ecology, 30*, 726–742. https://doi.org/10.1080/02757540.2014.917172

Charmantier, G. (1998). Ontogeny of osmoregulation in crustaceans: A review. *Invertebrate Reproduction & Development, 33*, 177–190. https://doi.org/10.1080/07924259.1998.9652630

Charmantier, G., & Anger, K. (2011). Ontogeny of osmoregulatory patterns in the South American shrimp *Macrobrachium amazonicum*: Loss of hypo-regulation in a land-locked population indicates phylogenetic separation from estuarine ancestors. *Journal of Experimental Marine Biology and Ecology, 396*, 89–98. https://doi.org/10.1016/j.jembe.2010.10.013

Clarke, R. J. (2009). Mechanism of allosteric effects of ATP on the kinetics of P-type ATPases. *European Biophysics Journal, 39*, 3–17. https://doi.org/10.1007/s00249-009-0407-3

Clausen, M. V., Hilbers, F., & Poulsen, H. (2017). The structure and function of the Na,K-ATPase isoforms in health and disease. *Frontiers in Physiology, 8*, 1–16. https://doi.org/10.3389/fphys.2017.00371

Claybrook, D. L. (1983). Nitrogen metabolism. In L. H. Mantel (Ed.), *The biology of crustacea: Internal anatomy and physiological regulation* (Vol. 5, pp. 163–213). Academic Press.

Coelho, P. A., & Ramos-Porto, M. R. (1985). Camarões de água doce do Brasil: Distribuição geográfica. *Revista Brasileira de Zoologia, 2*, 405–410. https://doi.org/10.1590/S0101-81751984000200014

Collins, M. P., & Forgac, M. (2018). Regulation of V-ATPase assembly in nutrient sensing and function of V-ATPases in breast cancer metastasis. *Frontiers in Physiology, 9*, 2012–2019. https://doi.org/10.3389/fphys.2018.00902

Compere, P., Wanson, S., Pequeux, A., et al. (1989). Ultrastructural changes in the gill epithelium of the green crab *Carcinus maenas* in relation to the external salinity. *Tissue & Cell, 21*, 299–318. https://doi.org/10.1016/0040-8166(89)90073-6

Costa, B. N. S., Almeida, H. P., da Silva, B. C. P., et al. (2020). *Macrobrachium amazonicum* (Crustacea, Decapoda) used to biomonitor mercury contamination in rivers. *Archives of Environmental Contamination and Toxicology, 78*, 245–253. https://doi.org/10.1007/s00244-019-00683-0

Cruz, B., Cunha, M., Bueno, A., et al. (2021). Natural diet of *Macrobrachium brasiliense* (Crustacea, Decapoda) in a Cerrado stream. *Iheringia. Série Zoologia, 111*, e2021003. https://doi.org/10.1590/1678-4766e2021003

Dammannagoda, L. K., Pavasovic, A., Prentis, P. J., et al. (2015). Expression and characterization of digestive enzyme genes from hepatopancreatic transcripts from redclaw crayfish (*Cherax quadricarinatus*). *Aquaculture Nutrition, 21*, 868–880. https://doi.org/10.1111/anu.12211

De Grave, S., & Fransen, C. H. J. M. (2011). Carideorum Catalogus: The recent species of the dendrobranchiate, stenopodidean, procarididean and caridean shrimps (Crustacea: Decapoda). *Zoologische Mededelingen, 85*, 195–588. http://www.vliz.be/en/imis?refid=210286

Dias, R., Pelicice, F. M., Tófoli, R., et al. (2021). Brazil's leading environmental agency and aquatic biodiversity threatened by federal decree. *Anais da Academia Brasileira de Ciências, 93*, e20210175. https://doi.org/10.1590/0001-3765202120210175

dos Anjos Guimarães, G., de Moraes, B. R., Ando, R. A., et al. (2023). Microplastic contamination in the freshwater shrimp *Macrobrachium amazonicum* in Itacoatiara, Amazonas, Brazil. *Environmental Monitoring and Assessment, 195*, 434. https://doi.org/10.1007/s10661-023-11019-w

Dutra, F. M., Forneck, S. C., Brazão, C. C., et al. (2016). Acute toxicity of ammonia to various life stages of the Amazon river prawn, *Macrobrachium amazonicum*, Heller, 1862. *Aquaculture, 453*, 104–109. https://doi.org/10.1016/j.aquaculture.2015.11.038

Dutra, F. M., Rönnau, M., Sponchiado, D., et al. (2017). Histological alterations in gills of *Macrobrachium amazonicum* juveniles exposed to ammonia and nitrite. *Aquatic Toxicology, 187*, 115–123. https://doi.org/10.1016/j.aquatox.2017.04.003

Fabri, L. M., Garçon, D. P., Moraes, C. M., et al. (2023). Kinetic characterization of the gill V(H^+)-ATPase of two hololimnetic population of the Amazon River Shrimp *Macrobrachium amazonicum* (Decapoda, Palaemonidae). *Comparative Biochemistry and Physiology B, 268*, 110880. https://doi.org/10.1016/j.cbpb.2023.110880

Fabri, L. M., Lucena, M. N., Garçon, D. P., et al. (2019). Kinetic characterization of the gill (Na^+, K^+)-ATPase in a hololimnetic population of the diadromous Amazon River shrimp

Macrobrachium amazonicum (Decapoda, Palaemonidae). *Comparative Biochemistry and Physiology, 227B*, 64–74. https://doi.org/10.1016/j.cbpb.2018.09.004

Fabri, L. M., Moraes, C. M., Costa, M. I. C., et al. (2022). Salinity-dependent modulation by protein kinases and the FXYD2 peptide of gill (Na^+, K^+)-ATPase activity in the freshwater shrimp *Macrobrachium amazonicum* (Decapoda, Palaemonidae). *Biochimica et Biophysica Acta (BBA)-Biomembranes, 1864*, 183982. https://doi.org/10.1016/j.bbamem.2022.183982

Fabri, L. M., Moraes, C. M., Calixto-Cunha, M., Almeida, A. C., Faleiros, R. O., Garçon, D. P., McNamara, J. C., Faria, S. C., & Leone, F. A. (2024). (Na+, K+)- ATPase kinetics in Macrobrachium pantanalense: highlighting intra- and interspecific variation within the Macrobrachium amazonicum complex. *Comparative Biochemistry and Physiology, 273B*, 110987. https://doi.org/10.1016/j.cbpb.2024.110987

Faleiros, R. O., Goldman, M. H. S., Furriel, R. P. M., et al. (2010). Differential adjustment in gill Na^+/K^+- and V-ATPase activities and transporter mRNA expression during osmoregulatory acclimation in the cinnamon shrimp *Macrobrachium amazonicum* (Decapoda, Palaemonidae). *The Journal of Experimental Biology, 213*, 3894–3905. https://doi.org/10.1242/jeb.046870

Foguesatto, K., Bastos, C. L. Q., Boyle, R. T., et al. (2019). Participation of Na^+/K^+-ATPase and aquaporins in the uptake of water during moult processes in the shrimp *Palaemon argentinus* (Nobili, 1901). *Journal of Comparative Physiology, 189B*, 523–535. https://doi.org/10.1007/s00360-019-01232-w

Foguesatto, K., Boyle, R. T., Rovani, M. T., et al. (2017). Aquaporin in different moult stages of a freshwater decapod crustacean: Expression and participation in muscle hydration control. *Comparative Biochemistry and Physiology, 208A*, 61–69. https://doi.org/10.1016/j.cbpa.2017.03.003

Foster, C., Amado, E. M., Souza, M. M., et al. (2010). Do osmoregulators have lower capacity of muscle water regulation than osmoconformers? A study on decapod crustaceans. *The Journal of Experimental Zoology, 313*, 80–94. https://doi.org/10.1002/jez.575

Freire, C. A., & McNamara, J. C. (1995). Fine structure of the gills of the freshwater shrimp *Macrobrachium olfersii* (Decapoda): Effect of acclimation to high salinity medium and evidence for involvement of the lamellar septum in ion uptake. *Journal of Crustacean Biology, 15*, 103–116. https://doi.org/10.1163/193724095X00622

Freire, C. A., Onken, H., & McNamara, J. C. (2008). A structure-function analysis of ion transport in crustacean gills and excretory organs. *Comparative Biochemistry and Physiology, 151A*, 272–304. https://doi.org/10.1016/j.cbpa.2007.05.008

Freire, C. A., Souza-Bastos, L. R., Amado, E. M., et al. (2013). Regulation of muscle hydration upon hypo- or hyper-osmotic shocks: Differences related to invasion of the freshwater habitat by decapod crustaceans. *The Journal of Experimental Zoology, 319*, 297283. https://doi.org/10.1002/jez.1793

Galvão, R. C. F., Holanda, I. B. B., De Carvalho, D. P., et al. (2018). Freshwater shrimps (*Macrobrachium depressimanum* and *Macrobrachium jelskii*) as biomonitors of Hg availability in the Madeira River Basin, Western Amazon. *Environmental Monitoring and Assessment, 190*, 77. https://doi.org/10.1007/s10661-018-6460-6

Garçon, D. P., Fabri, L. M., Moraes, C. M., et al. (2022). Effects of ammonia on gill (Na^+, K^+)-ATPase kinetics in a hololimnetic population of the Amazon River shrimp *Macrobrachium amazonicum*. *Aquatic Toxicology, 246*, 106144. https://doi.org/10.1016/j.aquatox.2022.106144

Garçon, D. P., Lucena, M. N., França, J. L., et al. (2011). Na/K-ATPase activity in the posterior gills of the blue crab, *Callinectes ornatus* (Decapoda, Brachyura): Modulation of ATP hydrolysis by the biogenic amines spermidine and spermine. *The Journal of Membrane Biology, 244*, 9–20. https://doi.org/10.1007/s00232-011-9391-5

Geering, K. (2006). FXYD proteins: New regulators of Na-K-ATPase. *American Journal of Physiology-Renal Physiology, 290*, F241–F250. https://doi.org/10.1152/ajprenal.00126.2005

Geering, K. (2008). Functional roles of Na,K-ATPase subunits. *Current Opinion in Nephrology and Hypertension, 17*, 526–532. https://doi.org/10.1097/MNH.0b013e3283036cbf

Gilles, R., & Péqueux, A. (1981). Cell volume regulation in crustaceans: Relationship between mechanisms for controlling the osmolality of extracellular and intracellular fluids. *The Journal of Experimental Biology, 215*, 351–362. https://doi.org/10.1002/jez.1402150312

Glendinning, S., Vosloo, A., & Morris, S. (2021). Ion regulation in a freshwater crab, *Potamonautes warreni*: The effects of trace metal exposure. *Aquatic Toxicology, 237*, 105885. https://doi.org/10.1016/j.aquatox.2021.105885

Gurjar, U. R., Xavier, M., Nayak, B. B., et al. (2021). Microplastics in shrimps: A study from the trawling grounds of northeastern part of Arabian Sea. *Environmental Science and Pollution Research, 28*, 48494–48504. https://doi.org/10.1007/s11356-021-14121-z

Hayd, L., & Anger, K. (2013). Reproductive and morphometric traits of *Macrobrachium amazonicum* (Decapoda: Palaemonidae) from the Pantanal, Brazil, suggests initial speciation. *Revista de Biología Tropical (International Journal of Tropical Biology), 61*, 39–57. https://doi.org/10.15517/rbt.v61i1.10881

Hayd, L. A., Lemos, D., & Valenti, W. C. (2014). Effects of ambient nitrite on Amazon River prawn, *Macrobrachium amazonicum*, larvae. *Journal of the World Aquaculture Society, 45*, 55–64. https://doi.org/10.1111/jwas.12071

Heinrich-Hirsch, B., Ahlers, J., & Peter, H. W. (1977). Inhibition of Na,K-ATPase from chick brain by polyamines. *Enzyme, 22*, 235–241. https://doi.org/10.1159/000458795

Heldt, A., Suita, S., Dutra, F. M., et al. (2019). Stable isotops as a method for analysis of the contribution of different ditary sources in the production of *Macrobrachium amazonicum. Latin American Journal of Aquatic Research, 47*, 282–229. https://doi.org/10.3856/vol47-issue2-fulltext-8

Henry, R. P., Garrelts, E. E., McCarty, M. M., et al. (2002). Differential induction of branchial carbonic anhydrase and NA(+)/K(+) ATPase activity in the euryhaline crab, *Carcinus maenas*, in response to low salinity exposure. *The Journal of Experimental Zoology, 292*, 595–603. https://doi.org/10.1002/jez.10075

Henry, R. P., & Wheatly, M. G. (1992). Interaction of respiration ion regulation, and acid-base balance in the everyday life of aquatic crustaceans. *American Zoologist, 32*, 407–416. https://www.jstor.org/stable/3883620

Hoang, N. T., & Kanemoto, K. (2021). Mapping the deforestation footprint of nations reveals growing threat to tropical forests. *Nature Ecology & Evolution, 5*, 845–853. https://doi.org/10.1038/s41559-021-01417-z

Hsieh, S. L., Wu, Y. C., Xu, R. Q., et al. (2021). Effect of polyethylene microplastics on oxidative stress and histopathology damages in *Litopenaeus vannamei. Environmental Pollution, 288*, 117800. https://doi.org/10.1016/j.envpol.2021.117800

Hu, K. J., & Leung, P. C. (2007). Food digestion by cathepsin L and digestion-related rapid cell differentiation in shrimp hepatopancreas. *Comparative Biochemistry and Physiology, 146B*, 69–80. https://doi.org/10.1016/j.cbpb.2006.09.010

Jalihal, D. R., Sankolli, K. N., & Shenoy, S. (1993). Evolution of larval developmental patterns and the process of freshwaterization in the prawn genus Macrobrachium Bate 1968 (Decapoda, Palemonidae). *Crustaceana, 6*, 365–376. https://doi.org/10.1163/156854093X00793

Jiang, K., Zhang, F., Zhang, D., et al. (2011). Identification of a trypsin gene from *Scylla paramamosain* and its expression profiling during larval development. *African Journal of Agricultural Research, 6*, 6613–6621. https://doi.org/10.5897/ajar11.784

Jung, H., Lyons, R. E., Hurwood, D. A., et al. (2013). Genes and growth performance in crustacean species: A review of relevant genomic studies in crustaceans and other taxa. *Reviews in Aquaculture, 5*, 77–110. https://doi.org/10.1111/raq.12005

Kaplan, J. H. (2002). Biochemistry of Na,K-ATPase. *Annual Review of Biochemistry, 71*, 511–535. https://doi.org/10.1146/annurev.biochem.71.102201.141218

Kirschner, L. B. (2004). The mechanism of sodium chloride uptake in hyperregulating aquatic animals. *The Journal of Experimental Biology, 207*, 1439–1452. https://doi.org/10.1242/jeb.00907

Kobayashi, M., Fujisaki, H., Sugawara, M., et al. (1999). The presence of an Na^+/spermine antiporter in the rat renal brush-border membrane. *The Journal of Pharmacy and Pharmacology, 51*, 279–284. https://doi.org/10.1211/0022357991772457

Lapola, D. M., Pinho, P., Barlow, J., et al. (2023). The drivers and impacts of Amazon Forest degradation. *Science, 379*, eabp8622. https://doi.org/10.1126/science.abp8622

Lee, C. E., Charmantier, G., & Lorin-Nebel, C. (2022). Mechanisms of Na^+ uptake from freshwater habitats in animals. *Frontiers in Physiology, 13*, 1–28. https://doi.org/10.3389/fphys.2022.1006113

Leone, F. A., Bezerra, T. M. S., Garçon, D. P., et al. (2014). Modulation y K^+ plus $NH_4{}^+$ of microsomal (Na^+, K^+)-ATPase activity in selected ontogenetic stages of the diadromous River Shrimp *Macrobrachium amazonicum* (Decapoda, Palaemonidae). *PLoS One, 9*, e89625. https://doi.org/10.1371/journal.pone.0089625

Leone, F. A., Fabri, L. M., Costa, M. I. C., et al. (2023). Differential effects of cobalt ions *in vitro* on gill (Na^+, K^+)-ATPase kinetics in the blue crab *Callinectes danae* (Decapoda, Brachyura). *Comparative Biochemistry and Physiology Part C: Toxicology & Pharmacology, 274*, 109757.

Leone, F. A., Furriel, R. P. M., McNamara, J. C., et al. (2005). (Na^+, K^+)-ATPase from crustacean gill microsomes: A molecular marker to evaluate adaptation to biotopes of different salinity. *Trends in Comparative Biochemistry & Physiology, 11*, 1–15.

Leone, F. A., Lucena, M. N., Garçon, D. P., et al. (2017). Gill ion transport ATPases and ammonia excretion in aquatic crustaceans. In D. Weihrauch & M. O'Donnell (Eds.), *Acid-base balance and nitrogen excretion in invertebrates* (pp. 61–107). Springer. https://doi.org/10.1007/978-3-319-39617-0_3

Leone, F. A., Masui, D. C., Bezerra, T. M. S., et al. (2012). Kinetic analysis of gill (Na^+, K^+)-ATPase activity in selected ontogenetic stages of the Amazon River shrimp, *Macrobrachium amazonicum* (Decapoda, Palaemonidae): Interactions at ATP- and cation-binding sites. *The Journal of Membrane Biology, 245*, 201–215. https://doi.org/10.1007/s00232-012-9431-9

Li, Y., Liu, Z., Jiang, Q., et al. (2022). Effects of nanoplastic on cell apoptosis and ion regulation in the gills of *Macrobrachium nipponense*. *Environmental Pollution, 300*, 118989. https://doi.org/10.1016/j.envpol.2022.118989

Liu, H. Y. L., Pan, L. Q., & Zheng, D. (2008a). Injection of biogenic amines modulates osmoregulation of *Litopenaeus vannamei*: Response of hemolymph osmotic pressure, ion concentration and osmolality effectors. *Comparative Biochemistry and Physiology, 151A*, 191–197. https://doi.org/10.1016/j.cbpa.2008.06.021

Liu, H. Y. L., Pan, L. Q., & Zheng, D. (2008b). Effects of salinity on biogenic amines, hemolymph osmotic pressure, and activity of gill's Na/K-ATPase in *Charybdis japonica* (Crustacea, Decapoda). *Journal of the World Aquaculture Society, 39*, 812–820. https://doi.org/10.1111/j.1749-7345.2008.00218.x

Lovett, D. L., & Watts, A. S. (1995). Changes in polyamine levels in response to acclimation salinity in gills of the blue-crab *Callinectes-sapidus* rathbun. *Comparative Biochemistry and Physiology, 110B*, 115–119. https://doi.org/10.1016/0305-0491(94)00133-F

Lucas-Solis, O., Moulatlet, G. M., Guamangallo, J., et al. (2021). Preliminary assessment of plastic litter and microplastic contamination in freshwater depositional areas: The case study of Puerto Misahualli, Ecuadorian Amazonia. *Bulletin of Environmental Contamination and Toxicology, 107*, 45–51. https://doi.org/10.1007/s00128-021-03138-2

Lucena, M. N., Garçon, D. P., Fontes, C. F. L., et al. (2017). Polyamines regulate phosphorylation–dephosphorylation kinetics in a crustacean gill (Na^+, K^+)-ATPase. *Molecular and Cellular Biochemistry, 429*, 187–198. https://doi.org/10.1007/s11010-017-2946-8

Lucena, M. N., Garçon, D. P., Fontes, C. F. L., et al. (2019). Dopamine binding directly up-regulates (Na^+, K^+)-ATPase activity in the gills of the freshwater shrimp *Macrobrachium amazonicum*. *Comparative Biochemistry and Physiology, 233A*, 39–47. https://doi.org/10.1016/j.cbpa.2019.03.018

Lucena, M. N., McNamara, J. C., & Leone, F. A. (2017). Gill (Na^+, K^+)-ATPase from the Amazon River shrimp, *Macrobrachium amazonicum* (Decapoda, Palaemonidae): Effect of exogenous biogenic amines on enzyme activity in juveniles and adults. *Hydrobiology, 789*, 59–76. https://doi.org/10.1007/s10750-016-2753-3

Lucena, M. N., Pinto, M. R., Garçon, D. P., et al. (2015). A kinetic characterization of the gill V(H^+)-ATPase in juvenile and adult *Macrobrachium amazonicum*, a diadromous palaemonid shrimp. *Comparative Biochemistry and Physiology, 181B*, 15–25. https://doi.org/10.1016/j.cbpb.2014.11.002

Lucu, Č., & Towle, D. W. (2003). $Na^{+}+K^{+}$-ATPase in gills of aquatic crustacea. *Comparative Biochemistry and Physiology, 135A*, 195–214. https://doi.org/10.1016/S1095-6433(03)00064-3

Maciel, C. R., & Valenti, W. C. (2009). Biology, fisheries, and aquaculture of the Amazon River prawn *Macrobrachium amazonicum*: A review. *Nauplius, 17*, 61–79.

Magalhães, C. (1985). The larval development of palaemonids from amazon region reared in the laboratory. 1. *Macrobrachium amazonicum* (Heller, 1862) (Crustacea, Decapoda). *Amazoniana-Limnologia et oecologia regionalis systemae fluminis amazonas, 9*, 247–274.

Magalhães, C., & Walker, I. (1988). Larval development and ecological distribution of central Amazonian palaemonid shrimps (Decapoda, Caridea). *Crustaceana, 55*, 279–292. https://doi.org/10.1163/156854088X00384

Maraschi, A. C., Marques, J. A., Costa, S. R., et al. (2022). Marine shrimps as biomonitors of the Fundão (Brazil) mine dam disaster: A multi-biomarker approach. *Environmental Pollution, 305*, 119245. https://doi.org/10.1016/j.envpol.2022.119245

Masui, D. C., Furriel, R. P. M., Silva, E. C. C., et al. (2005). Gill microsomal (Na^{+},K^{+})-ATPase from the blue crab *Callinectes danae*: Interactions at cationic sites. *The International Journal of Biochemistry & Cell Biology, 37*, 2521–2535. https://doi.org/10.1016/j.biocel.2005.06.004

Maytorena-Verdugo, C. I., Córdova-Murueta, J. H., & García-Carreño, F. L. (2017). Peptidase compensation in the digestive system of whiteleg shrimp *Penaeus vannamei* against dietary Kunitz type soybean trypsin inhibitor. *Aquaculture Nutrition, 23*, 1095–1103. https://doi.org/10.1111/anu.12477

McGrath, M. E. (1999). The lysosomal cysteine proteases. *Annual Review of Biophysics and Biomolecular Structure, 28*, 181–204. https://doi.org/10.1146/annurev.biophys.28.1.181

McNamara, J. C., & Faria, S. C. (2012). Evolution of osmoregulatory patterns and gill ion transport 1063 mechanisms in the decapod Crustacea: A review. *Journal of Comparative Physiology, 182B*, 997–1014. https://doi.org/10.1007/s00360-012-0665-8

McNamara, J. C., & Freire, C. A. (2022). Strategies of invertebrate osmoregulation: An evolutionary blueprint for transmuting into fresh water from the sea. *Integrative and Comparative Biology, 62*, 376–387. https://doi.org/10.1093/icb/icac081

McNamara, J. C., & Lima, A. G. (1997). The route of ion and water movements across the gill epithelium of the freshwater shrimp *Macrobrachium olfersii* (Decapoda, Palaemonidae): Evidence from ultrastructural changes induced by acclimation to saline media. *The Biological Bulletin, 192*, 321–331. https://doi.org/10.2307/1542725

McNamara, J. C., Moreira, G. S., & Moreira, P. S. (1983). The effect of salinity on respiratory metabolism, survival and moulting in the first zoea of *Macrobrachium amazonicum* (Heller, 1862) (Crustacea, Palaemonidae). *Hydrobiology, 101*, 239–242. https://doi.org/10.1007/BF00009880

McNamara, J. C., Moreira, G. S., & Souza, S. C. R. (1986). The effect of salinity on respiratory metabolism in selected ontogenetic stages of the freshwater shrimp *Macrobrachium olfersii* (Decapoda, Palaemonidae). *Comparative Biochemistry and Physiology, 83A*, 359–363. https://doi.org/10.1016/0300-9629(86)90590-6

McNamara, J. C., Rosa, J. C., Greene, L. J., et al. (2004). Free amino acid pools as effectors of osmotic adjustment in different tissues of the freshwater shrimp *Macrobrachium olfersii* (Crustacea, Decapoda) during long-term salinity acclimation. *Marine and Freshwater Behaviour and Physiology, 37*, 193–208. https://doi.org/10.1080/10236240400006208

McNamara, J. C., & Torres, A. H. (1999). Ultracytochemical location of Na^{+}/K^{+}-ATPase activity and effect of high salinity acclimation in gill and renal epithelia of the freshwater shrimp *Macrobrachium olfersii* (Crustacea, Decapoda). *The Journal of Experimental Zoology, 284*, 617–628.

Meireles, A. L., Valenti, W. C., & Mantelatto, F. L. (2013). Reproductive variability of the Amazon River prawn, *Macrobrachium amazonicum* (Caridea, Palaemonidae): Influence of life cycle on egg production. *Latin American Journal of Aquatic Research, 41*, 718–731. http://dx.doi.org/103856/vol41-issue4-fulltext-8

Melo, G. A. S. (2003). Famílias Atyidae, Palaemonidae, Sergestidae. In G. A. S. Melo (Ed.), *Manual de identificação dos Crustacea Decapoda de água doce do Brasil* (pp. 289–415). Edições Loyola.

Moller, T. H. (1978). Feeding behaviour of larval and pós-larval of *Macrobrachium rosenbergii* (de Man) (Crustacea: Palaemonidae). *Journal of Experimental Marine Biology and Ecology, 35*, 251–258. https://doi.org/10.1590/S0101-81751997000400003

Moraes-Riodades, P. M. C., & Valenti, W. C. (2004). Morphotypes in male Amazon River Prawns, *Macrobrachium amazonicum*. *Aquaculture, 236*, 297–307. https://doi.org/10.1016/j.aquaculture.2004.02.015

Moreira, G. S., McNamara, J. C., & Moreira, P. S. (1986). The effect of salinity on the upper thermal limits of survival and metamorphosis during larval development in *Macrobrachium amazonicum* (Heller) (Decapoda, palaemonidae). *Crustaceana, 50*, 231–238. https://doi.org/10.1163/156854086X00250

Morris, S., & Edwards, T. (1995). Control of osmoregulation via regulation of Na/K-ATPase activity in the amphibious purple shore crab, *Leptograpsus variegatus*. *Comparative Biochemistry and Physiology, 112C*, 129–136. https://doi.org/10.1016/0742-8413(95)02004-7

Morth, J. P., Pedersen, B. P., Toustrup-Jensen, M. S., et al. (2007). Crystal structure of the sodium-potassium pump. *Nature, 450*, 1043–1049. https://doi.org/10.1038/nature06419

Muhlia-Almazán, A., Sánchez-Paz, A., & García-Carreño, F. L. (2008). Invertebrate trypsins: A review. *Journal of Comparative Physiology, 178B*, 655–672. https://doi.org/10.1007/s00360-008-0263-y

Mulati, A. (2017). *Dieta Natural de Macrobrachium amazonicum (Heller, 1862) (Crustacea, Decapoda) no Oeste de Minas Gerais, Brasil* (p. 45). Dissertação de Mestrado - Universidade Federal de Lavras.

New, M. B. (2005). Freshwater prawn farming: Global status, recent research and a glance at the future. *Aquaculture Research, 36*, 210–230. https://doi.org/10.1111/j.1365-2109.2005.01237.x

Nogueira, C. S., Pantaleão, J. A. F., & Costa, R. C. (2020). Hybridization experiments between freshwater prawns *Macrobrachium amazonicum* and *M. pantanalense* (Decapoda: Palaemonidae), and the effects of geographical isolation. *Marine and Freshwater Research, 72*, 520–525. https://doi.org/10.1071/MF20226

Odinetz-Collart, O. (1988). Aspectos ecológicos do camarão *Macrobrachium amazonicum* (Heller, 1862) no baixo Tocatins (PA-Brasil). *Memoria de la Sociedad de Ciencias Naturales La Salle, 48*, 341–353.

Odinetz-Collart, O. (1991). Stratégie de reproduction de *Macrobrachium amazonicum* en Amazonie Centrale (Decapoda, Caridea, Palaemonidae). *Crustaceana, 61*, 253–270. https://doi.org/10.1163/156854091X00146

Palmgren, M. G., & Nissen, P. (2011). P-type ATPases. *Annual Review of Biophysics, 40*, 243–266. https://doi.org/10.1146/annurev.biophys.093008.131331

Pantaleão, J. A. F., Hirose, G. L., & Costa, R. C. (2012). Relative growth, morphological sexual maturity, and size of *Macrobrachium amazonicum* (Heller 1862) (Crustacea, Decapoda, Palaemonidae) in a population with an entirely freshwater life cycle. *Invertebrate Reproduction & Development, 56*, 37–41. https://doi.org/10.1080/07924259.2011.587276

Paschoal, L. R., & Zara, F. J. (2017). First record of intersexuality in the Amazon River shrimp *Macrobrachium amazonicum* (Heller, 1862) (Caridea: Palaemonidae). *The Journal of Crustacean Biology, 37*, 507–511. https://doi.org/10.1093/jcbiol/rux049

Pelicice, F. M., & Castello, L. (2021). A political tsunami hits Amazon conservation. *Aquatic Conservation: Marine and Freshwater Ecosystems, 31*, 1221–1229. https://doi.org/10.1002/aqc.3565

Péqueux, A. (1995). Osmotic regulation in crustaceans. *Journal of Crustacean Biology, 15*, 1–60. https://doi.org/10.2307/1549010

Péqueux, A., Le Bras, P., Cann-Moisan, C., et al. (2002). Polyamines, indolamines, and catecholamines in gills and haemolymph of the euryhaline crab, *Eriocheir sinensis*. Effects of high pressure and salinity. *Crustaceana, 75*, 567–578. https://doi.org/10.1163/156854002760095606

Perera, E., Rodríguez-Viera, L., Perdomo-Morales, R., et al. (2015). Trypsin isozymes in the lobster *Panulirus argus* (Latreille, 1804): From molecules to physiology. *Journal of Comparative Physiology, 185B*, 17–35. https://doi.org/10.1007/s00360-014-0851-y

Pierce, S. K. (1982). Invertebrate cell volume control mechanism: A coordinated use of intracellular amino acids and inorganic ions as osmotic solute. *The Biological Bulletin, 163*, 405–419. https://doi.org/10.2307/1541452

Pinto, M. R., McNamara, J. C., Almeida, E. A., et al. (2016). Effects of ammonia stress in the Amazon river shrimp *Macrobrachium amazonicum* (Decapoda, Palaemonidae). *Aquatic Toxicology, 170*, 13–23. https://doi.org/10.1016/j.aquatox.2015.10.021

Pirkmajer, S., & Chibalin, A. V. (2019). Hormonal regulation of Na^+-K^+-ATPase from the evolutionary perspective. *Current Topics in Membranes, 83*, 315–351. https://doi.org/10.1016/bs.ctm.2019.01.009

Post, R. L., Hegyvary, C., & Kune, S. (1972). Activation by adenosine triphosphate in the phosphorylation kinetics of sodium and potassium ion transport adenosine triphosphatase. *The Journal of Biological Chemistry, 247*, 6530–6540. https://doi.org/10.1016/S0021-9258(19)44725-x

Preto, B. L., Kimpara, J. M., Moraes-Valenti, P., et al. (2010). Population structure of pond-raised *Macrobrachium amazonicum* with different stocking and harvesting strategies. *Aquaculture, 307*, 206–211. https://doi.org/10.1016/j.aquaculture.2010.07.023

Queiroz, L. D., Abrunhosa, F. A., & Maciel, C. R. (2011). Ontogenesis and functional morphology of the digestive system of the freshwater prawn, *Macrobrachium amazonicum* (Decapoda: Palaemonidae). *Zoology, 28*, 395–402. https://doi.org/10.1590/S1984-46702011000300014

Rainbow, P. S. (2002). Trace metal concentrations in aquatic invertebrates: Why and so what? *Environmental Pollution, 120*, 497–507. https://doi.org/10.1016/S0269-7491(02)00238-5

Rico, A., Redondo-Hasselerharm, P. E., Vighi, M., et al. (2023). Large-scale monitoring and risk assessment of microplastics in the Amazon River. *Water Research, 232*, 119707. https://doi.org/10.1016/j.watres.2023.11970

Robinson, J. D., Leach, C. A., & Robinson, L. J. (1986). Cation sites, spermine, and the reaction of the (Na^+, K^+)-dependent ATPase. *Biochimica et Biophysica Acta, 856*, 536–544. https://doi.org/10.1016/0005-2736(86)90145-8

Rocha, C. P., Maciel, C. M. T., Valenti, W. C., et al. (2022). Prospection of putative genes for digestive enzymes based on functional genome of the hepatopancreas of Amazon river prawn. *Acta Scientiarum. Animal Sciences, 44*, e53894. https://doi.org/10.4025/actascianimsci.v44i1.53894

Rocha, C. P., Quadros, M. L. A., Maciel, M., et al. (2017). Morphological changes in the structure and function of the feeding appendages and foregut of the larvae and first juvenile of the freshwater prawn *Macrobrachium acanthurus*. *Journal of the Marine Biological Association of the United Kingdom, 98*, 713–720. https://doi.org/10.1017/S0025315416001855

Rocha, C. P., Souza, A. S., Maciel, M., et al. (2016). Development and functional morphology of the mouthparts and foregut in larvae and post-larvae of *Macrobrachium jelskii* (Decapoda: Palaemonidae). *Arthropod Structure & Development, 45*, 242–252. https://doi.org/10.1016/j.asd.2016.02.001

Sá, R. J. D., de Souza, A. G. S., dos Santos, J. E., et al. (2019). Indicadores biológicos de qualidade da água e as mudanças climáticas. *Multidisciplinary Reviews, 2*, e2019006–e2019006. https://doi.org/10.29327/multi.2019006

Saborowski, R. (2015). Nutrition and digestion. In E. S. Chang & M. Thiel (Eds.), *The natural history of the crustacea* (Physiology) (Vol. 4, pp. 285–319). University Press.

Sáez, A. G., Lozano, E., & Zaldívar-Riverón, A. (2009). Evolutionary history of Na,K-ATPases and their osmoregulatory role. *Genetica, 136*, 479–490. https://doi.org/10.1007/s10709-009-9356-0

Santos, L. C. F., Belli, N. M., Augusto, A., et al. (2007). Gill (Na^+,K^+)-ATPase in diadromous, freshwater palaemonid shrimps: Species-specific kinetic characteristics and α-subunit expression. *Comparative Biochemistry and Physiology, 148A*, 178–188. https://doi.org/10.1016/j.cbpa.2007.04.008

Santos, A., Hayd, L., & Anger, K. (2013). A new species of *Macrobrachium* Spence Bate, 1868 (Decapoda, Palaemonidae), *M. pantanalense*, from the Pantanal, Brazil. *Zootaxa, 3700*(4), 534–546. https://doi.org/10.11646/zootaxa.3700.4.2

Seiler, N., & Dezeure, F. (1990). Polyamine transport in mammalian cells. *The International Journal of Biochemistry, 22*, 211–218. https://doi.org/10.1016/0020-711x(90)90332-w

Siebers, D., Lucu, C., Sperling, K. R., & Eberlein, K. (1972). Kinetics of osmoregulation in the crab *Carcinus maenas*. *Marine Biology, 17*, 291–303. https://doi.org/10.1007/BF00366739

Silva, E. C. C., Masui, D. C., Furriel, R. P., et al. (2008). Regulation by the exogenous polyamine spermidine of Na/K-ATPase activity from the gills of the euryhaline swimming crab *Callinectes danae* (Brachyura, Portunidae). *Comparative Biochemistry and Physiology, 149B*, 622–629. https://doi.org/10.1016/j.cbpb.2007.12.010

Soares, M. P., Jesus, F., Almeida, A. R., et al. (2020). Effects of pH and nitrites on the toxicity of a cypermetrin-based pesticide to shrimps. *Chemosphere, 241*, 125089. https://doi.org/10.1016/j.chemosphere.2019.125089

Soegianto, A., Irawan, B., & Usman, N. (2013). Effects of sublethal copper concentrations on gills of white shrimp (*Litopenaeus vannamei*, Boone 1931). *Bulletin of Environmental Contamination and Toxicology, 91*, 630–634. https://doi.org/10.1007/s00128-013-1113-5

Stephens, A., Rojo, L., Araujo-Bernal, S., et al. (2012). Cathepsin B from the white shrimp *Litopenaeus vannamei*: cDNA sequence analysis, tissues-specific expression and biological activity. *Comparative Biochemistry and Physiology, 161B*, 32–40. https://doi.org/10.1016/j.cbpb.2011.09.004

Suami, R. B., Al Salah, D. M. M., Kabala, C. D., et al. (2019). Assessment of metal concentrations in oysters and shrimp from Atlantic Coast of The Democratic Republic of the Congo. *Heliyon, 5*, e03049. https://doi.org/10.1016/j.heliyon.2019.e03049

Suman, T. Y., Jia, P. P., Li, W. G., et al. (2020). Acute and chronic effects of polystyrene microplastics on brine shrimp: First evidence highlighting the molecular mechanism through transcriptome analysis. *Journal of Hazardous Materials, 400*, 123220. https://doi.org/10.1016/j.jhazmat.2020.123220

Therien, A. G., & Blostein, R. (2000). Mechanisms of sodium pump regulation. *American Journal of Physiology-Cell Physiology, 279*, C541–C566. https://doi.org/10.1152/ajpcell.2000.279.3.c541

Toei, M., Saum, R., & Forgac, M. (2010). Regulation and isoform function of the V-ATPases. *Biochemistry, 49*, 4715–4723. https://doi.org/10.1021/bi100397s

Towle, D. W. (1997). Molecular approaches to understanding salinity adaptation of estuarine animals. *American Zoologist, 37*, 575–584. https://www.jstor.org/stable/3884134

Valencia, D. M., & Campos, M. R. (2007). Freshwater prawns of the genus *Macrobrachium* Bate, 1968 (Crustacea: Decapoda: Palemonidae) of Colombia. *Zootaxa, 1456*, 1–44. https://doi.org/10.11646/zootaxa.1456.1.1

Vercesi, K., & Hayd, L. (2015). Avaliação do número de ovos em diferentes estágios de desenvolvimento embrionário de *Macrobrachium pantanalense*. *Boletim Instituto Pesca (Online), 41*, 655–663.

Vergamini, F., Pileggi, L. G., & Mantelatto, F. L. (2011). Genetic variability of the Amazon River prawn *Macrobrachium amazonicum* (Decapoda, Caridea, Palaemonidae). *Contributions to Zoology, 80*, 67–83.

Wang, Z., Tang, D., Guo, H., et al. (2020). Transcriptome analysis reveals potential genes involved in digestive enzyme function in a mudflat crab *Helice tientsinensis*. *Thalassas, 36*, 573–583. https://doi.org/10.1007/s41208-020-00212-0

Wang, Z., Tang, D., Shen, C., et al. (2022). Identification of genes involved in digestion from transcriptome of *Parasesarma pictum* and *Parasesarma affine* hepatopancreas. *Thalassas, 38*, 93–101. https://doi.org/10.1007/s41208-021-00296-2

Wei, J., Zhang, X., Yang, Y., et al. (2014). Comparative transcriptomic characterization of the early development in Pacific white shrimp *Litopenaeus vannamei*. *PLoS One, 9*, e106201. https://doi.org/10.1371/journal.pone.0106201

Weihrauch, D., Becker, W., Postel, U., et al. (1998). Active excretion of ammonia across the gills of the shore crab *Carcinus maenas* and its relation to osmoregulatory ion uptake. *Journal of Comparative Physiology, 168B*, 364–376. https://doi.org/10.1007/s003600050156

Weihrauch, D., Becker, W., Postel, U., et al. (1999). Potential of active excretion of ammonia in three different haline species of crabs. *Journal of Comparative Physiology, 169B*, 25–37. https://doi.org/10.1007/s003600050190

Weihrauch, D., Morris, S., & Towle, D. W. (2004). Ammonia excretion in aquatic terrestrial crabs. *The Journal of Experimental Biology, 207*, 4491–4504. https://doi.org/10.1242/jeb01308

Weiss, R., Anger, K., Hayd, L., et al. (2015). Interpreting genetic distances for species recognition: The case of *Macrobrachium amazonicum* Heller, 1862 and the recently described *M. pantanalense* Dos Santos, Hayd & Anger, 2013 (Decapoda, Palaemonidae) from Brazilian fresh waters. *Crustaceana, 88*, 1111–1126.

Wilder, M. N., Huong, D. T. T., Atmomarsono, M., et al. (2000). Characterization of Na/K--ATPase in *Macrobrachium rosenbergii* and the effects of changing salinity on enzymatic activity. *Comparative Biochemistry and Physiology, 125A*, 377–388. https://doi.org/10.1016/S1095-6433(00)00162-8

WoRMS. (2023). Caridea. Accessed through: World Register of Marine Species at: https://www.marinespecies.org/aphia.php?p=taxdetails&id=106674 on 2023-09-05.

Zanders, P., & Rodriguez, J. M. (1992). Effects of temperature and salinity stress on osmoionic regulation in adults and on oxygen consumption in larvae and adults of *Macrobrachium amazonicum* (Decapoda, Palaemonidae). *Comparative Biochemistry and Physiology, 101*, 505–509. https://doi.org/10.1016/0300-9629(92)90502-H

Effects of Temperature on Parameters of Aerobic Metabolism and the Capacity for Oxygen Transport in Fish Adapted to Different Climatic Zones

Daiani Kochhann, Tatjana Simčič,
Maria de Nazaré Paula-da-Silva, Anton Brancelj,
and Adalberto Luis Val

Abstract

Temperature stands out as one of the primary factors influencing physiological processes in ectothermic organisms. The impact of temperature increase on fish with diverse life histories varies significantly. Gaining knowledge about how fish respond to temperature elevation is crucial for understanding their susceptibility to climate change. Assessing aerobic metabolism provides valuable insights in this regard, and this chapter aims to evaluate the reliability of the electron transport system (ETS) activity assay by comparing physiological responses in fish from distinct climatic zones, specifically tropical and temperate regions. By conducting such comparisons, we can enhance our understanding of how fish from different climatic backgrounds may react to temperature increases and the potential implications for their adaptation to climate change.

D. Kochhann (✉)
Acaraú Valley State University, Sobral, CE, Brazil
e-mail: daiani_kochhann@uvanet.br

T. Simčič
National Institute of Biology, Ljubljana, Slovenia

M. de Nazaré Paula-da-Silva
Brazilian National Institute for Research of the Amazon, Manaus, AM, Brazil

A. Brancelj
National Institute of Biology, Ljubljana, Slovenia

University of Nova Gorica, Nova Gorica, Slovenia

A. L. Val
Laboratory of Ecophysiology and Molecular Evolution, Brazilian National Institute for Research of the Amazon (INPA), Manaus, Amazonas, Brazil

S. S. de Souza et al. (eds.), *The Future of Amazonian Aquatic Biota*,
https://doi.org/10.1007/978-3-031-66822-7_16

1 Introduction

Temperature is an essential environmental factor that profoundly impacts the survival of organisms, exerting its influence on a wide array of animal activities ranging from biochemical processes to behavior and reproduction. Consequently, it delineates the boundaries within which life can thrive. Remarkably, life flourishes across virtually all temperature ranges found on the Earth's surface. Fish species, in particular, inhabit diverse aquatic habitats, encompassing both marine and freshwater environments. These habitats span from frigid polar regions where temperatures can plummet to as low as −2 °C, to tropical areas characterized by average temperatures nearing 35 °C, and even exceeding 40 °C under the most extreme circumstances. For fish to inhabit this extensive temperature spectrum, they must exhibit adaptability in their life strategies. It is important to note that no single species can endure the entirety of the temperature range observed on Earth.

Throughout evolution, ectothermic animals, including fish, have undergone adaptations to align with the characteristic or optimal temperature range of their respective natural environments. Consequently, the thermal tolerance windows of different species will vary depending on the environmental temperatures they encounter (Pörtner, 2002). Fish that have adapted to colder temperatures necessitate specific adjustments to enhance enzymatic activities in low-temperature settings. Such adaptations can be achieved through various mechanisms, including alterations in membrane phospholipid composition, functional disparities among isozymes, and modifications in the quantity and functionality of mitochondria, among others (Hochachka & Somero, 1984; Johnston et al., 1998; St-Pierre et al., 2000). As temperature rises, the rate of enzymatic reactions also increases, leading to heightened enzymatic activities in thermally adapted animals (Angilletta Jr et al., 2010), albeit within the confines of their thermal tolerance thresholds. Extensive theoretical and empirical research has been conducted to explore how temperature variability influences animal performance in a changing world. Phenotypic plasticity emerges as a mechanism through which animals can sustain their performance across diverse environmental conditions (Araújo et al., 2017).

The extent of phenotypic plasticity in animals varies considerably. The concept of phenotypic plasticity suggests that the magnitude of temperature variation influences an animal's ability to acclimate and, consequently, sustain their species through reproduction. Animals inhabiting environments with higher levels of environmental parameter variation possess a greater potential to acclimate to a wider range of temperatures compared to those in areas with lower variability (e.g., eurytopic versus stenotopic species) (Angilletta Jr, 2009; Liefting et al., 2009; Masel et al., 2007; Morgan et al., 2022). However, empirical studies have revealed that the capacity for thermal acclimation is not necessarily linearly correlated with thermal variability across different latitudes (Angilletta Jr, 2009; Gunderson & Stillman, 2015; Seebacher et al., 2015). The ability to thermally acclimate, which refers to an animal's capability to modify its performance across different temperature ranges, showcases their capacity to cope with temperature variability and is regarded as a crucial skill in the face of global climate change (Rohr et al., 2018).

Climate change poses a significant physiological challenge, particularly for aquatic ectotherms, as their body temperature closely mirrors the environmental temperature and the aquatic environment restricts behavioral thermoregulation. The vulnerability of fish species to climate change has been a subject of research since the early 2000s (as discussed below), leading to the development of theories regarding the mechanisms that determine thermal tolerance. The oxygen-capacity limited thermal tolerance (OCLTT) framework finds its roots in the pioneering studies conducted by Fry nearly 70 years ago (Fry, 1947, 1971). Fry's conceptual work explores the interplay between temperature, oxygen, and metabolism, wherein temperature governs the pace of aerobic metabolism, which in turn is constrained by the availability of oxygen. The standard metabolic rate (SMR) represents the minimal energy required to sustain life, while the maximum metabolic rate (MMR) signifies peak aerobic performance. The difference between the two, known as the aerobic scope (AS), represents the energy available to support other vital activities such as reproduction and growth (Fry, 1947). OCLTT has been the subject of numerous studies and reviews since its proposal by Pörtner in the early 2000s (early reviews include Pörtner, 2001, 2002, 2010). Earlier investigations by Fry argued that at higher temperatures, the aerobic scope (AS) decreases due to the MMR reaching its maximum potential while the SMR continues to increase. The routine metabolic rate (RMR), which represents the rate of oxygen consumption during normal low-level behaviors, is commonly used as a substitute for SMR in studies examining AS in fish (Metcalfe et al., 2016).

Over the past few decades, empirical data have accumulated, indicating that oxygen concentration in water can indeed limit thermal tolerance in certain species. However, it appears that this is not the primary mechanism constraining thermal tolerance in all ectothermic organisms (Ern et al., 2023). Extensive research on the effects of climate change has focused on how it affects AS, with indications that reductions in AS resulting from limitations in oxygen transport capacity are the driving factors behind changes in the distribution and abundance of fish species already experiencing the effects of climate change (e.g., Pörtner & Farrell, 2008; Pörtner & Knust, 2007). Despite significant questioning of this framework (see, e.g., Clark et al., 2013; Ern et al., 2014; Schulte, 2015), investigating the impact of climate change on AS in species from different climatic regions can provide insights into why species exhibit diverse responses to temperature increases (Lefevre, 2016).

The objective of this chapter is to provide a comprehensive overview of empirical research conducted on the impact of elevated temperatures on parameters of aerobic metabolism in fish adapted to tropical and temperate regions. Our main focus is to compare the response of fish in tropical and temperate regions through empirical studies. Specifically, the effect of increasing temperature on metabolic rate and potential aerobic metabolism accessed using the ETS activity assay was evaluated. The goal is to evaluate the possibility of using a potential aerobic scope (PAS), using ETS activity and real respiration rate, as a marker of metabolic stress in fish. Additionally, we present novel data on oxygen-binding capacity through blood P_{50} measurements for two species. We hypothesized that tropical species would exhibit a greater metabolic stress response when confronted with rising environmental temperatures than temperate species.

2 Temperature Effects on Metabolism

Temperature exerts significant control over biochemical reactions in organisms, exerting a profound influence on biological systems at various levels, ranging from molecular processes to entire communities (McCullough et al., 2009). This understanding of temperature's impact on biological systems stems from the Arrhenius equation, which states that chemical reactions are accelerated by increases in temperature. The magnitude of this acceleration is captured by the temperature coefficient (Q10), which quantifies the sensitivity of chemical reactions to temperature changes for both terrestrial and aquatic organisms. In the case of aquatic ectothermic animals, such as fish, their body temperature is primarily regulated by the external environmental temperature, thereby dictating the majority of the organism's reactions.

In the field of physiological sciences, researchers have developed several techniques to investigate the metabolism of animals. These techniques encompass various measurements, including enzyme activity assessments related to cellular energy production (such as citrate synthase, cytochrome c oxidase, and lactate dehydrogenase), heat energy production, changes in metabolites, and measurements of oxygen consumption and/or carbon dioxide excretion. However, it is worth noting that some of these techniques have become outdated due to their inaccuracy or the complexity of the apparatus required for measurements, such as heat production and carbon dioxide excretion.

Given that the majority of energy production in animal bodies occurs through aerobic pathways, which rely on oxygen, the measurement of the enzyme activity associated with aerobic metabolism and oxygen consumption rates are pivotal parameters for comprehending the responses of organisms, including fish, to climate change. These measurements provide valuable insights into the metabolic adjustments and physiological adaptations of organisms in the face of changing environmental conditions. Given that the majority of energy production in animal bodies occurs through aerobic pathways, which rely on oxygen, the measurement of enzyme activity associated with aerobic metabolism and oxygen consumption rates are pivotal parameters for comprehending the responses of organisms, including fish, to climate change. These measurements provide valuable insights into the metabolic adjustments and physiological adaptations of organisms in the face of changing environmental conditions, including climate change.

Mitochondria play a crucial role in cellular energy generation and are highly susceptible to temperature influences. Within aerobic organisms, the electron transport system (ETS) comprises a chain of enzymes that transport electrons via dehydrogenases and cytochromes to oxygen in aerobic animals, which provides the energy that enables oxidative phosphorylation and results in a generation of the adenosine triphosphate (ATP) as the universal energy carrier within organisms (Cammen et al., 1990). ATP is synthesized from adenosine diphosphate (ADP) through the activity of the ETS, and it serves as the primary source of energy for all activities in animals. This energy enables the performance of essential functions required for survival, characterized as the standard metabolic rate (SMR), as well as activities associated with fitness, including growth and reproduction.

Adaptation for enhanced performance to extreme temperatures can potentially lead to reduced performance at lower temperatures due to biochemical constraints on enzyme structure, function, and membrane fluidity (Nati et al., 2016). As oxygen serves as the final acceptor in the respiratory chain, any increase in aerobic metabolism must be accompanied by a corresponding increase in oxygen consumption and subsequent delivery to the body tissues. The hypothesis is that the aerobic capacity of each organism has an optimal temperature and the limitation of aerobic capacity arises from constraints on the cardiorespiratory system's ability to enhance oxygen delivery to tissues (Pörtner & Farrell, 2008; Steinhausen et al., 2008; Munday et al., 2012). Fry's studies (Fry, 1947, 1971) proposed that an organism's maintenance costs, measured as standard, routine, and/or resting metabolic rates, would exponentially increase with temperature. This hypothesis was recently reviewed by Lefevre (2016) specifically for marine animals.

This imbalance between energetic costs and maximum energy production would consequently result in a decrease in the aerobic scope (AS), which represents the energy available for performances such as growth and reproduction. However, certain researchers and empirical data have challenged Fry's paradigm, suggesting that for certain species, the increase in routine metabolic rate due to rising temperatures may be less pronounced, and maximal metabolic rate may increase until very close to the species' critical thermal tolerance, thereby leading to an expansion of AS (Clark et al., 2013; Schulte, 2015). Furthermore, the length of the acclimation period and adjustments in mitochondrial density within an organism contribute to enhancing AS (Lannig et al., 2003), thereby assisting in the maintenance or even increase of fitness in response to higher temperatures.

The ongoing debate surrounding the effects of elevated temperatures on aerobic metabolism in general and on AS in particular, and the conflicting results obtained so far, studies that focus on this task are essential for enhancing our understanding of the physiological responses of organisms to increasing temperatures and their implications for ecological and evolutionary processes. In addition, the development of techniques that streamline data collection, particularly in remote or challenging environments, would be highly advantageous. The ability to gather data efficiently and accurately in such settings would enable broader and more comprehensive investigations. It would facilitate the examination of a wider range of species and habitats, contributing to a more comprehensive understanding of the impact of temperature on AS.

3 Temperature Effects on Electron Transport System (ETS) Activity

The main energy source for aerobic organisms is the respiratory chain inside mitochondria. Comprised of a series of enzymes, known as the electron transport system (ETS), it facilitates the transfer of electrons from dehydrogenases and cytochromes to oxygen, which provides the energy that allows oxidative phosphorylation and the consequent generation of adenosine triphosphate (ATP), which serves as the universal energy carrier within organisms (Cammen et al., 1990). Enzymatic measurements of ETS activity have been used as a tool for the estimation of the aerobic potential of animals, plants, and microorganisms since it was developed by Packard (1971). In this assay the substrates nicotinamide adenine dinucleotide (NADH) and nicotinamide adenine dinucleotide phosphate (NADPH) are added to activate the ETS enzymes. The indicator, the INT (salt (2-(4-iodophenyl)-3-(4-nitrophenyl)-5-phenyltetrazolium chloride) act as an artificial electron acceptor. When INT is reduced it forms formazan that can be measured spectrophotometrically. The rate of formazan production is used to calculate the oxygen consumption, typically expressed as the volume of oxygen consumed per unit weight of tissue.

It is important to note that the ETS kinetic assay is a measurement of the aerobic potential that would occur if enzymes work at their maximum capacity (Cammen et al., 1990) and not an actual measure of respiration. Nevertheless, from ETS activity measurements, it is possible to calculate actual respiration using models based on substrate concentrations, kinetic assays, or specific quotients related to respiration (Packard et al., 1996; Packard & Gómez, 2008; Packard & Christensen, 2004; Bondyale-Juez et al., 2017).

Using ETS activity as an indicator of potential maximum aerobic metabolism offers several advantages. Firstly, it has been shown that ETS activity remains relatively stable in the face of short-term environmental variations (Båmstedt, 1980; Cammen et al., 1990; Hernández-León & Ikeda, 2005; Herrera et al., 2011; Smith & Chong, 1982). This stability allows for consistent measurements and reduces the impact of immediate fluctuations in environmental factors on the assessment of aerobic potential. Secondly, ETS activity has the capacity to acclimate to long-term changes in metabolism (Smith & Chong, 1982).

This is particularly relevant in the context of increasing temperatures associated with climate change. As organisms acclimate to new thermal conditions, their metabolic processes, including ETS activity, can adjust to optimize energy production and utilization. Furthermore, the ability to freeze samples in liquid nitrogen and preserve their stability for later analysis is a significant advantage. This is particularly valuable in remote locations where conducting in-situ measurements may present logistical and technical challenges. For example, setting up a respirometry system for measuring maximum oxygen consumption rates can be complex and require access to electricity, which may not be feasible in remote areas. Additionally, measuring animals with a wide size range would require adaptations or multiple pieces of equipment in a respirometer setup, further complicating field measurements. Freezing samples for subsequent ETS activity analysis circumvents these challenges and allows for more convenient sample collection and transport.

We have accumulated knowledge about the relevance of the measurements of ETS through enzymatic assays especially in zooplankton (e.g., Hernández-León & Ikeda, 2005; Packard & Gómez, 2008; Simčič & Brancelj, 2004) and macrozoobenthos (Cammen et al., 1990; Bondyale-Juez et al., 2017; Herrera et al., 2011; Muskó et al., 1995; Simčič & Brancelj, 2003). However, the use of the technique developed by Packard and colleagues was much less applied in investigating fish aerobic potential (Table 1).

The Atlantic cod (*Gadus morhua* Linnaeus, 1758) is a species characterized by its ability to tolerate a wide range of low temperatures, making it a cold-eurytherm organism. Its temperature habitat range from 0 to 15 °C, as documented by Cohen et al. (1990). Lannig et al. (2003) conducted a study to investigate the ETS response of cod from three distinct populations to rising temperatures. Their findings revealed heightened ETS activity in the white muscle of fish acclimated to lower temperatures (2 °C), although the pattern was less evident in the liver of the sampled fish across the three populations. These results indicate that residing in cold environments incurs a significant energetic cost due to the observed eurythermy, a characteristic observed in numerous Arctic and sub-Arctic marine species. Such species often feed in relatively warmer surface water but seek rest in colder water closer to the seabed. Regrettably, the authors did not measure the whole animal oxygen consumption, precluding conclusive evidence regarding whether the observed increase in aerobic capacity among cod is accompanied by a corresponding rise in metabolic demand. A separate investigation examining ETS activity in seven freshwater fish species from temperate and subtropical regions in Alabama found that enzymatic activity reached its peak at temperatures around 33 °C or even higher (Horne et al., 2022). The maximum temperatures used in these experiments exceeded the natural temperature range for all the species studied and were close to the critical thermal maxima specific to each species. Moreover, in most of the species examined, ETS activity decreased when temperatures surpassed their critical thermal tolerance. However, it is worth noting that the marble trout (*Salmo marmoratus* Cuvier, 1829) of the Salmonidae family exhibited an increase in ETS activity during different developmental stages, even toward the upper limits of its temperature range (Simčič et al., 2015). Overall, the data on ETS activity in species from temperate and subtropical regions demonstrate a consistent pattern: ETS activity tends to increase as temperatures approach or exceed the species' natural temperature range. Notably, certain species, such as *Campostoma oligolepis* (Hubbs and Greene, 1935), and *Cottus carolinae* (Gill, 1861), display a continued increase in ETS activity even when exposed to temperatures exceeding their critical thermal maxima.

In contrast, tropical species exhibit a more diverse range of responses in terms of ETS activity. The redfin pyrrhulina (*Pyrrhulina* aff *brevis* Steindachner, 1876) displayed an increase in ETS activity as the temperature rose to 29.6 °C (Almeida-Silva et al., 2020), but no change was observed when the temperature further increased to 30.5 °C (Campos et al., 2019). For these studies, the low temperature was between 25 and 26.5 °C. Since the natural habitat temperature range for this species typically falls between 24 and 28 °C, it is possible that the study conducted by Campos et al. (2019) already captured the decline in the species' aerobic capacity, as the temperature used was 2.5 °C higher than the species' natural environmental temperature range. However, it is unfortunate that the authors only conducted measurements at two temperatures, limiting the available information on the fish's responses. Similar patterns were observed in the ETS activity of the Agassizi's dwarf cichlid (*Apistogramma agassizii* (Steindachner, 1875)). In the case of two

Table 1 Effects of temperature on electron transport system measurements (ETS) activity, routine metabolic rate (RMR), and the rate between ETS/RMR, a measurement of potential aerobic scope

Family	Species	Climate zone	Ambiental temperature (°C)	Interval of experimental temperatures (°C)	In vivo/In vitro	Period of fish acclimation	ETS activity	RMR	ETS/ RMR	Tissue analyzed	Reference
Gadidae	*Gadus morhua*	Temperate	0–15	04–15	In vitro		↓			WM	Lannig et al., 2003
Cottidae	*Cottus carolinae*	Temperate	0–30	11–44	In vitro		↑			WM	Horne et al., 2022
Leuciscidae	*Notropis baileyi*	Temperate	0–30	11–44	In vitro		↑			WM	Horne et al., 2022
Leuciscidae	*Semotilus atromaculatus*	Temperate	0–30	11–44	In vitro		↑			WM	Horne et al., 2022
Leuciscidae	*Phoxinus phoxinus*	Temperate	2–20	18–24	In vivo	96 h	↑	↑	=	WA	Unpublished data
Leuciscidae	*Alburnus alburnos*	Temperate	10–20	10–16	In vivo	10 days	↑	↑	=	WA	Unpublished data
Leuciscidae	*Campostoma oligolepis*	Subtropical	1–36	11–44	In vitro		↑			WM	Horne et al., 2022
Leuciscidae	*Cyprinella gibbsi*	Subtropical	1–36	11–44	In vitro		↑			WM	Horne et al., 2022
Leuciscidae	*Cyprinella venusta*	Subtropical	1–36	11–44	In vitro		↑			WM	Horne et al., 2022
Centrarchidae	*Lepomis macrochirus*	Subtropical	1–36	11–44	In vitro		↑			WM	Horne et al., 2022
Salmonidae	*Salmo marmoratus*	Temperate	2–15	10–16	In vivo	96 h	↑	↑	=	WA	Simčič et al., 2017

Lebiasinidae	*Pyrrhulina brevis*	Tropical	24–28	25 and 29.6	In vivo	15 days	↑	↑	= *	WA	Almeida-Silva et al., 2020
Lebiasinidae	*Pyrrhulina brevis*	Tropical	24–28	26.5 and 30.5	In vivo	30 days	=	↑	↓ *	WM	Campos et al., 2019
Cichlidae	*Apistogramma agassizi*	Tropical	26–29	26.5 and 30.5	In vivo	96 h	↓	↑	↓ *	WM	Campos et al., 2019
Cichlidae	*Apistogramma agassizi*	Tropical	26–29	26 and 29	In vivo	30 days	↑	=	↑	WA	Kochhann et al., 2015
Cichlidae	*Mesonauta festivus*	Tropical	25–34	28–34	In vivo	10 days	↑	↑	↑	WA	Unpublished data
Characidae	*Hyphessobrycon melazonatus*	Tropical	24–34	26.5 and 30.5	In vivo	30 days	=	↑	↓ *	WM	Campos et al., 2019
Characidae	*Paracheirodon axelrodi*	Tropical	23–27	26–32	In vivo	96 h	↑	↑	=	WA	Unpublished data

↑ means that an increase toward higher temperatures was observed, ↓ means that a decrease toward higher temperature was observed, and = means that there was no change between temperatures tested. WA = measurement of ETS activity was done in whole-body animal; WM = measurement of ETS activity in white muscle. *authors from the original study did not calculated the ETS/RMR ratio. In vivo = animals were experimentally acclimated to the tested temperature. In vitro*:* animals were sampled in the environmental temperature and ETS activity measurements were done in the tested temperature. Climate zone and environmental temperature of the species were from Fish Base (https://www.fishbase.se/search.php). Blank spaces: the measurement was not done

species, the cardinal tetra (*Paracheirodon axelrodi* (Schultzs, 1956)) and the flag cichlid (*Mesonauta festivus* (Heckel, 1840)), ETS activity increased at higher temperatures, even when it approached the upper limits of the species' natural temperature range (unpublished data). However, no statistical changes in ETS activity were observed in the tetra group member (*Hyphessobrycon melazonatus* Durbin, 1908) over the studied temperature interval (26.5 °C and 30.5 °C) (Campos et al., 2019). It is important to note that some studies assessed ETS activity in the entire animal, while in others, measurements were limited to white muscle. Since ETS activity varies across tissues, researchers should ideally opt, whenever possible, for ETS assays using whole-body homogenates.

Despite the differences in experimental approaches across studies, it becomes apparent that there is a consistent pattern in the response of ETS activity among subtropical and temperate zone fish species. Specifically, there is an increase in ETS activity at temperatures very close to or above the critical thermal maximum of the respective species. The only exception to this pattern is the cod (*G. morhua*), which is a cold-eurytherm species that underscores the costs associated with cold adaptation. For most temperate zones and subtropical species, it is expected that the aerobic capacity for energy production will only be affected by temperatures that fall outside the species' actual temperature range. This emphasizes the importance of conducting studies that explore a broader range of temperatures. Conversely, there is no clear pattern in the response of ETS activity among tropical fish to high temperatures, as differences in response are observed even within the same species in studies with slight temperature variations.

The observed differences in the ETS activity response between tropical and other species can potentially be explained by three factors. Firstly, studies involving tropical species were primarily conducted in vivo, with the species acclimated to higher temperatures for a minimum of 96 h, often extending to 10 days or more. In contrast, many studies that measured ETS activity using the kinetic method examined the response of enzymes to temperature in vitro, without acclimation. This difference in acclimation protocols could contribute to variations in the observed responses. Secondly, in two studies that measured ETS activity at different temperatures, carbon dioxide concentration in the water was also manipulated. The fish in these studies were acclimated to higher levels of carbon dioxide, resulting in slightly lower water pH (around 5.8 in water with high carbon dioxide and around 6.2 in water with actual carbon dioxide concentration) along with elevated temperature. However, it is highly unlikely that these small changes in pH would significantly influence the aerobic capacity of the species. Native to the Amazon River basin, these fish are adapted to acidic water conditions, as black water bodies in the region typically have a pH below 5 (Val & Wood, 2022). Lastly, most studies acclimated fish to only two different temperatures: a lower, average ambient temperature and a higher temperature. As a result, it becomes challenging to determine trends in ETS activity with temperature. It is crucial for future studies to investigate how ETS activity responds to a much broader range of temperature acclimations, allowing for a more comprehensive understanding of the relationship between temperature and ETS activity.

4 Temperature Effects on Energy Consumption

The routine metabolic rate (RMR) measurement is widely used to assess the impact of temperature on energy consumption in animals. RMR refers to the amount of energy required for biological processes and low-level routine activities at a given temperature to maintain an organism's normal functions and reflects an evolutionary optimization of energetic demands (Clarke, 2004; Clarke & Fraser, 2004; Rolfe & Brown, 1997). As RMR is temperature-dependent (Clark & Johnston, 2002; Pörtner, 2001, 2002; Pörtner et al., 2006), animals inhabiting higher-temperature environments exhibit higher energetic costs compared to those living in lower-temperature environments. This relationship between temperature and RMR highlights the influence of thermal conditions on an organism's energy expenditure. Indeed, studies have consistently shown that the routine metabolic rate (RMR) of aquatic ectotherms tends to increase as temperature rises. This relationship has been extensively reviewed in the scientific literature, as highlighted by studies such as Gillooly et al. (2001), Dell et al. (2011), Lefevre (2016), and Jørgensen et al. (2022). However, it is important to note that the magnitude of the change in metabolic rate can vary significantly among different species, populations, and even individuals within the same species (Cavieres & Sabat, 2008; Norin et al., 2016). Additionally, individual variation within a species can arise due to factors such as genetic differences, phenotypic plasticity, and acclimation or adaptation to specific environmental conditions.

A limited number of studies have simultaneously measured ETS activity and RMR in fish subjected to increasing temperatures. As the interest is evaluating the parallel measurements of ETS activity and RMR of fish facing increasing temperatures, a discussion of RMR from studies that also measured ETS activity through kinetic assays is presented here. Unfortunately, RMR measurements were not conducted in the majority of studies involving temperate zones or subtropical species (Table 1). In line with expectations, the studies that did measure RMR demonstrated an increase in metabolic rate with rising temperatures, with only one exception observed in the study involving the dwarf cichlid (*A. agassizii*). This exception could be attributed to the social context in which the species was maintained (for details see Kochhann et al., 2013). The increase in RMR at higher temperatures is often interpreted as a beneficial response, allowing animals to take advantage of warmer environmental conditions. This response may facilitate faster growth or earlier reproduction, as suggested by Careau et al. (2014). However, it is worth noting that thermally sensitive individuals can easily experience performance impairments with further temperature increases or prolonged exposure to high temperatures, as highlighted by Norin et al. (2016). The elevated energy expenditure associated with maintenance under higher temperatures could potentially hinder growth and reproduction. While some studies allowed sufficient time for fish acclimation (10 days or more), there was no indication of metabolic compensation occurring in response to higher temperatures for any of the species examined. Unfortunately, the studies did not assess growth rates, making it difficult to determine whether the observed increase in RMR had detrimental effects on the tested fish species.

5 Temperature Effects on ETS/RMR Ratio, as an Indicator of Potential Aerobic Scope

The potential rise in metabolic costs would not pose a significant detriment if it corresponds to an increase in energy production. One physiological mechanism that has been postulated to regulate species distribution and thermal tolerance is the restriction of aerobic scope (AS) observed under extreme temperatures, as indicated by the OCLTT hypothesis (Pörtner, 2001). Accumulated empirical data over the past decades provide substantial evidence that oxygen can constrain thermal tolerance in certain species, while in others, alternative mechanisms play a more pivotal role in limiting thermal tolerance (Ern et al., 2023). Additionally, a decline in aerobic scope (AS) is recognized as a factor influencing animal thermal performance, as lower temperatures restrict AS and consequently impact animal performance (Farrell, 2009; Fry, 1947). The diminished AS plays a crucial role in determining the geographic and/or habitat distribution of species (however, see Nati et al., 2016) as it hinders migration capabilities and impairs growth and reproductive performance, thereby exerting long-term effects on fitness.

A few studies investigated changes in the potential aerobic scope (PAS) measured as ETS/RMR ratio in fish subjected to either acclimation or acute exposure to elevated temperatures (Simčič et al., 2015, Campos et al., 2019, Almeida-Silva et al., 2020). However, the findings regarding the response pattern are inconclusive. For instance, in the case of temperate species such as marble trout (*S. marmoratus*) juveniles, no significant differences in the ETS/RMR ratio were observed following maintenance at temperatures ranging from 10 to 16 °C for 96 h (Simčič et al., 2015). Similarly, the Eurasian minnows (*P. phoxinus*) exhibited an absence of changes in the ETS/RMR ratio after being maintained for 96 h within a temperature range of 18–24 °C, as did the common bleak (*A. alburnus*) during a 10-day acclimation period spanning temperatures from 10–16 °C (Kochhann et al., unpublished data). In the case of tropical species, responses in PAS proved to be more unpredictable. Some species exhibited no change in PAS, as seen in the cardinal fish exposed to a temperature range of 26–32 °C for 96 h (Kochhann et al., unpublished data), and *P. brevis* acclimated for 15 days within a temperature range of 25–29.6 °C (Almeida-Silva et al. 2021). However, a decrease in PAS was observed in the dwarf cichlid (*A. agassizii*) acclimated for 30 days to temperatures of 26.5 and 30.5 °C (Campos et al., 2019), while an increase in PAS was noted in *M. festivus* during a 10-day acclimation period within a temperature range of 28–34 °C (Kochhann et al., unpublished data).

In the majority of species, both ETS and RMR exhibited concurrent increases, contributing to the observed trend of no change in PAS. This indicates that most species are capable of maintaining their aerobic capacity within a range of different temperatures, including those close to the maximum temperature found in their natural habitats, underscoring the high degree of metabolic adaptability. While a decrease in AS is generally considered detrimental, certain species can enhance their glycolytic power, as demonstrated in the case of the dwarf cichlid (Campos et al., 2019). On the other hand, the optimum performance that is supposed to occur on temperatures that optimize the AS is uncertain (Healy & Schulte, 2012). It is important to note that while RMR measures the actual energy required for routine activities, ETS activity represents the energy that would be available if the aerobic enzyme systems were functioning at maximum capacity, rather than the actual respiration rate (Cammen et al., 1990). Further investigations should explore the long-term effects of elevated temperatures on the ETS/RMR ratio to determine whether differences in PAS, as measured by these techniques, serve as reliable indicators of changes in optimal performance. Nonetheless, some evidence suggests that a higher PAS, as measured by these techniques, could serve as a predictor of improved performance in trout (Simčič et al., 2017).

To gain a more comprehensive understanding of the effects of future temperature increases in fish species, it is crucial to extend the upper experimental temperature limits beyond those currently experienced by the studied species. This approach will provide valuable insights into the impacts of elevated temperatures on fish physiology and behavior.

6 Temperature Effects on Oxygen-Binding Capacity

An elevation in aerobic metabolism necessitates a corresponding increase in oxygen uptake and subsequent delivery to tissues. This requirement becomes particularly critical in aquatic habitats, where rising temperatures can lead to a reduction in oxygen concentrations and availability for organisms. Consequently, at higher temperatures, even in well-oxygenated water, if gill-mediated oxygen uptake fails to match the heightened oxygen demand in tissues, oxygen levels within bodily fluids may decline (Pörtner, 2001).

To prevent tissue hypoxia, red blood cells need to adapt to environmental fluctuations and increased oxygen demands from target tissues (Val et al., 2016). The blood P_{50} value represents the oxygen quantity required to saturate 50% of hemoglobin within red blood cells, serving as an estimation of oxygen-binding capacity. The temperature sensitivity of blood P_{50} refers to how changes in environmental temperature can influence the oxygen affinities of blood (Kaufman et al., 2013). The aquatic environment poses a specific challenge for species because as temperature and metabolic demands increase, the available oxygen in water decreases (Holt & Jørgensen, 2015). So far, most studies did not include measurements of blood P_{50} together with aerobic metabolism. Limited unpublished data, involving only two species, are available for comparison, and they suggest an increase in blood oxygen affinities with rising temperatures (Table 2).

The increased blood oxygen affinities observed at higher temperatures play a crucial role in facilitating oxygen delivery to tissues, as evidenced in both tropical and temperate zone fish (Val et al., 2016). However, this adaptation also has a negative consequence, as a decrease in oxygen-binding capacity reduces the ability to bind oxygen at the gills due to the effects of higher temperatures (Kaufman et al., 2013). Although a decrease in oxygen-binding capacity occurs in both tropical and temperate zone species at elevated temperatures, the magnitudes of these decreases

Table 2 Partial oxygen pressure needed to saturate 50% of hemoglobin (blood P_{50}) in a temperate (*Alburnus alburnus* Linnaeus, 1758) and a tropical (*Mesonauta festivus* Heckel, 1840) fish species. Fish were acclimated for 10 days at assigned temperatures. Blood-O_2 dissociation curves were individually obtained using a Hemox analyzer (TCS Scientific, PA, USA). Blood P_{50} was determined at pH 7.0 and 7.4 at each respective temperature acclimation (in vitro temperature). We also measured blood P_{50} at a temperature that was 6 °C higher than acclimation temperatures

Acclimation temperature	pH 7.0	pH 7.4	In vitro temperature	pH 7.0	pH 7.4
Alburnus alburnus (Linnaeus, 1758)					
10 °C	20.0 ± 2.8	8.5 ± 3.2†	16 °C	29.5 ± 3.3	20.6 ± 2.5†
16 °C	32.6 ± 4.2*	24.6 ± 3.0†*	22 °C	30.9 ± 5.0	41.1 ± 5.7† ‡
Mesonauta festivus (Hackel, 1840)					
28 °C	32.2 ± 2.0	31.3 ± 2.9	34 °C	38.5 ± 2.7	46.8 ± 3.0†
34 °C	39.1 ± 3.7*	41.4 ± 5.6*	40 °C	46.2 ± 4.2 ‡	53.4 ± 3.6† ‡

*$p < 0.05$ between acclimation temperatures at the same pH
‡ $p < 0.05$ between in vitro temperatures at the same pH
†$p < 0.05$ between pHs at the same temperature

differ. In tropical species, the maximum increase in blood P_{50} reaches 1.3 times compared to animals under controlled temperature conditions at pH 7.4, whereas in temperate zone species, the increase can reach up to 3.0 times (Val et al., 2016).

The effects of pH on blood P_{50} were examined in both temperate zone and tropical species, with more pronounced effects observed in temperate zone species. An increase in blood P_{50} occurs when a decrease in pH facilitates O_2 delivery to tissues, which are normally more acidic because of a higher CO_2 concentration. A strong Bohr effect was observed in *A. alburnus* at 10 °C and 16 °C, resulting in a 2.3-fold increase in blood P_{50} at 10 °C and a 1.3-fold increase at 16 °C, compared to pH 7.4. However, an unexpected response was observed at 22 °C, where a decrease in blood P_{50} was noted at pH 7.0, indicating the need for further investigation. It is important to note that this measurement was conducted at a temperature higher than the fish's acclimation temperature, and 22 °C falls outside the typical temperature range of the *A. alburnus* habitat.

In the analysis of tropical fish, pH did not have a significant effect on blood P_{50} at 28 °C and 34 °C. However, an increase in P_{50} was observed at the highest tested temperatures of 34 °C and 40 °C. It is important to note that the discrepancy in results between measurements at 34 °C under acclimated conditions and at higher temperature exposure suggests that caution should be exercised when interpreting the results from higher temperature exposures, as acclimation of the entire organism to a higher temperature is a more realistic scenario. Furthermore, it is crucial to consider that our blood P_{50} data were obtained in a fully oxygenated environment. However, this is not reflective of natural habitats, particularly in tropical water bodies, which can experience naturally hypoxic or even anoxic conditions on a daily basis (Val & Wood, 2022).

7 Conclusions and Perspectives

The impact of climate change on animal physiology has become a prominent and highly discussed topic in recent years. Despite the increasing number of new studies and meta-analyses examining this issue, significant knowledge gaps still exist, particularly regarding the responses of tropical freshwater fish to further temperature increases. While some recent reviews have suggested a higher vulnerability of tropical fish compared to temperate fish (Campos et al., 2021; Nati et al., 2021), there is a substantial bias in the available data, with a limited number of Amazonian species having information on thermal tolerance and metabolism across different temperatures. Furthermore, some of the reviews have assessed vulnerability based on data from critical thermal maxima, which, while being a useful proxy for thermal tolerance, represents a measurement taken very close to the lethal temperature. It is evident from these studies that the majority of populations will not be able to sustain their routine activities over the long term when exposed to temperatures near their thermal maxima.

Considering the limitations of our data regarding metabolic changes, it appears that tropical species exhibit a more varied response to increased temperatures and may be more susceptible to further temperature increases. This finding aligns with a growing body of evidence that suggests fish species from tropical regions are more vulnerable to the impacts of climate change compared to temperate species. In our study, a decrease in potential aerobic scope (PAS) was observed solely in tropical species and occurred at temperatures beyond the range typically experienced in their natural distribution.

While the discussion on the use of aerobic scope (AS) as a reliable indicator of vulnerability to climate change is ongoing, our data suggest that measuring PAS using the ETS activity assay as a proxy for maximal aerobic capacity (keeping in mind that it represents metabolic potential) could be a valuable indicator of sensitivity to increased temperatures, especially when assays are done with whole-body homogenates. With global climate change resulting in rising temperatures worldwide, it is crucial to focus our studies on understanding species' capacity to acclimate to these new environmental conditions. Long-term measurements of ETS activity could aid in determining the long-term acclimation capacity of fish to higher temperatures. Future studies investigating how PAS changes with temperature in freshwater fish should aim to test a wide range of temperatures, with particular attention given to temperatures exceeding those currently observed in the natural range of the species. This approach will help us better understand the potential impacts of climate change as it brings about higher temperatures that may become a reality for these species. Additionally, studies trying to correlate maximum metabolic rate and ETS activity in fish under thermal metabolic stress are needed to validate the PAS as an indicator of sensitivity to increasing temperatures.

Another suggestion is to integrate data on oxygen-binding capacity changes that would occur in increased temperatures as, at least for some species, oxygen transport and delivery to tissues could be a factor limiting performance in environments with high temperatures (Ern et al., 2023). Studies that investigate changes in ETS activity integrating high temperature and low oxygen are also in need, particularly for tropical species that will face hypoxic waters more frequently. The time window is becoming short and only with an increase in these studies, we will be able to predict the impact of climate change on fish species.

References

Almeida-Silva, J., Campos, D., & Almeida-Val, V. (2020). Metabolic adjustment of Pyrrhulina aff. Brevis exposed to different climate change scenarios. *Journal of Thermal Biology, 92*, 102657. https://doi.org/10.1016/j.jtherbio.2020.102657

Angilletta, M. J., Jr. (2009). *Thermal adaptation: A theoretical and empirical synthesis*. Oxford University Press. https://doi.org/10.1093/acprof:oso/9780198570875.001.1

Angilletta, M. J., Jr., Huey, R. B., & Frazier, M. R. (2010). Thermodynamic effects on organismal performance: Is hotter better? *Physiological and Biochemical Zoology, 83*(2), 197–206. https://doi.org/10.1086/648567

Araújo, J. D., Ghelfi, A., & Val, A. L. (2017). *Triportheus albus* Cope, 1872 in the blackwater, clearwater, and whitewater of the Amazon: A case of phenotypic plasticity? *Frontiers in Genetics, 8*, 114. https://doi.org/10.3389/fgene.2017.00114

Båmstedt, U. (1980). ETS activity as an estimator of respiratory rate of zooplankton populations. The significance of variations in environmental factors. *Journal of Experimental Marine Biology and Ecology, 42*(3), 267–283. https://doi.org/10.1016/0022-0981(80)90181-1

Bondyale-Juez, D. R., Packard, T. T., Viera-Rodríguez, M. A., et al. (2017). Respiration: Comparison of the Winkler technique, O_2 electrodes, O_2 optodes and the respiratory electron transport system assay. *Marine Biology, 164*, 226. https://doi.org/10.1007/s00227-017-3271-1

Cammen, L. M., Corwin, S., & Christensen, J. P. (1990). Electron transport system (ETS) activity as a measure of benthic macrofaunal metabolism. *Marine Ecology Progress Series, 65*, 171–182.

Campos, D. F., Amanajás, R. D., Almeida-Val, V. M. F., & Val, A. L. (2021). Climate vulnerability of South American freshwater fish: Thermal tolerance and acclimation. *Journal of Experimental Zoology Part A: Ecological and Integrative Physiology, 335*(9–10), 723–734. https://doi.org/10.1002/jez.2452

Campos, D. F., Braz-Mota, S., Val, A. L., & Almeida-Val, V. M. F. (2019). Predicting thermal sensitivity of three Amazon fishes exposed to climate change scenarios. *Ecological Indicators, 101*, 533–540. https://doi.org/10.1016/j.ecolind.2019.01.051

Careau, V., Gifford, M. E., & Biro, P. A. (2014). Individual (co)variation inthermal reaction norms of standard and maximal metabolic rates inwild-caught slimy salamanders. *Functional Ecology, 28*, 1175–1186.

Cavieres, G., & Sabat, P. (2008). Geographic variation in the response to thermal acclimation in rufouscollared sparrows: Are physiological flexibility and environmental heterogeneity correlated? Funct. *Ecology, 22*, 509–515. https://doi.org/10.1111/j.1365-2435.2008.01382.x

Clark, A., & Johnston, N. M. (2002). Scaling of metabolic rate with body mass and temperature in teleost fish. *The Journal of Animal Ecology, 68*(5), 893–905.

Clark, T. D., Sandblom, E., & Jutfelt, F. (2013). Aerobic scope measurements of fishes in an era of climate change: Respirometry, relevance and recommendations. *The Journal of Experimental Biology, 216*, 2771–2782. https://doi.org/10.1242/jeb.084251

Clarke, A. (2004). Is there a universal temperature dependence of metabolism? *Functional Ecology, 18*, 252–256.

Clarke, A., & Fraser, K. P. P. (2004). Why does metabolism scale with temperature? *Functional Ecology, 18*(2), 243–251.

Cohen DM, Inada T, Iwamoto T, and Scialabba N (1990) FAO species catalogue. Vol. 10. Gadiform fishes of the world (order Gadiformes). An annotated and illustrated catalogue of cods, hakes, grenadiers and other gadiform fishes known to date. FAO Fisheries Synopsis. 125(10). : FAO. 442 p.

Dell, A. I., Pawar, S., & Savage, V. M. (2011). Systematic variation in the temperature dependence of physiological and ecological traits. *Proceedings of the National Academy of Sciences, 108*(26), 10591–10596. https://doi.org/10.1073/pnas.1015178108

Ern, R., Andreassen, A. H., & Jutfelt, F. (2023). Physiological mechanisms of acute upper thermal tolerance in fish. *Physiology, 38*(3), 141–158. https://doi.org/10.1152/physiol.00027.2022

Ern, R., Huong, D. T. T., Phuong, N. T., Wang, T., & Bayley, M. (2014). Oxygen delivery does not limit thermal tolerance in a tropical eurythermal crustacean. *The Journal of Experimental Biology, 217*, 809–814. https://doi.org/10.1242/jeb.094169

Farrell, A. P. (2009). Environment, antecedents and climate change: Lessons from the study of temperature physiology and river migration of salmonids. *Journal of Experimental Biology, 212*, 3771–3780. https://doi.org/10.1242/jeb.023671

Fry, F. E. J. (1947). Effects of the environment on animal activity. *Ontario Fisheries Research Laboratory Publication, Biology Series, 55*(68), 1–62.

Fry, F. E. J. (1971). The effect of environmental factorson the physiology of fish. In W. S. Hoar & D. J. Randall (Eds.), *Fish physiology* (Vol. 6, pp. 1–98). Academic Press.

Gillooly, J. F., Brown, J. H., West, G. B., Savage, V. M., & Charnov, E. L. (2001). Effects of size and temperature on metabolic rate. *Science, 293*(5538), 2248–2251. https://doi.org/10.1126/science.1061967

Gunderson, A. R., & Stillman, J. H. (2015). Plasticity in thermal tolerance has limited potential to buffer ectotherms from global warming. *Proceedings of the Royal Society B: Biological Sciences, 282*, 20150401.

Healy, T. M., & Schulte, P. M. (2012). Thermal acclimation is not necessary to maintain a wide thermal breadth of aerobic scope in the common killifish (*Fundulus heteroclitus*). *Physiological and Biochemical Zoology, 85*(2), 107–119. https://doi.org/10.1086/664584

Hernández-León, S., & Ikeda, T. (2005). Zooplankton respiration. In P. A. Del Giorgio & P. J. Williams (Eds.), *Respiration in aquatic systems* (pp. 57–82). Oxford University Press.

Herrera, A., Packard, T., Santana, A., & Gómez, M. (2011). Effect of starvation and feeding on respiratory metabolism in *Leptomysis lingvura* (GO Sars 1866). *Journal of Experimental Marine Biology and Ecology, 409*(1), 154–159. https://doi.org/10.1016/j.jembe.2011.08.016

Hochachka, P. W., & Somero, G. N. (1984). *Biochemical adaptation* (pp. 1–537). Princeton University Press.

Holt, R. E., & Jørgensen, C. (2015). Climate change in fish: Effects of respiratory constraints on optimal life history and behaviour. *Biology Letters, 11*(2), 1–4.

Horne, L. M., DeVries, D. R., Wright, R., Irwin, E., Staton, B. A., Abdelrahman, H. A., & Stoeckel, J. A. (2022). Thermal performance of the electron transport system complex III in seven Alabama fishes. *Journal of Experimental Zoology Part A: Ecological and Integrative Physiology, 339*(2), 153–162. https://doi.org/10.1002/jez.2667

Johnston, I. A., Calvo, J., Guderley, H., Fernandez, D., & Palmer, L. (1998). Latitudinal variation in the abundance and oxidative capacities of muscle mitochondria in perciform fishes. *The Journal of Experimental Biology, 201*, 1–12.

Jørgensen, L. B., Ørsted, M., Malte, H., Wang, T., & Overgaard, J. (2022). Extreme escalation of heat failure rates in ectotherms with global warming. *Nature, 611*(7934), 93–98. https://doi.org/10.1038/s41586-022-05334-4

Kaufman, R. C., et al. (2013). Effects of temperature on hardhead minnow (*Mylopharodoncon ocephalus*) blood-oxygen equilibria. *Environmental Biology of Fishes, 96*(12), 1389–1397.

Kochhann, D., Campos, D. F., & Val, A. L. (2015). Experimentally increased temperature and hypoxia affect stability of social hierarchy and metabolism of the Amazonian cichlid *Apistogramma agassizii*. *Comparative Biochemistry and Physiology Part A: Molecular and Integrative Physiology, 190*, 54–60. https://doi.org/10.1016/j.cbpa.2015.09.006

Lannig, G., Eckerle, L., Serendero, I., et al. (2003). Temperature adaptation in eurythermal cod (*Gadus morhua*): A comparison of mitochondrial enzyme capacities in boreal and Arctic populations. *Marine Biology, 142*, 589–599. https://doi.org/10.1007/s00227-002-0967-6

Lefevre, S. (2016). Are global warming and ocean acidification conspiring against marine ectotherms? A meta-analysis of the respiratory effects of elevated temperature, high CO_2 and their interaction. *Conservation Physiology, 4*(1), cow009. https://doi.org/10.1093/conphys/cow009

Liefting, M., Hoffmann, A. A., & Ellers, J. (2009). Plasticity versus environmental canalization: Population differences in thermal responses along a latitudinal gradient in *Drosophila serrata*. *Evolution, 63*, 1954–1963. https://doi.org/10.1111/j.1558-5646.2009.00683.x

Masel, J., King, O. D., & Maughan, H. (2007). The loss of adaptive plasticity during long periods of environmental stasis. *The American Naturalist, 169*, 38–46. https://doi.org/10.1086/510212

McCullough, D. A., et al. (2009). Reviews in fisheries science research in thermal biology: Burning questions for coldwater stream fishes. *Reviews in Fisheries Science, 17*(1), 90–115.

Metcalfe, N. B., Van Leeuwen, T. E., & Killen, S. S. (2016). Does individual variation in metabolic phenotype predict fish behaviour and performance? *Journal of Fish Biology, 88*(1), 298–321. 10.1111/jfb.12699.

Morgan, R., Andreassen, A. H., Åsheim, E. R., Finnøen, M. H., Dresler, G., Brembu, T., Loh, A., Miest, J. J., & Jutfelt, F. (2022). Reduced physiological plasticity in a fish adapted to stable temperatures. *Proceedings of the National Academy of Sciences, 119*(22), e2201919119. https://doi.org/10.1073/pnas.2201919119

Munday, P. L., McCormick, M. I., & Nilsson, G. E. (2012). Impact of global warming and rising CO_2 levels on coral reef fishes: what hope for the future? *The Journal of Experimental Biology.*; 15 November 2012, *215*(22), 3865–3873. https://doi.org/10.1242/jeb.074765

Muskó, I. B., G-Tóth, L., & Szábo, E. (1995). Respiration and respiratory electron transport system (ETS) activity of two amphipods: *Corophium curvispinum* G.O. Sars and *Gammarus fossarum* Koch. *Polskie Archiwum Hydrobiologii, 42*, 547–558.

Nati, J. J. H., Lindström, J., Halsey, L. G., & Killen, S. S. (2016). Is there a trade-off between peak performance and performance breadth across temperatures for aerobic scope in teleost fishes? *Biology Letters, 12*, 122016019120160191. https://doi.org/10.1098/rsbl.2016.0191

Nati, J. J., Svendsen, M. B., Marras, S., Killen, S. S., Steffensen, J. F., McKenzie, D. J., & Domenici, P. (2021). Intraspecific variation in thermal tolerance differs between tropical and temperate fishes. *Scientific Reports, 11*(1), 1–8. https://doi.org/10.1038/s41598-021-00695-8

Norin, T., Malte, H., & Clark, T. D. (2016). Differential plasticity of metabolic rate phenotypes in a tropical fish facing environmental change. *Functional Ecology, 30*(3), 369–378. https://doi.org/10.2307/48582145

Packard, T. T. (1971). The measurement of respiratory electron transport activity in marine phytoplankton. *Journal of Marine Research, 29*(3), 235–244. https://doi.org/10.1007/BF00393750

Packard, T. T., Berdalet, E., Blasco, D., Roy, S., St-Amand, L., Lagacé, B., Lee, K., & Gagné, J. (1996). Oxygen consumption in the marine bacterium pseudomonas nautical predicted from ETS activity and bisubstrate enzyme kinetics. *Journal of Plankton Research, 18*(10), 1819–1835.

Packard, T. T., & Christensen, J. P. (2004). Respiration and vertical carbon flux in the Gulf of Maine water column. *Journal of Marine Research, 62*(1), 93–115. https://doi.org/10.1357/00222400460744636

Packard, T. T., & Gómez, M. (2008). Exploring a first-principles-based model for zooplankton respiration. *ICES Journal of Marine Science, 65*(3), 371–378. https://doi.org/10.1093/icesjms/fsn003

Pörtner, H. (2001). Climate change and temperature-dependent biogeography: Oxygen limitation of thermal tolerance in animals. *Naturwissenschaften, 88*, 137–146. https://doi.org/10.1007/s001140100216

Pörtner, H. O. (2002). Climate variations and the physiological basis of temperature dependent biogeography: Systemic to molecular hierarchy of thermal tolerance in animals. *Comparative Biochemistry and Physiology Part A: Molecular & Integrative Physiology, 132*, 739–761.

Pörtner, H. O. (2010). Oxygen- and capacity-limitation of thermal tolerance: A matrixfor integrating climate-related stressor effects in marine ecosystems. *The Journal of Experimental Biology, 213*, 881–893.

Pörtner, H. O., & Farrell, A. P. (2008). Physiology and climate change. *Science, 322*, 690–692.

Pörtner, H. O., & Knust, R. (2007). Climate change affects marine fishes through the oxygen limitation of thermal tolerance. *Science, 315*, 95–97.

Pörtner, H. O., et al. (2006). Trade-offs in thermal adaptation: The need for a molecular to ecological integration. *Physiological and Biochemical Zoology, 79*(2), 295–313.

Rohr, J. R., Civitello, D. J., Cohen, J. M., Roznik, E. A., Sinervo, B., & Dell, A. I. (2018). The complex drivers of thermal acclimation and breadth in ectotherms. *Ecology Letters, 21*(9), 1425–1439. https://doi.org/10.1111/ele.13107

Rolfe, D. F. S., & Brown, G. C. (1997). Cellular energy metabolism and molecular origin of standard metabolic rate in mammals. *Physiological Reviews, 77*, 731–758.

Schulte, P. M. (2015). The effects of temperature on aerobic metabolism: Towards a mechanistic understanding of the responses of ectotherms to a changing environment. *The Journal of Experimental Biology, 218*, 1856–1866. https://doi.org/10.1242/jeb.118851

Seebacher, F., White, C. R., & Franklin, C. E. (2015). Physiological plasticity increases resilience of ectothermic animals to climate change. *Nature Climate Change, 5*, 61–66.

Simčič, T., & Brancelj, A. (2003). Estimation of the proportion of metabolically active mass in the amphipod *Gammarus fossarum*. *Freshwater Biology, 48*, 1093–1099.

Simčič, T., & Brancelj, A. (2004). Respiratory electron transport system (ETS) activity as an estimator of the thermal tolerance of two *Daphnia* hybrids. *Journal of Plankton Research, 26*, 525–534. https://doi.org/10.1093/plankt/fbh056

Simčič, T., Jesenšek, D., & Brancelj, A. (2015). Effects of increased temperature on metabolic activity and oxidative stress in the first life stages of marble trout (*Salmo marmoratus*). *Fish Physiology and Biochemistry, 41*(4), 1005–1014.

Simčič, T., Jesenšek, D., & Brancelj, A. (2017). Metabolic characteristics of early life history stages of native marble trout (*Salmo marmoratus*) and introduced brown trout (*Salmo trutta*) and their hybrids in the Soča River. *Ecology of Freshwater Fish, 26*(1), 141–149. https://doi.org/10.1111/eff.12264

Smith, J. C., & Chong, C. K. (1982). Body weight, activities of cytochrome oxidase and electron transport system in the liver of the American plaice *Hippoglossoides platessoides*. Can these activities serve as indicators of metabolism? *Marine Ecology Progress Series, 9*, 171–117.

St-Pierre, J., Brand, M. D., & Boutilier, R. G. (2000). Mitochondria as ATP consumers: Cellular treason in anoxia. *Proceedings of the National Academy of Sciences, 97*, 8670–8674. https://doi.org/10.1073/pnas.140093597

Steinhausen, M. F., Sandblom, E., Eliason, E. J., Verhille, C., & Farrell, A. P. (2008). The effect of acute temperature increases on the cardiorespiratory performance of resting and swimming sockeye salmon (Oncorhynchus nerka). *The Journal of Experimental Biology*.; 15 December 2008, *211*(24), 3915–3926. https://doi.org/10.1242/jeb.019281

Val, A. L., & Wood, C. M. (2022). Global change and physiological challenges for fish of the Amazon today and in the near future. *The Journal of Experimental Biology*. 15 May 2022, *225*(10) jeb216440. https://doi.org/10.1242/jeb.216440

Val, A. L., Paula-Silva, N., Almeida-Val, F., & V. M., & Wood, C. M. (2016). In vitro effects of increased temperature and decreased pH on blood oxygen affinity of 10 fish species of the Amazon. *Journal of Fish Biology, 89*(1), 264–279. https://doi.org/10.1111/jfb.13009

Limited Thermal Capacity of Amazonian Fishes to Survive in a Warming World

Angelina M. Dichiera, Ellen H. Jung, Muhammed A. Oyinlola, and Colin J. Brauner

Abstract

The Amazon basin contains more than 15% of the world's freshwater fishes, many of ecological and economical importance. Unfortunately, research suggests that tropical species, which have adapted to a thermally stable environment, are most at risk with a long-term increase above current temperatures, which is worrisome when the Amazon could experience an increase in air temperature of up to +6.5 °C within the next century. In this chapter, we review the thermal limits of Amazonian fishes and their capacity to acclimate to warm temperatures. Furthermore, we explore how this thermal tolerance data may be incorporated into prediction models to evaluate Amazonian fish's vulnerability to thermal stress and highlight the research that is needed to better understand their capacity to acclimate to a warming world.

A. M. Dichiera
Virginia Institute of Marine Science, College of William and Mary, Gloucester Point, VA, USA

E. H. Jung · C. J. Brauner (✉)
Department of Zoology, University of British Columbia, Vancouver, BC, Canada
e-mail: brauner@zoology.ubc.ca

M. A. Oyinlola
Changing Ocean Research Unit, Institute for the Oceans and Fisheries, The University of British Columbia, Vancouver, BC, Canada

S. S. de Souza et al. (eds.), *The Future of Amazonian Aquatic Biota*, https://doi.org/10.1007/978-3-031-66822-7_17

1 Importance of Amazonian Fishes and Risk of Temperature Change

The Amazon Basin is renowned for its biodiversity and species richness. Fishes of the Amazon River system make up roughly 15% of all freshwater fish species in the world, with >2700 species recorded and new species continuously discovered (Jézéquel et al., 2020). Beyond their ecological value, these fishes and their abundance and richness are intrinsic to the economic stability and health of many peoples of the Amazon Basin. However, tropical ectotherms are predicted to be particularly vulnerable to global climate change because they live in a thermally stable habitat with little fluctuation and are relatively sensitive to temperature change (Deutsch et al., 2008).

Near constant, warm temperatures are thought to have led to specialized thermal physiology (Ghalambor et al., 2006; Janzen, 1967; Payne & Smith, 2017) wherein tropical ectotherms have adapted to a narrow thermal niche and live close to their thermal maxima (Deutsch et al., 2008; Tewksbury et al., 2008). Currently, water temperatures of the Amazon River system have been reported to range from 16.4 to 39.3 °C, with an average temperature of 27.5 °C (Campos et al., 2021), but are rising at an unprecedented rate with climate change. Globally, temperatures have already risen 1.1 °C above pre-industrial levels (IPCC, 2023), accompanied by increased duration, frequency, and intensity of heatwaves. Specifically in the Amazon, air temperatures are currently increasing at a rate of 0.2–0.3 °C each decade, and in the worst-case scenario could increase up to +6.5 °C at the end of the century (Carvalho et al., 2020). A recent study indicates that the simultaneous change in hydrologic regimes (volume and flow rate) of Amazon River waters may buffer dramatic temperature increases; however, a moderate temperature increase of +0.5 °C is still expected (van Vliet et al., 2013). Furthermore, the heterogeneous nature of the Amazon River system will prompt variable responses to climate warming, wherein shallow floodplain lakes and forest streams (igarapés) will respond more dramatically than stable, deep river channels.

Aquatic ectotherms'' body temperature conforms to that of the environmental water temperature, such that changes in environmental temperature largely define their performance, fitness, and abundance. To deal with thermal stress, ectothermic organisms may display escape responses and migrate to favorable habitats. However, deforestation is increasing stream temperatures and limiting thermal refugia for much of the Amazon River system (Ilha et al., 2018). Instead, acclimatory or evolutionary responses must be utilized to cope with thermal stress. Unfortunately, many studies have demonstrated the relative inflexibility of upper thermal limits for ectotherms in general, providing little hope for evolutionary rescue (Addo-Bediako et al., 2000; Araújo et al., 2013; Bennett et al., 2021; Morgan et al., 2020; Sandblom et al., 2016). In other words, species' upper thermal limits may not evolve fast enough across generations to keep up with increasing temperatures (Morgan et al., 2020). Thus, understanding the capacity for acclimation is the best avenue to predicting how these fishes will respond to and/or survive in a warming world. Specifically in this chapter, we will address: (1) Are Amazonian fishes living close to their thermal limits? (2) Do they have the capacity to acclimate to warmer temperatures? (3) How can we use this data, and design future studies, to predict the future of Amazonian fishes in a warming world?

2 Are Amazonian Fishes Living Life on the (Thermal) Edge?

2.1 Measuring Thermal Limits

With the rising need for climate change research, studies characterizing thermal limitations and plasticity of ectothermic organisms have risen in recent years. Researchers have employed creative approaches to address these questions, and with the diversity of approaches comes an array of definitions for thermal limits. The most widely employed method to assess and compare animals' acute thermal tolerance in laboratories is critical thermal maximum or CT_{max} (first developed by Cowles & Bogert, 1944). Cost-effective, rapid, and repeatable, CT_{max} requires little more than tanks and heaters, which enables the use of this method even in remote field sites. Briefly, the CT_{max} method entails increasing the water temperature in a matter of minutes to hours (1–18 °C/h) until the fish reaches a predetermined endpoint. The endpoints are when either the locomotion or performance is negatively affected during the temperature ramping. Most researchers use loss of equilibrium (LOE) or the failure of the righting response as an endpoint as it is thought to be ecologically relevant without sacrificing the animal. At LOE, fish are unable to escape from acute critical conditions (e.g., cannot avoid predators or find shelter from thermal stress), eventually leading to mortality.

Because of its wide use, this metric of assessing upper thermal limits is what we will focus on for most of the chapter; however, it is important to highlight the other metrics of thermal limitations that may also be used in combination with, or in lieu of, CT_{max} measurements. Lethal thermal maxima (LT_{max}; or incipient lethal temperature, ILT) can be defined in a few ways depending on the study. Commonly, LT_{max} is similar in approach to the CT_{max} method; however, fishes are heated past their loss of equilibrium to mortality, wherein the fish can no longer recover. Some studies also identify the temperature at which 50% mortality occurs as the LT_{max} or ILT (Fry, 1967; Fry et al., 1942). In addition, chronic lethal maxima (CL_{max}) has been used to identify the temperature which incurred high morbidity >50% (Jung et al., 2020). While fewer studies have used this method, it has been important to demonstrate that mortality occurs at temperatures below CT_{max}. Though the difference between a species CT_{max} and CL_{max} is marginal, there is a significant correlation between the two metrics (Jung et al., 2020). Thus, understanding the relationship between a species' CT_{max} and CL_{max} may allow for predictability of more conservative and ecologically relevant measures of thermal tolerance (CL_{max}) from a very simple and standardized, nonlethal thermal tolerance assay (CT_{max}) that abound in the literature.

Within an experimental trial where thermal limits are measured, one can also measure agitation temperature (T_{ag}; McDonnell & Chapman, 2015). Similar to the avoidance behavior observed in hypoxia studies (Chapman & Mckenzie, 2009), fishes typically display a behavioral response prior to LOE, wherein they abruptly increase activity in their experimental tank to try to avoid or escape the increasing temperature. This behavioral metric requires more precise control of the experimental set-up, and the ability to record behavior during the trial with minimal stimulation from sources other than increasing temperature, and as such, is a method more prone to bias. However, this metric is growing in importance due to its ecological relevance and lends insight into the behavioral strategies fishes may use in their natural habitat in response to increasing temperature, setting a lower boundary for their thermal tolerance.

Conversely, researchers can also measure lower thermal limits (critical thermal minima, CT_{min}; first developed by Cowles & Bogert, 1944) in addition to CT_{max} to estimate the thermal window or difference between upper and lower thermal limits. The thermal window breadth provides insight into the ability of the organism to deal with a range of temperatures, whereas a narrow thermal window indicates the adaptation to a rather thermally stable environment and most likely fewer physiological tools to deal with varying environments. Similarly, researchers have used thermal polygons to understand how CT_{max} and CT_{min} change with acclimation temperature, where a larger polygon area indicates a better capacity for acclimation, and a smaller area indicates a poor capacity (reviewed by Beitinger & Bennett, 2000; first developed by Fry et al., 1942).

An additional measure to assess thermal limitations is through creating thermal performance curves (reviewed by Schulte et al., 2011). Performance can be any number of metrics associated with whole organism function (e.g., growth, metabolism, and swim performance). This time-intensive method requires measurements across a broad range of acute and/or acclimated temperatures to assess the upper and lower critical temperatures (T_{crit}) where performance declines below a sustainable range, and with this data also identify an optimal temperature (T_{opt}) where performance is maximized. These metrics are incredibly informative for truly understanding how temperature impacts physiological performance, but a recent study suggests that a fish's preferred temperature may not align with their T_{opt} (Ripley et al., 2022). Additional research is needed to understand what drives the choice of thermal habitat as opposed to what temperature the organism maximizes physiological performance.

2.2 What Causes Failure at Upper Thermal Limits?

Much of the recent thermal physiology literature is devoted to understanding what underlying mechanisms cause failure at upper thermal limits. Recently reviewed by Ern et al. (2023), oxygen limitation and neurological dysfunction are two of the most popular proposed mechanisms currently. Neurological dysfunction is receiving increased attention as methods continue to advance to experimentally test the link between the brain and thermal tolerance (e.g., Andreassen et al., 2022; Jutfelt et al., 2019). The main mechanism proposed here is a mismatch between the supply and demand of generating action potentials (Vornanen, 2016, 2020). No studies to our knowledge have tested if and how neurological dysfunction may underlie thermal limits in Amazonian fishes. Alternatively, the oxygen and capacity-limited thermal tolerance (OCLTT) theory posits that oxygen limitation is the ultimate limiting factor for thermal limitations in aquatic ectotherms, primarily stemming from a mismatch of oxygen supply and oxygen demand at high temperatures leading to cardiac failure (Pörtner, 2001). After this initial proposal, many species have been tested, and while some species do demonstrate that oxygen is the primary limiting factor to thermal performance (e.g., sockeye salmon; Eliason et al., 2011), some species do not (e.g., barramundi; Norin et al., 2014), limiting the universal applicability of the mechanism (Jutfelt et al., 2018).

Increasingly, more Amazonian species have been tested in the context of the OCLTT. Across five families of Amazonian fishes, Campos et al. (2018) found a positive linear relationship between CT_{max} and thermal factorial aerobic scope, a measure of a fish's capacity for aerobic functions limited by the maximum metabolic rate measured during a thermal ramp. This suggests that families with limited aerobic capacity to deal with warming also have lower thermal limits, in support of the OCLTT. However, there are exceptions such as the species from the family Rivulidae which have lower aerobic capacity with warming and high thermal tolerance. Jung et al. (2020) characterized the hypoxia and thermal tolerance of 37 species of Amazonian fishes in the field and lab, finding that thermally tolerant species are also hypoxia tolerant. This may indirectly support the OCLTT, if the mechanisms underlying thermal tolerance are similar to those underlying hypoxia tolerance. However, Jung et al. (2020) also directly tested how oxygen supply impacts thermal tolerance with exposures to hyperoxia. If oxygen is limiting thermal tolerance, then increased oxygen supply should significantly increase thermal tolerance. Out of the seven species tested, only four demonstrated increased CT_{max} with hyperoxia, demonstrating mixed support for the OCLTT once again. One of these species which did not increase CT_{max} with hyperoxia, *Colossoma macropomum*, previously demonstrated a decrease in survival and growth with warm acclimation, yet no change in aerobic performance (Lapointe et al., 2018), providing further evidence that the OCLTT may be species specific. While there may not be any "one-size-fits-all" rule for what physiological mechanism underlies thermal limitations for aquatic ectotherms, the field of experimental thermal physiology is developing new and exciting techniques to tackle this fundamental question, and more research is needed to clearly define how oxygen supply or neurological dysfunction affect thermal limits in Amazonian fishes.

2.3 Thermal Limits of Amazonian Fishes

In a recent review of South American fishes, Campos et al. (2021) showed that tropical freshwater fishes are living close to their maximum thermal limits. Most studies experimentally testing the upper thermal limits of Amazonian fishes have been conducted within the past decade and compared with the sheer number of species living in the Amazon River Basin, only a fraction of these fishes have been studied. We compiled the available data (to the best of our knowledge) for 68 species from 18 studies in Table 1, including CT_{max}, LT_{max}, and CL_{max} data, as well as the acclimation temperatures at which the upper thermal limits were measured. A majority of the data was provided by Jung et al. (2020) and Campos et al. (2018), where the authors examined the thermal tolerance of 37 species and 12 species, respectively.

The available thermal limitation data demonstrate the heterogeneity of responses to thermal stress. Importantly, fishes acclimated to 28–31 °C (the thermal regime used most frequently to study Amazonian fishes) have CT_{max} values ranging from 32.4 to 42.9 °C (average CT_{max} = 39.1 °C). This indicates some fish are at greater risk from warming than others. Climate change scenarios predict atmospheric temperatures of +6.5 °C pre-industrial levels in the worst-case scenario (Carvalho et al., 2020). If we predict water temperature increases will mirror that of atmospheric temperatures, we could expect Amazon River temperatures upwards of 36 °C in some regions at the end of this century. For fishes like *Hyphessobrychon melazonatus*, which exhibited a CT_{max} of 36.0 °C when acclimated to 30.5 °C (Campos et al., 2019), it is critical that we understand *if* and *how much* these vulnerable fish have the capacity to alter their upper thermal limits in response to long-term warming.

3 Capacity for Thermal Plasticity

3.1 Can Amazonian Fishes Increase Upper Thermal Limits through Acclimation?

With some Amazonian fishes living dangerously close to their thermal limits, the question becomes, is there hope that they can employ thermal plasticity to deal with a warming climate? Increasingly in conservation physiology, understanding the capacity of a population or species to acclimate to increased temperature is vital in predicting their ability to deal with a warming climate. Most ectotherms increase their upper thermal limits when acclimated to a warmer temperature for some time; however, the magnitude of this increase is variable. While there are a few ways to understand acclimation capacity for different physiological responses (Einum et al., 2019), acclimation response ratios (ARR) have become valuable metrics to compare populations and species CT_{max} response to increased acclimation temperatures. ARR is calculated as simply the difference between the CTmax values in fish acclimated to two acclimation temperatures (typically, control temperature and a warm acclimation temperature), divided by the difference between the acclimation temperatures:

$$ARR = \frac{CT_{max,2} - CT_{max,1}}{\Delta T}$$

Table 1 Critical and lethal thermal limits of 68 species of Amazonian fishes. Study sites were specified for the study by Jung et al. (2020) in the species name column (field: Anavilhanas Archipelago of the Rio Negro; INPA: the Instituto Nacional de Pesquisas da Amazônia in Manaus, Brazil; UBC: the University of British Columbia). Lethal thermal maxima (LT_{max}) values were defined based on individual studies (e.g., lethal temperature in response to slow fast ramp rates, or temperature where species displays 50% survival), and chronic lethal maxima (CL_{max}) were defined based on Jung et al. (2020) wherein the value represents the highest acclimation temperature the species withstood. Thermal safety margins were calculated as the difference between acclimation temperature and average CT_{max}, LT_{max}, or CL_{max}. Acclimation response ratios (ARRs) were calculated as the difference between the average acclimated CT_{max} and the average control CT_{max} divided by the difference between acclimation temperatures. **Bolded** ARR values represent the response between the highest acclimation temperature and control. All CT_{max} and LT_{max} values are reported as mean ± S.E.M. where S.E.M. was reported. *Italic* values indicate data extracted from figures using WebPlotDigitizer 4.1 (https://apps.automeris.io/wpd/) estimated to the nearest 0.1 °C

Species	CT_{max} (°C)	LT_{max} (°C)	CL_{max}(°C)	Acclimation temperature (°C)	Thermal safety margin (°C)	ARR	Study
Aequidens pallidus	38.38 ± 0.92			25	13.38		Campos et al. (2018)
Anablepsoides micropus	37.07 ± 0.37			25	12.07		Campos et al. (2018)
Ancistrus dolichopterus (field)	39.9 ± 0.3			30	9.9		Jung et al. (2020)
Apistogramma agassizii	40 ± 0.2			26.5	13.5		Campos et al. (2019)
	41.5 ± 0.2			30.5	11	**0.38**	Campos et al. (2019)
Apistogramma agassizii	40.9 ± 0.2			26.9	14		Kochhann et al. (2021)
Apistogramma borellii (INPA)	39.65 ± 0.51		33	28	11.65		Jung et al. (2020)
	40.47 ± 0.06		33	31	9.47	0.27	Jung et al. (2020)
	41.3 ± 0.32		33	33	8.3	**0.33**	Jung et al. (2020)
Apistogramma diplotaenia (field)	42.2 ± 0.2			30	12.2		Jung et al. (2020)
Apistogramma geophyra (INPA)	40.83 ± 0.15			28	12.83		Jung et al. (2020)
Apistogramma hippolytae	38.62 ± 0.37			25	13.62		Campos et al. (2018)
Apistogramma viejita (UBC)	41.01 ± 0.14		33	31	10.01		Jung et al. (2020)
	42.2 ± 0.09		33	33	9.2	**0.60**	Jung et al. (2020)
Astronotus ocellatus (INPA)	39.74 ± 0.62			28	11.74		Jung et al. (2020)
	42.21 ± 0.15			31	11.21	0.82	Jung et al. (2020)
	42.93 ± 0.04			33	9.93	**0.64**	Jung et al. (2020)

(continued)

Table 1 (continued)

Species	CT_{max} (°C)	LT_{max} (°C)	CL_{max}(°C)	Acclimation temperature (°C)	Thermal safety margin (°C)	ARR	Study
Astronotus ocellatus (large)		*40.8*		25	15.8		Gutierre et al. (2016)
		41.0		30	11.0		Gutierre et al. (2016)
Astronotus ocellatus (small)		*40.4*		25	15.4		Gutierre et al. (2016)
		40.5		30	10.5		Gutierre et al. (2016)
Astyanax bimaculatus	*38.2*			24	14.2		Chung and Méndez (1993)
	39			27	12	0.27	Chung and Méndez (1993)
	39.8			30	9.8	0.27	Chung and Méndez (1993)
	40.7			33	7.7	**0.28**	Chung and Méndez (1993)
Brycon amazonicus (INPA)	39.43 ± 0.17			28	11.43		Jung et al. (2020)
Brycon amazonicus	*40.6*			31	9.6		Lapointe et al. (2018)
	41.5			33	8.5		Lapointe et al. (2018)
	41.9			35	6.9		Lapointe et al. (2018)
Carnegiella strigata (INPA)	37.93 ± 0.10		31	28	9.93		Jung et al. (2020)
	39.63 ± 0.04		31	31	8.63	**0.57**	Jung et al. (2020)
Carnegiella strigata (UBC)	38.19 ± 0.10		31	31	7.19		Jung et al. (2020)
Characidium pteroides	35.16 ± 0.51			25	10.16		Campos et al. (2018)
Colossoma macropomum (INPA)	42.06 ± 0.06		>35	28	14.06		Jung et al. (2020)
	42.89 ± 0.10		>35	31	11.89	0.28	Jung et al. (2020)
	42.8 ± 0.11		>35	33	9.8	0.15	Jung et al. (2020)
	44.05 ± 0.09		>35	35	9.05	**0.28**	Jung et al. (2020)
Colossoma macropomum (field)	42.4 ± 0.1			30	12.4		Jung et al. (2020)
Colossoma macropomum	*42.1*			31	11.1		Lapointe et al. (2018)
	43.0			33	10.0	0.55	Lapointe et al. (2018)
	43.7			35	8.7	**0.40**	Lapointe et al. (2018)

Corydoras julii (UBC)	40.38 ± 0.15		33	31	9.38		Jung et al. (2020)
	40.74 ± 0.18		33	33	7.74	**0.18**	Jung et al. (2020)
Corydoras paleatus	33.91 ± 0.72			20	13.91		Yanar et al. (2019)
	35.81 ± 0.86			24	11.81	0.48	Yanar et al. (2019)
	37.51 ± 0.64			28	9.51	**0.45**	Yanar et al. (2019)
Corydoras paleatus	25.58 ± 1.21			7	18.58		Conte et al. (2023)
	26.18 ± 1.24			12	14.18	0.12	Conte et al. (2023)
	31.74 ± 0.19			17	14.74	0.62	Conte et al. (2023)
	34.71 ± 0.22		34.9 ± 0.52	22	12.71	0.61	Conte et al. (2023)
	37.65 ± 0.34			27	10.65	0.60	Conte et al. (2023)
	37.57 ± 0.27			32	5.57	**0.48**	Conte et al. (2023)
Corydoras pulcher (INPA)	37.49 ± 0.70			28	9.49		Jung et al. (2020)
Corydoras schwartzi (INPA)	38.74 ± 0.04		33	28	10.74		Jung et al. (2020)
	39.85 ± 0.10		33	31	8.85	0.37	Jung et al. (2020)
	41.01 ± 0.13		33	33	8.01	**0.45**	Jung et al. (2020)
Corydoras splendens (INPA)	39.00 ± 0.10			28	11.0		Jung et al. (2020)
Crenuchus spilurus	34.71 ± 0.5			25	9.71		Campos et al. (2018)
Exodon paradoxus (field)	38.0 ± 0.4			30	8.0		Jung et al. (2020)
Gasteropelecus maculatus	30.38 ± 0.29			22	8.38		Bohórquez (2016)
	32.41 ± 0.22			28	4.41	0.34	Bohórquez (2016)
	36.26 ± 0.19			32	4.26	**0.59**	Bohórquez (2016)
Geophagus brasiliensis		32.9		12.5	20.4		Rantin and Petersen (1985)
		33.0		15.0	18.0		Rantin and Petersen (1985)
		35.1		20.0	15.1		Rantin and Petersen (1985)
		36.5		25.0	11.5		Rantin and Petersen (1985)
		38.5		30.0	8.5		Rantin and Petersen (1985)
		38.5		32.5	6		Rantin and Petersen (1985)

(continued)

Table 1 (continued)

Species	CT_{max} (°C)	LT_{max} (°C)	CL_{max}(°C)	Acclimation temperature (°C)	Thermal safety margin (°C)	ARR	Study
Geophagus surinamensis (field)	38.0 ± 0.5			30	8.0		Jung et al. (2020)
Gymnotus anguillaris (field)	40.4 ± 0.1			30	10.4		Jung et al. (2020)
Hemigrammus rhodostomus (UBC)	40.41 ± 0.10		33	31	9.41		Jung et al. (2020)
	40.7 ± 0.12		33	33	7.7	**0.15**	Jung et al. (2020)
Hemigramus cf. *geisleri*	32.37 ± 0.85			25	7.37		Campos et al. (2018)
Hemiodus gracilis (INPA)	37.48 ± 0.4			28	9.48		Jung et al. (2020)
Hemiodus gracilis (field)	38.2 ± 0.4			30	8.2	**0.36**	Jung et al. (2020)
Heros severus (field)	41.4			30	11.4		Jung et al. (2020)
Hypostomus plecostomus	37.83 ± 0.73			20	17.83		Yanar et al. (2019)
	39.55 ± 0.95			24	15.55	0.43	Yanar et al. (2019)
	41.06 ± 0.74			28	13.06	**0.40**	Yanar et al. (2019)
Hypostomus watawata	*37*			24	13		Chung and Méndez (1993)
	37.8			27	10.8	0.27	Chung and Méndez (1993)
	38.6			30	8.6	0.27	Chung and Méndez (1993)
	38.9			33	5.9	**0.21**	Chung and Méndez (1993)
Hyphessobrycon erythrostigma (INPA)	39.13 ± 0.15		33	28	11.13		Jung et al. (2020)
	39.34 ± 0.20		33	31	8.34	0.07	Jung et al. (2020)
	40.04 ± 0.07		33	33	7.04	**0.18**	Jung et al. (2020)
Hyphessobrycon herbertaxelrodi	39.5 ± 0.21	35.9 ± 0.30		26	13.5		Cooper et al. (2019)
	40.6 ± 0.19			29	11.6	0.37	Cooper et al. (2019)
	41.0 ± 0.10			31	10.0	**0.30**	Cooper et al. (2019)
Hyphessobrycon megalopterus (field)	38.6 ± 0.7			30	8.6		Jung et al. (2020)
Hyphessobrycon melazonatus	32.60 ± 1.28			25	7.60		Campos et al. (2018)

Hyphessobrycon melazonatus	36 ± 0.5			26.5	9.5		Campos et al. (2019)
	36 ± 0.5			30.5	5.5	**0**	Campos et al. (2019)
Iguanodects geisleri	30.87 ± 0.63			25	5.87		Campos et al. (2018)
Laetacara fulvipinnis (INPA)	40.99 ± 0.26			28	12.99		Jung et al. (2020)
Leporinus fasciatus (field)	36.1 ± 0.5			30	6.1		Jung et al. (2020)
Metynnis hypsauchen (field)	40.7 ± 0.2			30	10.7		Jung et al. (2020)
Mesonauta insignis	41.3 ± 0.2			28.1	13.2		Kochhann et al. (2021)
Microcharacidium eleotrioides	35.21 ± 0.48			25	10.21		Campos et al. (2018)
Mikrogeophagus ramirezi (UBC)	41.10 ± 0.24		>35	31	10.10		Jung et al. (2020)
	42.63 ± 0.07		>35	33	9.63	0.77	Jung et al. (2020)
	43.95 ± 0.09		>35	35	8.95	**0.71**	Jung et al. (2020)
Myloplus asterias (field)	40.4 ± 0.2			30	10.4		Jung et al. (2020)
Nannostomus backfordi	35.50 ± 0.36			25	10.50		Campos et al. (2018)
Nannostomus eques (INPA)	39.09 ± 0.31		33	28	11.09		Jung et al. (2020)
	41.03 ± 0.04		33	31	10.03	0.65	Jung et al. (2020)
	41.68 ± 0.18		33	33	8.68	**0.52**	Jung et al. (2020)
Nannostomus marginatus	35.67 ± 1.07			25	10.67		Campos et al. (2018)
Nannostomus trifasciatus (UBC)	41.03 ± 0.19		33	31	10.03		Jung et al. (2020)
	42.7 ± 0.18		33	33	9.7	**0.84**	Jung et al. (2020)
Otocinclus spp. (INPA)	36.54 ± 0.26		31	28	8.54		Jung et al. (2020)
	38.9 ± 0.22		31	31	7.9	**0.79**	Jung et al. (2020)
Paracheirodon axelrodi	*38.1*			23	15.1		Campos et al. (2017)
	38.3			26	12.3	0.07	Campos et al. (2017)
	38.0			29	9	−0.02	Campos et al. (2017)
	39.9			32	7.9	**0.20**	Campos et al. (2017)

(continued)

Table 1 (continued)

Species	CT_{max} (°C)	LT_{max} (°C)	CL_{max}(°C)	Acclimation temperature (°C)	Thermal safety margin (°C)	ARR	Study
Paracheirodon axelrodi (INPA)	38.74 ± 0.06		33	28	10.74		Jung et al. (2020)
	39.85 ± 0.06		33	31	8.85	0.37	Jung et al. (2020)
	40.03 ± 0.17		33	33	7.03	**0.26**	Jung et al. (2020)
Paracheirodon innesi	38.5 ± 0.30	33.5 ± 0.26		26	12.5		Cooper et al. (2019)
	39.3 ± 0.22			29	10.3	0.27	Cooper et al. (2019)
	39.6 ± 0.13			31	8.6	**0.22**	Cooper et al. (2019)
Paracheirodon simulans	*39.0*			23	16		Campos et al. (2017)
	37.9			26	11.9	−0.37	Campos et al. (2017)
	41.7			29	12.7	0.45	Campos et al. (2017)
	41.9			32	9.9	**0.32**	Campos et al. (2017)
Piaractus brachypomus	*35.2*			16.5	18.7		Di Santo et al. (2018)
	37.8			20	17.8	0.74	Di Santo et al. (2018)
	40.3			25	15.3	0.60	Di Santo et al. (2018)
	41.9			30	11.9	0.50	Di Santo et al. (2018)
	43.9			35	8.9	**0.47**	Di Santo et al. (2018)
Piaractus mesoptamicus	39.6 ± 0.2			26	13.6		Blasco et al. (2020)
Pimelodellas chagresi	*38.2*			24	14.2		Chung and Méndez (1993)
	38.8			27	11.8	0.20	Chung and Méndez (1993)
	39.8			30	9.8	0.27	Chung and Méndez (1993)
	39.9			33	6.9	**0.19**	Chung and Méndez (1993)
Poecilia caucana	38.43 ± 0.64			20	18.43		Martínez et al. (2016)
	40.28 ± 0.92			25	15.28	0.37	Martínez et al. (2016)
	41.57 ± 0.27			28	13.57	**0.39**	Martínez et al. (2016)

Poecilia reticulata	39.71 ± 0.68			20	9.71		Yanar et al. (2019)
	40.88 ± 0.68			24	16.88	0.29	Yanar et al. (2019)
	41.80 ± 0.67			28	13.80	**0.26**	Yanar et al. (2019)
Poecilia reticulata (male)	38.95 ± 0.51	41.22 ± 0.34		24	14.95		Chung (2001)
	39.79 ± 0.47	41.69 ± 0.42		27	12.79	0.28	Chung (2001)
	39.98 ± 0.54	42.01 ± 0.50		30	9.98	0.17	Chung (2001)
	40.33 ± 0.25	42.55 ± 0.45		33	7.33	**0.15**	Chung (2001)
Poecilia reticulata (female)	39.48 ± 0.82	41.41 ± 0.31		24	15.48		Chung (2001)
	39.69 ± 0.51	42.02 ± 0.42		27	12.69	0.07	Chung (2001)
	39.85 ± 0.34	42.24 ± 0.41		30	9.85	0.06	Chung (2001)
	40.61 ± 0.28	42.86 ± 0.17		33	7.61	**0.13**	Chung (2001)
Potamorhina pristigaster (field)	38.3 ± 0.5			30	8.3		Jung et al. (2020)
Prochilodus scrofa (small)	33.9			15	18.9		Fernandes et al. (1995)
	36.7			20	16.7	0.56	Fernandes et al. (1995)
	38.7			25	13.7	0.48	Fernandes et al. (1995)
	40.3			30	10.3	0.43	Fernandes et al. (1995)
	42.0			35	7.0	**0.41**	Fernandes et al. (1995)
Prochilodus scrofa (large)	33.3			15	18.3		Fernandes et al. (1995)
	35.7			20	15.7	0.48	Fernandes et al. (1995)
	38.2			25	13.2	0.49	Fernandes et al. (1995)
	40.6			30	10.6	0.49	Fernandes et al. (1995)
	42.6			35	7.6	**0.47**	Fernandes et al. (1995)
Pterophyllum scalare	38.18 ± 0.57			20	18.18		Yanar et al. (2019)
	38.41 ± 0.40			24	14.41	0.06	Yanar et al. (2019)
	40.94 ± 0.47			28	12.94	**0.35**	Yanar et al. (2019)

(continued)

Table 1 (continued)

Species	CT_{max} (°C)	LT_{max} (°C)	CL_{max}(°C)	Acclimation temperature (°C)	Thermal safety margin (°C)	ARR	Study
Pterophyllum scalare (juvenile)	34.0			20	14		Pérez et al. (2003)
	35.9			24	11.9	0.48	Pérez et al. (2003)
	38.6			28	10.6	0.58	Pérez et al. (2003)
	38.8			32	6.8	**0.40**	Pérez et al. (2003)
Pterophyllum scalare (adults)	36.2			20	16.2		Pérez et al. (2003)
	36.5			24	12.5	0.08	Pérez et al. (2003)
	38.1			28	10.1	0.24	Pérez et al. (2003)
	41.2			32	9.2	**0.42**	Pérez et al. (2003)
Pygocentrus nattereri	32.62 ± 0.03			26	6.62		Divya and Ranjeet (2014)
	33.74 ± 0.05			28	5.74	0.56	Divya and Ranjeet (2014)
	34.16 ± 0.03			30	4.16	0.39	Divya and Ranjeet (2014)
	35.82 ± 0.04			32	3.82	0.53	Divya and Ranjeet (2014)
	37.12 ± 0.08			34	3.12	**0.56**	Divya and Ranjeet (2014)
Pyrrhulina aff brevis	36.21 ± 0.69			25	11.21		Campos et al. (2018)
Pyrrhullina brevis	39 ± 0.4			26.5	12.5		Campos et al. (2019)
	40 ± 0.2			30.5	9.5	**0.25**	Campos et al. (2019)
Satanoperca jurupari (field)	40.0 ± 0.5			30	10.0		Jung et al. (2020)
Satanoperca pappaterra (field)	42.8 ± 0.2			30	12.8		Jung et al. (2020)
Semaprochilodus insignis (field)	39.2 ± 0.3			30	9.2		Jung et al. (2020)
Serrasalmus rhombeus (field)	39.6 ± 0.6			30	9.6		Jung et al. (2020)
Sternopygus macrurus (field)	38.9 ± 0.4			30	8.9		Jung et al. (2020)
Uaru amphiacanthoides (field)	41.4			30	11.4		Jung et al. (2020)

This metric has become useful in meta-analyses to understand the capacity for thermal acclimation across ectothermic species, taxa, life stages, etc. (Gunderson & Stillman, 2015; Pottier et al., 2021, 2022). There is no sexual dimorphism when it comes to a fish's capacity to acclimate to high temperature (Pottier et al., 2021), but literature suggests differences in acclimation capacity at different life stages, and embryonic stages may be a critical window of vulnerability (Pottier et al., 2022). While aquatic ectotherms demonstrate two to three times the thermal plasticity of terrestrial taxa, they are still quite poor at acclimating to warming conditions (Gunderson & Stillman, 2015; Pottier et al., 2022). An ARR of 1 indicates complete compensation for increased thermal acclimation, and a commensurate increase in upper thermal limits with the increase in acclimation temperature. Freshwater ectotherms (fishes and crustaceans) have an ARR of 0.25 ± 0.02 (Gunderson & Stillman, 2015) up to a certain acclimation temperature, but heat tolerance does not completely track changes in environmental temperature (Gunderson & Stillman, 2015; Pottier et al., 2022).

However, these broad generalizations may not accurately capture the complexity of responses in Amazonian fishes, as there is a need for more thermal acclimation studies that specifically measure the effect of acclimation temperature on upper thermal limits. We calculated ARRs from 34 of the 68 species identified in Table 1. Where studies acclimated fish to multiple temperatures, we calculated ARRs using data for the control and the highest acclimation temperature. With this data we found an average ARR value of 0.38 ± 0.03, suggesting that Amazonian fishes are more capable of acclimating to warming than previously thought for freshwater ectotherms. Recently, Campos et al. (2021) also demonstrated that South American fishes had an average acclimation response of 0.34, and that tropical freshwater species were better able to compensate for warming with acclimation. However, as with the CT_{max} data, we found an incredible diversity of response ratios ranging from 0 (no acclimation capacity) to 0.84 (almost complete compensation). This is also evidenced by the wide range of ARRs reported by Campos et al. (2021) for Amazonian fishes, including the highest value reported for *Nannostomus trifasciatu* at 1.5 (overcompensation). The disparity in some of the ARR values reported by Campos et al. (2021) and here (Table 1) for the same species demonstrate how sensitive these values are based on the reported endpoint (LT_{max} vs. CT_{max}), species origin (e.g., differences in prior exposure), and even the precision (e.g., differences in rounding) with which we report CT_{max}. Importantly, this wide range of values highlights the need to use multiple characterizations to understand fishes' vulnerability to warming.

For example, *Gasteropelecus maculatus* demonstrates a high ARR of 0.59 when comparing their CT_{max} following acclimation to 22 °C and 32 °C (CT_{max} of 30.4 °C and 36.3 °C respectively; (Bohórquez, 2016). However, for all fish in Table 1, we also calculated the difference between the acclimation temperature and CT_{max}, here referred to as the thermal safety margin (TSM). Typically, the thermal safety margin (TSM) refers to the difference in temperature between an animal's thermal limit and their environment. For example, in wild fishes, the difference between the warm range boundary temperature (i.e., the average temperature of the warmest month at equatorward range limits) and optimal temperature for performance is smaller in tropical fishes compared to fishes at other latitudes, suggesting that tropical fishes are operating close to their thermal limits (Payne et al., 2016). Here we derived TSMs as another metric of thermal vulnerability using acclimation temperature (e.g., McKenzie et al., 2021). As such, *Gasteropelecus maculatus* has the lowest TSM of all 68 species we examined, with an average CT_{max} only 4.3 °C greater than the acclimation temperature (Bohórquez, 2016). This indicates that while the species may acclimate to warm temperatures well, they are at higher risk for rapid increases in temperature than other species that may have a poorer ARR but innately higher thermal tolerance.

Pyrrhullina brevis have an ARR of 0.25 because they demonstrated only a 1 °C improvement in their CT_{max} from an acclimation at 26.5–30.5 °C (Campos et al., 2019). Yet, they are more thermally tolerant than *G. maculatus* when considering their CT_{max} values (39 °C and 40 °C, respectively). On the other hand, *Hyphessobrychon melazonatus* demonstrates no capacity for thermal acclimation whatsoever (ARR = 0), and as previously mentioned, this species is living quite close to its upper thermal limit (Campos et al., 2019). While we can conclude this species is one of the most vulnerable to warming, the lack of pattern between CT_{max} and ARR seen in Amazonian fishes makes it difficult to predict vulnerability based on one metric alone.

3.2 Acclimation-Associated Mortality

Moreover, there are other considerations beyond both CT_{max} and ARR that may be more critical to explore in thermal tolerance and acclimation studies. In a study comparing the thermal sensitivity of equatorial fishes, Lapointe et al. (2018) found evidence that the 6 species studied increased their upper thermal limits when acclimated to a higher temperature and concluded that tropical species may not be as vulnerable as we think. Lapointe et al. (2018) quantified performance metrics (e.g., specific growth rate and metabolism) at warm acclimation temperatures, as well as CT_{max}. *Colossoma macropomum* increased CT_{max} by 1.6 °C with +4 °C acclimation from 31 °C to 35 °C (ARR = 0.4). Similarly, *Brycon amazonicus* increased CT_{max} by 1.3 °C with +4 °C acclimation from 31 °C to 35 °C (ARR = 0.325). With high upper thermal limits (>40 °C) and moderate acclimation capacity, we may predict both species will perform well in a warming world. However, strikingly, only 22% of *B. amazonicus* survived the 35 °C acclimation. Therefore, acclimation-associated mortality may be more indicative of thermal acclimation capacity than ARR alone, which is solely based on the surviving individuals' responses.

Jung et al. (2020) demonstrated that while CT_{max} may increase with warm acclimations, many Amazonian fishes simply do not survive short- or long-term acclimations to temperatures that are only a few degrees higher than current maximum river temperatures. In fact, only 2 of the 13 species the authors studied survived acclimation at 35 °C. Recently, Conte et al. (2023) used both CL_{max} and CT_{max} data to construct a complete thermal polygon for *Corydoras paleatus*. To our knowledge, thermal polygons have only been created for three Amazonian fishes, *Corydoras paleatus* (Conte et al., 2023), *Paracheirodon axelrodi*, and *Paracheirodon simulans* (Campos et al., 2017). With the addition of CL_{max} data to this thermal polygon, Conte et al. (2023) have provided a template for future researchers that relates chronic thermal limits to acute upper thermal limits as fishes respond to acclimation. These studies demonstrate the usefulness of the CL_{max} metric, and when coupled with CT_{max} and ARR data, enables us to better understand the sensitivity of Amazonian fishes. For example, *Mikrogeophagus ramirezi* represents one of the most tolerant species to warming (Jung et al., 2020), wherein their ability to tolerate short-term warm acclimation (CL_{max} > 35 °C) corroborates with their high acclimation capacity (ARR = 0.71) and high innate thermal tolerance (CT_{max} > 40 °C). *Colossoma macropomum* also demonstrates an ability to tolerate short-term warming (CL_{max} > 35 °C) and high innate thermal tolerance (CT_{max} > 40 °C) but low to moderate acclimation capacity as defined by their ARR (Jung et al., 2020; Lapointe et al., 2018). Here, it seems overall the species will be able to perform well in a warming environment, where the CL_{max} and CT_{max} provide more relevant information in combination with the ARR. In the opposite manner, *Otocinclus* spp. has one of the highest acclimation capacities (ARR = 0.79), where CT_{max} increased from 36.5 °C to 38.9 °C with a temperature increase of +3 °C (from 28 °C to 31 °C; Jung et al., 2020). However, this species also demonstrated high mortality (>50%) within 10 days at acclimation temperatures of 33 °C. Thus, its CL_{max} of 31 °C indicates the incredible sensitivity of this species to temperatures above the current river conditions. Thus, incorporating multiple metrics of thermal limitations and acclimation capacity is essential to predicting how fishes will cope with global warming.

4 Predictive Power of Thermal Limitation Data

For many researchers, the main objective of conducting thermal limitation studies is to use physiological data for predicting population-, species-, and even ecosystem-level responses to environmental change. However, translating lab-derived data to large-scale predictions is a difficult undertaking, often requiring a more substantial sample size than is typically found in any one physiological study. In an attempt to predict how Amazonian fishes will cope with global warming, we used the 18 studies in Table 1 to first determine if we can predict upper thermal limits based on environmental temperature, to extend our limited dataset. Second, because CL_{max} is a critically important metric that more fully informs us of species' vulnerability, we wanted to determine if there was a predictive relationship between CT_{max} and CL_{max}. With this information, we endeavored to create a predictive model using CT_{max} and CL_{max} as factors in assessing biodiversity changes across the Amazon River Basin with different climate change scenarios.

4.1 Methods

4.1.1 Can We Predict CT_{max} Based on Acclimation (Environmental) Temperature?

We explored the relationship between the critical thermal maximum (CT_{max}) of Amazon fish species and various environmental factors by applying a generalized additive model (GAM). The GAM framework allows for capturing non-linear associations between the response variable, thermal limit, and predictor variables. We obtained CT_{max} data for Amazonian fish species from the literature (Table 1). The data set included parameters latitude, acclimation temperature, acclimation duration, and warming rate as they were reported across all included studies and were found to be the best predictors for CT_{max}.

To model the nonlinear relationship between the response variable (CT_{max}) and the predictors, we employed a GAM using the "gam" function (Model 1). We utilized cubic regression spline basis functions ("bs = cr") to flexibly capture the potentially complex relationships between the predictors and the response variable. The number of knots ("k") was set appropriately to balance model flexibility and over-fitting. To account for residual dependencies, we incorporated an autoregressive process of order 1 (AR1) correlation structure with a correlation coefficient of 0.6873612. The GAM was fitted using the 'family' argument set as Gamma (link = inverse). The inverse link function was chosen to model the negatively skewed nature of the CT_{max} variable. This choice allows for capturing the appropriate relationship between the predictors and the CT_{max} of the Amazon fish species, accounting for the specific distributional characteristics of the response variable. We then developed the model with 65% of the data and tested the model with the remaining 35%. The fitted GAM model with the inverse link function is as follows:

$$\text{Model} = \text{gam}\left(\begin{array}{l} CT_{max} \sim s\left(\text{Latitude}, bs = "cr", k = 16\right) \\ +s\left(\text{Acclimation temperature}, bs = "cr", k = 17\right) \\ +s\left(\text{Acclimation duration}, bs = "cr", k = 12\right) \\ +s\left(\text{warming rate}, bs = "cr", k = 6\right), \text{correlation} \\ = \text{corAR1}\left(0.6873612, |\text{form} = \sim 1, |\text{Species}\right), \text{family} \\ = \text{Gamma}\left(\text{link} = \text{inverse}\right), \text{data} = \text{data} \end{array}\right) \quad (1)$$

We then assessed the importance of predictor variables in this GAM for predicting CT_{max} using the permutation importance to evaluate the relative influence of predictor variables on the model. The permutation importance method was calculated to determine the impact of each predictor variable on the model's deviance. We developed a function to compute the drop in deviance by permuting each variable and fitting the permuted model. The original deviance of the model was compared with the permuted deviance to quantify its importance. We further ranked the permutation importance in descending order to determine the relative importance of the predictor variables. Higher ranks indicated greater influence. We further developed an alternative model using the highest ranked variables. We then evaluate the two models using the Akaike information criterion (AIC). A lower AIC value indicates a better-fitting model, considering the balance between model complexity (number of parameters) and goodness of fit (log-likelihood).

4.1.2 Can We Predict Chronic Lethal Maxima from CT_{max} Data?

With the importance of CL_{max} in determining species' vulnerability (see Sect. 3.2), we examined the interspecific relationship between CT_{max} and CL_{max}. To accomplish this, we developed a generalized linear model (GLM; Model2) using data obtained from Jung et al. (2020). Using the developed model, we were able to predict CL_{max} values for other Amazon species in our database (Table 1) using CT_{max} and acclimation temperature as predictors, with the latter set at two different levels: 28 °C and 31 °C. The model is as follows:

$$CL_{max} = \beta 0 + \beta 1^{*} CT_{max} + \beta 2^{*} \text{ Acclimation temperature} \tag{2}$$

To carry out our projections, we first estimated the CT_{max} for these species using the GAM (Model 1) and then employed the GLM (Model 2) to project the CL_{max} values.

4.1.3 Can We Predict How Biodiversity in the Amazon River Basin Will Change with CT_{max} and CL_{max} Data?

To assess biodiversity using both critical thermal maximum (CT_{max}) data and presence data, we first determined species richness within the Amazon River Basin by analyzing presence data. We obtained species occurrence information from the Global Biodiversity Information Facility (https://www.gbif.org/). We extracted a total of 39 fish species based on our CT_{max} data. We then developed the georeferenced occurrence database over the Amazon on a 0.25 by 0.25 cell grid. Second, we assigned weights to each species based on their CT_{max} values and CL_{max}, which indicate their thermal tolerance using the following equations:

$$CL_{max}50 = \{1, \text{if } CL_{max} > CT_{max}, 0.5, \text{otherwise} \tag{3}$$

$$\text{Weight} = (1/\text{Thermal\_limit})^{*} CL_{max}50 \tag{4}$$

where $CL_{max}50$ represent 50% mortality based on Jung et al. (2020). This allows us to incorporate thermal tolerance as a factor in the assessment of biodiversity. Third, to calculate the weighted species richness, we summarized each species weight in each cell. The resulting measure provided a weighted species richness value that accounts for the thermal tolerance of each species and is indicated as the biodiversity thermal index (BTI). Lastly, we projected BTI under climate change scenarios. We used the combined climate change and socioeconomic scenarios SSP1–2.6 (1.3 °C increase by 2081–2100 relative to 1970–2000), SSP2-4.5 (4.0 °C increase) and SSP5–8.5 (6.5 °C increase) based on the projection of Carvalho et al. (2020). Also, we assumed the current mean temperature of the Amazon is 26 °C and 28 °C based on Carvalho et al. (2020) and Jung et al. (2020), respectively.

4.2 Predictive Models

4.2.1 Can We Predict CT_{max} Based on Acclimation (Environmental) Temperature?

Latitude, acclimation temperature, acclimation duration, and warming rate all had a significant effect on CT_{max}, and importantly, the resulting GAM could be used to predict CT_{max} as a response to these variables. The latitude smooth term has a significant effect of CT_{max} and shows a complex relationship that is not well-captured by a simple linear term (Fig. 1a; estimated degree of freedom [*df*] = 14.871, F = 15.92, $p < 2e\text{-}16$). The acclimation temperature smooth term also has a significant effect on CT_{max}, suggesting a linear relationship (Fig. 1b; $df = 1.000$, F = 131.66, $p < 2e\text{-}16$). The acclimation duration smooth term shows a significant effect on

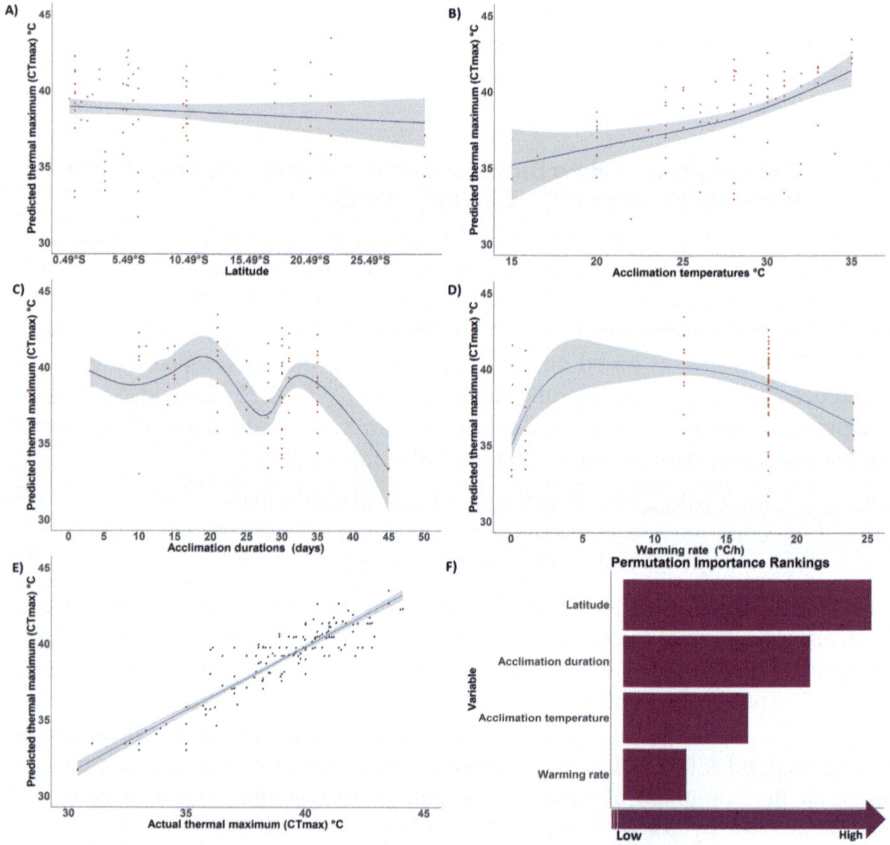

Fig. 1 Components of the GAM model fit for Amazonian fishes' critical thermal maximum (CT_{max}). The predicted critical thermal maximum (CT_{max}) vs (**a**) latitude, (**b**) acclimation temperature, (**c**) acclimation duration, (**d**) warming rate, and (**e**) predicted critical thermal maximum (CT_{max}) of 35% randomly selected data vs. laboratory-measured critical thermal maximum (CT_{max}) ($R^2 = 0.84$, $P < 0.001$). (**f**) Permutation importance rankings plot

CT_{max} with an edf of 3.678, indicating a nonlinear relationship (Fig. 1c; $df = 3.678$, $F = 15.87$, $p < 2e\text{-}16$). Similarly, the warming rate smooth term has a significant effect on CT_{max}, with an edf of 3.982, indicating a nonlinear relationship (Fig. 1d; $df = 3.982$, $F = 64.98$, $p < 2e\text{-}16$). The intercept term has a small standard error and an estimated value of 2.591e-02. Its high significance ($p < 2e\text{-}16$) implies a nonzero baseline CT_{max} even when all other variables are zero. Model 1 explains about 80.9% of the variation in the CT_{max} ($R^2 = 0.809$). The GCV score is 0.001261, indicating a good model fit with low generalization error. The deviance explained is 84.6% which indicates the model captures a substantial portion of the total deviance in the data (Fig. 1e).

We used the permutation importance technique to determine which predictor variables were most important in predicting CT_{max} in the GAM model. The rankings we obtained gave us valuable insights into the effects of each variable and helped us identify the most influential factors (Fig. 1f). Our results showed that latitude, acclimation temperature, and acclimation duration were the most important predictors of CT_{max}. We also developed an alternative GAM model without the warming rate, which differed from the main model in terms of estimated degrees of freedom, GCV scores, goodness-of-fit measures, and AIC values (main model = 541.46; alternative model = 699.40). Based on our analysis, the main model (Model 1) had higher degrees of freedom, lower GCV scores, higher R-squared (adjusted), and lower AIC values, indicating that it was a better fit for the data and provided a more accurate representation of the relationship between the predictors and the CT_{max} variable.

4.2.2 Can We Predict Chronic Lethal Maxima from CT_{max} Data?

Our analysis indicates that the coefficients in the GLM model represent the anticipated impact of certain factors on the response variable, CL_{max}. Specifically, the intercept coefficient is 11.9634, indicating that when both CT_{max} and acclimation temperature are at zero, the estimated CL_{max} is 11.9634. The CT_{max} coefficient is 0.8267, signifying that an increase of one unit in CT_{max} leads to an estimated rise of 0.8267 in CL_{max}, with other variables held constant. On the other hand, the acclimation temperature coefficient is −0.3982, meaning that an increase of one unit in acclimation temperature corresponds to an estimated decrease of 0.3982 in CL_{max}, while other factors are controlled. Our model suggests that CT_{max} and acclimation temperature have significant impacts on CL_{max}, and it fits the data reasonably well.

4.2.3 Can We Predict How Biodiversity in the Amazon River Basin Will Change with CT_{max} Data?

This study was undertaken to assess the potential changes in the biodiversity thermal index (BTI) across distinct latitudinal zones within the Amazon region, considering multiple climate change scenarios. The focus centered specifically on the assumed mean Amazon temperature of 26 °C and 28 °C. For the mean Amazon temperature of 26 °C (Figs. 2a,b,c and 3a), under the SSP1-2.6 scenario, characterized by moderate mitigation efforts, we found that all Amazon zones (Northern, Central, and Southern) would experience a minimal decrease in BTI, specifically by 1.3% (Figs. 2a and 3a). In contrast, the more severe SSP3-7.0 scenario (a higher

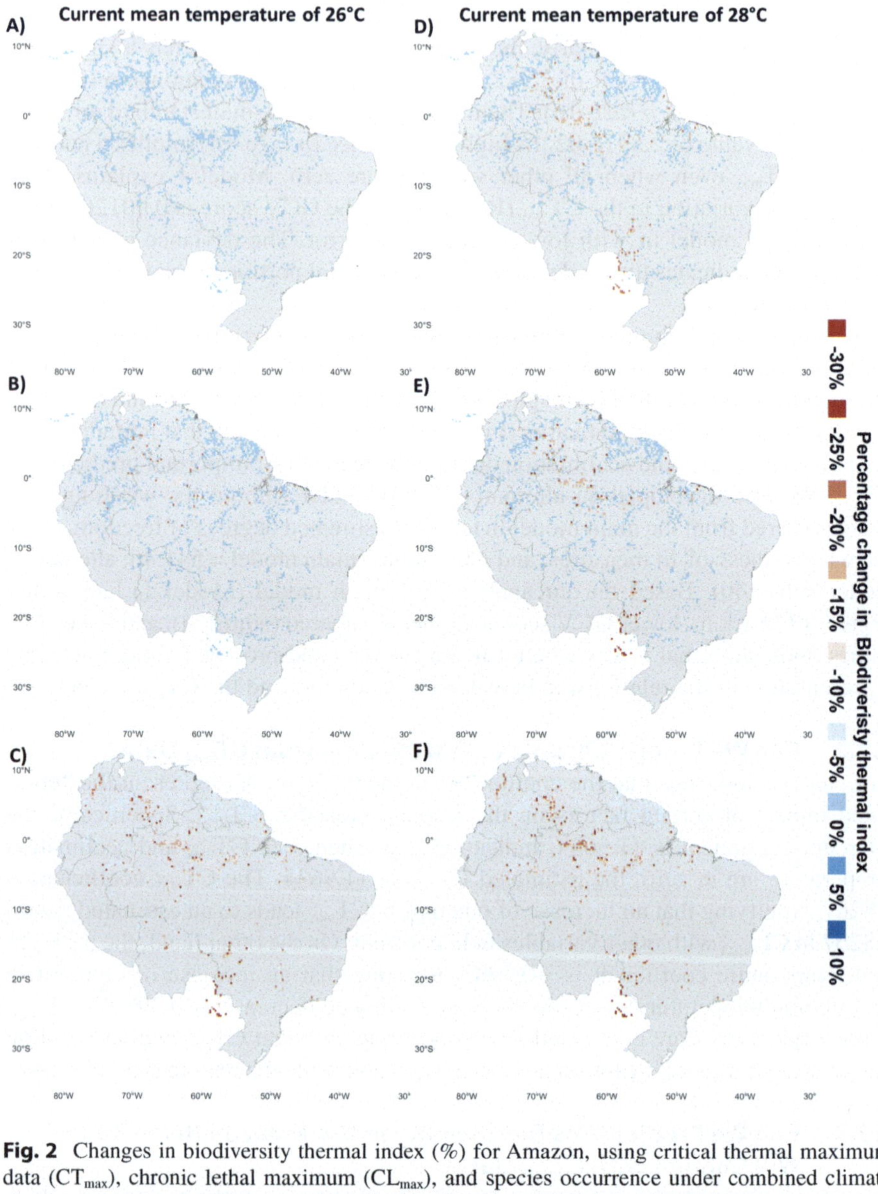

Fig. 2 Changes in biodiversity thermal index (%) for Amazon, using critical thermal maximum data (CT_{max}), chronic lethal maximum (CL_{max}), and species occurrence under combined climate change and socioeconomic scenarios (**a**,**d**: SSP1-2.6; **b**,**e**: SSP2-4.5; **c**,**f**: SSP5-8.5)

greenhouse gas emission trajectory) showed greater projected decreases in BTI. In the Northern Amazon, we estimated a decrease of 5.9 ± 5% (mean and standard deviation), while the Central and Southern Amazon regions were projected to experience a decrease of 6.9 ± 5% and 6.9 ± 7%, respectively (Figs. 2b and 3a). On the

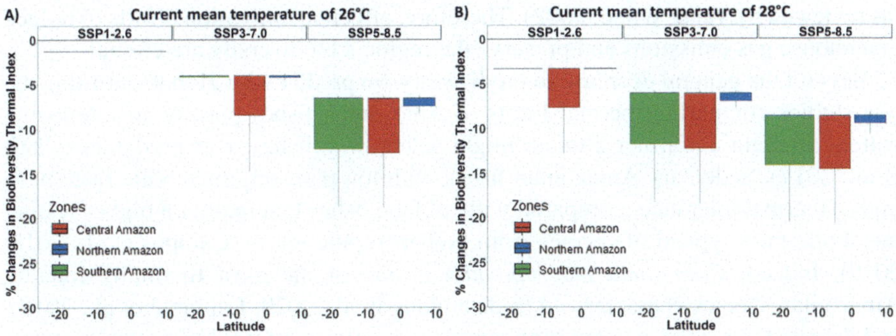

Fig. 3 Latitudinal changes in biodiversity thermal index (%) for Amazon regions; Northern Amazon (blue), Central Amazon (red), and Southern Amazon (green) under combined climate change and socio-economic scenarios, based on the current mean temperatures of (**a**) 26 °C and (**b**) 28 °C

other hand, under the high-emission SSP5-8.5 scenario, our results indicated that the BTI would decrease in the Amazon. We estimated a decrease of 8.9 ± 5%, 9.8 ± 5%, and 10.7 ± 7% in the Northern, Central, and Southern Amazon regions, respectively (Fig. 2c and 3c).

In the context of a mean Amazon temperature of 28 °C (Fig. 2d, e, f and 3b), under the SSP1-2.6 scenario, our analysis indicated a slight decline in BTI across all Amazon zones. The estimated reductions were 6.01 ± 4.96%, 5.08 ± 4.74%, and 6.15 ± 6.81% for the Central, Northern, and Southern Amazon, respectively (Fig. 2d and 3b). In contrast, the more severe SSP3-7.0 scenario projected greater decreases in BTI. Specifically, we estimated a decline of 9.28 ± 5.07%, 8.44 ± 5.13%, and 10.21 ± 7.31% in the Central, Northern, and Southern Amazon regions, respectively (Fig. 2e and 3b). Under the SSP5-8.5 scenario, the projected changes in BTI were even more alarming. Our analysis revealed estimated decreases of 11.70 ± 4.92%, 10.89 ± 4.97%, and 12.62 ± 7.09% in the Central, Northern, and Southern Amazon regions, respectively (Fig. 2f and 3b).

5 The Future of Amazonian Fishes in a Warming World

5.1 The Future of Amazonian Fishes Based on Current Data

Our analyses provide evidence that with different climate change scenarios, the physiological capacity of Amazonian fishes to tolerate higher temperatures may not be enough to maintain the incredible diversity we see today. We discovered that the BTI in the Amazon region is projected to decrease, which varies in severity across latitudinal zones and climate change scenarios. As one might expect, with higher emission scenarios, we find greater declines in BTI, and in particular, the Central Amazon will see the largest decrease across all scenarios. These projected changes in BTI highlight the vulnerability of the Amazon's biodiversity to climate change (as reviewed by Val & Wood, 2022). Therefore, effective mitigation efforts to reduce greenhouse gas emissions and preserve the region's biodiversity are crucial.

Beyond the general decrease in biodiversity we predict with global warming, the few studies comparing thermal limits of Amazonian fishes portray an interesting pattern wherein a warmer climate might select prey fishes over predators (Clark et al., 2017). Sedentary Amazonian fishes with lower resting metabolic rates have higher thermal tolerance, compared with athletic fishes that maintain higher resting metabolic rates typical of species from higher trophic levels (Campos et al., 2018, 2019). Indeed, *Colossoma macropomum* is one of the most thermally tolerant Amazonian species characterized to date (Jung et al., 2020; Lapointe et al., 2018), and despite their large size are omnivorous (sometimes referred to as the "vegetarian piranha"). Conversely, the red-bellied piranha (*Pygocentrus nattereri*) demonstrated an upper thermal limit of 37.1 °C when acclimated to 33 °C, indicating a relatively poor response to warmer temperatures (Divya & Ranjeet, 2014).

This idea that selection is stronger on prey than predators has been referred to as the "life-dinner principle" (Dawkins & Krebs, 1979)—if a predator loses, it misses a meal; but if prey loses, it dies. Thus, there is evidence that thermal sensitivity is correlated to aerobic and anaerobic metabolism (low basal aerobic metabolism and higher anaerobic capacity for escape response). Unfortunately, this pattern where a warming climate selects for fishes with "low-performance phenotypes" will have ramifications on community dynamics and ecosystem structure (Clark et al., 2017). There is evidence that aquatic omnivores increase herbivory with increased temperature (Zhang et al., 2020), and the relationship between temperature, feeding, and digestion has been used to explain the distribution of herbivorous fishes across latitudes (Floeter et al., 2005), specifically the range expansion of tropical herbivorous fishes (Vergés et al., 2014, 2016). However, a recent meta-analysis revealed herbivorous marine fishes absorb the same proportion of nutrients and energy from their food regardless of temperature, and in general have lower absorption efficiency than carnivorous fishes (Knight et al., 2021). While temperature is known to impact feeding and digestion (reviewed by Volkoff & Rønnestad, 2020), more research is needed to understand how feeding strategy (e.g., herbivory, omnivory, carnivory, etc.) relates to thermal limits, especially with regards to Amazonian fishes. Understanding how resource availability, competition, and predatory-prey interactions may impact thermal limits and optima is critical for predicting the ecological consequences of climate change (Litchman & Thomas, 2023; Woodward et al., 2010).

5.2 Research Needs for Predicting the Future of Amazonian Fishes

5.2.1 Refining and Increasing the Diversity of Methods

As with most laboratory-based experimental methods, there are caveats to using CT_{max} to predict the ecological impacts of a warming climate (reviewed in detail by Desforges et al. (2023)). For example, how appropriate is a rapid heating test to understanding responses over longer timescales, in more complex environmental scenarios? Certainly, this information is useful with the increased frequency, duration, and intensity of heatwaves brought about by global climate change, which are thought to exacerbate mortality (Jørgensen et al., 2022). Specifically, the *rate* of heatwave onset may be the leading cause of mass mortality (e.g., northern Red Sea in 2017; Genin et al., 2020). The rate of thermal ramping is indeed an issue to contend with for future studies, as the species' final endpoint (CT_{max}) vary in response. Some species demonstrate lower tolerance when warmed slowly (e.g., zebrafish; Åsheim et al., 2020), and when studying large-bodied fishes, faster ramping rates can artificially inflate CT_{max} values (e.g., Wheeler et al., 2022). Conversely, at least one species demonstrates an *increase* in CT_{max} with slower ramp rates (white sturgeon; Penman et al., 2023). A recent innovation combines CT_{max} methodology with a slower ramp rate (+1 °C per 30 min) during an aerobic swim trial, to determine what temperature the fish fatigue (CT_{swim}). Though not significantly different from CT_{max}, CT_{swim} was 2 °C lower—increasing the environmental relevancy of this method compared to traditional CT_{max} (Blasco et al., 2020). Few Amazonian fishes have been tested for CT_{swim}, but this method is particularly applicable for active Amazonian species where static CT_{max} methods may overestimate their upper thermal limits.

With the GAM, we determined that the warming rate is least predictive of CT_{max} in Amazonian fishes and showed an unusual nonlinear relationship where the fastest and slowest ramp rates predict similar CT_{max} values (Fig. 1d). However, only 5 warming rates have been used to study Amazonian fishes thus far; whether this is an accurate relationship remains to be tested. Similarly, acclimation duration, which is one of the best predictors of CT_{max} according to our model, demonstrates an unusual bimodal relationship with CT_{max} (Fig. 1c). Methodology studies may be incredibly useful to better understand how ramping rate and acclimation duration impact upper thermal limits for Amazonian fishes, creating predictive relationships between CT_{max}, ramping rates, and acclimation duration—which may be species- or taxa-specific—as a guide for researchers.

Though useful in inter- and intraspecific comparisons, CT_{max} alone does not truly capture a species' vulnerability, nor does ARR (calculated *using* CT_{max}) capture a species' acclimation capacity when considering the complexity of organismal responses to environmental challenges. Here we have highlighted the importance of including CL_{max} to understand species' vulnerability to warming. In their study, Jung et al. (2020) found a significant relationship between CL_{max} and CT_{max}, a trend further supported by the GLM presented here. This presents an exciting avenue for future work validating and refining this relationship with characterizations of additional Amazonian fishes and using this relationship to predict species' vulnerability when needed.

While we did not include thermal death time (TDT) in our models, the reintroduction of this model into physiological studies necessitates that we also consider how exposure duration and intensity impact an organisms' ability to tolerate heat (reviewed by Ørsted et al., 2022). A recent study demonstrated a clear relationship between thermal tolerance and the accumulation of thermal injury that eventually results in death (Jørgensen et al., 2021), which would be a useful model to incorporate in the future. By including the complex physiological considerations of heat injury and repair from heat stress, we can improve our predictions of habitat suitability and changes in biodiversity.

Furthermore, there is a need to explore the behavioral thermal limitations of Amazonian fishes. Few studies have measured agitation temperatures in these species, but the importance of this metric is highlighted in the study by Kochhann et al. (2021), comparing the upper thermal limits of two Amazonian species *Mesonauta insignus* and *Apistogramma agassizii*. The species' CT_{max} were not significantly different from each other (~41 °C). However, there was a 2 °C difference in agitation temperature (~35 °C and 37 °C, respectively), where *M. insignus* was significantly more sensitive to warming. While this metric requires more time and resources, it brings valuable insights into species' thermal sensitivity and how behavioral changes may be used to mitigate thermal stress.

5.2.2 Understanding the Impact of Multiple, co-Occurring Stressors

A limitation of our current approaches in thermal physiology is that with any laboratory-based experimental design, we are inevitably reducing the complexity that the experimental organism would encounter in the natural habitat. Global climate change will invariably impact Amazonian fishes with multiple stressors in combination with temperature, including decreased pH, increased ion levels, increased dissolved organic carbon, increased total suspended particles, higher carbon dioxide levels, and decreased oxygen levels (reviewed by Val & Wood, 2022). In addition, increased temperature is shown to compound the effects of contaminants in Amazonian fishes, e.g., microplastics accumulate faster in *Symphysodon aequifasciatus* (Wen et al., 2018) and copper toxicity increases in *Hoplosternum littorale* (Braz-Mota et al., 2017). More studies are needed to understand how increased temperature works in tandem with other climate change and anthropogenic stressors that will covary in the natural environment.

For example, the combined stressors of temperature and dissolved oxygen are one of many threats facing fishes in the Brazilian Amazon (Frederico et al., 2016). While a vast majority of Amazon Forest streams have high oxygen levels (Espírito-Santo et al., 2009), many regions experience seasonal cycles of flooding and drought. During drought seasons, fish assemblages are dominated by species with high thermal and hypoxia tolerances (Röpke et al., 2016). Increased temperature is predicted to reduce oxygen saturation levels of aquatic ecosystems, and more prominently impact ephemeral, shallow habitats such as the shallow palm swamps and floodplain lakes during the dry season. In addition, diurnal instances of oxygen supersaturation (hyperoxia) may increase in the day due to increased eutrophication, though less research has been devoted to this issue (Almeida-Val & Hochachka, 1995; Jung et al., 2020; Val & Wood, 2022). With these changes, we may see a strong selection specifically for the tolerant species that inhabit those harsh environments. There is evidence to suggest there is regional plasticity within species, wherein different physiological mechanisms may be employed to cope with thermal stress (Fé-Gonçalves et al., 2020). However, hydroelectric dams are causing decreased oxygen levels, as well as habitat fragmentation and loss of gene flow (Val et al., 2016). This loss of gene flow may further decrease the chances of rescue, limiting the in-migration of warm-adapted conspecifics (Somero, 2010), and little is known about the behavioral changes (e.g., migration) these fishes show in response to multiple, co-occurring stressors.

Jung et al. (2020) demonstrated that hypoxia tolerance and thermal tolerance are tightly correlated in Amazonian fishes, wherein species sensitive to warm temperatures were also sensitive to low oxygen. This is supported by work from Campos et al. (2017) showing two sympatric species of brilliant tetras are adapted to their thermal microhabitats, but the species with higher upper thermal limits has a greater hypoxia tolerance. However, Jung et al. (2020) demonstrated if fishes are acclimated to a higher temperature, this relationship between thermal tolerance and hypoxia tolerance ceases. Thermal tolerance increases with acclimation temperature, but fishes simultaneously become *more* sensitive to hypoxia with increased acclimation temperature. This complex, and worrying, relationship between thermal tolerance, hypoxia tolerance, and thermal acclimation warrants further investigation.

Furthermore, understanding how changing terrestrial and hydrological conditions will further impact the thermal stability of aquatic habitats is critically important to predicting how fish populations, distributions, and fisheries practices are affected (Feng et al., 2020). Fundamental changes in the Amazon's weather and landscape have resulted in increased forest fires (Alencar et al., 2004; Cochrane & Laurance, 2008), many of which are human-ignited (Laurance, 1998). In turn, habitat fragmentation resulting from deforestation further increases forest flammability (Alencar et al., 2015). These land-use changes are predicted to increase river discharge in deforested areas, increasing run-off, and prompting significant changes in water chemistry (Lima et al., 2014). Deforestation increases stream temperatures (up to 6 °C warmer) and while fish populations demonstrate local thermal adaptation, increased temperatures still negatively impact growth (Ilha et al., 2018). In addition, the regional hydrological cycles of flooding and drought seasons are becoming more variable and more intense (Gloor et al., 2015); for example, increased precipitation in the northwest and decreased precipitation in the south and southeast. Thus, the global effects of warming are compounded by the millions of hectares deforested every year in the Amazon River Basin. Economic activities and rapid expansion in infrastructure may therefore have more of an impact on species richness than global warming alone (Oberdorff et al., 2015).

Incorporating physiological data is critical to creating thermal habitat suitability models and predicting species' geographical distributions and abundance in response to global climate change (Wagner et al., 2023). Importantly, there is a need for studies from the Global South. In a recent meta-analysis synthesizing how fish distribution changes in response to global warming, Dahms and Killen (2023) identified a strong publication bias, wherein few studies originated from the most biodiverse regions of the world (South America, Southeast Asia, and Africa). Through increasing characterizations of Amazonian fishes' thermal limits and acclimation capacities, diversifying the methods used to understand their thermal responses, increasing the complexity of laboratory-based experiments to better reflect environmental conditions, and supporting and promoting the work conducted in this region, we can better predict the future of Amazonian fishes in a warming world.

References

Addo-Bediako, A., Chown, S. L., & Gaston, K. J. (2000). Thermal tolerance, climatic variability and latitude. *Proceedings of the Royal Society of London. Series B: Biological Sciences, 267*, 739–745.

Alencar, A. A., Brando, P. M., Asner, G. P., & Putz, F. E. (2015). Landscape fragmentation, severe drought, and the new Amazon forest fire regime. *Ecological Applications, 25*, 1493–1505.

Alencar, A. A., Solórzano, L. A., & Nepstad, D. C. (2004). Modeling forest understory fires in an eastern Amazonian landscape. *Ecological Applications, 14*, 139–149.

Almeida-Val, V. M. F., & Hochachka, P. W. (1995). Air-breathing fishes: Metabolic biochemistry of the first diving vertebrates. In *Biochemistry and molecular biology of fishes* (Vol. 5, pp. 45–55). Elsevier.

Andreassen, A. H., Hall, P., Khatibzadeh, P., Jutfelt, F., & Kermen, F. (2022). Brain dysfunction during warming is linked to oxygen limitation in larval zebrafish. *Proceedings of the National Academy of Sciences, 119*, e2207052119.

Araújo, M. B., Ferri-Yáñez, F., Bozinovic, F., Marquet, P. A., Valladares, F., & Chown, S. L. (2013). Heat freezes niche evolution. *Ecology Letters, 16*, 1206–1219.

Åsheim, E. R., Andreassen, A. H., Morgan, R., & Jutfelt, F. (2020). Rapid-warming tolerance correlates with tolerance to slow warming but not growth at non-optimal temperatures in zebrafish. *Journal of Experimental Biology, 223*, jeb229195.

Beitinger, T. L., & Bennett, W. A. (2000). Quantification of the role of acclimation temperature in temperature tolerance of fishes. *Environmental Biology of Fishes, 58*, 277–288.

Bennett, J. M., Sunday, J., Calosi, P., Villalobos, F., Martínez, B., Molina-Venegas, R., Araújo, M. B., Algar, A. C., Clusella-Trullas, S., & Hawkins, B. A. (2021). The evolution of critical thermal limits of life on earth. *Nature Communications, 12*, 1198.

Blasco, F. R., Esbaugh, A. J., Killen, S. S., Rantin, F. T., Taylor, E. W., & McKenzie, D. J. (2020). Using aerobic exercise to evaluate sub-lethal tolerance of acute warming in fishes. *Journal of Experimental Biology, 223*, jeb218602.

Bohórquez, B. (2016). Estimación del máximo crítico térmico para la palometa gasteropelecus maculatus (Characiformes: Gasteropelecidae). Una aproximación a su respuesta frente al cambio climático. Ph.D. Thesis, Universidad Distrital Francisco José de Caldas.

Braz-Mota, S., Fé, L. M. L., Delunardo, F. A. C., Sadauskas-Henrique, H., de Almeida-Val, V. M. F., & Val, A. L. (2017). Exposure to waterborne copper and high temperature induces the formation of reactive oxygen species and causes mortality in the Amazonian fish Hoplosternum littorale. *Hydrobiologia, 789*, 157–166.

Campos, D. F., Amanajás, R. D., Almeida-Val, V. M., & Val, A. L. (2021). Climate vulnerability of South American freshwater fish: Thermal tolerance and acclimation. *Journal of Experimental Zoology Part A: Ecological and Integrative Physiology, 335*, 723–734.

Campos, D., Braz-Mota, S., Val, A., & Almeida-Val, V. (2019). Predicting thermal sensitivity of three Amazon fishes exposed to climate change scenarios. *Ecological Indicators, 101*, 533–540.

Campos, D. F. D., Jesus, T. F., Kochhann, D., Heinrichs-Caldas, W., Coelho, M. M., & Almeida-Val, V. M. F. (2017). Metabolic rate and thermal tolerance in two congeneric Amazon fishes: Paracheirodon axelrodi Schultz, 1956 and Paracheirodon simulans Géry, 1963 (Characidae). *Hydrobiologia, 789*, 133–142.

Campos, D. F. D., Val, A. L., & Almeida-Val, V. M. F. (2018). The influence of lifestyle and swimming behavior on metabolic rate and thermal tolerance of twelve Amazon forest stream fish species. *Journal of Thermal Biology, 72*, 148–154.

Carvalho, S., Oliveira, A., Pedersen, J. S., Manhice, H., Lisboa, F., Norguet, J., de Wit, F., & Santos, F. D. (2020). A changing Amazon rainforest: Historical trends and future projections under post-Paris climate scenarios. *Global and Planetary Change, 195*, 103328.
Chapman, L. J., & Mckenzie, D. J. (2009). Behavioral responses and ecological consequences. In *Fish physiology* (Vol. 27, pp. 25–77). Elsevier.
Chung, K. S. (2001). Critical thermal maxima and acclimation rate of the tropical guppy Poecilla reticulata. *Hydrobiologia, 462*(1), 253–257. https://doi.org/10.1023/A:1013158904036
Chung, K. S., & Méndez, S. (1993). Tolerancia Térmica Comparativa en Algunos Peces Tropicales de Venezuela. *Ciencia, 1*(1), 1–7.
Clark, T. D., Messmer, V., Tobin, A. J., Hoey, A. S., & Pratchett, M. S. (2017). Rising temperatures may drive fishing-induced selection of low-performance phenotypes. *Scientific Reports, 7*, 40571.
Cochrane, M. A., & Laurance, W. F. (2008). Synergisms among fire, land use, and climate change in the Amazon. *Ambio: A Journal of the Human Environment, 37*, 522–527.
Conte, M., de Campos, D. F., & Eme, J. (2023). Effective practices for thermal tolerance polygon experiments using mottled catfish Corydoras paleatus. *Journal of Thermal Biology, 115*, 103616.
Cooper, C. J., Mueller, C. A., & Eme, J. (2019). Temperature tolerance and oxygen consumption of two South American tetras, Paracheirodon innessi and Hyphessobrycon herbertaxelrodi. *Journal of Thermal Biology, 86*, 102434. https://doi.org/10.1016/j.jtherbio.2019.102434
Cowles, R. B., & Bogert, C. M. (1944). A preliminary study of the thermal requirements of desert reptiles. *Bulletin of the AMNH, 83*, 5.
Dahms, C., & Killen, S. S. (2023). Temperature change effects on marine fish range shifts: A meta-analysis of ecological and methodological predictors. *Global Change Biology, 29*, 4459.
Dawkins, R., & Krebs, J. R. (1979). Arms races between and within species. *Proceedings of the Royal Society of London. Series B. Biological Sciences, 205*, 489–511.
Desforges, J. E., Birnie-Gauvin, K., Jutfelt, F., Gilmour, K. M., Eliason, E. J., Dressler, T. L., McKenzie, D. J., Bates, A. E., Lawrence, M. J., & Fangue, N. (2023). The ecological relevance of critical thermal maxima methodology (CTM) for fishes. *Journal of Fish Biology, 102*, 1000.
Deutsch, C. A., Tewksbury, J. J., Huey, R. B., Sheldon, K. S., Ghalambor, C. K., Haak, D. C., & Martin, P. R. (2008). Impacts of climate warming on terrestrial ectotherms across latitude. *Proceedings of the National Academy of Sciences, 105*, 6668–6672.
Di Santo, V., Jordan, H. L., Cooper, B., Currie, R. J., Beitinger, T. L., & Bennett, W. A. (2018). Thermal tolerance of the invasive red-bellied pacu and the risk of establishment in the United States. *Journal of Thermal Biology, 74*, 110–115. https://doi.org/10.1016/j.jtherbio.2018.03.015
Divya, P., & Ranjeet, K. (2014). Effect of temperature on oxygen consumption and ammonia excretion in red bellied piranha Pygocentrus nattereri kner, 1858. *Journal of Aquatic Biology and Fisheries, 2*, 97–104.
Einum, S., Ratikainen, I., Wright, J., Pélabon, C., Bech, C., Jutfelt, F., Stawski, C., & Burton, T. (2019). How to quantify thermal acclimation capacity? *Global Change Biology, 25*, 1893–1894.
Eliason, E. J., Clark, T. D., Hague, M. J., Hanson, L. M., Gallagher, Z. S., Jeffries, K. M., Gale, M. K., Patterson, D. A., Hinch, S. G., & Farrell, A. P. (2011). Differences in thermal tolerance among sockeye salmon populations. *Science, 332*, 109–112.
Ern, R., Andreassen, A. H., & Jutfelt, F. (2023). Physiological mechanisms of acute upper thermal tolerance in fish. *Physiology, 38*, 141–158.
Espírito-Santo, H. M. V., Magnusson, W. E., Zuanon, J., Mendonca, F. P., & Landeiro, V. L. (2009). Seasonal variation in the composition of fish assemblages in small Amazonian forest streams: Evidence for predictable changes. *Freshwater Biology, 54*, 536–548.
Fé-Gonçalves, L. M., Araújo, J. D. A., Santos, C. H. D. A. D., & Almeida-Val, V. M. F. d. (2020). Transcriptomic evidences of local thermal adaptation for the native fish Colossoma macropomum (Cuvier, 1818). *Genetics and Molecular Biology, 43*, e20190377.
Feng, D., Raoufi, R., Beighley, E., Melack, J. M., Goulding, M., Barthem, R. B., Venticinque, E., Cañas, C., Forsberg, B., & Sorribas, M. V. (2020). Future climate impacts on the hydrology

of headwater streams in the Amazon River basin: Implications for migratory goliath catfishes. *Hydrological Processes, 34*, 5402–5416.
Fernandes, M. N., Barrionuevo, W. R., & Rantin, F. T. (1995). Effects of thermal stress on respiratory responses to hypoxia of a South American Prochilodontid fish, Prochilodus scrofa. *Journal of Fish Biology, 46*(1), 123–133. https://doi.org/10.1111/j.1095-8649.1995.tb05951.x
Floeter, S. R., Behrens, M., Ferreira, C., Paddack, M., & Horn, M. (2005). Geographical gradients of marine herbivorous fishes: Patterns and processes. *Marine Biology, 147*, 1435–1447.
Frederico, R. G., Olden, J. D., & Zuanon, J. (2016). Climate change sensitivity of threatened, and largely unprotected, Amazonian fishes. *Aquatic Conservation: Marine and Freshwater Ecosystems, 26*, 91–102.
Fry, F. J. (1967). Responses of vertebrate poikilotherms to temperature. In *Thermobiology* (pp. 375–409).
Fry, F., Brett, J., & Clawson, G. (1942). Lethal limits of temperature for young goldfish. *Revue Canadienne de Biologie, 1*, 50–56.
Genin, A., Levy, L., Sharon, G., Raitsos, D. E., & Diamant, A. (2020). Rapid onsets of warming events trigger mass mortality of coral reef fish. *Proceedings of the National Academy of Sciences, 117*, 25378–25385.
Ghalambor, C. K., Huey, R. B., Martin, P. R., Tewksbury, J. J., & Wang, G. (2006). Are mountain passes higher in the tropics? Janzen's hypothesis revisited. *Integrative and Comparative Biology, 46*, 5–17.
Gloor, M., Barichivich, J., Ziv, G., Brienen, R., Schöngart, J., Peylin, P., Ladvocat Cintra, B. B., Feldpausch, T., Phillips, O., & Baker, J. (2015). Recent Amazon climate as background for possible ongoing and future changes of Amazon humid forests. *Global Biogeochemical Cycles, 29*, 1384–1399.
Gunderson, A. R., & Stillman, J. H. (2015). Plasticity in thermal tolerance has limited potential to buffer ectotherms from global warming. *Proceedings of the Royal Society B: Biological Sciences, 282*, 20150401.
Gutierre, S. M. M., Schofield, P. J., & Prodocimo, V. (2016). Salinity and temperature tolerance of an emergent alien species, the Amazon fish Astronotus ocellatus. *Hydrobiologia, 777*(1), 21–31. https://doi.org/10.1007/s10750-016-2740-8
Ilha, P., Schiesari, L., Yanagawa, F. I., Jankowski, K., & Navas, C. A. (2018). Deforestation and stream warming affect body size of Amazonian fishes. *PLoS One, 13*, e0196560.
IPCC. (2023). Climate change 2023: Synthesis report. A report of the intergovernmental panel on climate change. In Core Writing Team, H. Lee, & J. Romero (Eds.), *Contribution of working groups I, II and III to the sixth assessment report of the intergovernmental panel on climate change*. IPCC.
Janzen, D. H. (1967). Why mountain passes are higher in the tropics. *The American Naturalist, 101*, 233–249.
Jézéquel, C., Tedesco, P. A., Darwall, W., Dias, M. S., Frederico, R. G., Hidalgo, M., Hugueny, B., Maldonado-Ocampo, J., Martens, K., & Ortega, H. (2020). Freshwater fish diversity hotspots for conservation priorities in the Amazon Basin. *Conservation Biology, 34*, 956–965.
Jørgensen, L. B., Malte, H., Ørsted, M., Klahn, N. A., & Overgaard, J. (2021). A unifying model to estimate thermal tolerance limits in ectotherms across static, dynamic and fluctuating exposures to thermal stress. *Scientific Reports, 11*, 1–14.
Jørgensen, L. B., Ørsted, M., Malte, H., Wang, T., & Overgaard, J. (2022). Extreme escalation of heat failure rates in ectotherms with global warming. *Nature*, 1–6.
Jung, E. H., Brix, K. V., Richards, J. G., Val, A. L., & Brauner, C. J. (2020). Reduced hypoxia tolerance and survival at elevated temperatures may limit the ability of Amazonian fishes to survive in a warming world. *Science of the Total Environment, 748*, 141349.
Jutfelt, F., Norin, T., Ern, R., Overgaard, J., Wang, T., McKenzie, D. J., Lefevre, S., Nilsson, G. E., Metcalfe, N. B., & Hickey, A. J. (2018). Oxygen-and capacity-limited thermal tolerance: Blurring ecology and physiology. *Journal of Experimental Biology, 221*, jeb169615.

Jutfelt, F., Roche, D. G., Clark, T. D., Norin, T., Binning, S. A., Speers-Roesch, B., Amcoff, M., Morgan, R., Andreassen, A. H., & Sundin, J. (2019). Brain cooling marginally increases acute upper thermal tolerance in Atlantic cod. *Journal of Experimental Biology, 222*, jeb208249.

Knight, N. S., Guichard, F., & Altieri, A. H. (2021). A global meta-analysis of temperature effects on marine fishes' digestion across trophic groups. *Global Ecology and Biogeography, 30*, 795–810.

Kochhann, D., Sarmento, C. G., de Oliveira, J. C., Queiroz, H. L., Val, A. L., & Chapman, L. J. (2021). Take time to look at the fish: Behavioral response to acute thermal challenge in two Amazonian cichlids. *Journal of Experimental Zoology Part A: Ecological and Integrative Physiology, 335*, 735–744.

Lapointe, D., Cooperman, M. S., Chapman, L. J., Clark, T. D., Val, A. L., Ferreira, M. S., Balirwa, J. S., Mbabazi, D., Mwanja, M., & Chhom, L. (2018). Predicted impacts of climate warming on aerobic performance and upper thermal tolerance of six tropical freshwater fishes spanning three continents. *Conservation Physiology, 6*, coy056.

Laurance, W. F. (1998). A crisis in the making: Responses of Amazonian forests to land use and climate change. *Trends in Ecology & Evolution, 13*, 411–415.

Lima, L. S., Coe, M. T., Soares Filho, B. S., Cuadra, S. V., Dias, L. C., Costa, M. H., Lima, L. S., & Rodrigues, H. O. (2014). Feedbacks between deforestation, climate, and hydrology in the Southwestern Amazon: Implications for the provision of ecosystem services. *Landscape Ecology, 29*, 261–274.

Litchman, E., & Thomas, M. K. (2023). Are we underestimating the ecological and evolutionary effects of warming? Interactions with other environmental drivers may increase species vulnerability to high temperatures. *Oikos, 2023*, e09155.

McDonnell, L. H., & Chapman, L. J. (2015). At the edge of the thermal window: Effects of elevated temperature on the resting metabolism, hypoxia tolerance and upper critical thermal limit of a widespread African cichlid. *Conservation Physiology, 3*, cov050.

McKenzie, D. J., Zhang, Y., Eliason, E. J., Schulte, P. M., Claireaux, G., Blasco, F. R., Nati, J. J., & Farrell, A. P. (2021). Intraspecific variation in tolerance of warming in fishes. *Journal of Fish Biology, 98*, 1536–1555.

Morgan, R., Finnøen, M. H., Jensen, H., Pélabon, C., & Jutfelt, F. (2020). Low potential for evolutionary rescue from climate change in a tropical fish. *Proceedings of the National Academy of Sciences, 117*, 33365–33372.

Norin, T., Malte, H., & Clark, T. D. (2014). Aerobic scope does not predict the performance of a tropical eurythermal fish at elevated temperatures. *Journal of Experimental Biology, 217*, 244–251.

Oberdorff, T., Jézéquel, C., Campero, M., Carvajal-Vallejos, F., Cornu, J., Dias, M., Duponchelle, F., Maldonado-Ocampo, J., Ortega, H., & Renno, J. (2015). Opinion paper: How vulnerable are Amazonian freshwater fishes to ongoing climate change? *Journal of Applied Ichthyology, 31*, 4–9.

Ørsted, M., Jørgensen, L. B., & Overgaard, J. (2022). Finding the right thermal limit: A framework to reconcile ecological, physiological and methodological aspects of CTmax in ectotherms. *Journal of Experimental Biology, 225*, jeb244514.

Payne, N. L., & Smith, J. A. (2017). An alternative explanation for global trends in thermal tolerance. *Ecology Letters, 20*, 70–77.

Payne, N. L., Smith, J. A., van der Meulen, D. E., Taylor, M. D., Watanabe, Y. Y., Takahashi, A., Marzullo, T. A., Gray, C. A., Cadiou, G., & Suthers, I. M. (2016). Temperature dependence of fish performance in the wild: Links with species biogeography and physiological thermal tolerance. *Functional Ecology, 30*, 903–912.

Penman, R. J., Bugg, W., Rost-Komiya, B., Earhart, M. L., & Brauner, C. J. (2023). Slow heating rates increase thermal tolerance and alter mRNA HSP expression in juvenile white sturgeon (Acipenser transmontanus). *Journal of Thermal Biology, 115*, 103599.

Pérez, E., & Díaz, F., & Espina, S. (2003). Thermoregulatory behavior and critical thermal limits of the angelfish Pterophyllum scalare (Lichtenstein) (Pisces: Cichlidae). *Journal of Thermal Biology, 28*(8), 531–537. https://doi.org/10.1016/S0306-4565(03)00055-X

Pörtner, H. (2001). Climate change and temperature-dependent biogeography: Oxygen limitation of thermal tolerance in animals. *Naturwissenschaften, 88*, 137–146.

Pottier, P., Burke, S., Drobniak, S. M., Lagisz, M., & Nakagawa, S. (2021). Sexual (in) equality? A meta-analysis of sex differences in thermal acclimation capacity across ectotherms. *Functional Ecology, 35*, 2663–2678.

Pottier, P., Burke, S., Zhang, R. Y., Noble, D. W., Schwanz, L. E., Drobniak, S. M., & Nakagawa, S. (2022). Developmental plasticity in thermal tolerance: Ontogenetic variation, persistence, and future directions. *Ecology Letters, 25*, 2245–2268.

Rantin, F. T., & Petersen, J. A. (1985). Thermal tolerance of South American cichlid. *Geophagus brasiliensis. Revue d'hydrobiologie tropicale, 18*(3), 221–226.

Ripley, D. M., Quinn, F. A., Dickson, J., Arthur, J., & Shiels, H. A. (2022). Thermal preference does not align with optimal temperature for aerobic scope in zebrafish (Danio rerio). *Journal of Experimental Biology, 225*, jeb243774.

Röpke, C. P., Amadio, S. A., Winemiller, K. O., & Zuanon, J. (2016). Seasonal dynamics of the fish assemblage in a floodplain lake at the confluence of the Negro and Amazon Rivers. *Journal of Fish Biology, 89*, 194–212.

Sandblom, E., Clark, T. D., Gräns, A., Ekström, A., Brijs, J., Sundström, L. F., Odelström, A., Adill, A., Aho, T., & Jutfelt, F. (2016). Physiological constraints to climate warming in fish follow principles of plastic floors and concrete ceilings. *Nature Communications, 7*, 11447.

Schulte, P. M., Healy, T. M., & Fangue, N. A. (2011). Thermal performance curves, phenotypic plasticity, and the time scales of temperature exposure. *Integrative and Comparative Biology, 51*, 691–702.

Somero, G. (2010). The physiology of climate change: How potentials for acclimatization and genetic adaptation will determine 'winners' and 'losers'. *Journal of Experimental Biology, 213*, 912–920.

Tewksbury, J. J., Huey, R. B., & Deutsch, C. A. (2008). Putting the heat on tropical animals. *Science, 320*, 1296–1297.

Val, A. L., Fearnside, P. M., & Almeida-Val, V. (2016). Environmental disturbances and fishes in the Amazon. *Journal of Fish Biology, 89*, 192–193.

Val, A. L., & Wood, C. M. (2022). Global change and physiological challenges for fish of the Amazon today and in the near future. *Journal of Experimental Biology, 225*, jeb216440.

van Vliet, M. T., Franssen, W. H., Yearsley, J. R., Ludwig, F., Haddeland, I., Lettenmaier, D. P., & Kabat, P. (2013). Global river discharge and water temperature under climate change. *Global Environmental Change, 23*, 450–464.

Vergés, A., Doropoulos, C., Malcolm, H. A., Skye, M., Garcia-Pizá, M., Marzinelli, E. M., Campbell, A. H., Ballesteros, E., Hoey, A. S., & Vila-Concejo, A. (2016). Long-term empirical evidence of ocean warming leading to tropicalization of fish communities, increased herbivory, and loss of kelp. *Proceedings of the National Academy of Sciences, 113*, 13791–13796.

Vergés, A., Steinberg, P. D., Hay, M. E., Poore, A. G., Campbell, A. H., Ballesteros, E., Heck, K. L., Jr., Booth, D. J., Coleman, M. A., & Feary, D. A. (2014). The tropicalization of temperate marine ecosystems: Climate-mediated changes in herbivory and community phase shifts. *Proceedings of the Royal Society B: Biological Sciences, 281*, 20140846.

Volkoff, H., & Rønnestad, I. (2020). Effects of temperature on feeding and digestive processes in fish. *Temperature, 7*, 307–320.

Vornanen, M. (2016). The temperature dependence of electrical excitability in fish hearts. *Journal of Experimental Biology, 219*, 1941–1952.

Vornanen, M. (2020). Feeling the heat: Source–sink mismatch as a mechanism underlying the failure of thermal tolerance. *Journal of Experimental Biology, 223*, jeb225680.

Wagner, T., Schliep, E. M., North, J. S., Kundel, H., Custer, C. A., Ruzich, J. K., & Hansen, G. J. (2023). Predicting climate change impacts on poikilotherms using physiologically guided species abundance models. *Proceedings of the National Academy of Sciences, 120*, e2214199120.

Wen, B., Zhang, N., Jin, S.-R., Chen, Z.-Z., Gao, J.-Z., Liu, Y., Liu, H.-P., & Xu, Z. (2018). Microplastics have a more profound impact than elevated temperatures on the predatory performance, digestion and energy metabolism of an Amazonian cichlid. *Aquatic Toxicology, 195*, 67–76.

Wheeler, C. R., Lang, B. J., Mandelman, J. W., & Rummer, J. L. (2022). The upper thermal limit of epaulette sharks (Hemiscyllium ocellatum) is conserved across three life history stages, sex and body size. *Conservation Physiology, 10*, coac074.

Woodward, G., Perkins, D. M., & Brown, L. E. (2010). Climate change and freshwater ecosystems: Impacts across multiple levels of organization. *Philosophical Transactions of the Royal Society B: Biological Sciences, 365*, 2093–2106.

Yanar, M., Erdoğan, E., & Kumlu, M. (2019). Thermal tolerance of thirteen popular ornamental fish Species. *Aquaculture, 501*, 382–386. https://doi.org/10.1016/j.aquaculture.2018.11.041

Zhang, P., van Leeuwen, C. H., Bogers, D., Poelman, M., Xu, J., & Bakker, E. S. (2020). Ectothermic omnivores increase herbivory in response to rising temperature. *Oikos, 129*, 1028–1039.

Part V

Aquaculture as a Tool to Ensure Food Security in the Amazon

The Future of the Value Chain of the Pirarucu in the Brazilian Amazon

Maria Sylvia Macchione Saes, Elis Regina Monte Feitosa, Marcelo Nema Areco Paiva, and Jacques Marcovicth

Abstract

The Amazon biome, the world's largest tropical forest, faces challenges in biodiversity preservation, climate change mitigation, food security, and reducing inequalities. With the global population projected to reach nine billion by 2037, strategies reconciling forest preservation, climate change, and food security are crucial. The natural wealth of the Brazilian Amazon and indigenous knowledge present an opportunity to improve the quality of life in the region. However, despite the sparse population density, most municipalities in the Amazon have low human development levels. The *Arapaima gigas*, known as pirarucu or paiche, is vital for local communities, biodiversity, and has market potential as a protein source. Past policies aimed at commercial fishing led to overfishing, prompting efforts to protect the pirarucu population. This chapter maps two strategies for pirarucu exploitation: chain of management and fish culture. It explores challenges considering economic, social, and environmental dimensions aligned with the sustainable development goals (SDGs). Stakeholder collaboration among producers, fishermen, communities, and public/private sectors is crucial for structuring these chains, fostering social, environmental, and economic development. This collaboration drives a sustainable bioeconomy in the Amazon, addressing the challenges of the twenty-first century.

M. S. M. Saes (✉) · E. R. M. Feitosa · M. N. A. Paiva · J. Marcovicth
School of Economics, Management, Accounting and Actuarial Sciences, University of São Paulo (USP), São Paulo, Brazil
e-mail: ssaes@usp.br

S. S. de Souza et al. (eds.), *The Future of Amazonian Aquatic Biota*,
https://doi.org/10.1007/978-3-031-66822-7_18

1 Introduction

The twenty-first century brought a series of challenges that require attention and efficient solutions, that consider the preservation of biodiversity, confronting climate change, guaranteeing food security, and the reduction of social and geographical inequalities (SDGs, 2015). At the center of these issues is the Amazon biome, a natural heritage that represents the world's largest tropical forest and a broad diversity of species, playing a fundamental role in the storage of carbon dioxide. Because of this, it is crucial to find strategies that reconcile the dilemma of preserving the forest, mitigate the effects of climate change, and combat food insecurity, especially given that the world population is estimated to reach nine billion people by 2037 (UN, 2022).

The natural wealth of the Brazilian Amazon and knowledge of traditional people, indigenous, and riverside communities provide a great opportunity to guarantee quality of life, income, and well-being for its inhabitants (Guinato et al., 2022; Neves, 2022; Saes et al., 2023; Salomão et al., 2006). Despite all this potential, it is worrying to note that the majority of municipalities located in the biome have a low or medium municipal human development index (MHDI)[1] according to AtlasBR (2023)|; even with the sparse population density in the region, approximately 5.6 inhabitants per km^2, around 4.4 times smaller than the national mean (Amazonas, 2021). In this context, it is essential that actions aimed at the Amazon region are planned and implemented without neglecting the creation of incentives that promote productive sectors that generate well-being for the populations that inhabit it.

One of the central activities of the riverside communities is the exploitation of fish, among them the *Arapaima gigas*, otherwise known as the pirarucu in Brazil and paiche in Peru. This species is one of the largest freshwater fish in the world, reaching up to three meters in length and 300 kg in weight (Castello & Stewart, 2010; Imbiriba, 2001). As well as playing a fundamental role in the subsistence of local communities and the preservation of biodiversity, the fish has significant market potential and could become a significant alternative source of protein for the general public, having great potential for human consumption as its average meat yield is 57% of the animal.[2]

Since the middle of the last century, public policies have been implemented to create conditions for the commercialization of the species. Initially, the focus of these policies was to incentivize commercial fishing, with the main aim of generating development and income for the local population. However, the significant investments in the increase and modernization of the fishing fleet resulted in practices of overfishing.

---

[1] The Human Development Index (HDI) is calculated according to the mean of the criteria evaluated by the indicator: income per capita, number of years of school education received, and life expectancy.

[2] See for example, the tax incentives created by Law n. 221/1967, of the Brazilian Federal Government (see Lobato, 2009).

In 1976, with the aim of preserving and recovering the pirarucu population, the Superintendent of Fishing Development (Sudepe), which today is incorporated into the Brazilian Institute for Development of the Environment and Renewable Natural Resources (Ibama), published ordinance 15, declaring the species as facing extinction and prohibiting catching it for 6 months of the year (October to March) (CONAB, 2020).

However, in the face of increasing demand for fish, as well as the need to guarantee sources of income for riverside populations, the need to implement effective actions to promote commercial activity became evident. The economic exploitation of the species followed two distinct approaches. On the one hand, in 1998, Law no. 7679 was established, which set out norms for the authorization of commercial fishing in areas managed and authorized by Ibama and Institute for the Environmental Protection of Amazonas (IPAAM). On the other, Decree no. 34,100 of 23/10/2013 created a regulatory environment for pirarucu in captivity, establishing norms, and procedures for the regularization of fishery activity for the species in the state of Amazonas, through which it is possible fish year-round.[3]

The aim of this chapter is to map the two distinct strategies of pirarucu exploitation in Amazonia: the chain of management and of fish culture and their potential for the construction of sustainable bioeconomy in the region (see Saes et al., 2023). Our aim is to investigate the main challenges faced by these production chains, considering the economic, social, and environmental dimensions, aligned with targets set out in the sustainable development goals (SDGs). Finally, we intend to identify the propelling and inhibiting factors to sustainable development of this chain in the future.

2 Institutional Context of the Pirarucu Value Chain: Fish Culture and Management

Fishing activity is a traditional practice of the indigenous and riverside peoples of the Amazon region, benefitting from favorable natural conditions. Throughout precolonial times, fishing played a crucial role in the life of these communities, establishing it as a fundamental activity. In the state of Amazonas, for example, there are historical records that show its importance to indigenous peoples long before the arrival of the Europeans (Neves, 2022; Prestes-Carneiro et al., 2016; Santos, 2005; Santos & Santos, 2005).

The abundance of resources for fishing in the Amazon region has been a vital source of sustenance and subsistence for these communities throughout the centuries. Fishing plays an essential role not only in the provision of food, but also in the preservation of culture and traditional lifestyles of these populations (Veríssimo, 1895). Among the most valuable species, the pirarucu stands out for its economic, cultural, organizational, and social importance. Fishing for pirarucu is an activity handed down from generation to generation, beginning in childhood for local

[3] In the case of fish culture, fishing is called harvesting.

inhabitants, and can be done individually or collectively. Initially, the spear was used, but over time, nets became widely adopted because they allowed the fish to be caught without compromising the quality of the meat and skin. It is important to remember that the use of fishing equipment is regulated by local agreements, especially in managed lakes (da Hora, 2020; Silva, 2018).

The commercial exploitation of pirarucu began in the eighteenth century and became more widespread in the 1950s with the introduction of diesel engines. In the following decades, following the model of integration and development, the Brazilian state invested in the modernization of the fishing fleet, installation of ice factories, storage, and refrigeration facilities in the Amazon region. This policy contributed to the increase in pressure on stocks and overexploitation of the species (Castro & Mcgrath, 2001), resulting in the inclusion of pirarucu on the Cites list (Convention on International Trade in Endangered Species of Wild Fauna and Flora) in 1975 (Gonçalves et al., 2018).

From this moment, the federal government assumed responsibility for the management of this activity, from the drafting to implementation of shared management measures, including the setting of rules, monitoring, and oversight of fishing, according to some sustainability criteria, with the aim of conserving the population of fish (da Hora, 2020; Gonzaga da Silva et al., 2020).

However, the continuous fall in the population of pirarucu in the Amazon Basin brought Ibama to adopt more restrictive measures. In 1989, a minimum size for catch was established (150 cm), and from ordinance 480, of 03/03/1991,[4] the period of protection for reproduction was instituted between the 1st of December and 31st of May. During this period, fishermen receive insurance and complement their income with agricultural work, extractivism, or with small businesses. Despite these restrictions, due to the vast extension of the Amazon Basin and the logistical and manpower limitations of the agencies in charge of oversight, these measures have not been completely effective (WWF, 2016). In 1991, through a series of regulations, the capture and sale of pirarucu in the Amazon Basin was completely prohibited, except in cases of experimental and scientific fishing, or for fish that come from fisheries (Amaral et al., 2011; Viana et al., 2004). Chart 1 (Appendix 1) presents the main legal frameworks that refer to the pirarucu in the state of Amazonas.

Amid this backdrop, in 1998, the Mamirauá Institute of Sustainable Development (IDSM),[5] established the Fishing Trade Program, also known as the Fishing Management Program. The main aim of this program is to promote the conservation of fishing resources in the region through participative management, depending on the active collaboration of local communities (IDSM, 2023). This initiative

---

[4]Ordinance n° 480, of 03/03/1991, today revoked and substituted by the normative instruction Ibama no34/2004, regulated the annual prohibition of the exercise of fishing of pirarucu in the following locations and periods: the states of Amazonas, Pará, Acre and Amapá, in the period from first of December to 31st of May; state of Rondônia, in the period from first of November to the 30th of April; the state of Roraima, between the first of May to the 31st of August.

[5]IDSM is a social organization linked to the Ministry of Science, Technology and Innovation (MCTI), which aims to develop its activities through research programs, management of natural resources and social development (IDSM, 2023).

culminated in the implementation of sustainably managed areas for pirarucu, and was the first area of participative management officially recognized by Ibama in 1999, and was implemented with the support of IDSM in the mid Solimões region. Fishing was authorized in the area on an experimental basis.[6]

According to Gonçalves et al. (2018), the management project implemented on that occasion and the recovery of stocks through the establishment of extraction quotas. These quotas were determined based on a count of existing fish in the lake, allowing the removal of, at most, 30% of the adults counted (over 1.5 m in length), while the remaining 70% were left to ensure the reproduction of the species.

Furthermore, the project prioritized the interruption of activities during the reproductive period, also known as the closed season. The objective was to promote the reproduction and growth of the population of pirarucus. In parallel, procedures, principles, and directives for sustainable management were adopted, aiming to increase the generation of income for a collective of fishermen.

In 2005, Ibama/Amazonas published an ordinance that prohibited fishing of the pirarucu, permitting the catch and trade in pirarucu only from areas of management/fishing agreements or farmed (cultivated in captivity) (IN n° 01, of 1 June 2005).

In 2013, Decree no. 34,100 was established to discipline and regulate the rearing of pirarucus in captivity, allowing fish for all periods of the year. These areas of rearing require licensing from the environmental agency for developing activities, as provided in Conama Resolution no. 413/2009. However, the licensing process for these areas is bureaucratic and time consuming, as Embrapa's study (2016) showed. Some factors, such as the need to solicit a term of concession of use for the area in cases of fish culture in public waters, as well as the lack of clear norms, jurisprudence, and adequate enforcement complicate the licensing process. In this context, Law Project 9/2021 is in process. It proposes, among other issues, the dispensing with the need for a prior location visit from the environmental agency responsible for licensing for small businesses, with up to 5 hectares of water surface area in excavated tanks or 5000 cubic meters of water in net tanks, lined tanks, or suspended tanks (Brasil, 2023).

3 Fish Culture of Pirarucu

Despite the first experiences with raising pirarucu in captivity dating back to the 1940s (Ferreira, 2016), commercial production of the species in Brazil is relatively recent and was regulated by Law no. 11,959 of June 2009, which established a national policy of sustainable development of fishing and fish culture. With this law and its subsequent norms,[7] the federal government, through Ibama, shared the responsibility for the ordering of managing fishing in inland waters, allowing

[6] A quota of 120 fish was authorized in four communities, involving 42 fishermen (CONAB, 2020).

[7] For example: Decree n° 34,100 of 23rd of October 2013, which disciplines the raising of pirarucu (*Arapaima Gigas*) in fish culture in the State of Amazonas; Normative Directive 1, of 21st of December 2011 "establishes criteria and procedures for the registration of reproduction grounds of

production in tanks and year-round production, in registered establishments that have undergone visits from government officials. Industrial fishing, as defined in this Law (article 8), is that practiced by individuals or companies that involve professional fishermen, employed or in quota-share partnerships, using small, medium, or large vessels for commercial purposes.

This regulation allowed government incentives to be directed to large scale production of pirarucu, with the aim of turning it into a profitable economic activity for the northern region of Brazil[8] (Goularti Filho, 2016). According to a number of studies, pirarucu has characteristics that allow its effective commercialization, such as high productive efficiency, with a high food conversion rate (up to 10 kg/year), broad market acceptance, soft and boneless flesh, as well as high nutritional value (Campos-Silva et al., 2019; Castello & Stewart, 2010; Pedroza Filho et al., 2016).

Despite significant growth in fish culture in the North region, production is still concentrated in Rondônia, responsible for 80% of the total Brazilian production, as shown in Table 1. This is due to the installation of large private processing companies in this area, which explains its predominance in the sector. According to Pedroza Filho et al. (2016), the north region had 118 productive pirarucu fattening facilities, and 14 productive facilities for juvenile pirarucu in 2008.

Fish culture in captivity has shown a solution to preserve the natural stocks of pirarucu, reducing the pressure placed on the species by fishing. Furthermore, it presents benefits for food security because the fish can be harvested year-round and the environmental control of raising means that exposure to pathogens and contaminants harmful to human health is minimized. Another advantage of raising in captivity is the ability to adjust the size of pirarucu according to the demands of the market, allowing the slaughter of smaller fish (<1.5 m) (Mapa, 2020). However, development of fish culture of pirarucu has faced some technological challenges and limitations related to knowledge of the biology of the species applied to fish culture, such as low understanding of reproduction, high cost of fingerlings and feed, and susceptibility to diseases (see Lima, 2013).

Among the research organizations dedicated to study of the pirarucu in captivity, Embrapa—the Brazilian Company of Agricultural Research—plays a fundamental role. Since 1984, it has conducted tests of production in captivity in the Center for Agriculture and Livestock Research in the Humid Tropics (CPATU/Embrapa). Other institutions such as INPA (The National Institute of Research in the Amazon), the IIAP (Research Institute of the Peruvian Amazon), and the federal universities have contributed to supplying technological and scientific information on the raising of pirarucu, approaching themes such as sexing methods, conditions of water temperature, and types of tank that promote spawning, nutritional studies to

---

the pirarucu (*Arapaima gigas*) in the State of Rondônia, in order to regulate businesses on the management, reproduction in captivity, fattening and trade in their products."

[8] Law Decree no 221, of 28th February 1967, which conceded Import Tax exemption (II) and Industrialized Product Tax exemption (IPI) until 1972 for the following items: fishing vessels, equipment, machines, devices, instruments, and tools for fishing. It also exempted fishing companies from payment of Business Income Tax (IRPJ).

Table 1 Evolution of the production of pirarucu from fish culture (in tons)

	2013	2014	2015	2016	2017	2018	2019	2020	2021
Rondônia	1955.7	11129.7	7688.8	615.0	488.6	999.6	979.5	980.3	943.1
Acre	48.2	36.1	82.7	40.8	35.9	48.4	46.2	46.3	41.7
Amazonas		125.0	103.0	93.4	108.7	161.8	168.4	144.0	194.4
Pará	33.8	136.8	174.4	150.8	183.3	205.7	272.3	295.2	221.3
Amapá	3.0	3.5	4.5	5.2	5.9	6.4	6.9	6.5	7.1
Tocantins		93.6	97.8	146.8	192.3	205.8	205.8	202.4	227.4
Roraima	–	–	–	–	–	–	–	–	–
Others	260.34	238.169	237.31	214.49	244.6	204	213.51	211.2	224.2
Brazil	2301.0	11762.9	8388.5	1266.4	1259.3	1831.8	1892.7	1885.8	1859.2

Source: Brazilian Institute of Geography and Statistics—IBGE, 2022

evaluate the quality of flesh of animals raised in captivity, diet, and costs of different inputs to the production of feed, among others (INPA, 2021).

The production of pirarucu in captivity has also required investments in technology transfer to small producers and the adoption of good management practices.[9] Regarding dissemination and support for fish culture, a set of organizations with Emater—the State Company for Technical Support and Rural Outreach, which offers technical orientation for planning fish culture projects, applying for environmental licenses and construction of tanks. The State Secretary of Rural Production (Sepror), the National Service of Rural Learning (Senar), Sebrae (Small and Medium Businesses Support Service) have developed a project supplying assistance and specialized technological support in partnership with Embrapa. The Institute of Development of Sustainable Forest Agriculture in the State of Amazonas (Idam) has given support to this activity to enable local rural producers to implement the activity, and increase their income. The barriers to production are the need for the property to have electrical energy and year-round logistical access. Furthermore, the producers need to acquire fingerlings, feed, and medication, which make the operation expensive.

3.1 The Fish Culture Production Chain

The production of pirarucu in captivity is composed by three main segments: producers of fingerlings, producers of fish, and processing companies. The initial phase is fingerling, in which the fingerlings are captured and go through feeding training before being removed from their parents. This stage is conducted by producers, responsible for reproduction until the fish reaches 10–20 cm. The juvenile fish are then sent to the properties dedicated to commercial production, known as the rearing phase, which ends when the animals reach 0.5–1 kg. In this stage, fish are raised in smaller structures and are more protected from predators. The next phase is fattening, in which the fish are fed until they reach a weight between 10 kg and 13 kg (Imbiriba, 2001; Lima et al., 2015). According to Sousa (2017), when there is an ideal cost–benefit ratio between feed consumption and weight gain, this is a period of approximately 12 months for this process, considering the time between rearing and slaughter.

Figure 1 schematically illustrates the fish culture chain, formed of feed, ice, refrigeration companies, fishing boats, and fish farmers. In the input segment, the fish farmers need to acquire feed, juveniles and equipment, as well as having the infrastructure of earthen ponds, which can be excavated with an earth lining, excavated and covered with sheeting, net tanks or lined tanks. Net tanks are installed in streams, rivers, reservoirs, weirs, ponds, and lakes, where there is already a current of water, while other systems are supplied by natural sources, such as rivers, lakes,

[9] These practices include: correction of water quality, monitoring for parasitic diseases, evaluation of zootechnics performance, and disinfection of nurseries.

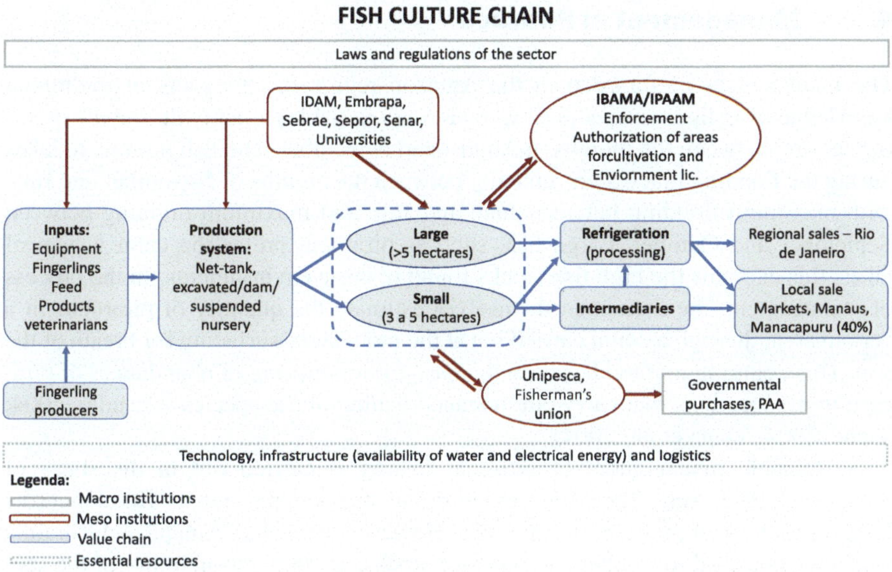

Fig. 1 Fish culture chain. Source: The authors

weirs, and/or ponds. Concrete tanks and vinyl-cover offer benefits in terms of cleaning, water quality, and pest control, while excavated tanks reproduce the natural habitat of the animal (Sousa, 2017).

Different to managed production, fish culture allows the concentration of production with the emergence of large producers who have farms with countless water surface and hire small producers who work in collaboration with the whole family. The growth of this activity created a new category of producer: the salaried producer, for the most part, or quota holder who has distinct characteristics from artisanal fishermen. While commercial artisanal fishing is carried out by autonomous fishermen or in a family economy, with its own means of production (a canoe with an outboard or tail motor) or through a partnership contract, commercial industrial fishing is practiced by individuals or companies, involving employed professional fishermen (registered with a fisherman's card), or in a quota-partnership regime (see Ferreira, 2016).

In general, salaried fishermen who work in industrial fishing in large farms produce a variety of species of fish, such as tambaqui, matrinxã among others, and work in other activities on the property, such as raising cattle or even agriculture (for example, soy and pineapple). This is a different situation to small producers who have two or three tanks and rely on family help and one or two jobs. These are closer to artisanal fishermen who are spread along the rivers and lakes and gain their income from production or extractivist collection of other products, depending on the region (Silva, 2009).

4 Management of Pirarucu

The management of pirarucu in the Amazon region has its roots in traditional knowledge of indigenous peoples, who arrived out fishing mainly during the draining season, between the months of August and November. The fish migrate to lakes during the flooding that occurs annually between the months of November and July, with maximum flooding between June and July and maximum draining between September and October. Large fish, such as pirarucu, prefer the calm waters of lakes, this explains the high fish stocks found in these environments. In the process of management, the fishermen themselves estimate the quantity of pirarucus in a certain area, through a count carried out at the moment of surfacing for breath of the fish. This counting method has been the base for monitoring of abundances of pirarucu in the locations that carry out managed fishing of the species (Castello, 2004; Viana, 2004; Viana et al., 2004).

Sustainable management of artisanal fishing is carried out in the states of Amazonas, Pará, Acre, Tocantins, and in initial stages in the state of Rondônia (the largest producer of pirarucu in captivity). However, states like Amapá and Roraima also have potential to implement managed stocks, given the availability of the species in the lakes of these regions. Despite the practice of management being identified in the states mentioned, it is still carried out in incipient form. Amazonas was a pioneer in the implementation of the activity and today has 34 areas with fishing agreements, ordered by the Secretary of the Environment (Sema), which benefit 5800 families and 10,200 fishermen (Amazonas, 2022).

The Sociobiodiversity Bulletin from CONAB publishes data on the production of pirarucu from extractivist fishing carried out in the state of Amazonas, through data gained from Ibama in the authorized areas. According to the data presented in the bulletin, in the period from 2011 to 2022, 548,110 fish were caught, with an average weight of 53 kg. In 11 years, there was a 317% increase in the authorization of fishing this species in the state. Table 2 presents these data.

It is important to highlight that the agreements for authorized fishing through Ibama are instruments that enable the formalization of management practices, protection of fishing stocks, organization of communities involved, as well as being an important tool to prevent conflict between fishermen, seeing that the local communities are involved and heard at all stages (awareness raising, construction of usage rules, environmental zoning of waterways, and validation of the fishing agreement), in order to take the interests of everyone into account (Amazonas, 2022).

To detail the activities of sustainable management of pirarucu, we used extractivist fishing in the Mid Solimões, AM region as a case study, especially the activities developed by the community located on the Mamirauá Sustainable Development Reserve (RDSM).

Table 2 Evolution of pirarucu caught from managed areas in Amazonas (quantity and mean weight)

	2011	2012	2013	2014	2015	2016	2017	2018	2019	2020	2021
Authorized	19,083	19,735	29,791	42,477	43,823	54,609	58,530	50,457	69,019	80,952	79,634
Caught	16,873	17,783	23,648	27,489	35,815	35,815	47,593	38,173	51,150	50,749	64,970
Mean weight (kg)	54	52	56	52	56	55	52	51	51	48	54
Total (ton)	911.1	924.7	1324.3	1429.4	2005.6	1969.8	2474.8	1946.8	2608.6	2435.9	3508.4

Source: CONAB, 2020

4.1 The Management of Pirarucu on the Mid Solimões: The Case of RDSM

Fishing in the Mid Solimões region occurred intensively during the 1960s and 70s, with the decline in the commercialization of rubber and jute in the region, fishing became an alternative productive activity. With the new producers who entered the market, fishing was incentivized by the government through the growth of the fishing fleet and the creation of refrigeration facilities in Amazonas, as already mentioned. These factors made fishing an important economic activity in the region (Gonçalves et al., 2018).

The high profitability of pirarucu in comparison with other species attracted many producers to this market. With the growth in the number of fishermen, and with government incentives, the stocks of this fish in the region were reduced, and the pirarucu became an overexploited resource. At the time, the local population saw the dwindling of the number of fish in the lakes, as well as a reduction in their average size. The incapability of the species to reproduce at a higher rate than the rate at which they were caught signaled the need for conscientious use of this species. Therefore in 1989, Ibama established the minimum catch size for pirarucu (150 cm), and in 1990 a closed season,[10] as was highlighted in item 2.

In 1996, the Mamirauá Sustainable Development Reserve (RDSM) was created by the state government of Amazonas, by way of State Decree no. 1619, aiming to protect biodiversity in the region, which is threatened by predatory exploitation, and conciliation with sustainable development of local communities through sustainable management of natural resources, scientific research, ecological tourism, and environmental education. In 1998, sustainable management of pirarucu began to be implemented in the region of RDSM. Since then, this activity enabled a 427% increase in the stocks of the species in the managed areas, as well as the composition of income of local populations, who have the fish resources that are fundamental for subsistence. Of all fishing activity, pirarucu represented 47% of the income derived from this activity in 1999 (Queiroz & Sardinha, 1999). Figure 2 shows the location of the Mamirauá Sustainable Development Reserve (RDSM).

Inside the RDSM, the population is divided into small population groupings called sites and communities. The sites are groups that have only one family tree, while communities can be made up of more than one family tree. Therefore, the communities have a more complex social organization because they house a greater number of individuals. In them schools, churches, centers, and community

[10] The closed season is a period of prohibition of fishing, or catching certain species in rivers, ponds, lakes, seas, or oceans. The main aim is to guarantee the preservation and reproduction of a species during the time in which they are most vulnerable, generally during the period of reproduction or growth. The closed season of the pirarucu in Amazonas is decreed annually by the state government and can vary according to climate conditions and the reproductive cycle of the species. In 2022, the closed season for pirarucu for Amazonas was established by the Environmental Protection Institute of Amazonas (IPAAM) through Ordinance n° 006/2022, published on the 31st of January 2022. The closed season of the pirarucu in the Amazonas region began on 15th of November 2022 and extended to the 15th of March 2023. During this period, catching pirarucu was prohibited, to guarantee the protection of the species and the preservation of fish stocks.

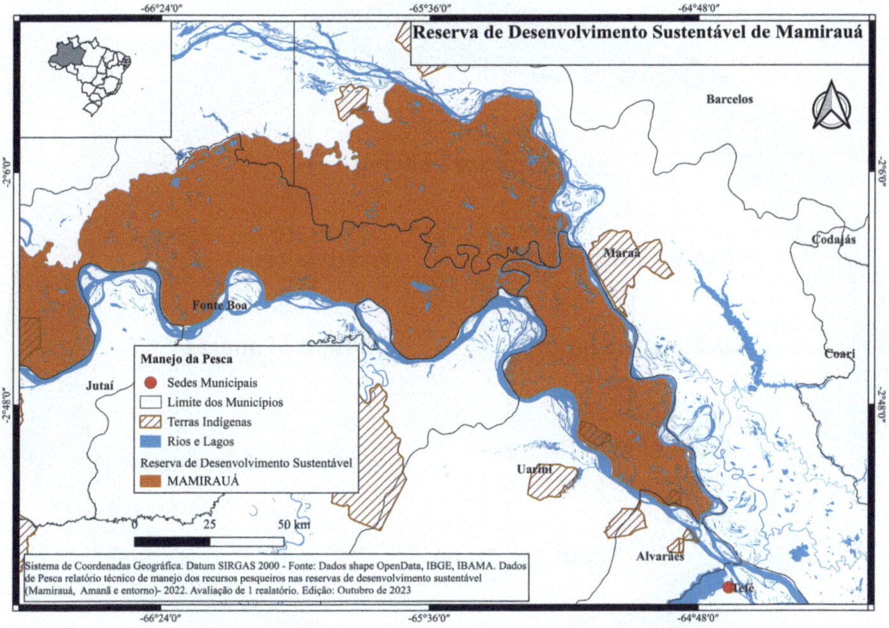

Fig. 2 Map of Brazil and the Mamirauá Sustainable Development Reserve (RDSM). Source: Bezerra (2005), p. 20

associations are found. In 2011, some demographic data collection was carried out of the reserve and its surroundings, finding a total population of 12,159 residents and users. The residents are the individuals who inhabit the RDSM, and the users are those who according to the National System of Conservation Units (SNUC) have rights to use the conservation unit's natural resources. In order to guarantee these rights, users must comply with the norms established by the Management Plan and the agreements formalized between associations, and participate in environmental protection plans.

The areas of management of pirarucu in the mid-Solimões are legally instituted by decisions from the Producers Associations, the Colony of Fishermen of Maraã-Z-32, the Union of Fishermen of Maraã, and the Colony of Fishermen of Alvarães-Z23, according to the management plan of RDSM. These decisions are deliberated by sectoral meetings and general assemblies, which reveals the need for judicial formalization of associations. On one hand, this increases the complexity of management of these associations and brings a high demand for organizational structure of these groups, but on the other, it allows access to new markets outside the state of Amazonas, willing to pay more for the fish.

With the areas defined, the planning of the fishing is done by fishermen together with the Mamirauá Institute of Sustainable Development (IDSM) and other technical support institutions. This participative planning is done through community meetings, stock counts, monitoring of fisheries, trade, and enforcement. The stock count is done by the fishermen themselves of the managed lakes. Because pirarucu

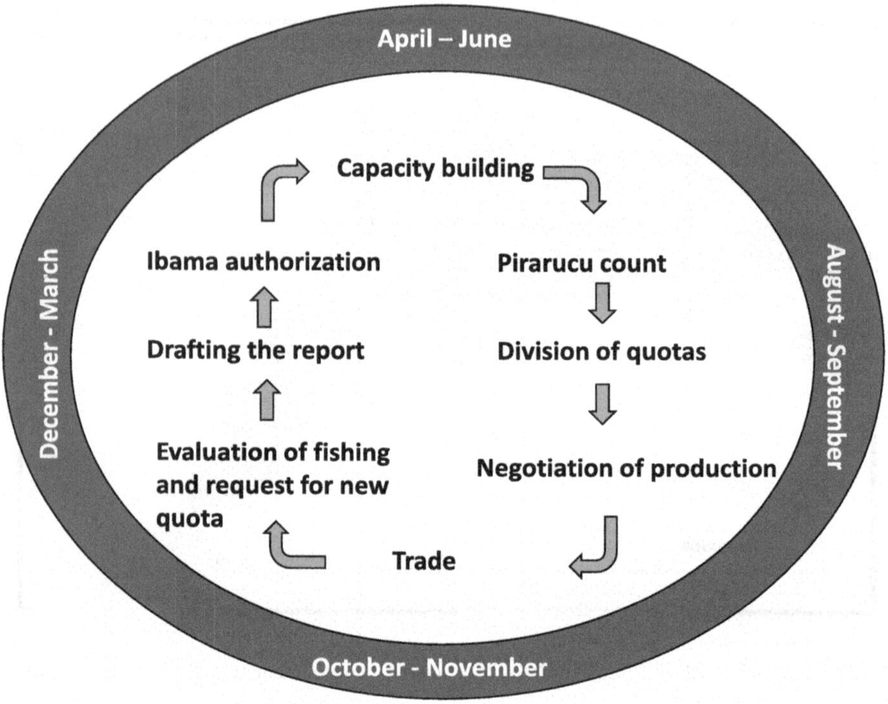

Fig. 3 Flow chart of the management activities of pirarucu throughout the year. Source: Amaral et al. (2011)

is a fish with lungs, the count is made when they surface for air ("rise") and then the number of fish is estimated. From this number, Ibama authorizes a quota of around 30% of the counted adult fish (fish measuring over 1.5 m). Figure 3 shows the management activities of the lakes throughout the year.

The division of quotas is done by the directors of the Associations, Unions, and Colonies, which follow rules previously established in councils. Each managed area formalizes its rules in a statute or internal regulations that follow the general guidelines required by Ibama (IDSM, 2014). Before the management program, the pirarucu was fished year-round, and its sale was negotiated by intermediaries at the ports of each community. This trader received the fish itself as payment for the sales and negotiated for small quantities. With the implementation of fishing management, the fishing began to be carried out collectively and the trade carried out through community associations of fishermen's colonies. Managed production became more noticeable and the intermediaries began to negotiate their purchases through the Trade Round. The final consumer could access the product in street markets at specific times in neighboring municipalities, which enabled higher prices for the fish (Amaral et al., 2011).

Participative management of fishing improved a number of indicators in the region, such as increasing the stocks of pirarucu and the income of the population. Furthermore, a greater number of fishermen joined the management projects (42 fishermen in 1999 to 1590 in 2017) and by including women in the production process, the management guaranteed greater gender equality in the region. Added to this is the increase in supply of legal pirarucu in the region (Gonçalves et al., 2018). Social technology developed at the site was recognized and won international prizes.[11]

4.2 The Production Chain of Management

The chain for pirarucu management involves local communities, fishermen, suppliers of inputs, associations, governmental and non-governmental organizations, processors, intermediaries, distributors, and consumers. The activity requires the use of inputs such as (i) vessels for the transport of the fish which, in some cases belong to the associations themselves or to the intermediaries; (ii) ice, to guarantee the preservation of the fish during transport; (iii) appliances for fishing, most importantly appropriate nets that avoid catching pirarucus smaller than the permitted size (1.5 m); and (iv) fuel used for transport and surveillance of the lakes, which represents a significant cost of production, around 70%.

Sustainable management involves the stages of: (i) monitoring the lakes year round to curb invaders who practice illegal fishing; (ii) planning and organization of the fishing timeline, counting the fish in each lake that make up part of the fishing agreement. This is needed to gain authorization from Ibama for the 30% of estimated stocks catch quota, and after, to decide how many fishermen will act in each area, the number of fish to be caught by each team, and how the income will be distributed that year; (iii) the fishing itself (catching and cleaning the fish, preparation of food for the team, etc.) (Amaral et al., 2011).

The production of the fish is directed by the fishermen's association, which carries out processing of part of the fish and sends the rest to the refrigeration facilities who buy the fish fresh. The fish is distributed through a number of channels, such as local markets, restaurants, and hotels, as well as local and regional distributors and a tannery that buys pirarucu skin to make leather. At this point in the chain, the consumption that happens is mostly for the local market. A series of public and nongovernmental intermediate organizations act in the chain, helping fishermen and women in the organization of collective actions, capacity building, and enabling market access with better prices (Saes et al., 2023). Figure 4 presents the chain described.

IDSM has an important strategic role in the mid-Solimões chain. It is an articulating link, which acts between licensing bodies and producers, enabling these relations and making sustainable management of pirarucu possible in the region. The institute carries out activities that prioritize interaction between traditional

[11] Best international experience of income Generation in flooded areas, by the Secretary of the Ramsar Convention, United Nations, in 2011. The project was a finalist in the national stage of the Energy Globe Award in 2015. In the same year it was a finalist in the Banco do Brazil Foundation Prize for Social Technology. In 2016, the project was also a finalist in the St. Andrews Award for the Environment, University of St. Andrews, in Scotland (Gonçalves et al., 2018).

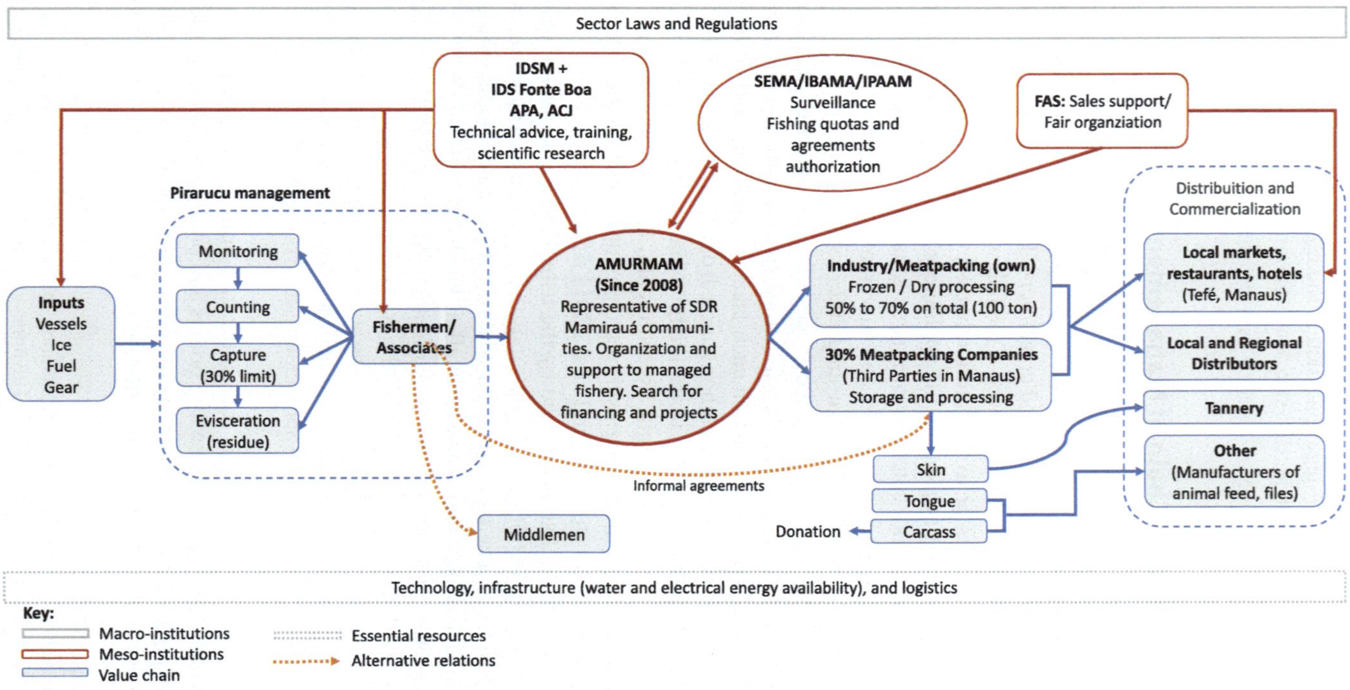

Fig. 4 Chain of the mid-Solimões. Source: Saes et al. (2023)

knowledge and knowledge derived from scientific research. This results in a conservation and sustainable social development experiment that currently serves as a model for replication in other areas of the Amazon (Amaral et al., 2011).

On the Mamirauá reserve, the Association of Residents and Users of the RDS Mamirauá Antônio Martins—Amurmam, created in 2008, has a fundamental function in the chain of the pirarucu, as well as representing all residents with governmental, environmental, land registry, and legal institutions in defense of their rights, the association is also responsible for the organization and support for managed fishing, as well as seeking financing to enable the activities in the region (Franco, 2019).

5 Challenges and Limitations for the Development of Fish Culture and Management Chain

To discuss the challenges and limitations of fish culture and management, information obtained from interviews with local actors and social organizations involved in sustainable management and fish culture of pirarucu in the region, as well as documental evidence (see Chart 2, Appendix 2).

Investigating the chain of pirarucu reveals a series of challenges that impact on sustainability and development. One of the main obstacles is related to predatory fishing that directly affects both the chain of managed fishing and of fish culture, given that fish that comes from illegal fishing is sold at extremely low prices, compromising the competitiveness of pirarucu on local markets.

Furthermore, illegal fishing compromises the ecosystem by not respecting the rates of reproduction of the species. Excessive fishing, especially during the breeding season, threatens the survival of the pirarucu. According to the president of the Macopani Sector Association, located in the RDSM, overfishing was exhausting the stock of pirarucu in the lakes in this area, which contributed to the decision taken to implement management by the communities, as he states:

> [...] we began to notice, to copy the others, because our region was vacante, and everyone came whenever They wanted. Then, we noticed that there was no longer enough fish to eat, for us to feed ourselves (President of the Macopani Sector Association, RDSM).

In the fish farmer's view, both illegal fishing and the competition from managed fishing is damaging, since the costs of production of raising in captivity are much higher. According to the producer from Canabrava do Norte, Mato Grosso:

> Fish from extractivism comes onto the market very cheaply. While we would have to be selling, like, let's talk on average, 22 reais per kilo, live weight, extractivist fishing is sending it for 8 reais per kilo of manta. I say manta, which is a fillet with the manta boneless. Then, it's a disordered business. There's no way we can compete with the price of extractivism (Fish farmer 1, Mato Grosso).

A lack of enforcement is also an important factor in the sustainable management of fishing, seeing as it incurs a series of costs associated with surveillance of the lakes, including expenditure on manpower and fuel. Monitoring the lakes is one of the costliest stages of management and is carried out 24 h per day with teams taking shifts. According to the president of the Macopani Sector Association, the cost is high because it is carried out year round,

> …there is the cost of gasoline, food, torches, batteries […] At night, noctural. At night the thieves … can come. They get close to… the fishing is done more at (President of the Macopani Sector Association, RDSM).

Another relevant issue is environmental degradation arising from constant pressure for deforestation, illegal mining, and climate change, directly affecting the reproduction and feeding environments of the pirarucu. Notwithstanding, a recent study on the levels of mercury in fish consumed in the Brazilian Amazon caused by artisanal mining, deforestation, and burning detected that over a fifth (21.3%) of the fish traded in the urban centers of six states (Acre, Amapá, Pará, Rondônia and Roraima) presented levels of contamination over the acceptable limit (greater than or equal to 0.5 μg/g) established by the World Health Organization (WHO), which threatens food security and human health (Basta et al., 2023).

In this sense, the fish culture chain is also affected, although cultivation in controlled environments reduces the interference from external factors, the water used in production originates from the rivers in the region. Furthermore, the high costs of feed reduce the profit margins of fish culture producers significantly. In some cases, they are not able to gain a price that covers the costs incurred during the process. For the fish farmer from Canabrava do Norte, for example, feed represents over 70% of the cost of production. The producer stated that there was no feed specifically developed to attend to the nutritional needs of the pirarucu, this input is substituted by feed developed for catfish, as highlighted in the interview:

> […] the thing about feed, it's like this, there is no specific feed in Brazil today, it doesn't exist. We use a feed that is used for catfish, mainly the fattening feed. So, it's not good for the fish! It's not good feed! […] To be honest, this feed that we have on the market, today, was made for a hybrid catfish, an improved (Fish farmer 1, Mato Grosso).

Furthermore, investment in installations for raising pirarucu in captivity is significant, mainly for small producers, because it requires tank infrastructure and nurseries that attend to the specific needs of the species, such as an adequate size, water quality, temperature, and controlled feeding. Brabo et al. (2022) estimated the cost of implementing fish culture with a hectare of excavated nurseries supplied by pump to be R$80,000 in the state of Pará in 2018, as well as the total cost of production at R$122,666.77, and the unit production cost at R$12.27/kg. Feed represented 50.7%, and the fingerlings accounted for 29.3% of the total. The authors concluded that raising pirarucu in excavated nurseries in the state of Pará, with feed as the only dietary source for the fish, does not represent a profitable investment because of the market prices. This difficulty is related by one of the fish farmers interviewed, although he was in Mato Grosso, he highlights that:

> The price of fish, today, is garbage, it almost doesn't even meet the cost of production, the margins are very fine. The fish has already fetched thirty, forty percent returns, today it's

giving, hmm, around them, fifteen percent. So, it's got a lot finer, the cost of inputs went up a lot and the price hasn't followed it (Fish farmer 2, Mato Grosso).

The logistical infrastructure in the Amazon is another challenge, mainly for managed areas, which are in remote and difficult to access regions. The high logistical costs to access these areas and the weakness in the refrigeration chain generates significant restrictions for transport and adequate storage of fish, often resulting in compromised quality, loss, and wastage in production. For the president of one of the associations located in mid Purús—AM, the difficulty transporting pirarucu is one of the main challenges, as he relates:

[...] the pirarucu, we work with it over a distance, from the communities to the municipal centers. And, when it finally reaches the refrigeration facility is very distant. From the communities until it arrives in the center in the municipality of Lábrea and then get more transport, to arrive here in Manaus, Manacapuru, where it is processed, this takes many days. This means we need to take great care, with ice, for it to not lose its quality. Yes, we need to undertake a lot of "articulation" with the vessels, which transport it, to make sure there is no delay and it arrives on time, there, for it not to lose quality when it arrives at the refrigeration facility to be processed. From the moment of slaughter to the time it is put on ice, three hours is recommended to not lose quality. If it is more than three hours, this slaughter time, before it goes on ice, it will already start to lose quality. (President of ATAMP, Médio Purús).

The fish culture chain in the region is also affected by this factor, which directly impacts on distribution and trade, complicating the arrival of products to consumer markers in acceptable condition.

There is also a series of problems with the fish culture chain that are not completely resolved. According to Lima et al. (2015), the production of the pirarucu in captivity has challenges such as low understanding of reproduction, which results in restrictions of supply, irregular supply, and high costs of fingerlings, little knowledge on the biology of the species, need for development of appropriate feed for nutritional demands and the high costs of them, a lack of technology for sexing the fish and high mortality rates, which results in low availability of fingerlings and consequently, high prices of these, among other factors (Ferreira, 2016; Lima et al., 2015).

Despite all of the challenges mentioned, the researcher from Embrapa Fish and Fish culture in Tocantins reported that the biggest obstacle is related to trade. As the data series of fish culture production (Table 1), the state of Rondônia significantly increased the quantity of pirarucu it produced between 2014 and 2015, because of the demand from a large firm that was established in the region, which aimed to export pirarucu from fish culture. After this period, the firm closed and production returned to levels below those verified in 2013. As she reports:

This, for us, was an alert, because we, for a long time said, that the biggest problem for producing pirarucu was a lack of fingerlings. And, after what happened in 2014 and 2015, we saw that this was not the main factor. The main factor has been, much more, in the issue of industry or in a better market for this species, which is a high value species. So the producers can't sell it easily, they can't sell it like the tambaqui, the whole fish, because it's a very large fish! (Researcher, Embrapa).

Added to the challenges of trade, issues around consumption involve consumer preferences and raising awareness of the importance of sustainability and the provenance of the products, which can be a limiting factor for market expansion, as well as valuing traditional knowledge and culture, which are intangible aspects in management activities. According to the supervisor of the Sustainable Amazon Fund (FAS) in Manaus, AM, one of the support projects developed together with the communities that work in extractivist fishing seeks to improve these aspects, as he reports:

> With this project we can, today, with this frontline with the association, make this production have more value and bring these environmental issues, bringing this work that the Community has developed. This more social work preserves more and at the same time takes a percentage. The work that has been developed in these areas is the increase of this population, you see that every year it grows. The issue is, of bringing this message, isn't it? Of you paying more for this product, for the Community to carry on there, doing its job, developing the work of conserving, preserving and at the same time, to be able to live (Supervisor, FAS).

These are just some of the challenges that need to be overcome for the chain of the pirarucu to develop sustainably in the Amazon region, generating income and wellbeing for the communities and producers who work in these areas. All of the challenges mentioned to a greater or lesser degree, affect the production chains of management and fish culture of the pirarucu and need solutions that reduce the resulting impacts. Mapping these factors is important to define strategies for the future of the chain. Chart 3 synthesizes of the main challenges identified and the way that they exert influence over chains.

6 A Sustainable Future for the Pirarucu Chain

The Organization for Economic Co-operation and Development (OECD) evaluates that, as a result especially of demographic evolution, fish consumption could reach 180 million tons in 2030, compared with 158 million tons in 2020. Given this tendency, fish culture is one of the forms of food production to grow most in the world. As previously highlighted, the pirarucu is a species that mainly inhabits the floodplains and rivers of the Amazon region, an ecosystem of great importance to the regulation of the global climate. These areas act as natural carbon stores, storing significant quantities of the gas in its trees, soil, and biomass. A sustainable chain for the pirarucu can contribute to the conservation of these ecosystems and, as a result, for the maintenance of the stock of carbon.

From the challenges identified, we analyze and propose a series of alternatives that stimulate a reflection on the future of the pirarucu chain in the Amazon. First, we highlight the coexistence and development of both chains; management and fish culture, are important for the sustainable development of this economic activity. It is evident that unorganized and predatory fishing create serious impacts on the ecosystem. In this sense, fish culture plays an important role in the conservation of the species, offering a sustainable alternative to predatory fishing. Second, we explore

Chart 3 Challenges for the chains of management and fish culture of the pirarucu in the Amazon region

Challenges in the pirarucu chain	Management	Fish culture
Overfishing/predatory fishing	●	
Illegal fishing	●	◒
Degradation of habitat caused by deforestation, illicit mining, mercury contamination, inadequate management, etc.	●	◒
High costs of surveillance of the lakes	●	
High cost of feed		●
Lack of logistical infrastructure	●	●
Low understanding of reproduction of the species		◒
Low understanding of the biology of the species	○	◒
Low competitiveness compared to other species	◒	◒
Low levels of consumption outside the Northern region	◒	◒
Lack of guarantee of adequate prices to cover the costs of production	◒	◒
High investment in installations		●
High mortality rate	◒	●
Low availability of fingerlings		●

Score: ● High influence; ◒ medium influence; ○ low influence
Source: The authors

alternatives that respect local conditions and natural jobs in the biome, which include traditional peoples, riverside communities, and small producers who are engaged in this activity, either through management or fish culture.

While the management chain presents a superior capacity for conserving and preserving the biome, due to the respect for natural cycles of reproduction of the ecosystem, it has challenges to become financially viable due to logistical issues and irregular demand. However, management creates additional value due to the positive externalities generated on the environment, which can be rewarded through carbon or ecotourism markets, as well as through certification of the product, making it a niche product. This niche market strategy could have a focus on valuing intangible attributes for consumers, such as those related to the development of the management activity itself, which brings together traditional knowledge, cultural aspects, conservation of lakes, and preservation of the forest. On the other hand, the fish culture chain has greater scalability. In the long term if technological limitations can be overcome, can cover the costs of production, becoming an important source of animal proteins that contributes to food security. As well as this, it is possible to unite small producers, in order to have a profitable activity during the closed season.

The possible solutions and an agenda for public policy that include the private sector and civil society, for the sustainable development of the management chain need to consider:

1. Adequate regulation and enforcement: adopt clear laws and regulations to manage the use of the lakes in forestry reserve areas to guarantee effective enforcement.
2. Education and awareness: educate to raise awareness among younger generations, local fishermen, and the general community about the importance of preservation of natural resources to promote responsible and sustainable fishing practices.
3. Monitoring the conditions of the lake: regularly evaluate the metrics and indicators of effectiveness of the measures adopted and guarantee that the fishing practices are sustainable in the long term.
4. Community participation: prioritize mechanisms that incentivize participation of the community in the management of the lake and forestry reserve to guarantee that needs and perspectives of the local community are considered when taking management decisions. For example, the implementation of fishing quotas helps to limit the number of fish taken from the lake. The defined quotas should incorporate the knowledge of the community. Taking the local specificities into account is essential for the success of collective action.
5. Differentiation strategy: valuing the characteristics of managed production through seals, certification of origin, and collective frameworks. Attributes that allow pricing that compensates the costs of production and is valued by consumers, in order for the benefits to be shared with consumers.
6. Strategies related to the ecosystem: structuring processes for the granting of carbon credit certificates, as well as making use of the landscape and extractivist fishing activity for ecotourism.

To increase the potential of the fish culture chain in the Amazon, the following measures are proposed:

1. Develop initiatives to strengthen the integration of partnerships between fish farmers, through associations and cooperatives.
2. Adoption of advanced techniques for raising in captivity, allied with scientific knowledge on the biology and behavior of the pirarucu, can contribute to the improvement of production and reduction of environmental impacts. For this reasons, supporting research organizations is essential.

3. Augment investment in research and development of technologies for sexing the fish to ensure adequate ratios of males to females, to preserve genetic diversity and reproductive capacity of the species.
4. Stimulate the use of technology for fish culture developed by Embrapa and revitalize existing nurseries, without the need to promote new deforestation.
5. Monitor and apply new technologies in order to minimize environmental damage, with more efficient use of water resources and optimization of the use of by-products allowing their recovery and use in agricultural activities, such as recirculating fish culture systems (RAS).

   Other factors can benefit both chains of production in the Amazon. Among them:

1. Infrastructure investment: supply roads, storage, and refrigeration, which promote trade and distribution of the pirarucu, as well as access to broader and more profitable markets.
2. Public policies and regulations: incentivize sustainable production of pirarucu, allied with effective regulations to ensure the protection of the species, the sustainability of the activity and rights of local communities.
3. Access to credit and technical assistance: build capacity of producers, improve management practices, and structure complementary chains such as the refrigeration facilities in the region.
4. Use and exploitation of residual products: encourage the production biogas, bio jewels, animal feed, and food for human consumption. Activities that generate additional income sources and reduce environmental impacts.
5. Build up capacity for the appropriate use of digitalization to ensure access to knowledge and improve levels of productivity, empowering innovation and reducing transaction costs bringing a better distribution of benefits.
6. Promote energy alternatives and efficiency in their use, with an emphasis on the reduction of fossil fuels, in management, in fish culture, in logistics, and in distribution to reach a greater level of sustainability.

These propositions cannot be underestimated given the magnitude of the importance of the Amazon biome. In general, it is clear that public policies and private incentives are necessary to propel both fish culture and managed fishing.

The two chains, as previously mentioned, are seen as potential driving forces for a sustainable bioeconomy in the Amazon. For this, articulation between a range of involved stakeholders, including producers, fishermen, communities, social organizations, public and private sectors, it is essential to reach adequate levels of structuring of the management and fish culture chains so that they reflect the potential for social, environmental, and economic development, as well as a sustainable future aligned with the challenges brought by the twenty-first century.

Acknowledgments National Council for Scientific and Technological Development (Conselho Nacional de Desenvolvimento Científico e Tecnológico—CNPQ, Grant No. 304233/2023-4).

Funding This research was funded by Fundação de Amparo à Pesquisa do Estado de São Paulo (FAPESP) and Fundação de Amparo à Pesquisa do Estado do Amazonas (FAPEAM) (Grant number 2020/08886-1).

Appendix 1

Chart 1 Key regulations related to the Pirarucu Supply Chain in the Amazon

Law n° 1.806, of 6th January 1953.	Created the Economic Valuing Plan of the Amazon as a "system of measures, services, enterprises and works, destined to increase the development of extractivist production and agricultural activity, mining, industrial and trade relations, in the sense of better social standards of life and economic wellbeing of the populations of the region and expansion of the wealth of the country." The plan was developed by the Superintendent of the Plan for Economic Valuing of the Amazon (SPVEA), created in 1953, and later the Superintendent for Amazon Development (Sudam), created in 1966.
Law-Decree n° 221/1967 (Fishing code)	Regulates fishing activity in Brazil seeking to promote industrialization in the sector.
Ordinance Ibama n° 466/1972	Prohibits the use of certain tools and forms of fishing in the basins of rivers in Amazonas, Araguaia and Tocantins. Practices involving dragnets, lances, and electricity were prohibited, as well as the use of corrals or pitfall traps, and methods that involve beating, poisons, or explosives. Additionally, the dimensions of permitted nets were subject to limitations, with a maximum length of 150 m and should be positioned at least 100 m apart. Furthermore, gillnets should have an opening of at most 70 mm, while castnets should have an opening of 50 mm.
Ordinance Ibama n° 15/1976	Prohibits the catching of pirarucu, annually, between October 1st and March 31st.
Constitution of 1988, article 195, paragraph 8°	Fishermen had their social rights levelled up to urban workers, giving them "special insurance" for Social Welfare, with all attendant rights: retirement, maternity leave, sick pay, coverage for accidents in the workplace, among others.
Ordinance no 480/1991	Regulates the annual prohibition of the exercise of pirarucu fishing in the Amazon Basin between December 1st and May 31st in the Araguaia and Tocantins rivers and their tributaries between October 1st and March 31st. A minimum catch size of 150 cm is established.
Ordinance Ibama no 14/1993	Limits trade in salted and dried fillets to 100 cm.
Ordinance Ibama n° 8/1996	Establishes general norms for fishing in the hydrographic basin of the Amazon River. Prohibits the catching, transport, and trade in pirarucus with a length of less than 150 cm.
Ordinance Ibama no 145-N/1998	Establishes norms for introduction, reintroduction, and transfer of fish, crustaceans, molluscs, and macrophytes for fish culture.
Law n° 7.679, of 23rd of November 1988	Prohibits fishing in certain periods, locations, as well as some methods, looking to protect aquatic flora and fauna. Establishes penalties, fines, and seizures for infractions, as well as criminalizing certain practices. Establishes that the Executive Power will regulate the necessary measures of fishing rules.
Portaria Ibama no 9/1999	Authorizes fishing in managed areas, establishing an annual quota of fish that can be fished in each community.

Decree 4.339 of 22/08/2002	Item 3 of the policy, titled "Sustainable Use of Components" establishes the aim to promote the sustainable use of the components of biodiversity, considering not only its economic value, but also environmental, social, and cultural value".
Normative Instruction Ibama no 29 12/2002	It establishes the criteria for the regulation of fishing agreements made by riverside communities, understanding that fishing agreements are "a set of specific measures resulting from consensual treaties between the various users and the management body of fishing resources in a given geographically defined area."
Law n° 10.779, 25th November 2003	Provides for the concession of an unemployment insurance benefit, during the closed season, to professional fishermen who carry out artisanal fishing activities. The benefit guarantees an income of one minimum monthly wage during the closed season, that is, while fishing activity is prohibited in order to preserve the species. "Professional fisherman who exercises his activity, exclusively and uninterruptedly, in an artisanal way, individually or in a family economy regime, during the closed fishing activity period for the preservation of the species."
Normative Instruction Ibama n° 34/2004	Establishes general norms for fishing of the pirarucu (*Arapaima gigas*) in the Amazon River Basin. Art. 2 prohibits "the capture, commercialization and transport of the pirarucu (*Arapaima gigas*), as follows I—in the States of Amazonas, Pará, Acre and Amapá, during the period from December 1st to May 31st; II—in the State of Rondônia, during the period from November 1st to April 30th; III—in the State of Roraima, during the period from March 1st to August 31st."; in art. 3 it states: "The pirarucu fishing areas, whose studies indicate a state of overfishing, may be closed at any time"; and in art. 5 "the following are excluded from this prohibition: I—specimens from duly registered fish culture, and accompanied by proof of origin; II—fishing of a scientific nature authorized by Ibama".
Normative Instruction Ibama n° 01/2005	Prohibits fishing and transportation, storage, and commercialization of pirarucu in the Amazon during the period from June 1st to November 30th. It also establishes criteria and procedures for pirarucu fishing under the management system and by fish farming.
Normative instruction MMA n° 24/2005	Prohibits, annually, in the hydrographic basin of the Araguaia-Tocantins rivers, the capture, transport, commercialization, and storage of pirarucu (*Arapaima gigas*), during the period from October 1st to March 31st.
Complementary state Law n° 53, 05/06/2007	It instituted the State System of Conservation Units and establishes criteria and norms for the creation, implementation, and management of Conservation Units in the state of Amazonas.
Ordinary Law no 3.245/2008	Establishes norms for the elaboration, artisanally, of edible products of animal origin and their commercialization in the State of Amazonas. The elaboration of edible products of fish origin, in an artisanal way, will be allowed only by rural producers who use raw material of their own production and/or acquired from third parties, as long as the sum of the two does not exceed 100 kg.
Normative Instruction Ibama-Acre n° 1/2008	Prohibits the capture, transport, storage, and commercialization of pirarucu in the state of Acre during the period from June 1st to November 30th. Production is allowed only in managed areas or by fish farming.

Resolution CEMA-AM N° 01/08—de 03 de julho de 2008	Establishes norms and procedures for the environmental regularization of tanks, nurseries, dams, small reservoirs, stream channels, and net tanks for fish culture in the state of Amazonas.
Normative instruction Ibama n° 19/2009	Establishes criteria and procedures for the regulation of Fishing Agreements of the Pantaleão, in the Conservation Unit of the Sustainable Development Reserve category (RDS Amanã, municipality of Maraã). It considers the categories: "1. breeding areas: destined exclusively for the reproduction of the species, where fishing is prohibited for an indefinite period; 2. maintenance areas: destined exclusively for the subsistence of the families living in the communities and of those enrolled in the Fishing Agreement, while they are in the surveillance activity of the Agreement area. 3) Commercial Use Areas: destined exclusively for commercial fishing carried out by the members enrolled in the Agreement. Commercial fishing must be done in a collective way and previously agreed upon among the Agreement participants, with entry approved in an Assembly among the communities and institutions participating in the Agreement."
Law n° 11.959, of 29 June 2009	Assigns competence to the states and the Federal District for regulating fishing in continental waters under their respective jurisdictions. It establishes guidelines for the access regime, total allowable catch, sustainable fishing effort, closed seasons, fishing seasons, capture sizes, prohibited areas, fishing gear and methods, environment support capacity, monitoring, control, and inspection of the activity, and reproduction protection and stock recomposition. It also considers the peculiarities and needs of artisanal, subsistence, and family fish culture fishermen. It also establishes rules for the classification of fishing as commercial (artisanal and industrial) and non-commercial (scientific, amateur, and subsistence). Rules are also defined for the classification of vessels and for the types of fish culture (commercial, scientific/demonstrative, environmental recomposition, and family fish culture).
Normative Instruction MPA/MMA no.10, 10th de June 2011	Art. 1 Approves the general norms and the organization of the permission system for fishing vessels for access and sustainable use of fishing resources, with the definition of fishing modalities, species to be caught and permitted areas of operation. Provides for the requirements for the Initial License for professional fishermen in artisanal fishing.
IN MPA n° 12/2011	Provides requirements for Initial License for professional fishermen in artisanal fishing.
Law n° 12.512, 14th October 2011 of the Federal Government—Green Grant	Program to support Environmental Conservation—Green Grant Program—which has the following objectives: I) to encourage the conservation of ecosystems, understood as their maintenance and sustainable use; II) to promote citizenship, improve living conditions and raise the income of the population living in extreme poverty who carry out activities to conserve natural resources in rural areas; and III) to encourage the participation of its beneficiaries in environmental, social, educational, technical, and professional training activities. The areas are predetermined by the Federal Government that, besides its own and specialized criteria, uses as a parameter the prioritization of reserves, verification of natural resources, extinction of species, and similar reasons.

Decree N° 34100 23/10/2013	Art. 1 This decree regulates the breeding of pirarucu (*Arapaima gigas*) in fish culture and establishes norms and procedures for the regularization of this activity in the State of Amazonas.
Normative Instruction 665/2014	It categorizes artisanal fishermen in three types: (1) exclusive: a fisherman who has fishing as their only professional activity or depends only on fishing to survive (receives Unemployment Insurance); (2) principal: a fisherman in which fishing is his main livelihood, but has another job. This worker is entitled to retire by the INSS, but cannot receive Unemployment Insurance; and (3) Subsidiary: a fisherman for whom fishing is not the main livelihood.
Decree n° 8.424 March 2015	Established that in order to be entitled to the benefits of the closed fishing season, the fisherman must be registered in the General Registry of Fishing Activity—RGP, hold a fishing license granted by the Ministry of Fisheries and Fish culture, and have made the payment of social security contributions in the last 12 months immediately prior to the requested benefit, or since the start of the closed fishing season.
Decree No. 36083 of 07/23/2015	State Regulation—Amazonas—Published in the Official Gazette on July 23, 2015, which regulates the managed fishing of pirarucu (*Arapaima* spp.) in the state of Amazonas.
Law No. 13,680, dated June 14, 2018.	ARTE Seal—With the aim of guiding adjustments in the regularization process of artisanal food products of animal origin, produced by family, artisanal, and community establishments associated with the artisanal fishing chain, particularly the *Arapaima gigas* (Pirarucu) species. The seal is a way to identify artisanal animal products, allowing their commercialization throughout Brazil.
Law No. 13,881, dated October 8, 2019.	Allows for the inclusion of extractive products of animal origin, including pirarucu, in the list of products covered by the Minimum Price Guarantee Policy.
July 13, 2020, the National Institute of Industrial Property (INPI) grants Geographical Indication (GI).	The National Institute of Industrial Property (INPI) has published the granting of the Geographical Indication (GI) for the Mamirauá denomination of origin for managed pirarucu. The registration, granted by the INPI, involves nine municipalities in the state of Amazonas (Alvarães, Fonte Boa, Japurá, Juruá, Jutaí, Maraã, Tefé, Tonantins, and Uarini). The delimited area encompasses the Tefé National Forest (Flona), Auatí-Paraná Extractive Reserve (Resex), and the Mamirauá (RDSM) and Amanã (RDSA) Sustainable Development Reserves.
Project of Law 9/2021	The proposal amends the law that establishes the National Policy for the Sustainable Development of Fish culture and Fisheries (Law 11,959 of 2009). It exempts small and medium-scale fish culture enterprises, classified according to specific conditions, from environmental licensing requirements.

Appendix 2

Chart 2 Report of the interviews with actors and organizations in the fish culture and management of pirarucu

Type of institution	Role of the interviewee	Company/Organization	Location	Date	Duration	Interview type
Fish culture (Private)	Fishing engineer	Agriculture and Fish culture	Rondônia	06/05/2021	1 h	On-line
Producer Association	1. Technical assistant 2. Technical coordinator 3. President of ATAMP	ASPROC ATAMP Memorial Chico Mendes	Amazonas Mid-Juruá Mid-Purús	14/08/2021	54 min	On-line
Tannery	Leather buyer	Tannery	Rio de Janeiro	6/010/2021	2 h	On-line
Processor: Refrigeration	Administrative manager (owner)	Refrigeration	Amazonas	09/12/2021	49 min	On-line
Producer Association	President of the Macopani Sector Association	Macopani Sector Producers Association	Juruá-AM RDSM	28/06/2022	15 min	In person
Producer Association	President of AMURMAM	Association of Residents and uders of RDSM Antonio Martins	Tefé-AM RDSM	28/06/2022	42 min	In person
Social Organization	Shared management of fishing resources	Mamirauá Sustainable Development Institute—IDSM	Tefé-AM RDSM	8/06/2022	48 min	In person
Social Organization	Project supervisor	Sustainable Amazon Foundation—FAS	Fonte Boa—AM	30/06/2022	1 h 22 min	In person
Research Center (Public)	Researcher	Embrapa Fishing and Fish culture	Tocantins	23/04/2023	44 min	On-line
Producer	Fish farmer (owner)	Individual fish farmer	Canabrava do Norte—MT	03/05/2023	50 min	On-line
Producer cooperative	President of the cooperative and fish farmer	Cooperative of fish farmers	Alta Floresta—MT	12/05/2023	1 h	On-line

References

Amaral, E., de Sousa, I. S., Gonçalves, A. C. T., Carvalho, G., Braga, R., Ronan, P., & Vilena, J. (2011). In IDSM/OS/MCTI (Ed.), *Manejo de pirarucus (Arapaima gigas) em lagos de várzea de uso compartilhado entre pescadores urbanos e ribeirinhos* (p. 50). Tefé.

Amazonas. (2021). Secretaria de Estado de Desenvolvimento Econômico, Ciência, Tecnologia e Inovação. Secretaria Executiva de Ciência, Tecnologia e Inovação. Nota Técnica: Um panorama sobre a Bioeconomia no Estado do Amazonas/Secretaria de Estado de Desenvolvimento Econômico, Ciência, Tecnologia e Inovação. Secretaria Executiva de Ciência, Tecnologia e Inovação. Governo do Amazonas. 7 p. http://www.sedecti.am.gov.br/

Amazonas. (2022). Amazonas ganhou três novos acordos de pesca ao longo de 2021. https://meioambiente.am.gov.br/amazonas-ganhou-tres-novos-acordos-de-pesca-ao-longo-de-2021/

AtlasBR. (2023). Atlas de Desenvolvimento Humano no Brasil. http://www.atlasbrasil.org.br/

Basta, P. C., et al. (2023). Análise regional dos níveis de mercúrio em peixes consumidos pela população da Amazonia brasileira: um alerta em saúde pública e uma ameaça à segurança alimentar. Nota técnica maio 2023. https://informe.ensp.fiocruz.br/assets/anexos/2441a041be660fb7575f8fe0bf6f8f34.PDF

Bezerra, N. P. (2005). Os ecoturistas estão chegando: aspectos da mudança social na RDS Mamirauá, AM. Dissertação de mestrado. Belém.

Brabo, et al. (2022). Criação de pirarucu em viveiros escavados no estado do Pará: um investimento rentável? *Informações Econômicas, SP, 52*, eie132020. https://doi.org/10.56468/1678-832X.eie1320.2022

Brasil. (2023). Projeto de Lei n° 9, de 2021. Dispensa o licenciamento ambiental aos empreendimentos aquícolas de pequeno e médio porte, assim classificados conforme condições que especifica. https://www25.senado.leg.br/web/atividade/materias/-/materia/146242.

Campos-Silva, J. V., Hawes, J. E., & Peres, C. A. (2019). Population recovery, seasonal site fidelity, and daily activity of pirarucu (*Arapaima spp.*) in an Amazonian floodplain mosaic. *Freshwater Biology, 64*(7), 1255–1264. https://doi.org/10.1111/fwb.13301

Castello, L. (2004). A method to count Pirarucu: Fishers, assessment, and management. *North American Journal of Fisheries Management, 24*(2), 379–389. https://doi.org/10.1577/M02-024.1

Castello, L., & Stewart, D. J. (2010). Assessing CITES non-detriment findings procedures for *Arapaima* in Brazil. *Journal of Applied Ichthyology, 26*(1), 49–56. https://doi.org/10.1111/j.1439-0426.2009.01355.x

Castro, F., & Mcgrath, D. (2001). O manejo comunitário de lagos na Amazônia. In: Parcerias Estratégicas. No. 12, setembro (pdf). https://www.academia.edu/download/49932660/Parcerias_Estrategicas_Castro_McGrath_2001.pdf

CONAB. (2020). Companhia Nacional de Abastecimento. Boletim da Sociobiodiversidade. 2020. PGMP_Bio Volume 4, Número 3 - 3° trimestre de 2020, pp. 47–52. https://www.conab.gov.br/info-agro/analises-do-mercado-agropecuario-e-extrativista/boletim-da-sociobiodiversidade.

da Hora, N. N. (2020). *Sistemas de co-manejo pesqueiro com enfoque no pirarucu (Arapaima spp.) em comunidades de várzea do Baixo Amazonas: análise institucional e custos de transação*. Universidade Federal do Oeste do Pará. https://repositorio.ufopa.edu.br/jspui/handle/123456789/351

Embrapa. (2016). *Discussão sobre a regularização da piscicultura brasileira: da produção à comercialização*. Embrapa Pesca e Aquicultura. https://ainfo.cnptia.embrapa.br/digital/bitstream/item/151862/1/CNPASA-2016-doc31.pdf

Ferreira, G. T. C. (2016). *Competitividade da cadeia produtiva do Arapaima gigas, o pirarucu da Amazônia brasileira*. Doctoral dissertation, Universidade de São Paulo. https://www.teses.usp.br/teses/disponiveis/12/12139/tde-19052016-114414/en.php

Franco, D. d. L. (2019). *Análise da cadeia produtiva do jacaré da Reserva de desenvolvimento sustentável Mamirauá, estado do Amazonas*. UFPE. https://doi.org/10.1590/s0102-33061991000200011

Gonçalves, A. C. T., Cunha, J. B. C. d., & Batista, J. d. S. (2018). *O gigante amazônico: manejo sustentável de pirarucu*. IDSM.

Gonzaga da Silva, N. M., Addor, F., Lianza, S., & Pereira, H. d. S. (2020). O debate sobre a tecnologia social na amazônia: a experiência do manejo participativo do pirarucu. *Revista Terceira Margem Amazônia, 4812*(6), 79–91. https://doi.org/10.36882/2525-4812.2020v6i14p79-91

Goularti Filho, A. (2016). Da Sudepe à criação da Secretaria Especial de Aquicultura e Pesca: as políticas públicas voltadas às atividades pesqueiras no Brasil. *Planejamento e Políticas Públicas, 49*. http://desafios2.ipea.gov.br/ppp/index.php/PPP/article/view/763

Guinato, R. B., do Nascimento, A. C. S., Pereira, H. C., da Silva, M. D. J., Corrêa, D. S. S., & Moura, E. A. F. (2022). Manejos sustentáveis de recursos naturais geram renda? Análise socioeconômica de iniciativas em áreas protegidas na amazônia central. *Contra Corrente: Revista do Programa de Pós-Graduação Interdisciplinar em Ciências Humanas, 18*, 82–102.

IBGE. (2022). *Pesquisa da Pecuária Municipal*. Produção da aquicultura, por tipo de produto. https://sidra.ibge.gov.br/tabela/3940

IDSM. (2014). *Plano de Gestão: Reserva de Desenvolvimento Sustentável Mamirauá—RDSM* (Vol. 1, 3rd ed.). IDSM. https://www.google.com/url?sa=t&rct=j&q=&esrc=s&source=web&cd=&ved=2ahUKEwi8zsr1z6X_AhWkHLkGHQy_D3wQFnoECAgQAQ&url=https%3A%2F%2Fwww.mamiraua.org.br%2Fdocumentos%2F968f2df0775ad27aaace7b5e1064dff9.pdf&usg=AOvVaw1WvfpD3yJi0ejmHMTUNnDS

IDSM. (2023). Programa de Manejo de Pesca. https://www.mamiraua.org.br/manejo-pesca

Imbiriba, E. P. (2001). Potencial de criação de pirarucu, *Arapaima gigas*, em cativeiro. *Acta Amazonica, 31*, 299–299.

INPA. Instituto Nacional de Pesquisas da Amazônia (2021). *Ipen, Embrapa e Inpa apresentam pesquisas sobre pirarucu em Encontro de Ideias*. Ministério da Ciência, Tecnologia e Inovações, 29 de novembro de 2021. https://www.gov.br/inpa/pt-br/assuntos/noticias/ipen-embrapa-e-inpa-apresentam-suas-pesquisas-sobre-especie-de-peixe-amazonico-em-encontro-de-ideias.

Lima. A. F. (2013). *Pirarucu da Amazônia - Ações de Pesquisa e Transferência de Tecnologias*. Embrapa Pesca e Aquicultura. Projeto 2013–2018. https://www.embrapa.br/busca-de-projetos/-/projeto/205513/pirarucu-da-amazonia%2D%2D-acoes-de-pesquisa-e-transferencia-de-tecnologias

Lima, F. L., et al. (2015). *Manejo de plantel de reprodutores de pirarucu*. Embrapa. https://www.embrapa.br/busca-de-publicacoes/-/publicacao/1040400/manejo-de-plantel-de-reprodutores-de-pirarucu.

Lobato, A. B. (2009). *A exploração pesqueira dos grandes bagres migradores na região amazônica e a responsabilidade do estado por dano ambiental transfronteiriço*. Dissertação de Mestrado em Direito Ambiental.

MAPA. (2020). *Guia Prático: Marcas coletivas para a comercialização de produtos da agricultura familiar*. Ministério da Agricultura, Pecuária e Abastecimento. https://www.gov.br/agricultura/pt-br/assuntos/sustentabilidade/indicacao-geografica/arquivos-publicacoes-ig/guia-pratico-marcas-coletivas-para-comercializacao-de-produtos-da-agricultura-familiar/view

Neves, E. G. (2022). *Sob os Tempos do Equinócio: oito mil anos de história na Amazônia Central*. Ubu Editora.

Pedroza Filho, M. X., Muñoz A. E. P., Rodrigues, A. P. O., Rezende, F. P., Lima, A., & Mataveli, F. M. (2016). Boletim Ativos da Aquicultura. Confederação da Agricultura e Pecuária do Brasil (CNA) e Embrapa Pesca e Aquicultura. Ano 2, Edição 8. Fevereiro de 2016. https://www.embrapa.br/pesca-e-aquicultura/busca-de-publicacoes/-/publicacao/busca/Peixe

Prestes-Carneiro, G., Béarez, P., Bailon, S., Py-Daniel, A. R., & Neves, E. G. (2016). Subsistence fishery at Hatahara (750-1230 CE), a pre-Columbian central Amazonian village. *Journal of Archaelogical Science—Reports, 08*, 454–462.

Queiroz, H. L., & Sardinha, A. D. (1999). A preservação e o uso sustentado do pirarucu de Mamirauá. In H. L. Queiroz & W. Cramptom (Eds.), *Estratégias para manejo de recursos pesqueiros em Mamirauá* (p. 208p). CNPQ/MCT.

Saes, M. S. M., Saes, B. M., Feitosa, E. R. M., Poschen, P., Val, A. L., & Marcovitch, J. (2023). When do supply chains strengthen biological and cultural diversity? Methods and indicators for the socio-biodiversity bioeconomy. *Sustainability, 15*(10), 8053.

Salomão, R. P., Rosa, N. A., Castilho, A., & Morais, K. A. C. (2006). Castanheira-do-brasil recuperando áreas degradadas e provendo alimento e renda para comunidades da Amazônia Setentrional. *Boletim do Museu Paraense Emílio Goeldi. Ciências Naturais, Belém, 1*(2), 65–78. maio-ago.

Santos, M. T. (2005). *A pesca na Região Amazônica* (p. 23). ProVárzea: Ibama. (ProVárzea. Cartilhas estudos estratégicos).

Santos, G. M., & Santos, A. C. M. (2005). Sustentabilidade da pesca na Amazônia. *Estudos Avançados, 19*(54), 165–182.

SDGs. (2015). Sustainable development goals. 2015. In *Transforming our world: The 2030 agenda for sustainable development*. United Nations.

Silva, C. D. D. (2009). *Pesca: classes sociais, territorialidades e trabalho em Manacapuru-AM*. p. 145 (Doctoral dissertation, Dissertação (Mestrado em Geografia). Programa de Pós-Graduação em Geografia da Universidade Federal do Amazonas, Manaus).

Silva, K. (2018). Histórias e Memórias Sobre a Pesca do Pirarucu no Auati-Paraná: do Barracão ao Manejo Comunitário. *Amazônica - Revista de Antropologia, 9*(1), 208. https://doi.org/10.18542/amazonica.v9i1.5489

Sousa, A. R. B. d. (2017). Piscicultura de pirarucu. In *Coleção Criar, 7*. Embrapa. https://www.embrapa.br/busca-de-publicacoes/publicacao/1087747/piscicultura-de-pirarucu

UN. (2022). *United Nations. Eight billion people, one humanity; Development milestone 'testament' to power of health and science*. UN News. Global perspective human stories. https://news.un.org/en/story/2022/11/1130417

Veríssimo, J. (1895). *A pesca na Amazônia (Monografia)*. Livraria Clássica de Alves & C.

Viana, J. P. (2004). A pesca no Médio Solimões. In M. L. Ruffino (Ed.), *A pesca e os recursos pesqueiros na Amazônia brasileira*. Ibama/Provárzea.

Viana, J. P., Damasceno, J. M. B., & Castello, L. (2004). Desarrollodel manejo pesquero comunitário en la Reserva de Mamirauá, Amazonas, Brasil. In C. Campos-Rozo & A. Ulloa (Eds.), *Fauna socializada. Tendencias en el manejo participativo de la fauna en América Latina*. Fundação Natura, MacArthur Foundation, Instituto Colombiano de Antropología e História. https://www.academia.edu/download/30495976/manejo.comunitario.mamiraua.espanhol.pdf

WWF. (2016). *Assessoria técnica à implementação da estratégia de desenvolvimento e fortalecimento da cadeia produtiva do pirarucu manejado em lagos naturais do Acre*. CON 00833/2016 - WWF Brasil. https://www.fundoamazonia.gov.br/export/sites/default/pt/.galleries/documentos/acervo-projetos-cartilhas-outros/WWF-Comercializacao-Pirarucu-estudo.pdf.

Impacts of Climate Change on the Parasite–Host Relationship in Freshwater Aquaculture

Gustavo da Silva Claudiano, Jefferson Yunis Aguinaga, Elizabeth Gusmão Affonso, and Bernardo Baldisserotto

Abstract

Extreme temperature events brought on by climate change are occurring more frequently and with greater intensity, which has an impact on all economic activities, including aquaculture. The chapter brings insights into the effect of climate change on the aquaculture industry as well as future research on the topic, exploring specific questions: What possible repercussions can climate change have on the parasite–host dynamic in aquaculture? How is the relationship between the host, biological agents, and environment? Climate change induces alterations in the physical–chemical parameters of the water, impacting all current ecological interactions and maybe limiting fish physiology and development throughout the whole aquaculture process. Through hosts and/or biological agents, fish health is predicted to be more impacted by climate change.

G. da Silva Claudiano (✉)
Federal University of Western Pará, Institute of Biodiversity and Forests, UFOPA, Santarém, Pará, Brazil
e-mail: gustavo.claudiano@ufopa.edu.br

J. Y. Aguinaga
Scientific University of the South, Marine Biology Faculty | IMARPE The Marine Institute of Peru, Lima, Peru

E. G. Affonso
Nilton Lins University, Graduate Program in Aquaculture UniNilton | INPA, Manaus, AM, Brazil

Department of Technology and Innovation, Brazilian National Amazonia Research Institute, INPA, Manaus, AM, Brazil

B. Baldisserotto
Federal University of Santa Maria, Health Sciences Center, Department of Physiology and Pharmacology, UFSM, Santa Maria, Rio Grande do Sul, Brazil

S. S. de Souza et al. (eds.), *The Future of Amazonian Aquatic Biota*,
https://doi.org/10.1007/978-3-031-66822-7_19

1 Introduction

Global aquaculture and fisheries are greatly impacted by the climate changes that have been witnessed over the past several decades everywhere in the world (Matvienko et al., 2020). Compared with marine systems, the effects of climate change on continental freshwater have not received as much attention; yet, it is anticipated to increase inland demand and drastically affect aquatic ecosystems (Fiorella et al., 2021). Farmed fish will be the most affected, as, unlike wild fish, they cannot migrate to areas with better water quality. This chapter will address how these modifications affect the way fish diseases evolve; we will employ the triad or epidemiological triangle model to understand the consequences of climate impact on biological agents (viruses, bacteria, parasites, and fungi), the host (fish), and the environment (fish farming)—diseases. We will also discuss how to treat and manage diseases that impact aquaculture, particularly those that damage fish that are valuable commercially in the Amazon.

2 Host–Parasite Relationships in Aquaculture

To comprehend how the parasite–host relationship will be impacted by climate change, we will employ the triad or epidemiological triangle model, one of the simplest and most traditional models for studying infectious and parasitic diseases. In a simplistic understanding, the triad consists of an agent (virus, bacteria, parasite, or other organisms that cause the disease), fish as the susceptible host, and fish farming as the environment that unites the host and agent. The combination of agent, host, and environmental factors is known as the "epidemiologic triad" among epidemiologists (Thrusfield et al., 2018). In this model, whether or not disease occurs in an individual depends on the interplay of these three factors.

The presence of the agent alone is not always sufficient to produce disease; different diseases require varied balances and interactions of these three components (host, agent, and environment); that is, the components can coexist in an environment without the disease occurring, for instance, from the opportunistic bacterium *Aeromonas hydrophila*, which causes hemorrhagic sepsis in tropical fish (Claudiano et al., 2019). This succession of events necessary for the disease to occur is called an epidemic process (Thrusfield et al., 2018). To trigger this process, an association between agent factors of the host and the environment is necessary; that is, any modification in any of the elements of the ecosystem results in adaptations of the other elements, which may be related to the disease development. Numerous factors, such as the organism's pathogenicity and virulence, dose, animal species, stress, age, host resistance, and others, affect whether exposure to an organism will result in disease (Fig. 1).

Improved comprehension of the intricate relationships is necessary to create conceptual models and potential processes that influence aquaculture productivity in a rapidly changing environment. The two main indicators of dramatic biological shifts are temperature and photoperiod, which can have a significant impact on

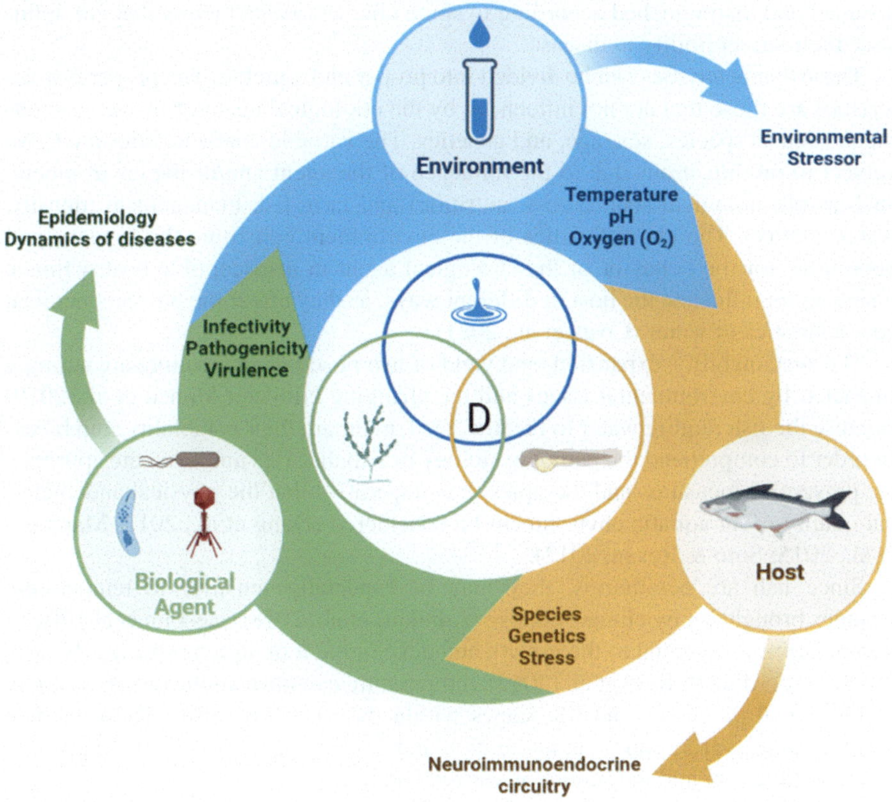

Fig. 1 Epidemiological triangle model applied to the disease triangle in the context of global warming—host–parasite relationships in aquaculture with the main characteristics and their effects on the agent, host, and environmental factors. *D* diseases

pathogen and animal behavior and influence how the three elements of the epidemiologic triad interact (Hueffer et al., 2011).

Climate change impacts aquaculture by raising temperature, decreasing pH, and dissolved oxygen levels in water (Singh, Ahmad, et al., 2021; Singh, Singh, et al., 2021). Changes in water quality can also impact fish behavior, immunophysiology, and parasites, which can impact the parasite–host interaction and perhaps lead to more frequent disease outbreaks (Bruneaux et al., 2017; Costa et al., 2021).

2.1 Host

Host is a living person or animal that, under natural circumstances, allows for the subsistence or accommodation of an infectious agent. In this chapter, we will consider the host as any fish capable of hosting in its organism a causal agent of a parasitic/infectious disease with which it can establish varied relationships. Hosts can be grouped and distinguished according to some characteristics/factors that can influence their susceptibility to disease.

These characteristics can be divided into proper and variable. The proper characteristics are those that are not influenced by the etiological agent or by the environment, such as species, sex, age, and genetics. The variable characteristics are those subject to modifications due to the influence of the agent and/or the environment, such as physiological state (stress, nutrition) and farm fish management (density, water quality). The characteristics of the environment constitute the fundamental conditions for the behavior of the etiological agent in a susceptible host. Climate changes can influence the host in different ways, as they affect the physicochemical characteristics of water (Costa et al., 2021).

The sustainability, expansion, and yield of fish raised in aquaculture are strongly impacted by environmental issues and the changing climate (Ahmed et al., 2019) because the fish require water to breathe, feed, maintain their osmolality, and breed. In order to comprehend the pathophysiology of fish diseases and to create appropriate preventive measures and therapies, it is imperative that the physical and chemical quality of an aquatic environment be considered (Dang et al., 2012; Marcusso et al., 2015; Soto & Revan, 2012).

Since fish are ectotherms, they may be especially sensitive to temperature changes brought on by climate change (Adhikari et al., 2018; Sae-Lim et al., 2017). Temperature is essential to the growth and development of aquatic species (Ngoan, 2018). According to Reid et al. (2019), temperature can have an impact on a variety of biological processes and life stages within the same organism. These include aerobic capacity, reproduction, maturation and development, growth rate, nutrition, immune function, and the timing of seasonal activities.

Increasing water temperatures will have different effects depending on the initial features of the environment. In southern Australia, for example, winter water temperatures are predicted to speed up the growth rate of four abalone species, while summer temperatures are predicted to increase juvenile mortality by a factor of ten (Russell et al., 2012). Hot weather in the tropics, like the Amazon region, can raise water temperatures over what many organisms can withstand. Since fish lack the ability to self-regulate their temperatures and frequently display a limited ability to avoid heat (Dupuis & Hann, 2009), the external temperature controls their body temperatures (Volkoff & Rønnestad, 2020). This is because their metabolic heat production and retaining mechanisms are insufficient to provide body warming (Prosser & Nelson, 1981).

According to Pol et al. (2017), temperature determines the rates of almost all biochemical reactions, and consequently, the speed of physiological processes. The standard metabolic rate increases with temperature (Glencross & Bermudes, 2010), and as a result, so does the maintenance requirement for protein and energy (Lupatsch & Kissil, 2005). In other words, feeding/growth rates rise gradually toward an optimal temperature and rapidly decrease if the temperature rises above this optimal temperature (Collins et al., 2020). Because of this, zootechnical performance depends on temperature and has a bell-shaped curve (Huey & Stevenson, 1979; Nati et al., 2016). Thus, fish exposed to temperatures below their optimum have food consumption limited to their basal metabolic rate, whereas fish exposed to temperatures above their optimum result in a greater energy deviation to obtain oxygen, contributing to the reduction of growth (Rebouças et al., 2014). This occurs because the connection between hemoglobin and oxygen becomes weaker, releasing oxygen more quickly; hence, a rise in the metabolic rate typically corresponds with a rise in temperature (oxygen consumption) (Ferreira et al., 2009; Lund & Tufts, 2003). The growth increases with temperature in fish and larvae, up to the point of adaptation of the species (de Oliveira & Val, 2017; Fey & Greszkiewicz, 2021), providing better zootechnical performance up to the thermal comfort range in some Brazilian tropical fish (Piana et al., 2003; Piedras et al., 2006; Tolussi et al., 2018). Nevertheless, the increased growth is related to the increase in feed consumption, evidenced by less weight gain and higher specific growth rate (de Oliveira & Val, 2017).

Furthermore, the feed conversion ratio (FCR, feed/gain), which quantifies how well feed is turned into meat, can be impacted by temperature (Reid et al., 2019). For some species, some little difference in temperature can produce important differences in feed conversion rate (Britz et al., 1997; Siikavuopio et al., 2012). In addition to the FCR, temperature can also influence gastric evacuation rate by accelerating the time passage of foods and the digestibility of specific nutrients, such as fatty acids or proteins (Hua & Bureau, 2009; Huguet et al., 2015). The temperature can influence the secretory activity of digestive juices (by affecting food ingestion), the motility of the gastrointestinal tract, the activity of digestive enzymes, the pace of digestion, and the rate of absorption (Kapoor et al., 1976; Volkoff & Rønnestad, 2020). Some studies indicate that warmer water may have an impact on gut transit time, as evidenced by an increase in yellowtail kingfish *Seriola lalandi* (Miegel et al., 2010) and *Pygocentrus nattereri* (Silva & Araujo-Lima, 2003) or a decrease in greenlip abalone *Haliotis laevigata* (Currie et al., 2015).

In summary, the high temperature or close to critical temperatures of the species in aquatic environments reduces the concentration of dissolved oxygen/hypoxia, leading to a reduction in anaerobic metabolism with a consequent reduction in the performance of the animals and thermal stress and/or stress by hypoxia.

Amazonian fish developed adaptations in response to environmental variables such as pH and temperature (Saint Paul & Soares, 1987). These adaptations involve behavioral changes due to fish migration to favorable environments and physiological adaptations, such as an increase in the number of allosteric modulators that affect the affinity of hemoglobin with oxygen (Val, 2000), added to ionoregulatory settings and mucus production (Aride et al., 2007; Kim et al., 2021). Despite the acquired adaptations, many fish species become vulnerable to environmental variations, which act as stressors inducing profound metabolic and physiological changes (Islam et al., 2022; Val et al., 2006). These physiological changes are more serious in farmed fish, as they are unable to migrate to a more favorable environment, due to the uniformity of the physicochemical variables and the limitations of the physical space of the culture tanks (Ali et al., 2003).

Fish species, genetics, age, developmental stages, physiological fitness, experimental model, and history of previous thermal exposure all affect how tolerant they are to heat (Bowden, 2008; Chadwick & McCormick, 2017; Islam et al., 2022; Sunday et al., 2012). The increase of temperature influences the comfort zone of tropical fish, affecting their homeostasis, and increasing susceptibility to bacterial or parasitic pathogens (Costa et al., 2021; Marcusso et al., 2015). For example, Nile tilapia raised at 28 °C increased plasma lysozyme and at 33 °C decreased it (Dominguez et al., 2005).

Acute or prolonged increase of temperature (Alfonso et al., 2021; Samaras et al., 2018), as well as other changes related to global warming such as acidification and changes in CO_2 concentration (Aride et al., 2007; Costa et al., 2021) induced stress, consequently greater susceptibility to diseases. Stress is not strictly bad, as physiological responses are important for a fish to adapt to the new reality, whether resulting from acute or chronic stressors (Heimbürge et al., 2019). In order to better adapt to an environmental challenge and reestablish equilibrium, the reaction works to induce a physiological and/or behavioral change (Alfonso et al., 2021). The environmental stressor (temperature, O_2, and pH) is perceived by different receptors and analyzed by the forebrain, mainly the hypothalamus and the telencephalon (Gorissen & Flik, 2016) and may initiate an alarm reaction with the activation of the hypothalamic–pituitary–interrenal (HPI) axis. As a result of this activation, glucocorticoids are released, which cause energy metabolism and neurological function to go into homeostatic mode (Alfonso et al., 2021). The anterior pituitary, the hypothalamus, and the interrenal system make up the three main parts of the HPI axis (Heimbürge et al., 2019).

Neuroendocrine activation has a biphasic character, consisting of two different neuroendocrine pathways, starting with the release of catecholamines and finally the release of cortisol. The first route is fast and transient, with neural stimulation of the chromaffin cells in the head kidney (brain–sympathetic–chromaffin cell axis), causing catecholamines to be released from reserves present in the chromaffin cells of the cephalic kidney (Alfonso et al., 2021; Bonga, 1997). In a longer way, the second route is the HPI axis. It starts with the activation of corticotropin-releasing factor (CRF), peptide that stimulates the production and secretion of adrenocorticotropic hormone (ACTH) into circulation, which activates cortisol synthesis by interrenal cells in the anterior kidney (Barton, 2002; Sadoul & Geffroy, 2019).

Catecholamines and cortisol (major corticosteroid in teleost) have secondary actions in various organs of the animal that improve the response to the stressor (Faught et al., 2020; Feidantsis et al., 2021). The objective of this neuroendocrine activation is to prepare the fish to the new reality, "fight-or-flight" response, being this reaction Seyle (1946) denominated general adaptation (SAG), with three distinct phases that can be divided as primary, secondary, or tertiary responses (Barton, 2002; Bonga, 1997; Fabbri & Moon, 2016; Faught et al., 2020; Schreck & Tort, 2016) (Fig. 2).

The primary response to stress starts with the neuroendocrine response, characterized by the release of catecholamines and corticosteroids; the latter is used as a marker of the primary response in fish, described earlier. Hormone release into the bloodstream and tissues, metabolic disturbance, energy mobilization, and hydromineral balance are examples of secondary stress reactions that help an animal survive

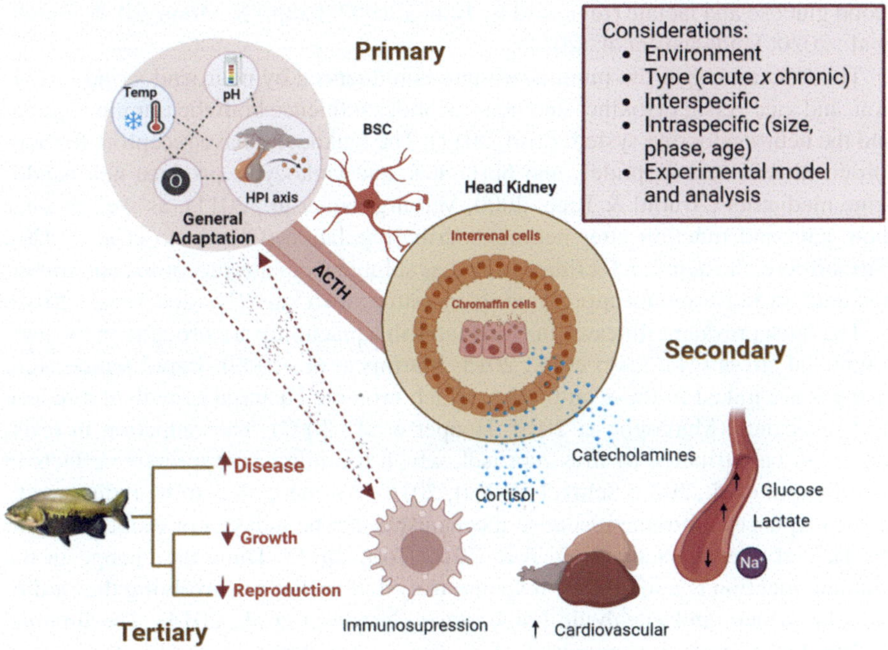

Fig. 2 Physiological responses to stress caused by climate change—water quality. Effects of the primary stress response by catecholamines/cortisol. Neuroendocrine activation has a biphasic character, with rapid and transient activation, with neural stimulation of the chromaffin cells in the head kidney (brain–sympathetic–chromaffin cell axis, BSC), and via the hypothalamic–pituitary–interrenal (HPI) axis with adrenocorticotropic hormone (ACTH) secretion into circulation. Secondary response—enhance cardiovascular frequency, mobilize energy (glucose, lactate, and Na^+) and immunosuppression. The tertiary responses are the result of prolonged stress, with energy diversion negatively impacting growth and reproduction and increasing susceptibility to disease. The immune system also regulates the central nervous system through cytokines secreted by leukocytes. All effects may vary according to the above considerations

a stressor danger (Alfonso et al., 2021; Islam et al., 2022). The implications of global warming in the secondary response of Amazonian fish need to be further studied given the higher metabolic demand of the warmer water climate (Costa et al., 2021; Windisch et al., 2014).

Energy substrate mobilization, which mostly consists of glucose, lactate, and amino acids, is caused by catecholamines and corticosteroids (Claudiano et al., 2020; Faught & Vijayan, 2016) (Fig. 2). The neuroendocrine response of fish to catecholamines occurs via cAMP-PKA signaling pathway plays (Feher, 2017) and cortisol by upregulating key genes (Claudiano et al., 2020; DiBattista et al., 2006; Vijayan et al., 2003). Although they have different mechanism of action, both induce glucose regulation by hepatic glycogen breakdown with redirection of energy reserves to provide energetic substrates that can be used by muscles and other tissues to facilitate behavioral and adaptive responses. We can include in this process of adaptation the increased heart rate, blood flow, vasodilation, and increase in blood glucose and lactate (reviewed by Bonga, 1997; Fabbri & Moon, 2016; Faught et al., 2020; Feidantsis et al., 2021).

The functioning of the immune system is influenced by neuroendocrine mediation, and vice versa; cytokines and neuropeptides influence both the immune system and the neuroendocrine system (Tort, 2011). The leukocytes have receptors for neurotransmitters, neuropeptides, and hormones, and leukocytes produce neuroendocrine mediators (Arnold & Rice, 2000; Verburg-van et al., 2011), as well as alter their state and function after neuroendocrine regulation (Claudiano et al., 2020). Furthermore, the neuroendocrine system has soluble mediator receptors, and immunostimulant and immunosuppression against stress are related (Nardocci et al., 2014).

The most prevalent illnesses in Brazilian fish aquaculture are brought on by environmental stress (Marcusso et al., 2015; Martins et al., 2009). Rapid temperature changes are linked to these outbreaks, which promote bacterial growth and reduce host resistance (Marcogliese, 2010; Hooper et al., 2007). The reduction in resistance can be attributed to stress/cortisol, which has immunosuppressive effects in fish (Faught et al., 2020; Schreck & Tort, 2016). As energy has to be redirected to stress responses, immune response mechanisms can be delayed or reduced due to the lack of energy (Schreck & Tort, 2016; Tort, 2011). Thus, the change in the immune function is a secondary response to systemic changes involving the cardiovascular system and metabolic homeostasis (Nardocci et al., 2014). The immune system, for example, is suppressed when survival is in threat due to elevated cortisol activity (Bonga, 1997; Faught et al., 2020; Nardocci et al., 2014). To counteract the widespread effect of the immune response, host cells can be protected by immunosuppressive mechanisms (Claudiano et al., 2020; Nardocci et al., 2014; Weyts et al., 1998).

Acute phase protein serum amyloid S (SAA) and proinflammatory cytokines including IL-1β, TNF-α, IL-12, and iNOS activity are all inhibited by cortisol when LPS is present (Fast et al., 2008; Huising et al., 2004; Saeij et al., 2003), as well as reduces the increase in vascular permeability and leukocyte extravasation in pacus, *P. mesopotamicus*, delaying leukocyte recruitment to the inflamed focus (Claudiano et al., 2013). In addition, cortisol inhibits chemotaxis and phagocytosis in goldfish (Shimon-Hophy, 2021) as well as proliferation of a fish macrophage cell line (Pagniello et al., 2002). Chronic cortisol stress decreased the expression of interferon (IFN) α-1 and heat shock proteins 70 (HSP70) and 90 (HSP90) (Gadan et al., 2012). In summary, the relationship between the immunological and neuroendocrine systems may be used to regulate stress-induced immunosuppression (Nardocci et al., 2014).

Both short-term and long-term temperature fluctuations outside of the ideal range have a detrimental impact on fish immune systems (Makrinos & Bowden, 2016; Wright & Cooper, 1981), but they may affect fish differently (Guo & Dixon, 2021). In acute and short-term stress, there is an immunostimulant response, and in chronic response, there is an immunosuppressive effect, a fact that explains the increased susceptibility of fish to viral, bacterial, and parasitic diseases (Costa et al., 2021; Guo & Dixon, 2021; Marcusso et al., 2015; Tort, 2011). Still, because acute and long-term temperature changes affect ion balance, they produce heart rhythm disturbances (Vornanen, 2016).

Transport, immunization, predation, fighting, and other discrete, short-term events are examples of acute stress. These events can cause an animal to die. However, the most prevalent type of stress in aquaculture is chronic stress, which can fluctuate in intensity but is often present at a low level (stocking density, socio-environmental interactions—dominance, infections, water quality, etc.). Fish release cortisol and catecholamines when they are abruptly exposed to high temperatures because they perceive it as a stress (Alfonso et al., 2021; LeBlanc et al., 2011).

It has been demonstrated in multiple teleost species that short-term, acute temperature changes can cause an increase in antibody levels in response to rising water temperatures (Bowden, 2008; Dominguez et al., 2004; Jung et al., 2012). The proteome of livers revealed an increase in a number of proteins, such as heat shock cognate protein (chaperoning), glutamine synthetase (ammonia metabolism), cofilin and beta-tubulin (cytoskeleton), protein-like fatty acid binding protein (lipid transport and antioxidant role), hemoglobin, and multiple proteins involved in carbohydrate metabolism. Numerous proteins were also involved in secondary and tertiary stress responses, including altered energy metabolism, osmotic regulation, and immune system responses (Ahmed et al., 2019).

Increasing temperature also induces increased lysozyme, IgM, and interleukin-1β (IL-1β) activity (Huang et al., 2015). Immediate responses of the innate humoral immunity in fish show high levels of lysozyme and C3 proteins, and upregulation of C5a (anaphylatoxin) (Demers & Bayne, 1997; Raida & Buchmann, 2007; Sunyer & Tort, 1995) also increases mucus immunity in fish, mainly IgM, IL-1β, hepcidin, transferrin, and acid/alkaline phosphatase after acute stress (Huang et al., 2011). On the other hand, reduction in temperature increases mortality in tilapias (Marcusso et al., 2015), which can be attributed to a reduction in innate immunity, verified by the decrease in the total leukocyte, respiratory burst activity, phagocytic index, and activities of the complement system (Ndong et al., 2007). Acute stress induces increased expression of proinflammatory chemical mediators such as TNF-α, IL-1β, IL-6, and IL-8 (Hoseini et al., 2019; Metz et al., 2006; Costa et al., 2021; Su et al., 2020), including downregulation of IL-10 anti-inflammatory mediator (Costa et al., 2021; Machado et al., 2020). The endocrine stress systems also influence the release of heat shock protein (HSPs) at the cellular level (Currie et al., 2008; Maloyan & Horowitz, 2002).

During acute heat stress, fish's antioxidant capability also significantly alters, with higher levels of superoxide dismutase, glutathione-S-transferase, catalase, lipid peroxidation, and cytochrome (Clotfelter et al., 2013; Kyprianou et al., 2010). In Amazonian fish, such as tambaqui, *Colossoma macropomum*, the extreme climate scenario shows a significant increase in the level of parasitism, which is accompanied by an inflammatory response, as well as by overexpression of proinflammatory cytokines such as IL-1β and downregulation of anti-inflammatory cytokine such as IL-10, with upregulation superoxide dismutase (Costa et al., 2021, Costa et al., 2021). The immunosuppressive effects of acute stress on innate defense, antioxidants, and modifications in chemical mediation that vary depending on the experimental model and species are some other contradictory findings (Bowden, 2008; Guo & Dixon, 2021; Makrinos & Bowden, 2016).

The production of fish outside their ideal temperature is a chronic stress that affect the homeostasis of the animals, with changes in the synthesis and release of neurotransmitters and cortisol that extend to secondary stress responses (Islam et al., 2022; Kim et al., 2019; Madaro et al., 2018; Vargas-Chacoff et al., 2019). Typically, elevated cortisol levels are seen concurrently with the secondary stress response (Gallant et al., 2017; Kandalski et al., 2018; Samaras et al., 2018; Steinhausen et al., 2008) appearing in peaks with varied time intervals (Alfonso et al., 2021; Benítez-Dorta et al., 2017) with negative feedback that affects cortisol production in fish (Benítez-Dorta et al., 2017; Mommsen et al., 1999). However, as observed by Alfonso et al. (2021), there have been documented instances of deviation from this overall trend (Davis, 2004; Samaras et al., 2018).

During chronic stress in liver tissues, there was downregulation in the levels of IL-8 and IL-6 (Philip & Vijayan, 2015). These results were confirmed in zebrafish, noting that after proinflammatory response in acute stress, there was upregulation for IL-8 and TNF-α in prolonged stress (Jiang et al., 2018). In rainbow trout, chronic stress induced significant downregulation in the expression of LyzII, IL-8, TNF-1α, IL-1β, and IFN-γ, accompanied by increased levels of cortisol, serum ACTH, and lactate. All genes and marker related to stress and immunity was downregulated, along with serum complement and bactericidal activity, IgM, and anti-protease activity, as well as specific antibody titers against *Aeromonas hydrophila* (Yarahmadi et al., 2016). Similar results were verified in turbot, *S. maximus*, indicating that chronic stress blocks the activities of metabolic and antioxidant enzymes, changes physiological parameters and immunosuppression (Liu et al., 2016). Overall, the data point to the fish immune system being mediated by the acute stress response being less suppressive than that of the chronic stress response (Guo & Dixon, 2021).

Thus, climate changes make aquatic organisms more susceptible to pathogens (Islam et al., 2022) because it induces tertiary responses to stress, negatively impacting growth and reducing reproductive and immune performances (Tort, 2011; Yada & Tort, 2016). Long-term changes, such as climate change, will alter not only the temperature, but also the pH and the oxygen levels in aquatic environments, affecting the physiology and immunity of the fish (Makrinos & Bowden, 2016). Global warming, together with an increase in atmospheric CO_2, decreases the pH of the water, resulting in water acidification (Chow et al., 1994; Feely et al., 2004).

The Amazon is home to three primary types of water: white (muddy, pH 6.2–7.2), black (pH 3.8–4.9), and clear (pH 4.5–7.8) (Aride et al., 2007). Fish health is impacted by daily and seasonal fluctuations in water pH levels (Matsuo & Val, 2002). Species have varied tolerance to extreme changes in pH, but all showed significant ion loss at low pH and the magnitude of ionic disturbance is related to the lethality to low pH (Costa et al., 2015).

Low pH reduces the ability of hemoglobin to absorb oxygen by inhibiting the excretion of H+ and CO_2 produced by metabolism, leading to acidemia (Fromm, 1980; Kim et al., 2021). Low pH (4.5) decreased phagocytic rates of leukocytes and plasma IgM levels (Nagae et al., 2001) and in vitro analysis it was observed that binding could eliminated in this pH range (Fernandes et al., 2004); immunosuppression was also verified by the decreasing lysozyme levels and antioxidant mechanisms in the serum at acidic pH (Ma et al., 2015; Yada et al., 2006). Fish under stress from low pH have morphological changes on their gills and skin histology, erythrocyte swelling, the formation of immature erythrocytes, reductions in total erythrocyte counts, hemoglobin, serum protein, and antioxidant mechanisms, all of which have an effect on survival (Costa et al., 2015; Das et al., 2006; Rodrigues et al., 2015; Shuangyao et al., 2018).

Both temperature and ionic concentration influence water oxygen levels (Ficke et al., 2007; Makrinos & Bowden, 2016). Dissolved oxygen (DO) that is biologically available enters the water column through the atmosphere, through motions like waves and waterfalls, and through photosynthetic activity (Kalff, 2000; Makrinos & Bowden, 2016; Stickney, 2000). Thus, DO levels may drop because of increased temperature—a consequence of increases in biological oxygen demand (Kalff, 2000), i.e., increased metabolism of algae, bacteria, and aquatic organisms, and temperature and oxygen solubility in water are inversely correlated (Ficke et al., 2007).

According to Martínez et al. (2020), one of the most important stressors in a fish farm is dissolved oxygen, which can cause hypoxia if the levels drop too far. Hypoxia is characterized by a decrease in oxygen concentration that has a negative impact on immunity, development, respiration, behavior, and metabolism in fish (Abdel-Tawwab et al., 2019; Jiang et al., 2017). Fish have primary, secondary, and tertiary stress reactions as a result of hypoxia, which can result in both acute and long-term stress (Bernier et al., 2012; Bernier & Craig, 2005; Martínez et al., 2020).

Depending on the species of fish, behavioral variations can be seen, such as fast swimming in a circle with its mouth open (Abdel-Tawwab et al., 2019), even the formation of a labial protuberance, as happens in tambaqui, verifying histological changes such as the submucosa of fibroblasts, loosely distributed, and the formation of spaces resembling alveoli, as well as numerous capillaries and lymphocytes (Alves et al., 2009). Certain species might either stay unmoving at the bottom of the tank in order to conserve energy or they might exhibit no particular behavior in an attempt to leave the hypoxic environment (Douxfils et al., 2012; Israeli & Kimmel, 1996; Wu, 2002). Anaerobic stress is linked to this rush/rest behavioral response and is characterized by elevated levels of cortisol, glucose, and lactate in the plasma (Cook & Herbert, 2012).

The fish gill is the primary location of gaseous exchange and the first organ to be affected during a hypoxic episode (Tiedke et al., 2015). Organs may develop necrotic or apoptotic lesions as a result of tissues receiving less oxygen (Geng, 2003; Van der Meer et al., 2005). Macroscopically, the gills show pale color (Mallya, 2007) and other pathological–anatomic alterations, such as decreases in the size of interlamellar cells and increases in the functional surface area of the gills in response to low levels of oxygen (Sollid et al., 2003). Hypoxia can even provoke necrosis, congestion, edema, hemorrhage, hyperplasia, and/or hypertrophy, i.e., the gill loses morphological structure and other organs such as liver, spleen, and anterior and posterior kidney can also be affected (Scott & Rogers, 1980).

The main and secondary stress responses are triggered by the DO decrease, which tries to adapt the body to hypoxic stress and restore equilibrium (Abdel-Tawwab & Wafeek, 2017; Gobi et al., 2018). Many species increase plasma cortisol levels, accompanied by increased erythrogram values and metabolism fueled by both glucose and lactate (Abdel-Tawwab et al., 2015; Affonso et al., 2002; Wells & Baldwin, 2006).

From an immunological perspective, hypoxia led to immunosuppression, which decreased defense against infections by pathogens (Di Marco et al., 2008; Evans et al., 2003). There is a drop of complement hemolytic activity (Welker & McNulty, 2007) and reduction in lysozyme activity, C3 components, reactive oxygen species, and bactericidal activity (Boleza et al., 2001; Ortuno et al., 2002), and the quantity of lymphocytes to the infectious focus was impaired (Schäfer et al., 2021). Chronic hypoxia increased plasma cortisol levels with downregulation in MHC II transcription in the kidney and increased IgM transcription in the spleen, indicating tissue-specific expression (Martínez et al., 2020).

As explained, climate changes can influence the host, as they impact the physicochemical parameters of the water, resulting in metabolic and functional organ changes, influencing health and zootechnical performance of the fish. Despite this, it is not possible to generalize the impact on aquaculture, due to the many factors involved: genetics, geographic location, species, type of production, and others. A fact that further studies need to investigate are the consequences of global warming on factors linked to the host (fish).

2.2 Biological Agents

In this chapter, we will consider the biological agent as the microorganism or parasite with potential infectious action on the host (fish) in general, including viruses, bacteria, fungi, and parasites (protozoa, helminths, and arthropods), capable of triggering the clinical signs and symptoms of a pathology. The characteristics of biological agents are morphology, infectivity, pathogenicity, virulence, immunogenicity, variability, resistance, and persistence (Thrusfield et al., 2018).

Temperature and pH are important factors that control the microbial and parasitic growth and pathogenesis (Adams & Moss, 2008; Hakalahti et al., 2006; Shapiro & Cowen, 2012; Silvestri et al., 2018). Microorganisms and parasites develop within a temperature range, which vary with the species (Khosravi et al., 2023). Due to its potential to impact critical steps in the pathogen's transmission chain, global warming may change the dynamics of illnesses (Khosravi et al., 2023), impacting outbreaks of infectious and parasitic diseases, particularly due to effects on transmission dynamics in aquatic organisms (Altizer et al., 2013; Harvell et al., 2002). Therefore, climate change affects fish farms by altering the parasite–host relationship, impacting the immunophysiology of fish and the characteristics of the biological agents (Costa et al., 2021; Marcusso et al., 2015; Peacock et al., 2022). Because of the intricacy of interactions between different species and the multiple interconnected ways that climate change might modify dynamics, it is difficult to forecast the outcome of this ecological relationship (Peacock et al., 2022). Through changes in the distribution, prevalence, and virulence of pathogens (bacteria, viruses, fungi, and parasites), climate change can raise the risk of disease (Altermatt et al., 2008; Tignat-Perrier et al., 2022).

Fish raised in farms typically exhibit little genetic diversity (down to single host lines or genotypes), are developed for high output, and are kept at extremely high stocking densities—many orders of magnitude higher than in the wild (Mennerat et al., 2010). The emergence of novel diseases can be triggered by these factors in conjunction with ecological changes. For example, intensive farming creates conditions for parasite growth and transmission that are significantly different from those experienced by parasites in wild host populations. This can lead to changes in selection on various traits, including biological agent characteristics (Mennerat et al., 2010).

Warmer winters may make it possible for year-round infection and several generations of parasites, which could lead to higher parasite loads due to greater opportunities for both direct and indirect transmission (Marcogliese, 2001). Raising the temperature can also improve the metabolism of parasites, producing more stages of transmission and increasing parasite fitness in a single outbreak. These two outcomes could lead to an increase in the prevalence of the disease (Lõhmus & Bjorklund, 2015).

As a result, the warmer water speeds up the metabolism of parasites, causing them to spread more quickly. If the ideal temperature for growth and transmission is surpassed, the prevalence of some parasites may also decline (Lõhmus & Bjorklund, 2015). For instance, warmer, drier weather throughout the period was connected with a fourfold increase in colonization rate by the metacommunity of three species, *Daphnia magna*, *Daphnia longispina*, and *Daphnia pulex* (Altermatt et al., 2008). In addition to these, cercarial stage, responsible for transmission to the intermediate host in the trematode cycle, has its higher infectivity rate with increasing temperature (Poulin, 2006; Thieltges & Rick, 2006), as verified in *Renicola roscovita*, *Acanthotrema tridactyla* and *Cyathocotylidae* gen. sp., aquatic trematode species (Khosravi et al., 2023; Montaudouin et al., 2016). In temperate places where parasite prevalence and transmission are strongly seasonal, elevated temperatures may also be advantageous to parasites if they prolong the time of transmission and reproduction (Hakalahti et al., 2006). Monogenean parasitism increased quickly and aggressively in Tambaqui exposed to the extreme climate scenario (Costa et al., 2021).

The analysis of the seasonal variation of parasites in tambaqui in the North region of Brazil verified that *Neoechinorhynchus buttnerae*, *Perulernaea gamitanae*, *Trichodina* sp., Monogenea, *Ergasilus* sp., and others presented greater diversity, richness (number of species in a community), and dominance in the rainy season (cooler), with evenness being greater in the dry season (warmer) (Medeiros, 2020).

Bacterial diseases common in many fish cultures are strongly impacted by temperature. There was an increase in incidence, disease severity, and salmon mortality due to *Flavobacterium columnare* in fingerlings in Finland through a period of

23 years (Pulkkinen et al., 2010), concomitant with high water temperature (Altermatt et al., 2008). Extended warm water periods can prolong the time between columnaris outbreaks because they lengthen *F. columnare* growth period (Pulkkinen et al., 2010). The pathogen *Streptococcus agalactiae* has been linked to outbreaks of septicemia and meningoencephalitis in Brazilian Nile tilapia farms, and it has been shown to increase mortality rates above 26 °C (Mian et al., 2008).

Temperature and antibiotic stress on bacteria can lead to a type of cross resistance, since there is evidence that links climate change to an increasing difficulty in treating illnesses due to increased antimicrobial resistance and biofilm development (Hauser et al., 2021; Lorenz & Wackernagel, 1994; MacFadden et al., 2018; Pepi & Focardi, 2021). The majority of aquaculture environments worldwide have high levels of the multiantibiotic resistance index (MAR), according to a meta-analysis of 460 articles. It also showed that rising temperatures are linked to higher mortality rates and MAR in the majority of aquatic organisms infected with major bacterial pathogens, including *Aeromonas* spp., *Edwardsiella* spp., *Flavobacterium* spp., *Streptococccus* spp., *Lactococcus* spp., and *Vibrio* spp. (Reverter et al., 2020). In addition to MAR expression, temperature affects a number of pathogenic characteristics, including adhesion, biofilm formation, motility, synthesis of flagellar components, and the generation of a quorum sensing signal (Cascarano et al., 2021). High bacteria growth and virulence factors expression of *Photobacterium damselae* subsp. *damselae*, which causes a disease characterized by granulomas in the liver and head kidney (Bordas et al., 1996; Shimada et al., 2014), have been related to elevate water temperature (Matanza & Osorio, 2018). Outbreaks of this pathogen in Cobia, *Rachycentron canadum*, have shown high resistance to the main antimicrobials used in Brazilian fish farms (Shimada et al., 2020).

Despite this, some pathogens showed the opposite response in elevated temperatures. If their optimal temperature for growth and transmission is exceeded, a lower prevalence of these pathogens can occur (Karvonen et al., 2010). Low water temperature (<22 °C) is one of the main predisposing factors for the occurrence of outbreaks in Brazilian farms caused by *Francisella noatunensis*, an important and emerging pathogen that has caused outbreaks in fish farms in numerous countries (Hsieh et al., 2007; Leal et al., 2014). When the water temperature drops below 22 °C, it causes a high death rate; when the temperature rises beyond 24 °C, it causes no mortality (Rodrigues et al., 2018).

Specific ideal temperatures are necessary for aquatic parasites and bacteria to complete their life cycles, and these temperatures can either raise or decrease the likelihood of illnesses (Marcogliese, 2001; Tignat-Perrier et al., 2022). As a result, climate change may alter host–parasite dynamics by altering opportunities for transmission and host susceptibility (Marcogliese, 2001; Pulkkinen et al., 2010). In summary, climate change can affect bacteria and parasites directly through the environment or indirectly through effects on host parameters, such as distribution, behavior, physiology, and mortality (Marcogliese, 2001; Tignat-Perrier et al., 2022).

2.3 Effects on Host–Parasite Dynamics

Climate change raises the danger to aquaculture by changing the distribution, prevalence, and virulence of infections and making hosts more susceptible to disease. Despite this, in current conditions it is difficult to predict the development of diseases on fish farms, as the culture environment is multistressor. The interactions of stressors are complex to predict, and climate changes will influence the parasite–host relationship, resulting in disease. Through a few examples of infections and co-infections linked to changes in water quality that result in high morbidity and mortality, this section aims to shed more light on the possibility of significant changes in the dynamics of infectious and parasitic diseases as a result of climate change.

These data, suggesting that variations in water temperature can impact not only the frequency, but also the diversity and dominance of parasite species such as those found in the work mentioned earlier, are *Streptococccus* spp., *Aeromonas* spp., *Perulernaea gamitanae*, *Edwardsiella* spp., *Flavobacterium* spp., *Lactococcus* spp., *Neoechinorhynchus buttnerae*, *Trichodina* sp., Monogenea, *Ergasilus* sp., and others.

Teleost physiology is temperature-dependent, meaning that variations in temperature will probably have a greater impact on the rate and timing of development in some species than in others. Therefore, during warm seasons, infections may become more prevalent than their hosts earlier due to wider thermal windows. As such, variable responses of the parasite and host to elevated temperatures might change the severity of the disease, causing it to worsen or not (Lõhmus & Bjorklund, 2015). The intensity and pathology of the parasites in infected host people may therefore rise in general for those parasites whose temperature increases do not exceed their fatal limits (Byers, 2020). As a result of interactions with other species in the ecosystem, temperature is likely to produce secondary effects that enhance impacts on the host and increase susceptibility to secondary diseases (Byers, 2020). Because immune system of fish is weakened by stressors including overcrowding, high temperatures, and osmotic stress, changes in global temperatures will also impact fish susceptibility to disease (Ficke et al., 2007).

In this context, for example, data from the research groups of the Federal University of Western Pará (UFOPA) and the National Institute for Amazonian Research (INPA), verified that juvenile tambaqui and matrinxã (*Brycon amazonicus*), reared in a recirculation aquaculture system (RAS) and biofloc system (BFS), respectively, with a drop in temperature at dawn in the winter period (temperatures below 22 °C), become more susceptible to infections by bacteria of the genus *Aeromonas* spp., more common of the species *Aeromonas hydrophila* and *Aeromonas jandaei*. In addition, co-infections with *Ichthyophthirius multifiliis* are frequent, in these conditions, mainly in tambaquis (data not published).

Stress due to temperature can either enhance host vulnerability and hence increase transmission or it can increase infected host death and thus decrease transmission. (Lafferty & Holt, 2003; Thomas & Blanford, 2003). This is part of the truth, especially in the environment, because in a culture system, a dead animal is a reservoir for microorganisms and parasites, even altering their pathogenicity (Pulkkinen et al., 2010). In this context, the increased temperature and CO_2 lead to a 7-day increase in parasitism by monogeneans, which decreased after 30 days but remained higher than that of the control group in tambaqui gills. This further highlights the increased stress and immune-related issues that accompany a swift and aggressive rise in parasitic infection brought on by the climate change scenario (Costa et al., 2021). Tambaquis exposed for 7 days to an environmental room at elevated temperature showed rapid increase of parasitism intensity and acute inflammation, related to the higher expression IL-1β and downregulated IL-10 anti-inflammatory gene, with tissue damage (Costa et al., 2021). Problems with water acidification increase the susceptibility of Amazonian fish, we report a case of a co-infection with trichodinids and *Aeromonas hydrophila* in pirarucu, *Arapaima gigas* from a commercial fish farm. The highest death rates have been observed at 24 °C and 32 °C. In Brazil, outbreaks of *S. agalactiae* infection have been documented in Nile tilapia when they are outside their thermal comfort zone (Marcusso et al., 2015).

According to Abdel-Tawwab et al. (2019), hypoxia alters fish innate and adaptive immune responses. Streptococcal infection caused a high death rate in tilapia (*Oreochromis* sp.) exposed to hypoxic circumstances and high temperatures. In addition, the stressed fish had lower serum levels of nonspecific immune parameters (adenosine deaminase, acetylcholinesterase, catalase, and superoxide dismutase). The head kidney of the stressed fish also showed lower levels of IgM and IgT, which may increase the likelihood of disease, and higher transcripts of inflammatory-related genes (TNF-α, IL-1β, and IL-10). The findings imply that fish homeostasis is impacted by displacement from the comfort zone, which heightens the fish's vulnerability to bacterial and parasite illnesses, including co-infections.

3 Conclusion

The impact of global warming on host–parasite dynamics is an increasingly important goal. The literature on climatic changes and the impact on the parasite–host relationship highlights the complexity of the interaction and the impact of changes in water quality on fish and on parasites and microorganisms. It is evident that understanding how fish will adapt to future climatic changes and how they will affect aquaculture and fisheries productivity are important.

Climate changes can influence the host, as they impact the water physicochemical parameters, resulting in metabolic and functional organ changes, influencing the health and zootechnical performance of the fish. Despite this, it is difficult to generalize the impact on aquaculture, due to the many factors involved: genetics, geographic location, species, type of production, and also because of the effect on parasites and bacteria. Climate change may modify host–parasite dynamics by altering possibilities for transmission and host susceptibility, either directly through the environment or indirectly through effects on host factors, such as distribution, behavior, physiology, and mortality. Therefore, more research is required to determine how the changing climate will affect various fish species, parasite species, and microorganisms, as well as how these will interact in real-world settings, in order to better understand how the parasite–host relationship will develop in the future. Furthermore, there aren't many studies on freshwater fish species, particularly those found in the Amazon, which emphasizes the need for more research to increase our knowledge of how the region is affected by global climate change.

References

Abdel-Tawwab, M., Hagras, A. E., Elbaghdady, H. M., & Monier, M. N. (2015). Effects of dissolved oxygen and fish size on Nile tilapia, *Oreochromis niloticus* (L.): Growth performance, whole-body composition, and innate immunity. *Aquaculture International, 23*, 1261–1274. https://doi.org/10.1007/s10499-015-9882-y

Abdel-Tawwab, M., Monier, M. N., Hoseinifar, S. H., & Faggio, C. (2019). Fish response to hypoxia stress: Growth, physiological, and immunological biomarkers. *Fish Physiology and Biochemistry, 45*(3), 997–1013. https://doi.org/10.1007/s10695-019-00614-9

Abdel-Tawwab, M., & Wafeek, M. (2017). Fluctuations in water temperature affected waterborne cadmium toxicity: Hematology, anaerobic glucose pathway, and oxidative stress status of Nile tilapia, *Oreochromis niloticus* (L.). *Aquaculture, 477*, 106–111. https://doi.org/10.1016/j.aquaculture.2017.05.007

Adams, M. R., & Moss, M. O. (2008). The microbiology of food preservation. In M. R. Adams & M. O. Moss (Eds.), *Food microbiology* (3rd ed., pp. 92–97). RSC Publishing.

Adhikari, S., Keshav, C. A., Barlaya, G., Rathod, R., Mandal, R. N., & Ikmail, S. (2018). Adaptation and mitigation strategies of climate change impact in freshwater aquaculture in some states of India. *Journal of Fisheries Sciences, 12*, 016–021.

Affonso, E. G., Polez, V. L., Correa, C. F., Mazon, A. F., Araujo, M. R., Moraes, G., & Rantin, F. T. (2002). Blood parameters and metabolites in the teleost fish *Colossoma macropomum* exposed to sulfide or hypoxia. *Comparative Biochemistry & Physiology (C), 133*, 375–382. https://doi.org/10.1016/S1532-0456(02)00127-8

Ahmed, N., Thompson, S., & Glaser, M. (2019). Global aquaculture productivity, environmental sustainability, and climate change adaptability. *Environmental Management, 63*, 159–172. https://doi.org/10.1007/s00267-018-1117-3

Alfonso, S., Gesto, M., & Sadoul, B. (2021). Temperature increase and its effects on fish stress physiology in the context of global warming. *Journal of Fish Biology, 98*(6), 1496–1508. https://doi.org/10.1111/jfb.14599

Ali, M., Nicieza, A., & Wootton, R. J. (2003). Compensatory growth in fishes: A response to growth depression. *Fish and Fisheries, 4*, 147–190. https://doi.org/10.1046/j.1467-2979.2003.00120.x

Altermatt, F., Pajunen, I. V., & Ebert, D. (2008). Climate change affects colonization dynamics in a metacommunity of three *Daphnia* species. *Global Change Biology, 14*, 1209–1220. https://doi.org/10.1111/j.1365-24862008.01588.x

Altizer, S., Ostfeld, R. S., Johnson, P. T., Kutz, S., & Harvell, C. D. (2013). Climate change and infectious diseases: From evidence to a predictive framework. *Science, 341*(6145), 514–519. https://doi.org/10.1126/science.1239401

Alves, M. I. M., Souza, P. C. G. d., Silva, P. R. R. d., & Mesquita, P. E. C. (2009). Estratégia Respiratória do Tambaqui, *Colossoma macropomum* (Cuvier, 1818), em Condições de Hipóxia. *Revista Científica De Produção Animal, 1*(2), 39. Recuperado de https://periodicos.ufpb.br/ojs/index.php/rcpa/article/view/42616

Aride, P. H. R., Roubach, R., & Val, A. L. (2007). Tolerance response of tambaqui *Colossoma macropomum* (Cuvier) to water pH. *Aquaculture Research, 38*, 588–594. https://doi.org/10.1111/j.1365-2109.2007.01693.x

Arnold, R. E., & Rice, C. D. (2000). Channel catfish, *Ictalurus punctatus*, leukocytes secrete immunoreactive adrenal corticotropin hormone (ACTH). *Fish Physiology and Biochemistry, 22*, 303e10. https://doi.org/10.1023/A:1007884903577

Barton, B. A. (2002). Stress in fishes: A diversity of responses with particular reference to changes in circulating corticosteroids. *Integrative and Comparative Biology, 42*, 517–525. https://doi.org/10.1093/icb/42.3.517

Benítez-Dorta, V., Caballero, M. J., Betancor, M. B., Manchado, M., Tort, L., Torrecillas, S., & Montero, D. (2017). Effects of thermal stress on the expression of glucocorticoid receptor complex linked genes in Senegalese sole (*Solea senegalensis*): Acute and adaptive stress responses. *General and Comparative Endocrinology, 252*, 173–185. https://doi.org/10.1016/j.ygcen.2017.06.022

Bernier, N. J., & Craig, P. M. (2005). CRF-related peptides contribute to stress response and regulation of appetite in hypoxic rainbow trout. *American Journal of Physiology-Regulatory, Integrative and Comparative Physiology, 289*, 982–990. https://doi.org/10.1152/ajpregu.00668.2004

Bernier, N. J., Gorissen, M., & Flik, G. (2012). Differential effects of chronic hypoxia and feed restriction on the expression of leptin and its receptor, food intake regulation and the endocrine stress response in common carp. *Journal of Experimental Biology, 215*, 2273–2282. https://doi.org/10.1242/jeb.066183

Boleza, K. A., Burnett, L. E., & Burnett, K. G. (2001). Hypercapnic hypoxia compromises bactericidal activity of fish anterior kidney cells against opportunistic environmental pathogens. *Fish & Shellfish Immunology, 11*, 593–610. https://doi.org/10.1006/fsim.2001.0339

Bonga, S. E. (1997). The stress response in fish. *Physiological Reviews, 77*, 591–625. https://doi.org/10.1152/physrev.1997.77.3.591

Bordas, M. A., Balebona, M. C., Zorrilla, I., Borrego, J. J., & Morinigo, M. A. (1996). Kinetics of adhesion of selected fish-pathogenic *vibrio* strains of skin mucus of gilt-head sea bream (*Sparus aurata* L.). *Applied and Environmental Microbiology, 62*, 3650–3654.

Bowden, T. J. (2008). Modulation of the immune system of fish by their environment. *Fish & Shellfish Immunology, 25*(4), 373–383. https://doi.org/10.1016/j.fsi.2008.03.017

Britz, P. J., Hecht, T., & Mangold, S. (1997). Effect of temperature on growth, feed consumption and nutritional indices of Haliotis midae fed a formulated diet. *Aquaculture, 152*(1–4), 191–203. https://doi.org/10.1016/S0044-8486(97)00002-1

Bruneaux, M., Visse, M., Gross, R., Pukk, L., Saks, L., & Vasemägi, A. (2017). Parasite infection and decreased thermal tolerance: Impact of proliferative kidney disease on a wild salmonid fish in the context of climate change. *Functional Ecology, 31*, 216–226. https://doi.org/10.1111/1365-2435.12701

Byers, J. E. (2020. Nov 24). Effects of climate change on parasites and disease in estuarine and nearshore environments. *PLoS Biol., 18*(11), e3000743. https://doi.org/10.1371/journal.pbio.3000743

Cascarano, M. C., Stavrakidis-Zachou, O., Mladineo, I., Thompson, K. D., Papandroulakis, N., & Katharios, P. (2021). Mediterranean aquaculture in a changing climate: Temperature effects on pathogens and diseases of three farmed fish species. *Pathogens, 10*, 1205. https://doi.org/10.3390/pathogens10091205

Chadwick, J. G., & McCormick, S. D. (2017). Upper thermal limits of growth in brook trout and their relationship to stress physiology. *Journal of Experimental Biology, 220*(21), 3976–3987. https://doi.org/10.1242/jeb.161224

Chow, P. S., Chen, T. W., & Teo, L. H. (1994). Physiological responses of the common clowfish *Amphiprion ocellaris* (Cuvier) to factors related to packaging and long-distance transport by air. *Aquaculture, 127*, 347–361. https://doi.org/10.1016/0044-8486(94)90237-2

Claudiano, G. S., Andrade, S. C. S., Souza, E. C., Yunis-Aguinaga, J., Coutinho, L. L., Moreira, D. K. T., Gonçalves, F. C., Mundim, A. V., Marzocchi-Machado, C. M., de Moraes, F. R., & Moraes, J. R. E. (2020). Role of neuroendocrine modulation and biochemistry in the sepsis in *Piaractus mesopotamicus*. *General and Comparative Endocrinology, 1*(288), 113338. https://doi.org/10.1016/j.ygcen.2019.113338

Claudiano, G., Petrillo, T. R., Manrique, W. G., Castro, M. P., Loureiro, B. A., Marcusso, P. F., & De Moraes, F. R. (2013). Acute aerocystitis in Piaractus mesopotamicus: Participation of eicosanoids and pro-inflammatory cytokines. *Fish & Shellfish Immunology, 34*(5), 1057–1062. https://doi.org/10.1016/j.fsi.2013.01.006

Claudiano, G. S., Yunis-Aguinaga, J., Marinho-Neto, F. A., Miranda, R. L., Martins, I. M., Otani, F. S., et al. (2019). Hematological and immune changes in Piaractus mesopotamicus in the sepsis induced by Aeromonas hydrophila. *Fish & Shellfish Immunology, 88*, 259–265. https://doi.org/10.1016/j.fsi.2019.01.044

Clotfelter, E. D., Lapidus, S. J. H., & Brown, A. C. (2013). The effects of temperature and dissolved oxygen on antioxidante defences and oxidative damage in the fathead minnow *Pimephales promelas*. *Journal of Fish Biology, 82*(3), 1086–1092. https://doi.org/10.1111/jfb.12050

Collins, C., Bresnan, E., Brown, L., Falconer, L., Guilder, J., Jones, L., Kennerley, A., Malham, S., Murray, A., & Stanley, M. (2020). Impacts of climate change on aquaculture. *Science Review, 2020*, 482–520. https://doi.org/10.14465/2020.arc21.aqu

Cook, D. G., & Herbert, N. A. (2012). The physiological and behavioural response of juvenile kingfish (*Seriola lalandi*) differs between escapable and inescapable progressive hypoxia. *Journal of Experimental Marine Biology and Ecology, 413*, 138–144. https://doi.org/10.1016/j.jembe.2011.12.006

Costa, J. C., de Souza, S. S., & Castro, J.d. (2021). Climate change affects the parasitism rate and impairs the regulation of genes related to oxidative stress and ionoregulation of *Colossoma macropomum*. *Scientific Reports, 11*, 22350. https://doi.org/10.1038/s41598-021-01830-1

Costa, S. T. D., Gressler, L. T., Sutili, F. J., Loebens, L., Lazzari, R., & Baldisserotto, B. (2015). Effect of humic acid on survival, ionoregulation and hematology of the silver catfish, *Rhamdia quelen* (Siluriformes: Heptapteridae), exposed to different pHs. *Zoologia (Curitiba), 32*, 215–224. https://doi.org/10.1590/S1984-46702015000300006

Currie, K. L., Lange, B., Herbert, E. W., Harris, J. O., & Stone, D. A. J. (2015). Gastrointestinal evacuation time, but not nutriente digestibility, of greenlip abalone, *Haliotis laevigata* Donovan, is affected by water temperature and age. *Aquaculture, 448*, 219–228. https://doi.org/10.1016/j.aquaculture.2015.01.037

Currie, S., Reddin, K., McGinn, P., McConnell, T., & Perry, S. F. (2008). β-Adrenergic stimulation enhances the heat-shock response in fish. *Physiological and Biochemical Zoology, 81*, 414–425. https://doi.org/10.1086/589095

Dang, V. T., Speck, P., & Benkendorff, K. (2012). Influence of elevated temperatures on the immune response of abalone, *Haliotis rubra*. *Fish & Shellfish Immunology, 32*, 732–740. https://doi.org/10.1016/j.fsi.2012.01.022

Das, P. C., Ayyappan, S., & Jena, J. K. (2006). Haematological changes in the three Indian major carps, *Catla catla* (Hamilton), *Labeo rohita* (Hamilton) and *Cirrhinus mrigala* (Hamilton) exposed to acidic and alkaline water pH. *Aquaculture, 256*, 80–87. https://doi.org/10.1016/j.aquaculture.2006.02.019

Davis, K. B. (2004). Temperature affects physiological stress responses to acute confinement in sunshine bass (*Morone chrysops* × *Morone saxatilis*). *Comparative Biochemistry and Physiology—A Molecular and Integrative Physiology, 139*, 433–440. https://doi.org/10.1016/j.cbpb.2004.09.012

de Oliveira, A. M., & Val, A. L. (2017). Effects of climate scenarios on the growth and physiology of the Amazonian fish tambaqui (*Colossoma macropomum*) (Characiformes: Serrasalmidae). *Hydrobiologia, 789*, 167–178. https://doi.org/10.1007/s10750-016-2926-0

Demers, N. E., & Bayne, C. J. (1997). The immediate effects of stress on hormones and plasma lysozyme in rainbow trout. *Developmental and Comparative Immunology, 21*(4), 363–373. https://doi.org/10.1016/S0145-305X(97)00009-8

Di Marco, P., Priori, A., Finoia, M. G., Massari, A., Mandich, A., & Marino, G. (2008). Physiological responses of European sea bass *Dicentrarchus labrax* to different stocking densities and acute stress challenge. *Aquaculture, 275*, 319–328. https://doi.org/10.1016/j.aquaculture.2007.12.012

DiBattista, J. D., Levesque, H. M., Moon, T. W., & Gilmour, K. M. (2006). Growth depression in socially subordinate rainbow trout *Oncorhynchus mykiss*: More than a fasting effect. *Physiological and Biochemical Zoology, 79*, 675–687. https://doi.org/10.1086/504612

Dominguez, M., Takemura, A., & Tsuchiya, M. (2005). Effects of changes in environmental factors on the non-specific immune response of Nile tilapia, *Oreochromis niloticus* L. *Aquaculture Research, 36*, 391e7. https://doi.org/10.1111/j.1365-2109.2005.01220.x

Dominguez, M., Takemura, A., Tsuchiya, M., & Nakamura, S. (2004). Impact of diferente environmental factors on the circulating immunoglobulin levels in the Nile tilapia. *Aquaculture, 241*, 491e500. https://doi.org/10.1016/j.aquaculture.2004.06.027

Douxfils, J., Deprez, M., Mandiki, S. N. M., Milla, S., Henrotte, E., Mathieu, C., Silvestre, F., Vandecan, M., Rougeot, C., Mélard, C., Dieu, M., Raes, M., & Kestemont, P. (2012). Physiologic al and proteomic responses to single and repeated hypoxia in juvenile Eurasian perch under domestication-clues to physiological acclimation and humoral immune modulations. *Fish & Shellfish Immunology, 33*, 1112–1122. https://doi.org/10.1016/j.fsi.2012.08.013

Dupuis, A. P., & Hann, B. J. (2009). Climate change, diapause termination and zooplankton population dynamics: An experimental and modelling approach. *Freshwater Biology, 54*, 221–235. https://doi.org/10.1111/j.1365-2427.2008.02103.x

Evans, J. J., Shoemaker, C. A., & Klesius, P. H. (2003). Effects of sublethal dissolved oxygen stress on blood glucose and susceptibility to *Streptococcus agalactiae* in Nile tilapia *Oreochromis niloticus*. *Journal of Aquatic Animal Health, 15*, 202–208. https://doi.org/10.1577/H03-024

Fabbri, E., & Moon, T. W. (2016). Adrenergic signaling in teleost fish liver, a challenging path. *Comparative Biochemistry and Physiology Part B: Biochemistry and Molecular Biology, 199*, 74–86. https://doi.org/10.1016/j.cbpb.2015.10.002

Fast, M. D., Hosoya, S., Johnson, S. C., & Afonso, L. O. (2008). Cortisol response and immunerelated effects of Atlantic salmon (*Salmo salar* Linnaeus) subjected to short- and long-term stress. *Fish & Shellfish Immunology, 24*, 194e204. https://doi.org/10.1016/j.fsi.2007.10.009

Faught, E., Hernandez-Perez, J., Wilson, J. M., & Vijayan, M. M. (2020). *Climate change and noninfectious fish disorders*. CAB International.

Faught, E., & Vijayan, M. M. (2016). Mechanisms of cortisol action in fish hepatocytes. *Comparative Biochemistry and Physiology Part B: Biochemistry and Molecular Biology, 199*, 136–145.

Feely, R. A., Sabine, C. L., Lee, K., Berelson, W., Kleypas, J., Fabry, V. J., & Millero, F. J. (2004). Impact of anthropogenic CO2 on the CaCO3 system in the oceans. *Science, 305*(5682), 362–366. https://doi.org/10.1126/science.1097329

Feher, J. (2017). 2.9 - ATP production I: Glycolysis. In J. Feher (Ed.), *Quantitative human physiology* (2nd ed., pp. 218–226). Academic Press. https://doi.org/10.1016/B978-0-12-800883-6.00020-3

Feidantsis, K., Pörtner, H. O., Giantsis, I. A., & Michaelidis, B. (2021). Advances in understanding the impacts of global warming on marine fishes farmed offshore: Sparus aurata as a case study. *Journal of Fish Biology, 98*, 1509–1523. https://doi.org/10.1111/jfb.14611

Fernandes, J. M., Kemp, G. D., & Smith, V. J. (2004). Two novel muramidases from skin mucosa of rainbow trout (*Oncorhynchus mykiss*). *Comparative Biochemistry and Physiology. Part B, Biochemistry & Molecular Biology, 138*, 53e64. https://doi.org/10.1016/j.cbpc.2004.02.004

Ferreira, P. D. M. F., Barbosa, J. M., Santos, E. L., Lima, M. R., & de Lima Cabral, G. A. (2009). Efeito da temperatura sobre a taxa metabólica da carpa-comum (Cyprinus carpio Linnaeus, 1758). *Revista Brasileira de Engenharia de Pesca, 4*(2), 1–10. https://doi.org/10.18817/repesca.v4i2.106

Fey, D. P., & Greszkiewicz, M. (2021). Effects of temperature on somatic growth, otolith growth, and uncoupling in the otolith to fish size relationship of larval northern pike, *Esox lucius* L. *Fisheries Research, 236*, 105843. https://doi.org/10.1016/j.fishres.2020.105843

Ficke, A. D., Myrick, C. A., & Hansen, L. J. (2007). Potential impacts of global climate change on freshwater fisheries. *Reviews in Fish Biology and Fisheries, 17*, 581–613. https://doi.org/10.1007/s11160-007-9059-5

Fiorella, K. J., Bageant, E. R., Schwartz, N. B., Thilsted, S. H., & Barrett, C. B. (2021). Fishers' response to temperature change reveals the importance of integrating human behavior in climate change analysis. *Science. Advances, 7*(18), eabc7425. https://doi.org/10.1126/sciadv.abc74

Fromm, P. O. (1980). A review of some physiological and toxicological responses of freshwater fish to acid stress. *Environmental Biology of Fishes, 5*, 79–93. https://doi.org/10.1007/BF00000954

Gadan, K., Marjara, I. S., Sundh, H., Sundell, K., & Evensen, O. (2012). Slow release cortisol implants result in impaired innate immune responses and higher infection prevalence following experimental challenge with infectious pancreatic necrosis virus in Atlantic salmon (*Salmo salar*) parr. *Fish & Shellfish Immunology, 32*, 637e44. https://doi.org/10.1016/j.fsi.2012.01.004

Gallant, M. J., LeBlanc, S., MacCormack, T. J., & Currie, S. (2017). Physiological responses to a short-term, environmentally realistic, acute heat dtress in Atlantic salmon, *Salmo salar*. *Facets, 2*, 330–341. https://doi.org/10.1139/facets-2016-0053

Geng, Y. J. (2003). Molecular mechanisms for cardiovascular stem cell apoptosis and growth in the hearts with atherosclerotic coronary disease and ischemic heart failure. *Annals of the New York Academy of Sciences, 1010*, 687–697. https://doi.org/10.1196/annals.1299.126

Glencross, B., & Bermudes, M. (2010). Effect of high water temperatures on the utilisation efficiencies of energy and protein by juvenile barramundi, Lates calcarifer. *Fisheries and Aquaculture Journal, 2010*, FAJ-14.

Gobi, N., Vaseeharan, B., Rekha, R., Vijayakumar, S., & Faggio, C. (2018). Bioaccumulation, cytotoxicity and oxidative stress of the acute exposure selenium in *Oreochromis mossambicus*. *Ecotoxicology and Environmental Safety, 162*, 147–159. https://doi.org/10.1016/j.ecoenv.2018.06.070

Gorissen, M., & Flik, G. (2016). The endocrinology of the stress response in fish: An adaptation-physiological view. *Fish Physiology, 35*, 74–111. https://doi.org/10.1016/B978-0-12-802728-8.00003-5

Guo, H., & Dixon, B. (2021). Understanding acute stress-mediated immunity in teleost fish. *Fish & Shellfish Immunology, 12*(2), 100010. https://doi.org/10.1016/j.fsirep.2021.100010

Hakalahti, T., Karvonen, A., & Valtonen, E. T. (2006). Climate warming and disease risks in temperate regions–Argulus coregoni and Diplostomum spathaceum as case studies. *Journal of Helminthology, 80*(2), 93–98. https://doi.org/10.1079/JOH2006351

Harvell, C. D., Mitchell, C. E., Ward, J. R., Altizer, S., Dobson, A. P., & Ostfed, R. S. (2002). Climate warming and disease risks for terrestrial and marine biota. *Science, 296*(5576), 2158–2162. https://doi.org/10.1126/science.1063699

Hauser, N., Conlon, K. C., Desai, A., & Kobziar, L. N. (2021). Climate change and infections on the move in North America. *Infection and Drug Resistance, 30*(14), 5711–5723. https://doi.org/10.2147/IDR.S305077

Heimbürge, S., Kanitz, E., & Otten, W. (2019). The use of hair cortisol for the assessment of stress in animals. *General and Comparative Endocrinology, 270*, 10–17. https://doi.org/10.1016/j.ygcen.2018.09.016

Hooper, C., Day, R., Slocombe, R., Handlinger, J., & Benkendorff, K. (2007). Stress and immune responses in abalone: Limitations in current knowledge and investigative methods based on other models. *Fish & Shellfish Immunology, 22*(4), 363–379. https://doi.org/10.1016/j.fsi.2006.06.009

Hoseini, S. M., Yousefi, M., Hoseinifar, S. H., & Van Doan, H. (2019). Cytokines' gene expression, humoral immune and biochemical responses of common carp (*Cyprinus carpio*, linnaeus, 1758) to transportation density and recovery in brackish water. *Aquaculture, 504*(2019), 13–21. https://doi.org/10.1016/j.aquaculture.2019.01.049

Hsieh, C., Wu, Z. B., & Tung, M. C. (2007). PCR and in situ hybridization for the detection and localization of a new pathogen *Francisella*-like bacterium (FLB) in ornamental cichlids. *Diseases of Aquatic Organisms, 75*, 29–36. https://doi.org/10.3354/dao075029

Hua, K., & Bureau, D. P. (2009). Development of a model to estimate digestible lipid content of salmonid fish feeds. *Aquaculture, 286*(3–4), 271–276. https://doi.org/10.1016/j.aquaculture.2008.09.028

Huang, Z. H., Ma, A. J., & Wang, X. A. (2011). The immune response of turbot, *Scophthalmus maximus* (L.), skin to high water temperature. *Journal of Fish Diseases, 34*, 619e627. https://doi.org/10.1111/j.1365-2761.2011.01275.x

Huang, Z., Ma, A., Wang, X. A., Lei, J., Li, W., & Wang, T. (2015). Interaction of temperature and salinity on the expression of immunity factors in diferente tissues of juvenile turbot *Scophthalmus maximus* based on response surface methodology. *Chinese Journal of Oceanology and Limnology, 33*, 28e36. https://doi.org/10.1007/s00343-015-4016-y

Hueffer, K., O'Hara, T. M., & Follmann, E. H. (2011). Adaptation of mammalian host-pathogen interactions in a changing arctic environment. *Acta Veterinaria Scandinavica, 53*, 17. https://doi.org/10.1186/1751-0147-53-17

Huey, R. B., & Stevenson, R. D. (1979). Integrating thermal physiology and ecology of ectotherms: A discussion of approaches. *American Zoologist, 19*(1), 357–366. https://doi.org/10.1093/icb/19.1.357

Huguet, C. T., Norambuena, F., Emery, J. A., Hermon, K., & Turchini, G. M. (2015). Dietary n-6/n-3 LC-PUFA ratio, temperature and time interactions on nutrients and fatty acids di - Gestibility in Atlantic salmon. *Aquaculture, 436*, 160–166. https://doi.org/10.1016/j.aquaculture.2014.11.011

Huising, M. O., Metz, J. R., Van, S. C., Taverne-Thiele, A. J., Hermsen, T., & Verburg-van Kemenade, B. M. (2004). Structural characterisation of a cyprinid (*Cyprinus carpio* L.) CRH, CRH-BP and CRH-R1, and the role of these proteins in the acute stress response. *Journal of Molecular Endocrinology, 32*, 627e48. https://doi.org/10.1677/jme.0.0320627

Islam, M. J., Kunzmann, A., & Slater, M. J. (2022). Responses of aquaculture fish to climate change-induced extreme temperatures: A review. *Journal of the World Aquaculture Society, 53*(2), 314–366. https://doi.org/10.1111/jwas.12853

Israeli, D., & Kimmel, E. (1996). Monitoring the behaviour of hypoxiastressed *Carassius auratus* using computer vision. *Aquacultural Engineering, 15*, 423–440. https://doi.org/10.1016/S0144-8609(96)01009-6

Jiang, J.-L., Mao, M.-G., Lü, H.-Q., Wen, S.-H., Sun, M.-L., & Liu, R. (2017). Digital gene expression analysis of takifugu rubripes brain after acute hypoxia exposure using next-generation sequencing. *Comparative Biochemistry & Physiology. Genomics Proteomics, 24*, 12–18. https://doi.org/10.1016/j.cbd.2017.05.003

Jiang, J., Shi, Y., Yu, R., Chen, L., & Zhao, X. (2018). Biological response of zebrafish after short-term exposure to azoxystrobin. *Chemosphere, 202*, 56–64. https://doi.org/10.1016/j.chemosphere.2018.03.055

Jung, T. S., del Castillo, C. S., Javaregowda, P. K., Dalvi, R. S., Nho, S. W., & Park, S. B. (2012). Seasonal variation and comparative analysis of non-specific humoral immune substances in the skin mucus of olive flounder (*Paralichthys olivaceus*). *Developmental and Comparative Immunology, 38*, 295e301. https://doi.org/10.1016/j.dci.2012.06.005

Kalff, J. (2000). *Limnology*. Prentice Hall.

Kandalski, P. K., de Souza, M. R. D. P., Herrerias, T., Machado, C., Zaleski, T., Forgati, M., & Donatti, L. (2018). Effects of short-term termal stress on the plasma biochemical profiles of two Antarctic *Nototheniid* species. *Reviews in Fish Biology and Fisheries, 28*, 925–940. https://doi.org/10.1007/s11160-018-9535-0

Kapoor, B. G., Smit, H., & Verighina, I. A. (1976). The alimentary canal and digestion in teleosts. In F. S. Russell & M. Yonge (Eds.), *Advances in marine biology* (Vol. 13, pp. 109–239). Academic Press.

Karvonen, A., Rintamaki, P., Jokela, J., & Valtonen, E. T. (2010). Increasing water temperature and disease risks in aquatic systems: Climate change increases the risk of some, but not all, diseases. *International Journal for Parasitology, 40*, 1483–1488. https://doi.org/10.1016/j.ijpara.2010.04.015

Khosravi, M., Díaz-Morales, D. M., Thieltges, D. W., Wahl, M., & Vajedsamiei, J. (2023). Thermal optima of cercarial emergence in trematodes from a marine high-temperature ecosystem, the Persian Gulf. *Scientific Reports, 13*(1), 4923. https://doi.org/10.1038/s41598-023-31670-0

Kim, J. H., Kim, S. K., & Hur, Y. B. (2019). Temperature-mediated changes in stress responses, acetylcholinesterase, and immune responses of juvenile olive flounder *Paralichthys olivaceus* in a bio-floc environment. *Aquaculture, 506*, 453–458. https://doi.org/10.1016/j.aquaculture.2019.03.045

Kim, J. H., Kim, S. R., Kim, S. K., & Kang, H. W. (2021). Effects of pH changes on blood physiology, antioxidant responses and Ig M of juvenile olive flounder, *Paralichthys olivaceus*. *Aquaculture Reports, 21*, 100790. https://doi.org/10.1016/j.aqrep.2021.100790

Kyprianou, T. D., Pörtner, H. O., Anestis, A., Kostoglou, B., Feidantsis, K., & Michaelidis, B. (2010). Metabolic and molecular stress responses of gilthead seam bream *Sparus aurata* during exposure to low ambient temperature: An analysis of mechanisms underlying the winter syndrome. *Journal of Comparative Physiology B, 180*(7), 1005–1018. https://doi.org/10.1007/s00360-010-0481-y

Lafferty, K. D., & Holt, R. D. (2003). How should environmental stress affect the population dynamics of disease? *Ecology Letters, 6*(7), 654–664. https://doi.org/10.1046/j.1461-0248.2003.00480.x

Leal, C. A., Tavares, G. C., & Figueiredo, H. C. (2014). Outbreaks and genetic diversity of Francisella noatunensis subsp orientalis isolated from farm-raised Nile tilapia (*Oreochromis niloticus*) in Brazil. *Genetics and Molecular Research, 13*(3), 5704–5712. https://doi.org/10.4238/2014

LeBlanc, S., Middleton, S., Gilmour, K. M., & Currie, S. (2011). Chronic social stress impairs thermal tolerance in the rainbow trout (Oncorhynchus mykiss). *Journal of Experimental Biology, 214*(10), 1721–1731. https://doi.org/10.1242/jeb.056135

Liu, B., Jia, R., Han, C., Huang, B., & Lei, J. L. (2016). Effects of stocking density on antioxidant status, metabolism and immune response in juvenile turbot (*Scophthalmus maximus*). *Comparative Biochemistry & Physiology, 190*, 1–8. https://doi.org/10.1016/j.cbpc.2016.07.007

Lõhmus, M., & Bjorklund, M. (2015). Climate change: What will it do to fish–parasite interactions. *Biological Journal of the Linnean Society, 116*, 397–411. https://doi.org/10.1111/bij.12584

Lorenz, M. G., & Wackernagel, W. (1994). Bacterial gene transfer by natural genetic transformation in the environment. *Microbiology Reviews, 58*(3), 563–602. https://doi.org/10.1128/mr.58.3.563-602.1994

Lund, S., & Tufts, B. (2003). The physiological effects ofheat stress and the role of heat shock proteins in rainbow trout (*Oncorhynchus*) red blood cells. *Fish Physiology and Biochemistry, 29*, 1–12. https://doi.org/10.1023/B:FISH.0000035891.54116.34

Lupatsch, I., & Kissil, G. W. (2005). Feed formulations based on energy and protein demands in white grouper (*Epinephelus aeneus*). *Aquaculture, 248*, 83–95. https://doi.org/10.1016/j.aquaculture.2005.03.004

Ma, J., Liu, Y., Niu, D., & Li, X. (2015). Effects of chlorpyrifos on the transcription of CYP3A cDNA, activity of acetylcholinesterase, and oxidative stress response of goldfish (*Carassius auratus*). *Environmental Toxicology, 30*(4), 422–429. https://doi.org/10.1002/tox.21918

MacFadden, D. R., McGough, S. F., Fisman, D., Santillana, M., & Brownstein, J. S. (2018). Antibiotic resistance increases with local temperature. *Nature Climate Change, 8*(6), 510–514. https://doi.org/10.1038/s41558-018-0161-6

Machado, M. G., Tavares, L. P., Souza, G. V. S., Queiroz-Junior, C. M., Ascenção, F. R., Lopes, M. E., & Sousa, L. P. (2020). The Annexin A1/FPR2 pathway controls the inflammatory response and bacterial dissemination in experimental pneumococcal pneumonia. *The FASEB Journal, 34*(2), 2749–2764. https://doi.org/10.1096/fj.201902172R

Madaro, A., Folkedal, O., Maiolo, S., Alvanopoulou, M., & Olsen, R. E. (2018). Effects of acclimation temperature on cortisol and oxygen consumption in Atlantic salmon (*Salmo salar*) post-smolt exposed to acute stress. *Aquaculture, 497*, 331–335. https://doi.org/10.1016/j.aquaculture.2018.07.056

Makrinos, D. L., & Bowden, T. J. (2016). Natural environmental impacts on teleost immune function. *Fish & Shellfish Immunology, 53*, 50–57. https://doi.org/10.1016/j.fsi.2016.03.008

Mallya, Y. J. (2007). *The effects of dissolved oxygen on fish growth in aquaculture* (p. 3). The United Nations University fisheries training programmer, Final project.

Maloyan, A., & Horowitz, M. (2002). β-Adrenergic signaling and thyroid hormones affect HSP72 expression during heat acclimation. *Journal of Applied Physiology, 93*, 107–115. https://doi.org/10.1152/japplphysiol.01122.2001

Marcogliese, D. J. (2001). Implications of climate change for parasitism of animals in the aquatic environment. *Canadian Journal of Zoology, 79*, 1331–1352. https://doi.org/10.1139/z01-067

Marcogliese, D. J. (2010). The impact of climate change on the parasites and infectious diseases of aquatic animals. *OIE Revue Scientifique et Technique, 27*, 467–484.

Marcusso, P. F., Aguinaga, J. Y., da Silva Claudiano, G., Eto, S. F., Fernandes, D. C., Mello, H., & de Moraes, F. R. (2015). Influence of temperature on Streptococcus agalactiae infection in Nile tilapia. *Brazilian Journal of Veterinary Research and Animal Science, 52*(1), 57–62. https://doi.org/10.11606/issn.1678-4456.v52i1p57-62

Martínez, D., De Lázaro, O., Cortés, P., Oyarzún-Salazar, R., Paschke, K., & Vargas-Chacoff, L. (2020). Hypoxia modulates the transcriptional immunological response in *Oncorhynchus kisutch*. *Fish & Shellfish Immunology, 106*, 1042–1051. https://doi.org/10.1016/j.fsi.2020.09.025

Martins, M. L., Miyazaki, D. M. Y., Tavares-Dias, M., Fenerick, J. R. J., Onaka, E. M., Bozzo, F. R., & Moraes, F. R. (2009). Characterization of the acute inflammatory response in the hybrid tambacu (*Piaractus mesopotamicus* male x *Colossoma macropomum* female) (Osteichthyes). *Brazilian Journal of Biology, 69*, 957–962. https://doi.org/10.1590/s1519-69842009000400026

Matanza, X. M., & Osorio, C. R. (2018). Transcriptome changes in response to temperature in the fish pathogen *Photobacterium damselae* subsp. *damselae*: Clues to understand the emergence of disease outbreaks at increased seawater temperatures. *PLoS One, 13*, e0210118. https://doi.org/10.1371/journal.pone.0210118

Matsuo, A. Y., & Val, A. L. (2002). Low pH and calcium efects on net Na1 and K1 fluxes in two catfish species from the Amazon River (Corydoras:Callichthyidae). *Brazilian Journal of Medical and Biological Research, 35*, 361–367. https://doi.org/10.1590/s0100-879x2002000300012

Matvienko, N., Levchenko, A., Danchuk, O., & Kvach, Y. (2020). Assessment of the occurrence of microorganisms and other fish parasites in the freshwater aquaculture of Ukraine in relation to the ambient temperature. *Acta Ichthyologica et Piscatoria, 50*(3), 333–348. https://doi.org/10.3750/AIEP/02979

Medeiros. (2020). Parasitos de tambaqui *Colossoma macropomum* (Cuvier, 1816) cultivados no estado de rondônia Simone Paiva. Tese em Português | Vetteses | ID: vtt-218693 Biblioteca responsável: BR68.1 Palotina 2020.

Mennerat, A., Nilsen, F., Ebert, D., & Skorping, A. (2010). Intensive farming: Evolutionary implications for parasites and pathogens. *Evolutionary Biology, 37*(2–3), 59–67. https://doi.org/10.1007/s11692-010-9089-0

Metz, J. R., Huising, M. O., Leon, K., Verburg-van Kemenade, B. M. L., & Flik, G. (2006). Central and peripheral interleukin-1 and interleukin-1 receptor I expression and their role in the acute stress response of common carp, *Cyprinus carpio* L. *Journal of Endocrinology, 191*(1), 25–35. https://doi.org/10.1677/joe.1.06640

Mian, G. F., Godoy, D. T., Leal, C. A., Yuhara, T. Y., Costa, G. M., & Figueiredo, H. C. (2008). Aspects of the natural history and virulence of *S. agalactiae* infection in Nile tilapia. *Veterinary Microbiology, 136*(1–2), 180–183. https://doi.org/10.1016/j.vetmic.2008.10.016

Miegel, R. P., Pain, S. J., van Wettere, W. H., Howarth, G. S., & Stone, D. A. J. (2010). Effect of water temperature on gut transit time, digestive enzyme activity and nutrient digestibility in yellowtail kingfish (*Seriola lalandi*). *Aquaculture, 308*, 145–151. https://doi.org/10.1016/j.aquaculture.2010.07.036

Mommsen, T. P., Vijayan, M. M., & Moon, T. W. (1999). Cortisol in teleosts: Dynamics, mechanisms of action, and metabolic regulation. *Reviews in Fish Biology and Fisheries, 9*, 211–268. https://doi.org/10.1023/A:1008924418720

Montaudouin, X., Blanchet, H., Desclaux-Marchand, C., Lavesque, N., & Bachelet, G. (2016). Cockle infection by Himasthla quissetensis–I. From cercariae emergence to metacercariae infection. *Journal of Sea Research, 113*, 99–107. https://doi.org/10.1016/j.seares.2015.02.008

Nagae, M., Ogawa, K., Kawahara, A., Yamaguchi, M., Nishimura, T., & Ito, F. (2001). Effect of acidification stress on endocrine and imune functions in carp, *Cyprinus carpio*. *Water Air and Soil Pollution, 130*, 893e8. https://doi.org/10.1023/A:1013855501445

Nardocci, G., Navarro, C., Cortes, P., Imarai, M., Montoya, M., Valenzuela, B., Jara, P., Acuna-Castillo, C., & Fernandez, R. (2014). Neuroendocrine mechanisms for immune system regulation during stress in fish. *Fish & Shellfish Immunology, 40*, 531–538. https://doi.org/10.1016/j.fsi.2014.08.001

Nati, J. J. H., Lindström, J., & Halsey, L. G. (2016). Is there a trade-off between peak performance and performance breadth across temperatures for aerobic scope in teleost fishes? *Biology Letters, 12*(9), 2016019. https://doi.org/10.1098/rsbl.2016.0191

Ndong, D., Chen, Y. Y., Lin, Y. H., Vaseeharan, B., & Chen, J. C. (2007). The immune response of tilapia *Oreochromis mossambicus* and its susceptibility to *Streptococcus iniae* under stress in low and high temperatures. *Fish & Shellfish Immunology, 22*(6), 686–694. https://doi.org/10.1016/j.fsi.2006.08.015

Ngoan, L. D. (2018). Effects of climate change in aquaculture: Case study in Thua Thien Hue Province, Vietnam. *Biomedical Journal of Scientific & Technical Research, 10*, 2018. https://doi.org/10.26717/BJSTR.2018.10.001892

Ortuno, J., Esteban, M. A., & Meseguer, J. (2002). Lack of effect of combining different stressors on innate immune responses of seabream (*Sparus aurata* L.). *Veterinary Immunology and Immunopathology, 84*(2002), 17–27. https://doi.org/10.1016/S0165-2427(01)00387-7

Pagniello, K. B., Bols, N. C., & Lee, L. E. (2002). Effect of corticosteroids on viability and proliferation of the rainbow trout monocyte/macrophage cell line, RTS11. *Fish & Shellfish Immunology, 13*, 199e214. https://doi.org/10.1006/fsim.2001.0395

Peacock, S. J., Kutz, S. J., Hoar, B. M., & Molnár, P. K. (2022). Behaviour is more important than thermal performance for an Arctic host–parasite system under climate change. *Royal Society Open Science, 9*(8), 220060. https://doi.org/10.1098/rsos.220060

Pepi, M., & Focardi, S. (2021). Antibiotic-resistant bacteria in aquaculture and climate change: A challenge for health in the Mediterranean area. *International Journal of Environmental Research and Public Health, 18*(11), 5723. https://doi.org/10.3390/ijerph18115723

Philip, A. M., & Vijayan, M. M. (2015). Stress-immune-growth interactions: Cortisol modulates suppressors of cytokine signaling and JAK/STAT pathway in rainbow trout liver. *PLoS One, 6*, 10. https://doi.org/10.1371/journal.pone.0129299

Piana, P. A., Baumgartner, G., & Gomes, L. C. (2003). Influência da temperatura sobre o desenvolvimento de juvenis de piapara (*Leporinus* cf. *obtusidens*). *Acta Scientiarum: Biological Sciences, 25*, 87–94.

Piedras, S. R. N., Moraes, P. R., & Pouey, J. L. O. F. (2006). Desempenho de juvenis de catfish (*Ictalurus punctatus*) em diferentes emperaturas. *Revista Brasileira de Agrociência, 12*, 367–370.

Pol, V., Flik, G., & Gorissen, M. (2017). Comparative physiology of energy metabolism: Fishing for endocrine signals in the early vertebrate pool. *Frontiers in Endocrinology, 2017*(8), 36. https://doi.org/10.3389/fendo.2017.00036

Poulin, R. (2006). Global warming and temperature-mediated increases in cercarial emergence in trematode parasites. *Parasitology, 132*, 143–151. https://doi.org/10.1017/S0031182005008693

Prosser, C. L., & Nelson, D. O. (1981). The role of nervous systems in temperature adaptation of poikilotherms. *Annual Review of Physiology, 43*(1), 281–300. https://doi.org/10.1146/annurev.ph.43.030181.001433

Pulkkinen, K., Suomalainen, L. R., Read, A. F., Ebert, D., Rintamäki, P., & Valtonen, E. T. (2010). Intensive fish farming and the evolution of pathogen virulence: The case of columnaris disease in Finland. *Proceedings of the Royal Society B: Biological Sciences, 277*(1681), 593–600. https://doi.org/10.1098/rspb.2009.1659

Raida, M. K., & Buchmann, K. (2007). Temperature-dependent expression of immune-relevant genes in rainbow trout following *Yersinia ruckeri* vaccination. *Diseases of Aquatic Organisms, 77*, 41–52. https://doi.org/10.3354/dao01808

Rebouças, P. M., Lima, L. R., Dias, I. F., & Barbosa, J. D. (2014). Influence of thermal oscillation in pisciculture water. *Journal of Animal Behaviour and Biometeorology, 2*(2), 35–42. https://doi.org/10.14269/2318-1265.v02n02a01

Reid, G. K., Gurney-Smith, H. J., Marcogliese, D. J., & Knowler, D. (2019). Climate change and aquaculture: Considering biological response and resources. *Aquaculture Environment Interactions, 11*, 569–602. https://doi.org/10.3354/aei00332

Reverter, M., Sarter, S., Caruso, D., et al. (2020). Aquaculture at the crossroads of global warming and antimicrobial resistance. *Nature Communications, 11*, 1870. https://doi.org/10.1038/s41467-020-15735-6

Rodrigues, M. V., Francisco, C. J., & David, G. S. (2018). Monitoring of *Francisella noatunensis* subsp. *orientalis* in farmed Nile tilapia (*Oreochromis niloticus*) in Brazil. *Aquaculture International, 26*, 127–138. https://doi.org/10.1007/s10499-017-0204-4

Rodrigues, R. V., Pedron, J. D. S., Romano, L. A., Tesser, M. B., & Sampaio, L. A. (2015). Acute responses of juvenile cobia *Rachycentron canadum* (Linnaeus 1766) to acid stress. *Aquaculture Research, 46*, 1241–1247. https://doi.org/10.1111/are.12282

Russell, B. D., Connell, S. D., Mellin, C., Brook, B. W., Burnell, O. W., & Fordham, D. A. (2012). Predicting the distribution of commercially important invertebrate stocks under future climate. *PLoS One, 7*, e46554. https://doi.org/10.1371/journal.pone.0046554

Sadoul, B., & Geffroy, B. (2019). Measuring cortisol, the major stress hormone in fishes. *Journal of Fish Biology, 94*, 540–555. https://doi.org/10.1111/jfb.13904

Saeij, J. P., Verburg-van Kemenade, L. B., van Muiswinkel, W. B., & Wiegertjes, G. F. (2003). Daily handling stress reduces resistance of carp to *Trypanoplasma borreli*: In vitro modulatory effects of cortisol on leukocyte function and apoptosis. *Developmental and Comparative Immunology, 27*, 233–245. https://doi.org/10.1016/S0145-305X(02)00093-9

Sae-Lim, P., Kause, A., Mulder, H. A., & Olesen, I. (2017). Breeding and genetics symposium: Climate change and selective breeding in aquaculture. *Journal of Animal Science, 95*, 1801–1812. https://doi.org/10.2527/jas2016.1066

Saint Paul, U., & Soares, M. G. (1987). Diurnal distribution and behavioral responses of fishes to extreme hypoxia in an Amazon floodplain lake. *Environmental Biology of Fishes, 20*(2), 91–104. https://doi.org/10.1007/BF00005289

Samaras, A., Papandroulakis, N., Lika, K., & Pavlidis, M. (2018). Water temperature modifies the acute stress response of European sea bass, *Dicentrarchus labrax* L. (1758). *Journal of Thermal Biology, 78*, 84–91. https://doi.org/10.1016/j.jtherbio.2018.09.006

Schäfer, N., Matoušek, J., Rebl, A., Stejskal, V., Brunner, R. M., Goldammer, T., Verleih, M., & Korytář, T. (2021). Effects of chronic hypoxia on the immune status of pikeperch (*Sander lucioperca* Linnaeus, 1758). *Biology (Basel), 10*(7), 649. https://doi.org/10.3390/biology10070649

Schreck, C. B., & Tort, L. (2016). The concept of stress in fish. In *Fish physiology* (Vol. 35, pp. 1–34). Academic Press.

Scott, A. L., & Rogers, W. A. (1980). Histological effects of prolonged sublethal hypoxia on channel catfish *Ictalurus punctatus* (Rafi nesque). *Journal of Fish Diseases, 3*, 305–316. https://doi.org/10.1111/j.1365-2761.1980.tb00401.x

Seyle, H. (1946). A síndrome geral da adaptação e as doenças da adaptação. *Jornal da Endocrinologia Clínica, 6*, 117–230.

Shapiro, R. S., & Cowen, L. E. (2012). Thermal control of microbial development and virulence: Molecular mechanisms of microbial temperature sensing. *MBio, 3*(5), 10–1128. https://doi.org/10.1128/mbio.00238-12

Shimada, M. T., Claudiano, G. S., Engracia Filho, J. R., Yunis, J., & Moraes, F. R. (2014). Hepatic Steatosis in cage-reared young cobia, *Rachycentron canadum* (Linnaeus, 1766), in Brazil. *Journal of Veterinary Science & Medical Diagnosis, 3*, 2. https://doi.org/10.4172/2325-9590.100013

Shimada, M. T., Yunis-Aguinaga, J., Cueva-Quiroz, V. A., Claudiano, G. S., Engracia Filho, J. R., Mouriño, J., Ruas de Moraes, F., & Engracia de Moraes, J. (2020). Isolation and characterization of pathology in case of massive mortality by *Photobacterium damselae* subsp.

piscicida in *Rachycentron canadum*. *Bioscience Journal, 36*, 1732–1741. https://doi.org/10.14393/BJ-v36n5a2020-41745

Shimon-Hophy, M. (2021). Avtalion RR. Influence of chronic stress on the mechanism of the cytotoxic system in common carp (Cyprinus carpio). *Immunology, 164*(2), 211–222. https://doi.org/10.1111/imm.13345.Epub. 2021 May 31.

Shuangyao, W., Zhiqiang, J., Mingguang, M., Shoukang, M., Yang, S., & Youzhen, S. (2018). Effects of seawater pH on survival, growth, energy budget and oxidative stress parameters of juvenile turbot *Scophthalmus maximus*. *Iranian Journal of Fisheries Sciences, 17*(4), 675–689. https://doi.org/10.22092/IJFS.2018.116814

Siikavuopio, S. I., James, P., Lysne, H., Sæther, B. S., Samuelsen, T. A., & Mortensen, A. (2012). Effects of size and temperature on growth and feed conversion of juvenile green sea urchin (Strongylocentrotus droebachiensis). *Aquaculture, 354*, 27–30. https://doi.org/10.1016/j.aquaculture.2012.04.036

Silva, E. C. S., & Araujo-Lima, C. A. (2003). Influence of food type and temperature on gastric evacuation of Pygocentrus nattereri and experimental conditions. *Acta Amazonica, 33*, 145–156. https://doi.org/10.1590/1809-4392200331156

Silvestri, A., Ferrari, E., Gozzi, S., Marchi, F., & Foschino, R. (2018). Determination of temperature dependent growth parameters in psychrotrophic pathogen bacteria and tentative use of mean kinetic temperature for the microbiological control of food. *Frontiers in Microbiology, 9*, 3023. https://doi.org/10.3389/fmicb.2018.03023

Singh, S. P., Ahmad, T., Sharma, J., & Chakrabarti, R. (2021). Effect of temperature on food consumption, immune system, antioxidant enzymes, and heat shock protein 70 of Channa punctata (Bloch, 1793). *Fish Physiology and Biochemistry, 47*, 79–91. https://doi.org/10.1007/s10695-020-00896-4

Singh, G., Singh, A., Singh, P., & Mishra, V. (2021). Chapter 4 - impact of climate change on freshwater ecosystem. In *Water conservation in the era of global climate change* (pp. 73–98). https://doi.org/10.1016/B978-0-12-820200-5.00017-8

Sollid, J., De Angelis, P., Gundersen, K., & Nilsson, G. E. (2003). Hypoxia induces adaptive and reversible gross morphological changes in crucian carp gills. *Journal of Experimental Biology, 206*(20), 3667–3673. https://doi.org/10.1242/jeb.00594

Soto, E., & Revan, F. (2012). Culturability and persistence of *Francisella noatunensis* subsp. *orientalis* (syn) in sea- and fresh water microcosms. *Microbial Ecology, 63*, 398–404. https://doi.org/10.1007/s00248-011-9932-6

Steinhausen, M. F., Sandblom, E., Eliason, E. J., Verhille, C., & Farrell, A. P. (2008). The effect of acute temperature increases on the cardiorespiratory performance of resting and swimming sockeye salmon (*Oncorhynchus nerka*). *Journal of Experimental Biology, 211*, 3915–3926. https://doi.org/10.1242/jeb.019281

Stickney, R. R. (2000). *Encyclopedia of aquaculture*. John Wiley and Sons.

Su, M., Zhang, R., Liu, N., & Zhang, J. (2020). Modulation of inflammatory response by cortisol in the kidney of spotted scat (scatophagus Argus) in vitro under different osmotic stresses. *Fish & Shellfish Immunology, 104*, 46–54. https://doi.org/10.1016/j.fsi.2020.05.060

Sunday, J. M., Bates, A. E., & Dulvy, N. K. (2012). Thermal tolerance and the global redistribution of animals. *Nature Climate Change, 2*(9), 686–690. https://doi.org/10.1038/nclimate1539

Sunyer, J. O., & Tort, L. (1995). Natural hemolytic and bactericidal activities of sea bream *Sparus aurata* serum are effected by the alternative complement pathway. *Veterinary Immunology and Immunopathology, 45*(3–4), 333–345. https://doi.org/10.1016/0165-2427(94)05430-Z

Thieltges, D. W., & Rick, J. (2006). Effect of temperature on emergence, survival and infectivity of cercariae of the marine trematode Renicola roscovita (Digenea: Renicolidae). *Diseases of Aquatic Organisms, 73*(1), 63–68. https://doi.org/10.3354/dao073063

Thomas, M. B., & Blanford, S. (2003). Thermal biology in insect-parasite interactions. *Trends in Ecology & Evolution, 18*(7), 344–350. https://doi.org/10.1016/S0169-5347(03)00069-7

Thrusfield, M., Christley, R., Brown, H., Diggle, P., French, N., Howe, K., Kelly, L., O'Connor, A., Sargeant, J., & Wood, H. (2018). *Veterinary epidemiology* (4th ed.). Wiley. ISBN:9781118280249. https://doi.org/10.1002/9781118280249

Tiedke, J., Borner, J., Beeck, H., Kwiatkowski, M., Schmidt, H., & Thiel, R. (2015). Evaluating the hypoxia response of Ruffe and Flounder Gills by a combined proteome and transcriptome approach. *PLoS One, 10*(8), e0135911. https://doi.org/10.1371/journal.pone.0135911

Tignat-Perrier, R., van de Water, J. A., Guillemain, D., Aurelle, D., Allemand, D., & Ferrier-Pagès, C. (2022). The effect of thermal stress on the physiology and bacterial communities of two key mediterranean gorgonians. *Applied and Environmental Microbiology., 88*(6), e0234021. https://doi.org/10.1128/aem.02340-21

Tolussi, C. E., Gomes, A. D., Ribeiro, C., Caneppele, D., Moreira, R., & Honji, R. (2018). Mobilization of energetic substrates in the endangered catfish *Steindachneridion parahybae* (Siluriformes: Pimelodidae): Changes in annual reproductive cycle in captivity. *Neotropical Ichthyology, 16*, e170120[1]. https://doi.org/10.1590/1982-0224-20170120

Tort, L. (2011). Stress and immune modulation in fish. *Developmental & Comparative Immunology, 35*(12), 1366–1375. https://doi.org/10.1016/j.dci.2011.07.002

Val, A. L. (2000). Organic phosphates in the red blood cells of fish. *Comparative Biochemistry and Physiology, 125*(A), 417–435. https://doi.org/10.1016/S1095-6433(00)00184-7

Val, A. L., Menezes, A. C. L., Ferreira, M. S., Silva, M. N. P., Araújo, R. M., & AlmeidaVal, V. M. F. (2006). Estresse em peixes: respostas integradas para a 126 sobrevivência e a adaptação. In A. T. Silva-Souza (Ed.), *Sanidade de Organismos Aquáticos no Brasil* (pp. 211–228). ABRAPOA.

Van der Meer, D. L., Van den Thillart, G. E., Witte, F., de Bakker, M. A., Besser, J., Richardson, M. K., Spaink, H. P., Leito, J. T., & Bagowski, C. P. (2005). Gene expression. Profiling of the long-term adaptive response to hypoxia in the gills of adult zebrafish. *American Journal of Physiology-Regulatory, Integrative and Comparative Physiology, 289*, R1512–R1519. https://doi.org/10.1152/ajpregu.00089.2005

Vargas-Chacoff, L., Muñoz, J. L. P., Ocampo, D., Paschke, K., & Navarro, J. M. (2019). The effect of alterations in salinity and temperature on neuroendocrine responses of the Antarctic fish *Harpagifer antarcticus*. *Comparative Biochemistry and Physiology -Part A: Molecular and Integrative Physiology, 235*, 131–137. https://doi.org/10.1016/j.cbpa.2019.05.029

Verburg-van, K. B., Ribeiro, C. M., & Chadzinska, M. (2011). Neuroendocrineimmune interaction in fish: Differential regulation of phagocyte activity by neuroendocrine factors. *General and Comparative Endocrinology, 172*, 31e8. https://doi.org/10.1016/j.ygcen.2011.01.004

Vijayan, M. M., Raptis, S., & Sathiyaa, R. (2003). Cortisol treatment affects glucocorticoid receptor and glucocorticoid-responsive genes in the liver of rainbow trout. *General and Comparative Endocrinology, 132*, 256–263. https://doi.org/10.1016/S0016-6480(03)00092-3

Volkoff, H., & Rønnestad, I. (2020). Effects of temperature on feeding and digestive processes in fish. *Temperature (Austin), 7*(4), 307–320. https://doi.org/10.1080/23328940.2020.1765950

Vornanen, M. (2016). The temperature dependence of electrical excitability in fish hearts. *Journal of Experimental Biology, 219*(13), 1941–1952. https://doi.org/10.1242/jeb.12843

Welker, T., & McNulty, S. (2007). Effect of sublethal hypoxia on the immune response and susceptibility of channel catfish, *Ictalurus punctatus*, to enteric septicemia. *Journal of the World Aquaculture Society, 38*(2007), 12–23. https://doi.org/10.1111/j.1749-7345.2006.00069.x

Wells, R. M. G., & Baldwin, J. (2006). Plasma lactate and glucose flushes following burst swimming in silver trevally (*Pseudocaranx dentex*: Carangidae) support the "releaser" hypothesis. *Comparative Biochemistry & Physiology, 143*, 347–352.

Weyts, F. A., Flik, G., & Verburg-van, K. B. (1998). Cortisol inhibits apoptosis in carp neutrophilic granulocytes. *Developmental and Comparative Immunology, 22*, 563–572. https://doi.org/10.1016/S0145-305X(98)00027-5

Windisch, H. S., Frickenhaus, S., John, U., Knust, R., Portner, H. O., & Lucassen, M. (2014). Stress response or beneficial temperature acclimation: Transcriptomic signature in Antarctic fish (*Pachycara brachycephalum*). *Molecular Ecology, 23*, 3469–3482. https://doi.org/10.1111/mec.12822

Wright, R. K., & Cooper, E. L. (1981). Temperature effects on ectotherm immune responses. *Developmental & Comparative Immunology, 5*, 117–122. https://doi.org/10.1016/0145-305X(81)90016-1

Wu, R. S. (2002). Hypoxia: From molecular responses to ecosystem responses. *Marine Pollution Bulletin, 45*, 35–45. https://doi.org/10.1016/S0025-326X(02)00061-9

Yada, T., Muto, K., Azuma, T., Fukamachi, S., Kaneko, T., & Hirano, T. (2006). Effects of acid water exposure on plasma cortisol, ion balance, and immune functions in the "cobalt" variant of rainbow trout. *Zoological Science, 23*, 707–713. https://doi.org/10.2108/zsj.23.707

Yada, T., & Tort, L. (2016). Stress and disease resistance: Immune system and Immunoendocrine interactions. *Fish Physiology, 35*, 365–403. https://doi.org/10.1016/B978-0-12-802728-8.00010-2

Yarahmadi, P., Miandare, H. K., Fayaz, S., & Caipang, C. M. (2016). Increased stocking density causes changes in expression of selected stress- and immune-related genes, humoral innate immune parameters and stress responses of rainbow trout (*Oncorhynchus mykiss*). *Fish & Shellfish Immunology, 48*, 43–53. https://doi.org/10.1016/j.fsi.2015.11.007

Conclusions and Perspectives

Samara Silva de Souza, Susana Braz-Mota,
and Adalberto Luis Val

This book presents a compilation of information about the megadiversity of the Amazon, including the responses of different groups of aquatic organisms to natural and anthropogenic environmental changes. The data collected in the chapters indicate that the Amazon biome has undergone significant changes due to human activity. This has led to large deforested areas, unprecedented fires, increased air and water temperatures, and hypoxic conditions that are already common in the region. Additionally, aquatic contamination by different groups of pollutants has been observed.

The information presented in this book suggests that the Amazonian aquatic biota can serve as an indicator of natural environmental changes as well as those caused by human actions. These changes can have drastic impacts on biodiversity, triggering a chain effect that compromises the entire food web and its interactions. The interactions between terrestrial and aquatic components are crucial for the functioning of Amazonian ecosystems. It is expected that climate change may alter the flow of energy and nutrients, potentially compromising the functioning of the food web and productivity, which could directly impact the entire ecological organization.

Increasing human pressure has caused significant changes in terrestrial and hydrological conditions, resulting in a cascade effect that ultimately leads to the vulnerability of the Amazon forest. Land use, deforestation, forest fires, and the use of pesticides can alter the climate and landscape of a region. These changes can, in turn, affect aquatic environments by increasing water flow and temperature, as well as altering hydrological cycles and chemical conditions of the water. Various factors contribute to climate change, and the chapters demonstrate that different organisms, such as symbiotic bacteria, plants, insects, crustaceans, fish, and amphibians, are already suffering from its impacts. This cascade of events will promote a decline in the assemblage of aquatic organisms, which are difficult to recover.

S. S. de Souza · S. Braz-Mota · A. L. Val
Laboratory of Ecophysiology and Molecular Evolution, Brazilian National Institute for Research of the Amazon (INPA), Manaus, Amazonas, Brazil

S. S. de Souza et al. (eds.), *The Future of Amazonian Aquatic Biota*,
https://doi.org/10.1007/978-3-031-66822-7

The presented information in this book contributes substantially to understanding the impacts of environmental changes, despite its limitations due to the size of the Amazon biome. However, further studies are needed to fully understand the various adaptations at all levels of biological organization. Additionally, new technologies must be developed to ensure the ecological webs of the biome function properly and to guarantee food security and the wellbeing of the population.

Additionally, it is crucial to enhance understanding of the species present in the region. This requires monitoring to gather data on the true Amazonian biodiversity, which is currently underestimated. First, it is necessary to gain knowledge and then take action towards the sustainable use of Amazonian resources. It is imperative to incorporate the data collected here into future projects aimed at mitigating or minimizing the effects of climate change on the Amazon. Understanding the Amazon is crucial for designing projects that aim to minimize the expected impacts on biodiversity. This will not only contribute to finding solutions to problems that affect the economy but also those that exacerbate social crises. Therefore, research and discoveries about the region should be included in government agendas.

The manufacturer's authorised representative in the EU is Springer Nature Customer Service Centre GmbH, Europaplatz 3, 69115 Heidelberg, Germany. If you have any concerns regarding our products, please contact ProductSafety@springernature.com

Printed and bound by CPI Group (UK) Ltd, Croydon, CR0 4YY
01/10/2026
02225910-0019